Resolving the Climate Change Crisis

Philip Lawn

Resolving the Climate Change Crisis

The Ecological Economics of Climate Change

Springer

Philip Lawn
Flinders Business School
Flinders University
Adelaide
Australia

ISBN 978-94-017-7501-4 ISBN 978-94-017-7502-1 (eBook)
DOI 10.1007/978-94-017-7502-1

Library of Congress Control Number: 2015956131

Printed on acid-free paper

This Springer imprint is published by SpringerNature
The registered company is Springer Science+Business Media B.V. Dordrecht

To my dad
Graham Charles Lawn
(20 March 1938–1 September 2010)

Preface

Many people regard climate change as an impending problem. I do not. Given the need to take immediate and drastic action to reduce greenhouse gas emissions, the threat posed by climate change has already reached a crisis point. However, a successful resolution to the crisis will require more than an effective emissions protocol. Unless the world's high-income nations begin the transition to a qualitatively improving steady-state economy, and low-income nations follow suit at some stage over the next 20–40 years, trying to avoid catastrophic climate change will be akin to putting a square peg in a round hole. To put it another way, if the world continues its predilection with continuous GDP growth, a well-designed emissions protocol will be as useless as the paper it is written on.

This book sets out why we must abandon the goal of continuous growth; how we can do so in a way that improves human well-being; what constitutes a safe atmospheric concentration of greenhouse gases; and what type of emissions protocol and emissions-trading framework is likely to achieve a desirable climate change outcome as well as promote the broader goal of sustainable development.

During the preparation of this book, a number of crucial United Nations climate change conferences were convened, including the highly publicised Copenhagen meeting in 2009. The initial aim of the Copenhagen conference was to establish a legally binding emissions protocol to take effect at the end of the first Kyoto commitment period in 2012. Sadly, no such accord emerged. At the 2012 meeting in Doha, an agreement was reached to further extend the Kyoto Protocol and to develop a new emissions protocol in Paris by 2015 to take effect in 2021. As promising as this sounds, deep emissions cuts must begin in 2016 if the rise in average global temperatures is to be restricted to 2 °C above pre-industrial levels. Thus, even if a strict new protocol is established, more must be done to kick-start the reduction in global greenhouse gas emissions during the 2016–2020 period.

Unfortunately, as I write this Preface, there is little sign that a new emissions protocol will achieve the emissions cuts necessary to prevent dangerous if not catastrophic climate change. Nor, given the need to quell GDP growth to realise greenhouse gas emissions targets, does an international agreement look like surfacing to deal adequately with population growth, biodiversity loss, and the rising rate of natural resource use—a sad reality attested by the abject failure of the 2012 United Nations Conference on Sustainable Development in Rio de Janeiro (Rio+20).

With the Paris conference fast approaching and the IPCC's Fifth Assessment Report having reconfirmed humankind's contribution to global warming, the world's leaders must take the 'bull by the horns' and put long-term concerns ahead of short-term political interests. Should they fail to do so, humanity not only faces a climatically tempestuous future, but one where human well-being, freedom in the liberal democratic tradition, and international peace will be gravely jeopardised.

Adelaide
April 2015

Philip Lawn

Contents

About the Author

Philip Lawn is an ecological economist at Flinders University in Adelaide, Australia, who completed his PhD at Griffith University in 1997. Since joining Flinders University in 1998, Philip has produced eight books, around 40 book chapters, and over 50 journal articles on sustainable development; green national accounting; international trade and the environment; ecological macroeconomics; and issues concerning the perceived conflict between environmental conservation and employment.

Philip is a member of the International Society for Ecological Economics (ISEE) and served on the executive of the Australia and New Zealand Society for Ecological Economics (ANZSEE)—a chapter of ISEE—from 2003 to 2008. During this period, Philip was involved in the organisation of three ANZSEE conferences and a pre-conference workshop on sustainable development indicators for the Asia-Pacific region in New Delhi in 2006.

In 2004, Philip became the inaugural editor of the International Journal of Environment, Workplace, and Employment (Inderscience) and currently serves on the Editorial Board of a number of academic journals. As an Associate Professor at the Flinders Business School, Philip provides advice to policy-makers and regularly offers his services as a speaker at public speaking forums.

List of Figures

List of Tables

List of Boxes

Part I
Ecological Economics, Climate Change, and Sustainable Development

Chapter 1
Ecological Economics and Climate Change

1.1 Introduction

The main thesis of this book is that the global warming observed over the past century is essentially the product of humankind's addiction to GDP growth and that catastrophic climate change will only be prevented if there is a global commitment to stabilise the atmospheric concentration of carbon dioxide-equivalent gases at no more than 450 parts-per-million (i.e., 450 ppm of CO_2-e). Because anthropogenic global warming is one of many global issues requiring urgent attention, most of which are inextricably linked, climate change must be resolved in conjunction with other sustainable development concerns (Beg et al. 2002; Halsnæs 2002; OECD 2004; Munasinghe and Swart 2005; Daly 2007; Lawn 2007; Sachs 2009). This not only means that greenhouse gas emissions have to be drastically reduced to achieve the 450 ppm target, it also means there is a need to: (i) decrease the global rate of natural resource use until it is ecologically sustainable; (ii) ensure population numbers are below eight billion people by 2100; (iii) reduce income and wealth disparities between and within countries; and (iv) encourage nations to make the transition to a qualitatively-improving steady-state economy. The latter is required in view of emerging evidence suggesting that the growth of many national economies is increasing social and environmental costs faster than it is increasing economic benefits. That is, growth is now reducing the per capita economic welfare of most high-GDP countries and an increasing number of low-GDP countries (e.g., China and Thailand) (Lawn and Clarke 2008).[1]

In order to prevent the concentration of greenhouse gases continuously rising or stabilising above 450 ppm of CO_2-e, a stringent emissions protocol encompassing all the world's nations is desperately needed (Chaps. 8–10). The centrepiece of the protocol must be an effective emissions-trading system. The type of emissions-trading system required and the reasons for its superiority over emissions taxes are

P. Lawn, *Resolving the Climate Change Crisis*, DOI 10.1007/978-94-017-7502-1_1

outlined in Chap. 7. Overall, for a global emissions-trading system to succeed, it must be built on the principles of ecological sustainability, distributional equity, and allocative efficiency. At the same time, the system needs to be accompanied by urgently required policies at the national level. A number of these policies are outlined in Chap. 3. In Chap. 10, it will be shown how some of these policies can be neatly dovetailed with a global emissions-trading system in order to maximise the chances of their success.

To support the book's central thesis, a sustainable emissions scenario is revealed in Chap. 4. Not only can the sustainable scenario identify the structural changes required to achieve a 450 ppm stabilisation target, it is able to highlight the implausibility of resolving the climate change crisis whilst trying to continuously grow the global economy. In addition, the sustainable scenario can provide valuable input when attempting to design an effective emissions protocol.

As suggested by the subtitle, this book involves an ecological economics approach to the current climate change crisis. Ecological economics differs from mainstream economics in many fundamental ways. In terms of directly addressing the climate change issue, mainstream economic approaches—including the well-known *Stern Review*—employ conventional benefit-cost analyses to inform their conclusions (Chap. 5). As will become evident, an ecological economics-motivated simulation and comparison of sustainable and growth-as-usual scenarios constitutes a major departure from the standard economic approach. This does not mean that ecological economists rule out the study of climate change-related benefits and costs as a valuable means of investigation. To the contrary, Chap. 6 reveals how an understanding of benefits and costs can provide useful information regarding least-cost ways of undertaking mitigation and adaptation measures. However, because of the primacy of ecological sustainability, ecological economists believe that any consideration of a desirable atmospheric concentration of greenhouse gases must be confined to a range of 'safe' or ecologically sustainable outcomes.

For most of the remainder of this chapter, the rationale behind a safe upper limit of 450 ppm of CO_2-e will be established. To do this, the basic principles of the Earth's natural greenhouse effect will be outlined. This will followed by a close examination of the climate change record over the past 542 million years, with a particular emphasis placed on the more accurate record of global temperatures over the last 800,000 years. It will then be explained why the global warming observed over the past century can be largely attributed to human activities; why anthropogenic global warming will almost certainly continue without appropriate human action; and how global warming is likely to impact on ecosystems, water supplies, food production, coastlines, and human health. As will become clear, it is by considering the potential impacts of future global warming that an upper limit on the safe concentration of greenhouse gases can best be determined.

1.2 Some Background Information on Climate Change Science

1.2.1 The Natural Greenhouse Effect

The Earth's climate is a complex, interactive system consisting of the atmosphere, the surface of the Earth, all living creatures, snow, ice, oceans, and other large bodies of water. The principal driver of the Earth's climate system is the Sun. The energy from the Sun radiates at very short wavelengths, predominantly in the visible or near-visible part of the spectrum. Averaged over the entire planet, the amount of radiant energy reaching the uppermost part of the Earth's atmosphere every second is 342 Watts per square metre (Wm^{-2}). Approximately 30 per cent of this energy is reflected back to space (107 Wm^{-2}). Of this, a further 77 Wm^{-2} is reflected by clouds and minute atmospheric particles broadly referred to as aerosols. The remaining 30 Wm^{-2} is reflected by light-coloured regions of the Earth's surface, such as ice, snow, and deserts (IPCC 2007b).

The energy not reflected back to space (235 Wm^{-2}) is absorbed by the Earth's surface and the Earth's atmosphere (168 and 67 Wm^{-2} respectively). To balance the incoming energy (i.e., to prevent the continual heating of the Earth's surface), the Earth must, on average, radiate the same amount of energy back to space. The Earth achieves this by emitting thermal energy in the form of long-wave radiation. Much of this thermal radiation is absorbed by the various greenhouse gases in the Earth's atmosphere—e.g., carbon dioxide (CO_2), methane (CH_4), and nitrous oxide (N_2O)—and is subsequently re-radiated back to Earth. It is via this process that greenhouse gases trap heat in the Earth's atmosphere. The blanketing of heat by greenhouse gases is known as the *natural greenhouse effect*. If not for this phenomenon, the Earth's average surface temperature would be 33 °C lower than its present value of around 14 °C.

To facilitate the natural greenhouse effect, greenhouse gases possess two crucial characteristics. On the one hand, they are transparent to the incoming shortwave energy not immediately reflected back to space. On the other hand, they absorb the long-wave, infra-red radiation through which the Earth returns energy to space (Flannery 2008). One notable exception is water vapour that exists in the form of clouds. Although clouds exert a blanketing effect similar to other greenhouse gases, they also reflect the Sun's radiant energy as it enters the atmosphere. Overall, clouds have a net cooling effect on the Earth's climate (IPCC 2007b).

There are many feedback mechanisms that either amplify or diminish the impact of shocks to the Earth's climate system. As examples of the latter, negative feedback mechanisms play the hemeostatic role of restoring the climate system to something approximating a dynamic equilibrium. Conversely, positive feedback mechanisms play their part in upsetting the dynamic equilibrium of the climate system. An example of a positive feedback mechanism is the so-called 'ice-albedo feedback', which occurs when a rise in the Earth's temperature reduces the surface coverage of ice and snow. Because this exposes dark land and water surfaces, it increases the Earth's absorption of heat. This, in turn, induces more warming and further melting, thus establishing a self-reinforcing cycle.[2] Unless countered by an offsetting negative feedback process, the warming triggered by the initial change in the climate system is amplified rather than suppressed. Not surprisingly, it is the difficulty associated with detecting, understanding, and quantifying feedback processes that makes climate change forecasting so very problematic (Houghton 1994; IPCC 2007b).

1.2.2 Climate Change Over the Course of Time

It is a widely recognised fact that the Earth's climate is constantly evolving (Houghton 1994; Fleming 1998; Flannery 2008). Changes in climate over time are due to the climate system's own internal dynamics and variations brought about by exogenous shocks. The latter are referred to as 'external forcings' and include volcanic eruptions, solar variations, and changes in the composition of the Earth's atmosphere.

Because the climate system is driven by the Sun's radiant energy, there are three main ways in which the radiative balance of the Earth can vary. Each has the capacity to alter average global temperatures. They are:

1. Changes in the incoming rate of solar radiation, which can be caused by variations in the Earth's orbit (Milankovitch cycles) or the solar output of the Sun.[3]
2. Changes in the proportion of incoming solar radiation reflected back to space, which can be caused by variations in cloud cover, atmospheric particles, and the Earth's coverage of ice, snow, and vegetation.
3. Changes in the proportion of the long-wave radiation absorbed in the atmosphere and subsequently re-radiated to Earth, which can be caused by variations in the concentration of greenhouse gases in the atmosphere.

Variations in all three factors have resulted in significant fluctuations over time in average global temperatures. Some temperature fluctuations are minor and short-lived. For example, the global cooling effects of the increase in atmospheric particles caused by the volcanic eruptions of Mt. Agung (1963), Mt. El Chichon (1982), and Mt. Pinatubo (1991) were rapidly corrected by negative feedback mechanisms present in the climate system. Another short-term influence on global temperatures is a heat transfer mechanism known as the El Niño phenomenon—a temporary change in climate caused by the intermittent warming of the tropical Pacific Ocean east of the International Dateline.[4]

Knowledge of much larger and long-term fluctuations in temperatures can be ascertained by analysing the palaeoclimate record of the Earth.[5] This record, going back 542 million years, is revealed in Fig. 1.1.[6] What is clearly evident from Fig. 1.1 is that pre-Quaternary climates (pre-2.6 million years ago) were generally much warmer than today—in some cases, up to 8 °C warmer (Cambrian period).[7]

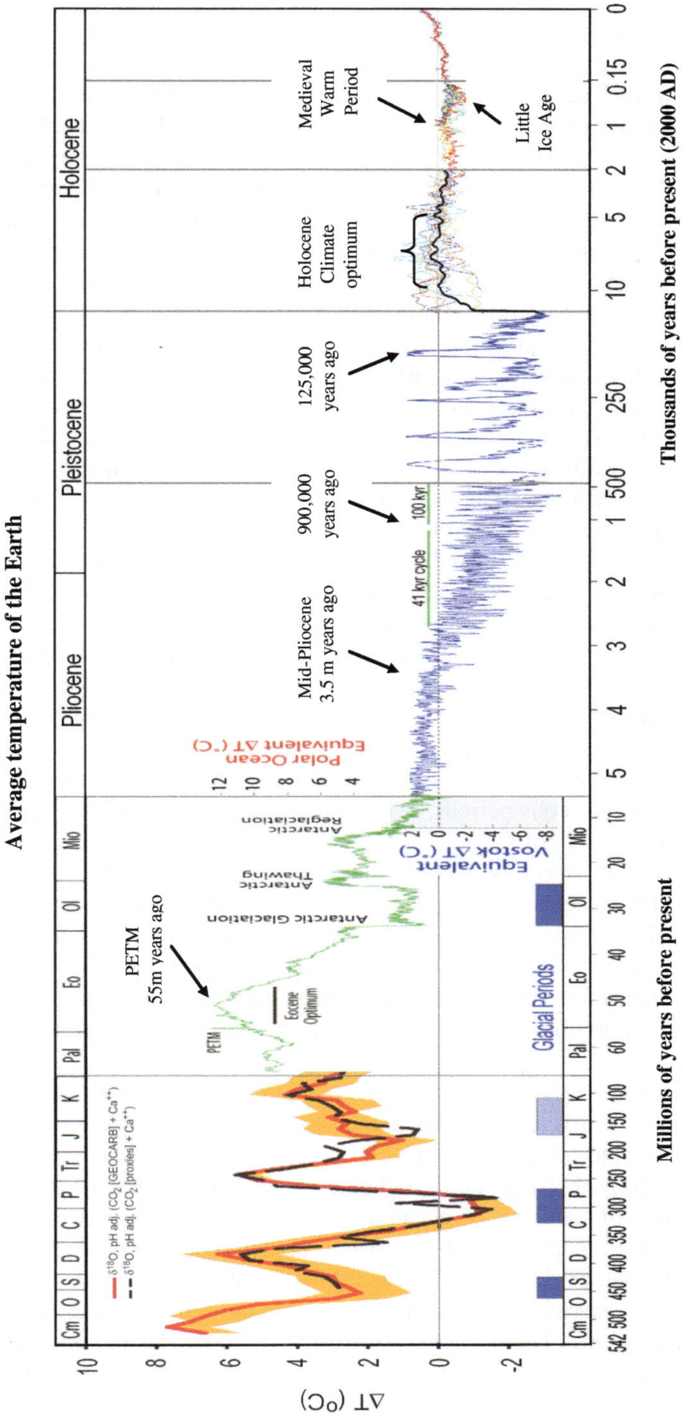

Fig. 1.1 Palaeotemperature record of the Earth. *Source* Modified version of http://en.wikipedia.org/wiki/File:All_palaeotemps.png. *Notes* Present refers to 2000 AD. The 'equivalent Vostok ΔT (°C)' and 'Polar Ocean Equivalent ΔT (°C)' temperature scales are polar, not global. They have been compressed to compensate. The temperature relativities are approximate

In most instances, these warmer climates were associated with very high CO_2 levels (IPCC 2007b). Having said this, considerable uncertainty surrounds the data on the greenhouse gas concentrations that existed more than one million years ago. Whereas data on more recent greenhouse concentration levels can be accessed via ice core samples, proxy indicators are required to obtain data on pre-Quaternary CO_2 levels. One of these proxies is based on the recognition that biological agents in soils and seawater have carbon isotope ratios distinct from the atmosphere (Cerling 1991; Freeman and Hayes 1992; Pagani et al. 2005). A second proxy involves the estimation of a boron isotope ratio (Pearson and Palmer 2000), while a third proxy is based on the empirical relationship between stomatal pores on tree leaves and the concentration of atmospheric CO_2 (McElwain and Chaloner 1995; Royer 2003). Although the use of these proxies has generated a wide range of reconstructed CO_2 values, their magnitudes are generally higher than the interglacial, pre-industrial CO_2 levels found in recently acquired ice core samples (IPCC 2007b). In other words, estimations of the CO_2 levels that prevailed during more distant times are invariably higher than the CO_2 levels present during the one million-year period leading up to the Industrial Revolution.

Further evidence of the general warmth of pre-Quaternary climates can also be gleaned from indicators of erstwhile continental ice. These indicators reveal that the Earth has been mostly ice-free during its geological history (IPCC 2007b). The two exceptions are the major expansion of Antarctic glaciation that began at the end of the Eocene epoch (35–40 million years ago) and the major glaciations that ushered in the Permian period around 300 million years ago. Both episodes coincided with low CO_2 concentration levels relative to adjacent periods. Conversely, the warmth of the Mesozoic era (65–250 million years ago), which included the Triassic, Jurassic, and Cretaceous periods, was greatly influenced by the presence of very high levels of atmospheric CO_2.[8]

One of the most abrupt warming events in the palaeoclimate record occurred at the interface of the Palaeocene and Eocene epochs, some 55 million years ago. Within a very short period of time, mean global temperatures rose by several degrees (Kennett and Scott 1991; Zachos et al. 2003; Tripati and Elderfield 2004). This event, often referred to as the Palaeocene-Eocene Thermal Maximum (PETM), endured for approximately 100,000 years (IPCC 2007b). The PETM is indicated by a sharp but brief temperature spike in Fig. 1.1.

What caused this unusually rapid rise in temperatures? Carbon isotopes in marine and continental records reveal that a large mass of carbon was released into the Earth's atmosphere and oceans. Not so well understood are the exact sources of this carbon (IPCC 2007b). Three main candidates stand out: (i) methane (CH_4) from the decomposition of clathrates on ocean floors; (ii) CO_2 from volcanic activity; and (iii) oxidation of sediments rich in organic matter (Dickens et al. 1997; Kurtz et al. 2003; Svensen et al. 2004). The PETM is the subject of intense study, not just because of the uncertainty surrounding its exact cause and its destructive impact on global ecosystems (Thomas 2003; Bowen et al. 2004; Harrington et al. 2004a), but because the quantity of carbon released at the time (around 1,000–2,000 Gigatonnes) is similar to the quantity of greenhouse gases expected to be released by humans during the 21st century (Dickens et al. 1997). Moreover, the period of recovery—that is, the period of time it took to reverse the PETM through natural sequestration processes—is in line with the forecast recovery period should humankind take the action needed to stabilise greenhouse gases at something near present concentration levels (IPCC 2007b). Overall, and despite great uncertainty regarding climate sensitivity during the PETM, enough is known for the event to serve as a conspicuous and salient example of the close relationship between extreme climate warming and a massive carbon release.[9]

Moving along the palaeoclimate record, the mid-point of the Pliocene epoch (3.5 million years ago) is the most recent time in the Earth's history when sustained global temperatures were substantially higher than at present. Around this time, temperatures were 2–3 °C above immediate pre-industrial levels (Haywood et al. 2000; Jiang et al. 2005). With respect to the current climate change issue, many observers believe there are two good reasons why the climate of the Mid-Pliocene constitutes a useful reference point. Firstly, given the hitherto 0.8 °C rise above pre-industrial temperatures and the predicted increase in global temperatures over the 21st century, the Earth's climate in 2100 is expected to be similar to the climate of the Mid-Pliocene (IPCC 2007b). Secondly, the geological arrangement of the continents and ocean basins during the Mid-Pliocene was not dissimilar to its present configuration. When all factors are considered, the average setting during one of the warmest periods during the Mid-Pliocene paints a cautionary picture—one in which the concentration of atmospheric CO_2 was typically between 360 and 400 ppm (Raymo and Rau 1992; Raymo et al. 1996); where sea levels were at least 15–25 metres above their current levels (Dowsett and Cronin 1990; Shackelton et al. 1995); and where snow and ice coverage was much less extensive than it is today (Guo et al. 2004). I say "cautionary picture" given that CO_2 levels, at around 400 ppm, are already at the upper end of the Mid-Pliocene concentration range. In addition, if a 15–25 metre sea level rise was to be replicated in the near future, huge areas of land would be inundated; some of the world's largest cities would be threatened; and hundreds of millions of people worldwide would be adversely affected (Stern 2007). The diminished ice coverage would also reduce the glacial meltwater that constitutes the principal water supply for millions of people in the Himalayas (e.g., western China and northern India) and the South American Andes.

Around 3 million years ago, the relative warmth of the Mid-Pliocene gradually gave way to colder temperatures. The colder climate led to the formation of the Arctic ice cap and the expansion of ice sheets in Antarctica. Although the climate record of this time—drawn from sediment cores from ocean floors and deep lake beds—is difficult to correlate, most agree that the colder temperatures were primarily driven by Milankovitch cycles that were amplified by changes in greenhouse gas concentrations (IPCC 2007b). In this sense, the variation in greenhouse gases acted as a positive feedback agent rather than a trigger of climate change. Towards the end of the Pliocene epoch, mean temperatures fell further and aridity increased. At the same time, the oscillations in temperature and aridity between the glacial and interglacial phases broadened significantly (see Fig. 1.1). Forests consequently declined and grasslands and savannas proliferated.

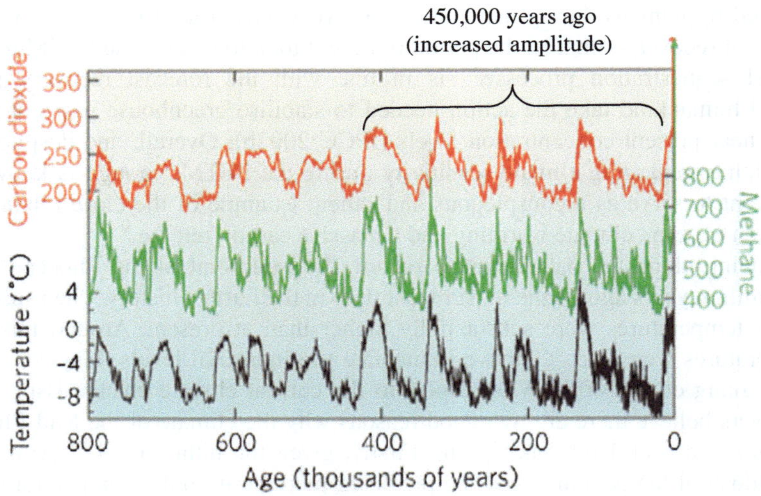

Fig. 1.2 The 800,000-year record of atmospheric carbon dioxide (*red* parts per million), methane (*green* parts per billion), and temperature (*black* relative to the average over the past millennium). *Source* Brook (2008)

Global temperatures continued to decrease with the onset of the Pleistocene epoch around 2.6 million years ago. Because of the very low average temperatures and the extent to which temperatures oscillated at this time, the climate of the Pleistocene epoch (11,500–2.6 million years ago) was characterised by repeated glacial cycles. At its peak, ice covered an estimated 30 per cent of the Earth's surface and continental glaciers often extended as far as the 40th parallel. This resulted in the locking up of huge volumes of water that lowered sea levels by around 100 metres.

Two intriguing changes occurred during the Pleistocene epoch (see Figs. 1.1 and 1.2). The first was the shift in the greenhouse gas-temperature cycle from 41,000 years to 100,000 years that occurred around 900,000 years ago. The second was the increase in the amplitude of the greenhouse gas-temperature cycle. Around 450,000 years ago, the interglacial peaks became warmer whereas the glacial troughs remained as cold as they had been at the end of the Pliocene epoch (Brook 2008).

The causes of both shifts are still under investigation. The most common explanation for the first shift is the possible reduction in the mean concentration of greenhouse gases around 900,000 years ago. Of course, if true, this raises the question as to what caused the decline in atmospheric greenhouse gas concentrations (Berger et al. 1999; Clark et al. 2006). This question is yet to be resolved. As for the second shift, Lüthi et al. (2008) have speculated that the full completion of the global CO_2 cycle may have somehow been extended to a period of 400,000–500,000 years.[10] As Brook (2008) highlights, this time-frame coincides with the larger 413,000 year change in the eccentricity of the Earth's orbit, a major determining factor of the Milankovitch cycle.[11] In both cases, more definitive answers are unlikely to emerge until a more extensive greenhouse gas-temperature record becomes available.[12] Having said this, it is undeniably clear from Fig. 1.2 that a tight coupling has existed over the past 800,000 years between global temperatures and the two key greenhouse gases of CO_2 and methane (CH_4). As previously mentioned, this does not imply that changes in atmospheric greenhouse gas concentrations were the driving force behind climate change (i.e., correlation does not imply causation). But it does suggest that greenhouse gases have continued to play a major positive feedback role in the Earth's climate system despite there being great uncertainty regarding the quantitative and mechanistic explanations of CO_2 variations (IPCC 2007b, 2014b).[13]

Returning to Fig. 1.1, it can be seen that the last great warming or interglacial period occurred around 125,000 years ago and that the end of the Pleistocene epoch marks the conclusion of the last glacial maximum.[14] During the last interglacial period—a short, warm period that was just 14,000 years in length—temperatures were 6 °C higher at the poles and 2 °C higher at the Equator. Global ice volumes were correspondingly low and oceans were up to 4–6 metres higher than at present (Plimer 2009).

Around 116,000 years ago, the warm conditions of the last interglacial abruptly ended. Evidence suggests that glaciation took hold within 400 years. Compared to present circumstances, air temperatures around the last glacial maximum were on average 5 °C colder and snowlines were 900 metres lower. Tropical sea surface temperatures were also 3 °C cooler. Once again, orbital factors were the principal driving forces behind the interglacial warming and the subsequent glacial cooling. Notwithstanding this, greenhouse gases continued to serve as a positive feedback influence on the Earth's climate.

During the final 1300 years of the Pleistocene epoch, just as the Earth was beginning to re-warm, a brief cold period emerged. Known as the Younger Dryas stadial (11,500–12,800 years ago), glacial conditions rapidly returned to the higher latitudes of the Northern Hemisphere (Houghton 1994).[15] Temperatures in parts of Greenland plunged to 15 °C below their present levels, wind strength increased, and the distribution of heat was significantly altered (Plimer 2009). Following the Younger Dryas event, which heralded the start of the current Holocene epoch, temperatures rapidly increased. Ice core records indicate that Arctic warming of around 7 °C occurred in just 50 years, with half of the increase transpiring in a 15-year period (Alley 2000).

Although questions remain as to whether the Younger Dryas event was uniformly global, there is no doubt that nothing of its type, nor the abrupt temperature increase that followed, has since been experienced. The prevailing theory suggests that the Younger Dryas was triggered by a shutdown of the North Atlantic thermohaline circulation—a large-scale oceanic circulation that transfers warm, low-density waters from the tropical North Atlantic to its colder, northern extremes. It its believed that this dramatic change was due to a freshwater influx into the North Atlantic Ocean caused by a breach in the glacial dam wall of Lake Agassiz in North America (Clark et al. 2002; Tarasov and Peltier 2005; IPCC 2007b). However, there are some observers who believe this theory is incomplete in that it fails to explain why cooling in the Southern Hemisphere preceded the Younger Dryas by as much as 1,000 years (Thompson et al. 2000). Clement and Cane (1999) have proposed that the abrupt temperature change may have been triggered by tropical ocean factors. In all, a consensus has yet to emerge as to what caused the Younger Dryas event or what brought about its sudden conclusion.

Following a gradual increase in global temperatures during the early stages of the Holocene epoch, temperatures peaked during a period known as the Holocene Climate Optimum (HCO). Spanning 4,000 years (5,000–9,000 BC), average global temperatures during the HCO were, despite its name, slightly below current levels. However, the lower latitudes were warmer than at present, with temperatures considerably higher near the North Pole and in Northwestern Europe.

Despite the relative warmth of the HCO, an abrupt cooling spell again took place about 8,200 years ago (6,200 BC). It would endure for 200–400 years. Milder than the Younger Dryas, the so-called 8.2 Kiloyear event was global in its extent, with a pronounced impact on sea levels (Alley and Ágústsdóttir 2005; Sarmaj-Korjonen 2007). It has been estimated that circum-North Atlantic temperatures fell by 1–5 °C (Rohling and Pälike 2005); atmospheric CO_2 concentrations declined by 25 ppm (Wagner et al. 2002); and sea levels decreased by as much as 14 metres. Similar to the popular theory behind the onset of the Younger Dryas, it is generally believed that the 8.2 Kiloyear event was caused by a slowdown of the North Atlantic thermohaline circulation—this time the consequence of a large meltwater pulse arising from the collapse of the Laurentide ice sheet in North America (Barber et al. 1999; Clarke et al. 2004; Ellison et al. 2006).

Not only was the climate colder during the 8.2 kiloyear event, it was markedly drier (Plimer 2009). East Africa, for example, experienced a prolonged and severe drought, whilst West Asia and Mesopotamia endured a 300-year aridification episode. It seems probable that the latter induced the development of irrigation which, for the first time, generated a consistent surplus of agricultural output. Many anthropologists believe this led to the earliest forms of urban civilisation.

Around 200 years into the 8.2 Kiloyear event, milder climatic factors reasserted themselves. By 5800 BC, the Earth's climate had returned to pre-event conditions. Ocean circulation changes appear to have been the main factor behind the end of the 8.2 Kiloyear event, although disagreement exists over the exact cause and nature of these changes (IPCC 2007b). Disagreement aside, the Younger Dryas and 8.2 Kiloyear events indicate how vulnerable the Earth's climate is to perturbations within the climate system, irrespective of their source (Steffensen et al. 2008; Schellnhuber 2008).

To this very day, minor fluctuations between cooler and warmer climes have continued. However, if one focuses on the past 2,000 years, three distinct climate regimes can be identified. They are: (i) the Medieval Warm Period (900–1300 AD); (ii) the Little Ice Age (1300–1850 AD); and (iii) the period of anthropogenic global warming (1850 AD to the present). During the Medieval Warm Period, temperatures in the Northern Hemisphere were, at times, higher than today. On occasions, agriculture thrived in the European Alps. Vineyards also existed in England and Germany in locations no longer conducive to grape growing (Plimer 2009). Because the North Atlantic Ocean was largely ice-free, Vikings were able to colonise Greenland, Iceland, and North America (Diamond 2005). The warmer conditions enabled the Vikings to grow barley and raise sheep in Greenland as well as establish vineyards in Newfoundland ('Vinland').

It was originally believed that the Medieval Warm Period was a global phenomenon. There is now increasing evidence to suggest it was not. In fact, global temperature records extracted from various sources reveal that the Earth was slightly cooler during the Medieval Warm Period than in the early and mid-20th century (Crowley and Lowery 2000; Bradley et al. 2003). As such, medieval warmth appears to have been restricted to areas neighbouring the North Atlantic Ocean (IPCC 2001b).

The Little Ice Age was not an ice age in the strict sense. It was a stadial or cool event within the current interglacial. During the Little Ice Age, agricultural production in Europe declined and human population numbers correspondingly fell (Lamb 1982; Fagan 2001). Although Mann et al. (1998) and Jones et al. (1998) have shown that the Little Ice Age was the Northern Hemisphere's coldest period during the last millennium, the cooling was modest with temperatures declining by less than 1 °C relative to late-20th century levels (Bradley and Jones 1993; Crowley and Lowery 2000). Not unlike the Medieval Warm Period, the Little Ice Age was predominantly felt in the North Atlantic region. The best explanation for the low North Atlantic temperatures appears to be an alteration in the patterns of atmospheric circulation—in particular, an enhanced easterly wind phase of the North Atlantic Oscillation (O'Brien et al. 1995).[16]

Another contributing factor to the low temperatures experienced during the Little Ice Age was the extensive volcanic activity that punctuated the period (Robock 1979). There are two ways that volcanic activity can reduce temperatures. Firstly, the resultant spread of ash in the upper atmosphere can block some of the incoming solar radiation. Secondly, when the sulphur oxide (SO_2) emitted by volcanic eruptions reaches the stratosphere, it converts to sulphuric acid particles. These particles reflect sunlight, thus further reducing the intensity of incoming solar radiation. One example of the cooling effect of volcanic activity during the Little Ice Age was the 1815 eruption of Mt. Tambora in Indonesia. It led to the 'year without a summer' in 1816, the coldest year on record in the USA.

It has also been suggested that the low intensity of solar radiation contributed to the low temperatures during the Little Ice Age. There were four periods during the Little Ice Age when the Sun emitted much less than its usual energy. These are referred to as the Wolf Minimum (1280–1340 AD), the Spörer Minimum (1450–1540 AD), the Maunder Minimum (1645–1715 AD), and the Dalton Minimum (1795–1825 AD) (Lamb 1982; Ribes and Nesme-Ribes 1993; Crowley 2000). Although these minima corresponded with cold spells, they also coincided with very warm temperatures, such as the extreme heat of 1685–86, which occurred during the Maunder Minimum (Plimer 2009). An expanding body of evidence suggests that the relationship between low solar activity and cool temperatures is not well understood (National Research Council 2005). It is therefore difficult to conclude that the variation in solar radiation greatly influenced temperatures during the Little Ice Age.

From a climate change perspective, the following best sums up the significance of the Medieval Warm Period and the Little Ice Age:

> [C]urrent evidence does not support globally synchronous periods of anomalous cold or warmth [….] and the conventional terms of 'Little Ice Age' and 'Medieval Warm Period' appear to have limited utility in describing trends in hemispheric or global mean temperature changes in past centuries (IPCC 2001b, p. 135).

1.2.3 Anthropogenic Global Warming—The Enhanced Greenhouse Effect

Of contemporary importance to humankind is the third identifiable period over the past 2,000 years—namely, the period since 1850. As Fig. 1.3 shows, average

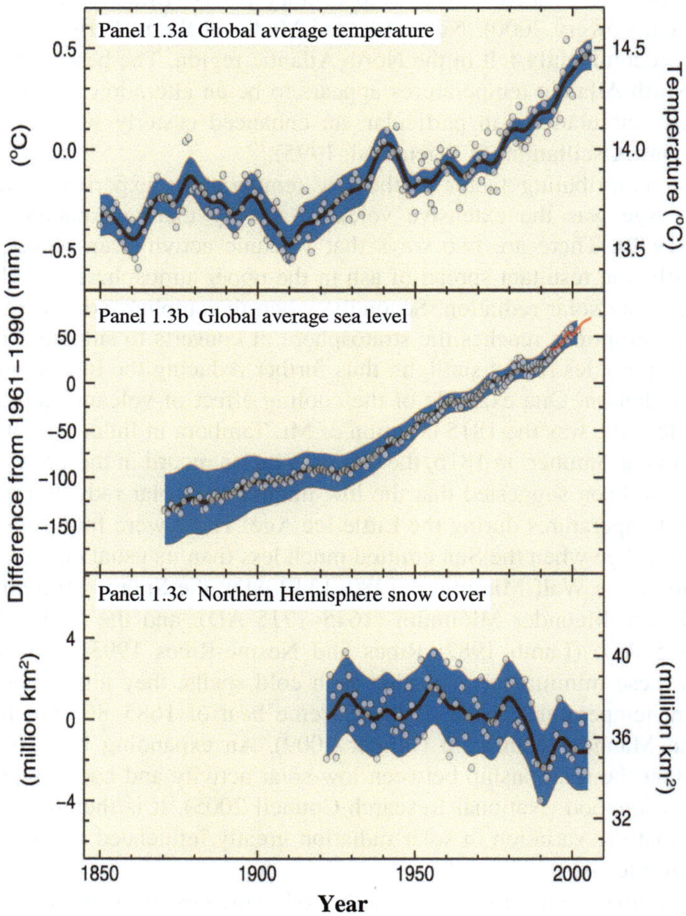

Fig. 1.3 Changes in temperatures, sea level, and Northern Hemisphere snow cover. *Source* IPCC (2007a, p. 31) (reproduced with kind permission from the IPCC)

global surface temperatures, particularly since 1950, have increased alarmingly (Panel 1.3a). Furthermore, sea levels have risen steadily, whilst the coverage of snow and ice in the Northern Hemisphere has receded (Panels 1.3b and 1.3c).

By recent historical standards, the latter part of the 20th century was unusually warm. Records indicate that the second half of the 20th century was the Northern Hemisphere's warmest 50-year period in the last 1,300 years. What's more, eleven of the twelve years from 1995–2006 (inclusive) ranked among the twelve warmest years in the instrumental record of global surface temperatures (IPCC 2007a).[17] Overall, since 1850, the average global surface temperature has increased by at least 0.8 °C and is likely to rise by a further 1 °C regardless of what action is taken to curb climate change (IPCC 2007b).

The dramatic recent rise in average global temperatures is further exemplified by Fig. 1.4. This figure is the same as Panel 1.3a except that linear trend-lines are added to reveal the extent of the temperature rise over four periods leading up to and including 2005. The 25, 50, 100, and 150-year trend-lines respectively correspond to the periods 1981–2005, 1956–2005, 1906–2005, and 1856–2005.[18] As can be seen from Fig. 1.4, the trend-lines get steeper as the periods leading up to 2005 shorten. This indicates that the average per decade increase in temperature has been escalating (0.045 °C for the 150-year period; 0.074 °C for the 100-year period; 0.128 °C for the 50-year period; and 0.177 °C for the 25-year period). Putting this into sharper perspective, the average rate of warming over the 1956–2005 period was 73 per cent faster than it was for the 100-year period from 1906 to 2005. In all, Fig. 1.4 indicates that the rate of temperature increase since 1850 has itself been rising.

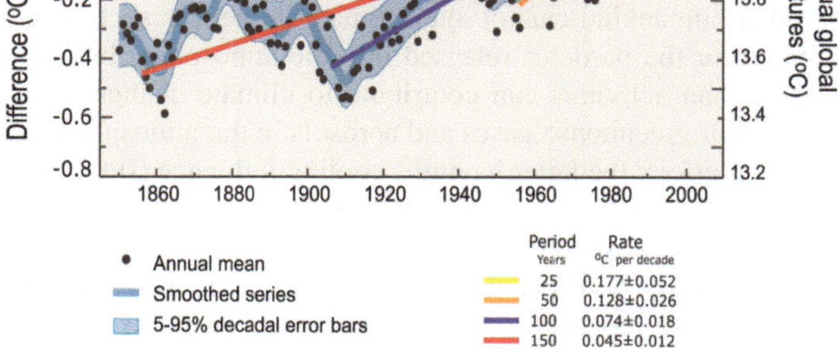

Fig. 1.4 Annual global mean observed temperatures (*black dots* and smoothed series) and linear trends for the last 25, 50, 100, and 150 years. *Source* IPCC (2007b, p. 253) (reproduced with kind permission from the IPCC)

Table 1.1 Concentration levels, global warming potentials, lifetimes, and radiative forcing of greenhouse gases

Gas type	1750 concentration (ppm)	2005 concentration (ppm)	100-year global warming potential	Atmospheric life-time (years)	2005 radiative forcing (Wm^{-2})
Carbon dioxide (CO_2)	278	379	1	~100	1.66
Methane (CH_4)	0.70	1.77	21	12	0.48
Nitrous oxide (N_2O)	0.27	0.32	310	114	0.16
CFCs total	0	8.68×10^{-10}	3,800–8,100	45–1,700	0.268
HCFCs total	0	2.02×10^{-10}	90–1,800	90–1,800	0.039
HFCs total	0	0.61×10^{-10}	140–11,700	1.3–17.9	0.010
PFCs total	0	0.77×10^{-10}	6,500–9,200	10,000–50,000	0.004
Sulfur hexafluoride (SF_6)	0	0.06×10^{-10}	23,900	3,200	0.003
Others	0	1.12×10^{-10}	Various	Various	0.01
Total long-life greenhouse gases	**278.97**	**381.09**			**2.63**
2005 CO_2-equivalent concentration of long-life greenhouse gases (ppm)					

Notes
- CFCs denotes chlorofluorocarbons
- HCFCs denotes hydrochlorofluorocarbons
- HFCs denotes hydrofluorocarbons
- PFCs denotes perfluorocarbons

own calculation

Source IPCC (2007b), Tables 2.1 and 2.14

As will soon be revealed, most of the increase in global temperatures since the mid-20th century can, unlike temperature changes in the past, be attributed to human activities (IPCC 2007a). It is for this reason that the current warming phase is often referred to as the *enhanced greenhouse effect*. Of course, in saying this, it is obvious that humankind cannot alter climate influences such as the intensity of solar radiation or the particles released into the atmosphere by volcanic activity. However, human activities can contribute to climate change by augmenting the concentration of greenhouse gases and aerosols in the atmosphere—the former having a warming effect; the latter having a cooling influence (IPCC 2007b). More so indirectly than directly, human activities can also increase the amount of water vapour in the Earth's atmosphere. This, too, can impact on the Earth's climate.

Table 1.1 and Fig. 1.5 reveal that the concentration of greenhouse gases in the Earth's atmosphere has grown dramatically since 1750. As at 2005, the concentration of carbon dioxide had increased from 278 ppm in 1750 to 379 ppm. It now (2014) stands at 398 ppm. Although the rise in the concentration of nitrous oxide over the 1750–2005 period was modest (0.27–0.32 ppm), the concentration

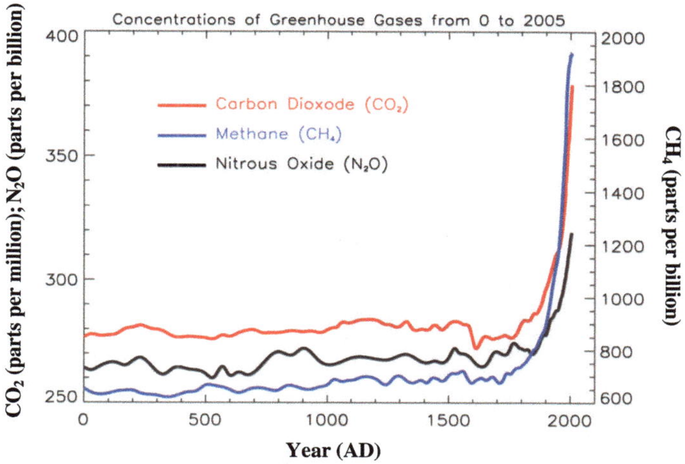

Fig. 1.5 Atmospheric concentrations of three crucial long-lived greenhouse gases over the last two millennia. *Source* IPCC (2007b, p. 135) (reproduced with kind permission from the IPCC)

of methane had more than doubled (0.70–1.77 ppm). Overall, the combined concentration of long-life greenhouse gases increased from 279.0 ppm in 1750 to 381.1 ppm by 2005—a rise of 37 per cent over the period.

The increase in the atmospheric concentration of greenhouse gases was almost entirely due to humankind's burning of fossil fuels, vegetation clearance, agricultural expansion, fertiliser use, and the growth in halocarbon production (Cunnold et al. 2002; Brovkin et al. 2004; Matthews et al. 2004; Marland et al. 2006; IPCC 2007b, 2014b). A question that often arises at this point is: To what extent is the observed change in global temperatures a consequence of the anthropogenic increase in greenhouse gas concentrations? In other words, is it possible that most if not all the recorded increase in global temperatures can be attributed to natural factors? To answer these questions, it is necessary to consider the 'radiative forcing' of the various climate-influencing factors and the extent to which these factors have changed since 1750.

Radiative forcing is a measure of how the energy balance of the Earth-atmosphere system is altered by climate-influencing factors. The term 'radiative' is used because climate-influencing factors alter the balance between the incoming and outgoing radiation within the Earth's atmosphere (IPCC 2007b). As explained earlier, it is the radiative balance within the Earth's atmosphere that determines global temperatures. The term 'forcing' is used in conjunction with 'radiative' to indicate how much and in what direction the Earth's radiative balance is being forced from its normal state.[19]

Radiative forcing is quantified as the "rate of energy change per unit area of the globe as measured at the top of the Earth's atmosphere". It is expressed in Watts per square metre (Wm^{-2}). When the radiative forcing from a climate-influencing factor or group of factors is positive, the energy of the Earth-atmosphere system

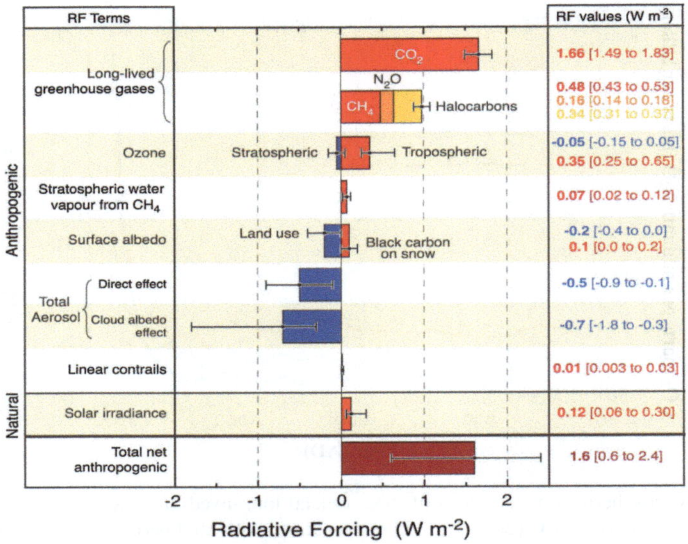

Fig. 1.6 Principal components of the radiative forcing (RF) of climate change. *Note* Values represent radiative forcings in 2005 relative to 1750 (start of industrial era). *Source* Truncated version of IPCC (2007a, Fig. 2.4, p. 39)

increases, thus leading to a warming of the system (IPCC 2007b). A negative value implies a reduction in energy and a cooling of the Earth-atmosphere system.

The importance of radiative forcing is illustrated in Table 1.1. Despite carbon dioxide being the most prevalent greenhouse gas in the atmosphere, its dominance is not proportionately reflected in its radiative forcing value of 1.66 Wm^{-2}. This is because the 100-year global warming potential of carbon dioxide is the least of all the long-life greenhouse gases.[20] For example, one tonne of sulphur hexafluoride has 23,900 times the global warming potential of one tonne of carbon dioxide.

Because of the different global warming potentials of each greenhouse gas, the combined warming potential of all greenhouse gases is often expressed in terms of its CO_2-equivalence. To obtain such a measure, the quantity of each greenhouse gas is multiplied by its global warming potential. The resultant value is converted to a quantity of carbon dioxide with the equivalent warming effect. Due to the more potent influence of non-CO_2 greenhouse gases, the total CO_2-equivalent concentration of all long-life greenhouse gases was 454.85 ppm in 2005 (see Table 1.1).[21] This is much higher than the physical concentration of all long-life greenhouse gases, which, to recall, was 381.1 ppm in 2005.[22] Table 1.1 also reveals that the radiative forcing of all long-life greenhouse gases in 2005 was 2.63 Wm^{-2}.

The relevant feature of the total radiative forcing between 1750 and 2005 is that most of it can be attributed to human activities. Figure 1.6 shows that the net anthropogenic impact on the Earth's climate since 1750 equated to a radiative forcing of 1.6 Wm^{-2}. Natural influences amounted to a mere 0.12 Wm^{-2}. This means that, since pre-industrial times, the human contribution to global warming exceeded natural forces by a factor of thirteen. It could easily have been more. The value of 1.6 Wm^{-2} is lower than the value of 2.63 Wm^{-2} in Table 1.1 because, as Fig. 1.6 indicates, the warming effect of all long-life greenhouse gases was tempered by land-use changes and the generation of aerosol particles. Indeed, the cooling effect of these two human-induced processes was such that the net effect of all anthropogenic forcing agents in 2005 was akin to a concentration of 374.91 ppm of CO_2-equivalent greenhouse gases.[23]

It is also important to note that the cooling effect of aerosols peaked in the 1950s and 1960s—a time when average global temperatures temporarily levelled off (IPCC 2007b).[24] If, as expected, the rising concentration of greenhouse gases continues to overwhelm the generation of aerosols, the warming influence of all anthropogenic forcing agents will continue to escalate.

I should point out that there is much confusion over what constitutes the most appropriate CO_2-equivalent concentration of greenhouse gases when referring to target stabilisation levels. We have already seen that there is a significant difference between the total CO_2-equivalent concentration of all long-life greenhouse gases (454.85 ppm in 2005) and that of all anthropogenic forcing agents (374.91 ppm in 2005). Because many forcing agents are short-lived, it is widely believed that the use of the lower CO_2-equivalent value leads to a drastic underestimation of the emissions cuts necessary to stabilise greenhouse gases at a specific target level. At the same time, the higher CO_2-equivalent value includes CFCs and HCFCs that are rapidly being phased out in line with obligations under the Montreal Protocol. For these reasons, Stern (2007) and others believe that the most relevant CO_2-equivalent concentration is that representing the radiative forcing of the six Kyoto Protocol gases (i.e., carbon dioxide (CO_2), methane (CH_4), nitrous oxide (N_2O), hydrofluorocarbons (HFCs), perfluorocarbons (PFCs), and sulphur hexafluoride (SF_6)). The combined radiative forcing of these six gases stood at 428.68 ppm of CO_2-e in 2005, although it was closer to 440 ppm in 2014.[25] Unless otherwise specified, all future references in this book to CO_2-equivalent concentrations of greenhouse gases will imply a reference to the radiative forcing of the six original Kyoto Protocol gases.[26]

Returning to the radiative forcing values in Fig. 1.6, these are computed by using climate change models to determine the warming and cooling influences of the individual radiative forcing components. A further question that often arises at this point is: How reliable are the climate change models? Like all models, climate change models are simplistic representations of reality. Nevertheless, they have become increasingly sophisticated and incorporate key features of the critical feedback mechanisms found within the climate system. This aside, one of the best ways to support a model is to conduct simulation exercises to replicate known outcomes. When these exercises are undertaken with just natural factors taken into account, climate change models are unable to reproduce the steep temperature rises observed in recent decades. However, as the pink-shaded series in Fig. 1.7 illustrates, climate change models are able to consistently simulate the observed 20th-century variations in temperatures *when they include human influences* (IPCC 2007b). By doing

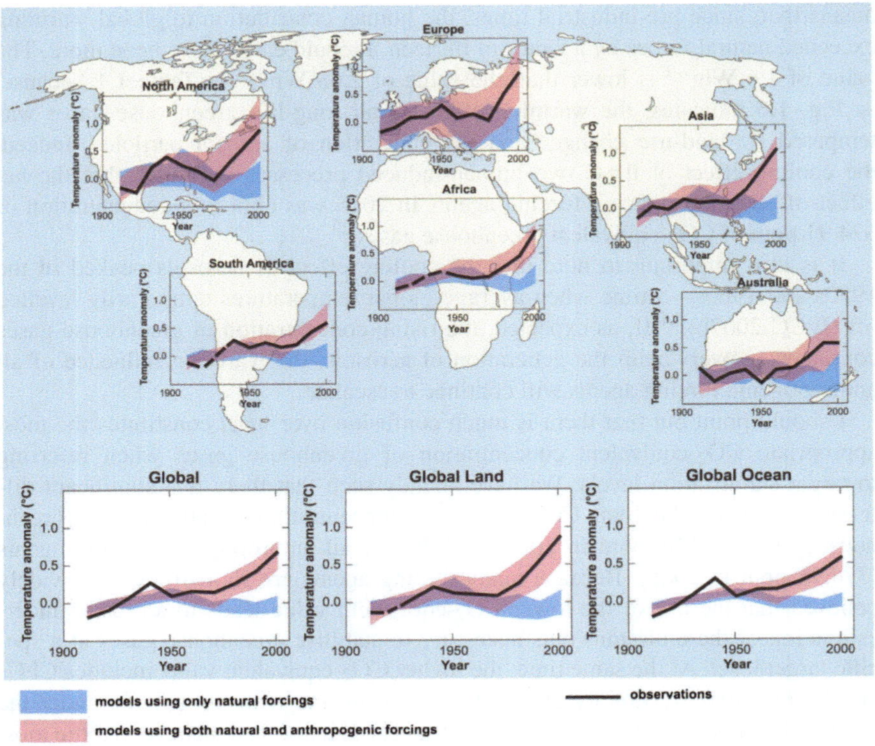

Fig. 1.7 Temperature changes relative to the 1901–1950 average. *Source* IPCC (2007b), FAQ 9.2, Fig. 1, p. 703 (reproduced with kind permission from the IPCC)

so, these models demonstrate that anthropogenic forcings have dominated all other potential causes of temperature rise during the past half-century.

There are two additional means of support for the explanatory power of climate change models. The first comes in the form of a consistent pattern of simulated results which indicate that more pronounced warming can be expected over land compared to oceans (see bottom of Fig. 1.7). This pattern of variation, which closely follows observed patterns of change, differs markedly from temperature changes that would normally be associated with natural forces, such as the El Niño phenomenon. The second is the similarity between climate change models and actual observations in terms of lower-atmospheric warming (troposphere) and upper-atmospheric cooling (stratosphere). It has been shown that if an increase in solar radiation had been mostly responsible for the recent rise in global temperatures, both the troposphere and stratosphere would have warmed. These and other considerations now provide us with the confidence that climate change models are rigorous and that anthropogenic influences have been the dominant source of the global warming observed over the past 50 years (IPCC 2007b).

In summary, climate change is an ongoing phenomenon that, until recently, had been exclusively caused by dynamic variations within the Earth's climate system or exogenous shocks of a natural cause. From time to time, a change in the Earth's climate has been initiated by a change in the atmospheric concentration of greenhouse gases. More often than not, however, the role of greenhouse gases has been confined to that of a positive feedback influence on the climate system.

The recent change in the Earth's climate system is unique insofar as the increase in the atmospheric concentration of greenhouse gases is almost entirely due to the proliferation of human industrial and agricultural activities since 1750. In this sense, humankind has elevated greenhouse gases to that of a primary trigger of climate change. Although the temperature increase over the past century has not been as abrupt as some of the temperature changes in the recent past (e.g., the Younger Dryas and 8.2 Kiloyear events), of particular concern is that greenhouse gas concentrations have risen well beyond the upper end of their range over the past 800,000 years. This means that a contemporary climate system parameter is now well outside its normal range of variation, thus pushing the climate system to a point where extreme changes are possible in the future (IPCC 2007b, 2014b).

Whilst uncertainties surround the explanation for the natural variation in CO_2 concentrations, the following seems certain: (i) despite short-term fluctuations, global temperatures have risen since 1850; (ii) an increase in atmospheric greenhouse concentrations can be closely associated with this global warming; (iii) the current rise in greenhouse gas concentrations has been facilitated by human activities; and (iv) the radiative forcing of human actions has far exceeded that of natural factors. There is, therefore, little doubt that the very nature and extent of humankind's recent endeavours have played a dominant role in the climate change that has occurred over the past century.

1.2.4 Climate Change Projections

Given humankind's impact on the Earth's climate so far, the next important issue of consideration is: What is the possible future impact of human activities on the Earth's climate? Ignoring the long-term implications of Milankovitch cycles, humankind's future impact will depend on anthropogenic greenhouse gas emissions, aerosol production, feedback effects, the influence of natural processes (which may accentuate or dampen humankind's influence), and the rate of greenhouse gas sequestration—the latter of which may be boosted by land-use changes (e.g., reafforestation) and technological advances (e.g., carbon capture-and-storage). Since these factors will ultimately determine the net effect of all anthropogenic forcing agents, considerable work has been undertaken to analyse the climate change impact of various emission pathways. Much of this work has been based on a set of emissions scenarios developed in 1996 by the Intergovernmental Panel on Climate Change (IPCC).[27] Originally aimed at providing input into the *IPCC Third Assessment Report*, the emissions scenarios consist of four scenario families designed to explore the climate-changing effect of a wide range of economic, demographic, and technological factors. The following are the storylines of each scenario family (IPCC 2000):

- A1 family: a rapid increase in Gross World Product (GWP); substantial reductions in per capita income disparities; a global population that peaks in the middle of the 21st century and declines thereafter; increased social and cultural interaction; and a rapid introduction of resource-efficient and pollution-reducing technologies.
- A2 family: a slower growth rate of GWP; a continuously increasing global population; less social and cultural interaction than A1; and a slow rate of technological advancement.
- B1 family: global convergence; the same peaking of the global population as A1; rapid technological change; an emphasis on equity; and a concerted shift towards a services-dominated economy.[28]
- B2 family: a moderate growth rate of GWP; a global population that continuously increases, but at a slower rate than A2; and a more localised response to economic, social, and environmental issues.

Of these four scenario families, the A1 family is divided into three emissions groups involving different energy paths. These are: (i) a fossil fuel-intensive path (A1FI); (ii) a path that leans heavily towards the use of non-fossil fuels (A1T); and (iii) a balanced energy path (A1B). Given the sub-division of the A1 family, there are, altogether, six scenario groups. Within each scenario group, there is a further sub-division of scenarios. Many of these scenarios share 'harmonised' assumptions regarding the rate of population growth, GWP, and final energy consumption. The remaining scenarios explore uncertainties in relation to the links between output growth, energy use, and greenhouse gas emissions. All up, the complete IPCC scenario set comprises 40 different emissions scenarios (IPCC 2000).

The extent to which the climate change projections differ amongst the various emissions scenarios depends on how far the projection extends into the future. Until around 2025, global warming of approximately 0.2 °C per decade can be expected irrespective of the emissions scenario (IPCC 2007a). Indeed, even if all greenhouse gases and aerosols remained at year 2000 concentration levels, warming of around 0.1 °C per decade could be expected over the next two decades. Beyond 2025, however, climate change projections depend increasingly upon the emissions scenario in question. Notwithstanding this, it appears certain that greenhouse gas emissions at or above current rates will induce much larger temperature rises than those observed during the 20th century (IPCC 2007b).

At the time of the release of the *IPCC Fourth Assessment Report* (IPCC 2007b, c, d), advances in climate change modelling had made it possible to provide *likely* climate change projections for each of the above emissions scenarios.[29] The best estimates and the *likely* range of temperature increases for all six scenario groups are revealed in Table 1.2.[30] They are also graphically illustrated in Fig. 1.8 along with 20th-century observations and the temperature projections of a constant year-2000 concentration of greenhouse gases.

Table 1.2 Projected global average temperature increase by end of the 21st century (temperature change at 2090–2099 relative to 1980–1999)

Emissions scenario	Year 2100 CO_2-e GHG concentration (ppm)	Best estimate (°C)	Likely range (°C)
Constant year 2000 concentrations	430	0.6	0.3–0.9
B1	600	1.8	1.1–2.9
A1T	700	2.4	1.4–3.8
B2	800	2.4	1.4–3.8
A1B	850	2.8	1.7–4.4
A2	1,250	3.4	2.0–5.4
A1FI	1,550	4.0	2.4–6.4

Notes
- ppm denotes parts per million
- GHG denotes greenhouse gas

Source Adapted from IPCC (2007a), Table 3.1, p. 45

In view of past projections, two things can be learned from Table 1.2. Firstly, the projected range of temperature increase of 1.1–6.4 °C (i.e., from the low end of the B1 range to the high end of the A1FI range) is broadly consistent with the temperature span disclosed in the *IPCC Third Assessment Report* (1.4–5.8 °C). That said, the upper ranges of the temperature projections are larger than the *Third Assessment Report*, essentially because more recent climate change models indicate stronger climate-carbon cycle feedbacks (IPCC 2007a). It is because of these stronger feedbacks that heatwaves, heavy precipitation events, and tropical cyclones (hurricanes and typhoons) are *very likely* to become more frequent and more intense during the 21st century.[31]

Secondly, estimated temperature increases are closely linked to the year-2100 concentration of greenhouse gases. The major exception is the A1T scenario group which, despite a lower greenhouse gas concentration in 2100 than the B2 group, is likely to produce a similar temperature increase by the end of the 21st century. The reason for this is that A1T emissions scenarios involve a higher rate of greenhouse gas emissions early in the century followed by a considerably lower rate much later. This is likely to induce an earlier positive feedback response within the Earth's climate system (see IPCC 2007a, Fig. 3.1). Only as the 22nd century proceeds will the radiative forcing and the corresponding temperature increase associated with the B2 scenario group exceed that of the A1T group.

The insights provided by Table 1.2 and Fig. 1.8 are confined to climate change projections at the global scale. Looking at the regional level, warming is expected to be greatest over land and most northern latitudes, and to be least over the Southern Ocean and parts of the North Atlantic Ocean (Stouffer 2004; Fyfe and Saenko 2005; Hazeleger 2005; Kunkel and Liang 2005; IPCC 2007a). In addition, a poleward shift of extra-tropical storm tracks is anticipated with significant impacts on wind, precipitation, and temperature patterns (Saenko et al. 2005; Stone and Fyfe 2005). For example, while precipitation in the high latitudes

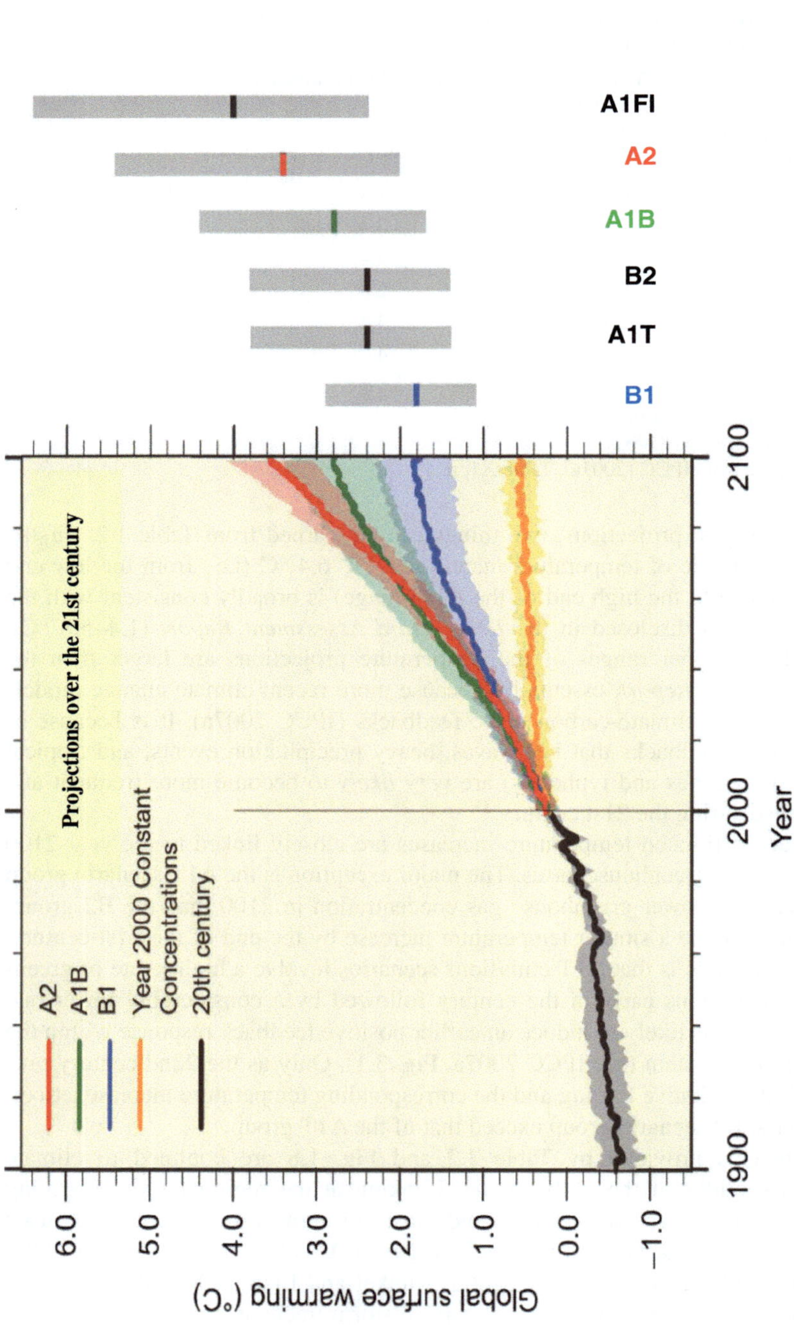

Fig. 1.8 Projected temperature changes for each emissions scenario relative to the 1980–1999 period. *Source* Adapted from IPCC (2007a), Fig. 3.2, p. 46 (reproduced in adapted form with kind permission from the IPCC)

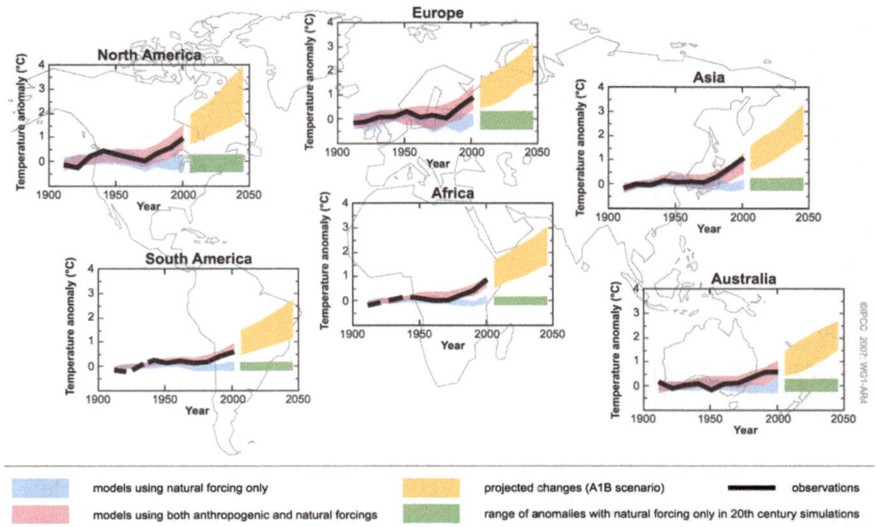

Fig. 1.9 Continental surface temperature anomalies: observations and projections. *Note* Anomalies based on temperature variations from the 1901–1950 temperature average. *Source* IPCC (2007b), Fig. TS.29, p. 75 (reproduced with kind permission from the IPCC)

is expected to increase, it is likely to decline in most sub-tropical land regions (Emori and Brown 2005; Wang 2005; Neelin et al. 2006; IPCC 2007a).

As a means of illustrating the likely disparate temperature increases across the globe, Fig. 1.9 summarises the projected temperature changes of six of the world's seven continents. Using A1B as a reference scenario, Fig. 1.9 shows that increases in temperature can be expected on all six continents. The figure also indicates that, up to 2050, temperature rises are likely to be strongest in North America, Europe, and Asia. The more moderate warming expected in Australia and South America can be partly explained by the projected lower rate of temperature increase in the Southern Ocean.

As useful as the IPCC emissions scenarios are, one of their major weaknesses is their inability to reveal the likely temperature changes for a range of stable concentration levels of greenhouse gases. This failure arises because, in the case of the A1B, A1FI, A2, and B2 scenario groups, concentration levels will still be rising sharply in the year 2100.

Knowledge of the probable temperature increase of specific stabilisation levels is important for three reasons. Firstly, global warming is expected to impact greatly on ecosystems, sea levels, agriculture, forestry, water supplies, human settlements, industry, and human health (IPCC 2007a, 2014c). Secondly, avoiding dangerous climate change will require the atmospheric concentration of greenhouse gases to be stabilised at or below certain levels. Finally, stabilising greenhouse gas concentrations at an appropriate level will necessitate adequate restrictions on future greenhouse gas emissions. Barring political obstacles, the latter is likely to dictate future greenhouse gas targets, emissions cuts, and associated policy measures (Stern 2007; Garnaut 2008; IPCC 2014b, c, d).

Table 1.3 Estimated temperature increase over pre-industrial temperatures for different stabilisation levels of CO_2-equivalent greenhouse gases

Equilibrium CO_2-equivalent (ppm)	Best estimate (°C)	Very likely above (°C)	Likely in the range (°C)
350	1.0	0.5	0.6–1.4
450	2.1	1.0	1.4–3.1
550	2.9	1.5	1.9–4.4
650	3.6	1.8	2.4–5.5
750	4.3	2.1	2.8–6.4
1,000	5.5	2.8	3.7–8.3
1,200	6.3	3.1	4.2–9.4

Note ppm denotes parts per million
Source IPCC (2007a), Table TS.5, p. 66

Best estimates of temperature increases and the likely temperature ranges for a variety of equilibrium greenhouse gas concentrations are provided in Table 1.3. What is evident from the table is that concentration levels above 450 ppm of CO_2-e are expected to increase global temperatures by more than 2 °C above pre-industrial (1750) levels. This is of particular concern given that, firstly, a 2 °C rise is considered by many to be the upper limit of a 'safe' temperature increase (Retallack 2005; Hamilton et al. 2005), and secondly, greenhouse gas concentrations are fast approaching the 450 ppm level.

What is not obvious from Table 1.3, but is of great significance, is that an upper bound temperature for each concentration level cannot be established. The reason for this is that feedback mechanisms within the Earth's climate system can change over time, either through ecosystem destruction or global warming itself, thus resulting in temperature rises well beyond estimated temperature ranges (IPCC 2007b). Moreover, as perturbations of the climate system progress, variations in climate feedbacks can occur unexpectedly. This can lead to abrupt climate changes of the kind described earlier in the chapter. Three possible triggers of abrupt climate change—the collapse of the West Antarctic Ice Sheet, the rapid loss of the Greenland Ice Sheet, and large-scale variations in ocean circulation systems (e.g., the North Atlantic thermohaline circulation)—are not expected to occur during the 21st century. However, should the perturbations continue into the 22nd century, the probability of their occurrence increases (IPCC 2007b). Moreover, the probability dramatically escalates in line with the severity of the perturbation—that is, as the stabilisation level of greenhouse gases and the associated temperature change rises. Clearly, the increasing risk of abrupt climate change must be considered when determining future emissions targets (Stern 2007; IPCC 2007b; Weitzman 2009).

1.2.5 The Potential Implications of Climate Change

Understanding the likely warming effect of different concentration levels of greenhouse gases is unquestionably important. Of equal importance, however, is a knowledge of the ecological, social, and economic implications of the various temperature rises, should they eventuate. From such an understanding, it is possible to make judgements as to what constitutes a 'safe' or 'ecologically sustainable' concentration level of greenhouse gases. Furthermore, it provides valuable insights into what might constitute a potentially desirable stabilisation target of greenhouse gases. I say "might" because it is only within the range of sustainable outcomes that ecological economists believe that comparisons between different stabilisation targets should be made. This is an entirely different approach to the mainstream economic position and one that will be explored in much greater detail in Chaps. 5 and 6.

It has already been mentioned that global warming will impact on ecosystems, sea levels, water supplies, industrial and agricultural production, and human health. In addition, these impacts will vary from region to region, both with regards to the nature and magnitude of the direct impacts and in terms of the capacity of different nations to adapt to whatever changes ensue (Brooks and Adger 2005; Berkhout et al. 2006; O'Brien et al. 2006; Tol and Yohe 2007; IPCC 2007c, Chaps. 9–16).

Table 1.4 summarises some of the estimated future impacts of climate change at the regional scale. It shows that a rise in water stress will become a major dilemma in Africa which, when coupled with declining agricultural and fisheries output, will increase malnutrition rates across the continent. Also of grave concern is Africa's high vulnerability to climate change and its limited institutional and economic capacity to adapt to a warmer world (AfDB et al. 2003).

Water stress will also be a future problem in Asia, particularly in areas reliant upon glacial meltwater. Although there is the potential for warming to increase crop productivity in various parts of Asia, a decline in agricultural output is expected in Central and Southern Asia. Another cause for concern is the forecast flooding of Asian megadeltas—an eventuation that would lead to the widespread damage of physical infrastructure and the displacement of many millions of people.

In the wealthier region of Australasia, climate change will have significant impacts in Australia in terms of water availability, species extinction, coral bleaching, and reduced agricultural output. In both Australia and New Zealand, cases of coastal flooding are likely to increase as a consequence of rising sea levels and more intense storms. Also expected is an increased rate of heat-related deaths.

The projected impact of warming in Europe is mixed. Like Australia and New Zealand, the health risk arising from more frequent and intense heatwaves will increase. As for agriculture, output is forecast to rise in some Northern regions, but decline in Southern Europe—the latter being a region that is also likely to suffer increased water shortages. Due to the high vulnerability of Europe's ecosystems, species extinction is expected to rise significantly, thus reducing Europe's already diminished biodiversity.

Latin America (Central and South America) is likely to experience a gradual change in vegetation cover as some tropical forests are naturally replaced by savannah woodlands. In other areas of Latin America, arid-land vegetation is expected to replace semi-arid flora. Because of changing rainfall patterns and disappearing glaciers, declining water availability will affect drinking water as well as reduce the water available for irrigation and the generation of hydro-electricity. In general, agricultural

Table 1.4 Expected 21st century climate change impacts on different regions of the world

Region	Water	Ecosystems	Food/forestry	Coasts	Health
Africa	75–250 million people exposed to increased water stress by 2020; 350–600 million people by 2050 (Ashton 2002; Arnell 2004)	Large impacts on terrestrial and aquatic ecosystems (Leemans and Eickhout 2004; Nkomo et al. 2006). Sub-Saharan species at risk of extinction (Thuiller et al. 2006; IPCC 2007c)	Severe reductions in agricultural and fisheries output (Mendelsohn et al. 2000a; Stige et al. 2006; Thornton et al. 2006)	Towards the end of the 21st century, rising sea levels will affect low-lying coastal areas with large populations. Further degradation of mangroves and coral reefs expected (Klein et al. 2002; Nicholls 2004)	Increased malnutrition due to food shortages (IPCC 2007c). Southward expansion of malaria into South Africa (Hartmann et al. 2002)
Asia	Himalayan glacial melting projected to increase the incidence of flooding and rock avalanches, yet decrease water supplies by 2030. Reduced freshwater availability in larger river basins of Central, South, East, and South-East Asia (Bou-Zield and El-Fadel 2002; Qin 2002; Hoanh et al. 2004; Batima et al. 2005; Stern 2007)	Increased pressure on ecosystems will compound the pressures caused by rapid urbanisation and industrialisation (Lu and Lu 2003; Callaghan et al. 2005; Ishigami et al. 2005; Malcolm et al. 2006)	By 2050, crop yields increase by up to 20 % in East and South-East Asia; decrease by up to 30 % in Central and South Asia (Parry et al. 1999; Murdiyarso 2000; Rosenzweig et al. 2001)	Coastal areas, especially heavily-populated megadeltas in South, East, and South-East Asia, will be subject to increased flooding from rising sea waters (Woodroffe et al. 2006; Li et al. 2004; Wassmann et al. 2004)	Increasing risk of hunger due to rising population and declining crop yields in Central and South Asia (McMichael et al. 2004)

(continued)

Table 1.4 (continued)

Region	Water	Ecosystems	Food/forestry	Coasts	Health
Australasia (Australia and New Zealand)	Reduced water supplies by 2030 caused by reduced precipitation and increased evaporation—most prevalent in southern/eastern Australia and Northland/eastern regions of New Zealand (Chiew et al. 2003; IPCC 2007c)	Significant loss of biodiversity by 2020 in ecologically rich sites (e.g., Great Barrier Reef, Kakadu wetlands, alpine areas, sub-Antarctic islands) (Halloy and Mark 2003; Williams et al. 2003; Jones et al. 2004; Frenot et al. 2005)	Reduced agricultural and forestry production by 2030—an exception being western and southern areas of New Zealand (Atwell et al. 2003; Howden et al. 2003; Luo et al. 2003; Howden and Jones 2004; Asseng et al. 2004)	Coasts adversely affected by rising sea levels and increased severity and frequency of storms and coastal flooding (MfE 2004; Voice et al. 2006; Pearman 2008)	Increased heat-related deaths (McMichael et al. 2003) and a rise in mosquito-borne infectious diseases in tropical Australia (e.g., Ross River fever and dengue fever) (Woodruff et al. 2006)
Europe	Increased incidence of flash flooding, particularly in Northern Europe. Reduced water availability in Southern Europe; high water stress in Central and Eastern Europe. Glacial retreat in mountainous areas (Krüger et al. 2002; Menzel and Bürger 2002; Eitzinger et al. 2003; Hock et al. 2005)	Fragmentation of ecosystems make it difficult for organisms to adapt to climate change (Thuiller et al. 2005). Extensive species loss in mountainous areas (up to 60 % species extinction by the end of the 21st century in some regions) (Walther 2004; Viner et al. 2006)	Decline in crop productivity in Southern Europe (Maracchi et al. 2005). Increased crop yields and forest growth rates in Northern Europe (Shiyatov et al. 2005; Olesen et al. 2007)	More frequent coastal flooding as a consequence of sea-level rise and storm surges (Hurrell et al. 2003; Meier et al. 2004)	Projected increase in health risks due to more frequent and more intense heatwaves (Casimiro and Calheiros 2002). Higher incidence of mortality and injury from wind storms and floods (Kirch et al. 2005)

(continued)

Table 1.4 (continued)

Region	Water	Ecosystems	Food/forestry	Coasts	Health
Latin America	Changing precipitation patterns and the disappearance of glaciers expected to affect water availability for human consumption, irrigation, and hydroelectricity generation (Ramírez and Brenes 2001; Arnell 2004; Vásquez 2004)	Increases in temperature and decreases in soil moisture to result in the gradual replacement of tropical forest by savannah in eastern Amazonia (Peterson et al. 2002; Miles et al. 2004). Semi-arid to be replaced by arid-land vegetation. Biodiversity losses in many tropical areas (Pounds et al. 1999; Nobre et al. 2005)	Rise in salinization and desertification of agricultural land in drier regions (IPCC 2007c). Productivity of some crops and livestock expected to decline (Parry et al. 2004; Warren et al. 2006)	Sea-level rise is projected to cause increased flooding in low-lying areas (Barros 2005; UCC 2005). Widespread damage to Mesoamerican coral reefs (IPCC 2007c)	Increased health risks due to the lack of economic, technological, and institutional capacity to adapt to climate change (UNEP 2003). A rise in the incidence of diarrhoea, malaria, and dengue fever (Kovats et al. 2005)
Small Islands	Changes in precipitation patterns increase the vulnerability of many small islands to water shortages (Arnell 2004).	Alien species colonise mid- and high-latitude islands (Smith et al. 2003; Frenot et al. 2005). Species extinction expected on Hawaiian and other Oceanic islands (Daehler 2005)	Agricultural losses of up to 5 % of GDP in high-terrain island nations and up to 20 % of GDP in low-terrain island nations (Ximena 1998; World Bank 2000; FAO 2004). Declines expected in fish stocks and harvest levels (Graham et al. 2006)	Rise in coastal inundation with consequent shoreline erosion and damage to infrastructure (Cowell and Kench 2001; Hay et al. 2003). Increasing rate of coral bleaching off the coasts of tropical islands (Sheppard 2003; Donner et al. 2005)	Tourism, a vital sector of many island nations' economies, greatly affected by climate change damages (e.g., sea-level rise and coral bleaching) (Becken 2005; Uyarra et al. 2005). Long-term viability of some atoll nations put at risk (Barnett and Adger 2003)

(continued)

Table 1.4 (continued)

Region	Water	Ecosystems	Food/forestry	Coasts	Health
North America	Warming in western mountain regions projected to reduce the winter snowpack, increase winter flooding, and decrease summer flows (Christensen et al. 2004; Merritt et al. 2005; Mote et al. 2005). Water supply stresses exacerbated by over-allocation of water resources (Postel and Richter 2003; Pulwarty et al. 2005)	70–120 % increase in the rate of forest fires in Canada (Brown et al. 2004; Flannigan et al. 2004). Coastal habitats adversely affected by rising sea levels and an increased intensity of tropical storms (Forbes et al. 2004). An overall decrease in vertebrate and tree species richness in the U.S. (Currie 2001)	5–20 % increase crop yield potential for temperature rises up to 2 °C, albeit with great variability among regions (Antle et al. 2004; Thomson et al. 2005). Increased crop failure where there is a strong reliance on irrigation water and where crops are grown near the warm end of their range (Polsky and Easterling 2001; Vasquez-Leon et al. 2002)	Coastal communities increasingly stressed by the interaction of climate change impacts (e.g., the increased intensity of tropical storms) and population growth (Titus 2002, 2005; Forbes et al. 2004; Kleinosky et al. 2006)	70 % increase in hazardous ozone days. A rise in heat-related deaths due to a 3–8 times increase in heat-wave days in some cities (IPCC 2007c)
Polar Regions	Reduced thickness and extent of glaciers and ice sheets (Rignot and Kanagaratnam 2006; IPCC 2007b). Warming, thawing, and decrease in the terrain underlain by permafrost (Sazonova et al. 2004; IPCC 2007a)	Changes to natural ecosystems to have detrimental impact on many species, including migratory birds, mammals, and higher predators (Wrona et al. 2005; Reist et al. 2006)	Decline in marine species and fish stocks that form part of the staple diet of indigenous Arctic communities (Nuttall et al. 2005; Wrona et al. 2005)	Increased coastal erosion and coastal reconfiguration (Forbes 2005; Instanes et al. 2005)	Reduced heating costs but increased stress caused by the need to adapt rapidly to climate change (Krupnik and Jolly 2002). Residents in some Arctic regions report respiratory problems associated with extreme summer days not previously experienced (Furgal et al. 2002)

output is expected to decline in many parts of Latin America, whilst rising sea levels are likely to inundate some coastal regions and damage Mesoamerican coral reefs.

In North America, the winter snowpack is expected to diminish in western mountain regions, thereby reducing summer stream flows. At the same time, forest fires will increase, particularly in Canada. Once again, rising sea levels will adversely affect coastal ecosystems—a problem expected to be exacerbated along the US south-east cost by the increasing intensity of tropical storms. Heat-related deaths are also projected to increase in many North American cities.

It is of little surprise that many of the world's small islands will be severely affected by the forecast rise in sea levels. Large-scale inundation of some islands, shoreline erosion, and damage to physical infrastructure is anticipated. In some instances, the viability of atoll nations (e.g., The Maldives) will be at risk. Very high agricultural losses are projected for low-terrain island nations, as are widespread declines in fish stocks. The tourism sectors of small island economies are also expected to be damaged by the adverse effects of climate change.

Finally, polar regions are likely to experience some of the most pronounced climate change impacts in the world. Projected changes include a reduction in the extent and thickness of glaciers and ice sheets and a widespread thawing of the permafrost. This will result in severe coastal erosion and, in some cases, reconfiguration of the coastline itself. The associated impact on polar ecosystems is expected to reduce polar biodiversity and fish populations. The latter is of particular concern given that seafood constitutes a major portion of the staple diet of many indigenous Arctic communities.

In terms of climate change impacts at the global scale, Fig. 1.10 reveals some of the major impacts expected should temperatures rise in accordance with various stabilisation levels of greenhouse gases. The top of Fig. 1.10 shows the estimated temperature increases and likely temperature ranges of four different greenhouse gas concentrations (450, 550, 650, and 750 ppm of CO_2-e). The remainder of Fig. 1.10 divides the major impacts into six categories to reflect the broader implications of climate change on water, ecosystems, food, coasts, and human health, plus likely singular events.

Given that temperatures have already risen by 0.8 °C above pre-industrial levels, Fig. 1.10 highlights the many impacts on the brink of emergence. With further temperature increases expected regardless of what mitigation measures are eventually implemented, it is clear that a number of significant impacts are likely to materialise in the not-too-distant future. Without trivialising the lesser impacts, Fig. 1.10 shows that the worst effects of climate change will occur once the rise in global temperatures exceeds 2 °C above pre-industrial levels (IARU 2009). For example, rises above the 2 °C mark will: (i) subject at least one billion more people to water shortages; (ii) increase the species at the risk of extinction to a dangerously high 20–30 per cent; (iii) extend the bleaching of corals from 'some' to 'most', with widespread coral mortality expected once the 3 °C mark is exceeded; (iv) render the terrestrial biosphere a net source of carbon emissions as opposed to a net carbon sink; (v) reduce the output of some cereals in the low latitudes and, should temperature rises exceed 3.5 °C, eventually reduce the output of all cereals; (vi) massively increase the number of additional people at risk of coastal flooding; and (vii) trigger several metres of sea-level rise.

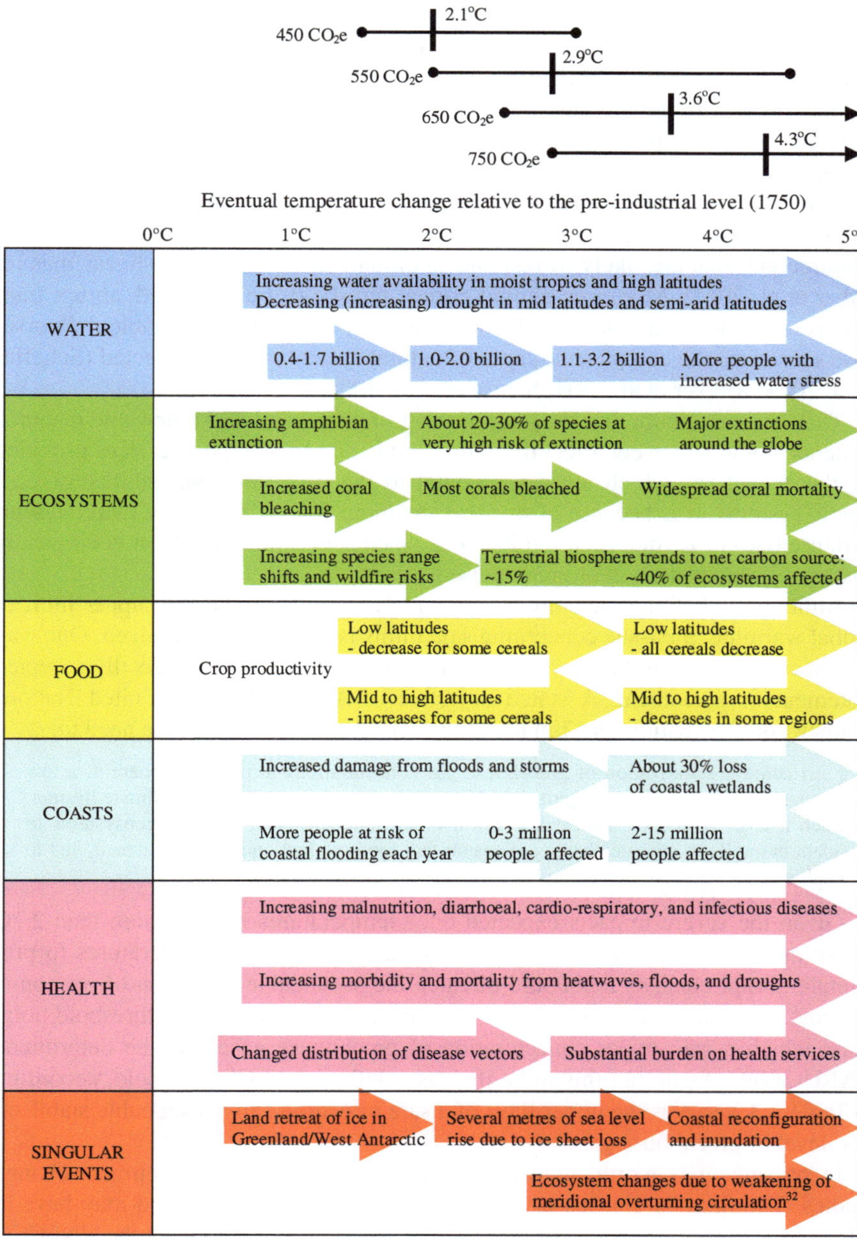

Fig. 1.10 Projected global impacts of temperature increases as they relate to different stabilisation levels of atmospheric greenhouse gases[32, 33]. *Source* Adapted from IPCC (2007c), Fig. 20.8, p. 828

Although it could be argued that, through adaptation, all but the most extreme impacts of climate change are surmountable, it needs to be recognised that the Earth has been greatly modified by human activities and that critical ecosystems are much fewer in number and highly fragmented. This significantly increases the difficulty with which fauna and flora can 'migrate' in response to climate system shocks. As a consequence, the capacity of the Earth to respond naturally to climate change has been greatly reduced, particularly when compared to past changes in the Earth's climate. Indeed, it has been shown that ecological and renewable resource systems are likely to respond to climate change in a non-linear manner (Aber et al. 2001). Moreover, should critical thresholds be surpassed, abrupt transitions to simplified and less productive natural systems seem inevitable. In worse-case scenarios, the complete collapse of natural systems can be expected (Scheffer et al. 2001; Rietker et al. 2004; Schröder et al. 2005).

There is little doubt that the undermining of ecological and renewable resource systems would adversely affect the source, sink, and life-support services provided by the ecosphere, which are fundamental to the long-run sustainability of economic systems. It is because of this that the more severe impacts of extreme temperature rises have important implications when determining a safe or ecologically sustainable concentration of greenhouse gases.

Although a 2 °C temperature rise is widely considered the safe upper limit of global warming, the issue concerning safe limits is by no means resolved. One way of objectively settling the issue is to define 'safe' in terms of a universally accepted statement or agreement. A stated objective of Article 2 of the United Nations Framework Convention on Climate Change (UNFCCC 1992)[34] is the need to:

> [....] achieve stabilization of greenhouse gas concentrations in the atmosphere at a low enough level to prevent dangerous anthropogenic interference with the climate system. Such a level should be achieved within a time-frame sufficient to allow ecosystems to adapt naturally to climate change, to ensure that food production is not threatened, and to enable economic development to proceed in a sustainable manner.

Given the severe impacts expected once temperatures rise by more than 2 °C above pre-industrial levels, plus the potential for increasing temperatures to precipitate abrupt changes that could threaten the sustainable operation of economic systems, it would seem that the 2 °C mark constitutes a reasonable threshold point upon which a safe upper concentration of greenhouse gases can be determined (IARU 2009). Assuming this to be the case, 450 ppm of CO_2-e would, according to Table 1.3, serve as the upper limit of a safe or ecologically sustainable stabilisation level of greenhouse gases.

In passing, it is worth highlighting the results of two recent climate change studies. The first, by Hansen et al. (2008), suggests that humankind may have to aim for a stabilisation target below 450 ppm of CO_2-e if it is to meet the UNFCCC objective. According to Hansen et al., a safe atmospheric concentration of CO_2 is probably in the order of 350 ppm (approximately 425 ppm of CO_2-e). The second study suggests that global warming of 2 °C above pre-industrial levels is best avoided by stabilising greenhouse gases at or below 400 ppm of CO_2-e

(Meinshausen 2006). Having said this, Meinshausen points to the fact that warming above 2 °C only becomes a *likely* prospect if greenhouse gases are stabilised at or above the 475 ppm level. In weighing up these two studies and the evidence revealed in the *IPCC Fourth Assessment Report*, it would seem that 450 ppm of CO_2-e constitutes the very upper end of what might be deemed a safe stabilisation level. Should future global emissions targets be based on a perceived 450 ppm safety level, they may have to be altered as new scientific evidence and information comes to hand. In all likelihood, any revision will be downwards.

1.2.6 The Scientific and Institutional Response to Climate Change

In Chap. 10, a proposal for a global emissions-trading system will be outlined. The system will in many ways need to reflect the institutional arrangements that exist at both the national and international levels. It is also likely that any new emissions-trading system will need to account for any future emissions targets, such as the obligations pertaining to a new protocol earmarked to replace the Kyoto Protocol in 2021. Given that a number of key climate change institutions are inextricably linked to the scientific community (e.g. the IPCC), then, as a basis for upcoming chapters in the book, it would seem pertinent to document the historical growth in the scientific understanding of climate change and the accompanying institutional response.

The idea that the earth's climate system involves energy balances was first postulated by Edme Mariotte in 1681 after he noted that the Sun's radiant heat could pass through glass and other transparent materials, but the heat radiated by other sources could not (Fleming 1998). This prompted Horace Benedict de Saussure, in the 1760s, to construct a heliothermometer—a glass-covered thermometer enclosed in a black, cork-lined box—to serve as an early practical demonstration of the greenhouse effect (IPCC 2007b). The conceptual advance that emerged from Saussure's experiments was the recognition that air itself is capable of absorbing thermal radiation. In 1824, and following on from Saussure's insights, Joseph Fourier outlined the importance of the atmosphere in trapping heat to warm the Earth, albeit Fourier had already used the term 'serre' (greenhouse) to describe the warming phenomenon in 1822 (Fleming 1998). C.S.M. Pouillit later supported Fourier's ideas in 1836 by declaring that the Earth's atmosphere exerts "unequal heat-absorbing actions" in the sense that the heat emitted from the Earth's surface differs to the heat received from the Sun (Fleming 1998). What was not understood at the time was the substance in the atmosphere responsible for the unequal absorption of heat.

An answer to this conundrum eventually emerged in 1859 when John Tyndall identified water vapour and CO_2 as heat-trapping gases in the atmosphere. Tyndall subsequently argued that variations in CO_2 had the potential to alter the Earth's climate (IPCC 2007b; Tyndall 1861). However, it was not until 1895 that Swedish electrochemist, Svante Arrhenius, made the first climate change predictions—namely, a doubling of the atmospheric concentration of CO_2 had the potential to increase global temperatures by as much as 5 °C, whilst a halving of CO_2 could induce an Ice Age (Arrhenius 1896).

As part of his overall theory, Arrhenius argued that past variations in atmospheric CO_2 could be largely attributed to volcanism (Fleming 1998). This aspect of Arrhenius' thesis was seriously challenged by T.C. Chamberlin. As a geologist, Chamberlin regarded the hypothesis of Arrhenius as overly simplistic and inconsistent with known geological events. Chamberlin had already developed theories on the global carbon cycle and its possible connection with CO_2 as an agent of climate change. In 1897, Chamberlin proposed that variations in the atmospheric concentration of CO_2—caused by environmental factors with the capacity to store and release carbon into the atmosphere—could, when combined with water vapour feedbacks, account for the advance and retreat of ice sheets and a range of other geological puzzles (Fleming 1998; Chamberlin 1897). Besides embodying a more sophisticated explanation for the global exchange of carbon, the model developed by Chamberlin was the first to include feedback mechanisms as a means of negating or amplifying climate change influences.

In 1899, and on the basis of emerging theoretical developments, Nils Eckholm predicted that the burning of pit coal would eventually double the atmospheric concentration of CO_2. This, Eckholm stressed, would increase the mean surface temperature of the Earth.[35]

Shortly later, the theory surrounding the warming potential of increased CO_2 was hotly disputed (Fleming 1998). At the turn of the 20th century, Knut Ångström (1900) claimed that more than sufficient CO_2 existed in the atmosphere to ensure maximum infra-red absorption. Consequently, any additional CO_2 would have no absorptive effect. William Humphries (1913), a long-time critic of climate change theories, used Ångström's results to argue that a doubling or halving of atmospheric CO_2 would leave unaltered the infra-red radiation absorbed by the Earth's atmosphere. As a result, variations in atmospheric CO_2 would have no marked impact on the average temperature of the Earth. Such expressions of doubts over the role of CO_2 were echoed by Charles Greely Abbot and F.E. Fowle, Jr., with their insistence that water vapour was the principal absorber of infra-red radiation (Fleming 1998).

Despite Chamberlin's dismissive reaction to Humphries, scepticism over the role of CO_2 continued to such an extent that, by 1929, G.C. Simpson (1929–30) claimed it was "now generally accepted that variations in carbon-dioxide in the atmosphere, even if they do occur, have no appreciable effect on the climate".

Nevertheless, by the 1930s, a prominent global warming trend had come to the attention of many climate scientists. Along with new measurements downplaying the cooling effect of water vapour, the rising temperatures prompted G.S. Callendar to re-evaluate the contribution of human activities to climate change (Fleming 1998). Two factors convinced Callendar, and others, that the increasing combustion of fossil fuels was responsible for the observed warming in the early part of the 20th century. The first was the human discharge of 150 billion tons of

CO_2 in the fifty years prior to the late-1930s that had underpinned a 6 per cent increase in the atmospheric concentration of CO_2 between 1900 and 1936. The second factor was the generation of a set of equations by Callendar (1938) linking greenhouse gases and global warming. Using the equations, Callendar calculated that a doubling of the atmospheric concentration of CO_2 would increase the mean global temperature by at least 2 °C, with considerably more warming likely to occur at the poles (IPCC 2007b).

It was also during the 1930s that Milutin Milankovitch outlined the theory that orbital changes—the Milankovitch cycles referred to earlier in the chapter—was the principal factor behind the oscillation between glacial and interglacial periods. As valuable as this theory became for explaining climate change over very long periods of time, it also revealed why the temperature changes experienced in the early part of the 20th century could not be strongly attributed to changes in the eccentricity of the Earth's orbit or to changes in the obliquity of the Earth's axis.

Until the 1940s, debates and concerns about climate change were largely confined to the scientific community. Upon revitalisation of the link between CO_2 and global warming and studies revealing that the thickness and extent of Arctic ice had greatly diminished between 1890 and 1940, the topic of climate change began to make its way into the public arena in the 1950s (Fleming 1998). During the same decade, Gilbert Plass (1956) integrated the understandings from various scientific disciplines to develop a more realistic model of radiative transfer. Plass used his more sophisticated theory to reinforce the findings of Callendar and to warn that humankind was, by increasing the atmospheric concentration of CO_2, conducting a large-scale experiment on the Earth's atmosphere with potentially dangerous repercussions (Fleming 1998).

Because the carbon cycle had emerged as a central component of many climate change theories, it became obvious that greater knowledge was required of the CO_2 exchange between the oceans and the atmosphere (IPCC 2007b). Roger Revelle and Hans Suess (1957) responded to the challenge by discovering that the world's oceans are unable to absorb all anthropogenically-generated CO_2 because of the extended time it takes for CO_2 to mix with deep ocean layers. Thus, given the rising rate of CO_2 emissions, Revelle and Suess concluded that CO_2 was likely to continue accumulating in the atmosphere. Revelle and Suess were soon supported by the first accurate measurements of atmospheric CO_2 in 1958 which showed that the atmospheric concentration of CO_2, at 315 ppm, was around 13 per cent higher than pre-industrial levels (Keeling 1960).[36]

Just as the link between increasing CO_2 and rising temperatures was gaining momentum, it emerged in the 1960s that global temperatures had been declining since the 1940s—in part, a consequence of rising aerosol production in North America, Europe, and Japan. This sparked new concerns about the prospect of global cooling, with the US Government and the United Nations contemplating the potential impact of a cooler world on agricultural output (Fleming 1998). As it turned out, air pollution regulations enacted throughout most of the industrialised world in the early-1970s quickly arrested the increase in aerosol production. Nonetheless, CO_2 emissions continued to rise. Recognising that the warming effect of the latter would eventually overwhelm the cooling effect of the former (see, again, Fig. 1.6), scientific opinion converged on global warming, not cooling, as the primary climate change risk of the 21st century. Support for this position was reinforced in the mid-1970s by, firstly, the most sophisticated model yet developed showing that a several degree increase in global temperatures would result from a doubling of atmospheric CO_2 (Manabe and Wetherald 1975), and secondly, by studies indicating that chlorofluorocarbons (CFCs) and methane also contribute positively to radiative forcing (Ramanathan 1975; Wang et al. 1976). Further support emerged in 1979 in the form a US National Academy of Sciences Report demonstrating that a doubling of CO_2 would lead to an increase in average global temperatures of 1.5–4.5 °C (NAS 1979). In the same year, the first World Climate Conference was held in Geneva. Concerns were voiced that "continued expansion of man's activities on Earth may cause significant extended [....] changes of climate" (WMO 1979, pp. 1–2).

As the 1980s begin, evidence revealed a return to a global warming trend with 1981 becoming the warmest year on the instrumental record. Reinforced by alarming reports published by the US National Academy of Sciences and the US Environmental Protection Agency, global warming entered the mainstream political arena in both North America and Europe. With climate change now prominent in mainstream political circles, Ramanathan et al. (1985) announced that global warming would occur at twice the previously expected rate because of rapid rises in methane levels and other trace greenhouse gases. This provoked a group of climatologists attending a 1985 climate change conference in Villach, Austria, to call upon governments around the world to establish international protocols to limit the rise in greenhouse gas emissions (Maunder 1992).

Despite the failure of governments to respond to the specific appeals of the scientific community, an important first step was taken in 1987 with the establishment of the Montreal Protocol—an environmental treaty that included binding restrictions on the generation of ozone-depleting gases, some of which also constituted greenhouse gases (Benedick 1991). A year later (1988), the United Nations established the Intergovernmental Panel on Climate Change (IPCC). The IPCC's task was to periodically generate reports detailing: (i) the causes and impacts of climate change; (ii) projections of future temperature changes; and (iii) the adaptation and mitigation options available to policy-makers. In the same year, James Hansen announced to the US Senate that "global warming had begun" and that the Earth was in for a rapid warming phase that could eventually degenerate into a runaway greenhouse effect.[37]

In 1990, the first IPCC report appeared. Highlighted in the report was evidence that average global temperatures had been rising and that global warming was likely to continue in the future (IPCC 1990).

The international response to climate change intensified in 1992 when, at the Earth Summit conference in Rio de Janeiro, the United Nations Framework Convention on Climate Change (UNFCCC) was signed by over 150 countries. Considered the most important climate change treaty yet created, the UNFCCC contained non-binding limits on greenhouse gas emissions as well as provisions to establish subsequent protocols as a means of institutionalising mandatory emission targets. The UNFCCC would eventually come into force on 21 March 1994.

In 1996, a second series of IPCC reports emerged outlining the strong detection of a human-enhanced greenhouse effect and the likelihood of serious global warming during the 21st century (IPCC 1996a, b, c). On the back of the IPCC reports, the 1997 Kyoto Protocol—the first mandatory emissions treaty to emerge from the UNFCCC—was successfully negotiated at the third 'Conference of the Parties' in Kyoto, Japan (COP-3). Under the Protocol, Annex I countries agreed to reduce their collective greenhouse gas emissions by 5 per cent relative to 1990 levels.[38] The applicable commitment period—that is, the period over which Annex I nations were required to meet their greenhouse gas targets—was set for 2008–2012. Although developing or non-Annex I countries were not required to cut emissions, they were encouraged to share in the common responsibility to reduce greenhouse gas emissions in the spirit of the Kyoto Protocol.

On 16 March 1998, the Kyoto Protocol opened for ratification and signature. As for year 1998 itself, it emerged as the warmest year ever recorded, enhanced by the most extreme El Niño event in recent memory.

At the dawn of the new millennium, studies revealed the importance of biological feedbacks in the carbon cycle that were likely to accelerate the pace of global warming (e.g., Cox et al. 2000). In 2001, a third published series of IPCC reports provided the strongest endorsement yet of humankind's contribution to climate change (IPCC 2001a, b, c, d). The reports bluntly stated that: (i) it was 'very likely' that most of the warming observed during the 20th century had been caused by the human generation of greenhouse gases; (ii) in the absence of corrective policies, temperatures were likely to rise by 1.4–5.8 °C over the course of the 21st century; and (iii) climate change surprises could eventually emerge with potentially devastating consequences.

In 2003, two events took place that many believe were a taste of things to come. The first was the weakening of West Antarctic and Greenland ice sheets that raised the prospect that sea levels could rise at a much faster rate than previously expected. The second was the most extreme heatwave in Europe in 500 years that resulted in 30,000 additional heat-related deaths (Larson 2003). With public concern about global warming on the rise, particularly in Europe, the Kyoto Protocol finally came into force on 16 February 2005 following its ratification by the Russian Federation in late-2004.[39] However, two major per capita emitters of greenhouse gases—the USA and Australia—would remain defiant by refusing to ratify the Kyoto Protocol. Globally, 2005 would become the second warmest year on record, a year punctuated by a number of severe tropical storms, including Hurricane Katrina.

In late-2006, the most publicised government-commissioned report on the economics of climate change was released. Soon to be widely known as the *Stern Review*, the report concluded that it would be necessary to invest at least one per cent of Gross World Product (GWP) every year to avoid climate change damage costs equivalent to the annual loss of 5–20 per cent of GWP (Stern 2007). The Stern conclusions were soon supported in 2007 with the publication of the fourth series of IPCC reports declaring that the cost of reducing emissions would be significantly less than the cost of climate change damages (IPCC 2007a). Also stressed by the IPCC was that "warming of the climate system is unequivocal" and that it was "very likely that global warming is the result of human activities".

2007 ended with a United Nations climate change conference in Bali, Indonesia, where the main objective was to put in place a roadmap with a view to establishing a legally-binding emissions agreement to take effect at the end of the Kyoto commitment period in 2012. It was hoped at the time that the successor treaty would be negotiated at a UNFCCC conference to be held in Copenhagen in December 2009. During the Bali conference, the European Community called for a peak in global emissions in 2020 with deep emission cuts thereafter. A number of countries, led by the USA, strongly opposed the strategy. The resulting compromise led to some criticism that the conference failed to achieve anything of significance (Carbon-info.org 2007). The Kyoto Protocol was finally ratified at the Bali conference by Australia, but not by the USA.

Despite a change in the US Government in 2008, the newly-elected President, Barack Obama, declined to ratify the Kyoto Protocol. Instead, President Obama stated that the USA would lead the world toward a new era of global co-operation on climate change. Included in a list of climate change promises were initiatives to reduce US emissions to 1990 levels by 2020 and to cut emissions by an additional 80 per cent by 2050 (Knowlton 2008).

In March 2009, the International Alliance of Research Universities convened an international scientific congress with the theme, *Climate Change: Global Risks, Challenges and Decisions*. The aims of the congress were twofold: (i) to coalesce the climate change knowledge released since the publication of the *IPCC Fourth Assessment Report* in 2007; and (ii) to produce a *Synthesis Report* (IARU 2009) to serve as scientific input into the UNFCCC conference to be held later in the year in Copenhagen.

Leading up to the Copenhagen conference, debate grew about the potential impact of future emissions cuts on the Gross Domestic Product (GDP) of nations. Some observers believed that the stimulatory measures being implemented across the world to deal with the global 'GDP' recession were taking precedence over the urgent need to reduce greenhouse gas emissions. A widespread fear emerged that the emissions reductions embodied in a Copenhagen protocol would be insufficient to combat the impact of human activities on the Earth's climate. Notwithstanding this, the Copenhagen conference took place with high expectations that an agreement would finally be struck that was both legally binding and genuinely effective at drastically reducing greenhouse gas emissions.

Despite the high hopes, the Copenhagen conference failed to deliver what had been anticipated. Many observers thus considered the conference to be a complete failure (Monbiot 2009; Harrabin 2009). With climate change negotiations "in disarray", a Copenhagen Accord was eventually drafted on the second-last day of the conference by the USA, China, India, Brazil, and South Africa. Although the US Government considered the Accord to be a "meaningful agreement", it was not immediately adopted by attendees. Consequently, the Accord failed to be unanimously passed. Whilst the Accord recognised the need to restrict temperature rises to no more than 2 °C above pre-industrial levels, the document itself contained no legally-binding commitments to reduce greenhouse gas emissions. Indeed, because the Accord did not include promised measures to cut global CO_2 emissions by 80 per cent by 2050, many countries—including Bolivia, Venezuela, Sudan, and Tuvalu—vehemently opposed the document. Despite the dissent, most countries eventually signed the Copenhagen Accord during 2010.

With the global 'GDP' recession continuing to deepen, public attention to climate change began to wane as did the sense of urgency amongst the world's politicians. The lack of genuine commitment to reducing emissions was reflected by the inability of the world's governments to broker a legally-binding emissions protocol at the United Nations climate change conference in 2010 (Cancún, Mexico). In what some people believe was an act of desperation, it was decided at the 2011 conference in Durban to further extend the Kyoto Protocol and to develop a new emissions protocol by 2015 to take effect in 2021. One consolation of the agreement, referred to as the Durban Platform, was that it included low-GDP countries for the first time (e.g., China and India) as well as the USA which had previously declined to ratify the Kyoto Protocol. In addition, the new protocol was to be legally binding. Although some commentators optimistically viewed the Durban Platform as a means of providing governments with more time to negotiate a lasting and effective emissions protocol, climatologists feared it would delay the urgent mitigation action required to avoid dangerous, if not catastrophic, climate change. The Durban Platform was reified at the 2012 conference in Doha in the form of amendments to the Kyoto Protocol—the most important being a pledge made by a number of Annex I countries to reduce their greenhouse gas emissions by at least 18 per cent below their 1990 levels during a second Kyoto commitment period that would span the years 2013–2020 (UNFCCC 2012).

In 2013, a new but alarming milestone was reached. Measurements taken at Mauna Loa, Hawaii revealed that the atmospheric concentration of CO_2 had reached 400 ppm.[40] Not only was this 41 % above immediate pre-industrial levels, it constituted the highest concentration of CO_2 at any time over the previous three million years (NOAA 2013). With the release of the IPCC's *Fifth Assessment Report* reconfirming humankind's contribution to global warming, the radiative forcing of the six original Kyoto Protocol gases stood at approximately 440 ppm of CO_2-e.

Finally, in late-2013, a United Nations conference was held in Warsaw to continue the negotiations towards the establishment of a new global agreement in Paris in 2015 (UNFCCC 2013a). During the conference, a number of key issues were raised. The first related to the financing of mitigation and adaptation measures and the transfer of technology from the world's richest to poorest nations. Concern was raised regarding the lack of capitalisation of the Green Climate Fund—an important element of the Copenhagen Accord—with only US$7.5 billion having been committed to the Fund by Annex I nations as of June 2013.[41] The second issue pertained to a new mechanism proposed during the conference to help the world's poorest nations cope with climate change-related losses and damages. Referred to as the Warsaw Mechanism, rich and poor nations hotly debated its terms and conditions with Annex I nations resisting any compulsion to explicitly compensate developing nations for weather-related natural disasters. The third issue related to emissions obligations post-2020. Whilst recognising the obligation to undertake most of the action to reduce global greenhouse gas emissions, Annex I nations stressed the need for developing countries to take on new responsibilities. This led to a widespread understanding that the Paris agreement would need to incorporate greenhouse gas targets for non-Annex I as well as Annex I nations.

Despite some positive developments, a number of observers believed that little progress was made at the Warsaw conference (e.g., Stern 2013). Worse still, many were left doubting whether an emissions protocol of the type needed to avoid catastrophic climate change would emerge from the Paris conference in 2015 (see Postscript on the outcome of the COP-20 conference in Lima in December 2014). Meanwhile, records at the end of 2013 indicated that average global temperatures had risen by 0.75 °C over the past century and that thirteen of the previous sixteen years were the warmest since 1850.

1.3 What Is Ecological Economics?

Because this book deals with climate change from an ecological economics perspective, it is important to say something about ecological economics itself. Ecological economics is a transdisciplinary paradigm that extends and integrates the study and management of 'nature's household' (ecosphere) and 'humankind's household' (economy).[42] A relatively new paradigm, ecological economics has emerged in response to the failure of mainstream economic paradigms to deal adequately with the coevolutionary interdependence of social, economic, and ecological systems. As such, ecological economics can be described as a means of bringing the false pre-analytical visions underpinning conventional economic assumptions into line with biophysical and existential realities (Lawn 2007).

Many attempts have been made to integrate the economy and the ecosphere. Two of note stand out. The first has involved the expansion of the economic domain until its boundaries coincide with those of the greater ecosphere. Daly (1991) refers to this as 'economic imperialism'—a chauvinistic exercise designed to bring all matter and energy under the regulating influence of market prices. Examples of economic imperialism include the imputation of shadow prices to value ecosystem services (contingent valuation), conventional benefit-cost analyses, and the assignment of unconditional property rights to permit the market exchange of open access resources. Mainstream economics is heavily based on the principle of economic imperialism.

The second popular attempt has involved the erasure of economic boundaries and the subsequent abandonment of any sort of economic valuation on the basis that the matter and energy flowing through the economy is governed by the same

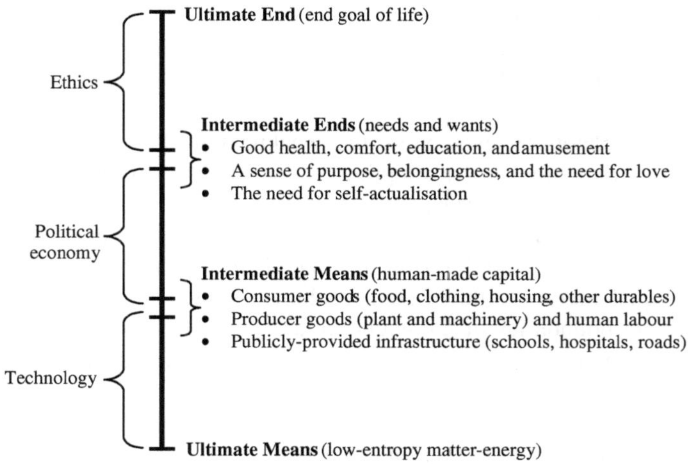

Fig. 1.11 Ends-means spectrum (adapted from Daly 1991, p. 19)

physical principles that reign throughout the containing ecosphere. Daly (1991) refers to this approach as 'ecological reductionism'—an exercise designed to explain all economic phenomena in terms of a biophysical metric. Examples of ecological reductionism include the 'maximum power principle' and the 'energy theory of value'. Although the principle of ecological reductionism is absent from mainstream economics, it plays a dominant role in the physical sciences.

There are a number of problems with these two integrating principles. In the first instance, economic imperialism reduces all physical and moral absolutes to relative and subjective values. In such a world, it is falsely contended that biophysical limits can be overcome by exchange relationships and the supposed anti-entropic nature of accumulated knowledge. Furthermore, whilst options still abound, no objective criteria exists for choosing between them. As a consequence, all efficient market outcomes are deemed equally desirable, which implies that notions of ecological sustainability and distributional equity are subjective concepts best resolved via the efficient allocation of scarce resources. After all, in a world devoid of objective value, public policy is feckless (Daly 2007).

Conversely, the problem with ecological reductionism is that by elevating chance, necessity, and physical determinism above everything else, it eliminates all scope for human will and purpose. In an ecological reductionist world, real alternatives do not exist. If they do, choices are made on our behalf by natural selection. Since human choice is essentially an illusion, then, like economic imperialism, there are no objective values. Once again, public policy is ineffectual except that it can be instituted to maximise the technical efficiency of resource use. Indeed, policies designed to maximise allocative efficiency are considered futile because allocative efficiency, which raises the spectre of choice, is a *non sequitur*. To ecological economists, the weakness of ecological reductionism as a means of policy setting is obvious—as much as it encourages policies to help reduce the resources consumed per unit of economic activity, it offers nothing in terms of what actions can be taken to ensure all newly produced goods possess the highest possible use value.[43]

Ecological economics overcomes the weaknesses of economic imperialism and ecological reductionism by adopting a 'dualistic' approach when integrating the economy and ecosphere (Daly 1991). It does this by recognising the significance of objective values; the existence of choice; the importance of adhering to biophysical constraints; and the need for relativism when choosing between alternatives of equal moral worth. In sum, ecological economics seeks to determine the appropriate boundaries of the economy and the limits to the effective use of markets. The importance of this second task is best illustrated by way of an ends-means spectrum (Fig. 1.11).

At the base of the ends-means spectrum is the *ultimate means*, which consists of low-entropy matter-energy in all its constituent forms (i.e., natural resources).[44] Low-entropy matter-energy constitutes the ultimate means because, as the fundamental, non-substitutable stuff of the universe, it is required to produce and maintain the stock of *intermediate means* (human-made capital).[45] Human-made capital is accumulated in order to satisfy *intermediate ends* (human needs and wants). At the apex of the ends-means spectrum is the *ultimate end* which Daly (1980, p. 9) describes as "intrinsically good" and something that "does not derive its goodness from any instrumental relation to some higher good". Although it may be impossible to agree on the ultimate end, we are compelled to recognise its existence if only because the prioritisation of goals requires an ordering or ethical guiding principle. It is with respect to the ultimate end that intermediate ends are ranked. More particularly, the attainment of the ultimate end requires the adequate satisfaction of intermediate ends in keeping with their hierarchical ranking.

Figure 1.11 reveals two vastly different relationships linking the intermediate and ultimate categories of the ends-means spectrum. The first involves the relationship between the ultimate end and intermediate ends. Since intermediate ends are ranked in accordance with a dogmatic belief in objective value, this relationship is an ethical one. The second involves the relationship between the ultimate means and intermediate means. Because the production and maintenance of human-made capital requires the initial extraction and subsequent transformation of low-entropy matter-energy, this relationship is a purely technological one. As for the central or intermediate segment of the ends-means spectrum, it does not involve a relationship but the political economic problem of valuing, allocating, and redistributing the intermediate means in ways that best serve the hierarchy of intermediate ends.

It is here, again, where the shortcomings of economic imperialism and ecological reductionism are exposed. In the case of economic imperialism, the economy and its associated markets are extended to embrace the entire ends-means spectrum. The ultimate end and ultimate means are consequently ignored or their existence denied. In terms of ecological reductionism, it is the segment linking the ultimate means and intermediate means that is extended to embrace the entire ends-means spectrum. Ignored on this occasion are the ultimate end and the segment linking the intermediate means and intermediate ends.

Ecological economics differs entirely. By recognising the significance of objective values, it understands the critical role that the ultimate end plays in ranking intermediate ends. It therefore leaves ethical considerations (e.g., what constitutes an equitable distribution of income and wealth) to non-economic institutions. Similarly, by recognising the need to adhere to biophysical constraints, ecological economics also leaves biophysical considerations (e.g., what constitutes an ecologically sustainable rate of resource use) to non-economic institutions. In the end, ecological economics confines market mechanisms to the intermediate segment of the ends-means spectrum, since it is here where relativism and exchange mechanisms become invaluable when choosing between ecologically sustainable alternatives of equal moral worth. It is the recognition of absolutes and the confinement of the market to the intermediate segment of the ends-means spectrum that makes the ecological economics approach uniquely dualistic.

Because of its dualistic approach, ecological economics differs from mainstream economics in two further ways. Firstly, instead of dealing with a particular issue or problem in isolation, ecological economists endeavour to tackle problems within an appropriate broader context. Secondly, ecological economists ask whether the problem under review is primarily the consequence of a much deeper problem, and whether it is possible to resolve the former (the symptoms) by successfully tackling the latter (the underlying cause). As it turns out, and despite the unique challenges that human-induced climate change presents, ecological economists believe the climate change crisis is the symptom of a more fundamental problem that is generally overlooked or emphatically denied. It is with this in mind that we turn to the final section of the chapter.

1.4 Putting the Climate Change Crisis into an Appropriate Context

For the purposes of this book, the climate change problem will be set within the broader context of sustainable development. I will have more to say about sustainable development in Chap. 2. For now, sustainable development can be described as a process that: (i) where possible, improves the total quality of life of each person; (ii) is characterised by an equitable distribution of income and wealth; and (iii) preserves the ecological carrying capacity of natural resource systems (Lawn 2000, 2007).

Why has sustainable development been chosen as the appropriate framing context for this book? To begin with, climate change has the potential to undermine the ecological integrity of resource systems that, as shown, is likely to have dramatic impacts on the quality of human life. Moreover, the resultant burdens are expected to fall disproportionately on the world's most disadvantaged people (Stern 2007; IPCC 2007c; Table 1.4). Unquestionably, unless action is taken to address the climate change issue, humankind's future sustainable development prospects will be adversely affected. But the sustainable development implications do not end here. It is equally important to ensure that any climate change response is also consistent with the goal of sustainable development. One must never lose sight of the fact that human-induced climate change is one of many sustainable development issues requiring urgent attention. The on-going problems of over-population, poverty, land degradation, water shortages, biodiversity loss, ecosystem destruction, and rising resource scarcity are already putting the sustainable development process at risk. These problems would remain unresolved even if an unlikely technological solution to global warming was unearthed tomorrow (Orr 2008). The climate change crisis must, as a consequence, be resolved in conjunction with other sustainable development concerns (Beg et al. 2002; Halsnæs 2002; OECD 2004; Munasinghe and Swart 2005; Daly 2007). This will require the implementation of a congruent set of policy initiatives as well as complementary forms of social and economic adjustments (Swart et al. 2003; Wilbanks 2003; Victor 2008; Jackson 2009).

I might add that, in view of the large emissions cuts required to deal with climate change, accounting for other sustainable development considerations will greatly reduce the difficulty associated with stabilising greenhouse gases at a specific concentration level. This will become clearer when the sustainable and growth-as-usual scenarios are revealed in Chap. 4.

Given the broader contextualisation of the climate change issue posed here, it is hardly surprising that the driving forces behind the emergence of other sustainable development concerns are similar to the forces behind the escalation in greenhouse gas emissions. This connection provides a clue as to what constitutes the underlying problem that must be addressed to resolve the climate change crisis. Contrary to mainstream opinion, the problem is humankind's predilection with the continuous growth of the economic subsystem. Apart from being a major cause for the rise in greenhouse gas emissions (IPCC 2007a), excessive growth has resulted in the global economy overshooting the ecosphere's long-run carrying capacity (Global Footprint Network 2008). At the same time, it is becoming clear that the economies of many nations have reached a point where the additional costs of further growth are exceeding the additional benefits that growth delivers (i.e., growth has become economically undesirable). Consequently, any congruent set of policies must facilitate the transition from a growth economy to a qualitatively-improving steady-state economy (Daly 1996, 2007; Lawn 2007).

What is a steady-state economy? I'll be outlining what a steady-state economy is in some detail in Chap. 2. Suffice to say now, a steady-state economy is an economic system made up of a constant magnitude or non-growing stock of physical goods (human-made capital).[46] It is also an economy comprised of a non-growing population of human beings. Hence, a qualitatively-improving steady-state economy is an economic system that doesn't physically grow but qualitatively progresses over time.

Despite the need for all nations to make the eventual transition to a steady-state economy, there is little doubt that low-GDP nations require more growth (some more than others). Critically, it must be a form of growth that is as efficient and equitable as possible. In direct contrast, the world's richest nations need to begin the transition immediately, not just to provide the 'ecological space' that would allow low-GDP countries to enjoy the benefits of further growth, but to significantly advance their own welfare interests. As we shall see later in the book, it is the variation in the rates at which different nations should make the transition to a steady-state economy that ought to dictate the extent and timing of the emissions cuts needed to stabilise greenhouse gases at a safe concentration level.

Annex 1A

Is the Earth Still Warming?—Yes It Is![47]

A lot of attention has recently been given to the notion that the Earth has stopped warming. One prominent IPCC member—Professor Mojib Latif—has gone so far as to suggest that average surface-air temperatures might fall over the next 10–20 years.[48] Given that average temperatures have not exceeded the peak year of 1998, a number of climate change sceptics believe this admission proves that anthropogenic global warming is not occurring. Such a claim by the climate change sceptics is totally misguided.

The basis for Latif's position is that a crucial heat transfer mechanism—the Pacific Decadal Oscillation (PDO)—has entered a negative phase. Heat transfer mechanisms regulate the storage of the heat trapped by greenhouse gases in the world's oceans, atmosphere, and various land surfaces.[49] It is generally believed that the PDO affects surface-air temperatures over much of the globe as well as other ocean oscillations that influence air temperatures elsewhere on the planet (Hansen et al. 2013; Kosaka and Xie 2013; Meehl et al. 2013). Although not conclusively proven, there is growing evidence to suggest that, during a positive PDO phase, a smaller than normal proportion of the heat trapped by greenhouse gases is stored in the world's oceans. This implies that a larger than normal proportion of the trapped heat ends up in the Earth's atmosphere (IPCC 2007b). Consequently, there is a tendency for average surface-air temperatures to be higher during a positive PDO phase than under prevailing circumstances. The opposite occurs during a negative PDO phase. Each phase lasts for thirty to forty years.

There have been three PDO phases over the past century. During the 1905–1945 positive phase, average surface-air temperatures rose by 0.4 °C; during the 1946–1976 negative phase, temperatures fell by 0.2 °C; and over the 1977–2007 positive phase, temperatures rose by 0.55 °C (equivalent to a 0.75 °C rise over the past century). It therefore seems reasonable to assume that the latest negative PDO phase will suppress any increase in average surface-air temperatures. However, many climatologists believe that the warming effect of rising greenhouse gases will overwhelm the cooling impact of the current negative PDO phase. They consequently believe that average surface-air temperatures will continue to rise over the next two to three decades.

There is considerable empirical evidence to support this warming position. Despite a clear upward trend in average surface-air temperatures during the previous two positive PDO phases, the entire 0.2 °C decline over the 1946–1976 negative phase can be attributed to temperature falls in the first two years of the period. That is, if 1946 and 1947 are omitted, the Earth's average surface-air temperature effectively plateaued, albeit annual temperatures continued to fluctuate. Moreover, the 1946–1976 negative PDO phase roughly coincided with the most prolific period of anthropogenic aerosol generation.[50] As explained in this chapter, aerosols have a cooling effect. For this reason, average surface-air temperatures should have fallen more dramatically during this period than they spectacularly rose during the previous two positive PDO phases. That they did not indicates that the warming effect of rising greenhouse gas concentrations was already sufficient to offset the cooling influence of the negative PDO phase.

Of greater significance is evidence produced by Murphy et al. (2009) showing that the Earth is rapidly warming. Although it is common to use trend changes in average surface-air temperatures as an indicator of climate change, it is more appropriate to examine variations in the Earth's *energy imbalance*. By energy imbalance, climatologists mean the difference between the heat emitted by the Earth back to space (heat lost) and the combined heat accumulated in the Earth's oceans, atmosphere, land, and ice.[51] Climatologists often use average surface-air temperatures as an indicator of climate change because, over a prolonged period, surface-air temperatures ultimately trend upwards in line with the accumulation of heat on Earth. The accumulated heat content of the Earth for the period 1950–2003 is revealed in Fig. 1.12.[52]

Figure 1.12 indicates that the Earth's accumulated heat content continued to increase beyond the peak year of 1998. The reason for the recent disparity in the trend changes in average surface-air temperatures and the Earth's accumulated heat content is that the heat storing capacities of land and the atmosphere are small compared to the heat storing capacity of the world's oceans.[53] Consequently, relatively small exchanges of heat between the atmosphere and the

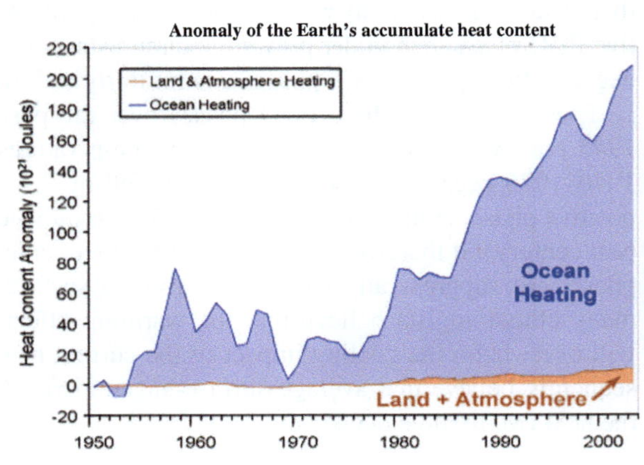

Fig. 1.12 Total heat content of the Earth, 1950–2003. *Source* www.skpeticalscience.com/global-cooling.htm (adapted from Murphy et al. 2009, Fig. 6b)

oceans can significantly alter average surface-air temperatures. This is no better exemplified than by the peak in average temperatures in 1998 that was caused by a massive El Niño-related transfer of heat from the Pacific Ocean to the atmosphere. Conversely, the failure in recent years for average surface-air temperatures to increase above the 1998 level has been the result of La Niña conditions combined with an alteration in the PDO cycle. Not only does Fig. 1.12 reveal the rise in the Earth's accumulated heat content, it exposes the extraordinary amount of warming that the Earth has recently experienced. Between 1970 and 2003, the Earth's accumulated heat content increased at an average rate of 6×10^{21} Joules or 190,000 Gigawatts per year.

As for the period since 2003, there is no equivalent time series of the Earth's accumulated heat content. Having said this, the next best thing exists in the form of a recent analysis of the ocean heat content down to a depth of 2,000 metres (von Schuckmann et al. 2009). Figure 1.13 reveals that the Earth's oceans continued to accumulate heat between 2003 and 2008. What's more, at 0.77 ± 0.11 Watts per square metre (Wm^{-2}), the heat absorbed during this period was by no means trivial.

There are, it would seem, three clear messages that emerge from the empirical evidence presented above and in this chapter. Firstly, given that average surface-air temperatures are an inherently noisy signal, we must avoid making climate change conclusions on the basis of short-term fluctuations of heat transfer mechanisms, such as the PDO and the El Niño/La Niña cycle. Secondly, even if Mojib Latif is correct and average temperatures fall slightly over the next decade or so, surface-air temperatures are likely to rise significantly in the future—particularly once the PDO cycle enters the next positive phase. An understanding of this second point is vital given that any short-term decline in average surface-air temperatures will almost certainly be used by climate change sceptics and opportunistic politicians to delay cuts in greenhouse gas emissions. Thirdly, global warming is undeniably with us and the rate of warming is likely to accelerate if no action is taken to limit the rise in the atmospheric concentration of greenhouse gases.

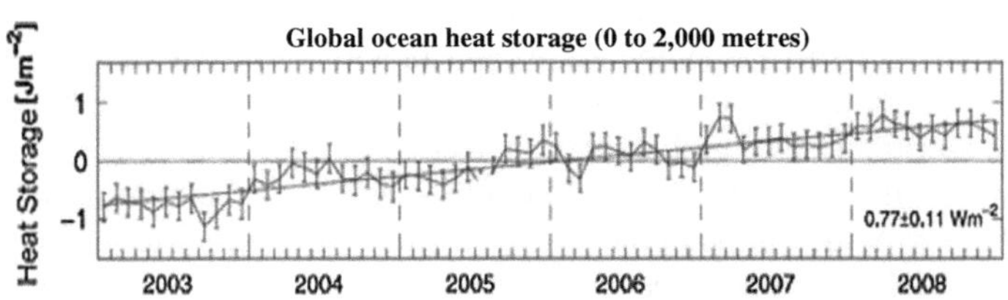

Fig. 1.13 Time series of global mean ocean heat storage (to a depth of 2,000 metres) measured in 10^8 Joules per square metre (Jm^{-2}). *Source* von Schuckmann et al. (2009)

Notes

1. For the purposes of this book, a high-GDP nation will imply a nation with a per capita GDP that is large by international standards. Conversely, a low-GDP nation will imply a nation with a low per capita GDP. The reason for using these terms instead of high-income and low-income nations is that GDP is a poor indicator of national income. I'll have more to say about this in Chap. 3.
2. Other positive feedback mechanisms include water vapour feedbacks (Soden and Held 2005); arctic methane releases (e.g., releases of methane from thawing permafrost) (Zimov et al. 2006); and reduced CO_2 absorption by the oceans (Buesseler et al. 2007).
3. Milankovitch cycles involve the regular and periodic changes in the parameters of the Earth's orbit around the sun. These cycles, which have little effect on global annual mean radiation, modify the seasonal and latitudinal distribution of the incoming solar radiation at the uppermost part of the Earth's atmosphere (IPCC 2007b; Berger 1977, 1978). There are three elements of the Milankovitch cycles: (i) changes in the *eccentricity* of the Earth's orbit due to variations in the minor axis of the ellipse; (ii) changes in the tilt or *obliquity* of the Earth's axis; and (iii) changes in the direction of the axis tilt at a given point of the Earth's orbit—referred to as *climate precession* (see IPCC 2007b, FAQ 6.1., Fig. 1, p. 449).

   Milankovitch cycles aside, the raw solar output of the Sun has gradually increased during the industrial era. This has resulted in a small positive radiative forcing since 1750 of around 0.25 Watts per square metre (IPCC 2007b, p. 136). On top of this, there is also a cyclical change in solar radiation of ± 0.1 % that follows an 11-year cycle.
4. The El Niño oceanic event involves the fluctuation of a global-scale tropical and subtropical surface pressure pattern called the Southern Oscillation. The combined atmosphere-ocean phenomenon is known as the El Niño-Southern Oscillation (ENSO). It is measured by way of a surface pressure anomaly between Darwin (Australia) and Tahiti plus prevailing sea temperatures in the central and eastern equatorial Pacific. The ESNO has a significant impact on the wind, sea surface temperature, and precipitation patterns in the tropical Pacific. It not only influences the climate of the Pacific region, but other parts of the world through 'global interconnections' (IPCC 2007b).

   The Pacific Decadal Oscillation (PDO) is another important heat transfer mechanism with the potential to cause short to medium-term fluctuations in average global temperatures (see Annex 1A).
5. Palaeoclimatology involves the study of ice sheets, tree rings, sediments, and rocks to determine the past state and fluctuations in the Earth's climate and its probable causes.
6. Figure 1.1 includes the time since the Precambrian super-eon.

7. Sources for the individual sections of Fig. 1.1 are:
   - 542–65 million years ago: Royer et al. (2004)
   - 65–5.5 million years ago: Zachos et al. (2001)
   - 5.5 million-420,000 years ago: Lisiecki and Raymo (2005)
   - 420,000–12,000 years ago: Petit et al. (1999)
   - 12,000–2,000 years ago: Image: Holocene Temperature Variations.png (various)
   - 2,000–150 years ago: Image: 2000 Year Temperature Comparison.png (various)
   - 150 years ago to present: Image: 2000 Year Temperature Comparison.png (various)
8. CO_2 levels peaked during the Cretaceous period around 100 million years ago.
9. Climate sensitivity refers to the equilibrium change in the mean global surface temperature following a doubling of the concentration of CO_2-equivalent gases in the atmosphere.
10. This possibility is based on the unusually low levels of CO_2 at the time of the glacial-interglacial cycles between 600,000 and 800,000 years ago (see Fig. 1.2).
11. The eccentricity of the Earth's orbit has two quasi-periodicities—one of 413,000 years and another of around 100,000 years (IPCC 2007b). The second is often regarded as the major factor behind the current 100,000 year greenhouse gas-temperature cycle.
12. Under the auspices of the International Partners in Ice Core Sciences (IPICS), efforts are currently underway to establish an unbroken 1.5 million year climate record to answer these and many other climate-related questions.
13. The change in levels, particularly over the past 800,000 years, can be attributed to a combination of processes in the atmosphere, oceans, in marine sediments, on land, as well as the dynamics of sea ice and ice sheets (Webb et al. 1997; Broeker and Henderson 1998; Archer et al. 2000; Sigman and Boyle 2000; Kohfeld et al. 2005). However, the quantitative and mechanistic explanation of CO_2 variations remains one of the major unsolved questions of climate change research.
14. The last glacial maximum peaked around 21,000 years ago (IPCC 2007b).
15. A stadial is a sub-division of a glacial stage. The Younger Dryas derives its name from the Arctic plant, the *dryas*, which is an early coloniser of Northern Hemisphere land following ice sheet recession.
16. The North Atlantic Oscillation consists of opposing variations in barometric pressure near Iceland and the Azores. It therefore corresponds to fluctuations in the strength of the main westerly winds across the Atlantic into Europe, and thus to fluctuations in the embedded cyclones with their associated frontal systems (IPCC 2007b).
17. More recent records indicate that average global temperatures had risen by 0.75 °C over the past century and that thirteen of the past sixteen years (1998–2013 inclusive) were the warmest since 1850.

18. The R-squared values for the trend-lines are 0.55 for 1856–2005; 0.73 for 1906–2005; 0.75 for 1956–2005; and 0.66 for 1981–2005.
19. By normal state, one is referring to the state of the Earth's radiative balance at a particular reference point—in this case, 1750 AD.
20. The global warming potential of each greenhouse gas depends on the intrinsic capability of a molecule in each gas type to absorb heat and the lifetime of each gas in the atmosphere.
21. Using the IPCC formula for radiative forcing, $454.85 = 278 \times \exp(2.634/5.35)$, where 278 represents the pre-industrial concentration of CO_2 in parts-per-million; 2.634 was the total radiative forcing of all long-life greenhouse gases in 2005; and 5.35 is a constant.
22. In effect, this means that if carbon dioxide was the only greenhouse gas in the atmosphere, it would have required a carbon dioxide concentration of 454.85 ppm in 2005 to have had the same warming potential as all greenhouse gases combined.
23. The value of $374.91 = 278 \times \exp(1.6/5.35)$, where 1.6 was the total radiative forcing of all anthropogenic agents in 2005.
24. In recent decades, the atmospheric concentration of aerosols has fallen in Europe, but has increased in Asia.
25. $428.68 = 278 \times \exp(2.317/5.35)$, where 2.317 was the total radiative forcing of the six Kyoto gases in 2005.
26. The 2012 Doha amendment to the Kyoto Protocol (COP-18) required a seventh greenhouse gas—Nitrogen trifluoride (NF_3)—to be included in greenhouse gas accounts for the purposes of setting greenhouse targets and assessing the performance of nations.
27. The Intergovernmental Panel on Climate Change or IPCC is a scientific, intergovernmental body that was established in 1988 by the World Meteorological Organization (WMO) and the United Nations Environment Programme (UNEP). One of the main tasks of the IPCC is to publish special reports on topic areas relevant to the implementation of the United Nations Framework Convention on Climate Change (UNFCCC). The IPCC itself does not conduct climate change research. It relies upon a broad spectrum of peer-reviewed and published scientific literature to make its assessments and to compile its special reports. Material published by the IPCC is widely considered to be authoritative.
28. Economic systems are usually divided into three sectors: (i) the primary sector, which includes agriculture and the resource-extractive industries; (ii) the secondary sector, which includes the manufacturing industries; and (iii) the tertiary sector, which includes service industries such as the health, education, recreation, hospitality, and life-style industries.
29. In the *IPCC Fourth Assessment Report* (IPCC 2007b), 'likely' is defined as something that has at least a 66 per cent chance of occurring.
30. The estimates for each family group are based on one chosen emissions scenario from each group. These emissions scenarios are referred to by the IPCC as 'illustrative' scenarios (IPCC 2000).

31. The term ‘very likely’ refers to a probability of more than 90 per cent.
32. The meridional overturning circulation (MOC) in the ocean is quantified by zonal sums of mass transports in depth or density layers. In the North Atlantic, away from sub-polar regions, the MOC is often identified with the thermohaline circulation. However, the MOC can also include shallower, wind-driven cells, such as those that occur in the upper ocean in the tropics and sub-tropics where warm waters moving poleward are transformed to slightly denser waters and subducted equatorward at deeper levels (IPCC 2007c).
33. The source figure is based on temperature variations from the 1980–1999 average. The temperature values in Fig. 1.10 have been adjusted so they are expressed in terms of variations from their pre-industrial (1750) values. As a means of illustration, a 2 °C increase above pre-industrial levels corresponds to a 1.4 °C increase above 1990–2000 levels or a 1.5 °C increase above the 1980–1999 average (IPCC 2007c, Box 19.2).
34. The UNFCCC is an international environmental treaty that was adopted on 9 May 1992 and later signed at the 1992 Earth Summit in Rio de Janeiro by more than 150 countries and the European Community. First entering into force on 21 March 1994, the UNFCCC contained non-binding, initial limits on greenhouse gases and provisions for subsequent updates, or protocols, to serve as mandatory emission targets. The first of these mandatory targets was the 1997 Kyoto Protocol.
35. Whilst the paper referring to this prediction was published in English in 1901 (Eckholm 1901), it first appeared in Swedish in 1899.
36. Although Keeling’s observations were the first accurate measures of atmospheric CO_2, measurements had been conducted, albeit with varying degrees of accuracy, since the beginning of the nineteenth century (Fleming 1998).
37. Hansen’s statement appeared in ‘Global warming has begun, expert tells Senate’, *New York Times*, June 24, 1988, p. 1.
38. Annex I countries are high-GDP, industrialised nations as defined under the UNFCCC. Most Annex I nations were required to reduce emissions by between five and eight per cent; the Russian Federation, Ukraine, and New Zealand were required to maintain emissions at 1990 levels; and Norway, Australia, and Iceland were permitted to increase emissions above 1990 levels by one, eight, and ten per cent respectively.
39. Russia’s ratification operationalised the Kyoto Protocol because of Article 25 which stipulates that the Protocol enters into force “on the ninetieth day after the date on which not less than 55 Parties to the Convention, incorporating Parties included in Annex I which accounted in total for at least 55 per cent of the total carbon dioxide emissions for 1990 of the Parties included in Annex I, have deposited their instruments of ratification, acceptance, approval or accession.” Although the ‘55 Parties’ clause was satisfied on 23 May 2002 upon ratification of the Kyoto Protocol by Iceland, it took ratification by the Russian Federation to satisfy the ‘55 %’ clause.

40. In actual fact, the 400 ppm level was first recorded in May 2012 at the National Oceanic and Atmospheric Administration's (NOAA) observatory in Barrow, Alaska. However, measurements taken at Mauna Loa, Hawaii are considered the 'benchmark' given that the station has, going back to 1958, the world's longest continuous record of atmospheric CO_2.
41. Annex I nations have promised to provide US$100 billion per year to the Green Climate Fund through to 2020.
42. Formally, ecological economics began as a distinct sub-discipline of economics following the creation of the International Society for Ecological Economics and the publication of the Society's journal, *Ecological Economics*, in 1989. For more on ecological economics, see Martinez-Alier (1987), Costanza et al. (1991), Daly and Farley (2004), Common and Stagl (2005), Lawn (2007), and Martinez-Alier and Røpke (2008).
43. By *use value*, I mean the service-yielding qualities of physical goods. This differs to the *exchange value* of a good, which is what a person must forego to obtain the good (its price). The aim of allocative efficiency is to maximise the use value generated from a given quantity of available resources.
    Technical efficiency (E) is a measure of the ratio of matter-energy embodied in the physical goods produced (Q) to the matter-energy embodied in the resources used to produced them (R). Hence, $E = Q/R$. Because of the first and second laws of thermodynamics, E must be less than a value of one. As will become clear in Chaps. 2 and 4, increases in technical efficiency enable a larger quantity of goods to be produced from a given resource flow. Technical efficiency can be regarded as one of a subset of factors that contribute to allocative efficiency. *Ceteris paribus*, an increase in the technical efficiency of production augments the use value generated from the allocation of a given resource flow. It therefore leads to greater allocative efficiency. However, a preoccupation with technical efficiency can impact negatively on other allocative factors (e.g., the choice of goods produced) and can thus lead to a reduction in allocative efficiency.
44. To understand what is meant by low-entropy and high-entropy matter-energy, one must know a little bit about the first and second laws of thermodynamics. The first law of thermodynamics is the *law of conservation of energy and matter*. It declares that energy and matter can never be created or destroyed. The second law is the *Entropy Law*. It declares that whenever energy is used in physical transformation processes, the amount of usable or 'available' energy always declines. Although the first law ensures the maintenance of a given quantity of energy and matter, the Entropy Law determines what proportion of it is usable. The Entropy Law is critical since, from a physical viewpoint, it is not the total quantity of matter-energy that is of primary concern, but the amount that exists in a readily available form.
    The best way to illustrate the relevance of these two laws is to provide a simple example. Consider a piece of coal. When it is burned, the matter-energy embodied within the coal is transformed into heat and ash. Whilst the first law ensures the total amount of matter-energy in the heat and ashes equals that previously

embodied in the piece of coal, the second law ensures that the usable quantity of matter-energy does not. In other words, the dispersed heat and ashes can no longer be used in a way similar to the original piece of coal. To make matters worse, any attempt to reconcentrate the dispersed matter-energy, which requires the input of additional energy, results in more usable energy being expended than that reconcentrated. Hence, all physical transformation processes involve an irrevocable loss of available energy or what is sometimes referred to as a 'net entropy deficit'. This enables one to understand the term *low entropy* and to distinguish it from *high entropy*. Low entropy refers to a highly ordered physical structure embodying energy and matter in a readily available form. Conversely, high entropy refers to a highly disordered and degraded physical structure embodying energy and matter that is, by itself, in an unusable or unavailable form. In all, the matter-energy used in economic processes can be considered a low-entropy resource whereas unusable by-products can be considered high-entropy wastes.

45. I am referring here to human-made capital in the Irving Fisher (1906) sense of all producer and consumer goods. Also included in this definition of human-made capital is the stock of public infrastructural assets.
46. There will naturally be some minor fluctuations either side of the steady physical quantity of goods, but the average quantity will effectively remained unchanged.
47. The central thesis of this annex is drawn from www.skpeticalscience.com/global-cooling.htm.
48. I should point out that Professor Latif, one of the world's leading climate modellers, is a strong believer in humankind's warming influence on the Earth's climate. Latif also believes that any decline in average surface-air temperatures will be temporary and that temperatures will again rise abruptly.
49. The North Atlantic Oscillation (NAO) is also an important heat transfer mechanism. However, its influence on global air temperatures appears to be much less significant than the PDO. Another heat transfer mechanism that receives a lot of publicity is the El Niño Southern Oscillation (ENSO). Whilst the ENSO is very influential, its impact is felt in terms of annual temperature variations. The El Niño phase of the ENSO tends to increase air temperatures, whilst a La Niña event has a cooling effect.
50. The anthropogenic generation of aerosols was at its greatest between 1950 and 1985.
51. The accumulated heat content of the Earth does not include the heat contained within the Earth itself.
52. In order to make their calculations of the Earth's total heat content, Murphy et al. firstly used data of the heat content of the upper 700 metres of ocean depth (Domingues et al. 2008; Ishii and Kimoto 2009; Levitus et al. 2009). They then added the heat content data of deeper ocean waters down to a depth of 3,000 metres (Levitus et al. 2000; Köhl et al. 2007; Köhl and Stammer 2008). To compute the heat content of the atmosphere, Murphy et al. used the surface temperature record plus the heat capacity of the troposphere. Finally, the authors added the heat content of land and ice (IPCC 2007b).
53. 'Land + atmosphere' in Fig. 1.12 also includes the heat absorbed by ice.

Chapter 2
Sustainable Development and Climate Change

2.1 Defining Sustainable Development

Given my desire to situate the climate change problem within a sustainable development context, this chapter begins by establishing a broad definition of sustainable development. The chapter continues with an elucidation of the ecological and economic limits to growth and the rationale behind the steady-state economy as a means of achieving the sustainable development goal. What will emerge is not only the backdrop for the remainder of the book, but the basis for the policies and reform measures to be outlined in Chap. 3, including those closely associated with climate change mitigation.

The concept of sustainable development first gained notoriety following the release of the Brundtland Report by the World Commission on Environment and Development in the 1980s (WCED 1987).[1] However, it was not until the 1992 Earth Summit in Rio de Janeiro and the widespread promotion of the United Nations' Agenda 21 that sustainable development became firmly established as a desirable policy objective. Despite its general acceptance, sustainable development continues to mean different things to different people. There are multiple reasons for this. Firstly, the concept of sustainable development is used in many contexts, for different purposes, and by people from varying cultural backgrounds and disciplinary schools of thought. Secondly, the sustainable development concept has rapidly evolved over a relatively short period of time. Finally, debates about sustainable development have been influenced by a wide range of underlying views on the relationship between human beings, economic systems, and the natural environment of which they are a part. Consequently, there are various opinions as to how sustainable development should be measured and what is required to move toward the sustainable development goal.

Ecological economists believe that any definition of sustainable development must be premised on a concrete representation of the economic process.

P. Lawn, *Resolving the Climate Change Crisis*, DOI 10.1007/978-94-017-7502-1_2

Unfortunately, most interpretations of sustainable development reflect the pre-occupation that many observers have with an atomistic-mechanistic world-view and their subsequent failure to recognise the coevolutionary nature of economic, social, and ecological change (Norgaard 1988; Mulder and van den Bergh 2001).

2.1.1 The Coevolutionary World-View

Coevolution is a term used to describe the evolving relationships and feedback responses typically associated with two or more interdependent systems. Coevolution takes place when at least one feedback loop is altered by within-system activity that, in turn, initiates an ongoing and reciprocal process of change (Norgaard 1985, 1988, 1994). A coevolutionary world-view offers a more realistic interpretation of the many critical relationships that bind together the various systems that make up the global system, including the Earth's climate system.

There are a number of basic features of the coevolutionary world-view worthy of elaboration. Firstly, the coevolutionary paradigm begins from the premise that the Earth is a system comprised of closely interacting and interdependent subsystems. Secondly, it recognises that the Earth and its constituent systems are dissipative structures in the sense that the Earth exists as an open system with respect to energy (a solar gradient), whereas the Earth's constituent subsystems exist as open systems with respect to energy, matter, and information.[2] Thirdly, because each system is connected to and dependent in some way on all other systems, everything evolves together over time. Fourthly, and given its complexity, the global system is far richer than the sum of its parts. Fifthly, coevolution is characterised by path-dependency—a proclivity of systems to be inextricably related to their past characteristics and to thus display a strong sense of structural inertia (David 1985; Arthur 1989). Sixthly, and despite the previous point, the coevolutionary world-view regards disequilibria and change as the rule rather than the exception. Finally, the coevolutionary world-view is based on a principle of system embeddedness that is sometimes referred to as the *logos* of nature. Metaphorically, logos is a term used to embrace the natural order of the universe. By acknowledging the logos of the global system, the coevolutionary world-view recognises, firstly, that the world is characterised by self-organisation (Capra 1982). Secondly, it recognises that systems exist at varying levels of complexity and are characteristically stratified and multi-levelled (Laszlo 1972). The logos of the global system and the embedded relationship between the three major spheres of influence—the economy, society, and ecosphere—are illustrated by way of Fig. 2.1.

In Fig. 2.1, the three major spheres of influence represent different systems at varying degrees of complexity. Each system can be considered a *holon* insofar as all systems exhibit the independent and autonomous properties of wholes and the dependent properties of parts.[3] In consequence, each sphere consists of smaller parts whilst simultaneously acting as the part of a larger whole (i.e., the economy constitutes a component of society whilst society constitutes a component of the

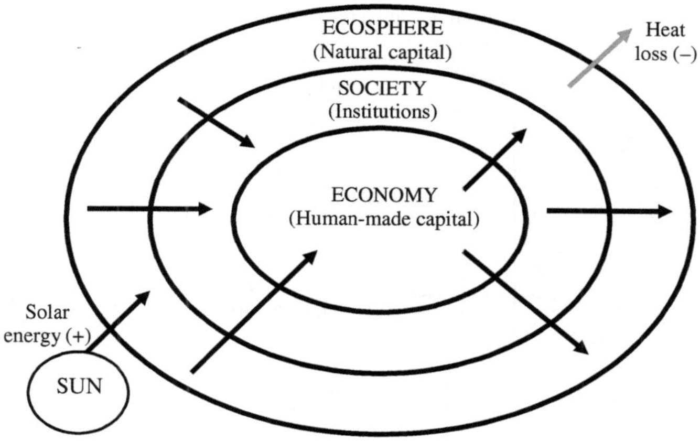

Fig. 2.1 A coevolutionary depiction of the interdependent relationship between the economy, society, and the ecosphere

ecosphere). In this sense, Fig. 2.1 represents society as the interface between the economy and the larger ecosphere, thus highlighting the crucial role played by institutions and social capital in promoting stable human behaviour in the face of indeterminacy, novelty, and surprise (Hodgson 1988; Faber et al. 1992).

2.1.2 The Linear Throughput Representation of the Economic Process

In order to convey the coevolutionary world-view in greater detail, consider the linear throughput representation of the economic process in Fig. 2.2. In keeping with the coevolutionary paradigm, the linear throughput model: (i) depicts the economy as a subsystem of society that, in turn, is depicted as a subsystem of the ecosphere; (ii) recognises the ongoing exchange of matter, energy, and information between the three major spheres of influence and all constituent subsystems; and (iii) acknowledges the evolving relationships and feedback responses typically associated with coevolutionary change.

Although the linear model comprises a multitude of individual elements, each element can be classified into five elemental categories. The first elemental category, *natural capital*, consists of mineral ores, fossil fuels, soil, forests, fisheries, rivers, oceans, lakes, wetlands, ecosystems, and the Earth's climate system. It is because natural capital is the only source of low-entropy resources, the ultimate waste-assimilating sink, and the sole provider of life-support services that natural capital constitutes the original source of all economic activity.

Human-made capital is the second elemental category and, in the Fisherian tradition (Fisher 1906), includes all human-made goods (i.e., producer goods

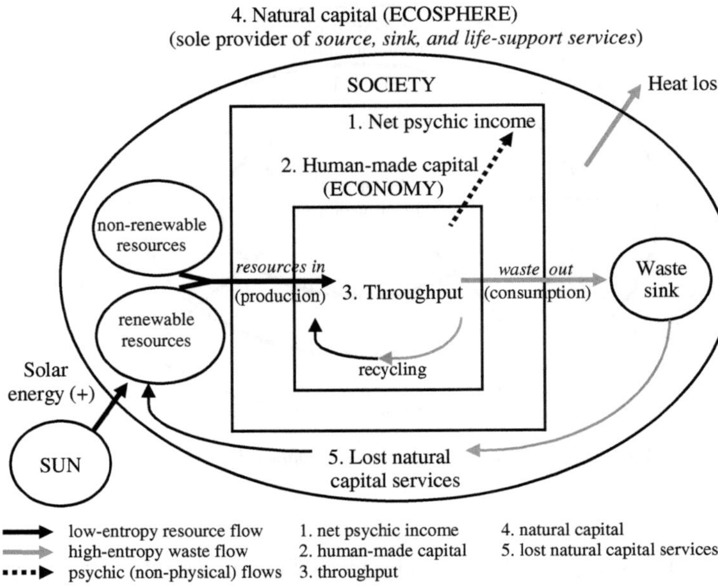

Fig. 2.2 A linear throughput depiction of the economic process

and consumer goods) plus human labour. Human-made capital is accumulated to increase human well-being beyond the level provided by natural capital alone.[4]

The third elemental category is the entropic *throughput* of matter-energy—the input into the economy of low-entropy resources and the subsequent output of high-entropy wastes. The throughput is the physical intermediary connecting natural capital and human-made capital.[5]

The fourth important elemental category is a psychic rather than physical category. Contrary to some opinions, human well-being depends not on the rate of production and consumption, but on what Fisher (1906) described as 'psychic income'.[6] Psychic income is more commonly referred to by economists as 'utility'. As the true benefit of all economic activity, psychic income has four main sources. The first source is the utility that emanates from the consumption and use of human-made capital. The second source of psychic income is derived from being engaged in production activities (e.g., the enjoyment and self-worth obtained from paid employment). A third source of psychic income comes from non-economic pursuits, such as time spent with family and friends, volunteer work, and leisure activities. The final source of psychic income is received from the natural environment in terms of its direct benefits, aesthetic qualities, and existence values. Of course, this final source of psychic income is not directly generated by economic activities. If anything, economic activities destroy rather than enhance these values. It is therefore better that these values be taken as a given and their subsequent destruction be counted as an opportunity cost of the economic process.

This last point reminds us that not all economic activity enhances the psychic enjoyment of life. Consumption of some portion of human-made capital can reduce psychic income if consumers make bad spending choices or if needs and wants have been inappropriately ranked. In addition, while benefits can be enjoyed by individuals at work, production activities are often unpleasant. Unpleasant things that lower one's psychic enjoyment of life represent the 'psychic outgo' of economic activity. It is by subtracting psychic outgo from psychic income that one arrives at the fourth elemental category—*net psychic income*.

In many ways, net psychic income constitutes the 'uncancelled benefit' of economic activity (Daly 1979). Why is this so? Imagine tracing the economic process from natural capital to its final psychic conclusion. Every intermediate transaction involves the cancelling out of a receipt and expenditure of the same magnitude (i.e., the seller receives what the buyer pays). Once a physical good is in the possession of the final consumer, there is no further exchange and no further cancelling out of transactions. Apart from the good itself, what remains at the end of the process is the uncancelled exchange value of the psychic income that the ultimate consumer expects to gain from consuming the good plus any psychic disbenefits and other costs associated with the good's production. Note, therefore, that if the costs are subtracted from the good's final selling price, the difference constitutes the 'use value' added to low-entropy matter-energy during the production process. Presumably the difference is positive otherwise the economic process has been a pointless exercise.

The fifth and final elemental category is the cost of *lost natural capital services* and arises because, in obtaining the throughput to produce and maintain human-made capital, natural capital must be manipulated and exploited both as a source of low-entropy and as a high-entropy waste-absorbing sink. Perrings (1986) has shown that no matter how benignly human beings conduct their exploitative activities, the resultant disarrangement of matter-energy and the inevitable coevolutionary feedback responses have deleterious impacts on the natural environment. Consequently, human beings must accept some loss of the ecosphere's source, sink, and life-support services as the low-entropy resources provided by natural capital are transformed into physical goods and then returned to the ecosphere, once the goods have been consumed, as high-entropy wastes.

In a similar way to net psychic income, lost natural capital services constitute the 'uncancelled cost' of economic activity (Daly 1979; Lawn and Sanders 1999). Why? Imagine, this time, tracing the economic process from its psychic conclusion back to natural capital. Once again, all transactions cancel out. What remains on this occasion is the opportunity cost of resource use or, more definitively, the uncancelled exchange value of any natural capital services sacrificed in obtaining the throughput of matter-energy required to fuel the economic process.[7]

In sum, the linear throughput model illustrates the following. Natural capital provides the throughput of matter-energy that is needed to produce and maintain the stock of human-made capital. Human-made capital is needed to enjoy a level of net psychic income greater than what would be experienced if the economic process did not take place. Finally, in manipulating and exploiting natural capital to obtain the throughput of matter-energy, the three critical services provided by natural capital are, to some degree, unavoidably sacrificed.

2.1.3 Ecological Constraints, Human Needs, and Intragenerational Equity

The above discussion places us in an ideal position to reflect on three aspects central to defining and achieving sustainable development. The first aspect is the importance of ecological factors and the need to adhere to ecological constraints to achieve ecological sustainability. The second and third aspects concern the hierarchy of human needs and the principle of intragenerational equity. As we shall see, both have a powerful influence on the 'development' side of the sustainable development coin.

2.1.3.1 Ecological Factors and Constraints

It was mentioned above that the throughput of matter-energy is the physical intermediary connecting natural capital and human-made capital. It was also pointed out that natural capital constitutes the original source of all economic activity. Given the obvious role that natural capital plays in achieving ecological sustainability, one must ask the following questions:

- How much natural capital is required to ensure the ecological sustainability objective is not recklessly put at risk?
- Should natural capital maintenance be a necessary sustainability tenet, what rules-of-thumb must humankind adhere to in order to prevent the decline in the quantity and quality of natural capital stocks?

I will endeavour to answer the first question by beginning with a consideration of production possibilities. Ever since Hicks (1946) defined income as the maximum amount that can be consumed in the present without compromising the ability to consume the same amount in the future, it has been widely recognised that sustaining the production and consumption of a particular quantity of physical output requires the maintenance of income-generating capital. However, a hot debate has longed raged over what form the capital should take. Whilst some observers believe that natural capital and human-made capital must be individually maintained (strong sustainability), others believe it is sufficient to maintain an appropriately combined stock of both forms of capital (weak sustainability). In the end, the most appropriate action depends on whether human-made capital and the technology embodied in it can adequately substitute for the low-entropy matter-energy and other crucial services provided by natural capital. If it cannot, it is necessary to follow the approach advocated by the proponents of strong sustainability—namely, 'keep natural capital intact'.

It is undeniably true that advances in the technology embodied in human-made capital can for some time reduce the incoming resource flow required to produce a given physical quantity of goods. However, for three related reasons, this does not amount to substitution. Firstly, technological progress only reduces the high-entropy waste generated in the transformation of natural capital to human-made capital. It does not allow human-made capital to "take the place of" natural capital. Secondly, because of the first and second laws of thermodynamics, there is a limit to how much production waste can be reduced by technological progress. This is because 100 per cent technical efficiency is physically impossible; there can never be 100 per cent recycling of matter; and energy cannot be recycled at all.[8] Thirdly, a value of one or more for the elasticity of substitution between human-made and natural capital is necessary to demonstrate the long-run substitutability of the former for the latter. Disconcertingly, recent studies have shown that the value of the elasticity of substitution derived from a production function obeying the first and second laws of thermodynamics is always less than one (Lawn 2007).

Thus, all things considered, human-made capital and natural capital are complements not substitutes. Consequently, the production of a given quantity of physical goods requires a minimum, irreducible rate of resource use and, therefore, a minimum quantity of resource-providing natural capital (Meadows et al. 1972; Pearce et al. 1989; Folke et al. 1994; Daly 1996; Lawn 2007). It is for this reason that ecological economists believe the strong sustainability approach to capital maintenance is necessary to sustain the economic process.

However, before a satisfactory answer can be given to the first of the above questions, it is necessary to consider the minimum amount of natural capital required to ensure ecological sustainability. It is at this point that we must go beyond production possibilities and turn our attention to the life-support function of natural capital.

The ability of natural capital to provide life-support services exists because the ecosphere, as a far-from-thermodynamic-equilibrium system characterised by a range of biogeochemical clocks and essential feedback mechanisms, has developed the self-organisational capacity to regulate the conditions necessary for life. There has, unfortunately, been a growing tendency for human beings to take the conditions for life for granted—a consequence of technological optimism and a growing detachment most people have from the vagaries of the natural world. In particular, two falsely held beliefs have emerged. The first is a widely held conviction that the Earth's current uniqueness for life was preordained. This is patently untrue, since, as Blum (1962) has explained, had any one of an infinite number of past events occurred only marginally differently, the evolution of living organisms on Earth might never have eventuated.

Secondly, it is widely believed that organic evolution is confined to living organisms responding to exogenously determined environmental factors. However, it is now transparently clear that 'fitness' is a byproduct of the coevolutionary relationship that exists between the ecosphere and its constituent species. Indeed, the ecosphere is as uniquely suited to existing species as are the latter to the ambient characteristics of the ecosphere. Hence, according to Blum (1962, p. 61), it is “impossible to treat the environment as a separable aspect of the problem of organic evolution; it becomes an integral part thereof.” Clearly, just as current environmental conditions were not preordained, the environmental conditions of the future will always be strongly influenced by the evolution of constituent species and, in particular, the actions of recalcitrant specimens.

An awareness of the above brings to bear a critical point. Although human intervention can never ensure the Earth remains eternally fit for human habitability, humankind does have the capacity to bring about a premature change in its prevailing comfortable state. Many people believe that anthropogenic global warming is just one of many signs of a radical change in the planet’s comfortable conditions. Nonetheless, there are some observers who argue that these events, if occurring, are of no great concern since they are little more than symptoms of a benign coevolutionary adjustment brought on by the eccentricities of humankind. That is, any malady caused by human activity will be short-lived because whatever may threaten the human habitability of the planet will induce the evolution of a new and more comfortable environmental state. For such observers, humankind is potentially immune from the consequences of its own actions.

Nothing, however, could be further from the truth. The quasi-immortality of the ecosphere prevails because of the informal association that exists between the global system and its constituent species. But quasi-immortality in no way extends to any particular species. Indeed, historical evidence indicates a tendency for the global system to correct ecological imbalances in ways that are invariably unpleasant for incumbent species. Hence, while the Earth has revealed itself to be immune to the emergence of wayward species (e.g., oxygen-bearers in the past), individual species—including human beings—are in no way immune from the consequences of their own collective folly. We can therefore conclude that the quantity of natural capital needed to ensure ecological sustainability is likely to greatly exceed the quantity needed for production purposes alone. Of course, this still leaves the first of the above questions unanswered.

Deeper insight into the minimum required natural capital can be gained by considering what bestows natural capital with the unique capacity to support life. Is it the quantity of natural capital or is it some particular aspect of it? Lovelock leaves us in no doubt by emphasising that a minimum number and complexity of species are required to establish, develop, and maintain the Earth’s biogeochemical clocks and essential feedback mechanisms. To wit:

> The presence of a sufficient array of living organisms on a planet is needed for the regulation of the environment. Where there is incomplete occupation, the ineluctable forces of physical or chemical evolution would soon render it uninhabitable (Lovelock 1988, p. 63).

It is, therefore, a combination of the interactions and interdependencies between the various species, the diversity of species, and the complexity of ecological systems—in all, the *biodiversity* present in natural capital—that underpins its life-supporting function. This doesn’t mean that the quantity of natural capital is unimportant. The quantity is vital if only because the biodiversity needed to maintain the Earth’s habitable status requires a full, not partial, occupation by living organisms. But the quantity of natural capital, itself, should never be equated with biodiversity.

If the sheer magnitude of natural capital is an inadequate indication of the effectiveness with which it can support life, what is the minimum level of biodiversity needed to maintain the ecosphere's life-support function? Unfortunately, this is not known, although there is general agreement that some semblance of a biodiversity threshold does exist. What we do know is that in the same way biodiversity begets greater biodiversity, so do diminutions in biodiversity beget further diminutions (Norton 1986). It is also known that the present rate of species extinction is far exceeding the rate of speciation—indeed, so much that biodiversity has, on any relevant time scale, become a non-renewable resource (Daily and Ehrlich 1992).

Given that a rise in the global rate of extinction increases the vulnerability of human beings to its own extinction, a sensible risk-averse strategy for humankind to adopt is a rigid adherence to a biodiversity 'line in the sand'.[9] Ehrlich (1993) provides a hint as to where this line should be drawn by pointing out that humankind knows enough about the value of biodiversity to operate on the principle that "all reductions in biodiversity should be avoided because of the potential threats to ecosystem functioning and its life-support role". As a corollary of Ehrlich's dictum, a line should be drawn at the currently existing level of biodiversity, especially given the magnitude of the loss of biodiversity over the past century. Conscious efforts should also be made to preserve remnant vegetation and important ecosystems.

We are now in a position to answer the second of the above questions—that is, what sustainability precepts should humankind follow to prevent the decline in the quantity and quality of natural capital stocks? There are essentially four fundamental rules-of-thumb or precepts requiring adherence:

1. The rate of renewable resource extraction should not exceed the regeneration rate of renewable resource stocks.
2. The depletion of non-renewable resources should be offset by the cultivation of renewable resource substitutes.
3. The rate of high-entropy waste generation should not exceed the ecosphere's waste-assimilative capacity. This third precept is of great relevance to the phenomenon of anthropocentric global warming.
4. Native vegetation and critical ecosystems should be preserved, rehabilitated, and/or restored. In addition, future exploitation of natural capital should be confined to areas already strongly modified by previous human activities.

2.1.3.2 Human Needs and the Principle of Intragenerational Equity

It has already been explained that human well-being depends critically on the psychic enjoyment of life. Despite having a good sense of what contributes directly to net psychic income, it is worth contemplating the extent to which each contributing factor is likely to advance the human condition. Although this will differ from culture to culture, a greater understanding can be gained by considering Maslow's (1954) hierarchy of human needs (Fig. 2.3).

Maslow's hierarchical framework is underpinned by a rigorous psychological theory of human motivation in which human needs are classified into five distinct categories. Beginning with lower-order needs and ascending through to higher-order needs, the five categories of human needs are:

- Basic *physiological needs*—i.e., one's basic need for food, clothing, and shelter.
- *Safety needs*—i.e., the need for physical and mental security; freedom from fear, anxiety, and chaos; and the need for stability, dependency, and protection.
- The need for *belongingness and love*—i.e., the need for affectionate relationships with people in general; the hunger for contact and intimacy; the desire for a sense of place in one's family or peer group; and the need to avoid the pangs of loneliness, ostracism, rejection, and rootlessness.
- The need for *esteem*—i.e., the desire for strength, achievement, adequacy, mastery, and competence; the need for independence and freedom; and the desire for recognition, attention, importance, dignity, and appreciation.
- *Self-actualisation needs*—i.e., the desire for self-fulfilment or, equivalently, the desire to become fully actualised in what one is capable of becoming.

By organising human needs into a hierarchy of relative prepotency, Maslow's needs hierarchy represents the human personality as an integrated whole in which every part, level, and dimension is interdependent. The framework itself indicates that human endeavour is initially devoted to the satisfaction of basic physiological needs. Once satisfied, the greater part of one's endeavours shifts to the next tier of

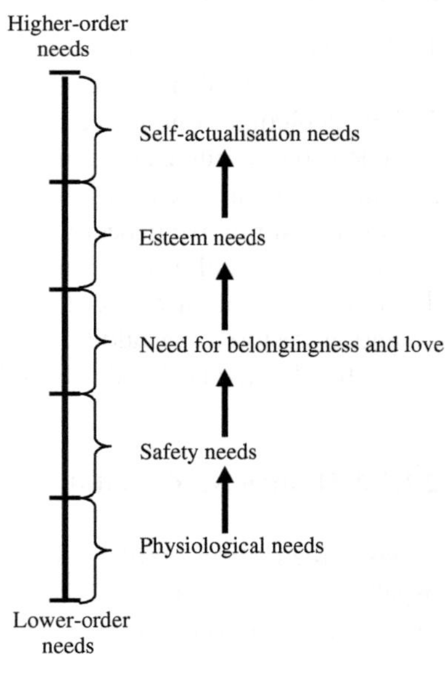

Fig. 2.3 Maslow's needs hierarchy (Maslow 1954)

human needs. The process ultimately culminates with the need for self-actualisation, which, according to Maslow, is the most creative and rewarding phase of the human development process. Overall, Maslow's needs hierarchy highlights that a healthy human existence requires the satisfaction of emerging higher-order needs as well as basic physiological needs. Achievement of this desirable human state has previously been described as a healthy 'existential balance' (Weisskopf 1973).

It is important to realise that the economic process need not always operate in a manner consistent with the adequate satisfaction of emerging higher-order needs. There are two reasons why. Firstly, physiological need satisfaction (such as being well fed) has no enduring qualities. Hence, the satisfaction of lower-order needs requires frequent engagement in physiological need-satisfying activities (such as eating regularly). Secondly, if higher-order needs are being inadequately satisfied, it is still possible to obtain a sense of equilibrium—albeit an unhealthy, unbalanced one—by engaging in more physiological need-satisfying pursuits (such as increased consumption). Because physiological need satisfaction quickly evaporates, the desire for more consumption requires greater production, which in turn reduces the time available to satisfy their higher-order needs. The increased lack of psychological need satisfaction further intensifies the desire for higher rates of production and consumption. Eventually, an illusionary need for continued growth becomes self-perpetuating. In a coevolutionary world characterised by path-dependency, a growth addiction can arise even though it may be contrary to the betterment of the human condition. This growth addiction is commonly referred to as 'consumerism' or the 'treadmill of production' (Schnaiberg 1980).

What does this all mean in terms of the human developmental process? To begin with, it is patently obvious that increasing the supply of intermediate means along one level of need at the expense of needs on a different level disturbs the balance of human existence (Kenny 1999). Thus, once lower-order needs have been satisfied, it follows that attempts to expand the stock of human-made capital should largely cease once it begins to impede the satisfaction of higher-order needs. Moreover, it suggests that human development demands, at the very least, a widespread concern for posterity and an ongoing need to address intragenerational inequities and injustices. Clearly, this means upholding various universal rights and privileges, one of which must be the eradication of absolute poverty. Not only does poverty alleviation ensure the satisfaction of basic physiological needs, it provides the foundation upon which higher-order needs can be subsequently satisfied.

One of the best ways to uphold the principle of intragenerational equity is to ensure access to paid employment for anyone who seeks it. Although unemployed people in high-GDP countries are rarely deprived of basic lower-order needs, they are often deprived of their safety and esteem needs. In almost all instances, they are starved of the means required to satisfy their self-actualisation needs. This invariably leads to alienation, disillusionment, depression, and the increased likelihood of committing a serious crime (Fryer 1995; Feather 1997; Sen 1997; Theodossiou 1998; Harvey 2000; Watts and Mitchell 2000; Biddle 2001; Layard 2005). Unemployment also results in the loss of human capital skills and the depreciation of a nation's productive capacity (Mitchell and Muysken 2008). There is little doubt, therefore, that the negative side-effects of unemployment constitute a welfare-reducing cost that impacts on society generally, not just on the unemployed. For this reason, the cost of unemployment should be included in aggregate measures of economic welfare. Moreover, unemployment should be given more consideration by policy-makers than it is at present. Indeed, since it can be argued that access to paid employment is a basic human right (Burgess and Mitchell 1998; Lawn 2009), full employment should be viewed as an obligatory objective of any nation aiming to achieve sustainable development.

Having said this, there is the potential for the full employment objective to conflict with the goal of ecological sustainability (Lawn 2009). Under the current institutional arrangements in virtually every country, there is a well-established link between real GDP and employment levels. This link compels governments to expand the economy to prevent the rise in unemployment.[10] Because, as we shall see, the continued expansion of the economy is both undesirable and ecologically unsustainable, it will become increasingly important to discover ways to sever the GDP-employment nexus so that full employment can be achieved without the perceived need for continued growth.[11]

2.1.4 A Broad Definition of Sustainable Development

Taking account of the aforementioned, the following will serve as our broad definition of sustainable development: "A nation is achieving sustainable development if it undergoes a coevolutionary process that improves the total quality of life of every citizen, both now and into the future, while ensuring its rate of resource use does not exceed the regenerative and waste-assimilative capacities of the natural environment. It is also a nation that ensures the survival of the biosphere and all its evolving processes while recognising, to some extent, the intrinsic value of sentient non-human beings."

Despite this definition being open to individual interpretation, its strength lies in its emphasis on: (i) the quality of human life, not simply the material standard of living; (ii) the welfare of *all* people, present and future; and (iii) the need to preserve the ecosphere upon which all welfare-related activities depend—a critical factor given the complementary relationship between human-made capital and natural capital. The emphasis on these three aspects suggests that sustainable development can be defined more narrowly as *non-declining economic welfare*, which, at the very least, requires natural capital to be kept intact (Lawn 2007). Provided we do not stray from the central tenets of the broad definition above, this narrow view of sustainable development can be of great practical use in that it is possible to measure natural capital and a nation's aggregate economic welfare. As we shall see, this makes it feasible to determine whether a nation is operating sustainably and whether the economic welfare being generated by its economic activities is rising over time.

2.2 Sustainable Development and the Ecological and Economic Limits to Growth

2.2.1 *Economic and Uneconomic Growth*

In view of the above definition of sustainable development, two further questions naturally emerge:

- Firstly, how big can the economy get before the throughput of matter-energy required to maintain it can no longer be ecologically sustained?
- Secondly, how big can the economy get before the additional costs of growth begin to exceed the additional benefits, at which point the economic welfare generated by a physically expanding economy begins to decline?

Unbeknown to many, the answers to these questions are not the same. This is because the first question relates to a physical scale of the economy that ought to be avoided at all costs whereas the second question relates to a physical scale we would be better off avoiding even if the long-term consequences of reaching it are not ecologically catastrophic. As we shall see, the desirable or optimal scale of the economy (the answer to the second question) is considerably smaller than the economy's maximum sustainable scale (the answer to the first question). This has important consequences for how humankind should deal with climate change.

To answer the above questions, the two elemental categories of net psychic income (uncancelled benefits) and lost natural capital services (uncancelled costs) can be diagrammatically presented to demonstrate the economic and ecological impacts of a growing economy. Consider Fig. 2.4 where we shall ignore efficiency-increasing technological progress for the moment and assume that all technological advances are of the throughput-increasing kind. Throughput-increasing technological progress enables a nation to augment the rate of resource throughput that, in turn, allows it to physically expand its economy. In Fig. 2.4, the

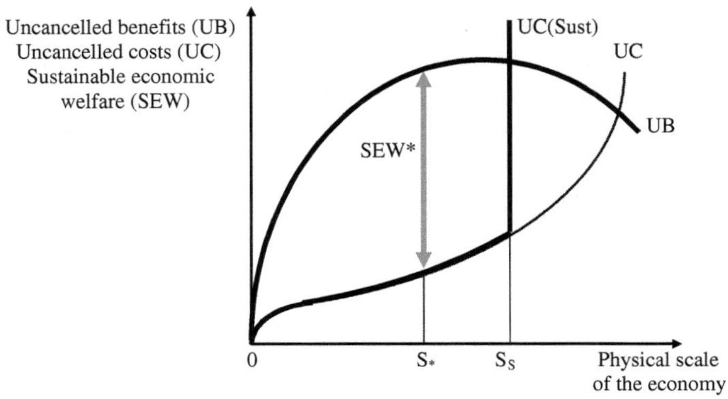

Fig. 2.4 The sustainable economic welfare generated by a growing economy

physical expansion of the economy is represented by a rightward movement along the horizontal axis. The uncancelled benefits and uncancelled costs are respectively represented by the UB and UC curves.

In keeping with the principle of diminishing marginal benefits, we can expect the uncancelled benefits associated with a growing economy to increase at a declining rate. Conversely, due to the principle of increasing marginal costs, we can expect the uncancelled costs to rise at an increasing rate.[12] The shapes of the UB and UC curves in Fig. 2.4 reflect these two standard economic principles. From a sustainability perspective, the UC curve is vertical at the point where the economy reaches its *maximum sustainable scale* of S_S (i.e., the largest physical scale of the economy consistent with a sustainable rate of resource throughput). As Fig. 2.4 shows, growth of the economy up to a physical scale of S_S is ecologically sustainable. Although growth beyond S_S is technically feasible for a short period of time (i.e., by drawing down stocks of natural capital), it is ecologically unsustainable in the long-run.

From an economic perspective, matters differ considerably. The economic welfare generated by a growing economy is measured by the vertical distance between the UB and UC curves. Figure 2.4 indicates that growth up to a physical scale of S_* increases benefits faster than costs. As such, it increases a nation's economic welfare and thus constitutes a form of 'economic' growth. However, growth beyond S_* reduces economic welfare. That is, physical expansion beyond the *optimal macroeconomic scale* (i.e., where sustainable economic welfare is maximised) increases costs faster than benefits. It therefore constitutes a form of 'uneconomic' growth and ought to be avoided. The critical message here is that growth beyond the optimal scale becomes economically undesirable even though the physical expansion of the economy between S_* and S_S is ecologically sustainable. In all, Fig. 2.4 demonstrates that an economic limit to growth (S_*) is likely to precede an ecological limit to growth (S_S).

2.2.2 Efficiency-Increasing Technological Progress

The previous analysis was somewhat over-simplified in that it ignored the possibility of efficiency-increasing technological progress. Technological advances of the efficiency-increasing kind bring about an upward shift of the UB curve or a downward/rightward shift of the UC curve. To explain how, we can arrange the uncancelled benefits and uncancelled costs to arrive at a macro measure of efficiency—sometimes referred to as 'ecological economic efficiency' (EEE):

$$\text{EEE} = \frac{\textit{Uncancelled benefits}}{\textit{Uncancelled costs}} = \frac{\textit{Net psychic income}}{\textit{Lost natural capital services}} \qquad (2.1)$$

For a given physical scale of the economy, an increase in the EEE ratio indicates an improvement in the efficiency with which natural capital and the low-entropy resources it provides are transformed into service-yielding human-made capital. A multitude of factors can contribute to an increase in the EEE ratio.

To demonstrate how, the EEE ratio can be decomposed to reveal the following four eco-efficiency ratios (Daly 1996):

$$\text{EEE} = \frac{\text{NPY}}{\text{LNCS}} = \overset{\text{Ratio 1}}{\frac{\text{NPY}}{\text{HMK}}} \times \overset{\text{Ratio 2}}{\frac{\text{HMK}}{\text{RT}}} \times \overset{\text{Ratio 3}}{\frac{\text{RT}}{\text{NK}}} \times \overset{\text{Ratio 4}}{\frac{\text{NK}}{\text{LNCS}}} \tag{2.2}$$

where:

- EEE = ecological economic efficiency
- NPY = net psychic income
- LNCS = lost natural capital services
- HMK = human-made capital
- RT = resource throughput
- NK = natural capital.

The order in which the four eco-efficiency ratios are presented in Eq. (2.2) is in keeping with the nature of the economic process—that is, net psychic income is enjoyed as a consequence of the creation and maintenance of human-made capital (Ratio 1); the maintenance of human-made capital requires the continued throughput of matter-energy (Ratio 2); the throughput of matter-energy is made possible thanks to the three instrumental services provided by natural capital (Ratio 3); and, in exploiting natural capital, the three instrumental services provided by natural capital are to some degree sacrificed (Ratio 4). Each eco-efficiency ratio represents a different form of efficiency pertaining to a particular sub-problem contained within the larger problem of achieving sustainable development. The four eco-efficiency ratios will now be individually explained along with the implications they have for the UB and UC curves.

2.2.2.1 Beneficial Shifts of the Uncancelled Benefits (UB) Curve

Ratio 1 is a measure of the *service efficiency* of human-made capital. It increases whenever a given amount of human-made capital yields a higher level of net psychic income. An increase in Ratio 1 causes the UB curve to shift upwards (see Fig. 2.5). This can be achieved by improving the technical design of all newly produced goods, altering the composition of final output (i.e., producing a greater proportion of goods with higher service-yielding qualities), or by advancing the means by which human beings organise themselves in the course of producing and maintaining the stock of human-made capital. The latter is important because it can reduce such things as the disutility of labour, the cost of commuting, and the cost of unemployment. A beneficial shift in the UB curve can also be achieved by redistributing income from the low marginal service or psychic income uses of the rich to the higher marginal service uses of the poor (Robinson 1962).[13]

Figure 2.5 illustrates what happens to sustainable economic welfare when the UB curve shifts upwards. Because an increase in Ratio 1 augments the net psychic

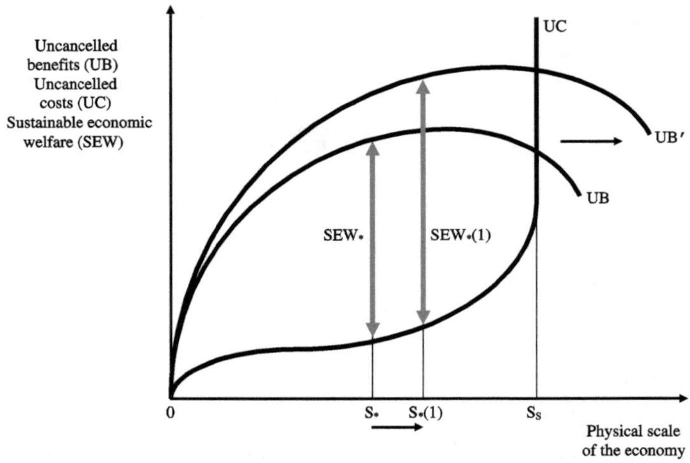

Fig. 2.5 An increase in sustainable economic welfare brought about by an upward shift of the UB curve

income yielded by a given amount of human-made capital, the UB curve shifts up to UB′. As a consequence, sustainable economic welfare is no longer maximised at the prevailing scale of S*. In the circumstances depicted in Fig. 2.5, it is economically desirable to expand the economy to the new optimal scale of S*(1).[14] In the process, sustainable economic welfare increases to SEW*(1).

2.2.2.2 Beneficial Shifts of the Uncancelled Costs (UC) Curve

Shifts of the UC curve arise as a consequence of changes in Ratios 2, 3, and 4. Ratio 2 is a measure of the *maintenance efficiency* of human-made capital. It increases whenever a given physical magnitude of human-made capital can be maintained by a lower rate of resource throughput. This can be achieved via any one of the following advances: (i) an increase in the technical efficiency of production; (ii) increased rates of material recycling; (iii) greater product durability; and (iv) improved operational efficiency (Lawn 2000). An increase in Ratio 2 causes the UC curve to shift downwards and to the right by enabling a given physical scale of the economy to be sustained by a reduced rate of resource throughput. This lessens the natural capital that needs to be exploited, including the Earth's greenhouse gas-absorbing sinks, which correspondingly results in the loss of fewer natural capital services.

Ratio 3 is a measure of the *growth efficiency* or productivity of natural capital. This form of efficiency is increased whenever a given amount of natural capital can sustainably yield more low-entropy resources and/or assimilate a greater quantity of high-entropy waste. Better management of natural resource systems and the preservation of critical ecosystems can lead to a more productive stock of natural capital. From a climate change perspective, an increase in the productivity of natural capital would be reflected by an increased capacity of the ecosphere to sequester greenhouse gases. An increase in Ratio 3 leads to a downward and rightward shift of the UC curve because a more productive stock of natural capital reduces the quantity of natural capital that must be exploited to obtain the throughput of matter-energy needed to sustain the economy at a given physical scale. This allows an economy of a particular scale to be sustained at the expense of fewer natural capital services.

Ratio 4 is a measure of the *exploitative efficiency* of natural capital. An increase in Ratio 4 occurs whenever there is a reduction in the natural capital services lost from directly exploiting a given quantity of natural capital. Once again, advances of this nature allow an economy of a particular physical scale to be sustained at the expense of fewer natural capital services. Increases in Ratio 4 can be obtained through the development and execution of more sensitive resource-extraction techniques, such as the use of underground rather than open-cut mining practices.

Figure 2.6 illustrates what happens following a beneficial shift of the UC curve. Because increases in Ratios 2, 3, and 4 reduce the uncancelled cost of maintaining an economy at a given physical scale, the UC curve shifts down and out to UC′. In doing so, the maximum sustainable scale of the economy increases from S_S to $S_S(1)$. At the same time, it becomes economically desirable to grow the economy to $S_*(2)$. Expansion to the new optimum increases sustainable economic welfare to $SEW_*(2)$.

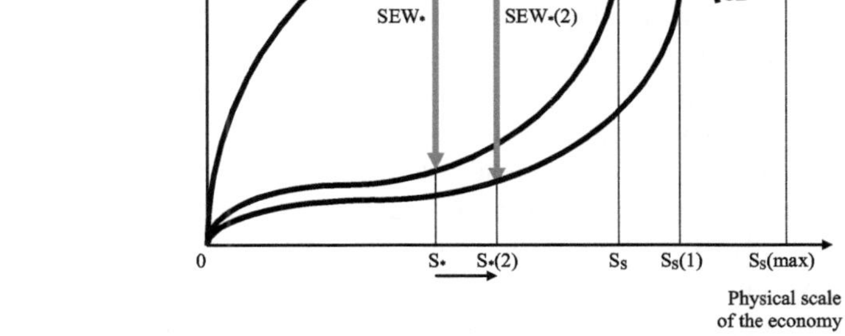

Fig. 2.6 A change in sustainable economic welfare brought about by a rightward/downward shift of the UC curve

2.2.2.3 Limits to the Beneficial Shifts of the UB and UC Curves and the Necessity of the Steady-State Economy

There is considerable debate surrounding how much and for how long humankind can beneficially shift the UB and UC curves. Because of biophysical constraints, there are many observers who correctly point out that humankind's ability to shift the UC curve is ultimately limited. As previously stressed, the first and second laws of thermodynamics preclude the 100 per cent technical efficiency of production and the 100 per cent recycling rate of waste materials. Both laws also forbid any recycling of waste energy. Consequently, both laws limit increases in Ratio 2. On top of this, it is impossible to continuously increase the productivity of natural capital (Ratio 3) and prevent the loss of at least some of the ecosphere's three instrumental services (Ratio 4). In view of these constraints, it follows that an upper limit must exist on the maximum sustainable scale of economic systems. This ultimate ecological limit is represented in Fig. 2.6 by the dotted line at the physical scale of S_S(max).

What about the UB curve? Is there a limit on its upward adjustment? This is a more complex issue because net psychic income is not subject to the same physical laws as physical goods. Having said this, the following should be borne in mind. Firstly, because human-made capital is required to experience the welfare generated by human activities, net psychic income always has a physical foundation. Hence, growth in the physical basis of human well-being always remains biophysically limited. Secondly, there is a probable limit on the human capacity to experience a sense of psychic well-being. A human being can only be so happy, contented, and fulfilled. Finally, efforts to increase human well-being by altering the composition of what humans consume are severely restricted. Thus, even if the information provided by cyberspace can offer wonderful welfare-increasing opportunities, a hungry person requires a meal to be fed, not a recipe. The same person also requires physical shelter, clothes, and heating/cooling to remain dry and comfortable. No amount of downloaded images or information can directly supply these physiological requirements.

Although there is, as a consequence, a clear limit on the human capacity to experience a sense of psychic well-being, there are good reasons to believe that the potential to shift the UB curve is far from exhausted. Moreover, because of impending ecological limits to growth (i.e., severe limits to the beneficial shift of the UC curve), achieving sustainable development will require all nations to eventually make the transition to a steady-state economy—preferably settling somewhere near the optimal scale. Given this steady-state imperative, it is clear that the goal of physically expanding the economy (growth) must give way to an emphasis on qualitative improvement (development). This means that all nations will need to shift their focus towards qualitatively advancing the stock of physical goods; ensuring the stock is more equitably distributed; minimising the rate of resource throughput; and reorganising the production process to increase job satisfaction and reduce any associated social costs. Provided these advances can be made, the necessity of the steady-state economy should not be a cause for concern. In fact, unlike growing the economy beyond its ecological and economic limits, there is no reason why a qualitatively-improving steady-state economy should not deliver increasing levels of sustainable economic welfare for many years to come.

2.3 Empirical Evidence of the Ecological and Economic Limits to Growth

2.3.1 Ecological Limits to Growth: Ecological Footprint Versus Biocapacity

As important as it is to recognise the inevitability of a steady-state economy, it is equally important to know where a nation's economy is in relation to its maximum sustainable scale (S_S) and optimal scale (S_*). Without this information, it is impossible to ascertain when to initiate the transition to a steady-state economy. Furthermore, it is impossible to know whether the transition will require a nation to gradually slow the growth of its economy (i.e., decelerate towards a larger economy) or reduce its physical scale (i.e., settle at a smaller economy).

A number of indicators have been developed to determine whether economic systems are nearing or have surpassed their ecological limit (Vitousek et al. 1986; Yale Center for Environmental Law and Policy et al. 2005; Pearce and Atkinson 1993; Wackernagel et al. 1999). Despite its conservative nature, many people believe that the ecological footprint is the best indicator of ecological limits so far established.[15] A country's ecological footprint represents the area of land *required* to generate the resources, absorb the wastes, and provide the critical ecosystem services needed to sustain economic activity at its current level (Wackernagel and Rees 1996). To determine if a nation's economy has exceeded its maximum sustainable scale, the ecological footprint is compared to its biocapacity. A nation's biocapacity is indicated by the quantity of land *available* to generate an ongoing supply of resources, absorb wastes, and maintain critical ecosystem services. Ecological unsustainability (ecological deficit) occurs if a nation's ecological footprint exceeds its biocapacity.

Table 2.1 reveals that, in 2005, 78 of 143 surveyed nations had ecological footprints in excess of their biocapacities (Global Footprint Network 2008). Some observers have suggested that ecological deficits are not a problem given that countries with ecological surpluses can aid deficit nations by exporting their surplus resources and/or by importing deficit countries' surplus wastes. However, Table 2.1 shows that the global ecological footprint exceeded the Earth's biocapacity by an amount equal to 0.6 global hectares per person. It also shows that 1.3 Earths are required to sustain the rate of global consumption at 2005 levels (Global Footprint Network 2008).[16] Since this situation cannot continue indefinitely, it follows that the ecological surpluses enjoyed by some nations are insufficient to 'finance' the combined ecological deficits of the remainder. Overall, it is

Table 2.1 Ecological footprint (EF) and biocapacity (2005)

	Ecological footprint *(gha per person)*	Bio-capacity *(gha per person)*	Ecological deficit (−) surplus (+) *(gha per person)*		Ecological footprint *(gha per person)*	Bio-capacity *(gha per person)*	Ecological deficit (−) surplus (+) *(gha per person)*		Ecological footprint *(gha per person)*	Bio-capacity *(gha per person)*	Ecological deficit (−) surplus (+) *(gha per person)*
World	2.7	2.1	−0.6								
Africa											
Algeria	1.7	0.9	−0.7	Gabon	1.3	25.0	23.7	Namibia	3.7	9.0	5.3
Angola	0.9	3.2	2.3	Gambia	1.2	1.2	0.0	Niger	1.6	1.8	0.2
Benin	1.0	1.5	0.5	Ghana	1.5	1.2	−0.3	Nigeria	1.3	1.0	−0.4
Botswana	3.6	8.5	4.8	Guinea	1.3	3.0	1.8	Rwanda	0.8	0.5	−0.3
Burkina Faso	2.0	1.6	−0.4	Guinea-Bissau	0.9	3.4	2.5	Senegal	1.4	1.5	0.2
Burundi	0.8	0.7	−0.1	Kenya	1.1	1.2	0.1	Sierra Leone	0.8	1.0	0.2
Cameroon	1.3	3.1	1.8	Lesotho	1.1	1.1	0.0	Somalia	1.4	1.4	0.0
Cent Afr Rep	1.6	9.4	7.8	Liberia	0.9	2.5	1.6	South Africa	2.1	2.2	0.1
Chad	1.7	3.0	1.3	Libya	4.3	1.0	−3.3	Sudan	2.4	2.8	0.4
Congo	0.5	13.9	13.3	Madagascar	1.1	3.7	2.7	Swaziland	0.7	1.7	0.9
Congo Dem Rep	0.6	4.2	3.6	Malawi	0.5	0.5	0.0	Tanzania	1.1	1.2	0.1
Cote d'Ivoire	0.9	2.2	1.3	Mali	1.6	2.6	0.9	Tunisia	1.8	1.1	−0.6
Egypt	1.7	0.4	−1.3	Mauritania	1.9	6.4	4.5	Uganda	1.4	0.9	−0.4
Eritrea	1.1	2.1	0.9	Morocco	1.1	0.7	−0.4	Zambia	0.8	2.9	2.1
Ethiopia	1.4	1.0	−0.3	Mozambique	0.9	3.4	2.5	Zimbabwe	1.1	0.7	−0.4
Asia-Pacific											
Australia	7.8	15.4	7.6	Korea DPRP	1.6	0.6	−0.9	New Zealand	7.7	14.1	6.4
Bangladesh	0.6	0.3	−0.3	Korea Republic	3.7	0.7	−3.0	Pakistan	0.8	0.4	−0.4
Cambodia	0.9	0.9	0.0	Laos	1.1	2.3	1.3	Papua NG	1.7	4.4	2.8
China	2.1	0.9	−1.2	Malaysia	2.4	2.7	0.3	Philippines	0.9	0.5	−0.3
India	0.9	0.4	−0.5	Mongolia	3.5	14.6	11.2	Sri Lanka	1.0	0.4	−0.6
Indonesia	0.9	1.4	0.4	Myanmar	1.1	1.5	0.4	Thailand	2.1	1.0	−1.2
Japan	4.9	0.6	−4.3	Nepal	0.8	0.4	−0.4	Vietnam	1.3	0.8	−0.5

(continued)

Table 2.1 (continued)

	Ecological footprint *(gha per person)*	Bio-capacity *(gha per person)*	Ecological deficit (−) surplus (+) *(gha per person)*		Ecological footprint *(gha per person)*	Bio-capacity *(gha per person)*	Ecological deficit (−) surplus (+) *(gha per person)*		Ecological footprint *(gha per person)*	Bio-capacity *(gha per person)*	Ecological deficit (−) surplus (+) *(gha per person)*
Latin America											
Argentina	2.5	8.1	5.7	Dominican Rep	1.5	0.8	−0.7	Mexico	3.4	1.7	−1.7
Bolivia	2.1	15.7	13.6	Ecuador	2.2	2.1	−0.1	Nicaragua	2.0	3.3	1.2
Brazil	2.4	7.3	4.9	El Salvador	1.6	0.7	−0.9	Panama	3.2	3.5	0.3
Chile	3.0	4.1	1.1	Guatemala	1.5	1.3	−0.2	Paraguay	3.2	9.7	6.5
Colombia	1.8	3.9	2.1	Haiti	0.5	0.3	−0.3	Peru	1.6	4.0	2.5
Costa Rica	2.3	1.8	−0.4	Honduras	1.8	1.9	0.1	Uruguay	5.5	10.5	5.0
Cuba	1.8	1.1	−0.7	Jamaica	1.1	0.6	−0.5	Venezuela	2.8	3.2	0.3
North America											
Canada	7.1	20.0	13.0	USA	9.4	5.0	−4.8				
M-E & Cent Asia											
Afghanistan	0.5	0.7	0.3	Israel	4.6	0.4	−4.2	Syria	2.1	0.8	−1.2
Armenia	1.4	0.8	−0.6	Jordan	1.7	0.3	−1.4	Turkey	2.7	1.7	−1.1
Azerbaijan	2.2	1.0	−1.1	Kazakhstan	3.4	4.3	0.9	Turkmenistan	3.9	3.7	−0.2
Georgia	1.1	1.8	0.7	Kuwait	8.9	0.5	−8.4	UAE	9.5	1.1	−8.4
Iran	2.7	1.4	−1.3	Lebanon	3.1	0.4	−2.7	Uzbekistan	1.8	1.0	−0.8
Iraq	1.3	0.3	−1.1	Saudi Arabia	2.6	1.3	−1.4	Yemen	0.9	0.6	−0.3

(continued)

Table 2.1 (continued)

	Ecological footprint *(gha per person)*	Bio-capacity *(gha per person)*	Ecological deficit (−) surplus (+) *(gha per person)*		Ecological footprint *(gha per person)*	Bio-capacity *(gha per person)*	Ecological deficit (−) surplus (+) *(gha per person)*		Ecological footprint *(gha per person)*	Bio-capacity *(gha per person)*	Ecological deficit (−) surplus (+) *(gha per person)*
EU25											
Austria	5.0	2.9	−2.1	Germany	4.2	1.9	−2.3	Netherlands	4.4	1.1	−3.3
Belg. and L/bourg	5.1	1.1	−4.0	Greece	5.9	1.7	−4.2	Poland	4.0	2.1	−1.9
Czech Republic	5.4	2.7	−2.6	Hungary	3.5	2.8	−0.7	Portugal	4.4	1.2	−3.2
Denmark	8.0	5.7	−2.3	Ireland	6.3	4.3	−2.0	Slovakia	3.3	2.8	−0.5
Estonia	6.4	9.1	2.7	Italy	4.8	1.2	−3.5	Spain	5.7	1.3	−4.4
Finland	5.2	11.7	6.5	Latvia	3.5	7.0	3.5	Sweden	5.1	10.0	4.9
France	4.9	3.0	−1.9	Lithuania	3.2	4.2	1.0	UK	5.3	1.6	−3.7
Rest of Europe											
Albania	2.2	1.2	−1.0	Croatia	3.2	2.2	−1.0	Russia	3.7	8.1	4.4
Belarus	3.9	3.4	−0.4	Macedonia	4.6	1.4	−3.2	Serbia and	2.6	1.6	−1.0
Bosnia Herzeg.	2.9	2.0	−0.9	Norway	6.9	6.1	−0.8	Mont.	5.0	1.3	−3.7
Bulgaria	2.7	2.8	0.1	Romania	2.9	2.3	−0.6	Switzerland	2.7	2.4	−0.3
								Ukraine			

Notes
- Totals may not add up due to rounding
- gha denotes global hectares

Source Global Footprint Network (2006)

apparent that the economies of most nations, as well as the global economy as a whole, have surpassed their maximum sustainable scale.

2.3.2 *Economic Limits to Growth: The Genuine Progress Indicator*

Unlike biophysical indicators, less work has been undertaken to determine whether economic systems have exceeded their optimal scale. However, an indicator has recently emerged that incorporates around twenty-five benefit and cost items of the economic, social, and environmental variety. Known as the Genuine Progress Indicator (GPI), this indicator involves subtracting the costs of economic activity from the benefits it generates to obtain a macroeconomic estimate of economic welfare.[17] From a diagrammatical perspective, the GPI is equivalent to the vertical distance between the UB and UC curves in Fig. 2.4.

GPI studies have been predominantly conducted on high-GDP nations. A recent project involving seven countries in the Asia-Pacific region has boosted the number of GPI studies of poor nations (Lawn and Clarke 2008). The results of the Asia-Pacific study are very illuminating and I shall refer to the significance of them soon. For now, consider Fig. 2.7 which reveals the results of GPI studies conducted on six wealthy nations in the 1990s. In all six cases, the GPI initially rises in unison with real GDP. A point is then reached where the GPI decreases or plateaus, although the timing of this turning point differs between nations. What doesn't alter significantly is that the rise in the GPI ceases once a nation's per capita GDP reaches Int$15,000 to Int$20,000 (2004 prices).[18]

Although real GDP is not strictly an indicator of the physical scale of a nation's economy, the initial decline in the GPI within this per capita GDP range suggests that all six countries have surpassed their optimal scale (S_* in Fig. 2.4). This implies they have all exceeded their economic limit to growth. Disturbingly, the GPI results of other wealthy nations reveal a similar pattern (see Diefenbacher 1994; Moffatt and Wilson 1994; Rosenberg and Oegema 1995; Jackson et al. 1997; Stockammer et al. 1997; Guenno and Tiezzi 1998; Makino 2008).[19]

The uniform trend displayed in Fig. 2.7 was first recognised in the mid-1990s. It led Max-Neef (1995) to put forward a 'threshold hypothesis'—namely, when a nation's per capita GDP exceeds a critical threshold, one can expect its per capita economic welfare to plateau or decline. As disconcerting as this was for countries at or beyond the threshold, the hypothesis offered comfort to the world's poorest nations. That is, with the per capita GDP of poor countries well below Int$15,000, the theory suggested that a positive relationship should exist for some time between the growth of their economies and the economic welfare they generate. Unfortunately, this does not appear to be occurring. Recent GPI studies on China and Thailand indicate that the per capita GPI of both countries has already begun to fall—China's per capita GPI peaking in 2002 (Wen et al. 2008); Thailand's peaking in 2001 (Clarke and Shaw 2008). Crucially, these declines took place when the per

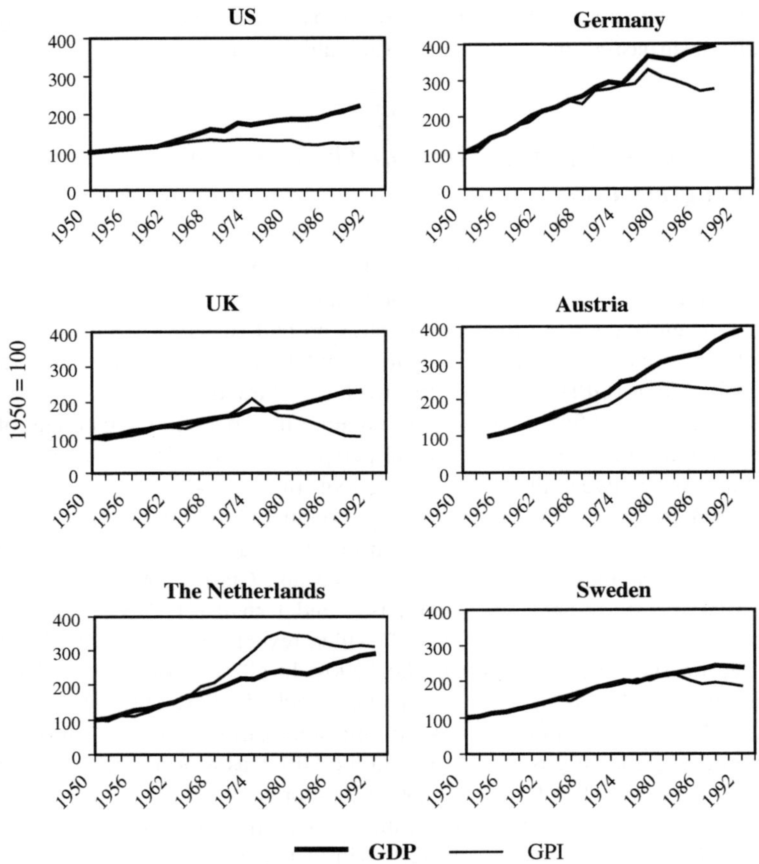

Fig. 2.7 The GPI and GDP for the USA and a range of European countries. *Source* Jackson and Stymne (1996)

capita GDP for China and Thailand was just $Int4926 and $Int7373 respectively (Lawn and Clarke 2008). In addition, the per capita GPI of China and Thailand peaked well short of the value currently being enjoyed by high-GDP nations.

In an endeavour to examine the relationship between growth and economic welfare in the Asia-Pacific region, Lawn and Clarke (2008) plotted the annual per capita GPI values of the seven countries included in the Asia-Pacific GPI study against their corresponding per capita GDP values. Figure 2.8 is the result of this comparison. It reveals that the three wealthy countries—Australia, New Zealand, and Japan—have all reached a threshold level of per capita GDP (Lawn 2008a; Forgie et al. 2008; Makino 2008). The figure also shows that China and Thailand have reached an apparent GDP threshold. Alarmingly, Fig. 2.8 suggests that the later a nation begins to rapidly expand its economy, the lower is its per capita GDP when its

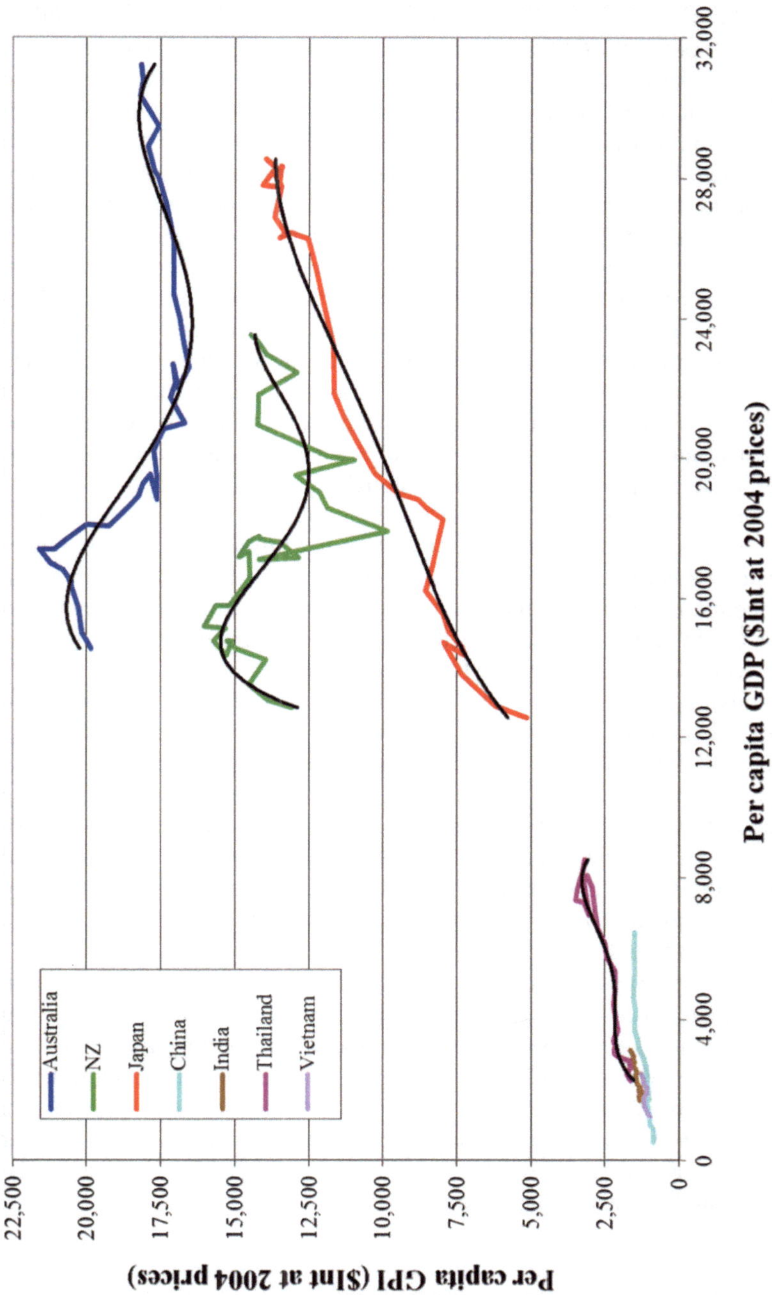

Fig. 2.8 Per capita GPI versus per capita GDP of selected Asia-Pacific countries. *Source* Lawn and Clarke (2008)

per capita GPI begins to decline. This is no better exemplified than by the tunnelling of each country's per capita GPI-GDP curve below that of its growth predecessor.

Lawn and Clarke (2008) believe the phenomenon revealed in Fig. 2.8 can be largely attributed to three factors. The first is the low level of consumption in poor nations relative to domestic production—very much the consequence of export-led growth strategies. This policy, which the Chinese Government is presently reconsidering, reduces the consumption benefits that poor nations enjoy from their productive endeavours.[20] The second factor is the migration of manufacturing operations to countries with low wages and feeble environmental regulations. Although this has helped poor countries to increase their real GDP, it has left many of them bearing a disproportionately large share of the world's social and environmental costs. Finally, as growth 'late-comers', low-GDP nations are attempting to expand their economies in a world replete with human beings and human-made capital, yet one with much less natural capital and fewer pristine ecosystems. Consequently, the marginal cost of an increment of GDP growth is far higher than it was when the world's high-GDP countries underwent their initial expansion phase.

It is because of the above factors that Lawn and Clarke (2008) have extended Max-Neef's theory to propose a *contracting threshold hypothesis*. The hypothesis is essentially this: As the economies of the world collectively expand in a globalised economic environment, there is a contraction over time in the threshold level of per capita GDP. As such, growth late-comers face the prospect of never attaining the economic welfare enjoyed by the early growth-movers.

Despite this new hypothesis, Lawn and Clarke still believe it is possible for poor nations to experience higher levels of economic welfare. However, they argue that progress will only occur if an extension can be made to the threshold at which the per capita GPI of the world's poor countries begins to decline. This, according to Lawn and Clarke (2008), will necessitate dramatic policy changes on the part of the world's low-GDP countries. Just as importantly, it will require rich nations to cease growing their economies in order to provide the 'ecological space' that poor nations need to enjoy a phase of welfare-increasing growth before they, too, must make the transition to a steady-state economy.

Should this self-imposed check on growth be a cause for concern for rich nations? Not at all, since many rich countries already need to reduce the physical scale of their economies to advance the economic welfare enjoyed by their own citizens. As will be revealed in Chap. 4, this is likely to be very important in terms of resolving the climate change crisis. Economic downsizing or de-growth (see Martinez-Alier 2009) will not only render it easier for rich nations to reduce their own greenhouse gas emissions, it will allow some poor countries to increase emissions as they complete their economic development process.

Notes

1. The WCED (1987, p. 43) defined sustainable development as ".... development that meets the needs of the present without compromising the ability of future generations to meet their own needs."

2. In the natural world, information exists as genetic information coded in the DNA molecule. In the anthropocentric world, information exists as knowledge encoded in various institutions and organisations, and on physical objects, such as books and computer disks.
3. A holon is a term made popular by Arthur Koestler. See Capra (1982), p. 303.
4. Human-made capital constitutes the intermediate means depicted in Fig. 1.11.
5. The throughput is equivalent to the ultimate means depicted in Fig. 1.11.
6. Fisher (1906) described psychic income as the subjective satisfaction that emerges in the stream of human consciousness from consumption and other benefit-yielding endeavours.
7. There are two things worthy of note here. Firstly, uncancelled costs are often undervalued because many natural capital values escape market valuation. Secondly, uncancelled costs should reflect the higher of two distinct classes of opportunity costs. The first is the cost of transforming a unit of low-entropy resources into a physical good in terms of the next best alternative good forgone. For example, if a unit of resource X is used to produce good A, it cannot be used to produce good B. The second class of opportunity cost involves the reduced capacity of natural capital to provide a flow of resources required to produce future goods. For example, if the extraction of a unit of resource X reduces the capacity of natural capital to provide a unit of X over time, which it will if X is a non-renewable resource or if a renewable resource is unsustainably harvested, a unit of X will be unavailable to produce goods of any type in the future. It is the larger of the two classes of opportunity costs that constitutes the true uncancelled costs of economic activity.
8. The technical efficiency of production (E) can be written as the ratio of energy-matter embodied in physical goods (Q) to the energy-matter embodied in the low-entropy resources used to produce them (R). That is, $E = Q/R$. While the value of E can be increased via technological progress, the first and second laws of thermodynamics dictate that E must be less than a value of one.
9. The concept of a 'line in the sand' is very similar to Ciriacy-Wantrup's (1952) notion of a 'safe minimum standard'. See, also, Crowards (1998).
10. It is invariably argued that a growth rate of 2–3 per cent of real GDP is required just to prevent the unemployment rate from rising.
11. Some of the ways to sever the GDP-employment nexus are outlined in Lawn (2009).
12. Why does the principle of increasing marginal costs apply to the entire economy? Firstly, it is customary for nations to extract the more readily available resources first and be left with the more complicated task of extracting lower quality resources later. Secondly, the cost of the undesirable ecological feedbacks associated with each incremental disruption of natural capital increases as the economy expands relative to a finite natural environment.
13. Having said this, there is a limit on the capacity for income redistribution to increase Ratio 1 because, at some point, excessive redistribution is likely to adversely dilute the incentive structure built into a market-based economy.

14. Importantly, the new optimal scale need not necessarily be physically larger as a consequence of the UB curve shifting upwards. The size of the optimal scale will depend upon the nature of the UB and UC curves—in particular, the extent to which the shape of the UB curve alters following its upwards shift.
15. There are a number of critics of the ecological footprint (e.g., see *Ecological Economics* (2000), volume 32(3); and Fiala (2008)). A major weakness of the ecological footprint is the use of land area as a numeraire for sustainability. There is no doubt that if a nation was compelled to generate its entire resource flow in the form of renewable resources, land area and fertility would constitute critical limiting factors. However, there are other factors that also impinge on a nation's renewable resources, such as water availability (Patterson 2006). Because factors other than land area can restrict the generation of renewable resources, ecological footprint studies almost certainly overestimate a nation's biocapacity (and underestimate ecological deficits). As such, the ecological footprint can be regarded as a conservative indicator, which is all the more concerning given the ecological footprint estimates revealed in Table 2.1. It is worth pointing out that, following the work of Lenzen and Murray (2001), a number of improvements have been made to ecological footprint estimates to better account for additional limiting factors.
16. 1.3 Earths = 2.7 hectares ÷ 2.1 hectares. A value of 1.3 is only possible because: (i) non-renewable resources constitute a major portion of all resource use at present, and (ii) resource limits can be exceeded in the short-run by liquidating natural capital stocks.
17. The GPI was originally labelled an Index of Sustainable Economic Welfare (Daly and Cobb 1989). To view the items typically used to calculate the GPI, see Lawn (2007).
18. These figures are based on international dollars (Int$). An international dollar is a fictitious monetary unit which equilibrates the purchasing power that a nation's currency has over its own GDP with that of the US dollar over America's GDP. Two people living in different countries earning the same international dollar-valued income would be able to purchase an equivalent basket of goods and services.
19. Although the per capita GPI of some wealthy countries has recovered slightly since the early-1990s, in virtually every case it has failed to reach its earlier peak value.
20. It is also unnecessary because, if the central government is the monopoly owner and issuer of the nation's currency, it can always purchase the difference between exports and imports and provide the same goods to its citizens rather than have them enjoyed by foreigners. Better still, the government can redirect the resources involved to provide more schools, hospitals, and other critical infrastructure with public goods characteristics. In doing this, the nation's real output would not change, but the goods consumed or used would increase, thereby raising the per capita economic welfare of the nation (Lawn 2011).

Chapter 3
Policies to Achieve Sustainable Development

3.1 Achieving Sustainable Development: Policy Goals and Instruments

In this chapter, a number of policies are outlined to achieve sustainable development. Readers may ask why it is necessary to detail a range of sustainable development policies when so few of them deal directly with the climate change crisis. There is a simple explanation. As already intimated, anthropogenic global warming cannot be resolved independently of the sustainable development goal. Thus, any policy that assists in achieving sustainable development will be indirectly helping to resolve the climate change crisis.

In the previous chapter, the structural changes and forms of technological progress required to achieve sustainable development were elucidated. The aim of the policy solutions outlined in this chapter is to facilitate and induce these changes. Many of the recommended policies are unconventional, largely because the eventual transition to a qualitatively-improving steady-state economy, which is necessary to achieve sustainable development, is itself an unconventional proposal. Consequently, standard policy prescriptions will not suffice. This is not to say that all standard policies should be abandoned. However, almost all that remain will require cosmetic changes.

Before I detail any policies at length, I would first like to say something about policy goals and instruments. Despite the multi-dimensional nature of sustainable development, ecological economists believe that achieving the condition requires the resolution of three major policy goals. They are[1]:

1. *Ecological sustainability*—ensuring the natural resource throughput required to maintain the economy remains within the regenerative and waste assimilative capacities of the natural environment (ecosphere).

P. Lawn, *Resolving the Climate Change Crisis*, DOI 10.1007/978-94-017-7502-1_3

2. *Distributional equity*—ensuring the distribution of income and wealth is equitable both within and across nations.
3. *Allocative efficiency*—ensuring the natural resources entering the economy are allocated to their best possible use, which, if achieved, would maximise a jurisdiction's economic welfare.

An understanding of these three policy goals is best appreciated via an alternative representation of the linear throughput model revealed in Chap. 2. Consider Fig. 3.1, where the economy is portrayed as a subsystem of the ecosphere and therefore a system dependent upon the source, sink, and life-support services provided by natural capital. The source and sink functions are respectively represented by the arrows labelled 'Resources IN' and 'Wastes OUT'. The unidirectional flow of matter-energy represented by these two arrows constitutes the throughput that nourishes the economic subsystem. As we have seen, for the economy to remain sustainable in the long-run, the throughput of matter-energy must not exceed the ecosphere's regenerative and waste assimilative capacities. It is worth mentioning that because anthropogenic global warming is a consequence of the excessive generation of greenhouse gases, the phenomenon is largely the product of humankind's failure to resolve the first of the above policy goals.

The arrows branching off from 'Resources IN' to the various physical goods/structures represent the allocation or relative division of the incoming resource flow to alternative product uses. The allocation of the incoming resource flow is efficient if the highest level of economic welfare (use value) is experienced from the consumption/use of the various goods produced.[2]

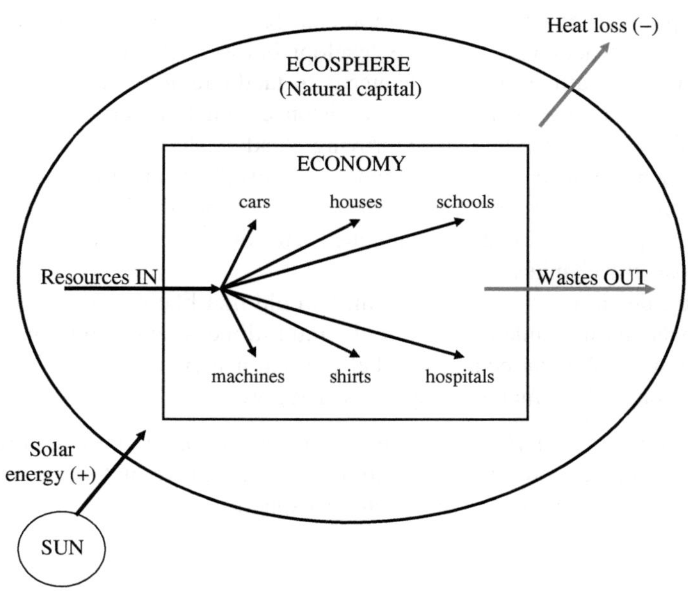

Fig. 3.1 The difference between resource throughput and resource allocation

Exactly who enjoys the benefits of existing and newly-produced goods depends on the distribution of income and wealth. Although it is infeasible for each person to possess an equal share of all physical goods, as we saw from Maslow's needs hierarchy, it is desirable and indeed ethically 'proper' to ensure that no person possesses fewer goods than what is deemed necessary to live a decent life. Ecological economists also believe that a society should limit the income differential between its richest and poorest citizens. Apart from the ethical dimension, severe inequities can lead to social upheaval that can greatly impede the sustainable development process. As we shall see in Part III, the pressure to cut global greenhouse gas emissions at a time when per capita emissions vary enormously across nations means that equity considerations will undoubtedly be at the core of future global emissions treaties.

Identifying policy goals is one thing. Achieving them through the application of an appropriate policy instrument is another. Mainstream economists believe that markets can simultaneously achieve the policy goals of sustainable scale and efficient allocation.[3] That is, by utilising markets to generate prices that reflect the scarcity of various resource types, waste sinks, and ecosystem services, mainstream economists believe that the resultant efficiency of resource allocation can ensure a sustainable rate of resource use.

Ecological economists disagree for two reasons. Firstly, they point out that no two separate policy goals can be solved via the application of a single policy instrument (Tinbergen 1952). In other words, it is impossible to 'kill' two independent policy 'birds' with one policy 'stone' (Daly 1992; Lawn 2007). Secondly, market prices—and this includes tax-adjusted market prices—cannot reflect the *absolute* scarcity of natural resources. At best, they can only reflect the *relative* scarcity of different resource types—e.g., how scarce coal is relative to oil (Norgaard 1990; Bishop 1993; Reynolds 1999; Lawn 2007; Daly 2008). The fact that markets are very good at reflecting relative scarcities is what makes them effective allocation mechanisms. However, it is because the sustainability goal pertains, in large part, to absolute resource scarcities, not relative scarcities, that markets are unable to ensure a sustainable rate of resource throughput.

This critical failure of markets is now revealing itself in the form of a phenomenon referred to as the 'Jevons' Paradox'. Despite the often stultifying effect of globalisation forces on governments (to be explained later), there are many instances where governments attempt to exploit the efficiency benefits of markets to achieve environmental outcomes. In many of these cases, governments manipulate the market prices of resources and/or wastes by using Pigouvian taxes to internalise previously unaccounted for environmental costs (Pigou 1932; Hoerner and Bosquet 2001; Schöb 2005; Lawn 2007).[4] On most occasions, the policy results in a more efficient allocation of resources. This, in turn, reduces the environmental impact *per unit of economic activity*. However, because markets fail to quantitatively limit the rate of resource throughput, the efficiency increases are almost always overwhelmed by the scale effect of rising economic activity (IPCC 2007d; Lawn 2009).

One of the driving forces behind the Jevons' Paradox is the tendency for increases in efficiency, by lowering the resource cost per unit of economic activity, to give the false impression that less frugality is needed to achieve sustainability (Daly 2008). This leads to an increase in the quantity of resources expended, which invariably exceeds the quantity of resources initially saved from the efficiency advances (Brookes 2000). Consequently, the aggregate rate of resource throughput ends up rising rather than diminishing (Ayres 2005; Haberl et al. 2006; Polimeni 2008).

Exacerbating the situation is the fact that market decisions regarding the current use of natural resources are made entirely by currently existing people who have a natural tendency to discount the future ramifications of their present actions. This leaves future generations—the people who will suffer most in a resource-poor and substantially warmer world—unable to partake in the current resource-bidding process. Thus, market decisions involving the future manifestation of significant costs are always biased against future generations.

All up, there is nothing that markets can do to prevent the stock of natural capital from declining. As we shall see in Chaps. 6 and 7, this glaring weakness of markets has significant implications for benefit-cost assessments and the use of taxes to reduce greenhouse gas emissions.

Given the inability of markets to ensure ecological sustainability, ecological economists believe the resolution to the three sustainable development goals requires the application of the following policy instruments[5]:

- *Quantitative restrictions* (caps or quotas) on the rate of resource throughput to achieve ecological sustainability.
- *Transfer systems* to achieve a just distribution of income and wealth.
- *Relative prices* determined by interacting demand and supply forces to achieve allocative efficiency.

Not only does sustainable development require the application of specific policy instruments but, as Daly (1992) has shown, the policy goals must be resolved in the order listed above. For example, it makes no sense to resolve the allocation problem first and then make the necessary adjustments to ensure the incoming resource flow is ecologically sustainable and equitably distributed. Because allocation involves the relative division, through exchange, of the incoming resource flow to alternative product uses, it is too late to adjust the physical volume of the resource flow should it be unsustainable. Similarly, since an individual's command over the allocation of the incoming resource flow depends on the ability to pay for the means required to satisfy needs and wants, it is too late to adjust the distribution of the incoming resource flow among alternative people, following its allocation, should it be inequitable.

Of course, ensuring a just distribution prior to the allocation process does not guarantee a just distribution following it. Hence, there is always the need for some further redistribution. But redistribution, following allocation, is much less disruptive and market-distorting if the distribution of the incoming resource flow is equitable to begin with. Above all, the policy goals of ecological sustainability and distributional equity must be resolved prior to the efficiency goal. For this reason, the sustainability and equity goals must be resolved: (i) prior to the market allocation of the incoming resource flow; (ii) with respect to ecological and ethical criteria respectively, not economic criteria; and (iii) through the use of institutional arrangements that lie outside the domain of the market—ideally in the democratic spirit of Hardin's "mutual coercion mutually agreed upon" (Hardin 1968).

Importantly, resolving the sustainability and equity goals prior to the efficiency goal internalises ecological and distributive *limits*, not just costs, into market prices. This not only paves the way for markets to facilitate a macroeconomic adjustment to the optimal scale, it ensures, once the optimal scale has been reached, that the sustainable incoming resource flow is allocated in ways that can further increase a nation's economic welfare.

It is with this in mind that many of the policies required to achieve sustainable development will now be outlined. Although the exact nature of the policies and the timing of their implementation will differ from country to country, the differences will largely depend on whether one is referring to a high-GDP or a low-GDP nation. This is because, as indicated, many low-GDP nations need to benefit from further growth (i.e., growth up to the optimal scale), whereas high-GDP nations need to immediately begin the transition to a steady-state economy, with many requiring a phase of economic downsizing or de-growth.

3.2 Ecological Tax Reform

Despite its shortcomings, which I shall deal with shortly, ecological tax reform should constitute the centrepiece of any national policy programme to achieve sustainable development. Ecological tax reform involves shifting the tax base away from the value added in production (i.e., away from the wages earned by labour and the income generated from the use of producer goods) onto depletion and pollution activities (Daly 1996; Lawn 2006a).

Because a reduction in the tax impost on labour and producer goods rewards value-adding in production, ecological tax reform encourages production excellence that, in turn, helps to increase the service efficiency of human-made capital (Ratio 1). Service efficiency is also potentially boosted because the concomitant increase in hourly wage rates[6] can induce workers located on the backward-bending section of the labour supply curve to increase their leisure time.[7] Not only would this reduce many social costs (e.g., the cost of family breakdown), it would promote job-sharing and lower the unemployment rate without the need for more production (growth).

By taxing depletion and pollution, ecological tax reform internalises ecological costs by pricing the scarce but previously unpriced contribution of nature. As alluded to in Chap. 2, this can increase the efficiency of resource allocation by creating an immediate incentive for producers to reduce resource wastage and a long-term incentive for producers to develop and install resource-saving and pollution-reducing technologies (i.e., increase Ratio 2). In the process, this can reduce the environmental impact per unit of economic activity. Moreover, it can reduce the quantity of greenhouse gases emitted per dollar of real GDP produced.

Ecological tax reform is invariably promoted on a platform of tax-revenue neutrality. By this I mean that advocates of ecological tax reform usually argue for tax changes that result in the revenue raised from depletion/pollution taxes equalling the reduced revenue obtained from lower income taxes.[8] This ensures that the government's total tax take remains unchanged. Despite the political attractiveness of this approach, there is no reason why the return of depletion/pollution tax revenues must be equally shared across the income spectrum. Indeed, the opportunity exists for governments to redistribute a large proportion of the tax revenue to the poor, thus enabling ecological tax reform to play a vital role in achieving distributional equity.[9] Redistribution is best conducted by increasing the tax-free income threshold and/or by offering the largest tax cuts to people on low incomes. In view of the ever-present problem of unemployment, distributional equity can also be served by using some of the funds to part-finance a Job Guarantee programme (more on this soon).

How practicable is ecological tax reform for the world's poorer nations? There is no doubt that the potential exists for depletion and pollution taxes to impede much-needed industrial development in impoverished countries. This is not to say that industries in poor nations should be permitted to wantonly deplete and pollute. Rather, it may be more efficacious to offer tax rebates or subsidies to encourage existing and emerging industries to take up resource-saving and pollution-reducing technologies. Either way, since the growth required by many impoverished nations must be as clean and efficient as possible, it is equally critical for governments in these countries to introduce appropriate incentives (rebates) and/or disincentives (taxes) to reduce the resource intensity of industrial practices.

3.3 Achieving Ecological Sustainability

3.3.1 Cap-Auction-Trade Systems

One of the major shortcomings of ecological tax reform is that it cannot, in the form presented above, guarantee ecological sustainability. This is because tax-adjusted market prices cannot reflect the absolute scarcity of natural resources. Nor is there anything specific about ecological tax reform—such as quantitative restrictions on the rate of resource throughput—to prevent the Jevons' Paradox from eventuating. This shortcoming does not mean that ecological tax reform should be rejected. It simply means that an ecological tax reform package must incorporate additional policy measures to overcome its inherent weaknesses.

The first of the required policies is a comprehensive cap-auction-trade system to encompass all major renewable resources and various forms of waste. In the case of renewable resources, a purpose-designed government authority should be established to cap the harvest rate of each major resource type. Within the cap, a number of permits would be created and subsequently auctioned to the highest bidders.[10] For obvious reasons, the number of permits pertaining to each resource type would be restricted to the regeneration rate of the relevant resource. Importantly, the limit on the number of permits would impose an across-the-board restriction on the rate of renewable resource extraction that would ensure adherence to sustainability precept # 1 from Chap. 2.

Once bought, the possession of a permit would grant an individual or entity the right to purchase a portion of the permissible resource flow from resource sellers. For example, a single permit might confer its possessor the right to purchase one cubic metre of unprocessed timber. If so, a resource buyer would need to acquire ten permits in order to obtain ten cubic metres of raw timber. For each cubic metre of timber purchased, resource buyers would automatically acquit one permit.

The auctioning process would be undertaken periodically to allow the government authority to vary the number permits in line with novel ecological changes. For a specific timber species, this may be every three to five years. For water, where drought in some locations can rapidly affect water supplies, it may be necessary to auction permits annually. Clearly, the life of a permit would be limited to the period between auctions—e.g., three years for timber; one year for water.[11] Expired permits would be non-redeemable. To maintain competitive markets, a limit would be placed on the number of permits any individual or firm could purchase. Permits can be resold to other individuals or firms provided the buyers are not already in possession of the maximum quota of permits.

To maximise the effectiveness of a cap-auction-trade system, specific arrangements would be devised to cater for the different geographical regions and jurisdictions where particular resources are located. The regional aspect is important because particular resource types (e.g., timber species) grow at different rates depending on their geographical location. In addition, the failure to consider regional effects can lead to an entire quota of a particular timber or fish species being sourced from one location. This has the potential to devastate a local fish population or timber reserve. Finally, since ecological sustainability requires the future exploitation of natural capital to be confined to locations already strongly modified by previous human activities (second part of sustainability precept # 4), cap-auction-trade systems with regional requirements in mind can help reduce or prohibit resource extraction from sensitive and hitherto low-impacted areas.

A key feature of a cap-auction-trade system is the higher price that resource buyers must pay to secure a portion of the renewable resources extracted for use. As per normal, resource buyers would pay the usual amount charged by resource suppliers when selling renewable resources. However, resource buyers would also be required to pay a 'premium' to obtain resource-use permits. The magnitude of the premium would be determined by interacting demand-side and throughput-limited supply-side forces in the permit market. As such, the premium would be equivalent to an absolute scarcity tax that would reflect the market internalisation of ecological limits, not just ecological costs. This last aspect of a cap-auction-trade system is crucial because it ensures the existence of a policy instrument to facilitate the efficient allocation of the incoming resource flow. It also generates revenue for redistribution purposes. Thus, a cap-auction-trade system would achieve the efficiency and distributional benefits of a conventional ecological tax reform package but go one step further and guarantee a sustainable rate of renewable resource use—a desideratum obtained from incorporating the three sustainable development policy instruments into the one institutional mechanism (the cap-auction-trade system) and addressing them in the appropriate sequence.[12]

A cap-auction-trade system would also serve as an effective means for controlling pollution levels. With pollution, caps would be imposed to ensure the discharged quantities of various forms of waste remained within the ecosphere's waste assimilative capacity, thus ensuring adherence to sustainability precept # 3. Not unlike renewable resources, permits would be created in accordance with the cap and auctioned to the highest bidders.

It is important to recognise that pollution caps need not be applied directly to pollution itself. As Daly (2008) has highlighted, the first law of thermodynamics ensures that the quantity of waste matter-energy generated by the economic process equals the quantity of matter-energy entering the economy in the form of natural resources. Hence, it is possible to limit certain forms of pollution by capping the extraction of the very resources that, when used, become the wastes that must be regulated. Indeed, because many resource-extraction activities are more spatially concentrated than pollution activities, it is often more effective to regulate pollution levels by capping resource-extraction rates than regulating pollution activities directly. This is particularly so when end-of-line polluters are numerous and widely dispersed (e.g., car-owners), but the number of resource-extractors is small (e.g., oil companies). In these circumstances, capping resource extraction can simplify the monitoring of pollution levels as well as streamline the permit-auctioning process by eliminating the need for end-of-line polluters to engage in the permit market. Capping resource extraction also has the advantage of inducing greater efficiency at each upstream stage of the economic process.

Having said this, any advantage gained from capping resource use to limit pollution levels depends largely on the type of waste under consideration. In the case of greenhouse gas emissions, the sheer variety of gases and their sources (e.g., carbon-based fuels, agriculture, and deforestation) suggests that achieving a safe greenhouse gas target will require a cap-auction-trade system aimed predominantly at regulating emissions levels rather than the resource-use activities generating the emissions. The complexities associated with carbon offsets and sequestration possibilities also make the targeting of greenhouse gas emissions more appealing.

Before I move on, it is worth highlighting two additional benefits of cap-auction-trade systems. It is often claimed that cap-auction-trade systems are more complicated than depletion/pollution taxes. In a logistical sense, this is probably true since, with taxes, there is no need to establish a government authority to conduct permit auctions.[13] From a practical sense, I believe cap-auction-trade systems are simpler insofar as bureaucrats only need to estimate the sustainable rate of renewable resource use and waste generation. They do not need to calculate permit prices—the equivalent of absolute scarcity taxes—since these are determined by demand and supply forces in the various permit markets.

If taxes are imposed, it is necessary for bureaucrats to estimate sustainability targets and then calculate the tax rates deemed necessary to achieve them. Assuming that the former can be correctly estimated, why would anyone believe that bureaucrats can determine extraction/pollution charges more accurately than permit markets?[14] Trying to control the entropic rate of throughput with the use of extraction/pollution taxes virtually guarantees bureaucratic error, especially given that fluctuating demand-side forces will require tax rates to be constantly adjusted to achieve throughput targets. Permit prices, on the other hand, would adjust automatically. And while permit prices can also be incorrect, the margin of error is likely to be considerably smaller given that markets are much better at collecting and interpreting masses of piecemeal information than bureaucrats. More importantly, with cap-auction-trade systems in place, adherence to sustainability precepts # 1 and 3 is guaranteed. In sum, incorrect permit prices mean that sustainability is achieved but maximum allocative efficiency is not; incorrect tax rates mean that neither sustainability nor maximum efficiency is achieved.

Secondly, as I will explain in later chapters, dealing successfully with the climate change crisis will require the introduction of a global cap-auction-trade system for greenhouse gas emissions. For the system to succeed, much of the institutional framework required to support it will need to exist at the international level. Nevertheless, to be truly effective, international institutions will need to work closely with nation-level authorities. Should cap-auction-trade systems be in place to deal with renewable resources and various forms of pollution, the necessary government authorities and nation-level institutions would already exist to successfully administer the likely regulations, procedures, and monitoring processes embodied in a global emissions-trading system. This would enable a global system to neatly dovetail with a nation's sustainable development policies. The same would not occur if a system of extraction/pollution taxes was in place.

3.3.2 Dealing with Non-renewable Resources

Although cap-auction-trade systems ensure adherence to sustainability precepts # 1 and 3, they do not guarantee adherence to precepts # 2 and 4. Hence, they only go part of the way towards achieving ecological sustainability. The reason why cap-auction-trade systems fail with respect to precept # 2 is obvious—placing caps on the extraction of non-renewable resources does not prevent their exhaustion.

To keep natural capital intact, some of the proceeds from depletion activities must be reinvested to cultivate renewable resource replacements. A useful formula exists to assist the reinvestment process (El Serafy 1989). The formula is predicated on the Hicksian principle that sustainable income requires a non-renewable resource earmarked for depletion to be converted into a perpetual income stream (Hicks 1946). From a strong sustainability perspective, this requires some of the earnings from the sale of a finite series of non-renewable resource extractions to be invested in such a way as to ensure the annual depletion rate of the non-renewable resource is equal to the sustainable harvest rate of a substitute renewable resource. To achieve this, so-called 'income' and 'capital' components of the finite series of non-renewable resource extractions must be calculated. Once obtained, the capital component represents a 'user cost'—in effect, the quantity of low-entropy matter-energy that needs to be invested each year during the depletion phase of the non-renewable resource to ensure renewable matter-energy of an equivalent magnitude can be perpetually harvested. A variation of El Serafy's user cost formula is given below:

$$R_{NR}/R_{\underline{NR}} = \frac{1}{(1+r)^{n+1}} \tag{3.1}$$

where:

- R_{NR} = the quantity of an extracted non-renewable resource available each year for production purposes;
- $R_{\underline{NR}}$ = the total quantity extracted each year of the non-renewable resource;
- r = the natural regeneration rate of the cultivated renewable resource asset;
- n = the number of years to fully exhaust the non-renewable resource.

Provided $R_{\underline{NR}}$, r and n are known, it is possible to calculate R_{NR} by rearranging Eq. (3.1) as follows:

$$R_{NR} = R_{\underline{NR}}\left[1 - \frac{1}{(1+r)^{n+1}}\right] \tag{3.2}$$

Once R_{NR} has been calculated, the quantity of the extracted non-renewable resource that must be invested (R_{INV}) to establish a renewable resource asset (N_R) becomes:

$$R_{INV} = R_{\underline{NR}} - R_{NR} \tag{3.3}$$

To simplistically illustrate the transition process, consider a hypothetical situation where a non-renewable resource consists of 1,000 units of low-entropy matter-energy (i.e., $N_{NR} = 1{,}000$); the resource is scheduled to be fully exhausted over nine years (i.e., $n = 9$); there are ten extractions in total, beginning with the first extraction at time zero (i.e., $R_{\underline{NR}} = N_{NR}/10 = 100$); and the regeneration rate of the substitute renewable resource is 5 per cent (i.e., $r = 0.05$). Using Eqs. (3.2) and (3.3), one obtains the following values: $R_{NR} = 38.61$ and $R_{INV} = 61.39$. The depletion and reinvestment processes are presented in Table 3.1 (Note: R = the total low-entropy matter-energy extracted in a particular year and R_R = the annual quantity of low-entropy matter-energy that eventually flows from the cultivated renewable resource asset).

Table 3.1 shows the non-renewable resource (N_{NR}) diminishing at the rate of 100 units of low-entropy matter-energy per year for nine years. The fourth column (R_{NR}) indicates that 38.61 units of low-entropy matter-energy are available each

Table 3.1 Conversion of a non-renewable resource into a sustainable flow of renewable resources

Time	N_{NR}	$R_{\underline{NR}}$	R_{NR}	R_{INV}	N_R	r	$N_R(1+r)-R_R$	R_R	$R=R_{NR}+R_R$
0	1000	100	38.61	0	0	0.05	0	0	38.61
1	900	100	38.61	61.39	61.39	0.05	64.46	0	38.61
2	800	100	38.61	61.39	125.85	0.05	132.14	0	38.61
3	700	100	38.61	61.39	193.54	0.05	203.21	0	38.61
4	600	100	38.61	61.39	264.60	0.05	277.83	0	38.61
5	500	100	38.61	61.39	339.23	0.05	356.19	0	38.61
6	400	100	38.61	61.39	417.58	0.05	438.46	0	38.61
7	300	100	38.61	61.39	499.85	0.05	524.84	0	38.61
8	200	100	38.61	61.39	586.23	0.05	615.54	0	38.61
9	100	100	38.61	61.39	676.94	0.05	710.78	0	38.61
10	0	0	0	61.39	772.17	0.05	772.17	38.61	38.61
∞	0	0	0	0	772.17	0.05	772.17	38.61	38.61

year for production purposes. The remaining 61.39 units (R_{INV}) must be annually invested to build up the renewable resource (N_R) which grows over the nine-year period at the rate of 5 per cent per year. By the end of the ninth year (beginning of the tenth year), 772.17 units of renewable matter-energy have been cultivated. It is now possible, on a permanent basis, to provide an annual flow of 38.61 units of low-entropy matter-energy for production purposes—an amount exactly equal to the annual quantity that was made available for production purposes during the depletion phase of the non-renewable resource (Note: $38.61 = 772.17 \times 0.05$). In effect, the stock of resource-providing natural capital has been kept intact.[15]

There are, however, two things worth bearing in mind. Firstly, there are limits to how much renewable natural capital can be cultivated at any point in time and over time. Although this limitation does not restrict how much non-renewable natural capital can be exploited over time, if we wish to adhere to sustainability precept # 2, it will undoubtedly constrain the rate at which non-renewable natural capital can be exploited at a single point in time. Hence, the discretion over the scheduled rate of exhaustion of a non-renewable resource will be restricted by the maximum amount of renewable natural capital that can be cultivated in each investment period. If, in the hypothetical scenario depicted in Table 3.1, it is not possible to cultivate 61.39 units of additional renewable matter-energy per year, it will be necessary to lengthen the exhaustion schedule beyond nine years. This will subsequently reduce the amount of non-renewable resources available for production purposes in each year during the depletion phase.[16]

Secondly, renewable resource substitutes do not exist for some non-renewable resources. We cannot, therefore, apply the El Serafy formula in such instances. Since many non-substitutable resources have very useful properties, a sensible course of action would involve the maximum extension of their availability and efforts to overcome any long-run reliance upon them well before their eventual depletion. To accomplish this, cap-auction-trade systems should again be employed. In these circumstances, the number of permits auctioned each year would be determined on the basis of estimated stock levels, projected future discoveries, and the period over which the continued availability of the resource is deemed to be intergenerationally just. Given the restrictions that cap-auction-trade systems would place on the annual consumption of non-substitutable resources, it is reasonable to believe that the prices of these resources would rise relative to the prices of renewable and substitutable non-renewable resources. In doing so, cap-auction-trade systems would encourage the wider use of the latter category of resources and assist a nation to wean itself off non-substitutable resources. By also promoting better resource management, a cap-auction-trade system would also boost the productivity of renewable resources (i.e., facilitate increases in Ratio 3).

Returning to our attention to non-renewable resources with renewable resource substitutes, how might a variation of the El Serafy formula be operationalised to ensure adherence to sustainability precept # 2? One possible solution is to compel resource liquidators to establish 'capital replacement' accounts in the same way business-managers in many countries are required to establish a superannuation fund for employees. This could be accomplished through changes in taxation and accounting legislation. Ideally, the legislative changes would include a strict schedule of discount rates and average mine lives that would be applied when calculating the set-aside component for each non-renewable resource type. The capital replacement accounts would be held by government-approved resource management companies whose task it would be to establish renewable replacement assets on behalf of the non-renewable resource liquidators.

3.3.3 The Sustainable Use of Agricultural Land

Agricultural land is not directly harvested but exploited for its propagating properties. Thus, unlike a flow of timber that is sustained by ensuring the harvest rate from a forest/plantation does not exceed its capacity to regenerate, the sustainable use of agricultural land cannot be achieved via controls on resource flows into the economy. Short of having to directly regulate all agricultural activities, which would clearly be untenable, the sustainable use of agricultural land is best be achieved by encouraging farmers to adopt sustainable land use practices. This is best facilitated by a policy mix that places the practical or stewardship responsibility of land management on farmers and most of the onus for funding sustainable land-use practices on the government. Having the financial responsibility rest predominantly with the government is entirely legitimate given that the condition of 'sustainability' is essentially a *public good* and therefore requires government intervention to be achieved.[17]

The first major component of a sustainable land use policy would be the use of subsidies and substantial tax rebates to assist farmers to adopt sustainable land use practices. The second policy component would be the levying of financial penalties on farmers who fail to fulfill their stewardship responsibility. The extent of the penalty would depend on the degree of land degradation and would be paid by farmers through the tax system.

How would the penalty be determined and meted out? Allowing for environmental factors that can detrimentally affect the condition of agricultural land (e.g., drought), representatives of a government authority (e.g., Department of Primary Industries) would conduct random inspections of farms to assess their overall condition. Should it be clear that the productive capacity of a farmer's land has been maintained, the farmer in question would not incur a penalty. However, worst-case offenders would incur both the highest penalty and something akin to a 'yellow card'. Three yellow cards would equate to a 'red card' and compulsory acquisition of the farm. The farmer would receive the market value of the property less any outstanding penalties and would be barred from engaging in agricultural activities for a specified period, not unlike the manner in which professionals and tradespeople are barred for engaging in negligent or sub-standard practices.

For obvious reasons, the government acquisition of properties would not be a trivially inexpensive exercise, although much of the cost would be recouped upon the eventual resale of the land. To finance the net cost of such a proposal, a government should establish an Environmental Trust Fund. The Fund would be chiefly financed by way of revenue raised by cap-auction-trade systems and other environmental taxes, including the penalties imposed on farmers for poorly managing agricultural land. Although the Environmental Trust Fund would be used to achieve a range of environmental objectives, some of the funds contained within it would be allocated to acquire the properties of negligent farmers.

It is often shown that the dire financial situation of a farmer is the main factor contributing to unsustainable land practices—i.e., where farmers, in their efforts to remain financially viable, over-extend the productive capacity of their land. I do not believe that the penalty system advocated above would compound such a problem. Nonetheless, to further assist farmers in this matter, Environmental Trust Funds should be used to enable struggling farmers to exit the agricultural industry and resettle and gain employment elsewhere or obtain qualifications in a new field of endeavor where employment can be found locally. Where a marginal farming region is particularly at risk, the Environmental Trust Fund would be used to invest in a community-level project to establish a replacement industry. A similar approach could also be applied in the case of the forestry, mining, and irrigation industries. Whilst many would question the possible high cost of this restructuring process, it is my belief that the cost would be much less than the cost of failing to adopt a proactive adjustment policy.

3.3.4 Native Vegetation Clearance Controls

It was explained above how regional-based cap-auction-trade systems can assist in limiting resource exploitation in areas yet to be significantly modified by human activities. Unfortunately, cap-auction-trade systems cannot ensure the preservation of native vegetation and critical ecosystems, particularly on privately-owned land, which is necessary to satisfy the first part of sustainability precept # 4. It has been conservatively estimated that around twenty per cent of a nation's land area should be preserved as habitat for wildlife conservation (Wilson 2002).[18] To maximise the ecological values of protected areas, it has also been suggested that additional land needs to be set aside to serve as native vegetation refuges and vegetation corridors to connect critical ecosystems.

In order to satisfy the first part of sustainability precept # 4, future land clearance needs to be kept to a minimum or be entirely prohibited. To achieve this end, explicit and strict controls over native vegetation clearance must be imposed. A policy of this nature has already been introduced in the Australian state of South Australia in the form of the Native Vegetation Clearance Act (1990). Since its enactment, wholesale land clearance within the state has ceased. There are two major features of the Act. Firstly, land-owners require permission to clear native vegetation, which is often denied. Secondly, unsuccessful applicants are provided with funds to fence off native vegetation and manage it sustainably. As for public land, valuable parcels of remnant vegetation and critical ecosystems should be encompassed within newly established National Parks to meet the twenty per cent 'bottom line' recommended by ecologists.

One of the weaknesses of the Native Vegetation Clearance Act is its failure to compensate land-owners for the potential loss of agricultural production or any other forgone mode of production. On pure equity grounds, compensation payments should be distributed by governments to farmers. These payments could be drawn from the previously mentioned Environmental Trust Fund. Moreover, because native vegetation can sequester carbon, it should be possible for farmers to receive and sell carbon credits for revegetating sections of their land.

An Environmental Trust Fund could be used to assist unviable farmers, following lack of government approval to clear land, to exit the agricultural industry at minimal personal cost. Compensation could also be extended to other industries often affected by land-clearance controls (e.g., forestry and mining).

As is widely understood, the benefits of preserving native vegetation extend well beyond national boundaries. Of particular concern is that many of the world's critical ecosystems (e.g., rainforests and wetlands) are located in impoverished nations where further GDP growth is required. These ecosystems generate considerable non-direct use benefits for all nations, yet are likely to come under intense pressure in coming decades. Despite the global benefits generated by ecosystems, preservation denies the host country many direct use benefits. In addition, poor nations lack the revenue sources to establish Environmental Trust Funds. To promote ecosystem preservation in poor nations, some of the revenue raised from a global emissions-trading system should be siphoned off to set up Environmental Trust Funds on their behalf (more on this in Chap. 9). Aid money from rich nations should also be directed in this manner. Trust funds could then be used to compensate low-GDP countries for the direct use benefits foregone as a consequence of ecosystem preservation—sometimes referred to as 'payments for ecosystem services' (Pagiola et al. 2002; Daly and Farley 2004; Engel et al. 2008; Corbera et al. 2009). The funds could then be redistributed by the recipient country to the citizens most affected by the lack of direct ecosystem access and assist them in their management or guardianship role. The former could take the form of direct compensation or the establishment of a substitute industry (e.g., tourism to replace logging).

In terms of climate change policy, one of the important services provided by native vegetation is its capacity to sequester large quantities of carbon. Indeed, greenhouse gas emissions from deforestation/land clearance accounts for close to 20 per cent of annual global emissions (Garnaut 2008; Stern 2009; Gillenwater and Seres 2011). To reverse this trend, reforestation and the regrowth of native vegetation can be encouraged by allowing land-owners to earn carbon credits or offsets that could be subsequently sold in emissions-trading markets. Should this be widely implemented, land-owners would become holistic land managers and would rightly receive income for adequately providing a range of land-use services, not simply income from the generation of agricultural products—an imbalance that has long induced land-owners to sacrifice many critical ecosystems services with public goods features to maximise, albeit in the short-term in many instances, agricultural outputs. Altogether, the establishment of Environmental Trust Funds and saleable carbon credits would provide two important nation-level institutions through which a global emissions-trading system could be effectively integrated.

3.3.5 Population Stabilisation

We have seen that the rate of resource throughput must remain within the ecosphere's regenerative and waste assimilative capacities to achieve ecological sustainability. Irrespective of what underlying forces determine a nation's rate of resource use, the aggregate rate of resource throughput depends ultimately on the per capita rate of throughput and a nation's population. That is:

$$\text{Rate of resource throughput} = \text{per capita throughput} \times \text{population} \quad (3.4)$$

As important as it is to focus on the per capita rate of resource throughput, it is equally important to focus on a nation's population (O'Connor and Lines 2008). Given that a minimum per capita level of resource consumption is required to provide the basic necessities of life, let alone a decent existence, quelling population growth may well be the most significant factor. Putting debates aside, human population numbers must eventually be stabilised to achieve ecological sustainability.

In view of the demographic momentum associated with a rising population, it makes obvious sense to move as rapidly as possible towards a stable human population. Even in the case of poor nations, which require a further phase of real GDP growth, population stabilisation is best addressed immediately, particularly when so many possess the world's highest population growth rates. Reliance upon the demographic transition alone—that is, on increases in per capita GDP to help lower fertility rates—will be insufficient to stabilise human population numbers at the speed required.

Exactly what policies should be introduced to stabilise a nation's population numbers will depend chiefly on whether the source of the population increase is a high domestic fertility rate or large-scale net immigration. In the case of low-GDP nations with high fertility rates, failure in the past to deal effectively with population growth can be principally attributed to the inadequate delivery of contraception and family planning programmes. Also at fault are gross inadequacies in health-related systems and a chronic shortage of equipment and qualified personnel (National Commission on Population 2000).

There are many reasons why the delivery of contraception and family planning programmes has not reached desired levels. One of the more critical factors is funding. In India, for example, only 50 per cent of budgetary outlays designated for population stabilisation are directly allocated to stabilisation activities, including the procurement of equipment and supplies (National Commission on Population 2000). However, in most instances, the problem is the insufficient fiscal capacity to finance the full range of programmes desperately needed. It is therefore incumbent upon the international community—especially wealthy nations—to assist in the financing of these measures. As it is, funding of these measures is in the best interests of high-GDP countries given that an excessive population in an increasingly affluent Asia and Africa would have significant global consequences, not the least being the increased difficulty of achieving a safe atmospheric concentration of greenhouse gases.

Another important factor behind the unsuccessful impact of contraception and family planning programmes in many low-GDP countries is the lack of women's rights. Women must be suitably empowered and legally protected to feel sufficiently secure to use the available contraception. They must also be afforded greater economic independence given that a lack of it reduces choice and increases the power of economic providers (men). Unquestionably, more must be done in many countries to increase the rights and independence of women and by the international community at large to pressure lagging nations in this regard.

Along with augmenting the welfare contribution of consumption, a more equitable distribution of income would also help reduce population growth. Evidence indicates that providing adequate incomes to the poor can go a long way towards lowering a nation's fertility rate (Kuznets 1974; Peterson 1975; Pakrasi and Halder 1981; Todaro 1994; Daly 1996). For example, whilst increases in per capita GDP have reduced fertility rates in many low-GDP countries, fertility rates continue to remain high amongst the very poor—in some cases enough to prevent overall fertility rates from rapidly falling (Daly 1996).[19] This suggests that a more equitable distribution of income would greatly assist in reducing the total fertility rate to at least something that would suppress the high rates of population growth found in many countries.

As for the rich nations with high net immigration at the root of their population growth (e.g., USA, Canada, and Australia), achieving a stable population is not possible without addressing the contentious issue of immigration. To what extent immigration must be restricted depends upon the desired population target, the natural rate of population increase (births minus deaths), and emigration numbers.

Demography is a more complex science than many people realise (Ehrlich and Ehrlich 1990; Yaukey et al. 2007; O'Connor and Lines 2008). Few are aware that, even with zero net immigration, a fertility rate below one per person (i.e., below the replacement rate) does not imply an immediate decline in a nation's population numbers. Assuming that a nation's fertility rate has fallen from having once been above the replacement rate, its population will not begin to shrink until the first generation with a below-replacement fertility rate passes on, since only then is there fewer people in each lower-age cohort than the previous generation.

If we therefore assume that a nation wishes to stabilise its population numbers and the fertility rate of its reproductive adults is below the replacement level, the permissible level of net immigration would be determined by the excess of deaths over births. Because this disparity is likely to be relatively insignificant, then, unless a nation's emigration rate is very high, the annual immigration intake would have to be kept very low to maintain a steady population. For high-GDP nations wishing to meet humanitarian goals, this would leave little room for immigration beyond the admission of political and economic refugees, and what is likely to become an increasing number of ecological and climate change refugees.

A lower immigration intake of predominantly needy people ought not to be considered undesirable. As things stand, the immigration policy of many wealthy nations includes a significant intake of highly-skilled migrants from low-GDP nations that, whilst beneficial to the host country, is often detrimental to source countries. It is also unjust to give immigration preference to wealthy people who desire a more amenable climate or lifestyle change over people in genuine need.

I would like to finish off by putting forward a population stabilisation policy that, although controversial, may be of great value in reducing the very high population growth rates present in some countries. The policy involves the introduction of transferable birth licences—a scheme first proposed by Boulding (1964) and revisited by Heer (1975) and Daly (1991). The first aspect of the scheme involves granting each person the right of reproductive replacement. This right would exist in the form of a freely allotted birth licence. Limiting each person to one licence would immediately reduce a nation's fertility rate to the replacement rate. Because the licences are freely issued, each couple would be able to produce and raise two children at normal cost. The licences would be transferable by sale or gift so that, donation aside, couples wanting more than two children could only do so by purchasing a licence in a market best administered by a government authority.[20] The buying and selling of licences would be restricted to the adult population (e.g., people aged eighteen years and above), although there would be no age-based rule governing the forfeiting of one's freely allotted licence.[21]

For countries with an excessive population, the scheme could be amended to reduce the fertility rate below the replacement rate. For instance, each person could be granted half a licence. This would mean that two licences would be required to produce one child, which would effectively translate to one freely allotted licence per couple. A scheme with this degree of austerity would be similar to the population policy that exists in China, except for one critical difference—the scheme would invoke the use of the market to enable willing and financially-able couples to have two or more children, thus combining macro-control with micro-variability (Daly 1991).[22]

Herein lies the beauty of the scheme. By limiting the number of birth licences, the fertility rate can be adjusted to achieve a population target consistent with the ecosphere's carrying capacity. By distributing reproductive rights on the basis of equality, the scheme is fair and just. Finally, by permitting the buying and selling of licences, the scheme facilitates a reallocation of reproductive rights in conformity with the child-rearing preferences of the population and their ability to pay. Overall, sustainability is promoted by the number of allotted licences; distributional equity is promoted by the means of initial distribution; and allocative efficiency is achieved by installing a market for birth licences. Hence, like cap-auction-trade systems, the scheme incorporates the separate policy instruments required to resolve the three sustainable development goals and institutes them in the necessary order.

Most people regard the transferability of reproductive rights—which assigns a monetary value to procreation—as morally repugnant. Moreover, they argue that the scheme is advantageous to the rich and, despite claims to the contrary, is unjust. Such opposition overlooks four facts. Firstly, although the scheme permits the transferability of birth licences, the right to procreate up to the imposed limit is free and equally provided to both rich and poor. Secondly, no-one is compelled to donate or sell his or her licence. Thirdly, although the scheme advantages the rich in that they are more able to afford an additional licence, the rich always have a purchasing-power advantage, the extent of which depends on the income disparity between rich and poor, not on the scheme itself (Daly 1991). Finally, should the rich have more children than the poor, the exchange of money for birth licences would contribute to bridging the income gap between rich and poor.

A more creditable objection to the scheme relates to the issue of enforcement. How should violators be penalised? For obvious reasons, the penalty for violation must be sufficiently severe to minimise the number of violators. On the other hand, if the penalty is excessive, it is possible that the scheme would fail to gain the support required to be instituted in the first instance. A penalty that is likely to be least resisted is a procreation levy. The levy would be set at a given amount above the going market price for a birth licence. Setting a procreation levy in this manner would limit the number of violations by ensuring that the cost of a licence is always less than the cost of violating the scheme. Why produce a child when not in possession of a birth licence if the penalty exceeds the cost of purchasing a licence in order to have a child legally?

Because some offenders would be unable to pay the procreation levy in full, they would be required to pay the levy over time through the tax system (e.g., by increasing the marginal tax rate on an offender's income). Of course, in very poor nations, some violators may never be in a position to pay off the procreation levy, irrespective of the payment mechanism. Indeed, making them pay a penalty could further plunge them into destitution. Again, this reinforces the importance of an equitable distribution of income and, furthermore, the need to ensure a decent minimum income for those at the bottom of the income spectrum.

Ultimately, the success of a transferable birth licencing scheme presupposes a number of things: (i) its broad community acceptance; (ii) an equitable distribution of income and wealth; and (iii) accessibility to contraception and family planning programmes. There are many other aspects of the scheme that would need to be ironed out prior to its inception. However, given the demographic success of China's population policy, the scheme offers the potential to achieve the same level of success but with greater flexibility and minimal loss of individual freedom.[23] For these reasons, the scheme should be given serious consideration in countries with high fertility rates.

3.4 Achieving Distributional Equity

3.4.1 Minimum Income

Because distributional equity requires, at the very least, no person receiving less income than what is deemed necessary to live a decent life, a minimum income of sorts must be provided to all citizens. There are some who believe that a minimum income should be supplied in the form of a basic income guarantee or negative income tax[24]—in effect, an unconditional transfer payment to each and every citizen from the central government (Baetz 1972; Van Parijs 1991, 2000, 2004; Gintis 1997; A. Atkinson 1995; Clark and Kavanagh 1996; Lord 2003; Widerquist and Lewis 2009). There are others who, because of the potential inflationary and disincentive effects of a basic income and the important contribution that work makes towards satisfying the full spectrum of human needs, believe that a minimum income should be provided via a system of guaranteed employment (Mitchell 1998; Wray 1998; Watts and Mitchell 2000; Mitchell and Mosler 2002; Mitchell and Wray 2005; Mitchell and Muysken 2008).

Commonly referred to as a Job Guarantee (Mitchell and Muysken 2008), a system of guaranteed employment would involve the central government acting as an employer-of-last-resort to provide work for anyone unable to secure paid employment in the private sector or conventional public sector. Although there would be no limit on how long a person could remain employed within the scheme, Job Guarantee jobs would be designed to provide temporary employment to avoid having people unemployed and consequently reliant upon on an inadequate dole payment (or, in many poor countries, devoid of an income altogether). Job Guarantee workers would: (i) engage in a form of work similar to their usual or qualified form of employment, thereby minimising the depreciation of their human capital; (ii) if required, undertake training and/or further education; and (iii) produce goods and services predominantly of the public goods variety. In keeping with business-cycle fluctuations, spending by the central government on the Job Guarantee would increase as the number of private-sector jobs declined, but decrease in line with a private-sector recovery (Mitchell and Mosler 2002).

An important feature of the scheme would be the payment of a minimum hourly wage to all Job Guarantee employees. This would guarantee the receipt of the minimum income required to meet the basic needs of full-time Job Guarantee employees and their dependents. As an added bonus, the minimum hourly wage would establish a wage floor for the entire economy insofar as private-sector employers would be deterred from paying workers a lower wage rate than that provided by the Job Guarantee (Mitchell and Muysken 2008; Tcherneva 2009).[25]

Although a Job Guarantee would, by its very nature, ensure full employment, two major concerns are usually expressed about the scheme. They are: (i) would the Job Guarantee ensure price stability?; and (ii) is it possible that the Job Guarantee would harm the private sector so much that it would result in an absurdly large number of people employed in the Job Guarantee scheme?

As to price stability, there are three ways that the Job Guarantee scheme can stifle excessive inflation. Firstly, because Job Guarantee workers are hired from the 'bottom' of the labour pool—i.e., are paid a minimum living wage—the government would avoid all wage-increasing competition with the private sector. Secondly, by ensuring the monetary value of the goods and services generated by the scheme approximates the Job Guarantee wage bill (i.e., by ensuring the goods and services produced are useful), the extra claims on real wealth would be matched by the increase in real wealth, thereby avoiding any domestic devaluation of the currency. Thirdly, because the Job Guarantee involves just enough additional government spending to attain full employment—no more, no less—it would avoid the inflationary problems associated with the indiscriminate Keynesian pump-priming of the 1960s and 1970s.

The second and third factors are crucial given that much of the criticism directed at the Job Guarantee centres on the belief that it would lead to a destabilising rate of price inflation. Mainstream positions on inflation are grounded on the concept of a non-accelerating inflation rate of unemployment (NAIRU). The NAIRU is a specific unemployment rate accompanied by a constant inflation rate.[26] Mainstream economists argue that an increase in real GDP that reduces the unemployment rate below the NAIRU leads to non-productivity-related wage rises and an accelerated rate of price inflation (Dornbusch and Fischer 1990).

Whilst evidence supports the inflation-quelling impact of a NAIRU policy approach, Job Guarantee advocates have shown that the Job Guarantee scheme incorporates an inflation-control mechanism similar to the NAIRU. Mitchell and Muysken (2008) refer to this mechanism as the NAIBER—a 'non-accelerating inflation buffer employment ratio'. It works in the following manner. Assume that a NAIRU policy is being employed and exists at a 6 per cent unemployment rate.[27] The Job Guarantee is then introduced to eliminate all but frictional unemployment. Because the scheme necessitates an increase in government spending, it boosts aggregate demand and real GDP. This, in turn, leads to demand-pull inflation, which reduces real incomes and dampens private-sector activity. As people spill out of private-sector employment and into the Job Guarantee, the percentage of the labour force employed in the private sector declines whilst the percentage employed at the minimum wage under the Job Guarantee increases. The ratio of

Job Guarantee workers to private-sector employees rises until the inflation rate stabilises (i.e., until the NAIBER is attained). Thus, through the use of its unique spending powers, a central government is able to achieve full employment in a non-inflationary manner (Wray 1998; Mitchell and Muysken 2008).

There are a number of other positive aspects associated with the Job Guarantee worth considering. Firstly, unlike unemployed labour, Job Guarantee workers are able to retain and acquire new and existing skills. This can increase the combined productivity of the entire labour force and boost the minimum (floor) wage. Secondly, since Job Guarantee workers maintain their employability, they represent a more credible threat to disaffected private-sector employees than the unemployed. Presumably the NAIBER would serve as a more effective inflation-control mechanism than the NAIRU (Mitchell 2000). Thirdly, because the combined labour force would be more productive under a Job Guarantee scheme, the NAIBER would decline over time and fall below the NAIRU in the long-run. Hence, apart from achieving full employment, the Job Guarantee would eventually increase the percentage of the labour force receiving a wage above the minimum level. Consequently, any increase in the number of people living on the floor wage would be short-lived and constitute a small price to pay to ensure the availability of paid employment for anyone who desires it.

The NAIBER mechanism aside, I believe an alternative source of inflationary pressure would arise if a Job Guarantee scheme was introduced in tandem with the cap-auction-trade systems outlined earlier in the chapter. Inflationary pressure would emerge within the economy because the associated increase in aggregate demand would force resource buyers to pay a high price for the restricted number of resource permits. Because this would raise resource prices, it would increase production costs and inflate the prices of final goods and services. In turn, this would reduce real incomes and lower the equilibrium output level. Thus, in the short-run, one might expect the number of Job Guarantee employees to initially be greater than it would under a typical NAIBER scenario (i.e., where just a Job Guarantee scheme was in place). Consequently, what I would call an 'ecologically sustainable' NAIBER, or ESNAIBER, would be higher than the NAIBER.

What about the long-run? The much higher price paid for resources would almost certainly induce a much greater rate of resource-saving technological progress. By increasing Ratio 2 (the maintenance efficiency of human-made capital), this would allow higher levels of real GDP to be obtained from a sustainable rate of resource throughput. In doing so, it would dramatically reduce the inflationary pressure generated by the introduction of a Job Guarantee. The lower inflationary pressure would keep interest rates low and encourage producers to adopt the best available 'green' technologies.[28] It is therefore highly probable that the ESNAIBER would be lower than the NAIBER in the long-run that, as just explained, would be lower than the NAIRU. As a result, there are likely to be fewer people employed by the Job Guarantee scheme under an ESNAIBER policy than would be unemployed people under a NAIRU policy approach. Above all, concerns regarding the possibility of a large number of people ending up as Job Guarantee employees are unfounded.

I mentioned earlier that there is a strong institutional link between real GDP and employment levels that must be severed to prevent the full employment objective conflicting with the goal of ecological sustainability. One of the great benefits of introducing a Job Guarantee scheme together with cap-auction-trade systems is that they would both assist in severing the GDP-employment nexus. In doing so, they would enable full employment and ecological sustainability to be simultaneously achieved. Importantly, should a nation's economy be larger than its maximum sustainable scale, full employment would not result from an increase in real GDP brought on by the rise in government expenditure.[29] Full employment would instead be achieved by having the Job Guarantee ration paid work to the extent required to eliminate unemployment and by having cap-auction-trade systems perform their function of keeping the scale of the economy within the sustainable carrying capacity of the ecosphere. This beneficial 'rationing' role of the Job Guarantee is not what its advocates envisaged when they first conceived of the scheme. Their aim was to boost aggregate demand sufficiently to close the unemployment gap. But in an ecologically 'full' world where the economies of most countries already appear to have exceeded their maximum sustainable scale, and where numerous more are about to do likewise, the employment-rationing role of the Job Guarantee might prove to be its most important practical function.

Advocates of a basic income guarantee—many of which do not support the Job Guarantee—claim that an unconditional transfer payment to each citizen would also weaken the GDP-employment nexus. Thus, they also claim that a basic income guarantee would reduce the environmental impact of providing a universal minimum income. More than this, they argue that, by severing the link between income and work, a basic income guarantee would go further than the Job Guarantee and afford individuals the freedom from work exigency (Gintis 1997).

There is no doubt that the introduction of a Job Guarantee scheme would make it difficult for many people to avoid paid work. Nonetheless, it also true that if a basic income guarantee triggered a massive withdrawal of labour from formal labour markets, the people continuing to engage in paid work would effectively end up 'paying' for the non-work of the people exiting the labour market. Hence, the freedom enjoyed by one person from not having to engage in paid work would become the source of another person's alienation (Cowling et al. 2006).

The other major concern regarding a large-scale withdrawal of labour is the possibility that it could precipitate a hyper-inflationary episode.[30] Whether this occurs depends largely on whether the withdrawal of labour from formal labour markets is 'real' or 'artificial'. One can identify three main sources of a genuine withdrawal of labour. They are:

1. Increased labour productivity. Improvements in labour productivity lead to higher real wages that allow people to reduce the number of hours they work.
2. Increased labour market flexibility. Flexible labour markets enable people who would like to reduce their work hours, but presently cannot, to do so.
3. Government cash payments that reflect the contribution that non-paid work makes to the social product (e.g., non-paid household work, child rearing, and volunteer work).

Why would the latter be an example of a real labour supply withdrawal? Because the government cash payment would not only reflect the contribution that one makes towards the nation's real income,[31] thereby ensuring that the value of any withdrawn labour is matched by a real demand-side outcome, it would ensure that those who continued to work would not be subsidising the non-work of those who have exited the labour market.

Of these three sources of genuine labour supply withdrawal, it is the latter which is most relevant to the basic income guarantee. To what extent the basic income guarantee would induce a real or artificial labour supply withdrawal depends on how much the government transfer payment exceeds a level of remuneration approximating the non-paid work contribution made by the *average* citizen towards the social product. It is important to focus on the average citizen because it would be too complex to determine the non-paid work contribution of each person and remunerate them accordingly. It would also be administratively simpler to provide a basic income guarantee on a universal basis.

Clearly, if a basic income guarantee was introduced and set at the basic living wage—as its proponents advocate—it would far exceed the average person's non-paid work contribution to the social product and precipitate a large 'artificial' withdrawal of labour. However, a government cash payment set at a value equivalent to the unemployment benefit paid in most wealthy countries (approximately 40 per cent of the minimum income) would be very close to the mark.[32] If so, the lesser cash payment would induce little in the way of an artificial withdrawal of labour and suppress any inflationary pressure that a fully-fledged basic income guarantee would otherwise generate. Should the potential for an artificial labour supply withdrawal still exist, it could be minimised by increasing the range of fractional employment options, thereby allowing people to supply the portion of their labour not covered by the smaller cash payment.

Making available a universal cash payment equal to around 40 per cent of a minimum income would deliver one further benefit. It would address the distortive impact on worker incentives that has emerged because of the failure of most governments to remunerate non-paid work. In much the same way that many observers fear that government cash payments would induce an artificial withdrawal of labour, the non-payment of household and volunteer work has long induced an artificial influx of reluctant workers (e.g., stay-at-home parents) into formal labour markets. Not only has this placed enormous pressure on families and other social institutions, it has increased the full employment level of real GDP (i.e., it has increased the real GDP required to achieve full employment). In doing so, it has boosted the rate of resource throughput required to fully engage all people in meaningful activities. The '40 per cent' cash payment proposed here would correct this socially destructive labour market distortion.[33]

In the end, I believe that a combination of a Job Guarantee scheme and universal cash payments approximating the average person's non-paid work contribution to the social product is the best way to provide a minimum income. Whether the combination of the two should be instituted in the form recommended here is a matter for further investigation, although it certainly should not constitute a reason to delay the introduction of a minimum income.

At this point in time, it would be unrealistic to expect poor nations to deliver a minimum income as generous as that advocated above. But it would not be unreasonable to expect the welfare system of a low-GDP country to guarantee its citizens access to essential goods and services. This could be achieved by providing basic foodstuffs and public housing for all citizens in dire need. In many low-GDP countries, foodstuffs for the very poor are supplied by public distribution systems (e.g., India). Unfortunately, only a small fraction of all subsidised food usually reaches the people being targeted (Shah 2004). Failed public distribution systems should therefore be replaced by a system of food stamps. According to Panagariya (2002) a well-designed system of food stamps would not only increase the quantity of food reaching the very needy, it would minimise the cost of collection, storage, and distribution of essential foodstuffs, as well as reduce the corruption that often plagues public distribution systems.

Finally, as important as any welfare assistance is in alleviating poverty, it deals largely with the symptoms and not the underlying causes of gross income disparities. Nothing does more to liberate the poor from debilitating levels of poverty than the opportunity to participate in economic life. To achieve this, there must, in all countries, be universal access to adequate education and health services. In low-GDP nations, this is likely to come at a significant cost which, again, suggests the need for financial assistance from the world's richer nations. The great advantage of 'enabling' education policies is that they lessen the future reliance of the poor on government handouts. This can reduce the financial burden that a welfare system imposes on the governments of low-GDP nations as well as quickly obviate their need for external financial assistance.

3.4.2 Maximum Income/Limiting the Range of Income Inequality

Because sustainable development requires the eventual transition to a steady-state economy, which means restricting the growth of a nation's real GDP, a maximum limit must be placed on the incomes of the rich. Of course, some would argue that the limit on real GDP would place a natural limit on income levels. This is true except that it would not prevent the emergence of an unjust gap between a nation's poorest and richest citizens. As such, it would not prevent the majority of a nation's citizens living on a minimum income and a small percentage of citizens earning absurdly high incomes. As explained earlier, distributional equity is as much about limiting the range of income inequality as it is about ensuring a minimum income for the very poorest in society. In some countries (e.g., USA), the order-of-magnitude difference in the incomes of the rich and poor is as high as 500. This difference has steadily grown over the past fifty years and is not confined to wealthy nations (Daly 2008).

To institute a maximum income limit, it would be necessary to impose a 100 per cent marginal tax rate on incomes beyond a certain income threshold.[34] It is probably best to initially apply the 100 per cent tax rate to incomes well above the desired maximum and gradually reduce the income threshold over time.

As difficult as it would be to determine an appropriate income threshold, it would seem reasonable to use as a reference point the annual salary of a nation's President, Chancellor, or Prime Minister. After all, no occupation entails greater responsibility or importance than that of an elected Head of State or leader of a national government. From my Australian perspective, the order-of-magnitude difference between the Prime Minister's salary and the annual income of an elderly Australian reliant upon the old-age pension is a factor of around 20.[35]

There would, no doubt, be many who would suggest that a maximum income limit of any kind is undesirable because it would stifle incentive. Furthermore, given my earlier recommendation that income tax rates should be lowered to reward value-adding, others might argue that the two policies are inconsistent. For two reasons, I don't believe this is the case. Firstly, marginal tax rates can still be cut for incomes below the maximum threshold level. Secondly, I have identified the salary of a nation's leader as a possible threshold income because it is likely to approximate the point where any additional income amounts to an *economic rent*. An economic rent equals the difference between the amount paid to a factor of production and the minimum payment required to have the factor supplied in a factor market (Fischer et al. 1988). Because economic rents emerge largely as a consequence of a rise in the scarcity of a particular production factor, not because of any increase in the quantity and/or quality of the goods or services it yields, economic rents constitute a form of 'unearned' income. On the basis of fairness alone, all economic rents should be taxed at 100 per cent. Just as importantly, however, taxing economic rents does not alter the supply of the relevant production factor since the factor is still paid the minimum amount required for it to be supplied (i.e., the tax is non-distorting). Therefore, a 100 per cent marginal tax rate on incomes above the maximum level would not reduce individual incentive.

Another overlooked benefit of a 100 per cent tax on economic rents is that the converse—i.e., the untaxed retention of economic rents—promotes unproductive forms of investment and asset price bubbles. Unproductive investment is promoted in the sense that the retention of economic rents increases the financial claims on real wealth of those who earn the rents without having generated a commensurate rise in real wealth (George 1879). Why go to the bother of investing in productive capital when it is easier to purchase an asset and effortlessly gain by waiting for its increasing scarcity to inflate its exchange value? Buy now, sell later, and increase one's financial claims on real wealth without ever contributing to the latter's maintenance, let alone improvement or expansion. Provided there are more people purchasing economic rent-earning assets than there are selling them, asset prices boom. Eventually, as a disproportionate number of asset-owners attempt to convert their assets to more desirable forms of real wealth, the number of sellers exceeds the number of buyers, asset prices collapse, and macroeconomic instability ensues.

Interestingly, it is because of the unearned income feature of economic rents that they also constitute a means by which financial claims on real wealth are redistributed, usually from the poor to the rich, since it is the rich who are best positioned to purchase economic rent-earning assets. This serves to further justify the confiscation of economic rents. In all, allowing the retention of economic rents defeats the purpose of a wealth-creating market economy, undermines the sustainable development process, and runs counter to every economic argument supporting the superiority of a market economy over a command economy.

3.4.3 Foreign Aid

As indicated previously, distributional equity should apply across countries as well as within countries. To accomplish equity internationally, it will be necessary for all countries to gravitate towards a similar per capita GDP. Although domestic policies are likely to contribute towards this goal, especially if they are designed to facilitate the transition to a steady-state economy (i.e., an optimal per capita GDP), international institutions, protocols, and treaties also have a major role to play. For example, a new greenhouse gas emissions protocol not only provides an opportunity to achieve a safe atmospheric concentration of greenhouse gases, but a means of attaining an equal per capita share of emissions across the globe, which would go a long way towards promoting international equity.

There are many other equity determinants and impediments requiring attention, not the least being the enormous advantage that wealthy nations have over their poorer cousins in terms of spending power, human capital, technology, and general productive capacity. In addition, it has been shown that high-GDP countries have enjoyed the advantage of having initially grown their economies when the marginal cost of growth was considerably lower than at present. In recognition of this privileged position, high-GDP countries should increase the financial aid they provide to the world's poorest nations. At present, average spending on foreign aid per wealthy nation is 0.3 per cent of their real GDP (OECD Development Statistics Online). This is woefully inadequate. A foreign aid rate of at least 0.7 per cent of real GDP would be more appropriate and considerably more just.[36]

Apart from the areas of assistance already recommended, such as the establishment of Environmental Trust Funds; payments for ecosystem services; and the financing of population stabilisation measures, there are three additional areas where increased aid money should be directed. In the first instance, aid money should be used to provide basic goods and services to people suffering from extreme poverty, famine, war, and natural disasters. Although wealthy countries already administer ongoing aid programmes to deal with these circumstances, an increase in financial aid would vastly alleviate the suffering that occurs during and following such events. Secondly, aid money should be distributed to low-GDP nations to boost their investment in natural capital—in particular, reforestation programmes; wetland restoration and rehabilitation projects; and sustainable land management schemes. Thirdly, even if the atmospheric concentration of greenhouse gases is stabilised at a safe level, further increases in average global temperatures will have a negative impact on the world's ecosystems and economies. Most wealthy countries have the capacity to adapt to these changes. Besides their greater vulnerability, few low-GDP countries have the same adaptive capacity. Aid money should therefore be provided to help disadvantaged nations adapt to the inevitable changes in temperature and rainfall patterns.

Finally, on top of the aid money, rich countries should establish a transfer mechanism to assist low-GDP countries in the take-up of resource-saving and pollution-reducing technologies, including greenhouse gas-abating technologies. Such a mechanism should include funds drawn from the sale of emissions permits within a global emissions-trading system. It should also include a compulsory funding commitment embodied in a future global protocol. I'll have more to say about such a commitment in Chap. 9.

3.5 Achieving Allocative Efficiency

3.5.1 Appropriately Defined Property Rights

Ecological sustainability and distributional equity are the bedrock goals for achieving sustainable development. To take the next step and efficiently allocate the incoming resource flow, a number of market-related conditions must be satisfied. The first of these is the right of property ownership. Property rights are a legal set of non-price rules which govern the way that property owners can utilise, exploit, and sell the goods or assets in their possession. A clearly defined and enforceable set of property rights is essential to the efficient and effective operation of markets.

In most wealthy nations, the legislative mechanisms required to define, institutionalise, and enforce property rights are well established. The same cannot be said of many countries with command economies or economies in transition. These countries would be well advised to install institutional mechanisms similar to those found in nations with well-established market economies. This aside, problems universally exist in terms of the *exclusivity* of property ownership.[37] It is here where policy reform is required. I do not wish to examine all possible areas of concern. However, there are three areas worth focusing on for the purposes of this book: (i) knowledge and information; (ii) property rights and externalities; and (iii) the impact of sunk costs on the contestability of markets. I will have something to say about knowledge and information now, and externalities and sunk costs later.

Knowledge and information exhibit public goods characteristics. Their accessibility increases the likely emergence of efficiency-increasing technological progress (i.e., technology that can increase all four eco-efficiency ratios referred to in Chap. 2). In most countries, certain forms of knowledge and information are subject to exclusive private ownership under patent and copyright laws. Commonly referred to as 'intellectual property rights', the rationale for their existence is to grant short-term monopoly power to the generators of new knowledge to reward them for their creative endeavours. It is argued that, without such a reward, there would be insufficient incentive to encourage the private sector to generate new knowledge. Whilst there is considerable truth in this argument, it overlooks the fact that patents, by often depriving the rest of society of knowledge and other useful information, can impede the next major technological advance—perhaps more so than what the exclusivity of knowledge encourages.

The increase in efficiency-increasing knowledge is best facilitated, not by intellectual property rights, but by 'intellectual royalty rights'. If introduced, intellectual royalty rights would—for a limited period—grant the generators of new knowledge the right to a small royalty payment each time 'their' knowledge is accessed by external parties.[38] At all times, knowledge-generators would be denied exclusive use of their newly-created knowledge. Whilst the receipt of royalty payments would provide a financial incentive to generate new knowledge, non-exclusivity would ensure that any new knowledge capable of furthering the public good would not be confined to those wanting to use it for private gain alone.

There are, of course, some weaknesses associated with a system of intellectual royalty rights. Firstly, not all new knowledge has direct commercial value. For this reason, the creation and dissemination of many forms of knowledge requires government funding. In terms of generating new knowledge, governments must adequately fund the research activities of public research institutes and offer prizes, grants, and bonuses to the private sector (Daly 1996). In terms of the dissemination of knowledge, the key reports and information generated by government departments and statistical agencies should be accessible to all citizens.[39] Governments also need to introduce laws that ensure adequate product-labelling, whilst governments themselves should actively participate in prominent forms of information-disseminating media, such as television, radio, and the internet. Without government-owned broadcasting agencies, 'uneconomical' forms of information are unlikely to receive adequate media coverage. Given the importance of the media, governments also need to introduce laws that guarantee a diversity of private media ownership. Diversity of ownership ensures the dissemination of a wide range of opinions and the availability of a broad range of information.

Finally, the extent to which a system of intellectual royalty rights provides open access to knowledge and information depends on how many people or parties are able to afford the royalty fees. This depends on the magnitude of the fees and the financial resources of the poorest members in society. Once again, the latter can be addressed by ensuring there is an equitable distribution of income and wealth. It can also be addressed by setting royalty fees in a way that balances the need to provide firms with an incentive to generate new knowledge and the need to ensure the poor have the capacity to pay royalty fees. Unfortunately, this can still leave people or entities located in poor nations unable to afford the new knowledge generated in high-GDP countries.

I earlier recommended the establishment of a transfer mechanism to assist poor nations to uptake new technology. This mechanism should also be used to subsidise the royalty fees that low-GDP nations would be required to pay to access new knowledge and information. As we shall see in Chap. 9, this will be important in terms of assisting low-GDP countries to access low-emissions and carbon-sequestration technologies.

3.5.2 Externalities

The efficiency of resource allocation also depends on how well market failures are ameliorated. It has already been highlighted that markets are unable to deal adequately with the goals of ecological sustainability and distributional equity. Unfortunately, markets also have a propensity to fail with respect to the efficiency goal. However, unlike the sustainability and equity goals, which cannot be resolved by markets regardless of matter how much they are manipulated, markets remain the best mechanisms to allocate the incoming resource flow. The failure of markets on occasions to deal adequately with the efficiency goal means that governments should intervene to improve their allocative function, not reject them outright.

There are five main ways in which markets fail to efficiently allocate the incoming resource flow. They include:

- imperfect information[40];
- externalities;
- public goods;
- imperfect competition;
- natural monopolies.

In a strict economic sense, an *externality* occurs when: (i) the production or consumption activity of one party impacts beneficially or detrimentally upon an external party or parties; and (ii) no reward is granted to, or compensation is paid by, the party undertaking the activity. Externalities can be positive or negative. In the former instance, the activity involves a spillover benefit; in the latter instance, the activity involves a spillover cost. As perplexing as it may seem, both positive and negative externalities are socially undesirable in the sense that the unfettered allocation of resources to externality-related activities reduces a society's economic welfare. For example, in the case of negative externalities, the failure to penalise transgressors for the spillover costs of their actions leads to a divergence between the private and social costs of their actions. Because private benefits and costs are relied upon when economic agents make market decisions, the existence of negative externalities results in the over-allocation of the incoming resource flow towards activities that generate spillover costs. Conversely, in the case of positive externalities, the failure to reward benefactors for the spillover benefits of their actions leads to a divergence between the private and social benefits of their activities. On these occasions, there tends to be an under-allocation of the incoming resource flow towards activities that generate spillover benefits. In sum, positive externalities result in too little of a good thing, whereas negative externalities result in too much of a bad thing.

The policies required to deal with externalities have long been debated by economists (Kahn 2005). Despite differences in opinion as to what mechanisms should be employed to adequately deal with them, the objective is essentially the same—'internalise' all spillover benefits and costs to ensure private benefits and costs coincide with social benefits and costs. To achieve this, it is necessary to fully reward benefactors for the spillover benefits they generate and to fully penalise transgressors for the spillover costs they impose.

One simple means of internalising spillover benefits and costs is to introduce a system of Pigouvian taxes and subsidies (Pigou 1932). In the case of Pigouvian taxes, if a government authority estimates that each tonne of a particular pollutant imposes a $20 spillover cost on the rest of society (e.g., clean-up costs, deteriorating health, and lost production elsewhere in the economy), the polluters should be charged a $20 tax for every tonne of the pollution they generate. As for Pigouvian subsidies, if it has been estimated that each ream of recycled-paper generates a $5 spillover benefit to the rest of society (e.g., carbon sequestration and ecosystem service benefits from logging fewer trees), the recyclers should be granted a $5 subsidy for every ream of recycled-paper they produce. Thus, by increasing the private cost of polluting, a Pigouvian tax ensures that the adjusted profit-maximising decisions of polluters and their associated resource demands coincide with the maximisation of society's economic welfare. Similarly, by increasing the profits associated with producing recycled-paper, a Pigouvian subsidy ensures that the adjusted decisions and resource demands of recyclers also coincide with the maximisation of society's economic welfare.

Although Pigouvian taxes and subsidies seem an obvious solution to externalities, there are some who believe they are unnecessary and harmful. In the tradition of Coase (1960), these observers argue that, in the presence of well-defined property rights, agreements will be struck between the generators of spillover effects and the parties impacted by them that will bring forth allocatively efficient outcomes. For example, in the pollution scenario described above, if the property rights are vested with the polluters (i.e., if polluters are granted the inalienable right to determine pollution levels), it is argued that the pollution-affected parties would offer the polluters a payment to reduce the pollution they generate until the allocatively efficient level of pollution is reached.

The basis of this argument is that the marginal spillover costs of pollution generally increase as pollution levels rise, whilst the marginal benefits (profits) of the polluters typically decline. Hence, initially, it is less costly for pollution-affected parties to offer a payment to persuade polluters to reduce their pollution levels than to incur the spillover cost of the polluters' desired level of pollution. At the same time, it is more profitable for the polluters to accept the payment on offer than to produce and sell the output associated with their originally desired level of pollution. This process continues until the amount paid and received by the opposing parties equals the marginal spillover costs and the marginal benefits of pollution. The process ceases at this point because it is no longer beneficial for the pollution-affected parties to pay the polluters to further reduce their pollution levels.[41] Importantly, once this point has been reached, the total pollution generated comes to rest at the social welfare-maximising level. So, too, does the pollution price. Thus, an allocatively efficient outcome is achieved without the need for a Pigouvian tax.

Because the advocates of the Coasian position believe that allocatively efficient outcomes can naturally emerge through private negotiation, they contend that government intervention through the imposition of Pigouvian taxes and subsidies is superfluous. Indeed, from an efficiency perspective, they believe that government intervention should be confined to adequately vesting the property rights to the generators of the spillover effects or the parties impacted by them.[42]

In addition, the advocates of the Coasian position believe that Coasian outcomes are preferable to Pigouvian taxes and subsidies because the former are more likely to achieve allocative efficiency. This is because the latter rely on bureaucrats being able to accurately estimate the taxes and subsidies required to realise allocatively efficient outcomes. Moreover, bureaucrats must regularly modify tax rates and subsidies as market conditions vary. On top of this, Coasian supporters claim that a bureaucratic department is likely to absorb more resources than the resources expended by the negotiating private-sector stakeholders.

Given the additional absorption of resources and the likely bureaucratic errors associated with Pigouvian taxes and subsidies—sometimes collectively referred to as 'government failure'—Coasian supporters believe that Pigouvian taxes and subsidies are counterproductive. What's more, some go further and suggest that if the cost of government failure is sufficiently large, it is possible for a Pigouvian-adjusted outcome to be less desirable than the outcome (externality) that existed prior to the government intervention.

As elegant and simplistic as Coasian solutions appear, their success depends on a number of crucial factors. Firstly, they can only achieve allocatively efficient outcomes if the spillover-related activities involve *private goods* rather than public goods—that is, if the consumption or use of the goods in question is potentially excludable and rival. Secondly, the number of parties involved must be relatively few in number and readily identifiable. Thirdly, the source of the spillover effect must be easily known to the impacted parties. For this reason, the spillover effect must almost always be non-pervasive, which often rules out the use of Coasian solutions to resolve many pollution-related externalities, including the emission of greenhouse gases. Finally, the transaction costs associated with a potential bargaining process must be negligible. Transaction costs constitute the time and resources expended to successfully negotiate an allocatively efficient agreement. If transaction costs are high (e.g., if successful negotiation involves exorbitant legal costs), it may be less costly for the relevant parties to accept the initial outcome. If so, the prevailing outcome will be allocatively inefficient.

This last point is important given that transaction costs often exceed the potential cost of government failure. Should this be the case, it is better to employ Pigouvian taxes and subsidies to deal with an externality than rely on Coasian solutions.

Having said all this, both Coasian and Pigouvian solutions fail with respect to the sustainability goal. This is because both mechanisms, at best, can only facilitate increases in allocative efficiency, and allocative efficiency does not, as previously explained, guarantee ecological sustainability. Thus, if the negative externality in question involves the generation of pollution, the likely pollution level must be well within the absorptive capacity of the natural environment for the application of Coasian and Pigouvian solutions to be seriously considered. Given that we now live in an ecologically 'full' world where resource extraction rates and pollution levels are often near or beyond ecologically sustainable limits, Coasian and Pigouvian solutions are unlikely to be the preferred policy choice in most instances where the externality has an ecological connection. For this reason, cap-auction-trade systems are likely to be the most effective institutional mechanisms when dealing with most ecologically-related externalities, not only because they ensure an ecologically sustainable rate of resource throughput, but because the initial auctioning and subsequent trading of depletion/pollution permits internalises whatever spillover costs are involved.[43] Thus, cap-auction-trade systems are also able to achieve allocatively efficient outcomes. As alluded to earlier, this has considerable implications when choosing between Pigouvian taxes and an emissions-trading system to reduce global greenhouse gas emissions. I'll have more to say about this in Chap. 7.

In the end, I suggest the use of the following when dealing with externalities to achieve allocative efficiency:

- *Property rights (Coasian) solutions*. Most applicable where the externality in question involves private rather than public goods; where only a small number of easily identifiable parties are affected; where the spillover effects are non-pervasive; and where the transaction costs of a potential Coasian bargaining process are negligible.
- *Pigouvian taxes and subsidies*. The preferred solution in circumstances where the spillover effects are pervasive; where the number of affected parties is large and diverse; where the identification of affected parties is difficult; where the transaction costs of a potential Coasian bargaining process are prohibitively high; and where a negative externality involves resource extraction rates and pollution levels that are well within the ecosphere's regenerative and waste assimilative capacities. Pigouvian remedies are also preferred in circumstances where Coasian solutions fail to adequately compensate injured parties or give rise to crucial environmental rehabilitation responses. Taxes, for instance, allow governments to compensate injured parties and, if necessary, rehabilitate the natural environment. However, the choice between Pigouvian and Coasian solutions must be considered in light of the fact that Pigouvian solutions are considerably more complex to implement because they require bureaucrats to accurately calculate taxes and subsidies, which requires the prior identification and accurate measurement of spillover benefits and costs.

- *Cap-auction-trade systems*. Most preferred in circumstances where the spillover costs of certain activities pose a long-term threat to ecological sustainability. By automatically internalising previously unpriced spillover costs, cap-auction-trade systems ameliorate negative externalities in much the same way as Pigouvian taxes but without the need for bureaucrats to calculate the correct tax rate. Unlike Pigouvian and Coasian solutions, the caps imposed by these systems on resource extraction and pollution generation limit the throughput of matter-energy to the ecologically sustainable rate.

3.5.3 Public Goods

Public goods, which I have referred to a number of times, are distinguishable from private goods in terms of the non-rivalry and the non-excludability of their consumption or use. By non-rivalry, I mean a situation where one person's consumption or use of a good does not affect another person's consumption or use of the same good. Non-excludability implies that the private owner of a public good is unable to prevent someone from consuming or using the good should the consumer or user refuse to pay to access the good. An example of a public good is a lighthouse. A mariner's navigational use of a lighthouse does not affect another mariner's use of the same lighthouse. Hence, there is non-rivalry of use. In addition, the private owner of the lighthouse cannot exclude mariners from enjoying the benefits of the lighthouse if they refuse to pay for lighthouse services. Even if there is a strong demand for lighthouse services, the propensity of people to 'free-ride' combined with the inability of the lighthouse owner to force users to pay for lighthouse services prevents the owner from earning sufficient revenue to cover construction and maintenance costs, let alone make a profit. Consequently, the lighthouse is unlikely to be built. Left to the private sector, too little of the incoming resource flow is ultimately allocated to provide public goods in the quantities necessary to maximise society's economic welfare.[44]

Public goods can exist in human-made forms (e.g., lighthouses) or in natural forms (e.g., atmosphere). Because we are endowed with naturally-occurring public goods, the inadequate provision of public goods generally applies to human-made public goods. Exceptions include cultivated natural capital (e.g., timber plantations) and overexploited and degraded forms of natural capital. In many cases, government intervention is required to re-establish and restore natural capital stocks to critical levels.

More often than not, there is a tendency for naturally-occurring public goods to be abused, although the likelihood of abuse usually depends on the nature of the ownership and management of the public goods. For example, in increasingly rare circumstances, the stakeholders of a naturally-occurring public good are able to jointly own and manage the good in a sustainable and efficient manner without the need for government intervention. In these instances, naturally-occurring public goods are referred to as *common-property resources*. However, in most contemporary circumstances, the number and diversity of stakeholders plus the sheer complexity and cost of managing naturally-occurring public goods preclude a common-property arrangement. Since it is not possible for the private sector to exclude the abusers of naturally-occurring public goods from accessing them, nor possible for the private sector to compel users to pay for their use, the exploitation of naturally-occurring public goods invariably resembles a 'free-for-all'. In these circumstances, naturally-occurring public goods are described as *open-access resources*. In order to deal with the potential overexploitation of open-access resources, government ownership is required. For reasons previously explained, cap-auction-trade systems should be introduced to promote the sustainable and efficient use of naturally-occurring public goods, unless they are goods with high conservation values. Naturally-occurring public goods of this variety are best preserved in National Parks or a similar public reserve.

Returning to human-made public goods, the inability of the market to provide them in sufficient quantities has efficiency implications that go well beyond the initial failure to satisfy a community's demands. Many human-made public goods include much of society's physical infrastructure, such as schools, universities, technical colleges, hospitals, roads, bridges, energy-supply systems, railway networks, and ports. These all play a significant role in facilitating the development and uptake of resource-saving and pollution-reducing technologies, including the renewable energy and carbon sequestration technologies needed to resolve the climate change problem. Clearly, to address the problems associated with public goods, governments must use their taxing, spending, and legislative powers to satisfy the community demand for human-made public goods; to establish an efficiency-increasing and low-carbon infrastructure (i.e., by gradually replacing the existing high-throughput and carbon-intensive infrastructure); and to maintain natural capital stocks at critical levels.

3.5.4 Sunk Costs and the Contestability of Markets

Market power provides those who possess it with the capacity to restrict the supply of a particular good and raise prices without fear of reprisal. This allows the possessors of market power to earn economic profits well above the 'normal' or efficient level, thus reducing the economic welfare obtainable from the supply of the relevant good.[45]

The conventional structure-conduct-performance paradigm assumes that market power is a function of the number of firms or agents operating in a particular market. For example, where a market is characterised by many buyers and sellers, it is normally assumed that the market is highly competitive and that market outcomes are very efficient (i.e., normal economic profits are earned by sellers). Conversely, where there are very few sellers in a particular market, it is assumed that incumbent firms possess the market power to prevent the entry of new firms.

This permits incumbents to earn above-normal economic profits to the detriment of society's economic welfare. The structure-conduct-performance paradigm thus suggests that a highly concentrated or imperfectly competitive market is less efficient than a market with many sellers.

One of the weaknesses of the structure-conduct-performance paradigm is that it overlooks the fact that concentrated markets can be desirable in circumstances where economies of scale prevail. Economies of scale exist when the average cost of production declines as an individual operation expands. In industries where economies of scale predominate, the efficient number of firms is usually quite small (e.g., the automotive industry). Contrary to what is implied by the structure-conduct-performance paradigm, breaking up an industry of this type can be undesirable because the existence of small-scale firms would hugely inflate the average cost of production. Thus, even if more intensive competition could reduce the mark-up of price above average cost, the much higher average cost of production would result in a higher market price of the relevant product.

It is commonly believed that economies of scale confer market power to market incumbents. However, market power is ultimately determined by the prevalence of sunk costs. A sunk cost is a cost that, once incurred, cannot be recovered. Markets characterised by a lack of sunk costs are deemed 'contestable' and relatively free of market power because the absence of sunk costs enables a market entrant to recover its up-front outlays should it exit the market (Baumol et al. 1982).[46] This reduces entrant risk and enhances the likelihood of market entry. In consequence, it increases potential competition. It has been shown that potential competition can discipline market incumbents as much as actual competition by allowing hit-and-run entrants to share in transient profit opportunities created by the over-pricing strategies of incumbent firms, which dampens the initial incentive to raise prices (Bailey 1981). Hence, contestable markets have the capacity to generate outcomes as allocatively efficient as those generated by traditional competitive markets.

From a policy perspective, the great advantage of the contestability approach to competition policy is that it overcomes the need to break up industries with significant economies of scale. Policy-makers can instead focus on increasing potential competition by introducing policies to detach sunk-cost facilities from the serving or utilising firms (i.e., industry incumbents). There are two possible ways to achieve this. Firstly, sunk costs could be borne by a government instrumentality, as is normally the case with highways, railways, and airports (Bailey 1981). The sunk-cost facilities could then be leased to the highest-bidding firms on a transferable basis on the presumption that the highest bidders are the most efficient producers. The transferability of the leases would prevent the leases from becoming a sunk cost. Secondly, the government could mandate that sunk costs be borne by a consortium. Consortiums already share in the provision of international broadcasting satellites and, in some countries, the provision of telecommunications networks. The sunk-cost facilities could again be leased to utilising firms in the manner described above.

There are, unfortunately, three major problems with the above policy. In the first instance, it can reduce the incentive for utilising firms to improve upon existing sunk-cost facilities, which is necessary to increase eco-efficiency Ratios 1 and 2. Why make advances to sunk-cost facilities if you do not have exclusive ownership of them? To overcome this shortcoming, royalty rather than property rights should be conferred to those who supply improved sunk-cost facilities. By granting royalty rather than property rights, the superior producer goods would remain accessible to those who wish to utilise them—thereby ensuring the contestability status of markets—whilst the supplier of the superior facilities would receive an ongoing but limited pecuniary reward for their efficiency-increasing efforts.

In the second instance, sunk-cost facilities cannot always be detached from the serving or utilising firms. Thirdly, in many capital-intensive industries, it is often too expensive or too risky for governments or a private consortium to bear the cost of their provision. Since it is under these two circumstances that the greatest potential exists for market incumbents to exploit market power, anti-trust legislation should be enacted to: (i) prohibit anti-competitive behaviour by market incumbents; and (ii) limit mergers where they are likely to substantially lessen competition, albeit this must be weighed up against the potential efficiency benefits of having large firms exploit their economies of scale.

In most industrialised nations, anti-trust legislation exists in some form, although it is often not rigorously applied. To ensure the success of anti-trust legislation, a well-funded and independent statutory authority is required to scrutinise market participants; to investigate and prosecute serious breaches of the legislation; and to assess merger applications. Since prevention is better than cure, penalties must be sufficiently large to deter anti-competitive behaviour.

In many low-GDP countries, detaching sunk costs from utilising firms is all but implausible. Hence, the opportunities to increase market contestability are very limited. Countries in this position must rely entirely upon anti-trust legislation, which, for no other reason, is well worth enacting given that a thriving small-business sector constitutes the foundation of every successful market economy.

3.5.5 Market Deregulation and the Privatisation of Public Assets

The recent global trend towards market deregulation and the corporatisation and privatisation of government instrumentalities reflects the strong desire of policy-makers to facilitate the efficiency-increasing benefits of greater competition and private ownership (Brown 1996).[47] Whilst many people assume that market deregulation and privatisation always lead to increases in the efficient operation of markets, there are a number of aspects worthy of consideration. Firstly, most of the policy measures designed to increase the level of potential competition have focused upon the freeing up of highly regulated markets. Although these measures have successfully exposed legalised monopolies and duopolies to market forces, they have failed to deal adequately with sunk cost considerations. Consequently, many deregulatory policies have left markets vulnerable to abuses of market power.

Secondly, the quest for increased efficiency has led to a dramatic rise in the privatisation of government instrumentalities in the belief that privately-owned business enterprises operate at lower cost than their publicly-owned counterparts. Despite some apparent evidence to support this belief (Hutchinson 1991; Vining and Boardman 1992; Clare and Johnston 1992; and Megginson et al. 1994), sufficient doubt exists to conclude that privatisation itself cannot guarantee efficiency gains (Bishop and Thompson 1992; Hall and Lobina 2005). This doubt exists largely because efficiency is often empirically evaluated in terms of profitability or sales volume per employee/asset. Yet, as some observers have pointed out, private enterprises can be expected to outperform government instrumentalities in these two areas because the public sector, by its very nature, has multiple objectives of which profit maximisation is rarely one of them (Bishop and Thompson 1992; Brown 1996; Quiggin 2002). Hence, most of the efficiency comparisons between private and public enterprises are invariably weighted in favour of private enterprises.

Not surprisingly, in circumstances where private enterprise-based practices are incorporated into the operational objectives of government instrumentalities, of which profitability and sales volume usually become important considerations, operational efficiency dramatically improves (EPAC 1995). What does this indicate? It indicates that a properly conceived and executed form of market deregulation (i.e., one where sunk costs are also appropriately dealt with) is likely to have far greater efficiency-increasing benefits than the mere transfer of assets from public to private ownership (Bishop and Thompson 1992).

Finally, even if it can be demonstrated that the privatisation and/or corporatisation of a particular government instrumentality can enhance its operational efficiency, it must firstly be demonstrated that the instrumentality in question is not a natural monopoly or a public-goods provider. Should either or both be the case, the private ownership or the introduction of private enterprise-based objectives would almost certainly lead to inflated market prices or the under-provision of the public goods. Only government ownership of natural monopolies and the government provision of public goods can prevent a market failure from arising. Clearly, to improve the operational efficiency of public instrumentalities in such instances, an alternative means to privatisation is required. Introducing incentives for public sector employees, such as rewards and bonuses for improvements in both productivity and efficiency are likely to be of considerable assistance in this regard.

3.5.6 Flexible Labour Markets and Adequate Skills Formation

Achieving allocative efficiency does not simply require the appropriate allocation of the incoming natural resource flow. It requires the adequate supply and appropriate allocation of the resource-transforming agents of the production process—labour and producer goods. There is little point in efficiently allocating natural resources if inappropriate quantities and forms of producer goods and labour power are available to transform the incoming resource flow into useful goods and services.

In wealthy nations, ensuring labour power is adequately available and efficiently allocated is more problematic than with producer goods. This is because institutional factors in most high-GDP countries generate wage relativities that are 'stickier' than the relative price differences of producer goods. The conditions of employment also tend to be very stable. Consequently, labour markets are unable to respond rapidly to changing employer demands for different forms of labour and desired variations in production schedules and associated workplace arrangements.

It has been shown that a lack of production and workplace flexibility can hinder efficiency advances and technological progress at the enterprise level (Blandy and Brummitt 1990). Furthermore, the lack of labour market flexibility can impede the ability of workers to reduce their work hours that could help reduce a nation's need to increase real GDP to achieve full employment.

There are, of course, good reasons for having in place some degree of relative wage stickiness and stable workplace arrangements. The employment of labour involves the employment of human beings, not machines, and wildly fluctuating wages and the lack of minimum working conditions can have a detrimental impact on workers' well-being and their productivity. Clearly, an appropriate balance must be struck between the efficiency benefits of relative wage and workplace malleability and the welfare and productivity benefits from having certainty of pay and secure conditions of employment.

Without prescribing specific labour market policies, one way of attaining this balance is to set award-wage brackets for different occupations and periodically update them to reflect changing market conditions. As for matching the supply and demand for different forms of human capital, this might be improved by better identifying and forecasting the future skill requirements of a nation's economy. A similar approach could also be used to determine the appropriate allocation of resources for educational and training needs.

At the enterprise level, improving operational efficiency requires a new management approach to 'internal' labour markets. Internal labour markets exist when employees within an organisation remain insulated from the 'external' market forces that normally influence wage rates and the demand and supply for labour (Norris 1989). Empirical evidence suggests that internal labour market arrangements that best promote increases in operational efficiency are those with incentive-based means of remuneration, such as profit-sharing arrangements (Weitzman 1984; Estrin 1986; Blandy and Brummitt 1990). Enterprises with these types of arrangements are usually characterised by harmonious workplace relationships and high levels of employee commitment, motivation, and self-responsibility—factors that help drive innovation and improvements in the quality of final goods and services (i.e., increases in Ratios 1 and 2). For this reason, profit-sharing and other incentive-based workplace arrangements should be more actively encouraged.

3.6 Monetary Reform

In a fiat-currency economy, money constitutes the fundamental means by which individuals gain market access to real goods and services. Should a nation's money supply increase at a faster rate than the availability of goods and services, claims on real wealth eventually outgrow claimable real wealth. This leads to price inflation and/or the failure of many individuals to service their outstanding debts (debt repudiation). Inflation and debt repudiation can potentially destabilise market economies, as evidenced by the recent global financial crisis—a case of too much liquidity, not a lack of it.

To avoid macroeconomic instability in the presence of an expanding money supply, it is necessary to expand real wealth. However, because continued growth is ecologically unsustainable, this so-called requirement exacerbates the impending ecological crisis. Thus, promoting the growth of economic systems to avoid macroeconomic instability of a financial origin sets a nation up to eventually experience macroeconomic instability of an ecological origin. In the end, one form of potential instability is substituted by another more insidious and irrecoverable form of instability.

A major factor in the growth of a nation's money supply is the ability of private banks to create money out of nothing and lend it at interest.[48] As the money-creating capacity of banks has increased over time, the emphasis of market activities has shifted from commodity circulation to capitalist circulation and ultimately to debt circulation (Daly 1996). What was previously the dominant form of market activity, simple *commodity circulation* involved the use of money as a medium of exchange. One began with a physical good or commodity (C), exchanged it for money (M), and then used the money to obtain a physical commodity of greater use value (C^1). The chain of transactions applicable to commodity circulation could thus be represented by $C \rightarrow M \rightarrow C^1$.

As the creation of money by private banks expanded, an increasing proportion of all market activities involved a more complex form of *capitalist circulation.* In these circumstances, one began with money (M), exchanged it for a commodity or used it to fund the generation of a new commodity (C), and finally exchanged the commodity for money (M^1). The chain of transactions could now be represented by $M \rightarrow C \rightarrow M^1$.

The important change in emphasis as the world moved from commodity circulation to capitalist circulation was that many physical commodities became little more than intermediaries in a process designed to increase one's financial claim on real wealth.[49] This did not always lead to macroeconomic problems. Indeed, provided the increased claims on real wealth were commensurate with the contribution made towards the augmentation of real wealth and/or the increase in its use value, the capitalist mode of circulation was able to be non-inflationary and desirable. However, if many of the increased claims on real wealth arose from the purchase and later sale of economic rent-earning assets and did not reflect the generation of new wealth, the potential for high rates of inflation and other forms of macroeconomic instability remained an ever-present possibility.

As many would know, the financial sector has become increasingly imaginative in terms of its creation of credit facilities and interest-bearing paper. Because of it, a growing proportion of market activities now involve *debt circulation*. With debt circulation, one begins with money that is often borrowed (M), exchanges it for an alternative financial asset, and later on exchanges the financial asset for money (M^1). The chain of transactions is therefore simplified to $M \rightarrow M^1$, meaning that physical commodities no longer feature in the exchange process. In these circumstances, the expansion of financial claims on real wealth over and above available real wealth is unambiguous. It is also treacherously destabilising.

Some of the policies outlined in this chapter could deal adequately with the problems created by the capitalist mode of circulation. For example, maximum income limits and the scarcity taxes embodied in cap-auction-trade systems would suppress some of the inflationary effects of economic rent-seeking behaviour. In addition, the policies aimed at achieving ecological sustainability would prevent any ecological stress emerging from the need to expand real wealth to keep pace with a growing money supply. However, none of these policies can avert the potential macroeconomic instability caused by the growing disconnect between real wealth and the growing claims on real wealth induced by debt circulation.

To minimise the macroeconomic instability of debt circulation, there is a need for central governments to introduce two monetary reforms. In the first instance, a central government must increase its control over the nation's money supply. This can be achieved by substantially reducing the ability of private banks to create money out of nothing and to lend it at interest. At present, commercial banks in most countries are subject to miniscule reserve requirements, thus giving them enormous power to create and destroy money.

To enable central governments to increase their control over the money supply, some observers have suggested that all quasi-bank financial institutions should be treated like commercial banks and that the entire financial sector should be subject to a 100 per cent reserve requirement (Soddy 1926; Fisher 1935; Daly 2008). Introduced gradually over time, these changes would eventually abolish the money-creating powers of the financial sector. What's more, banks would be restricted to earning profits by financial intermediation—i.e., by lending already-existing money deposited by savers and charging a higher interest rate to borrowers than what they pay to depositors—and by providing chequing, safe-keeping, and other basic financial services.

Many observers would respond by arguing that the majority of bank loans advanced on the back of bank-created money are based on judicious assessments of the capacity of borrowers to service the loans. Furthermore, they would argue that the sheer magnitude of the task required to appropriately manipulate the money supply would render absolute central bank control of a nation's money supply inefficient at best, and woefully ineffective at worst. Given these incontrovertible claims, it would seem that private banks are as well positioned as central governments to determine the appropriate size of a nation's money supply.

I beg to differ. The financial sector cannot be relied upon to determine the appropriate size of the money supply if only because prudence on the part of banks at the micro level (i.e., rational lending policies) need not correspond with collective prudence at the macro level. Having said this, I believe there is an efficiency justification for giving banks limited money-creating powers. Thus, a balance is required between the need for central government control to prevent the undesirable expansion and contraction of a nation's money supply and the desirability of allowing privately-owned banks to efficiently create some of the nation's money.

I therefore recommend the setting of money supply targets by central banks with caps imposed on the quantity of new money that the banking sector can create. The caps would be gradually tightened and the monetary base increased until a desirable mix is reached between government-created and bank-created money.[50] Periodically, banks would be required to bid for 'money-creating' permits. As per the caps, limits would be set on the total number of permits sold. To facilitate competition within the financial sector, limits would be placed on the number of permits held by an individual bank. Much of the cost of permits would be met by the seigniorage[51] enjoyed by banks upon acquiring the permits. In this sense, this first reform measure would involve the application of a cap-auction-trade system to assist in the management of money creation.

A second monetary reform measure is needed because caps on bank-created money do not eliminate the problems caused by compound interest. Consider the following example. I own a timber plantation consisting of 1,000 cubic metres of timber. At its current state of maturity, it regenerates at the rate of 10 per cent per annum. I can therefore harvest 100 cubic metres of timber indefinitely. If we assume no inflation and a constant price of $10 per cubic metre of timber, my plantation is worth $10,000. I am able to derive an income from timber harvesting of $1,000 per year (assume no harvesting costs). Of course, I need not sell the entire $1,000 worth of harvested timber each year. I can stockpile timber and sell more than a $1,000 worth of timber in a later year. Despite this option being available to me, my income is effectively $1,000 per year. Moreover, my financial claim on real wealth reflects my contribution to the annual flow of claimable real wealth.

I then decide to sell my timber plantation for $10,000 and deposit the $10,000 in an interest-bearing account. The account offers an interest rate of 10 per cent per annum. It therefore generates an annual interest income of $1,000. So long as I spend the $1,000 each year, my financial claim on real wealth is no different to when I owned the timber plantation. Moreover, assuming that the new plantation owner is harvesting timber at the same rate as I had previously, my annual claim on real wealth is still matched by the annual flow of claimable real wealth.

What, however, if I decide not to spend the $1,000 of annual interest income and instead roll it back into the interest-bearing account? We shall assume that the new plantation owner is stockpiling all harvested timber. At the end of the first year, I will have $11,000 ($10,000 + $1,000 interest). At the end of the second year, I will have $12,100 ($11,000 + $1,100 interest). If I decide to withdraw and spend the $2,100 of interest earned over two years, my claim on real wealth will be $100 greater than if I had spent my interest income at the end of each of the two years (i.e., $100 = $2,100 − $2,000). In the meantime, the new plantation owner has stockpiled, at most, $2,000 worth of timber. I say 'at most' because stockpiled timber degenerates and its commercial value consequently diminishes over time. Crucially, by allowing my interest income to compound, my $2,100 of spending power exceeds the value of the real wealth accumulated over the two-year period by the new plantation owner.

Why is there now a disconnect between my financial claims on real wealth and claimable real wealth? Unlike real wealth, which can only generate an income equivalent to earning a simple rate of interest, money left to accumulate in an interest-bearing account can increase exponentially. By convention only, the latter is able to expand in a manner that is totally inconsistent with biophysical realities.

To ensure the financial domain conforms to the biophysical realm, only simple-interest payments from interest-bearing accounts and other financial assets should be permitted. To impose a restriction of this kind, I recommend a second reform measure that would include the following basic features. Firstly, simple interest would be payable in the form of 'simple-interest dollars'. Secondly, simple-interest dollars would exist electronically in a specially designed account where only the conversion of simple-interest dollars into real wealth, not financial assets, would be permitted. Thirdly, simple-interest dollars would have a limited life of, say, one year on the basis that simple-interest income should reflect the simple interest annually generated by a stock of real wealth. Consequently, simple-interest dollars would be electronically confiscated if not spent within a year of their receipt. This last feature would ensure that the spending of simple-interest income on real wealth roughly coincides with the length of time it takes for existing real wealth to generate new real wealth.

3.7 National Accounting Reforms

To provide improved policy guidance and reveal whether a nation is moving towards the sustainable development goal, better indicators are required. Many government policies are based on the indicators present in the system of national accounts. These indicators provide a misleading and incomplete representation of a nation's sustainable development performance. There is, therefore, an urgent need to reform the system of national accounts, which should start with the modification of many existing economic indicators. The national accounting system should also be broadened to include indicators of the social and environmental variety.

3.7.1 National Accounting Reform # 1—National Income

The first required reform is a better measure of national income. As revealed earlier in the chapter, a nation's annual income is correctly defined as the quantity of goods and services that a nation can consume in a given year and be capable of consuming as many goods and services in the following year and beyond (Hicks 1946; Daly 1996). For a nation to sustain its consumption, it must, at the very least, maintain its productive capacity. To do this, it must set aside some portion of its annual product to: (i) replace depreciated producer goods and critical infrastructure; (ii) rehabilitate its citizens and economy from the negative effects of past and present economic activities; (iii) defend its citizens and economy from the negative effects of future economic activities; and (iv) maintain the natural capital needed to sustain the ecosphere's source, sink, and life-support services that are required to sustain the economic process. Clearly, to sustain its consumption over time, a nation cannot consume its entire annual output of goods and services.

In conventional terms, real GDP is regarded as a nation's real income. Real GDP is a constant-price measure of the goods and services annually produced by domestically-located factors of production. Since a nation wanting to sustain its consumption cannot consume its entire output of goods and services, real GDP overstates a nation's real income. To obtain a meaningful measure of national income—sometimes referred to as Sustainable Net Domestic Product (SNDP)—the following subtractions should be made to real GDP in the system of national accounts:

$$\text{SNDP} = \text{real GDP} - \text{DHK} - \text{DNC} - \text{DRA} \tag{3.5}$$

where:

- SNDP = Sustainable Net Domestic Product;
- DHK = depreciation of human-made capital (producer goods);
- DNK = depletion of natural capital;
- DRA = defensive and remedial activities.

3.7.2 National Accounting Reform # 2—Sustainable Economic Welfare

The second required reform is the inclusion of an indicator of sustainable economic welfare. This would enable a government to determine where the nation's economy is in relation to its optimal macroeconomic scale (S_* in Fig. 2.4). As already explained, a nation's economic welfare constitutes the difference between the uncancelled benefits and uncancelled costs of economic activity. For a variety of reasons, GDP is also a poor indicator of a nation's economic welfare. Firstly, many costs that ought to be subtracted when calculating economic welfare (e.g., the cost of resource depletion and the cost of crime) are added in the calculation of GDP. Secondly, GDP overlooks many benefits (e.g., the value of non-paid work) and the welfare implications of a change in the distribution of income. Finally, the calculation of GDP includes defensive and remedial activities that do not directly increase economic welfare.

I should point out that although a measure of Sustainable Net Domestic Product (SNDP) is a better indicator of a nation's real income than real GDP, it too is not an adequate indicator of a nation's economic welfare. This is because some of the benefit and cost factors overlooked in the calculation of GDP are also overlooked in the calculation of SNDP (see Lawn 2006b).

In Chap. 2, the Genuine Progress Indicator (GPI) was introduced as a viable indicator of sustainable economic welfare.[52] The GPI is the best indicator of its type because it is the only indicator that explicitly identifies, values, and compares the major benefits and costs of economic activity. The GPI, or something similar, urgently needs to be incorporated into the system of national accounts.

There are, however, some people who have criticised the GPI in terms of its theoretical underpinnings and the choice of benefit and cost items used in its calculation (Atkinson 1995; Neumayer 1999a, 2000; Harris 2007). Much has been done to address these concerns, although the need to refine the GPI remains (Lawn 2003, 2005, 2007, 2008b; Clarke and Lawn 2008).

Perhaps the greatest weakness of the GPI is the lack of a consistent set of valuation methods to estimate the value of some benefit and cost items. The problem of inconsistency also extends to the choice of items. For example, in some studies, the imputed value of leisure time is treated as a benefit item (e.g., Lawn and Sanders 1999; Lawn 2007); in others, the value of lost leisure time is deducted as a cost item (e.g., Redefining Progress 1995). However, in some GPI studies, there is no recognition whatsoever of leisure/lost leisure time (e.g., Daly and Cobb 1989; Stockhammer et al. 1997).

Most people are aware of the United Nations System of National Accounts (SNA). The SNA sets out the standardised methods by which GDP and other macroeconomic indicators are calculated. A consistent set of valuation methods and procedures is also required to calculate the GPI. This would not only boost its credibility, it would increase the meaningfulness of international GPI comparisons.

3.7.3 National Accounting Reform # 3—Ecological Economic Efficiency

Once a nation has begun the transition to a steady-state economy, increasing its sustainable economic welfare depends, not on augmenting the rate of resource throughput, but on advances in efficiency-increasing technological progress. To reveal whether and by how much the latter is being achieved, a third reform measure is required—namely, the calculation of a nation's ecological economic efficiency (EEE) and its four eco-efficiency ratios (Ratios 1–4). In order to do this, it is first necessary to estimate the values of the five elemental categories of the economic process. To recall, they are net psychic income (uncancelled benefits); the cost of lost natural capital services (uncancelled costs); human-made capital; resource throughput; and natural capital. A separate account should be established for each category and be included in the system of national accounts.

The accounts for the first two categories are relatively easy to compile in the sense that the items included in them would be the same items used to calculate the GPI. Indeed, the value of the GPI is the difference between the total value of the net psychic income account and the total value of the lost natural capital services account (Lawn and Sanders 1999). The difficulty associated with the construction of the remaining three accounts varies. The human-made capital account, which would include durable consumer goods, privately and publicly-owned dwellings, producer goods, and human labour, is a reasonably straightforward account to compile. This is because, in most countries, the values of these items and the relevant price deflators are readily accessible from national statistical agencies. Compiling the natural capital account is more problematic in that the physical and monetary values of the relevant items are not easy to obtain. From my own experience in compiling a natural capital account for Australia (Lawn 2000), the values of some items must be estimated by the researcher or be obtained from individual and often one-off reports. In addition, the need to make assumptions about the monetary value of various forms of natural capital involves a great deal of subjectivity. Clearly, before it would be feasible to include a natural capital account in the system of national accounts, these shortcomings would need to be remedied, as would the need for a standardised set of items and valuation methods.

The throughput account is the most difficult of all the accounts to compile, not simply because it is difficult to record the various resources entering the economy and the wastes exiting it, but because it is impossible to aggregate them into a single index. Since economic activity cannot proceed without the use of energy, energy consumption is a good proxy for resource throughput. In my own work, I have used energy consumption as the basis for the establishment of a throughput account (Lawn 2000, 2007). Although an account of this type does not reveal the respective resource inputs and waste outputs associated with a nation's economic activities, it permits the calculation of Ratios 2 and 3 (see Eq. 2.2).

Should energy consumption be adopted as the prime indicator of resource throughput, there is still good reason for a nation to publish a satellite account to reveal the quantities of the various resources it has used and the various wastes it has generated (e.g., cubic metres of timber logged and tonnes of greenhouse gases emitted). By analysing satellite accounts over time, it would be possible to determine trend changes in the use rates of certain resource types and the rates at which particular wastes are being created and released.

3.7.4 National Accounting Reform # 4—Biophysical Indicators

Although the GPI provides an indication of where a nation's economy is in relation to its optimal scale, an entirely different indicator is required to ascertain where the economy is in relation to its maximum sustainable scale (S_S in Fig. 2.4).

As suggested earlier, there are a range of indicators with the capacity to fulfil this role. Without going into any great detail, they include:

- the Human Appropriation of Net Primary Production (HANPP), which is an indicator designed to measure the product of photosynthesis appropriated by humankind for its own purposes (Vitousek et al. 1986; Haberl et al. 2007). Proponents regard the HANPP as a means of measuring the scale of human activities relative to the supporting ecosphere.
- an Environmental Sustainability Index, which is a composite indicator aimed at capturing five key components of ecological sustainability (Yale Center for Environmental Law and Policy et al. 2005).
- Material-Flow Accounting, which involves input-output matrices to measure the throughput of materials used in the economic process (Perman et al. 2003).
- comparison of a nation's ecological footprint and biocapacity, where ecological sustainability requires the latter to exceed the former (Wackernagel et al. 1999).

To recall, it was the ecological footprint that was previously used as an indicator of maximum sustainable scale (see Table 2.1). Notwithstanding some of its weaknesses, I believe the ecological footprint is the best of the sustainability indicators outlined above, in part due to the intractable shortcomings of its rival indicators. For example, the Environmental Sustainability Index includes social variables which unduly flatter wealthy, high resource-consuming countries; there are considerable measurement uncertainties surrounding the HANNP; and Material-Flow Accounts are not amenable to time-series comparisons. Two further aspects work strongly in favour of the ecological footprint. Firstly, it incorporates a sustainability measuring stick insofar as it permits comparisons between 'actual consumption levels' (ecological footprint) and 'maximum sustainable consumption levels' (biocapacity). Secondly, a measure of a nation's ecological footprint includes the renewable resources that must be cultivated to replace declining non-renewable resource stocks. This is a significant advance over other indicators.

A major weakness of the ecological footprint is its use of land area as a numéraire for sustainability. Although land area and fertility constitute critical resource-limiting factors, there are other factors that impinge upon the supply of natural resources—for example, climate and water availability (Patterson 2006). Australia is a good case in point. According to Table 2.1, Australia had a per capita ecological surplus in 2005 of 7.6 hectares. At the best of times, a large proportion of Australia must cope with chronic water shortages. Recent droughts have significantly reduced water allocations to irrigators in the Murray-Darling Basin. This has severely reduced agricultural output from what is normally regarded as Australia's 'bread basket'. Should the severity and frequency of droughts increase in future years, as many climatologists predict, Australia's biocapacity is likely to decline significantly even if the area of land available for resource generation remains unchanged (Note: the arable land area can diminish due to urban sprawl).

In recent years, a number of refinements have been made to ecological footprint estimates to better account for resource-limiting factors (Lenzen and Murray 2001; Global Footprint Network 2008). These refinements have raised the ecological footprint to a standard that now warrants its inclusion in the national accounts as a vital biophysical indicator of maximum sustainable scale.

3.7.5 National Accounting Reform # 5—Social Capital Indicators

Sustainable development requires a nourishing association or coevolutionary linkage between the economy and the ecosphere that supports it. The quality of this linkage depends critically on the social sphere or the social capital at the interface of the economy and the ecosphere (see Fig. 2.1) (Hodgson 1988; Putman 1993, 2000). Social capital refers to the civic interaction of citizens and the networks of co-operation and solidarity that represent the self-governance of community life. Crucial to the stock of social capital are such elements as social cohesion, trust, and reciprocity (Franke 2005; ABS 2004).

Many attempts have been made by academic researchers, international organisations, and national statistical agencies to measure social capital (see Onyx and Bullen 1998; Grootaert and van Bastelaer 2001; ABS 2004, 2006; Zukewich and Norris 2005). It would not be unfair to say that social capital indicators are still in the embryonic stage of development. Most attempts have so far focused on measuring social participation in clubs, societies, religious organisations, and not-for-profit entities and the extent to which they offer support beyond the market and the government provision of goods and services.

In measuring social capital, it is important to distinguish between social networks, the determinants of social capital, and its spillover effects (ABS 2004). The distinction is necessary given that the spillover effects of social capital are largely reflected in other indicators (e.g., the GPI and some of its social cost items). Hence, there is little need for its specific measurement. Conversely, measuring social networks, in all its forms, can serve as a useful means of tracking changes in social capital, whilst measuring the determinants of social capital can provide insights into why stocks of social capital are quantitatively and qualitatively changing over time. This, in turn, can highlight the driving forces behind the enhancement or otherwise of the interactions between the economy and the ecosphere that are fundamental to achieving sustainable development. Despite the considerable work that needs to be undertaken to establish worthwhile indicators of social capital, improvements and refinements should eventually bring about indicators sufficient in quality to be included in the system of national accounts.

3.8 International Reforms

As important as it is for individual nations to introduce policies to achieve sustainable development, it is also crucial for nations to take collective action at the international level. International action is required for three main reasons. Firstly, climate change is a global problem and cannot be resolved by individual nations acting alone. Secondly, it is within the international domain that a global emissions-trading system must be dovetailed with a range of national and international policies to achieve sustainable development. Thirdly, virtually every nation is now an integral part of the global economy. International market forces can, in a 'globalised' world, discourage the introduction of the policies required to achieve a more sustainable, just, and efficient global economy.

The need for a global emissions-trading system to reflect the required paths of the world's rich and poor nations will be addressed in Part III of the book. For now, my focus is on international product and resource markets and the need for institutional reforms at the international level.

As just mentioned, one of the obstacles to the introduction of desirable domestic policies is the globalised nature of the international economy. By globalisation, I mean the integration of the world's national economies into one single economy through free trade and free capital mobility. Largely originating from former President Nixon's decision to sever the link between gold and the US dollar in 1971, globalisation involves the erasure of economic boundaries and the subsequent ability of corporations to bypass many national institutions and laws designed to serve useful social purposes. In a globalised world, the fundamental unit of concern is the corporation and the individual consumer (Daly and Cobb 1989; Røpke 1994; Daly 1996; Lawn 2007, 2013).

In stark contrast to globalisation is the concept of 'internationalisation'. Internationalisation refers to a global economy where national economies exist as separate and autonomous entities tied together in recognition of the potential value of international trade, treaties, and alliances. Internationalisation was a feature of the global economy in the two-and-a-half decades following the formation of the Bretton-Woods system in 1944. In an internationalist world, national institutions and laws impinge on economic activities for the purposes for which they are intended. Accordingly, the fundamental unit of concern is the nation state. In addition, the people residing within each nation are viewed as a community of citizens rather than a collection of individual consumers (Daly 1996).

The most significant difference between globalisation and internationalisation is the mobility of international capital and its implications for international trade. In the former circumstance, capital can be freely moved from one international location to another. In the latter circumstance, the mobility of international capital is severely restricted. What does this mean for international trade? In a globalised economy, international trade is governed by the principle of *absolute advantage*. Absolute advantage is where the terms of international trade are dictated by absolute rather than relative profitability. As such, decisions regarding the most appropriate production location are primarily based on where the absolute cost of production is lowest. This differs from an internationalist situation where international trade is governed by the principle of *comparative advantage*. In the case of comparative advantage, it is relative profitability that matters. In these circumstances, privately-owned firms, limited in their capacity to shift their production location internationally, are forced to specialise in the production of goods where the absolute cost of doing so domestically may be high, but where the relative cost of doing so must be low.

To understand how the latter situation might occur, one must go back to the basic premise underlying the rationale for international trade. Early in the 19th century, David Ricardo (1817) pointed out that the comparative advantage argument for free trade rests entirely on the immobility of capital. For instance, the principle of comparative advantage can never operate within the confines of a national economy because capital is always free to move to locations offering the most profitable investment opportunities—that is, where goods can be produced at the lowest absolute cost. Hence, at the intranational level, investment and the allocation of resources are always governed by absolute rather than relative profitability. Ricardo promoted free trade because, in 1817, the international mobility of capital was severely limited.

Should it matter that international trade is now governed by a different principle? After all, no national economy has been brought to ruin because intranational trade is governed by the principle of absolute advantage. For a number of good reasons, yes. First, intranationally, all production and exchange activities are subject to basically the same non-price rules, including any national policies regarding the rate of resource use, the distribution of income and wealth, and the efficiency of resource allocation. Consequently, no single producer can gain an unfair advantage from paying an equivalent form of work a significantly lower wage, by polluting when and where another producer cannot, or by paying a much lower rate of tax.[53] To gain a competitive advantage, producers must be genuinely more efficient than their nearest competitors. The same, however, cannot be said of the international market. This is because the international market is not a formally instituted market in the sense of being a collective set of social and cultural institutions within which a large number of commodity exchanges between buyer and seller take place (Hodgson 1988). Indeed, because social and cultural institutions rarely exist beyond national boundaries, commodity exchanges between international buyers and sellers take place in a domain largely free of institutional constraints. Consequently, the 'price-determining parameters' of the national markets where domestic production takes place are, for many countries, grossly incommensurate with those of the global market.[54]

To some extent, this is not a bad thing. On the positive side, it is desirable for price signals to reflect variations in economic efficiency. If a domestic producer is inefficient because a foreign producer is better at producing a similar commodity, the variation in prices should ensure the survival of the latter and the demise of the former. On the negative side, it is undesirable to have domestic producers ceasing their operations because of an inability to compete with a foreign producer subject to much weaker social and environmental standards. Yet industrial flight is precisely what an unfettered globalised market promotes because the free mobility of capital allows nationally instituted non-price rules and any explicit policy of cost internalisation to be avoided by transnational corporations (Daly 1993, 2013). Furthermore, because the price-determining parameters of the global market often come to rest at the lowest common denominator, the competitive pressure to lower the cost of production often leads to the erosion of environmental and social standards at the national level (Daly and Cobb 1989; Rees 2013). This so-called 'race to the bottom' has been exacerbated by World Trade Organisation (WTO) Articles that render governments powerless to introduce compensating tariffs that might otherwise offset any cost advantage enjoyed by foreign producers subject to weaker production-related standards. Pressure in Australia to reduce the minimum wage, to allow mineral exploration in National Parks, and to lower tax rates in line with taxation regimes of its nearest Asian neighbours is symptomatic of the degenerative impact of a global free trade environment governed by the law of absolute advantage. It is the pressure to reduce standards that often constitutes the political obstacle to the introduction of sustainable development policies, including climate change initiatives.

Secondly, since highly mobile capital will generally flow to locations with an absolute advantage in production, the potential for large trade imbalances to emerge is significantly high. The same does not occur when capital is effectively immobile because the level of international lending and borrowing required to accumulate large foreign debts is precluded. The growth in foreign debts has not only forced many low-GDP nations to deplete their natural capital to service their indebtedness, it has often led to the undesirable restructuring of economies as a means of securing loans from such institutions as the International Monetary Fund (IMF). This invariably results in significant hardship to the economically disadvantaged residing in the affected countries.[55]

Finally, there are many who believe that globalisation encourages poor nations to earn additional export income that can be used to invest in human-made capital and new technology. This investment, it is argued, allows poor nations to establish the necessary productive capacity to catch up with richer nations. In view of what has been said so far, the benefits of increased productive capacity are of little value if it involves having to attract capital by maintaining absurdly low wages, poor working conditions, and weak environmental standards. Yet it is the erosion of social and environmental standards—the bedrock of any beneficial increase in productive capacity—that is rapidly becoming an undesirable by-product of globalisation itself. Globalisation, it seems, facilitates the emergence of opposing forces rather than complementary beneficial outcomes.

These and other long-term problems associated with export-oriented globalisation have convinced internationalists that all nations, not simply impoverished nations, should focus on import-replacement policies. Let me say upfront that an import-replacement policy is not, as some believe, 'anti-trade'. Nor does it require the imposition of tariffs and quotas to protect inefficient and under-performing industries. Import replacement is where a country increases the efficiency of production to such an extent that it is able to produce a variety of goods at a lower cost than it previously cost to import them. Thus, instead of earning an additional $1billion from the production and exportation of more wheat, a country might reduce import spending on cars by $1billion by becoming more efficient at automobile production. Clearly, from a balance of trade perspective, nothing is lost by switching from an export emphasis to import replacement. But there is much to be gained. In the first instance, a successful import-replacement policy leaves a country producing a greater variety of goods. This not only increases a country's self-sufficiency, it reduces its exposure to volatile global market forces.[56] Next, the production of a greater range of goods renders a country less reliant on exports as a source of income, which in turn renders it increasingly free not to trade. Lastly, over-specialisation in the quest for higher export income has brought with it rural underdevelopment, urban overpopulation, and the loss and destruction of once self-reliant communities. An import-replacement policy would do much to overturn these undesirable trends.

There are usually two responses to the criticism directed towards the globalisation of the international economy. Firstly, there is little evidence of widespread industrial flight arising from disparate wages, working conditions, and environmental standards. Secondly, globalisation has helped drag many people out of abject poverty.

To the first point, a number of studies have been undertaken to verify or repudiate the theory that capital moves to locations with weaker social and environmental standards—otherwise known as the 'pollution-haven hypothesis'. The majority of these studies support the position that differences in labour costs account for at least some industrial flight (Leonard 1988; Hodge 1995; Garrod 1998; Ratnayake and Wydeveld 1998). However, almost all studies lead to the conclusion that environmental stringency has virtually no impact on the choice of production location (Dean 1992; Pearce and Warford 1993; Jaffe et al. 1995; Garrod 1998). The reason for this, it seems, is that the cost of adjusting to environmental standards is minimal for all but a few highly pollutive industries and that avoiding such costs through relocation is almost always absorbed by the cost of relocation itself (Leonard 1988; Stevens 1993).

For some observers, the lack of conclusive statistical evidence means the verdict is still out on whether variations in environmental standards cause industrial flight (Hodge 1995; Field 1998; Ratnayake and Wydeveld 1998). As I see it, the weakest aspect of the empirical studies so far undertaken is that they concentrate solely on the relocation of existing firms and industrials from high-GDP to low-GDP countries. They have not considered three other potential manifestations of industrial flight, namely: (i) how many new industries have emerged in low-GDP countries where, if not for the disparities in standards between rich and poor nations, most would have emerged in high-GDP countries?; (ii) to what extent is the low cost of adjusting to strict environmental standards due to standards in wealthy countries falling short of what is required to meet sustainability and equity requirements, in which case if standards were suitably tightened, the cost differential would be significant enough to induce the relocation of capital?; and (iii) how much has the threat of offshore relocation discouraged the introduction of more exacting environmental standards in high-GDP countries or, worse still, has led to the dilution of existing standards?[57] Until these questions have been adequately answered, the apparent lack of any mass relocation of existing industries from rich to poor nations cannot be used to disclaim the pollution haven hypothesis.[58]

To the second point, most of the people purportedly dragged from poverty by globalisation have gone from earning US$1 a day to a mere US$2 a day. Although US$2 is better than US$1, it has been gained at the expense of environmental degradation, social dislocation, and longer working hours. These constitute welfare losses but are not highlighted by the globalisation advocates. Nor are they reflected in a nation's GDP, although they have played a significant role in the recent decline in the per capita GPI of many countries, including China and Thailand (see Fig. 2.8). Some globalisation supporters concede this fact but argue that there is no alternative way for poor nations to progress. It is either globalisation or regress. This claim is nonsense. The incomes of people in poor countries can be lifted without the degenerative effects of globalisation, just as incomes and environmental and social standards were raised in today's high-GDP countries during the immediate post-World War 2 period (Lawn and Clarke 2008).

3.8.1 *The WTO and Compensating Tariffs*

To avoid the degenerative effects of globalisation, the world does not need countries disengaging from the global economy. What the world needs is a return to fair and balanced trade and the restoration of comparative advantage as the principle governing international trade. Also required is a mechanism to cancel out the non-efficiency-related cost advantages enjoyed by the producers of internationally traded goods.

One potential solution is to permit countries with similar wages, tax regimes, and environmental standards to freely trade with each other but impose 'compensating' tariffs on countries with lower standards. To work effectively, the tariff would need to reflect the undesirable cost advantage arising from disparities in standards, not from genuine differences in the efficiency of production.[59] Because the abuse of compensating tariffs has the potential to trigger a degenerative tariff war (see Bhagwati and Mavroidis 2007), a system of compensating tariffs would need to be overseen by an international organisation. An important responsibility of the chosen organisation would include the approval or otherwise of compensating tariff applications. The WTO would seem a likely candidate, although success in this regard would require the WTO to radically revamp its anti-tariff attitude regarding international trade matters.

In many ways, the WTO ought not to be antagonistic towards compensating tariffs. As it is, Article XX of the General Agreement on Tariffs and Trade (GATT), which serves as the basis for the WTO's assessment of international trade matters, permits the imposition of compensating tariffs on environmental grounds, albeit under strict and narrowly-defined circumstances. Moreover, much of the WTO's rhetoric on international trade centres on its potential efficiency benefits. As things currently stand, international trade in the presence of externalised social and environmental costs leads to a grossly inefficient allocation of global resources. Introducing a system of compensating tariffs would do much to eradicate this undesirable feature of globalisation.

Understandably, antagonism towards a system of compensating tariffs is likely to come from those who would view it as a means by which rich nations can maintain their wealth advantage over the world's poorest countries. There is no doubt that a system of this nature would make it more difficult for many poor nations to export their wares to high-GDP countries. However, this difficulty may not be undesirable. Indeed, it may help to close the gap between rich and poor. Consider the following response by Daly when asked about the possible implications of internalising social and environmental standards into the prices of Third World goods:

> Granted this makes it harder for poor countries to export — so does a decent minimum wage and the existence of free labour unions and the outlawing of child labour within the poor country. In my view it is not all bad to make it harder for poor countries to export to the US. It means that instead of planting all their land in bananas or fancy fruits and flowers for export, the poor country might have to plant more rice and beans for its own citizens. And to sell the rice and beans to its own citizens, it will have to worry about their purchasing power — about domestic jobs and decent wages, and the distribution of income within their country. And they might worry a bit less about cutting wages and social benefits in order to be more competitive in the global market, as they must do in the export-led model of development to which the IMF and WTO are so committed. Admittedly, less export revenue will be available to buy expensive toys for the elite, but even that might not be all that bad. Maybe they will begin to invest some of their surplus in their own country.[60]

In other words, a system of compensating tariffs would force policy-makers in the world's low-GDP nations to treat the issue of domestic spending power more seriously. In turn, it would compel policy-makers to implement measures to improve the distribution of income and wealth within their own nations. Furthermore, to compete in the global market, low-GDP countries would need to become genuinely more efficient, not increasingly attractive pollution havens and/or cheap labour locales. Because this would gradually close the productivity gap between the world's rich and poor countries, it would help to bridge income disparities across the globe.

3.8.2 A New System of Foreign Exchange Management

To restore balanced trade and the principle of comparative advantage, it is necessary to restrict the international mobility of financial capital. Such a restriction reduces industrial flight insofar as the mobility of financial capital is necessary to gain from the international movement of human-made capital. If one cannot move the profits generated elsewhere back home, there is little incentive to relocate production in the first place. Importantly, capital immobility does not prevent goods from being traded and consumed globally.

Ultimately, a restriction on the international mobility of capital forces the owners of capital to consider what goods should be produced domestically and where, domestically, it is best to produce them. The former consideration is dictated by the principle of comparative advantage, whilst the latter is dictated, in the domestic context, by the principle of absolute advantage.

Unfortunately, a direct macro constraint on the international flow of financial capital impedes the beneficial adjustment of exchange rates. There are, however, two practical means of overcoming this predicament. The first would involve the formation of John Maynard Keynes' concept of an International Clearing Union (ICU). Keynes' plan, developed during the 1944 Bretton Woods conference, was to have all international trade measured in terms of a neutral unit of international currency called the 'bancor'. Whereas exporting would accrue bancors, importing would involve the expenditure of bancors. Each nation's bancor account would be tied to its currency through a fixed, albeit adjustable, exchange rate. By charging a penalty fee on excess bancors (trade surplus) and overdrawn bancor accounts (trade deficit), Keynes believed that countries would be induced to balance their trade on the current account.

The second potential means of restoring the principle of comparative advantage entails the creation of a lesser known IMPEX system of foreign exchange management (Iggulden 1996). Operated at the national level, the IMPEX system would involve the creation of an IMPEX facility in each country to deal with all international transactions. The facility, which would come under the supervision of a country's central bank, would operate under the following five rules:

1. Every international transaction must pass through the IMPEX facility.
2. All foreign currency must be exchanged for IMPEX dollars ($IMP).
3. The purchase of foreign currency requires the possession of IMPEX dollars.
4. No spending on imports is permitted unless there is sufficient 'earned' foreign exchange available on the day (held in the form of IMPEX dollars) and only if importers are willing to pay the price demanded by the possessors of IMPEX dollars.
5. The buying and selling of the IMPEX dollars of any particular country is only open to the citizens of that country.

The exchange process in each country would operate as follows. When foreign currency enters a country from export sales, the exporter would receive IMPEX dollars based on the exchange rate between the domestic currency and the foreign currency earned. For instance, if an Australian exporter earned $US100 and the going exchange rate between an Australian and American dollar was $US1.00 = $Aus1.20, the Australian would receive $IMP120 from the IMPEX facility. The possessor of the $IMP120 would then be free to purchase another foreign currency in order to import foreign goods or, if he or she had no importing intentions, to sell the IMPEX dollars to an Australian who does. The Australian would not be permitted to sell the IMPEX dollars to a foreign national. Should the Australian exporter want immediate conversion of the American dollars to

Australian dollars, the earned IMPEX dollars would remain available for purchase by potential Australian importers, but would be held by the IMPEX facility.

In the day-to-day buying and selling of IMPEX dollars, an IMPEX rate would fluctuate relative to the domestic currency. For example, let's assume that the going IMPEX rate in Australia was 1.40. Should the amount of IMPEX dollars demanded by Australians for import purposes increase relative to the IMPEX dollars earned, the IMPEX rate would appreciate (i.e., rise above 1.40). On the other hand, if the demand for IMPEX dollars fell relative to its supply, the IMPEX rate would depreciate (i.e., fall below 1.40).

What, then, would Australians have to do if they required \$US100 to import American goods? Unless they were already in possession of IMPEX dollars, they would firstly be required to purchase \$IMP120 as per the going exchange rate between the Australian and American dollar (\$US1.00 = \$Aus1.20). Secondly, to purchase \$IMP120, they would be required to part with \$Aus168 as per the going IMPEX rate of 1.40 (i.e., \$Aus168 = \$US100 × 1.20 × 1.40). Thus, in order for Australians to import \$US100 worth of American goods, it would cost Australians \$Aus168, not \$Aus120 as per usual. Clearly, there would be two currency markets in place—one being the traditional foreign exchange market; the other being the domestic IMPEX market.

In the former market, exchange rates would continue to fluctuate as per normal because although the IMPEX system would ensure balanced trade, it would still allow a country to have trade imbalances with individual countries. For example, Australians might choose to import more American goods and fewer Japanese goods, thereby leading to a deterioration in the terms of trade with the US but an improvement in the terms of trade with Japan. Since, in these circumstances, Australians would demand more American dollars and less Japanese Yen then, ceteris paribus, the Australian dollar should depreciate relative to the former and appreciate relative to the latter.

Operated along these lines, the IMPEX system would deliver three important benefits. To begin with, by balancing trade, it would ensure the absence of unserviceably large foreign debts. Secondly, it would eliminate the destabilising impact of currency speculation. Thirdly, the IMPEX system would restore the principle of comparative advantage by eradicating much of the competitive (absolute) disadvantage that domestic producers suffer when strict environmental and social standards increase the cost of domestic production. In doing so, it would lessen the need for compensating tariffs, although, for efficiency reasons, it would not eliminate the need for them altogether.

How would the IMPEX system deliver the latter benefit? Let's assume that a national government introduces stricter environmental and workplace standards and this increases the cost of domestic production. This would increase the prices of all domestic goods relative to foreign goods. One would expect exports to decrease; the quantity of earned foreign exchange to fall; and available IMPEX dollars to decline. On the import side, one would expect an increased demand for imported goods and a corresponding rise in the domestic demand for IMPEX dollars. The increasing demand and falling supply of IMPEX dollars would inflate the value of IMPEX dollars and increase the cost of imports. To a significant extent, this would offset the price advantage enjoyed by the foreign producers subject to much weaker standards. In other words, the IMPEX system would go a long way towards internalising domestic environmental and social standards into the price of 'advantaged' foreign goods.[61]

I should add that an IMPEX system of foreign exchange management need not be as rigid as outlined here. It is possible for the IMPEX facility to make available a small quantity of 'unearned' IMPEX dollars. This would permit small trade imbalances; allow for limited overseas borrowing/lending that could be of value to poor nations requiring an injection of overseas investment funds; and incorporate some degree of flexibility into the system. Flexibility is required given that a certain quantity of unearned IMPEX dollars must be issued to activate the system. After all, if all nations introduced the IMPEX system and were forced to export before they could import goods and services, a stalemate would ensue that would only be resolvable by enabling at least one country to import without having first exported. Only the availability of unearned IMPEX dollars could overcome the stalemate.

Given the potential for foreign debts to become economically and ecologically unserviceable, there is a need for strict controls on the quantity of unearned IMPEX dollars issued by an IMPEX facility. Ideally, unearned IMPEX dollars would cease to be issued once a nation's accumulated foreign debt reached a small fraction of its sustainable income (e.g., 2 or 3 per cent of Sustainable Net Domestic Product).[62] A pre-assigned limit would also need to be imposed on the quantity of foreign IMPEX dollars made available for sale to overseas investors. This would prevent future interest payments exceeding a nation's capacity to service them.

Of the two systems outlined above, I prefer the IMPEX system of exchange rate management. With the bancor system, the fees paid on excess and overdrawn bancor accounts merely encourage countries to balance imports with exports. They do not guarantee balanced trade. Conversely, with the IMPEX system, a means exists by which the magnitude of permissible trade imbalances can be explicitly regulated. There is also greater plasticity with the IMPEX system than the bancor system, since the exchange rates of national currencies and IMPEX dollars remain fully flexible.

3.8.3 Redirecting the Goals and Functions of Bretton Woods Institutions

In order to satisfy the needs of an internationalist arrangement, the Bretton Woods institutions of the World Bank, IMF, and World Trade Organisation (WTO) must operate in accordance with the charter upon which they were conceived.[63] I have already explained the possible new role of the WTO in relation to compensating tariffs. In the case of the IMF, it must significantly limit the creation and allocation of special drawing rights to reduce the expansion of international liquidity. In addition, the IMF must alter the conditionality of its loans to allow poor countries to service their existing foreign debts in ways that do not cause short-term hardship and long-term impoverishment.

The IMF was originally created to administer the international monetary system. Assuming that an IMPEX system of foreign exchange management was introduced, the IMF is ideally placed to oversee its operation. Working in conjunction with the central banks of each participating country, the IMF would monitor the issuing of IMPEX dollars and ensure the trade in IMPEX dollars complied with the system's five critical rules.

As for the World Bank, a greater proportion of its investment capital should be distributed to help low-GDP countries acquire much needed human-made capital and technology. In countries where environmental degradation is severe, funds should also be used to rehabilitate and restore natural capital stocks. Not only would this raise the productive capacity of needy nations—a primary reason for the World Bank's creation—it would facilitate the narrowing of the technology and income gaps between rich and poor countries. Bridging the technology gap would enable low-GDP countries to compete with wealthy nations on genuine efficiency grounds without the need for low wages, poor working conditions, and weak environmental standards. This would increase the profits that are internally generated and retained, which, presumably, would boost domestic investment levels and strengthen a nation's self-sufficiency.

Notes

1. There are, of course, many policy goals applicable to achieving sustainable development. The three policy goals listed here are best thought of as all-encompassing categories where: (i) the first goal captures the ecological dimension of sustainable development; (ii) the second goal captures the social and ethical dimension; and (iii) the third goal captures the economic dimension. For example, the second goal of distributional equity would include more than just considerations about the distribution of income and wealth. It would also include political and social justice concerns.
2. Whether a good is directly 'consumed' or 'used' depends on whether it is a durable or a non-durable good. Non-durable goods such as food, beverages, and petrol are consumed. Durable goods such as cars, televisions, and clothes are used. Whilst durable goods are not directly consumed, they depreciate over time.
3. Mainstream economists acknowledge that markets cannot ensure distributional equity.

4. The term 'Pigouvian taxes' has been applied to the internalisation of environmental externalities into market prices since the work of Pigou (1932).
5. These are broad-based terms to describe the policy instruments required to resolve the three policy goals. See Daly (1991, 1996) and Lawn (2000, 2007) for more detail.
6. The increase in hourly wage rates is dependent upon workers being able to share in the increased profits that arise from the production of better quality goods.
7. At low real wage rates, the labour supply curve for an individual is typically upward sloping insofar as an individual worker will generally be willing to work more hours as their real wage rate rises. However, once real wage rates and consumption levels are very high, the marginal benefit of additional work (i.e., the marginal benefit of consumption) becomes quite small. At some point—that is, once a particular real wage rate is reached—the marginal benefit of leisure exceeds the marginal benefit of consumption. It is above this real wage rate that an individual's labour supply curve is downward sloping or backward bending in the sense that a rise in the real wage will induce the worker to increase their leisure time (and reduce their work time).
8. In actual fact, taxes received by currency-issuing central governments do not increase their spending power. As a consequence, currency-issuing central governments do not technically earn revenue from taxation. Taxation involves the destruction of private-sector spending power. Following the imposition of a tax, private-sector banks accounts held at the nation's central bank are debited. However, despite there being a double-book 'T' account entry, there is no crediting of a government account. Central-government spending involves the creation of high-powered money (net financial assets) out of nothing and its insertion into the economy via the central government's spending. Assuming the government has hired labour and/or purchased goods and services generated by the private sector, private-sector accounts are credited but there is no actual debiting of a government account. In reality, therefore, a central government surplus constitutes a 'net financial asset-destroying fiscal stance' and a deficit constitutes a 'net financial asset-injecting fiscal stance'. Thus, if a currency-issuing central government achieves tax-revenue neutrality following the alteration of tax rates, it is merely leaving unchanged the difference between the net financial assets it is injecting into the economy and those it is destroying. However, for the sake of exposition, I shall continue to refer to the destruction of net financial assets from taxation as equivalent to raising tax revenue. The issue of tax revenue and government spending is better left to an alternative forum. For more, see Mitchell and Muysken (2008).
9. Redistribution is also important given that depletion/pollution taxes do not discriminate between rich and poor and are therefore regressive.
10. Why the highest bidders? Presumably those who are willing to pay the highest price to obtain a permit are those able to add the most value to the resources they wish to secure (i.e., are able to sell the end product made from the resources at the highest price). Hence, selling to the highest bidders

is likely to better facilitate qualitative improvements in all newly produced goods. Of course, the capacity to pay a higher price for a permit is also a key factor. Being rich is likely to confer an advantage when bidding for permits. This conceded, the extent of the advantage of the rich—which applies to the purchase of all things—is a matter of distribution, which ought to be dealt with by ensuring an equitable distribution of income and wealth. It might also be an issue of market power. This can be dealt with by ensuring all markets are suitably 'contestable' (Baumol et al. 1982; Lawn 2000) and by limiting the number of permits that any individual or entity can purchase.

11. Short permit life-spans would also massively reduce the incentive for people to engage in the speculative buying and selling of permits.
12. By this I mean, firstly, the incorporation of quantitative limits to achieve ecological sustainability; secondly, the public capture of scarcity rents and their redistribution to achieve distributional equity; and thirdly, the internalisation of ecological limits into renewable resource prices to facilitate allocative efficiency.
13. In both cases, it is necessary to police resource buyers and sellers to prevent abuse of the system. Hence, there is no logistical advantage here.
14. Of course, cap-auction-trade systems are not immune from bureaucratic error. Whether they assist in achieving ecological sustainability depends largely on how accurately the sustainable rate of throughput is estimated. If an overestimation is made and too many permits are auctioned off, the incoming resource flow will exceed the maximum sustainable rate. Should depletion/pollution taxes be preferred, any overestimation of the maximum sustainable rate of throughput will result in insufficient tax rates. Hence, the possibility of error does not amount to an argument against cap-auction-trade systems. To avoid any problems that might emerge from incorrectly estimating the maximum sustainable rate of throughput, it would be wise to adopt a 'precautionary' approach and limit the incoming resource flow and the number of permits sold to, say, 75 per cent of the estimated maximum sustainable rate.
15. In doing so, one satisfies the condition of 'strong sustainability'. See Lawn (2007).
16. As a means of illustration, if 50 units per year is the maximum quantity of renewable matter-energy cultivatable from an investment of the non-renewable resource in question, the sustainable exhaustion schedule increases to 19 years (20 extractions) and the annual amount of non-renewable matter-energy available for production purposes falls from 38.61 to 31.16 units of low-entropy matter-energy.
17. Public goods exhibit two main characteristics: (i) non-rivalry in consumption/use; and (ii) non-excludability in consumption/use. It is because of these features that it is often financially unviable for the private sector to provide public goods in sufficient quantities (DeSerpa 1988). This explains why the public goods problem constitutes a form of market failure and why there is a need for governments to provide public goods.
18. Wilson believes it should be more in the region of 50 per cent.

19. This in many ways supports the proposition that a lack of economic independence of women—often the poorest citizens in many countries—reduces the capacity of women to reduce birth rates.
20. Having a government authority administer the market for transferable birth licences would provide a reputable clearing-house for unsold licences. This would enable the number of outstanding licences to be tallied and guarantee sellers a fair price for their licence. It would also ensure that people are not forced to sell their licence under duress. Meanwhile, a government authority could provide counselling services and 'cooling off' periods (say, one week) to ensure individuals are making rational and calculated decisions.
21. People below the age of eighteen would not be encouraged to use their licences and, in many countries, would be below the age of consent. However, teenage pregnancies do occur and licences would be forfeited if such people went through with the birth of the child.
22. Under the scheme, child rearing would not be the exclusive preserve of heterosexual couples, or couples at all, for that matter. Surrogacy, donated sperm, and in vitro-fertilisation would enable homosexual couples and single people to make use of their freely allotted birth licence and any purchased licences.
23. Because of the so-called 'one-child' policy first adopted in China in 1978, China's 2008 population was estimated to be 300–400 million people less than what it would have been without the policy. Criticism has been directed at the one-child policy because of the abuses associated with the preference many Chinese have for a boy over a girl. This is a separate human rights issue that cannot be attributed directly to the one-child policy.
24. A negative income tax involves a person receiving a universal cash payment from the central government (demogrant) that exceeds what he or she pays in the form of income tax. The amount of a negative income tax therefore constitutes the difference between the cash payment and the total income tax paid. For example, if the minimum cash payment is \$10,000 per year and the marginal tax rate on all income is 40 per cent, then a person with an income of \$20,000 per year would have an annual tax bill of \$8,000 (i.e., \$20,000 × 0.40). They would therefore retain \$12,000 per year, meaning they would receive a negative income tax of \$2,000 (i.e., \$10,000 − \$8,000). For a person earning \$25,000, their tax bill of \$10,000 (i.e., \$25,000 × 0.40) would be equal to the minimum cash payment. Clearly, only people with incomes above \$25,000 per year would pay a positive income tax. For example, for a person earning \$50,000, their tax bill is \$20,000 (i.e., \$50,000 × 0.40), and their positive income tax is \$10,000 (i.e., \$20,000 − \$10,000).
25. Although it is not the aim of the Job Guarantee scheme to compete against the private sector for labour, workers are free to quit their private-sector job, if dissatisfied, and accept a Job Guarantee occupation. This allows the government to indirectly impose on the private sector a minimum hourly wage and minimum conditions of employment.
26. The NAIRU differs to the natural rate of unemployment. The natural rate arises as a consequence of people moving between jobs or because of an imbalance

between the location and skills of the unemployed and those of the jobs on offer. It is a rate of unemployment independent of the rate of price inflation. The NAIRU, on the other hand, exists entirely in the context of a non-accelerating rate of price inflation. It is therefore possible for the NAIRU to exist at an unemployment rate where structural unemployment is quite prevalent. In most cases, the NAIRU is slightly higher than the natural rate of unemployment.

27. The NAIRU is typically around 5–6 per cent of the labour force in most countries.
28. The higher is the interest rate, the greater is the opportunity cost of employing cleaner production techniques. See Lawn (2007, Chap. 13).
29. An increase in aggregate demand is still permissible, even with cap-and-trade systems in place, if a nation's current rate of resource use is well within ecologically sustainable limits.
30. Hyperinflation would result if the financial claims on real goods and services remained largely unchanged, but, as a consequence of the withdrawal of labour, the quantity of goods and services for sale dramatically declined. The same purchasing power would be chasing fewer goods and services, thereby devaluing the currency (inflation).
31. By a nation's real income, I mean all real output irrespective of whether it is included in a measure of a nation's real GDP.
32. This would not apply to old-age pensioners and people on disability-support payments. These people would still receive the full payment.
33. One would prefer to see traditional non-paid work remain unpaid on the basis that it constitutes an integral part of a nation's social capital. But if market forces have the propensity to deplete social capital (Hirsch 1977; Daly and Cobb 1989), the preservation and replenishment of social capital may require non-pecuniary assets to be valued in the same way as other assets. If so, the case for a modified basic income guarantee is further enhanced.
34. A 100 per cent marginal tax rate on income beyond a certain income threshold is not as unrealistic as many people believe. In fact, recent moves by G20 leaders to cap executive salaries indicates that income caps are politically feasible and socially acceptable.
35. This takes into account the concessions received by old-aged pensioners with respect to transport, health, rental assistance, and utility bills.
36. 0.7 per cent of real GDP was the figure agreed to by wealthy donor countries when making pledges as part of their Monterrey (2002) and Gleneagles (2005) aid commitments.
37. Two other key areas related to property rights—namely, *universality* and the *transferability* of property—are generally well covered by property rights legislation.
38. The size of royalty payments would be determined by a government department dealing with patent and copyright applications. The department would set the payments at a level which provides the creators of knowledge an adequate financial return on their investment without pricing prospective users of the knowledge out of the market.

39. By altering human behaviour, the information provided by governments can sometimes reduce the initial emergence of inefficiencies and externalities (Kennedy et al. 1994). Indeed, in some cases, government-provided information can act as a form of 'moral suasion'—a case where governments convince citizens they have a moral obligation to refrain from behaving in ways that are detrimental to society (Kahn 2005).
40. I have already explained how the potential information problem should be dealt with by governments.
41. Nor, at this point, is it beneficial for polluters to receive a payment to further reduce their pollution levels.
42. In both cases, allocatively efficient outcomes ensue. See Perman et al. (2003) for a diagrammatic explanation.
43. What's more, the ongoing trade in depletion/pollution permits ensures that the embedded scarcity tax varies in accordance with changing market forces. It is therefore a low-cost means of achieving allocative efficiency.
44. In other words, without government intervention, the macro-allocation of the incoming resource flow to the public sector is inevitably insufficient, thus resulting in the provision of too few public goods. Macro-allocation is a relatively new term used by ecological economists to distinguish between the resource flow respectively allocated to the private and public sectors (Daly and Farley 2004).
45. A normal economic profit is one where the accounting profit earned (i.e., revenue less explicit costs) is equal to implicit costs. Implicit costs constitute the foregone value of the next best use of time and capital.
46. For a detailed and rigorous explanation of contestable markets and the importance of sunk costs, see Baumol et al. (1982).
47. Privatisation involves the sale and subsequent transfer of public assets to the private sector. Corporatisation, on the other hand, involves the introduction of private enterprise-based objectives and practices to the operations of government instrumentalities.
48. Private banks cannot exactly create money out of nothing since the creation of something out of nothing is forbidden by the first law of thermodynamics. All money has some physical dimension, even in cases when its creation involves a simple increase in the credit value of a deposit account on a computer hard-drive. The term 'created out of nothing' in the context of money is therefore designed to indicate that banks can create money without the need for reserve money to fully support it.
49. The increase in the financial claims on real wealth are denoted as $\Delta M = M^1 - M$.
50. The increase in the monetary base would require the issuance of more non-interest-bearing fiat money by the central government.
51. Seigniorage is the difference between the monetary value of a money token and the cost of producing and maintaining the token. Seigniorage provides a money-creating entity considerable power to obtain real resources and/or goods and services. For example, if an entity can create $100 of spending power, it can claim $100 of real wealth. If the creation of the $100 only

requires the entity to expend (forego) $1 of real resources, the entity can obtain $99 of real wealth simply through its power to create money. This entity would enjoy $99 of seigniorage.

52. In earlier work, the GPI was referred to as an Index of Sustainable Economic Welfare (ISEW). It has also been labelled a Sustainable Net Benefit Index (SNBI) (Lawn and Sanders 1999; Lawn 2000).
53. While there are often disparities between the non-price rules of different states or provinces in a given country, they are usually much smaller in magnitude than the disparities between different nations.
54. Price-determining parameters are the various economic, social, and environmental factors that constitute the institutional context within which a particular market operates. As such, these parameters influence or 'determine' the market price for different goods and services. Examples include natural capital services, human know-how, cultural norms and beliefs, as well as individual tastes and preferences (d'Arge 1994).
55. There are, however, a number of countries with very large foreign debts that appear quite serviceable. According to Pitchford (1990), most foreign debts are of little concern since many are the result of numerous rational arrangements established between domestic borrowers and foreign lenders. While this may be so, one must be careful not to fall victim to the fallacy of composition. Micro rationality can lead to macro irrationality if transactions between individuals and entities across international borders are incommensurate with the social and environmental standards of the countries in which they reside.
56. Self-sufficiency was promoted as a desirable national goal in the United Nations Report on the World Summit on Sustainable Development held in Johannesburg in 2002.
57. This last question is particularly pertinent to the climate change issue. In many high-GDP countries, the concern about the loss of industries and associated jobs has led many governments to avoid introducing an emissions-trading system or emissions tax.
58. More recent work by Cole and Fredriksson (2013) has provided statistical support for the proposition that the impact of environmental standards on foreign direct investment is deterring governments from introducing stringent environmental regulations.
59. Compensating tariffs would be protectionist insofar as they would protect hard-won social and environmental standards. They would not be protectionist in the sense of shielding genuinely inefficient industries. If, following the imposition of a compensating tariff, the price of a Third World good remained lower than in a high-GDP country, the compensating tariff would not prevent the producer in the rich country from being competed out of existence. In other words, the compensating tariff would not preserve a genuinely inefficient producer in a high-GDP country.
60. Taken from an internet seminar on Herman Daly's book, *Beyond Growth* (http://csf.colorado.edu/seminars/ daly97/proceedings).

61. It would not internalise standards entirely because the appreciation of domestic IMPEX dollars would affect the willingness to import all foreign goods, not simply the 'advantaged' foreign goods. Hence, the need for compensating tariffs remains.
62. A 2 to 3 per cent rate is consistent with the regeneration rate of most renewable resource stocks—in effect, the interest rate generated by the natural capital that all nations are ultimately reliant upon.
63. The WTO emerged out of the General Agreement on Tariffs and Trade (GATT), an original Bretton Woods institution.

Chapter 4
A Sustainable Versus a Growth-as-Usual Scenario

4.1 Introduction

My aim in this chapter is to simulate a 'sustainable' (steady-state) scenario and a 'growth-as-usual' scenario and compare the results. There are three good reasons for performing this exercise. Firstly, it reveals the structural changes required to achieve a safe emissions target consistent with the broader goal of sustainable development. Secondly, it enables one to assess and compare the plausibility of a conventional emissions scenario and what many would regard as a more radical emissions scenario. As we shall see, the radical scenario—the sustainable scenario—is entirely feasible but the growth-as-usual scenario is not. Finally, consideration of the plausibility of both scenarios can provide valuable input when designing a global emissions protocol and the various policy institutions that would be embodied in it, including a global emissions-trading system.

I should point out that the simulation exercises conducted in this chapter constitute a major departure from most economic approaches to the climate change problem. Almost all mainstream approaches involve the use of a conventional benefit-cost analysis where, invariably, a so-called optimal concentration of greenhouse gases is determined by equating the marginal benefits and marginal costs of climate change mitigation. The approach adopted in this chapter is strictly target-based.[1] By this I mean that a number of desirable targets are pre-determined and the structural adjustments required to achieve them are then simulated, which requires assumptions to be made about renewable resource potentials, carbon sequestration possibilities, and the rate of technological progress. In this sense, the pre-determined targets serve as the guiding parameters for the simulation exercises. This stands in direct contrast to a conventional benefit-cost analysis approach where no desirable targets or outcomes are initially assumed.

P. Lawn, *Resolving the Climate Change Crisis*, DOI 10.1007/978-94-017-7502-1_4

4.2 A Sustainable Emissions Scenario

4.2.1 Setting the Guiding Parameters for a Sustainable Emissions Scenario

This first major section of the chapter focuses on the simulation and plausibility of a sustainable emissions scenario. In order to establish the guiding parameters for such a scenario, it is first necessary to resolve the following:

- What constitutes the dividing line between a sustainable and unsustainable economy? In other words, what constitutes the maximum sustainable scale of national economies and the global economy as a whole (S_S in Fig. 2.4)?
- Given that sustainable development requires gravitation towards the optimal macroeconomic scale (i.e., where sustainable economic welfare is maximised), what level of economic activity approximates the optimum (S_* in Fig. 2.4)?

Answers to the first dot point are crucial because, first and foremost, the objective of the international community must be an emissions trajectory consistent with ecological sustainability. To achieve this, humankind must not only avoid a potentially catastrophic concentration of greenhouse gases in the Earth's atmosphere, it must ensure that the aggregate rate of resource use is ecologically sustainable. Since the total rate of resource use is very much a function of human population numbers, it will also be necessary, as explained in Chap. 3, to stabilise the human population at a level consistent with the ecosphere's long-run carrying capacity.

The answer to the second dot point is crucial because, once national economies are operating sustainably, further economic adjustment is required to deliver higher levels of economic welfare. However, to know how far and in what direction the adjustment should take, it is necessary to determine the per capita real GDP that approximates an optimal macroeconomic scale. Once known, the optimum can become the nation's long-term macroeconomic target.

4.2.2 Ecological Sustainability: Avoiding Catastrophic Climate Change

In Chap. 1, it was argued that a safe concentration of greenhouse gases is one that will limit global warming to no more than a 2 °C rise above pre-industrial levels. To ensure this, it was argued that the concentration of greenhouse gases must not exceed 450 ppm of CO_2-e. If we therefore assume that a stabilisation level of 450 ppm should be the aim of a post-Kyoto emissions protocol, the question that arises is: What emissions trajectory should we follow to achieve a 450 ppm target? This question is an important one because the concentration of greenhouse gases at any point in time is primarily a function of cumulative emissions levels. As a consequence, a multitude of emissions pathways can lead to stabilisation at a desired greenhouse gas target (IPCC 2000, 2007a, b, c; den Elzen and Meinshausen 2006; Stern 2007; Anderson and Bows 2008).

In order to select one of the many emissions trajectories that would satisfy the 450 ppm target, the following should be borne in mind. Firstly, the shallower are the initial reductions in greenhouse gas emissions, the deeper the cuts must be in future years to stabilise at the desired concentration level. Secondly, because of the path-dependent nature of economic systems (David 1985; Arthur 1989), deep initial reductions in greenhouse gas emissions are likely to increase the ease and reduce the relative cost of having to make future emissions reductions. Thirdly, although it is possible to overshoot a stabilisation target by a considerable margin and later achieve it through a combination of massive emissions reductions and high rates of carbon sequestration, the potential warming effect of climate inertia suggests that excessive overshoot trajectories are best avoided (Meinshausen 2006; Stern 2007; Hansen et al. 2008).[2] A desirable emissions trajectory must therefore involve deep initial emissions cuts (Stern 2007; IPCC 2007a; Garnaut 2008). A smooth rate of decline in greenhouse gas emissions is also likely to be more desirable than a trajectory interspersed with abrupt emissions reductions.

In many ways, little additional effort has been made at recent UNFCCC meetings to quell rising emissions levels. This is because global emissions have been significantly determined by the obligations pertaining to the Kyoto Protocol, which were extended at the 2012 conference in Doha (COP-18). Emissions have also been strongly influenced by the underlying growth rate of the global economy. Given the reductions in emissions required by many Annex I nations to meet their Kyoto obligations plus the impact of the global 'GDP' recession, which has endured in many countries, the growth in global emissions has remained subdued in recent years. Emissions levels beyond 2015 will be largely determined by whatever emissions protocol emerges from a future UNFCCC conference and when or whether nations begin the transition to a steady-state economy.

Let's assume a slight rise in global emissions between 2010 and 2015—the consequence of existing Kyoto obligations and the fact that concerted efforts to further limit global emissions will not come into effect until after 2015. Should emissions peak around 2015, Anderson and Bows (2008) have shown that stabilising greenhouse gases at 450 ppm of CO_2-e without any major overshoot[3] will require a 4 per cent annual reduction in global emissions.[4] Of this, there will need to be a 6.5 per cent annual reduction in process-related CO_2 emissions.[5]

The Anderson and Bows trajectory is more stringent than what is generally considered necessary to achieve a 450 ppm target.[6] There are two reasons for this. Firstly, new evidence has emerged regarding the impact of temperature changes on carbon-cycle feedbacks and its effect on the absorptive capacity of natural carbon sinks (Cox et al. 2006; Jones et al. 2006; Friedlingstein et al. 2006; Canadell et al. 2007; Le Quéré et al. 2007). The updated evidence indicates that a rise in global mean temperatures will reduce the ecosphere's capacity to store carbon emissions much more than previously anticipated. Secondly, recent empirical data reveals that greenhouse gas emissions have risen faster than the rates assumed in most studies (Raupach et al. 2007). Because of these two factors, much deeper emissions cuts will be required to achieve a specific stabilisation target.

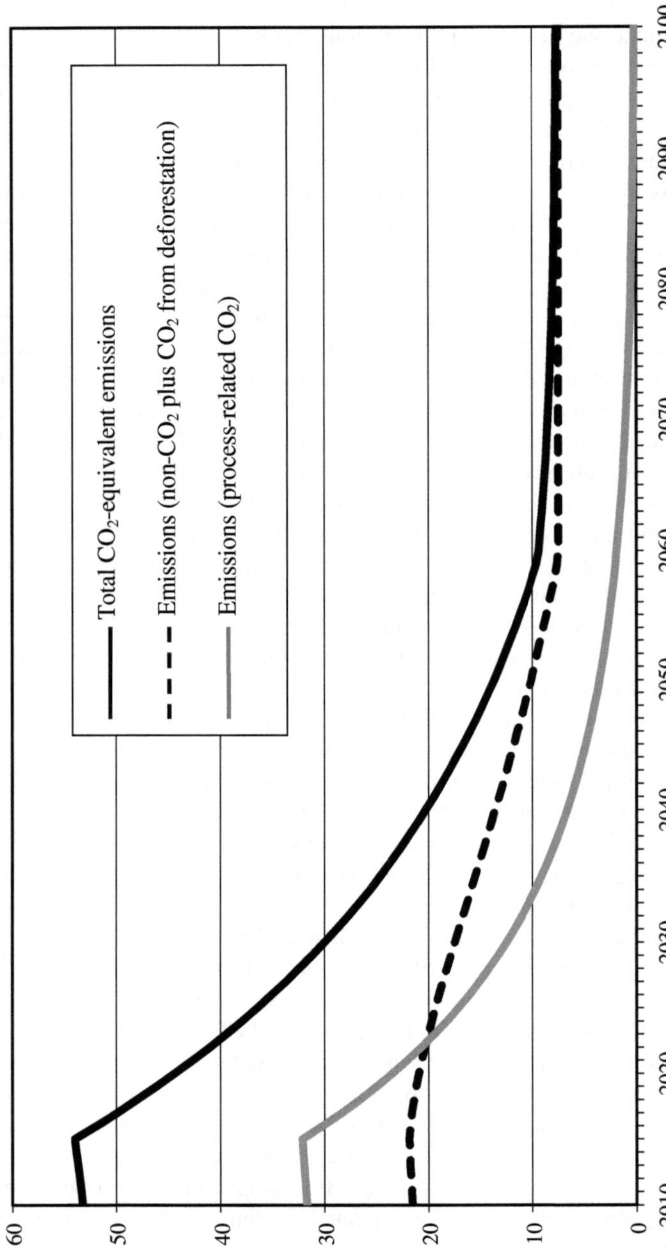

Fig. 4.1 CO_2-equivalent emissions to stabilise the atmospheric concentration of greenhouse gases at 450 ppm of CO_2-e, 2010–2100. *Source* Adapted from Anderson and Bows (2008)

The emissions trajectory recommended by Anderson and Bows (2008) to achieve a 450 ppm stabilisation target is illustrated in Fig. 4.1. The annual greenhouse gas emissions applicable to this trajectory will serve as our first sustainability parameter. Of particular note is that annual process-related CO_2 emissions must be reduced to negligible levels by the end of the 21st century. Unquestionably, the achievement of such a target will require all countries to be included in any new emissions protocol. It will also require high-GDP nations, as major beneficiaries of past emissions, to undertake massive emissions reductions in the years up to 2050.

4.2.3 Ecological Sustainability: A Sustainable Population and a Sustainable Rate of Resource Use

To ensure that the rate of resource use is ecologically sustainable, it will be assumed that the global ecological footprint is an appropriate indicator of aggregate resource use. Clearly, to operate sustainably, the global ecological footprint must be no greater than the Earth's biocapacity. For precautionary reasons, it will be assumed that ecological sustainability requires humankind to reduce its ecological footprint to 90 per cent of the Earth's biocapacity.[7] The 90 per cent factor will thus serve as our second guiding sustainability parameter.

The third and final sustainability parameter—a maximum sustainable human population—is a difficult parameter to estimate because it depends on the average per capita resource consumption of the future. This, in turn, depends on: (i) the per capita real GDP that approximates an optimal macroeconomic scale; (ii) the rate of resource-saving and pollution-reducing technological progress; and (iii) whether nations are willing to make the transition to a steady-state economy and, if so, whether they are willing to stabilise their economies at the estimated optimal scale.

Ideally, humankind would bring the global ecological footprint into line with the Earth's biocapacity before 2050 and further reduce the footprint to satisfy the 90 per cent precautionary factor soon after. Assuming a rapid rate of technological progress and the willingness of all nations to adjust to an optimum per capita GDP, achieving a sustainable human population will require population numbers to peak by around 2060 and be no more than 8 billion by the end of the century. Achieving this ambitious population target will require, where needed, the introduction of population stabilisation programmes described in Chap. 3.

4.2.4 Increasing Economic Welfare: Moving to the Optimal Macroeconomic Scale

We are now left with the task of determining the optimal per capita GDP. Genuine Progress Indicator (GPI) studies have indicated that the sustainable economic welfare of high-GDP nations has ceased to rise once per capita GDP reaches a level between Int$15,000 and Int$20,000 (2004 prices). At first blush, this suggests that the optimum is somewhere within in this range. However, it was pointed out in in Chaps. 2 and 3 that the economic welfare of some poorer nations has started to decline at lower levels of per capita GDP (e.g., China and Thailand). This suggests that the optimum could now be much less than Int$15,000.

In a world actively seeking to achieve sustainable development, there is good reason to discount the second factor when determining the optimal per capita GDP. As explained in Chap. 2, the early decline in the economic welfare of poor nations ought to be reversible if high-GDP countries, by reducing the physical scale of their own economies, provide the ecological space for low-GDP nations to enjoy an extended phase of welfare-increasing growth. Since our main interest lies in a sustainable emissions scenario, we shall assume that high-GDP countries will make the transition to a smaller, albeit qualitatively superior economy, which will allow the economic welfare of the world's poor nations to rise as their per capita GDP increases.

With the second factor excluded, where in the Int$15,000–$20,000 range might the optimum lie? There are two key influences worth considering. Firstly, the global ecological footprint is presently 50 per cent larger than it was when high-GDP nations reached their per capita welfare peak in the 1970s and 1980s (Global Footprint Network 2008). Even if there is a gradual decline in humankind's ecological impact, the large footprint will guarantee that the marginal cost of global economic activity remains high for some time to come, thus placing considerable downward pressure on the GPI. Secondly, the world's population is much larger than it was and will inevitably rise regardless of the measures taken to quell its growth. Consequently, I believe that Int$15,000 is the probable per capita optimum.[8] It will therefore serve as our welfare-maximising parameter.[9]

Before moving on, it is important to say a few things about a per capita optimum of Int$15,000 and its efficacy as a macroeconomic target. For high-GDP nations, the target implies a need to reduce per capita consumption to levels reminiscent of the 1970s/80s. Many would see this as a self-imposed recession that would lead to widespread suffering not seen since the 1930s Great Depression. There is no doubt that, in a growth-dependent economy, a sharp fall in GDP leads to considerable hardship.[10] Nonetheless, we have seen that excessive GDP levels can also be detrimental (see Figs. 2.7 and 2.8). The fact that an increase in hardship can be minimised by augmenting GDP when a nation is in possession of a growth-dependent economy does not imply that GDP growth beyond the optimal scale is desirable.

What needs to be understood is that an increase in hardship need not occur if the decline in GDP constitutes a necessary step in the transition to a qualitatively-improving steady-state economy. Through the implementation of appropriate policies, it should be possible to increase the service-yielding quality of most goods; improve the equity of wealth ownership; augment the productivity of labour; and reduce many social costs. At the same time, it should also be possible to increase the technical efficiency of production; improve the recycling rates of waste materials; augment the productivity of natural capital; and develop more sensitive resource-extraction techniques. If achieved, these advances would significantly increase the GPI of wealthy countries. The GPI would also be given a boost by reductions in real output given that a physically smaller economy would further reduce the uncancelled costs of economic activity (lost natural capital services). Thus, by the time high-GDP nations would be stabilising their per capita GDP at an optimal value of Int$15,000, the economic welfare being delivered would be considerably higher than the welfare once generated by the same per capita GDP in the 1970s/80s.

Mainstream commentators would argue that GDP growth combined with qualitative change would produce the best of all welfare outcomes. There is little doubt that a qualitative emphasis would generate a better outcome than a policy focused on growth alone. But the economies of high-income nations would still be larger than their optimal scale. Hence, the marginal benefits of qualitatively better growth would still be less than the additional costs associated with an excessively large economy. Furthermore, if rich countries undertake more growth, most will have economies that are considerably larger than their maximum sustainable scale. The failure to recognise this and the fact that, at some point, sustainable economic welfare can only be augmented by operating a qualitatively-improving steady-state economy reflects a widespread misunderstanding of the disparate welfare-generating potentials of the steady-state alternative and that of a shrinking growth-dependent economy or an economy that has grown well beyond its optimal scale.

4.2.5 Assumptions

We are now in a position to simulate the sustainable emissions scenario. In what follows, annual global emissions will adhere to the trajectory prescribed by Anderson and Bows (2008) to ensure greenhouse gas emissions stabilise at 450 ppm of CO_2-e. As for real Gross World Product (GWP), it will vary in accordance with physical production possibilities; the need to reduce the global ecological footprint to at least 90 per cent of the Earth's biocapacity; and the desire to settle at an optimal per capita GWP of Int$15,000 (2004 prices).

Before the sustainable emissions scenarios can be presented and assessed, it is first necessary to outline the assumptions employed. In sum, the following has been assumed:

- Annual greenhouse gas emissions begin in 2010 at 53.2 Gigatonnes of CO_2-e ($GtCO_2$-e) and peak in 2015 at 54.0 $GtCO_2$-e.[11] The peak comprises non-CO_2 and land use-related emissions of 21.9 $GtCO_2$-e and process-related CO_2 emissions of 32.1 $GtCO_2$-e. Non-CO_2 and land use emissions eventually fall to a floor level of 7.5 $GtCO_2$-e (Anderson and Bows 2008).
- The global population in 2010 is 6.92 billion. Due to population stabilisation programmes, the annual population is thereafter determined by the following equation:

$$Pop(t) = \frac{PPop(t)}{1.002^{(t-t_0)}} \tag{4.1}$$

where *Pop(t)* = population in year *t*; *PPop(t)* = projected population in year *t* (without stabilisation measures); $t_0 = 2010$.

- In 2010, the ecological footprint is 18.22 billion global hectares and the Earth's biocapacity is 13.29 billion global hectares. The footprint/biocapacity ratio in 2010 is 1.37 (Global Footprint Network 2008).
- The biocapacity of each global hectare increases annually in line with advances in the productivity of natural capital at the rate described by the following equation:

$$BC(t) = BC(2010) \times \left[1 + \exp\left(-7.6 - \frac{t - 2011}{100}\right)\right] \quad (4.2)$$

where $BC(t)$ = per hectare biocapacity in year t; $BC(2010)$ = per hectare biocapacity in year 2010; t = any year from 2011 onwards.

- The total number of bioactive hectares declines by 0.1 per cent in any year where the biocapacity is exceeded by the ecological footprint (i.e., where there is an ecological deficit). No downward adjustment is made to global biocapacity to account for the likely impact of rising global temperatures.[12]
- The annual resource flow used for production purposes changes in direct proportion to the change in the global ecological footprint.
- Real GWP is a function of the annual resource flow and the technical efficiency of production. The production function for real GWP is:

$$GWP = \left[1 - \exp\left(-\beta . K_H\right)\right] \times R \quad (4.3)$$

where K_H is a combined factor of producer goods and labour; β = the state of resource-saving technology; R = the annual resource flow (reflected by the annual change in the global ecological footprint); $\left[1 - \exp\left(-\beta . K_H\right)\right]$ = the technical efficiency of production (E); and $0 < E < 1$.

- $\beta . K_H$ constitutes the human-made production factor and is augmented at the rate of 2.3 per cent per annum.
- In 2010, real GWP is Int$61.0 trillion (2004 prices); $E = 0.25$; and $\beta . K_H = 0.29$.
- To begin the transition to ecological sustainability, the global ecological footprint is reduced at the rate of 1 per cent per year until the footprint/biocapacity ratio falls to 0.9.
- Upon reducing the footprint/biocapacity ratio to 0.9, the reduction in the ecological footprint is momentarily suspended to minimise the remaining time it takes to reach the optimal per capita GWP of Int$15,000.
- Upon reaching the optimal per capita GWP, a 1 per cent annual decline in the ecological footprint is resumed. By reducing the ecological footprint, the uncancelled costs of economic activity are further lowered. This progressively increases the sustainable economic welfare enjoyed at the optimal macroeconomic scale.

Let me just say a few things about some of the above assumptions. Firstly, the 0.1 per cent reduction in the total number of bioactive hectares in any year where the ecological footprint exceeds the Earth's biocapacity reflects the fact that an ecological deficit diminishes the stock of natural capital and, in doing so, reduces the number of bioactive hectares available to generate a flow of resources, absorb wastes, and provide life-support services.

Secondly, the annual increase in the productivity of each global hectare diminishes over time. As per Eq. (4.2), the biocapacity of each global hectare increases by 0.05 per cent in 2011 and diminishes thereafter. By 2100, the annual increase declines to 0.02 per cent.

Thirdly, the starting value of $E = 0.25$ in 2010 is based on a potential factor-four increase in the technical efficiency of production (Weiszacker et al. 1998). Given this assumption, it follows from the production function equation that the starting value for the combined human-made factor of production, $\beta.K_H$, is 0.29 (i.e., $[1 - \exp(-0.29)] = 0.25$).

Fourthly, the production function for real GWP (Eq. 4.3) is an example of a Bergstrom production function (Ayres and Miller 1980; Lawn 2007). It is a non-conventional production function designed to obey the first and second laws of thermodynamics. As such, it embodies a number of key features. To begin with, it treats low-entropy resources as the only true input of the production process—the 'material' cause of production. At the same time, it treats producer goods and labour, not as inputs *per se*, but as the resource-transforming agents of the production process. That is, the Bergstrom production function treats producer goods and labour as the 'efficient' cause of production (Daly 1996, 2008).

Next, despite allowing producer goods to adequately substitute for labour, and vice versa, the Bergstrom production function does not permit producer goods and labour, as the combined human-made factor of production ($\beta.K_H$), to substitute for low-entropy resource inputs. Thus, as producer goods and labour are augmented and the technology/know-how embodied within them improves, the Bergstrom function permits more output to be produced from a given input of resources. But it does so *only* because it allows for a greater proportion of the matter-energy embodied in resource inputs to make its way into final goods. Hence, augmentation of the human-made factor of production merely reduces the waste immediately generated in the production process.[13] This is not equivalent to human-made capital taking the place of low-entropy resources, which would be required for human-made capital to constitute a genuine resource substitute.

Finally, the Bergstrom production function ensures that the technical efficiency of production (E) is always less than 100 per cent (Ayres and Miller 1980; Ayres and van den Bergh 2005; Lawn 2007). It therefore represents the limit to which production waste can be reduced, thus imposing an upper limit on the quantity of output that can be produced from a given quantity of low-entropy resource inputs.

4.2.6 Results

The sustainable emissions scenario has been simulated for the 2010–2100 period. Table 4.1 reveals the end-of-decade values of each major variable and the values pertaining to the years in which key milestones were attained during the simulation period. The five columns in the left-hand segment of Table 4.1 reveal the changes in total and component greenhouse gas emissions, population, and per capita emissions. The values in the emissions column reflect the emissions trajectory needed to stabilise greenhouse gas concentrations at the 450 ppm target.[14] The population column shows the global population peaking at 8.55 billion in 2062, but falling to 7.95 billion by the end of the century (Fig. 4.2). Due to the large decrease in total greenhouse gas emissions over the simulation period, per capita global emissions decline from 7.69 tonnes per person in 2010 to 0.96 tonnes per person in 2100 (Fig. 4.2). This constitutes an 87.5 per cent decrease over the simulation period or an average annual decline of 2.3 per cent.

The five columns in the central segment of Table 4.1 reveal the ecologically-related variables. The third column shows the footprint/biocapacity ratio falling from 1.37 in 2010 to 1.0 by 2044 and then to 0.9 by 2054. The ecological footprint is subsequently maintained at 90 per cent of the Earth's biocapacity until the per capita GWP reaches Int$15,003 in 2084. The ecological footprint is then reduced to 10.13 billion global hectares by 2100 which lowers the footprint/biocapacity ratio to 0.76 (Fig. 4.3). Over the entire simulation period, the global ecological footprint declines by 44.4 per cent (average decrease of 0.7 per cent per annum). The per capita ecological footprint, by falling from 2.63 to 1.27 global hectares, declines by 51.6 per cent over the simulation period (average decrease of 0.8 per cent per annum).

The economic adjustment variables are revealed in the right-hand segment of Table 4.1. The first column shows real GWP increasing from Int$61.0 trillion in 2010 to Int$125.2 trillion in 2084, and then declines slightly to Int$119.3 trillion by 2100 (Fig. 4.4). Per capita GWP (second column) rises from its 2010 value of Int$8,811 to its optimal value of Int$15,003 in 2084. Beyond 2084, per capita GWP remains at the optimal value (Fig. 4.4).

The third, fourth, and fifth columns in the right-hand segment are efficiency variables. These variables are of critical concern since they provide insight into the technical plausibility of the sustainable scenario. As the fifth column shows, the technical efficiency of production rises from its initial value of 0.25 in 2010 to 0.89 in 2100. The third column, which reveals the real GWP/emissions ratio (i.e., the real output per tonne of greenhouse gas emissions), begins in 2010 at Int$1,145 per tonne and increases to Int$15,588 per tonne by 2100 (Fig. 4.5). This corresponds to a 13.6-factor increase over the simulation period. Finally, the real GWP/footprint ratio in the fourth column rises from Int$3,347 per global hectare to Int$11,781 per global hectare (Fig. 4.5).

Table 4.1 A sustainable scenario with an optimal scale of the world economy at $Int15,000 per capita (2004 prices)

Year	Greenhouse gas trajectory to stabilise at 450 CO_2-e ppm					Ecological sustainability trajectory (Ecological footprint ≤ biocapacity)					Economic adjustment—pc Int$ 15,000 equates to optimal scale				
	Total emissions ($GtCO_2$-e)	Non-CO_2 & land use emissions ($GtCO_2$-e)	Energy & process emissions ($GtCO_2$-e)	Pop[n] (millions)	Per capita CO_2-e emissions (t/person)	Ecological footprint (EF) (mill. of gha)	Bio capacity (BC) (mill. of gha)	Earths required (EF/BC)	Per capita EF (gha)	Per capita BC (gha)	GWP (bill. of Int$)	Per capita GWP (Int$)	GWP/ emissions (mt$/ tonne)	GWP/ gha (Int$/ ha)	Technical efficiency
2010	53.2	21.6	31.7	6,922.0	7.69	18,220.2	13,294.3	1.37	2.63	1.92	60,987.5	8,811	1,145	3,347	0.25
2015	**54.0**	**21.9**	**32.1**	7,222.6	7.48	17,335.9	13,260.5	1.31	2.40	1.84	63,954.6	8,855	1,184	3,689	0.28
2020	44.4	20.9	23.5	7,515.4	5.91	16,494.5	13,225.2	1.25	2.19	1.76	66,941.3	8,907	1,508	4,058	0.30
2030	30.0	17.5	12.5	7,991.9	3.75	14,932.3	13,142.7	1.14	1.87	1.64	72,873.5	9,118	2,430	4,880	0.36
2040	20.3	13.6	6.7	8,310.2	2.44	13,518.0	13,070.7	1.03	1.63	1.57	78,543.4	9,451	3,877	5,810	0.43
2044	17.3	12.1	5.2	8,398.0	2.06	12,990.5	13,037.6	**1.00**	1.55	1.55	80,671.8	9,606	4,659	6,210	0.46
2050	13.7	10.1	3.5	8,485.3	1.61	12,237.7	13,064.8	0.94	1.44	1.54	83,647.2	9,858	6,113	6,835	0.51
2054	11.7	8.9	2.8	8,519.2	1.37	11,760.2	13,082.0	0.90	1.38	1.54	85,452.0	10,031	7,305	7,266	0.54
2060	9.4	7.5	1.9	8,544.6	1.10	11,796.0	13,106.7	**0.90**	1.38	1.53	93,518.5	10,945	9,960	7,928	0.59
2062	9.2	7.5	1.7	**8,546.3**	1.07	11,803.2	13,114.6	0.90	1.38	1.53	96,209.2	11,257	10,496	8,151	0.61
2070	8.5	7.5	1.0	8,518.9	1.00	11,830.3	13,144.8	0.90	1.39	1.54	107,011.7	12,562	12,580	9,046	0.68
2080	8.0	7.5	0.5	8,408.8	0.96	11,861.4	13,179.3	0.90	1.41	1.57	120,163.1	14,290	14,953	10,131	0.76
2084	7.9	7.5	0.4	8,341.8	0.95	11,873.0	13,192.2	0.90	1.42	1.58	125,151.7	**15,003**	15,808	10,541	0.79
2090	7.8	7.5	0.3	8,217.6	0.95	11,184.9	13,210.6	0.85	1.36	1.61	123,288.3	15,003	15,835	11,023	0.83
2100	7.7	7.5	0.2	7,950.8	0.96	10,125.5	13,239.0	0.76	1.27	1.67	119,285.2	15,003	15,588	11,781	0.89
% change	−85.6 %	−65.2 %	−99.5 %	14.9 %	−87.5 %	−44.4 %	−0.4 %	−44.2 %	−51.6 %	−13.3 %	95.6 %	70.3 %	1260.8 %	252.0 %	256.9 %
Ave. p.a.	−2.1 %	−1.2 %	−5.8 %	0.2 %	−2.3 %	−0.7 %	0.0 %	−0.6 %	−0.8 %	−0.2 %	0.7 %	0.6 %	2.9 %	1.4 %	1.4 %

Milestone years

2010: Policies implemented to reduce CO_2-e emissions to the levels required to stabilise the atmospheric concentration of greenhouse gases at 450 ppm

2015: Annual emissions levels peak

2044: World economy at maximum sustainable scale or one Earth (i.e. ecological footprint equals biocapacity)

2054: Scale of world economy at safe level of 0.9 Earths

2062: World population peaks

2084: World economy settles at optimal macroeconomic scale of approximately Int$15,000 per person

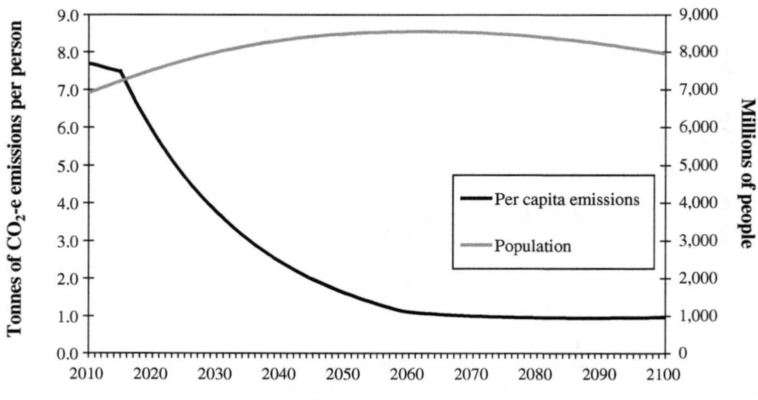

Fig. 4.2 Population and per capita emissions—World, 2010–2100

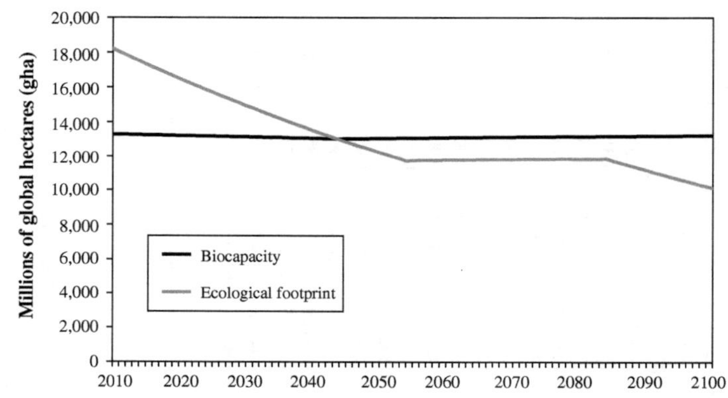

Fig. 4.3 Ecological footprint versus biocapacity—World, 2010–2100

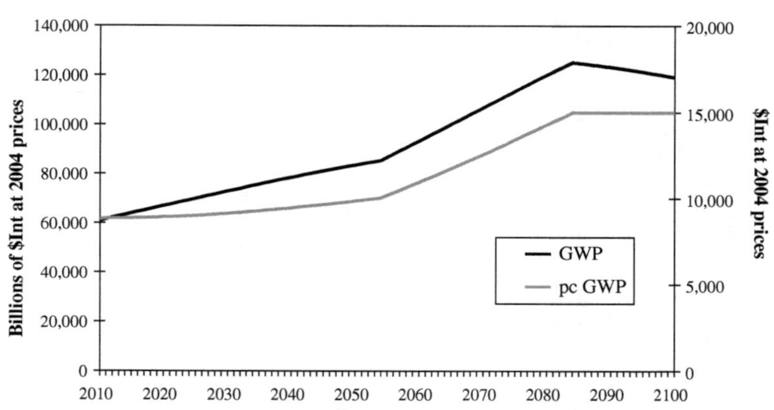

Fig. 4.4 Real GWP and per capita GWP—World, 2010–2100

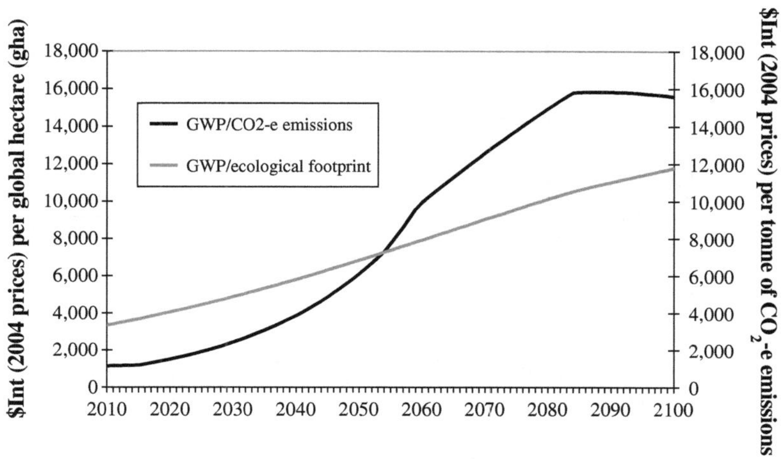

Fig. 4.5 Required footprint and emissions efficiency—World, 2010–2100

4.3 An Assessment of the Sustainable Emissions Scenario

4.3.1 Evolution of the Global Economy

How might the global economy evolve in the context of this sustainable emissions scenario? One way of illustrating this possibility is to represent the change in the scale of the global economy in terms of the UB and UC curves presented in Figs. 2.4–2.6. Figure 4.6 shows the probable shift in the UB and UC curves of the global economy over the simulation period should appropriate policies be introduced to facilitate advances in efficiency-increasing technology. In 2010, it can be seen that generating a real GWP of Int$61 trillion requires an ecological footprint equivalent to 1.37 Earths. In other words, the global economy far exceeds its maximum sustainable scale at 2010 technology levels. The global GPI is also much lower than its optimal value.

By 2044, qualitative improvements in the stock of physical goods shift the UB curve upwards. Over the same time, greater technical efficiency, higher rates of materials recycling, increased product durability, and a more productive stock of natural capital shift the UC curve downwards and to the right. In the process, the maximum sustainable scale of the global economy increases to a real GWP of Int$81 trillion. The global economy now operates at the maximum sustainable scale (1 Earth). Although the GPI increases between 2010 and 2044, the scale of the global economy does not maximise the GPI.

Further efficiency-increasing progress between 2044 and 2054 continues to shift the UB and UC curves. Although the scale of the global economy expands (up from Int$81 trillion in 2044 to Int$85 trillion by 2054), it is now smaller than

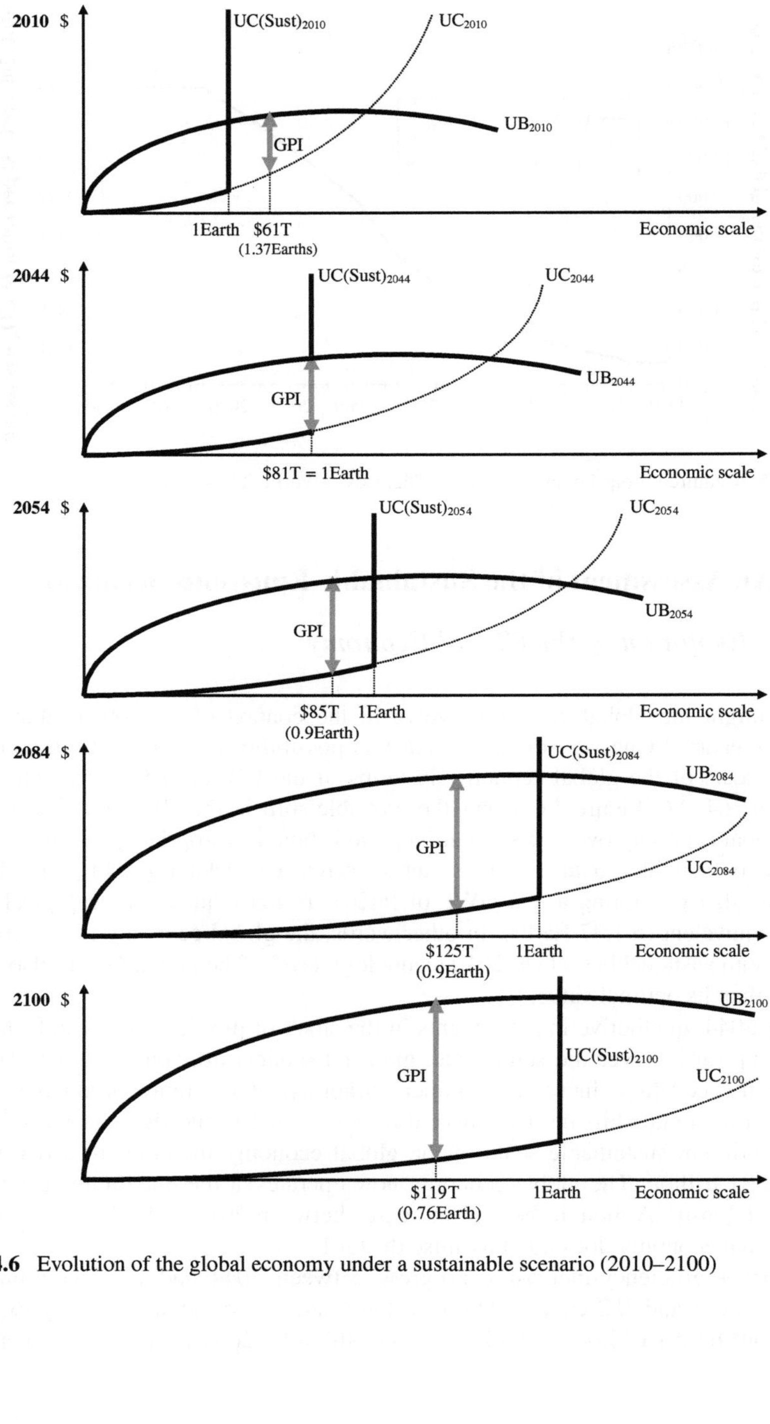

Fig. 4.6 Evolution of the global economy under a sustainable scenario (2010–2100)

the maximum sustainable scale at 2054 technology levels (0.9 Earths). The GPI also rises but has not yet been maximised.

Between 2054 and 2084 the UB and UC curves continue to shift in a desirable direction. With the global economy operating within safe ecological limits, advances in technical efficiency are directed towards reaching the optimal per capita GWP of Int$15,000 as rapidly as possible. By 2084, real GWP has increased to Int$125 trillion, the optimum has been attained (per capita GWP of Int$15,003), and the GPI has been maximised.

Beyond 2084, positive shifts of the UB and UC curves are exploited to boost economic welfare without the need for further economic expansion. Indeed, the scale of the global economy decreases slightly to Int$119 trillion by 2100. In the process, humankind's ecological footprint declines to 76 per cent of the Earth's biocapacity. Over the same period of time, the GPI increases.

There are two points about this description worth highlighting. Firstly, the optimal per capita GWP of Int$15,003 is an *average* value. Thus, when the global optimum is first attained in 2084, disparities in the per capita GPI of nations are still likely to exist. To minimise any welfare discrepancies, the international redistribution of income, wealth, and technology will be of critical importance, particularly in the post-2084 period when the average per capita GWP is stabilised at the optimal value. However, no amount of redistribution can be expected to overcome the welfare disparities caused by an inept government, corruption, institutional inadequacies, and failed population stabilisation policies. Clearly, disparities in the per capita economic welfare of nations are unlikely to be entirely eliminated.

Secondly, since it is impossible to know the initial positions of the UB and UC curves, or how much they are likely to shift over time, the changing value of the global GPI cannot be known with any great precision.[15] Moreover, we cannot be certain that the UB and UC curves will shift in the desirable directions assumed in Fig. 4.6. This will depend on the nature of global and national policies and future rates of efficiency-increasing technological progress. In this sense, Fig. 4.6 does not definitively indicate how the global economy will evolve. Rather, it forms an approximate image of how the global economy ought to evolve in order to achieve sustainable development.

4.3.2 The Plausibility of the Sustainable Emissions Scenario

Although the scenario just presented meets the three sustainability parameters and the one welfare-maximising target, doubts surrounding its plausibility rest mostly with the real GWP/emissions ratio and the technical efficiency ratio. This is because the two ratios must increase over the remainder of the century by factors of 13.6 and 3.5 respectively to generate the real GWP indicated in Table 4.1 (i.e., they must increase at average rates of 2.9 per cent and 1.4 per cent per annum).[16] Should the ratios fail to increase sufficiently, it will be impossible to achieve the optimal per capita GWP of Int$15,003 by 2084, if ever.

There is no denying that it will be difficult to increase the technical efficiency ratio at the rate assumed in this scenario. Having said this, the 1.4 per cent annual increase required over the remainder of the century is only slightly greater than the projected annual increase of 1 per cent (Nakicenovic et al. 2006; IPCC 2014d). As things stand, modern production processes and transport systems are extremely resource-wasteful. In addition, the recycling of waste materials falls well short of potential recycling rates (Weiszacker et al. 1998). Provided appropriate policies and incentives are introduced to stimulate efficiency-increasing technological progress (e.g., ecological tax reform), the rise in technical efficiency assumed in this scenario would appear feasible (IEA 2006; Jolley 2006c; IPCC 2014d).

More problematic is the required rate of increase in the real GWP/emissions ratio. At 2.9 per cent per annum, it is well above the 1.2 per cent historical rate of increase (IPCC 2007d). There are, nonetheless, encouraging results emerging from interventionist emissions scenarios aimed specifically at reducing the carbon intensity of economic activity. Some of these simulation studies forecast a decline in the carbon intensity of real GWP of around 2.5 per cent per annum (IPCC 2000; Riahi and Roehrl 2001; Nakicenovic et al. 2005; Riahi et al. 2006). Whilst this is below the 2.9 per cent rate needed, I believe the policies outlined in Chap. 3, together with the emissions-trading system to be revealed in Chap. 10, would be far more conducive to efficiency improvements, carbon-reducing technological progress, and the transition away from carbon-based fuels than the policy measures assumed in these interventionist scenarios. Should this be the case, and should the policies be enough to bridge the 0.4 per cent gap, the required increase in the real GWP/emissions ratio should be comfortably achieved.

Another important aspect requiring serious consideration is whether it will be possible to shift to renewable energy at the rates and levels required to satisfy this or any other sustainable emissions scenario. Such an assessment is not easy to make. There is, for example, considerable debate over the future prospects of: (i) renewable energy; (ii) nuclear energy as a relatively safe, low-carbon, non-renewable energy source; and (iii) carbon capture-and-storage technology (geosequestration) as a means of reducing the rate at which fossil fuel use must be decreased. The most optimistic estimates suggest that the energy demands of the sustainable emissions scenario just presented should be easily met (Pacala and Socolow 2004; IPCC 2007c, 2011; MacKay 2009).

Because of the possible inaccuracy of the optimistic claims, it makes more sense to focus on the pessimistic estimates. After all, should they exceed the necessary energy requirements, there is little cause for concern (see Table 4.2). A number of pessimistic outlooks have been articulated in relation to the development, expansion, and integration of wind and solar energy, biofuels, hydro-electricity, nuclear energy, hydrogen conversion, and geosequestration technologies (Trainer 1995, 2005, 2007, 2008, 2011; Davy and Coppin 2003; Bossel 2003, 2004; Torvanger et al. 2004; Hendricks et al. 2004; Jolley 2006b; Augenstein and Benemann 2007; Storm van Leeuwin and Smith 2008). Taken together, these viewpoints indicate the need to augment nuclear energy as a means of overcoming initial energy shortfalls (Lovelock 2006). However, due to safety and international security concerns, it

Table 4.2 Energy potential to meet energy needs and CO_2-e emissions caps (sustainable scenario)—World, 2010–2100

Year	Energy trajectory (EJ)	Energy & process emissions ($GtCO_2$-e)	Permissible non-renew. energy (EJ)	Nuclear energy (EJ)	Geosequest. (EJ)	Renewable energy							Excess renewable capacity (%)	Renewable % of total energy (%)
						Total required (EJ)	Potential renewable							
							Wind (EJ)	Solar (EJ)	Biomass (EJ)	Hydro (EJ)	Remainder (EJ)	Total (EJ)		
	A	*B*	*C*	*D*	*E*	*F*	*G*	*H*	*I*	*J*	*K*	*L*	*M*	*N*
2010	495.0	31.7	405.9	29.2	0.0	89.1	1.5	0.3	64.4	14.9	8.1	89.1	100.0 %	18.0 %
2020	448.1	23.5	326.4	36.3	15.0	121.7	16.6	26.4	75.8	18.6	12.1	150.8	123.9 %	27.2 %
2030	405.7	12.5	200.8	43.4	16.8	204.9	31.7	52.6	87.2	22.3	16.9	213.3	104.1 %	50.5 %
2040	367.3	6.7	132.7	50.4	14.2	234.5	46.9	78.7	98.6	26.0	22.0	276.1	117.7 %	63.9 %
2050	332.5	3.5	96.8	57.5	10.7	235.6	62.0	110.0	110.0	29.7	26.4	338.1	143.5 %	70.9 %
2060	320.5	1.9	75.7	59.7	8.0	244.8	65.1	115.5	115.5	31.2	29.1	356.3	145.6 %	76.4 %
2070	321.4	1.0	34.2	21.4	6.0	287.2	67.7	120.1	120.1	32.4	30.5	370.9	129.2 %	89.3 %
2080	322.2	0.5	15.0	5.4	4.5	307.2	69.7	123.7	123.7	33.4	31.7	382.3	124.5 %	95.3 %
2090	303.9	0.3	9.5	3.4	3.5	294.4	71.1	126.2	126.2	34.1	32.7	390.3	132.6 %	96.9 %
2100	275.1	0.2	6.1	1.9	2.8	269.0	71.8	127.5	127.5	34.4	33.3	394.5	146.6 %	97.8 %

Sources
A: Trainer (2007); IEA (2009); own estimates
B: Anderson and Bows (2008)
C: Own estimates
D: Trainer (2007), Storm van Leeuwen and Smith (2008), IEA (2009), own estimates
E: Torvanger et al. (2004), Hendricks et al. (2004), Trainer (2007), own estimates
G: Davy and Coppin (2003), Stern (2007), Trainer (2007), Global Wind Energy Council (2008)
H: Hayden (2003), Stern (2007), Trainer (2007), IEA (2009), own estimates
I: Stern (2007), Trainer (2007), Augenstaein and Benemann (2007), REN21 (2009), own estimates
J: Trainer (2007); REN21 (2009); own estimates
K: Bossel (2003, 2004), Trainer (2007), IEA (2009), own estimates
L: Sum of columns G–K
Note Values may not sum precisely due to rounding errors

would be desirable to minimise the use of nuclear energy once renewable energy has reached sufficient supply capacity.

In order to assess the plausibility of the sustainable emissions scenario, it is first necessary to convert the global ecological footprint from Table 4.1 to an equivalent level of energy use. It is then necessary to make assumptions regarding the potential to expand the different energy sources as well as assumptions about the greenhouse gas emissions associated with their use. Using the above references as a guide, Table 4.2 and Fig. 4.7 reveal: (i) the total energy required over the simulation period (measured in Exajoules (EJ)); (ii) the change in the composition of energy use; and (iii) a lower-bound estimate of the world's renewable energy potential. As column *C* shows, the permissible use of non-renewable energy—the major contributor to process-related CO_2 emissions—decreases dramatically over the simulation period from 405.9 EJ in 2010 to 6.1 EJ by 2100. Conversely, the use of nuclear energy (column *D*) rises rapidly between 2010 and 2050 (29.2 EJ to 57.5 EJ); increases marginally between 2050 and 2060 (57.5 EJ to 59.7 EJ); and is reduced to negligible levels by 2100 (1.9 EJ in 2100).[17]

Except for a minimal rise between 2040 and 2050, renewable energy requirements (column *F*) increase steadily between 2010 and 2080 (89.1 EJ to 307.2 EJ). Beyond 2080, the reduction in total energy demand reduces renewable energy requirements to 269.0 EJ by 2100. As for the lower-bound estimate of the world's renewable energy potential (column *L*), it rises throughout the simulation period (up from 89.1 EJ in 2010 to 394.5 by 2100). Nevertheless, the rate of increase slows in the second half of the century due to diminishing returns caused by power integration difficulties, lengthening transmission distances, a lack of suitable locations (wind, wave, and solar), increasing siltation of dams (hydro-electricity), and declining soil fertility and water shortages (biofuels).

Critically, the lower-bound estimate of the world's renewable energy potential remains above the estimated renewable energy requirements for the entire

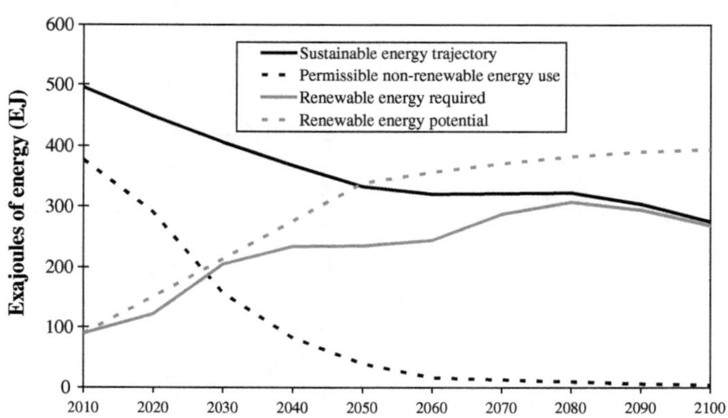

Fig. 4.7 World energy trajectory in a sustainable scenario with greenhouse gas stabilisation at 450 ppm of CO_2-e, 2010–2100

simulation period.[18] Column M and Fig. 4.7 show the extent to which the former exceeds the latter. In all but 2030, the world's renewable energy potential at the end of each decade is at least 17.7 per cent higher than its renewable energy requirements. In 2030, the excess capacity is just 4.1 per cent. Provided this constitutes a sufficiently secure buffer, it should be possible to generate the renewable energy needed to satisfy the sustainable emissions scenario. Importantly, too, the energy sources considered necessary to maintain base-load energy needs never fall below 78.7 per cent of total energy requirements.[19] Given that this assessment is based on a pessimistic renewable energy outlook, the sustainable scenario presented in this chapter appears eminently plausible.

4.4 A Growth-as-Usual Scenario

4.4.1 Assumptions

Having revealed and assessed the plausibility of a sustainable emissions scenario, a growth-as-usual scenario will now be simulated. Some of the assumptions applied in the sustainable emissions scenario are also used in the growth-as-usual scenario. The most obvious similarity is the greenhouse gas emissions trajectory. Since the aim remains one of stabilising the atmospheric concentration of greenhouse gases at 450 ppm of CO_2-e, the emissions trajectory is identical to the sustainable emissions scenario.

Where the growth-as-usual scenario differs to the sustainable scenario is in terms of real GWP. Being a growth-as-usual scenario, it is assumed that real GWP rises at a rate consistent with IPCC projections (IPCC 2000, 2007c). It is also assumed that there is no aim to reduce the global ecological footprint below the Earth's biocapacity, nor any desire to have per capita GWP settle at a prescribed optimal level. Also different in this growth-as-usual scenario is the change in the world's population. Because of an assumed lack of population stabilisation measures, it is assumed that the world's population varies in accordance with projections made by the Population Division of the United Nations Department of Economic and Social Affairs (www.earthtrends.wri.org).

Some of the additional assumptions employed in the growth-as-usual scenario include:

- All 2010 values are the same as the sustainable emissions scenario.
- Real GWP increases at an annual rate of 2.3 per cent. This is consistent with the median growth rate predicted by the IPCC (2000, Table 2–3).
- The production function for real GWP is the same as the sustainable emissions scenario (Eq. (4.3)). The human-made production factor ($\beta.K_H$) is augmented at the rate of 2.3 per cent per annum.
- The global ecological footprint varies in proportion to the resource throughput required to increase real GWP by 2.3 per cent per annum.

- The biocapacity of each global hectare increases in line with advances in the productivity of natural capital (i.e., at the rate described by Eq. (4.2)).
- The total number of bioactive hectares declines by 0.1 per cent in any year where the biocapacity is exceeded by the ecological footprint (i.e., where there is an ecological deficit). Not unlike the sustainable emissions scenario, no downward adjustment is made to biocapacity for the likely impact of rising global temperatures.

4.4.2 Results

Like the sustainable emissions scenario, the growth-as-usual scenario is also simulated for the 2010–2100 period. The end-of-decade values of each major variable and the values pertaining to one milestone year are revealed in Table 4.3. The values of all but one of the five columns in the left-hand segment of the table are the same as the sustainable emissions scenario. The one exception—the population column—shows the global population rising to a peak of 9.67 billion by 2082 and then falling to 9.52 billion by 2100.[20]

The ecologically-related variables in the central segment of Table 4.3 are considerably different to the sustainable emissions scenario. The first column shows the global ecological footprint rising throughout the designated simulation period—the result of having to constantly increase the rate of throughput to boost real GWP by 2.3 per cent per annum. By 2100, the ecological footprint is 39.52 billion global hectares. This constitutes a 116.9 per cent rise over the simulation period or an average rate of increase of 0.9 per cent per annum. Because the Earth's biocapacity (second column) is always exceeded by the ecological footprint, the biocapacity declines in every year between 2010 and 2100. By 2100, the Earth's biocapacity is 12.52 billion global hectares—a 5.8 per cent decrease over the simulation period. The increase in the global ecological footprint and a comparison between the footprints pertaining to the growth-as-usual and sustainable emissions scenarios are diagrammatically presented in Fig. 4.8.

Unlike the sustainable emissions scenario, the footprint/biocapacity ratio (third column) increases throughout the simulation period. Beginning at 1.37 in 2010, the ratio rises to 3.16 by 2100. On a per capita basis, the ecological footprint increases from 2.63 to 4.15 hectares (57.8 per cent increase). Conversely, the per capita biocapacity falls from 1.92 to 1.32 hectares (31.5 per cent decrease).

The first of the economic adjustment variables in the right-hand segment of Table 4.3 shows real GWP growing at 2.3 per cent per annum. By 2100, real GWP is Int$472.1 trillion—a 7.7-factor increase over the simulation period. Per capita GWP (second column) rises from its 2010 value of Int$8,811 to Int$49,607 by 2100. It therefore grows well beyond the optimal value of Int$15,000 (Fig. 4.9).

Of the efficiency variables, the technical efficiency of production (fifth column) rises from its initial value of 0.25 in 2010 to 0.89 by 2100. This is the same rate

Table 4.3 A growth-as-usual scenario with no population stabilisation policy (2004 prices)

Year	Greenhouse gas trajectory to stabilise at 450 CO_2-e ppm					Growth-as-usual trajectory					Growth-as-usual trajectory—GWP increases at the rate of 2.3 % p.a				
	Total emissions ($GtCO_2$-e)	Non-CO_2 & land use emissions ($GtCO_2$-e)	Energy & process emissions ($GtCO_2$-e)	Pop^n (millions)	Per capita CO_2-e emissions (t/person)	Ecological footprint (EF) (mill. of gha)	Bio capacity (BC) (mill. of gha)	Earths required (EF/BC)	Per capita EF (gha)	Per capita BC (gha)	GWP (bill. of Int$)	Per capita GWP (Int$)	GWP/ emissions (Int$/ tonne)	GWP/ gha (Int$/ ha)	Techical efficiency
2010	53.2	21.6	31.7	6,922.0	7.69	18,220.2	13,294.3	1.37	2.63	1.92	60,987.5	8,811	1,145	3,347	0.25
2015	**54.0**	**21.9**	**32.1**	7,295.1	7.40	18,522.3	13,260.5	1.40	2.54	1.82	68,331.2	9367	1,265	3,689	0.28
2020	44.4	20.9	23.5	7,667.1	5.79	18,864.4	13,225.2	1.43	2.46	1.72	76,559.2	9985	1,725	4,058	0.30
2030	30.0	17.5	12.5	8,317.7	3.60	19,692.9	13,150.4	1.50	2.37	1.58	96,106.7	11,554	3,205	4,880	0.36
2040	20.3	13.6	6.7	8,823.5	2.30	20,764.1	13,070.7	1.59	2.35	1.48	120,645.1	13,673	5,956	5,810	0.43
2050	13.7	10.1	3.5	9,191.3	1.49	22,157.1	12,986.6	1.71	2.41	1.41	151,448.9	16,477	11,067	6,835	0.51
2060	9.4	7.5	1.9	9,442.3	0.99	23,980.7	12,898.8	1.86	2.54	1.37	190,117.7	20,135	20,248	7,928	0.59
2070	8.5	7.5	1.0	9,603.9	0.89	26,384.1	12,880.8	2.05	2.75	1.34	238,659.6	24,850	28,056	9,046	0.68
2080	8.0	7.5	0.5	9,671.1	0.83	29,573.2	12,807.6	2.31	3.06	1.32	299,595.4	30,978	37,281	10,131	0.76
2082	8.0	7.5	0.5	**9,673.0**	0.82	30,328.5	12,713.5	2.39	3.14	1.31	313,535.3	32,413	39,326	10,338	0.77
2090	7.8	7.5	0.3	9,641.9	0.81	33,828.4	12,617.0	2.68	3.51	1.31	376,089.8	39,006	48,305	11,118	0.83
2100	7.7	7.5	0.2	9,517.1	0.80	39,524.2	12,518.4	3.16	4.15	1.32	472,115.1	49,607	61,697	11,945	0.89
% change	−85.6 %	−65.2 %	−99.5 %	37.5 %	−89.5 %	116.9 %	−5.8 %	130.4 %	57.8 %	−31.5 %	674.1 %	463.0 %	5286.1 %	256.9 %	256.9 %
Ave. p.a.	−2.1 %	−1.2 %	−5.8 %	0.4 %	−2.5 %	0.9 %	−0.1 %	0.9 %	0.5 %	−0.4 %	2.3 %	1.9 %	4.5 %	1.4 %	1.4 %

Milestone years

2010: Policies implemented to reduce CO_2-e emissions to the levels required to stabilise the atmospheric concentration of greenhouse gases at 450 ppm

2015: Annual emissions levels peak

2082: World population peaks

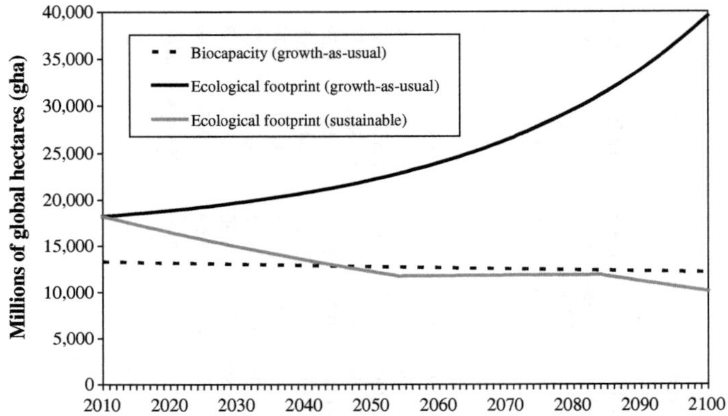

Fig. 4.8 Ecological footprint versus biocapacity—World, 2010–2100

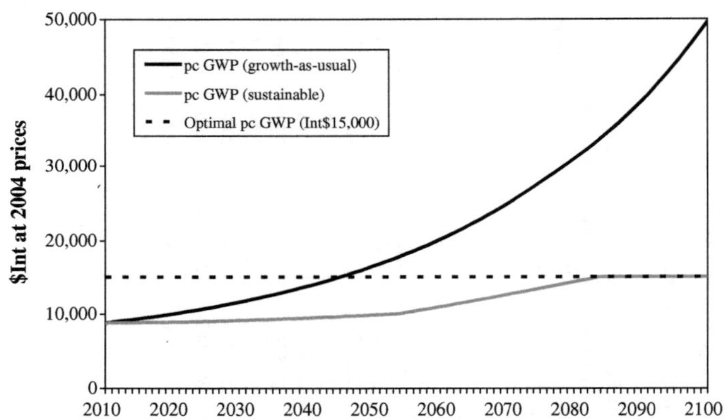

Fig. 4.9 Per capita GWP—World, 2010–2100

of increase as experienced in the sustainable emissions scenario.[21] As a consequence, the real GWP/footprint ratio in the fourth column also rises by the same rate as the sustainable scenario (i.e., up from Int$3,347 per global hectare in 2010 to Int$11,945 per global hectare by 2100). I should point out that the identical change in the real GWP/footprint ratio also explains why the rapid rate of growth in the growth-as-usual scenario results in a large and ever-increasing ecological footprint. Finally, the real GWP/emissions ratio (column three) increases from Int$1,145 to Int$61,697 per tonne of greenhouse gas emissions. This constitutes a 53.9-factor rise over the simulation period or an average rate of increase of 4.5 per cent per annum. This is around four times higher than the 13.6-factor rise pertaining to the sustainable emissions scenario (Fig. 4.10).

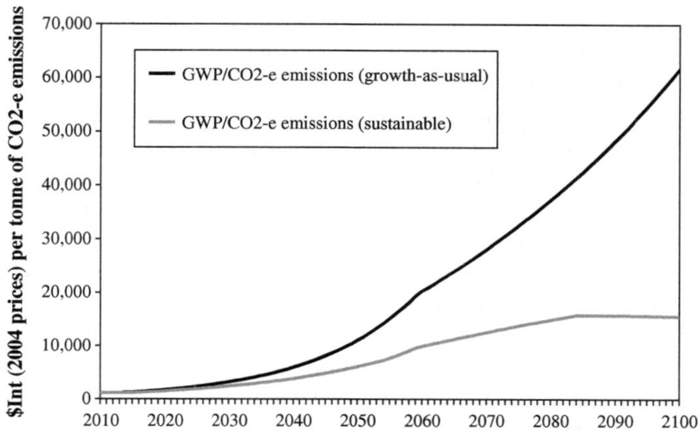

Fig. 4.10 Required emissions efficiency—World, 2010–2100

4.4.3 Plausibility of the Growth-as-Usual Scenario

Although, by adhering to a 450 ppm stabilisation target, the growth-as-usual scenario satisfies the first sustainability parameter, there are many reasons why it can be dismissed as a plausible means of achieving sustainable development. Firstly, with a footprint/biocapacity ratio rising to 3.16 by 2100, the scenario is grossly unsustainable. In fact, it is questionable whether a footprint/biocapacity ratio of this magnitude would be attainable without triggering a catastrophic collapse of the world's ecosystems and renewable resource stocks. In view of the complementarity that exists between human-made capital and natural capital, it is doubtful whether the real GWP levels assumed in this scenario—particularly those towards the end of the century—are in any way achievable.

Secondly, even allowing for the most optimistic predictions regarding the development and implementation of geosequestration technologies, it is unlikely that a 53.9-factor increase in the real GWP/emissions ratio could be achieved.[22] Whilst a massive increase in nuclear energy could greatly assist in this regard, it would be insufficient to bridge the gap (Trainer 2007; Storm van Leeuwin and Smith 2008). In any case, the nuclear industry is dependent upon non-renewable resources and is itself unsustainable in the long-run.

Thirdly, although many of the optimistic estimates concerning renewable energy supplies appear to be sufficient to power a global economy producing a real GWP of Int$472.1 trillion in 2100 (1,073.8 EJ), these estimates include such developments as the expansion of biofuels and hydro-electricity (see, for example, IPCC 2011).[23] They are consequently based on the capacity to further increase humankind's ecological footprint. As explained above, the ecological footprint/biocapacity ratio will be manifestly unsustainable by the year 2100. The optimistic estimates are therefore wildly over-inflated. As for the lower-bound estimates of the world's renewable energy potential (Table 4.2), it is demonstrably clear that it would be impossible to generate the renewable energy needed to power the global economy, let alone generate it in an ecologically sustainable manner (Note: the lower-bound estimate of the world's renewable energy potential for 2010 is 394.5 EJ which is considerably less than the required 1,073.8 EJ).[24]

Finally, even if a real GWP of Int$472.1 trillion could be produced in 2100, the associated per capita GWP of Int$49,607 would be well above the optimal value of Int$15,000. Thus, as much as average consumption levels would be very high in a growth-as-usual scenario, the additional benefits generated would be far outweighed by the additional costs, especially given that the projected rise in the ecological footprint/biocapacity ratio suggests there would be a massive future increase in the uncancelled costs of economic activity. It is also highly likely that a growing proportion of the rising real GWP would be comprised of defensive and remedial activities (Scheraga et al. 1993; Leipert 1986; Fankhauser and Tol 2005; Spash 2007; Lawn and Clarke 2008). As explained in Chap. 3, these activities do not directly increase well-being but merely minimise the current and future welfare losses associated with excessive growth (Leipert 1986). Combined with the high cost of lost natural capital services, the economic welfare experienced by the average global citizen is likely to be much lower than the welfare experienced in a sustainable emissions scenario. Overall, the growth-as-usual scenario is both infeasible and undesirable.

Annex 4A: Energy Requirements and Energy Potentials of the Sustainable Emissions Scenario

In this chapter included a sustainable emissions scenario and an assessment of global energy requirements and potentials. The simulated exercise and subsequent energy assessment indicated that the sustainable emissions scenario is both desirable and feasible. The energy requirements and potentials were presented in Table 4.2, where the energy assessment involved a *lower-bound* estimate of the world's renewable energy potential. With this in mind, this annex outlines the assumptions and methods used to calculate the values included in Table 4.2.

Column A: *Energy trajectory*

- Column A presents the total energy required to generate the Gross World Product (GWP) pertaining to the sustainable emissions scenario.
- The initial value of 495.0 Exajoules (EJ) corresponds to the estimated energy requirements of the global economy in 2010 (IEA 2011).[25]
- The remaining values were calculated by adjusting the 2010 energy requirements in line with projected changes in the global ecological footprint. Thus, for any year (t), global energy requirements were calculated using the following formula:

$$Energy\ requirements_t\ (EJ) = \frac{Ecological\ footprint_t}{18,220.2} \times 495.0 \quad \text{(A4.1)}$$

where *Ecological footprint* is measured in global hectares (gha) and 18,220.2 gha represents the ecological footprint of the global economy in 2010.

Column B: *Energy and process emissions*

- Column B reveals the permissible level of greenhouse gas emissions generated from the use of energy and industrial processes.
- The values are drawn directly from Table 4.1 and are based on Anderson and Bows (2008). The values reflect the reduction in energy and process emissions required to stabilise greenhouse gases at 450 ppm of CO_2-e.

Column C: *Permissible non-renewable energy use*

- Column C shows the maximum amount of non-renewable energy that can be used given the cut in greenhouse gas emissions required to stabilise at 450 ppm of CO_2-e.
- The initial value of 405.9 EJ equals the 2010 total energy requirement of 495.0 EJ less the renewable supply of energy (89.1 EJ). The renewable energy supply in 2010 was 18 per cent of the total energy supply (REN21 2011).
- The remaining values equal the sum of the permissible use of carbon-based fuels and nuclear-based energy. Included in these values is the additional use of carbon-based fuels made possible through the application of geosequestration technologies (*column E*).
- Beyond 2050, it is assumed that:
    - Non-renewable energy (nuclear and non-nuclear) is exploited up to the limits imposed by the need to make emissions reductions to achieve a 450 ppm stabilisation target.
    - Carbon-based fuels are required for, but limited to, transport needs, which constitute:
        - 5 per cent of total energy requirements in 2060
        - 4 per cent of total energy requirements in 2070
        - 3 per cent of total energy requirements in 2080
        - 2 per cent of total energy requirements in 2090
        - 1.5 per cent of total energy requirements in 2100.
    - Nuclear energy is only used to make up for any shortfall in non-renewable energy use (i.e., the difference between the allowable use of non-renewable energy and the use of carbon-based fuels for transport purposes).

Column D: *Nuclear energy*

- Column D reveals the quantity of nuclear energy required to reduce the world's reliance on carbon-based fuels and to overcome the limited rate with which renewable energy can be augmented.
- The initial value of 29.2 EJ reflects the use of nuclear energy in 2010 (IEA 2011).[26]

- Between 2010 and 2050, it is assumed that nuclear energy can be boosted at a quantity of 7.1 EJ per decade, thus reaching 57.5 EJ by 2050 (note: 57.5 EJ is half the maximum eventual supply assumed by Stern (2007)).
- Beyond 2050, it is assumed that nuclear energy is exploited only as a means of making up for any shortfall in non-renewable energy use as a consequence of restricting carbon-based fuels to transport purposes. It is because of the increase in renewable energy potential and the decrease in total energy requirements that nuclear energy use eventually decreases from a peak of 59.7 EJ in 2060 to negligible levels by 2010 (1.9 EJ).
- Throughout the period, it is assumed that the carbon released from nuclear energy use is 25 per cent of the carbon released from the use of carbon-based fuels. The release of carbon in the use of nuclear energy is the consequence of having to utilise carbon-based fuels to extract and transport uranium and to fuel the nuclear enrichment and waste treatment processes.

Column E: *Geosequestration*

- Column E constitutes the increase in the permissible use of non-renewable energy made possible by the deployment of geosequestration technologies. In other words, the energy value in this column reflects the additional quantity of non-renewable resources that can be used for production purposes as a consequence of geologically sequestering some of the carbon emitted from the use of carbon-based fuels.
- The additional use of non-renewable energy in 2010 is assumed to be 0.01 EJ.
- Beyond 2010, the geosequestration values equal the difference between the permissible use of non-renewable energy from deploying geosequestration technologies and that which would be permissible if geosequestration technologies did not exist.
- Up to 2050, it is assumed that the rate of carbon sequestration will rise steadily to reach 20 per cent of all carbon emissions. Beyond 2050, it is assumed that the rate of carbon sequestration will rise at a more intense rate as the use of carbon-based fuels rapidly diminishes (Torvanger et al. 2004; Hendricks et al. 2004; Trainer 2007; IPCC 2014d).
- The geosequestration rates between 2010 and 2050 are assumed to be:
    - 5 per cent of greenhouse gas emissions between 2010 and 2020
    - 10 per cent of greenhouse gas emissions between 2020 and 2030
    - 15 per cent of greenhouse gas emissions between 2030 and 2040
    - 20 per cent of greenhouse gas emissions between 2040 and 2050.
- Given the fall in the use of carbon-based fuels, geosequestration rates between 2060 and 2100 are assumed to be:
    - 26 per cent of greenhouse gas emissions between 2050 and 2060
    - 33 per cent of greenhouse gas emissions between 2060 and 2070
    - 41 per cent of greenhouse gas emissions between 2070 and 2080
    - 50 per cent of greenhouse gas emissions between 2080 and 2090
    - 60 per cent of greenhouse gas emissions between 2090 and 2100.

- The assumed rate of increase in geosequestration between 2010 and 2050 is based on the projected growth in the use of geosequestration technologies (Stern 2007). Between 2060 and 2100, the assumed rate of increase in geosequestration is based on a scaling up to a 60 per cent geosequestration rate by 2100. The 60 per cent rate is premised on an 80 per cent extraction rate of greenhouse gases applied to around 75 per cent of all greenhouse gas emissions (Stern 2007; Trainer 2007).

Column F: *Total renewable energy requirements*

- Column F is the quantity of renewable energy required to meet the total energy needs of the global economy given the limits imposed on non-renewable energy use by the necessary reductions in greenhouse gas emissions.
- Renewable energy requirements are equal to the total energy requirements (*column A*) less the permissible use of non-renewable energy (nuclear and non-nuclear) (*column C*).

Column G: *Wind energy*

- The wind energy supply capacity for 2010 was estimated at 1.5 EJ (REN21 2011; Global Wind Energy Council 2011).
- Between 2010 and 2050, the assumed growth in wind energy is based on a straight-line increase in the quantity of wind energy expected in 2010 (1.5 EJ) and Stern's (2007) reported estimate of 62.0 EJ by 2050 (i.e., this assumes an average rate of increase of 15.1 EJ per decade) (Davy and Coppin 2003; Trainer 2007).
- Beyond 2050, the assumed increase in harnessed wind energy is:
    - 5 per cent increase between 2050 and 2060
    - 4 per cent increase between 2060 and 2070
    - 3 per cent increase between 2070 and 2080
    - 2 per cent increase between 2080 and 2090
    - 1 per cent increase between 2090 and 2100.
- The decline in the rate of increase beyond 2050 reflects inevitable diminishing returns due to a lack of suitable wind turbine sites, seasonal factors, energy storage limitations, and transmission and integration inefficiencies (IPCC 2011; REN21 2014)
- The 2100 value of 71.8 EJ per year of wind energy potential is marginally higher than the IPCC's (2011) lower-bound estimate of 70 EJ per year.

Column H: *Solar energy*

- The solar energy supply for 2010 was 0.3 EJ (IEA 2011).
- The assumed growth in solar energy between 2010 and 2050 is based on a straight-line increase in the quantity of solar energy anticipated for 2010 (0.3 EJ) and Stern's (2007) reported estimate of 110.0 EJ by 2050 (i.e., this assumes an average rate of increase of 27.4 EJ per decade).
- Beyond 2050, the assumed percentage increase in solar energy is the same as for wind energy.

- The decline in the rate of increase in solar energy beyond 2050 reflects inevitable diminishing returns caused by similar factors to those limiting the generation of wind energy (REN21 2014).

Column I: *Energy from biomass*

- Column I includes the energy sourced from biological material that can be used as a transport fuel or for industrial purposes. It excludes organic material chemically transformed by geological processes (e.g., oil and coal).
- The energy generated from biomass for 2010 was 64.4 EJ or around 72 per cent of the 2010 total renewable energy supply (REN21 2011).
- The assumed growth in biomass energy between 2010 and 2050 is based on a straight-line increase in the quantity of biomass energy anticipated for 2010 (64.4 EJ) and Stern's (2007) reported estimate of 110.0 EJ by 2050 (i.e., 11.4 EJ per decade).
- Beyond 2050, the assumed percentage increase in biomass energy is the same as for wind and solar energy.
- The declining rate of increase in biomass energy beyond 2050 reflects diminishing returns caused by land and water shortages.
- The 2050 value of 110 EJ per year of biomass energy potential (see Table 4.2) is marginally higher than the IPCC's (2011) lower-bound estimate of 100 EJ per year.

Column J: *Hydro-electricity*

- The energy generated from hydro-electricity for 2010 was 14.9 EJ or around 16.7 per cent of the 2010 total renewable energy supply (REN21 2011).
- Between 2010 and 2050, the assumed growth in hydro-electricity is based on a straight-line increase in the quantity of hydro-electricity expected in both 2010 (14.9 EJ) and 2050 (29.7 EJ) (i.e., 3.7 EJ per decade).
- Beyond 2050, the assumed percentage increase in hydro-electricity is the same as for wind, solar, and biomass energy.
- The decline in the rate of increase beyond 2050 again reflects diminishing returns caused by a shortage of suitable sites, competing water needs, and the growing siltation of dams (REN21 2014).

Column K: *Remaining sources of renewable energy*

- Column K includes all remaining renewable energy sources (e.g., tidal and geothermal).
- The 2010 supply of energy from remaining renewable sources is equal to nine per cent of all renewable energy supplied, which amounts to 8.1 EJ.
- Between 2010 and 2050, the rate of increase in the remaining sources of renewable energy is assumed to be:

  - 50 per cent increase between 2010 and 2020
  - 40 per cent increase between 2020 and 2030
  - 30 per cent increase between 2030 and 2040
  - 20 per cent increase between 2040 and 2050.

- Beyond 2050, the assumed rate of increase in the remaining sources of renewable energy is assumed to be:

  - 10 per cent increase between 2050 and 2060
  - 5 per cent increase between 2060 and 2070
  - 4 per cent increase between 2070 and 2080
  - 3 per cent increase between 2080 and 2090
  - 2 per cent increase between 2090 and 2100.

- The high rates of increase reflect the enormous potential for expansion and the small scale with which many of the remaining renewable energy sources have so far been exploited.
- Once again, the decline in the rate of increase beyond 2050 reflects diminishing returns from a wide range of limiting factors (Trainer 2007; IPCC 2011).

Column L: *Total renewable energy potential*

- Column L is the sum of all potential sources of renewable energy.

Column M: *Excess renewable energy capacity*

- Column M represents, in percentage terms, the excess of total renewable energy capacity over total renewable energy requirements (*Column L/Column F* × *100 %*).

Column N: *Renewable energy as a percentage of total energy requirements*

- Column N represents renewable energy requirements as a percentage of total energy requirements (*Column F/Column A* × *100 %*).

Notes

1. A target-based approach is described by the IPCC (2007d) as one that optimises policy strategies in response to assumptions regarding the likes of greenhouse gas emissions targets, technological change, and climate change impacts.
2. Given current and projected emissions levels and the fact that the atmospheric concentration of greenhouse gases stood at 440 ppm of CO_2-e as of early-2014, the 450 ppm stabilisation target will inevitably be exceeded. Hence, an overshoot trajectory is unavoidable. Indeed, evidence suggests that stabilisation at 450 ppm is likely to involve a peak concentration level of around 500 ppm by 2050 (den Elzen and Meinshausen 2006). Therefore, what

I am arguing against is a dramatic overshoot trajectory (i.e., in the order of 100 ppm), which would be impossible to avoid if deep emissions cuts do not begin soon.

3. As will be explained in Chap. 6, avoiding major overshoot amounts to preventing the atmospheric concentration of greenhouse gases exceeding 500 ppm of CO_2-e before stabilising at 450 ppm.
4. A 4 per cent annual reduction in global CO_2-e (Kyoto-gas) emissions beyond 2015 would amount to total greenhouse gas emissions of approximately 1,270 Gigatonnes of CO_2-e ($GtCO_2$-e) between 2015 and 2050 and 1,490 $GtCO_2$-e between 2015 and 2100. Both constitute emissions totals consistent with limiting the probability of average temperatures rising by more than 2 °C to below 50 per cent (Meinshausen et al. 2009).
5. Greenhouse gas emissions can be classified as either: (i) process-related CO_2 emissions; or (ii) non-CO_2 and land use-related CO_2 emissions (e.g., methane and nitrous oxide). Process-related CO_2 emissions include CO_2 emissions from energy sources, industrial processes, solvent and other product use, waste, and sundry activities.
6. Another potential benefit of choosing a more stringent emissions trajectory is that a 450 ppm target may prove to be too high. As it is, stabilisation at 450 ppm still leaves open the possibility of average global temperatures rising by 3 °C (see Table 1.3). Even if the Anderson and Bows trajectory involves much deeper emissions cuts than is required to stabilise at 450 ppm of CO_2-e, it might well equate to the emissions cuts needed to achieve a lower, yet more appropriate, stabilisation level.
7. The 'precautionary principle' is enshrined in Article 3.3 of the United Nations Framework Convention on Climate Change (UNFCCC 1992).
8. As a means of support, surveys over time of self-reported 'happiness' suggest that individual happiness ceases to be related to per capita income once a nation's per capita GDP reaches around US$15,000 (Layard 2005).
9. A recent paper by Kubiszewski el al. (2013), of which I was a co-author, indicates that the per capita GPI of the global economy peaked in 1978 when global per capita GDP reached approximately US$7,000 (at 2005 prices). This would suggest that a per capita GDP of Int$15,000 exceeds the likely per capita optimum at the global level. It should, however, be remembered that the proposed Int$15,000 per capita optimum assumes that all nations will eventually move to a steady-state economy and ultimately gravitate towards the Int$15,000 value. Because high-GDP countries would need to reduce their real GDP to achieve this goal, it would, as explained in Chap. 2, provide the ecological space for the world's poorest nations to enjoy a period of welfare-increasing GDP growth, unlike the present situation where excessive GDP growth by the world's wealthy nations has increased the marginal cost of growth for the world's poorest countries. Hence, the per capita US$7,000 identified in Kubiszewski el al. (2013) must be seen as the legacy of a suboptimal growth path of the global economy and not an indication of the 'true' global optimum.

10. A growth-dependent economy is one that is structurally designed to grow. As such, it incorporates a growth-imperative irrespective of whether the growth is desirable. There are many reasons as to why economies become 'growth-reliant'. However, the path-dependent nature of economic systems and its associated structural inertia plays a significant role.
11. The assumed 53.2 $GtCO_2$-e of global emissions for 2010 falls within the estimated range of anthropogenic greenhouse gas emissions in 2010 (49 $\pm$ 4.5 $GtCO_2$-e) (IPCC 2014). The assumed 54.0 $GtCO_2$-e of global emissions for 2015 prior to major emissions cuts is based on an emissions projection made by the World Resources Institute (WRI, *Earthtrends*, www.earthtrends.wri.org). The assumption of 54.0 $GtCO_2$-e is also similar to the global greenhouse gas emissions projected for 2015 by the UNEP (2011).
12. Even though the 450 ppm stabilisation target is assumed to be 'safe', average global temperatures will still rise to around 2 °C above the pre-industrial average. This implies a further rise of around 1.2 °C, which will undoubtedly have some negative impacts on global biocapacity (see Table 1.4 and Fig. 1.10). The impact of climate change on the ecosphere and its broader implications for the total output of the economy are discussed at length in Chap. 6.
13. I say immediate production waste because, as a consequence of the first and second laws of thermodynamics, all matter-energy embodied in the resources entering the economy eventually exits the economy as waste (see Figs. 2.2 and 2.9).
14. It is worth noting that the need to cut global emissions reductions to 44.4 $GtCO_2$-e by 2020 is consistent with a recent UNEP (2011) study of emissions pathways.
15. The lack of data at the global level precludes a precise calculation of a global GPI.
16. The order-of-magnitude increases of 13.6 and 3.5 are implied by the 1,260.8 per cent and 252.0 per cent values at the bottom of the real GWP/emissions and technical efficiency columns in Table 4.1.
17. Remember, these energy needs are based on pessimistic estimates of renewable energy development. A more rapid rate of increase in renewable energy capacity would significantly reduce the need for nuclear energy expansion.
18. Renewable energy potential and renewable energy requirements are the same for 2010 because it is assumed that the amount of energy available was the same amount used.
19. Base-load energy sources are energy sources that are always available to meet base-load energy requirements. Included in such sources are non-renewable energy resources, one-half of wind power, one-quarter of solar-electricity, hydro-electricity, biomass, and one-half of 'remaining' renewable energy sources (e.g., tidal and geo-thermal). Based on Table 4.2, in 2080, the availability of these resources—assuming the world remains on track to achieve a 450 ppm stabilisation target—constitutes 78.7 per cent of total global energy requirements. By 2100, this ratio increases to 91.8 per cent. These ratios are more than sufficient to meet base-load energy needs.

20. Based on estimates by the UN Department of Economic and Social Affairs (Population Division) (WRI, *Earthtrends*, www.earthtrends.wri.org).
21. It might be argued that a much faster rate of growth in the growth-as-usual scenario would lead to a larger annual increase in the human-made production factor ($\beta.K_H$) and hence to a faster rate of increase in the technical efficiency of production. Although more producer goods would be generated in the growth-as-usual scenario, less of them are likely to be of the efficiency-increasing kind. Indeed, more are likely to be of the throughput-increasing variety. This is because there would be none of the explicit resource constraints or additional policies imposed in the sustainable scenario that would presumably induce much greater levels of efficiency-increasing progress. Hence, whilst K_H would increase more in the growth-as-usual scenario, β is likely to rise at a much slower rate.
22. A similar factor increase would be required in the 450 ppm scenario envisaged by Stern (2007), which Stern claims would be unachievable with current and foreseeable technologies.
23. The 1,073.8 Exajoules required in 2100 is equal to the 2010 energy requirement of 495.0 EJ multiplied by the assumed factor increase in the global ecological footprint over the 2010-2100 period (i.e., 1,073.8 = 495.0 × [39,524.2 gha/18,220.2 gha]). 1,073.8 Exajoules falls within the estimated the range of 900 and 1,350 Exajoules.
24. Although some studies indicate the technical potential for renewable energy as a whole to be around 2.5 times as large as 2007 global primary energy demand (e.g., IPCC 2011 and Fischedick et al. 2011), it is also widely acknowledged that various factors—in particular, the infrastructure required to accommodate variable output levels and transmit renewable electricity to load centres—are likely to significantly limit the deployment of individual renewable energy technologies before absolute technical resource limits are reached. In other words, whilst, technically, there is great potential to generate renewable energy, the ability to utilise all that can be generated is and always will be severely restricted.
25. 495 Exajoules = 523 Quadrillion Btu ÷ 1.055 (Note: 1 Btu = 1.055×10^{-15} Exajoules).
26. 29.2 Exajoules = 495 Exajoules × 0.059 (Note: The consumption of nuclear energy constituted 5.9 per cent of total global energy consumption in 2010).

Part II
Alternative Perspectives of the Climate Change Crisis

Chapter 5
A Mainstream Economic Perspective of the Climate Change Crisis

5.1 Introduction

How does the ecological economics perspective in Part I compare with the mainstream economic position on climate change? Mainstream economists have been working on the climate change problem since the early-1990s. More recent mainstream work includes the well-known reviews by Stern (2007) and Garnaut (2008), and the studies conducted by McKibbin and Wilcoxen (2002), van Kooten (2004), Fankhauser and Tol (2005), Metz and Vuuren (2006), Tol and Yohe (2006b), Nordhaus (2007b), Weitzman (2007), and Beinhocker et al. (2008).

In a general sense, the mainstream economic approach amounts to determining how much we should cut greenhouse gas emissions to reduce the negative impacts of anthropogenic global warming.[1] More specifically, it involves ascertaining the most 'economically efficient' emissions trajectory that humankind should travel along over time. To make such a determination, mainstream economists advocate the deployment of a benefit-cost analysis, which, in its simplest form, involves comparing the benefits and costs of different emissions trajectories. However, many mainstream economists go much further and argue that a benefit-cost analysis should be conducted in the form of an optimisation exercise, such as those conducted in the past by Nordhaus (1991, 1993), Peck and Teisberg (1992, 1994), Maddison (1995), Nordhaus and Yang (1996), Gupta and Bhandari (1999), MacCracken et al. (1999), Manne and Richels (2001), Tol (2001, 2002a, b), McKibbin and Wilcoxen (2002), Azar and Lindgren (2003), Fankhauser and Tol (2005), and den Elzen and Meinshausen (2006).

P. Lawn, *Resolving the Climate Change Crisis*, DOI 10.1007/978-94-017-7502-1_5

5.2 The Stern Review: An Example of a Mainstream Economic Approach to the Climate Change Crisis

Despite mainstream agreement regarding the use of a benefit-cost analysis to determine the extent and timing of emissions reductions, there is considerable disagreement about how the benefits and costs should be estimated. To demonstrate the level of disagreement that exists, I shall focus on the approach and subsequent conclusions of the *Stern Review* (Stern 2007), since much can be learned from its close scrutiny. Before I do, let me begin by refuting any suggestion that the *Stern Review* is a non-mainstream economic document.

There are many commentators who believe that the *Stern Review* does not qualify as a mainstream approach to the climate change crisis because it places excessive weight on ethics and the risks associated with potentially catastrophic climate change. According to these commentators, it is a consequence of this approach that Stern's recommended stabilisation target of 550 ppm of CO_2-e is much lower than many mainstream suggested targets, albeit it is considerably higher than the upper 450 ppm limit recommended in this book. As we shall see, mainstream economists have accused Stern of manipulating accepted assumptions and conventional methods to arrive at an excessively stringent mitigation strategy. In my opinion, this accusation is false and I shall return to the debate shortly.

What should not be debatable is the fact that as much as the *Stern Review* goes beyond most studies to stress the importance of risk, uncertainty, and ethics, Stern continues to deal with these issues within the context of a mainstream benefit-cost framework. Amongst other things, the *Stern Review* internalises the risk of catastrophic climate change and its potential implications for future generations via the use of a low discount rate. It does not adopt the ecological economics position of avoiding unacceptably risky outcomes and confining choices over the atmospheric concentration of greenhouse gases to a range of 'safe' or ecologically sustainable alternatives. Moreover, Stern assumes that strong GDP growth up to 2100 is both desirable and ecologically sustainable; that climate change mitigation and damage costs are best represented as GDP losses; that human-made capital is a near perfect substitute for natural capital; and that the nexus between GDP growth and greenhouse gas emissions can be severed. It is for these reasons that the *Stern Review* can be considered, at best, a novel variation on the mainstream economic approach to the climate change problem (Spash 2007).

5.2.1 Stern's General Conclusions

As pointed out in Chap. 1, the main conclusion of the *Stern Review* is that at least one per cent of Gross World Product (GWP) must be invested each year to prevent climate change damage costs equivalent to the annual loss of 5–20 per cent of GWP, now and forever (Stern 2007, p. xv).[2] Viewed this way, the forgone one per cent of

GWP constitutes the cost of mitigation, whereas the avoided loss of 5–20 per cent of GWP constitutes the benefit. I should point out that as a consequence of faster than expected climate change (IPCC 2007a), Stern has revised the cost of mitigation from 1 to 2 per cent of GWP (Jowit and Wintour 2008). Nevertheless, because the benefits of mitigation far exceed the estimated costs, Stern has argued for immediate and decisive action to stabilise the atmospheric concentration of greenhouse gases at no more than 550 ppm of CO_2-e. All up, Stern's suggested action amounts to cutting global emissions by at least 25 per cent by 2050 (Stern 2007, p. xvi).

Despite acknowledging that the increase in greenhouse gas emissions has been largely driven by the rise in GWP, Stern does not believe that a choice must be made between averting climate change and rising real output. To the contrary, Stern believes that new energy technologies and production processes should enable nations to 'decarbonise' their economies sufficiently to achieve climate stabilisation without having to sacrifice GDP growth (Stern 2007, 2009). In this sense, the Stern position on climate change mitigation resembles the growth-as-usual scenario presented but rejected in the previous chapter.

5.2.2 Stern's General Methodology

To arrive at his conclusions, Stern employed 'bottom-up' and 'top-down' approaches to determine the cost of climate change mitigation (Stern 2007). A bottom-up method involves the estimation of what it costs to adopt conservation measures, geosequestration technologies, and low-emission energy resources to achieve a specific stabilisation target. Conversely, a top-down method involves the use of macroeconomic models to simulate the demand and supply for various resources and likely feedbacks between emissions levels and output growth. These models are utilised to compare the future paths of key macroeconomic variables under a no-mitigation or base scenario and alternative mitigation scenarios. The cost of a given mitigation policy constitutes the difference between the base scenario and the policy-induced scenario. It was by weighing up the results from bottom-up and top-down methods that Stern estimated the mitigation cost—in effect, the cost of stabilising greenhouse gases at 550 ppm of CO_2-e—at one per cent of GWP.

As for estimating the benefit of climate change mitigation—which equates to the avoided cost of climate change damages—Stern relied heavily upon an integrated assessment model referred to as PAGE2002 (Hope 2006).[3] Informed by the integrated assessment literature and Nordhaus and Boyer (2000), Stern began his assessment by using the PAGE2002 model to estimate the value of market damages plus the 'willingness-to-pay' to avoid the catastrophic effects of abrupt climate change. The model was then used to estimate the non-market impacts of inaction. Finally, Stern added high climate change sensitivities and feedbacks into the overall calculus. By combining all the estimated damage costs, Stern arrived at an average reduction of 14.4 per cent in *per capita consumption equivalents* or a total damage cost comparable to a 5–20 per cent decline in GWP.[4] When converted to the 'social cost of carbon', the cost of inaction amounted to US$85 per tonne of CO_2 or US$312 per tonne of carbon.[5] Upon finally comparing the benefits and costs of climate change mitigation, Stern estimated the global net benefits from shifting to an emissions pathway that is consistent with a 550 ppm stabilisation target at around US$2.5 trillion (at 2006 prices). Stern went further to claim that the net benefits of climate change mitigation would increase over time.

It is worth explaining what is meant by the social cost of carbon, since it will become particularly relevant in Chaps. 6 and 7. The social cost of carbon equals the present value of the extended impact of climate change caused by the discharge of one additional tonne of carbon today. In other words, the social cost of carbon represents the marginal damage cost of today's carbon emissions. Although the social cost of carbon can be expressed in terms of dollars per tonne of carbon or dollars per tonne of carbon dioxide, it is usually expressed as the price of carbon emissions. However, given the radiative forcing potential of all greenhouse gases, the term 'social cost of carbon' can also be used as a metaphor for the social cost of *all* greenhouse gas emissions. In this case, the social cost of carbon is best viewed as the price of all CO_2-equivalent greenhouse gas emissions. For the sake of convenience, the social cost of carbon will henceforth be referred to in this manner.

5.2.3 Stern's Choice of Discount Rate

What is not understood by many non-economists is that the result of a benefit-cost analysis depends heavily on the *present value* estimates of monetary values. To put this another way, the calculation of the net benefits in any conventional benefit-cost analysis entails the prior discounting of all future benefits and costs. Although the application of the discounting procedure does not reduce the raw magnitude of any future benefit or cost, it does reduce how much we value, *in the present*, a benefit or cost we envisage enjoying or incurring in the future. For example, at an annual discount rate of 5 per cent, the present value of a $1 benefit/cost expected in 50 years' time is just 8.7 cents.

Since the seminal work of Ramsey (1928), it has generally been accepted that the following formula should be employed when determining the most appropriate discount rate for use in a benefit-cost analysis[6]:

$$r = \delta + \eta g \tag{5.1}$$

where $r =$ the annual discount rate; $\delta =$ the pure rate of time preference and reflects how much we value the consumption of something today as opposed to consumption of the very same thing in the future; $\eta =$ the elasticity of marginal utility (consumption elasticity) and represents the aversion to economic inequality among different generations; and $g =$ the average rate of growth in per capita consumption.

In the case of the *Stern Review*, present value calculations were made of the growth-as-usual pathway by applying an annual discount rate of 1.4 per cent.

A discount rate of this magnitude implies that a $1 benefit/cost expected in 50 years' time has a present value of 49.9 cents. Stern attained the discount rate of 1.4 per cent by assuming: (i) a pure rate of time preference (δ) of 0.1 per cent; (ii) a consumption elasticity (η) of one; and (iii) a growth rate of per capita consumption (g) of 1.3 per cent per annum (i.e., $1.4\ \% = 0.1\ \% + [1 \times 1.3\ \%]$). Stern's (2007) rationale for using these values was that:

- δ of 0.1 per cent reflects the widespread view that it is unethical to regard a cost borne by future generations as any less important than a similar cost borne by the current generation. The miniscule value for δ reflects the outside possibility that future generations of human beings may not exist due to an unavoidable catastrophe.
- η of one assumes that if the income of person A is twice that of person B, an extra dollar received by both individuals will increase the utility of A by half as much as it increases the utility of B.[7] According to Stern, an assumed value of one best balances the rate of inter-temporal inequality aversion and risk aversion.
- g of 1.3 per cent represents most of the projected rates of increase in per capita GWP over the coming century.[8]

5.3 Mainstream Criticisms of the Stern Review

In view of the media attention given to the *Stern Review* and its potential policy influence, economists have had a great deal to say about Stern's conclusions and recommended mitigation strategy. Overall, most mainstream economists are critical of the *Stern Review*. The various criticisms can be classified into five categories. For instructive reasons, these will now be outlined and discussed.

5.3.1 Stern Underestimates the Cost of Climate Change Mitigation

As mentioned above, the conclusions in the *Stern Review* depend considerably upon the estimated benefits and costs of climate change mitigation. Many mainstream economists believe that Stern has vastly underestimated the cost of mitigation, which has biased his recommended course of action towards deeper than necessary emissions reductions over the next two to three decades. According to Weitzman (2007), the underestimation is in part due to Stern's consistent leaning towards assumptions and formulations that emphasise optimistically-low mitigation costs. Tol and Yohe (2006a) agree with Weitzman, yet are intrigued as to why Stern's estimated mitigation cost is so low given that his team did little more than review the prevailing mitigation cost literature and re-run existing integrated assessment models (e.g., PAGE2002).[9]

Although baffled, Tol and Yohe (2006a) believe the underestimation of the mitigation costs can be explained by a number of factors. Firstly, Stern relies heavily upon the mitigation cost estimates included in a commissioned report produced by Anderson (2006). These estimates are far more optimistic than those found in previous studies. Secondly, Stern underplays the uncertainty associated with emissions reduction costs despite emphasising uncertainties on the damages side of their benefit-cost calculations. Had Stern given equal consideration to the uncertainties pertaining to mitigation measures, then, according to Tol and Yohe, the estimated cost of climate change mitigation would have been substantially higher. As mentioned above, new evidence regarding the speed of climate change has already prompted Stern to revise the annual mitigation cost from 1 to 2 per cent of GWP. However, the IPCC has gone further and suggested that the annual cost of mitigation could be 5 per cent of GWP, or more (IPCC 2007d).

Thirdly, Stern largely ignores the economy-wide impact of rising energy costs that would inevitably follow the introduction of measures to reduce greenhouse gas emissions (e.g., a carbon tax or an emissions-trading system). Fourthly, Stern overlooks the cost of having to rapidly replace the existing stock of human-made capital with low-emissions capital—potentially a major factor in mitigation cost calculations. Finally, Stern limits the estimation of mitigation costs to 2050. Thus, by truncating the time horizon over which the mitigation costs are calculated, Stern ignores numerous downstream costs that, according to the Energy Modeling Forum (Weyant et al. 2006), are likely to increase the expected annual loss of GWP to 2.2 per cent by 2050 and 6.4 per cent by 2100 (Note: compare this to the average annual loss of 1 per cent of GWP initially estimated by Stern for the period up to 2050).

5.3.2 *Stern Overestimates the Cost of Climate Change Damages*

Many mainstream commentators believe that Stern has also overestimated the cost of climate change damages and that this, too, has influenced Stern's recommended course of action. Once again, Weitzman (2007) and Tol and Yohe (2006a) have led the way, albeit with Byatt et al. (2006), Nordhaus (2007a), and Mendelsohn (2007) in support, by arguing that Stern consistently selected the most pessimistic damage estimates when calculating damage costs. However, Tol and Yohe go further by claiming that Stern has misleadingly treated all damages from climate change as the potential benefit of climate change mitigation. Given Stern's recommended stabilisation target of around 550 ppm of CO_2-e, Tol and Yohe point out that the associated level of greenhouse gas abatement would merely slow, rather than prevent, all damages from climate change. As such, the cost of inevitable damage must be subtracted from the total damage costs to accurately estimate the benefit of climate change mitigation.

Tol and Yohe's criticism does not stop here. They also argue that Stern has double-counted the risk of climate change. How? According to Tol and Yohe, Stern accounts for 'catastrophic risk' in his estimates of the willingness-to-pay to avoid extreme climate change outcomes and again in the integrated assessment exercises in which 'catastrophic risk' is incorporated as an uncertainty parameter. By double-counting risk, Tol and Yohe believe an additional factor has contributed to Stern's high estimated damage costs. I should point out that Stern refutes this suggestion and I shall return to this dispute shortly.

In another critical paper, Yohe (2006) highlights the distortionary impact of the methods used by Stern to incorporate high climate sensitivities and feedbacks into the *Review*'s damage cost estimates. Yohe begins by pointing out that Stern rightly accounts for high climate sensitivities by assuming a 20 per cent likelihood of large economic losses should global temperatures rise by 5 °C above pre-industrial levels. However, Yohe then reveals how Stern accounts for climate change sensitivities by extending the distribution for potential temperature increases by 0.4 °C through to 2100. In doing so, Stern succeeds in anchoring an accelerated rate of global warming for every emissions path even though feedback processes can, as complexity theory informs us, generate outcomes that are characteristically nonlinear (Charney 1979; Faber and Proops 1990; IPCC 2007b; Hansen et al. 2008).

In the end, critics of Stern point to what they consider to be 'proof in the pudding'—namely, Stern's estimated social cost of inaction (US$312 per tonne of carbon) which is 8–10 times larger than the social cost suggested by standard economic models (IPCC 2007a; Nordhaus 2007c).[10] Of course, it could be argued that Stern's very high damage cost estimate is essentially the result of the low discount rate applied in the *Stern Review*. However, when the discount rate used by Stern (1.4 per cent) is applied to a widely respected DICE model developed by Nordhaus (1993, 2007c), the social cost of inaction increases to just US$159 per tonne of carbon, which is little more than one-half of Stern's estimate.[11] According to most mainstream economists, this indicates that factors beyond the choice of discount rate, such as the extreme and pessimistic assumptions outlined above, are largely to blame for Stern's high estimated damage cost.

5.3.3 The Discount Rate Used by Stern Is Too Low

As alluded to, the choice of discount rate plays a critical role in determining the present value estimates of the benefits and costs of climate change mitigation. Almost all mainstream economists believe that the 1.4 per cent discount rate used by Stern is too low and a key reason behind Stern's high damage cost estimates (e.g., Yohe 2006; Tol and Yohe 2006a; Arrow 2007; Mendelsohn 2007; Dasgupta 2007, Nordhaus 2007a, c; Weitzman 2007). The major criticism of the 1.4 per cent discount rate stems from the mainstream view that discount rates should reflect real rates of return on capital, which are typically in the order of 4–6 per cent per annum. As such, mainstream economists believe that Stern's discount rate is at odds with market rates of return on capital and observed savings rates (Nordhaus 2007c).

With respect to savings rates, an optimal consumption trajectory requires a society to save a constant amount (s) of permanent income. If r equals the real rate of return on capital and g constitutes the implied balanced growth rate, then $g = s \times r$ and Eq. (5.1) becomes:

$$r = \delta + \eta(s \times r) \tag{5.2}$$

Rearranging (5.2) yields:

$$s = \frac{r - \delta}{\eta r} \tag{5.3}$$

In view of Stern's use of $\delta = 0.1\ \%$, $\eta = 1$, and $g = 1.3$ per cent, Stern's assumed savings rate is 93 per cent (i.e., $0.929 = [1.4\ \% - 0.1\ \%]/1.4\ \%$). According to Dasgupta (2007), a savings rate of this magnitude is incongruous since it far exceeds the 15 per cent savings rate observed in most countries. Worse still, it absurdly implies that the current generation is willing to deprive itself almost entirely of consumption to augment the consumption of future generations. Concurring with Dasgupta, Weitzman (2007) accuses Stern of ignoring market-based observations and behavioural inferences by treating them as largely irrelevant to the issue of long-run discounting. Although Weitzman concedes that valid ethical reasons may exist for placing high weights on the welfare of future generations, Weitzman considers Stern's use of a low discount rate, and the extreme parameter values employed to arrive at it, as an unconvincing means of justifying the call for strong emissions reductions.

Exactly how much of an influence does the 1.4 per cent discount rate have on Stern's conclusions? Nordhaus (2007c) provides a clue by drawing attention to Stern's claim that the cost of inaction could result in a 20 per cent cut in per capita consumption, "now and forever". As Nordhaus stresses, the substantial losses 'now' do not equate to substantial losses 'today'. Indeed, Stern's estimate of the actual output losses 'today' is essentially zero. Nordhaus illustrates this by focusing on the *Stern Review*'s high-climate change scenario. In this scenario, the mean losses of GWP are 0.4 per cent in 2060, 2.9 per cent in 2100, and 13.8 per cent in 2200. Yet this is reported as a loss in current per capita consumption of 14.4 per cent.

How, asks Nordhaus, do damages averaging around 1 per cent of GWP over the next century equate to a 14.4 per cent loss of GWP, "now and forever"? The conversion is made possible by applying a discount rate that is low enough for the relatively small damage costs expected over the next two centuries to be overwhelmed by the high damage costs expected beyond 2200. Indeed, as Nordhaus highlights, more than half of the estimated damages "now and forever" occur after 2800. On the other hand, if a discount rate is used that is in keeping with the real returns on capital and observed savings rates, the cost of inaction falls dramatically—by at least one-half according to Nordhaus's DICE model.

Given the criticism directed towards Stern, it is instructive to consider what some mainstream economists believe is a more appropriate discount rate, including what constitutes a more realistic set of parameter values. Because of uncertainty, Weitzman (2007) believes that a plausible value for η, which also represents a coefficient of relative risk aversion, should be somewhere between 1 and 4, preferably 2. As for δ, the pure rate of time preference, Weitzman argues that it must be significantly greater than zero to ensure the discount rate is consistent with market behaviour and observed savings rates. In the end, Weitzman opts for a 'trio of twos'—namely, $\delta = 2$ per cent, $\eta = 2$, and $g = 2$ per cent, which results in a discount rate of 6 per cent and an optimal savings rate of 33 per cent.[12] It would be remiss of me to conceal that Weitzman favours the application of a declining discount rate that would approach the Stern rate of 1.4 per cent over time.

For different reasons, Nordhaus believes that a more appropriate pure rate of time preference is 3 per cent, although he accepts a value of 1 for the elasticity of marginal utility. Like many mainstream economists, Nordhaus argues for a 3 per cent pure rate of time preference on the basis that it more accurately reflects the inherent discounting behaviour of individuals. If we accept Stern's value of 1.3 per cent for g, Nordhaus's preferred parameter values produce a discount rate of 4.3 per cent and an optimal savings rate of 30 per cent.[13]

Contrary to Nordhaus, Dasgupta (2007) has little problem with the 0.1 per cent value for the pure rate of time preference, but finds Stern's value of $\eta = 1$ deeply unsatisfactory. This is because an assumed value of $\eta = 1$ is equivalent to saying that the distribution of well-being is not very important and, furthermore, that the current generation should invest heavily in mitigation measures even though future generations are likely to be considerably richer. From past experience, Dasgupta explains why values between 2 and 4 for the elasticity of marginal utility are ethically more satisfactory. Dasgupta ultimately opts for a value of $\eta = 3$. Again, if we adopt Stern's value of 1.3 per cent for g, Dasgupta's preferred parameter values generate a discount rate of 4 per cent and an optimal savings rate of 32.5 per cent.[14]

As can be seen, the discount rates assumed by Weitzman, Nordhaus, and Dasgupta (6, 4.3, and 4 per cent respectively) are considerably higher than the discount rate assumed by Stern and much closer to the average 4–6 per cent rate of return on capital. Although the optimal savings rates of 33, 30, and 32.5 per cent are double the typically observed savings rates of 15 per cent, they are well below the extreme rate implied by Stern. More importantly, they are savings rates that many mainstream economists believe are necessary to combat climate change in the most efficient and equitable manner.

5.3.4 The Stern Review Does not Constitute a Proper Benefit-Cost Analysis

I mentioned earlier that mainstream economists believe that a benefit-cost analysis should be conducted in the form of an optimisation exercise. That is, to conduct a comprehensive benefit-cost analysis, it is necessary to equate the marginal benefits and marginal costs of various emissions trajectories in order to determine the economically 'most efficient' trajectory (Nordhaus 2007c). I might add that because the benefits of mitigation are measured in terms of avoided climate change damages, an efficient emissions trajectory can also be viewed as one that minimises the sum of total mitigation costs and total damage costs.

Figure 5.1 reveals two mainstream representations of an efficient emissions level at time t_0. In the top panel, the Total Damage Cost curve is upward sloping to indicate that the higher is the level of greenhouse gases emitted at t_0, the higher are the damage costs of climate change, both at t_0 and beyond (Note: future damage costs are important because of the lengthy time that greenhouse gases remain in the atmosphere (see Table 1.1)). The Total Damage Cost curve is relatively flat because the additional damage cost of higher emissions at t_0 is minor compared to the impact of a larger stock of greenhouse gases. Although downward sloping, the Total Mitigation Cost curve is best viewed in terms of the cost associated with reducing greenhouse gas emissions at t_0. The Total Mitigation Cost curve appears in the form presented in Fig. 5.1 because, as mitigation levels are increased (i.e., as emissions levels are reduced), the additional mitigation effort required to further

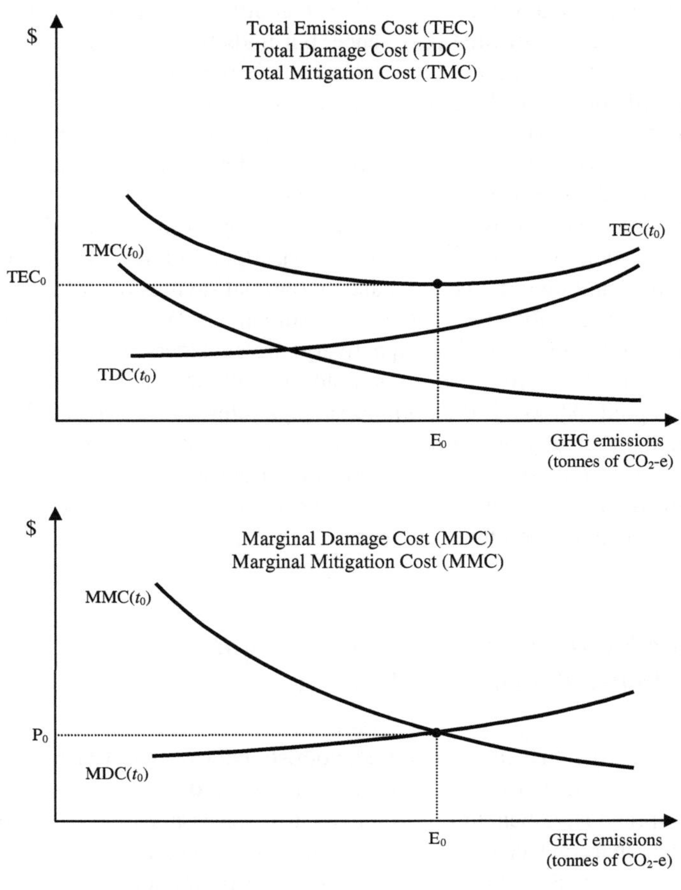

Fig. 5.1 An efficient level of greenhouse gas emissions at time t_0

reduce emissions rises. As such, an incremental reduction in greenhouse gas emissions is more costly to achieve. The Total Emissions Cost curve represents the sum of the total damage and total mitigation costs at different emissions levels. Taken together, the efficient emissions level at t_0 is E_0, which occurs where the Total Emissions Cost curve is at its lowest point (i.e., where the total cost of current emissions is minimised). At E_0, the total emissions cost is TEC_0.

The lower panel of Fig. 5.1 shows the Marginal Damage Cost curve and Marginal Mitigation Cost curve. Whereas the former curve represents the present value cost of emitting one additional tonne of greenhouse gases, the latter curve represents the present value cost of reducing greenhouse gas emissions by one tonne.[15] The Marginal Damage Cost curve is upward sloping to indicate that total damages rise at an increasing rate as emissions levels escalate. Given the nature of the Total Mitigation Cost curve depicted in the top panel, the Marginal Mitigation Cost curve is downward sloping with respect to increasing greenhouse gas emissions or, equivalently, upward sloping with respect to decreasing emissions. The efficient level of greenhouse gas emissions is represented by the intersection of the Marginal Damage Cost and Marginal Mitigation Cost curves. Correspondingly, the efficient price of greenhouse gas emissions (i.e., the social cost of carbon) is P_0 per tonne of CO_2-equivalent greenhouse gases.

It is important to recognise that the positions of the Total Damage Cost, Marginal Damage Cost, and Total Emissions Cost curves depend not only on past and present emissions levels, but on future emissions levels. Because greenhouse gases remain in the atmosphere for a very long time, higher future emissions increase the climate change impact of the greenhouse gases emitted today. Should the emissions levels beyond t_0 be greater than the levels assumed in Fig. 5.1, the positions of the above-mentioned curves would be higher than those shown in the figure. Moreover, the efficient emissions level at t_0 would be less than E_0 and the efficient price of greenhouse gas emissions would be higher than P_0.

Whilst Fig. 5.1 shows the efficient emissions level at t_0, Fig. 5.2 goes further to reveal the mainstream concept of an efficient emissions trajectory. As can be seen in the top panel of Fig. 5.2, the Total Emissions Cost curve can be expected to shift upwards at least until the atmospheric concentration of greenhouse gases stabilises. The reason for this is that total emissions costs will continue to rise as the concentration of greenhouse gases escalates. Why? As was explained in Chap. 1, renewable resources of various kinds, ecosystems, and physical infrastructure become increasingly vulnerable to global warming as the atmospheric concentration of greenhouse gases increases. This, in turn, increases the damage costs associated with any given emissions level. Furthermore, since ecosystem and renewable resource degradation leads to greater resource scarcity and higher resource prices, an escalating concentration of greenhouse gases also increases the cost of maintaining greenhouse gas emissions at a particular level.

Consider time t_0 again. The Total Emissions Cost curve is represented in the top panel of Fig. 5.2 by $TEC(t_0)$. As with Fig. 5.1, the efficient emissions level is E_0. Assuming that E_0 exceeds the Earth's greenhouse gas-absorbing capacity (A), the atmospheric concentration of greenhouse gases rises. This causes the Total

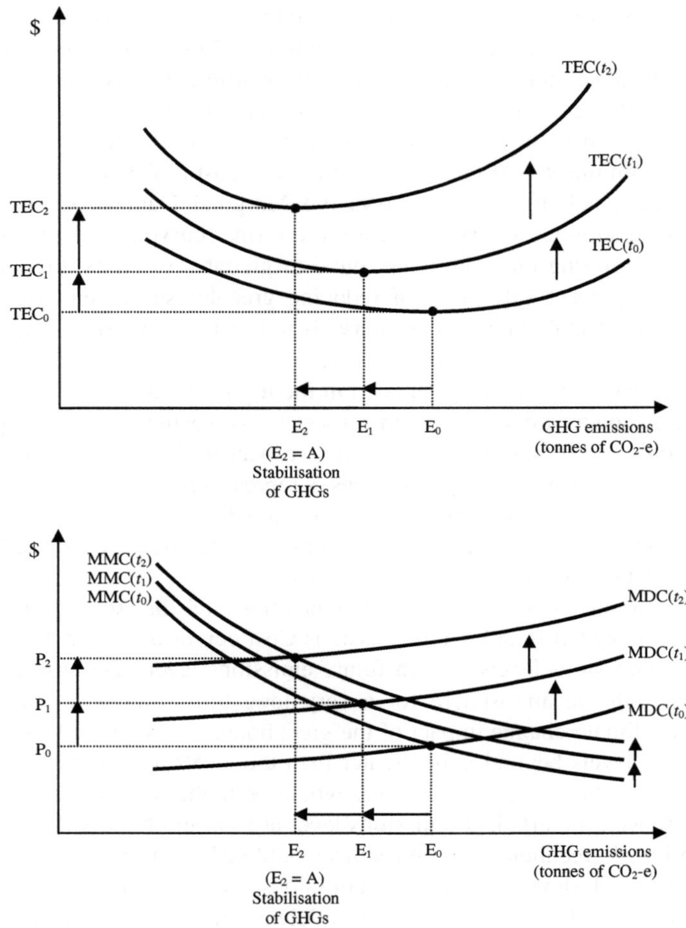

Fig. 5.2 An efficient emissions trajectory leading to an eventual stabilisation of greenhouse gases at time t_2

Emissions Cost curve at time t_1 to shift upwards to TEC(t_1). Accordingly, the efficient emissions level decreases to E_1. Assuming that E_1 also exceeds the Earth's greenhouse gas-absorbing capacity, the Total Emissions Cost curve shifts upwards to TEC(t_2), thus resulting in a new efficient emissions level of E_2. Since E_2 is equal to the Earth's greenhouse gas-absorbing capacity (i.e., $E_2 = A$), the concentration of greenhouse gases stabilises. Provided emissions levels beyond t_2 remain at E_2 and the stabilised concentration level does not increase climate change damages further, the Total Emissions Cost curve remains stationary at TEC_2.[16] Overall, the efficient emissions trajectory depicted in Fig. 5.2 involves a reduction in greenhouse gas emissions from time t_0 to t_2 and an increase in total emissions costs from TEC_0 to TEC_2.

As for the lower panel, both the Marginal Damage Cost and Marginal Mitigation Cost curves shift upwards for reasons just given. Assuming that the Marginal Damage Cost curve rises more than the Marginal Mitigation Cost curve (to be explained in Chap. 6), it can be seen that the shifting intersection of the two curves corresponds to a lower emissions level (E_0 to E_2) and a rise in the efficient price of greenhouse gas emissions (P_0 to P_2) (i.e., an increase in the social cost of carbon).

I should point out that the dynamics represented in Fig. 5.2 are based on the assumption that many factors with the capacity to influence total mitigation and damage costs—such as technological progress, learning-by-doing, the co-benefits of mitigation, and changes in resource scarcity caused by factors other than global warming—remain constant over time. As we shall see in Chap. 6, these factors are likely to play a more significant role in determining mitigation and damage costs than what has been acknowledged in past climate change studies, including the *Stern Review*. Hence, they are likely to appreciably alter what might be considered a desirable mitigation strategy.

In view of the mainstream perspective of an efficient emissions trajectory just outlined, Mendelsohn (2008) and Tol and Yohe (2006a) believe that the benefit-cost analysis conducted by Stern is seriously incomplete insofar as it narrowly compares the cost of inaction with the cost of stabilising emissions at one particular concentration level (550 ppm of CO_2-e). Thus, rather than equating marginal benefits and marginal costs in the manner illustrated in Figs. 5.1 and 5.2, Stern's approach amounts to little more than a comparison between the total benefits and costs of one, aggressive, near-term, mitigation strategy.

Does this really matter? Potentially, yes. Let us assume that Stern is correct and $US2.5 trillion of net benefits can be expected from stabilising greenhouse gas emissions at 550 ppm of CO_2-e, and by achieving a 550 ppm target via deep, early, emissions reductions.[17] How are we to know that higher net benefits would not emerge from an alternative stabilisation target and emissions trajectory? According to mainstream economists, we cannot, unless we explore benefit-cost comparisons over a range of alternative emissions scenarios. Because Stern fails to do this, mainstream economists believe it is highly probable that Stern's recommended course of action is well wide of the desirable mark.

In fact, according to Tol and Yohe (2007b), Stern's recommended course of action is, by deduction, sub-optimal. Tol and Yohe's reasoning is this. Assume that an optimal emissions strategy has been implemented. Since an efficient emissions trajectory would be characterised by the minimisation of discounted mitigation costs, Tol and Yohe believe we should expect the marginal cost of mitigation to rise over time as measures are taken to reduce greenhouse gas emissions. Indeed, Tol and Yohe believe that marginal mitigation costs should rise in keeping with the increase over time of the social cost of carbon. Yet, according to Tol and Yohe (2007b), the results borrowed by Stern from Anderson (2006) suggest that marginal mitigation costs are likely to decline over time. In other words, the variation over time of the marginal mitigation costs used in Stern's analysis is inconsistent with the mainstream understanding of an efficient emissions trajectory. As such, Tol and Yohe (2007b) believe that Stern's mitigation strategy cannot be an optimal one.

5.3.5 Stern's Conclusions Are Incommensurate with His Economic Analysis

Virtually all mainstream economists believe there is a need to reduce greenhouse gas emissions. Furthermore, a number of them agree with Stern's call to stabilise greenhouse gas concentrations at around 550 ppm of CO_2-e. Where they differ from Stern is that they believe that the great majority of the required emissions reductions should occur later rather than sooner.

Although Stern emphasises the importance of the physical implications of climate change, including the risks associated with different greenhouse gas concentration levels (see Stern 2007, Chaps. 3–5), Stern still relies upon the results of his economic analysis to justify the *Review*'s recommended course of action. This is no better exemplified than by the 'Summary of Conclusions', which is dominated by economic-based arguments and frequent references to the benefits and costs of climate change mitigation.

Given that Stern's economic results lie well outside the range of results reported in most mainstream economic studies, it would seem odd that Stern's stabilisation target is much the same as that recommended by most mainstream economists, even if the recommended pathway to its achievement is significantly different. Many mainstream economists agree. Tol and Yohe (2006a), for example, believe that in view of the benefits and costs used in the *Stern Review*, Stern should have recommended a far more stringent emissions reduction strategy and a much lower stabilisation target. Weitzman (2007), who agrees with Stern's stabilisation target but not with his economic analysis, argues that Stern's recommended course of action is so incommensurate with the economic evidence presented that Stern is effectively "right for all the wrong reasons".

Many advocates of the need for urgent climate change mitigation might conceivably argue that a desirable conclusion is what ultimately matters and that if the potential risk of a climate change catastrophe justifies a course of action similar to that recommended by Stern, all is well (assuming that a stabilisation target of around 550 ppm is desirable, which I believe it is not). Tol and Yohe (2006a) disagree. They argue that Stern's flawed economic analysis plays into the hands of climate change sceptics by providing them with ammunition to fire at the proponents of climate change action. Thus, instead of being a voice of reason, Tol and Yohe believe that the *Stern Review* is a counter-productive document.

5.4 Stern's Response to the Mainstream Criticisms of the Stern Review

Given the above, Stern and his colleagues have spent considerable time and effort responding to the many mainstream criticisms of the *Stern Review* (e.g., Dietz et al. 2007a, b, c; Dietz and Stern 2008, 2009).[18] Although Stern has addressed each of the above criticisms individually, the main thrust of Stern's rejoinder is that most of the criticisms display a gross misinterpretation of the *Review's* results and conclusions (Dietz et al. 2007b). According to Stern, this is best exemplified by the undue attention paid to the aggregate modelling exercise conducted in Chap. 6 of the *Review*. As important as aggregate studies can be in explaining the logic surrounding key climate change issues, Stern has stressed that the main purpose of the modelling exercise was to supplement the more pertinent statements expressed in the *Review* regarding risks and ethics and why consideration of these factors, alone, justifies stringent, up-front emissions cuts.

In defending the *Stern Review*, Stern points out that the methodology he adopted reflects the sound policy-making approach developed long ago by James Meade (1952). Widely employed in the 1960s and 1970s, Meade's philosophy involves a method of appraisal that not only addresses the specific problem at hand, but accounts for all stakeholders (whether they be rich or poor or present or future generations); the likely timing of benefits and costs; the development path under consideration; and the potential impact of extreme, albeit low-risk outcomes. Regrettably, in Stern's opinion, Meade's policy-setting philosophy has all but been abandoned. It is for this reason that Stern believes that mainstream economists have dangerously elevated aggregate models to the centre stage of policy considerations when it is clear that decision-making requires an understanding of the intricacies of policy-setting in a risky, imperfect world characterised by price distortions, irreversibility, and limited information (Stern 2009).

Furthermore, Stern believes that most criticisms of the *Stern Review* reveal the reluctance of mainstream economists to contemplate ethical considerations—a result of poor training, ignorance, and a conviction among mainstream economists that ethics is something economists should strenuously avoid (Stern 2009). Given the intergenerational nature of the climate change problem, Stern is adamant that ethical considerations cannot be avoided. Yet it is because mainstream economists shun any examination of ethics that, according to Stern, their restricted approach compounds the very problems that an over-emphasis on formal climate change models creates.

Ultimately, Stern believes that sound climate change policy requires all the evidence pertaining to climate change to be assembled in a structured way. Despite the informational value of formal economic modelling, Stern warns that these models unavoidably omit key elements and crucial matters of concern. Thus, to confine climate change analyses to formal economic models, one runs the risk of drawing conclusions on the basis of modelling assumptions often chosen for mere analytical convenience (Dietz and Stern 2008). Having said this, Stern believes that the economic analysis conducted in the *Stern Review* is sufficient to support strong action on climate change because, unlike most mainstream studies, it adequately accounts for risks and uncertainty and incorporates the latest climate change science and mitigation possibilities (Dietz et al. 2007b). It is with the importance of risks, ethics, and up-to-date science in mind that we now turn to Stern's response to the specific criticisms of the *Stern Review* outlined earlier in the chapter.

5.4.1 Stern Underestimates the Cost of Climate Change Mitigation—Stern's Response

As revealed above, mainstream economists believe that Stern has underestimated the cost of climate change mitigation by basing mitigation cost estimates on optimistic assumptions and forecast methods. Stern vehemently disagrees with this assessment by stressing that his mitigation cost estimates are methodologically robust and do not rest on any one particular assumption or modelling approach (Dietz et al. 2007b). Moreover, Stern has argued that his mitigation cost estimates are exclusively drawn from a comprehensive and up-to-date sweep of the mitigation cost literature. To support his claims, Stern points out that the mitigation cost estimates revealed in the *Stern Review* have been verified by reputable organisations, such as the International Energy Agency (IEA 2006).

Besides Stern's general response, there are many detailed reasons why Stern believes the various criticisms of his mitigation cost estimates are unfounded. To begin with, Stern believes that many of his critics have failed to sufficiently account for probable future rates of technological innovation and the likely deployment of mitigation technologies. In doing so, critics have downplayed the likely emergence of substitution possibilities and options that, if exploited, would significantly lower mitigation costs (Dietz et al. 2007b). In fact, according Stern, the many pessimistic assumptions about technological innovation and substitution possibilities, which have led to very high mitigation cost estimates, are entirely at odds with past rates of technological progress. Conversely, Stern believes that his assumptions regarding technological change are very conservative by historical standards (Anderson 2006; Dietz and Stern 2008).

Secondly, and this relates to Stern's first point, most mainstream economists have ignored the facilitating role of future policy changes. In most studies, it has been assumed that policy-induced price signals and institutional mechanisms, including increased carbon trading opportunities, will remain largely unchanged into the future. This, of course, is highly improbable. Instead, forthcoming policy measures will almost certainly spur on mitigation efforts that, through learning-by-doing and economies of scale, are likely to exert considerable downward pressure on mitigation costs (Dietz et al. 2007b).

Thirdly, Stern believes that most modelling exercises have overlooked an important link between primary energy costs and mitigations costs. To recall, the cost of climate change mitigation is measured in terms of the cost associated with meeting a particular emissions target. Many observers believe that a major mitigation cost will be the rise in the cost of primary energy. Acknowledging this, Stern points out that primary energy costs in most industrialised countries currently amount to around 3–4 per cent of GDP (a little higher in most industrialising nations). This means that if mitigation strategies increase the cost of primary fuels by no more than 25 per cent, as Stern and others predict, mitigation costs are unlikely to rise above a maximum of one per cent of GDP (Dietz et al. 2007b). Whilst mitigation costs are likely to be higher in poorer countries, Stern believes that the gradual uptake of new technologies and a change in the composition of national output should bridge the mitigation cost gap between rich and poor nations. Stern does concede that a shift towards biofuels could dramatically increase the cost of primary energy; however, Stern believes that energy-saving forms of technological progress should keep the cost of energy to a manageable level, thus keeping a lid on the cost of climate change mitigation (Stern 2009).

Finally, since the cost of mitigation equals the cost of achieving a particular greenhouse stabilisation target, then, as Stern highlights, any co-benefits associated with reducing greenhouse gas emissions must be subtracted to obtain the 'true' cost of mitigation. At various stages during the *Stern Review* (in particular, Chap. 12), Stern outlines a range of benefits that would emerge as a consequence of measures taken to reduce greenhouse gas emissions. These include soil conservation benefits and averted landslides from avoided deforestation; the preservation of ecosystem services; less waste and the reduced generation of many hazardous forms of pollution; energy-cost savings from improved building design and the supply of more energy-efficient appliances and modes of transport; and a reduced reliance on non-renewable energy sources (Stern 2007). All of these co-benefits, plus more, are often overlooked in mitigation cost studies, thus contributing further to mitigation cost estimates well above those revealed in the *Stern Review*.

I mentioned earlier that, owing to new climate change information, Stern has revised his estimated cost of climate change mitigation from one to two per cent of GWP. Given this upward revision, there are some critics who would argue that their criticism of Stern has been wholly confirmed. This is not so, since the basis for Stern's revision would apply to all mitigation cost studies, not just the study conducted by Stern. Consequently, any relative difference between Stern's cost estimates and those of his critics remains. Also remaining intact is the validity of Stern's response—although, as we shall see, there are non-mainstream reasons for believing that Stern has underestimated the cost of climate change mitigation.

5.4.2 Stern Overestimates the Cost of Climate Change Damages—Stern's Response

Of all the mainstream criticisms levelled at the *Stern Review*, the majority have been directed at the alleged overestimation of the cost of climate change damages. Stern and his colleagues have accordingly devoted most of their response efforts defending this category of costs. Much of the argument put forward by critics to explain Stern's overestimation of damage costs have centred on Stern's supposed application of an unjustifiably low discount rate.

Before we consider Stern's response to the discount rate allegation, it is worth highlighting a number of other reasons why Stern believes the criticisms of his damage cost estimates are groundless. To begin with, Stern argues that his critics are wrong to claim his assessment of damages has been based on selective pickings from the climate change literature. To the contrary, Stern stresses that his damage cost estimates are founded on a broad appreciation of damage cost studies and an acknowledgment of the latest climate change science (Dietz et al. 2007b).

Secondly, and more importantly, Stern argues that most of the Integrated Assessment Models (IAMs) used to calculate damage costs are restricted to a narrow set of the most measurable impacts of climate change. Very few IAMs include the impacts of large-scale, discontinuous changes to the climate system that are likely to have a significant and non-marginal impact on future economic activity (Dietz et al. 2007a). Moreover, almost no IAMs have been extended to incorporate 'socially contingent' impacts, such as the potential cost of the rise in international conflict and ecologically-induced migration. As a consequence, almost all IAMs are incapable of accounting for the high damage costs that are likely to emerge if global temperatures rapidly escalate. According to Stern, it is for this and other reasons that the losses of welfare-equivalent consumption generated by most IAMs are just a few per cent of GWP. On the other hand, when large-scale, discontinuous climate change impacts are incorporated into climate change models—as Stern performs in the *Stern Review*—comparable losses increase to at least 10 per cent of GWP (Dietz et al. 2007a).

Throughout Stern's many responses to his critics, Stern refers to three additional reasons for the inadequate nature of most IAMs. Firstly, Stern believes that most IAMs downplay the potential for rising greenhouse gas emissions to deplete and degrade the stock of natural capital essential for human development (Dietz et al. 2007a; Dietz and Stern 2008). Secondly, by assuming that increases in the consumption of human-made goods can compensate for the loss of critical natural capital services, most IAMs overlook the detrimental impacts that would arise from the irreversible loss of non-substitutable natural capital (Dietz et al. 2007a).[19] Thirdly, although most IAMs involve the use of Monte Carlo procedures that allow climate change impacts to be modelled probabilistically, invariably the links between greenhouse gas emissions and the economic impacts of global warming are inadequately parameterised.[20] According to Stern, this arises because most modelling strategies involve a 'best guess' of each parameter rather than a full estimate of the uncertainty surrounding the various links. For example, it is recognised by mainstream economists that a meaningful valuation of relative climate change risks requires a proper application of *expected-utility analysis* (Dietz et al. 2007a). In applying this form of analysis, most researchers use the Monte Carlo procedure to generate a probability distribution of consumption in order to calculate the discounted utility of *mean* expected utility. Stern believes this is an incomplete application of the 'economics of risk' because it fails to preserve important information about the probability distribution of consumption that is required to value the probability distribution of social welfare. Given this blatant shortcoming, Stern argues that modellers should calculate the discounted utility of each possible climate change outcome prior to making each Monte Carlo draw (Dietz et al. 2007a). According to Stern, this approach would ensure retention of the critical information embodied in the probability distribution. Not surprisingly, this alternative approach constituted a core element of the welfare assessment performed in the *Stern Review*. Moreover, it played a key role in the disparity between Stern's estimate of climate change damages and those revealed in other mainstream studies.

Finally, Stern is at pains to stress that the impact function used to link damage costs to global mean temperatures was formulated following a close examination of a wide variety of optimistic to pessimistic climate change 'stories' (Dietz et al. 2007b). As such, Stern believes his raw damage costs lie at the centre of a range of impact studies and do not constitute an outlier.[21] What's more, because the PAGE2002 model employed by Stern omits many climate change impacts, Stern admits he has probably underestimated the full cost of climate change damages.

Before moving on, it was pointed out earlier that Tol and Yohe (2006a) believe that Stern's high damage costs can be partly attributed to Stern having double-counted the risk of catastrophic climate change. To a large extent, Tol and Yohe's accusation has been prompted by Stern's broader-than-usual treatment of uncertainty and the incorporation of potentially large-scale climate change impacts. Upon close scrutiny of Stern's expected-utility analysis—which includes the specification of a high-climate change scenario to account for recent quantitative modelling of positive natural feedbacks—it is clear that Stern has not double-counted the risk of climate change. Stern has simply added an extreme, albeit low-probability climate change pathway in order to generate a probability distribution capable of providing a better assessment of the potential impacts of climate change. As Stern emphasises, this constitutes a more comprehensive treatment of risk, not a doubling up of risk. In all, Stern believes it is the majority researchers that are at fault since, by failing to incorporate the potential for extreme climate change impacts, they are only 'part-counting' risk (Dietz et al. 2007b).

5.4.3 The Discount Rate Used by Stern Is Too Low—Stern's Response

As mentioned above, a great deal of the criticism levelled at Stern's climate change modelling has centred on Stern's choice of discount rate. As far as Stern is concerned, this criticism reflects a gross undervaluation of risk and the ethical considerations surrounding the climate change issue as well as an incorrect understanding of how Stern estimated the cost of climate change damages.

In an effort to defend his 1.4 per cent discount rate, Stern believes that the choice of rate should be made on the basis of what most people would deem as morally acceptable. If we consider the discount rate parameters of the Ramsey formula, then, beginning with the pure rate of time preference (δ), Stern stresses that its value must be understood in terms of how we ethically discriminate by birth date (Dietz and Stern 2008). Given that all generations should be regarded as people of equal moral worth, Stern believes that δ should be approximately zero per cent.[22] In other words, if all generations are more or less equally well off, there is no reason why a climate change cost borne by future generations should be valued any less than the same cost borne today. According to Stern, arguing any differently amounts to a position that most people would find morally unacceptable (see, also, Broome 1992; Cline 1992). Yet this is the exact position taken by those who believe Stern's choice of discount rate is too low.

With regard to the elasticity of marginal utility (η), Stern explains how it also constitutes an ethical parameter, since its value represents an assumed aversion to economic inequality both within and between generations. In defending his choice of $\eta = 1$, Stern begins by arguing that a value of $\eta = 0$—which would assume that the marginal utility of an extra dollar of income is the same for rich and poor individuals—is seriously problematic (Dietz and Stern 2008). As Stern stresses, although a value of $\eta = 0$ is used in many benefit-cost analyses, few people would accept that an extra dollar of income would provide a millionaire with as much additional utility as it would for a pauper.[23]

As for $\eta = 2$, Stern invites readers to conduct a 'leaky bucket' experiment with respect to the redistribution of income. In general, the aim of a redistribution policy is to increase the net welfare of society. This is successfully achieved so long as the increase in the utility of the poor exceeds the decrease in the utility of the rich. With this is mind, the central question is this: How much 'redistributed' income can be lost from administrative and other related processes without compromising the aim of increasing the net welfare of society? As Stern highlights, when critics argue that η should equal 2, they are effectively saying that if person A has five times the income of person B, then taking one dollar from A and giving it to B will produce a net social welfare gain even if 95 per cent of the redistributed income is lost during the redistribution process (Note: this assumes that the marginal utility of an extra dollar received by person B is 25 times more than it is for person A). With a value of $\eta = 1$ (i.e., where the marginal utility of an extra dollar received by person B is assumed to be five times more than for person A), a loss of up to 80 per cent is tolerable. Given that many observers consider an 80 per cent loss to be extremely high, Stern believes it is difficult to justify that $\eta = 1$ is too low (Dietz et al. 2007b; Dietz and Stern 2008). Since it can be concluded that $\eta = 2$ is too high and $\eta = 0$ is unrealistic, Stern believes that a value of $\eta = 1$ is more than acceptable.

A key factor underlying the mainstream criticism of Stern's 1.4 per cent discount rate pertains to the relative merits of employing a 'prescriptive' or 'descriptive' approach to discounting. The latter approach, which is preferred by most mainstream economists, involves the assumption that the ethical preferences of society can be adequately revealed by current market behaviour. For this reason, many mainstream economists believe that discount rates should reflect real rates of return on capital, which, as previously mentioned, are usually much higher than Stern's annual discount rate of 1.4 per cent.

For a variety of reasons, Stern vehemently disagrees with this logic. Firstly, Stern stresses that discount rates are essentially marginal concepts designed to deal with small changes around a specified development path. It therefore matters little at the macro level if a relatively high discount rate is used to assess an individual project because a nation continues on much the same development path regardless of whether the project proceeds. However, in the case of climate change, where changes are likely to be severe and non-marginal, the choice of mitigation strategy is likely to have a profound impact on the future development path of a nation and the world generally (Stern 2009). As such, we cannot use rates of return on capital to assess climate change policy in the same way that we assess individual projects.

Secondly, Stern is quick to remind us that most market interactions involve aggregated private decisions that take account of short-term and medium-term benefits and costs. They do not involve collective decisions concerning actions with potentially severe, long-term ramifications (Dietz et al. 2007b). Hence, no financial market or other market of substance can be expected to reveal how a generation of people, when faced with the possibility of inflicting major costs on future generations, should conduct itself (Stern 2009). Nor, therefore, can any market provide the signals upon which to base a discount rate for assessing climate change costs and the future direction of climate change policy.

Thirdly, even if we employ a descriptive approach to discounting, one discovers that the recommended discount rate range of 4–6 per cent far exceeds the long-term, inflation-adjusted, low-risk interest rates on consumption and other loans. For instance, real long-term rates on government bonds are often in the order of 1.5 per cent over fifty years—a disparity in rates largely brought about by varying patterns of risk and capital market imperfections (Stern 2009). Upon careful consideration of this disparity, Stern believes that a riskless real return of 1.5 per cent on consumption loans reveals more relevant discount information than long-term investment returns of around 4–6 per cent. The reason for this is that an assessment of climate change mitigation strategies must involve a comparison of pathways using a measure of social welfare expressed in terms of the discounted utility of *consumption* over the indefinite future. It should therefore be our preferences over consumption possibilities, not returns on private investment decisions, which guide climate change policy. Moreover, the riskless real return is of greater relevance because uncertainty is best treated by explicitly incorporating it within climate change models. Uncertainty should not, according to Stern, be separately incorporated into the analysis via the discount rate (Stern 2009).

The above aside, Stern questions whether it is possible to confidently say that investment returns on private capital will remain in the 4–6 per cent range. If some of the expected climate change impacts materialise, long-term returns on all forms of capital could fall to much lower levels than those enjoyed in the past (Dietz et al. 2007b). In fact, ironically, it is the higher discount rate recommended by most mainstream economists that is likely to result in a comparatively higher concentration of greenhouse gases, greater climate change damage, and a reduced possibility of long-term investment returns remaining in the 4–6 per cent range.

Fourthly, as referred to earlier, mainstream economists normally assume that increases in the consumption of human-made goods can compensate for any loss of the natural capital services caused by accelerated climate change. Apart from the obvious fact that consumer goods cannot offset the degradation of critical forms of natural capital, this view fails to account for the likely change in the relative prices of various goods and services over the coming century (Sterner and Persson 2008; Stern 2009). As Stern explains, if we invest in human-made capital as the primary means of generating goods and services and ignore the impact of climate change on natural capital, the balance between the flow of environmental and human-made goods and services will, if only in the meantime, tip towards the latter. As a consequence, the prices of environmental goods are likely to increase relative to the prices of human-made goods. Should they do so, the future cost of more expensive environmental goods will almost certainly exceed the cost of climate change mitigation, thus reflecting a deficient mitigation policy (Stern 2009).

How does this flaw in mainstream thinking relate to Stern's criticism of the descriptive approach to discounting? Since most people recognise the critical nature of many natural capital services, there is, as Stern and others believe, a good reason to believe that society has a lower discount rate for environmental goods than for consumption goods (Quiggin 2008; Stern 2009). To therefore apply an across-the-board discount rate based on perceived rates of return on human-made capital would almost certainly result in the undervaluation of critical natural capital services and an inadequately weak mitigation strategy. All things considered, Stern believes that a descriptive approach to discounting is seriously defective and that a normative or prescriptive approach is required.

Finally, and very importantly, there is a serious misunderstanding of how Stern applied discount rates when estimating the present value of future climate change costs. What is not widely recognised is that Stern's 1.4 per cent discount rate was calibrated on a growth-as-usual pathway unaffected by climate change—that is, on a pathway where per capita consumption (g) was assumed to be increasing at an average rate of 1.3 per cent per annum. The 1.4 per cent rate was not applied to pathways affected by climate change.

How did Stern deal with climate change-affected pathways? It was mentioned earlier that Stern believes modellers should calculate the discounted utility of each possible climate change outcome prior to probabilistically modelling climate change impacts via the use of a Monte Carlo procedure (Dietz et al. 2007a). Imagine, then, a potential outcome where climate change damages reduce the increase in per capita consumption to a rate of just 0.5 per cent per annum. To discount the future utility of such an outcome, Stern altered the value of g to 0.5 per cent. Thus, with values of $\eta = 1$ and $\delta = 0.1$ per cent employed, Stern assumed a discount rate of 0.6 per cent (i.e., $0.6\ \% = 0.1\ \% + [1 \times 0.5\ \%]$).

In view of the wide range of per capita consumption values existing across a given probability distribution, each possible consumption path possesses a unique set of discount factors and hence a unique discount rate. In recognition of this, Stern applied a different discount rate for each potential climate change outcome. By doing so, Stern effectively applied what might be referred to as an *endogenous* discount rate (Dietz et al. 2007a). In direct contrast, most researchers make the mistake of applying a fixed *exogenous* discount rate. That is, they employ a fixed value for g in the Ramsey formula even though the consumption growth rate varies with each potential climate change outcome. As Stern explains, this latter approach leads to an underestimation of the cost of climate change damages (Dietz et al. 2007a). Thus, putting aside debates about the appropriate values of δ and η, Stern believes it is not his own discount rate that is too low, but the discount rates recommended by most mainstream economists that are too high.

I would just like to conclude this sub-section with one last point. Despite all the controversy surrounding Stern's choice of discount rate, Stern believes his critics have placed too much emphasis on the issue of discounting. To reinforce this conviction, Dietz et al. (2007a) have conducted a comprehensive sensitivity analysis of Stern's damage cost estimates. Dietz et al. begin their analysis by varying the elasticity of marginal utility (η) between 1 and 3 and the pure rate of time preference (δ) between 0.1 and 1.5 per cent. What Dietz et al. reveal is that increases in δ significantly reduce the cost of climate change. For example, in a baseline climate change scenario, with δ increased from 0.1 to 1.5 per cent, the mean cost of climate change falls from 11.1 to 3.3 per cent of per capita consumption equivalents. In the case of a high climate change scenario, the same increase in δ reduces the mean cost of climate change from 14.4 to 4.2 per cent of per capita consumption equivalents Dietz et al. (2007a).

As for the elasticity of marginal utility, Dietz et al. (2007a) discover that an increase in η has complex effects on the estimated cost of climate change. It is well understood that a rise in η implies an increased aversion to risk and inter-temporal inequality. What isn't understood is that the discounting impact of the elasticity of marginal utility varies as the severity of climate change damages rises. To illustrate how, Dietz et al. begin by showing that, with $\delta = 0.1$ per cent, an increase in η from 1 to 3 in the baseline climate change scenario causes the mean cost of climate change to fall from 11.1 to 1.3 per cent of per capita consumption equivalents. Dietz et al. then show that the same continuous decline does not occur in the high climate change scenario. For example, while the mean cost of climate change falls from 14.4 to 7.4 per cent of per capita consumption equivalents as η is increased from 1 to 2, it rises back up to 13.2 per cent as η is increased from 2 to 3 (Dietz et al. 2007a). The reason for this turn-around in costs is that when higher values of η are applied to more extreme climate change scenarios, the aversion to risk outweighs the aversion to inter-temporal inequality. In other words, as η increases, we become increasingly averse to the high climate change scenarios where future consumption is likely to be significantly affected.

To further demonstrate the alternating influence of η on the welfare cost of climate change, Dietz et al. increase the damage function exponent (γ) embodied in the PAGE2002 model from a value of 1–3. The damage function exponent is the main exponent determining the modelled impact of global warming on consumption-equivalent welfare. What Dietz et al. show is that, with $\delta = 0.1$ per cent and $\eta = 1$, an increase in γ from 1 to 3 boosts the mean cost of climate change from 5.4 to 33.3 per cent of per capita consumption equivalents. However, with $\eta = 3$, the same rise in γ increases the mean cost of climate change from 0.9 per cent to a massive 51.9 per cent of per capita consumption equivalents (Dietz et al. 2007a). Thus, a high value for η, which produces a large discount rate, can, under particular circumstances, generate very high climate change costs.

Overall, Dietz et al.'s sensitivity analysis makes two things abundantly clear. Firstly, it is wrong to assert that the low values ascribed to δ and η, for which Stern has been highly criticised, are the major factors behind Stern's high damage cost estimates. Secondly, Stern's treatment of risk and uncertainty and incorporation of the latest climate change science had at least as much of an influence on the *Stern Review's* as Stern's novel application of an endogenous discount rate—something most economists have overlooked.

5.4.4 The Stern Review Does not Constitute a Proper Benefit-Cost Analysis—Stern's Response

To recall, the mainstream view of a benefit-cost analysis involves estimating the 'optimal' level of greenhouse gas emissions, which is determined by equating the marginal damage cost of climate change and the marginal cost of mitigation. According to mainstream economists, performing this exercise requires a thorough investigation of benefit-cost comparisons over a wide range of alternative emissions scenarios. As previously mentioned, Stern has been heavily criticised for confining his analysis to a comparison between the cost of inaction and the cost of stabilising emissions at one particular concentration level—namely, 550 ppm of CO_2-e.

In response, Stern has argued that since the conclusions and policy recommendations in the *Stern Review* did not rest on the results of a formal optimisation exercise, it made little sense to conduct a benefit-cost analysis of the type advocated by mainstream economists. Moreover, as much as Stern has acknowledged that economic modelling can quantify the economic consequences of unabated climate change and can illuminate the implications of value judgements made with regards to time preference, risk, and inequality aversion, Stern has also been at pains to stress that aggregate modelling studies should, for the following reasons, be treated with great circumspection (Dietz et al. 2007b).

Firstly, there are limits to information, multiple ethical objectives, and unrepresented economic agents (i.e., future generations) that drive a wedge between economic theory and reality. What's more, this gap is magnified in relation to climate change policy because the causes and consequences of climate change are global; the risks and uncertainties are pervasive; and the impacts are long-term, persistent, and potentially irreversible and non-marginal.

Secondly, since economic models represent an aggregated attempt to express the impacts of climate change in terms of forgone future consumption, they are highly simplistic and restricted to a relatively narrow set of the most measurable climate change impacts. Consequently, economic models have an inbuilt tendency to underestimate expected damages.

Thirdly, as mentioned, economic models inadequately represent the uncertainty surrounding the impact of climate change. It is for this reason, according to Stern, that aggregate modelling studies suppress most of what is interesting and troubling about climate change (Dietz et al. 2007b; Dietz and Stern 2008).

Given the aforementioned, one might ask why Stern bothered to formally model the economic impacts of climate change. Stern did so, not because formal economic models provide precise estimates of climate change damages, but because economic models, when appropriately calibrated, can provide what Stern refers to as a "canvas on which debates about intergenerational fairness, the distribution of wealth, and the management of risk and uncertainty can be painted" (Dietz et al. 2007a, p. 323).

Thus, in sum, Stern believes it is both misleading and dangerous to base climate change policy on exercises designed to minimise the present value of the sum total of damage and mitigation costs (Dietz and Stern 2008). Instead, Stern is of the view that economic modelling is better confined to the less ambitious task of determining whether the costs of strong and immediate mitigation action are greater or less than the costs of climate change damages under a business-as-usual approach (Dietz et al. 2007b). Stern makes no apologies for having made the latter assessment—an assessment that led Stern to conclude that a stabilisation target above 550 ppm of CO_2-e is too risky and that a target below 450 ppm is too costly.

Finally, as for the claim by Tol and Yohe (2007b) that Stern's recommended course of action is sub-optimal, Stern disagrees. To recall, Tol and Yohe argue that an optimal emissions strategy requires the marginal costs of mitigation to rise over time as measures are taken to reduce greenhouse gas emissions. Yet, according to Tol and Yohe, the marginal mitigation costs borrowed by Stern from Anderson (2006) decline over time. In response, Stern believes that Tol and Yohe have confused marginal costs with average costs, since it is not marginal mitigation costs but average mitigation costs which should be falling (Dietz et al. 2007b).[24] In order to forestall further criticism, Stern has stressed that declining average mitigation costs are entirely consistent with the academic literature on the evolution of climate change costs.

5.4.5 Stern's Conclusions Are Incommensurate with His Economic Analysis—Stern's Response

Although Stern's recommended stabilisation level of around 550 ppm of CO_2-e is similar to the target advocated by many mainstream economists, it is Stern's conviction that deep emissions cuts are required now, rather than later, that sets him apart from most of his contemporaries. This said, it is Stern's high damage cost and low mitigation cost estimates which have led mainstream economists to argue that Stern should have recommended a lower stabilisation target and a more stringent emissions reduction strategy to achieve it.

In the time since the *Stern Review* was released, Stern has repeatedly indicated that he did not rule out a lower stabilisation target than 550 ppm of CO_2-e. Indeed, Stern is quick to remind people that he essentially advocated a target range between 450 and 550 ppm of CO_2-e (Stern 2007, p. xvi). Because Stern's recommended target is more ambitious than what many economists recognise, Stern believes his target is wholly commensurate with his economic analysis.

Even so, Stern stresses that a disproportionate amount of attention has been paid to his economic modelling when it is clear that the climate change crisis raises issues that cannot be resolved through economic analyses alone. As such, and since more than mere cost comparisons lie beneath the conclusions of the *Stern Review*, Stern believes it matters little that his recommended emissions strategy appears somewhat inconsistent with his modelling results.

5.5 Support and Further Criticisms of the Stern Review

5.5.1 Mainstream Support for the Stern Review

Despite the widespread criticism of the *Stern Review*, a number of mainstream economists have openly supported Stern by endorsing many of the previously-outlined points made by Stern (e.g., Solow, Mirrlees, Arrow, and Stiglitz).[25] Arrow, for example, believes that even if there is a good reason to heavily discount uncertainty and future climate change costs, Stern's fundamental conclusion that we should take immediate and sharp action to reduce greenhouse gases is entirely justified. More specifically, Ackerman (2007) asserts that, in terms of the benefit and cost comparisons of climate change, Stern's economic reasoning is watertight, which is evidenced by the unsuccessful attempts by critics to argue otherwise. Ackerman also believes that Stern's novel treatment of uncertainty and risk has failed to receive the attention it deserves. Where economists have taken note, they have generally addressed both aspects misleadingly and incorrectly. Reflecting the views of most supportive economists, Ackerman remarks:

> [T]he Stern Review takes us a long way toward understanding the economics of climate change, posing many big questions and answering some of them quite well. The arguments against conventional, high discount rates and the massive review of sectoral estimates of damages and mitigation costs will be hard to improve on, except in issues of detail (Ackerman 2007, p. 24).

Ackerman also touches on another crucial issue raised by many of Stern's supporters—namely, the false belief that the discount rate used in relation to climate change should reflect market rates of return on capital. Like Stern, Ackerman (2007) believes that the descriptive approach to discounting is erroneous and grounded in abstract theories of perfect markets and the false view that people have a consistent attitude towards short-term investment options and the non-marginal impacts of long-term policy decisions, such as policies related to climate change mitigation.

In defending Stern's choice of a low discount rate, Quiggin (2008) raises the *equity premium puzzle* and the concept of *hyperbolic discounting* to further invalidate the descriptive approach to discounting. The equity premium puzzle refers to the notion that, for plausible values of η (from the Ramsey formula), the difference between the real bond rate and the rate of return on equities or stocks should be no more than half a percentage point. However, historically, annual real returns on US government bonds have been around 1 per cent whilst annual returns on US stocks have averaged around 7 per cent. Hence, the real bond rate should be higher than what it is and the rate of return on equities should be much lower (Mehra and Prescott 1985). Ultimately, the large discrepancy between bond and equity rates can only be resolved by assuming the presence of a large risk premium, which means having to assume a high aversion to risk and therefore very high values for η. Yet, as Quiggin stresses, a high value for η implies a high discount rate, which only exacerbates the equity premium puzzle.

As for hyperbolic discounting, Quiggin (2008) points to a substantial body of literature on expected utility theory and behavioural economics which consistently reveals that people apply a high discount rate to trades between the present and the near future, but a low discount rate for trades between the near and the far future. Furthermore, it has been shown that an individual's choice of discount rate is also affected by whether the trade involves something with minimal spillover impacts or something with potentially pervasive consequences. Invariably, individuals attach a lower discount rate to circumstances involving potentially widespread, risky, and deleterious impacts, such as those expected from climate change (e.g., Kahneman and Tversky 1979). Taken together, the failure of researchers to account for hyperbolic discounting and the equity premium puzzle, which has prompted the use of discount rates equal to average rates of return on capital rather than a riskless bond rate between 1 and 2 per cent, has led to untenable conclusions and undesirable policy prescriptions.

I earlier referred to the implications of discount rates on the assumed savings rate of an optimal consumption trajectory (Eq. 5.3). One of the mainstream criticisms of Stern's 1.4 per cent discount rate was that it assumes a ludicrously high savings rate of over 90 per cent. To recall, savings rates in most countries are in the vicinity of 15 per cent. DeLong (2006) has come to the aid of Stern by demonstrating that the derivation of a very high savings rate is grounded on the implausible assumption that there is no future technological progress. Should technological progress continue at current rates, as DeLong and others predict, Stern's discount rate implies an entirely believable savings rate and one wholly consistent with a 1.3 per cent growth rate in per capita consumption.

Whilst on rates of return and discount rates, I might take this opportunity to question whether real rates of return on investment capital have actually been in the 4–6 per cent range, as is generally assumed. It has already been suggested that rates of return of around 4–6 per cent may not be achievable in a world afflicted by extreme climate change. It is my view that a significant proportion of the seemingly healthy rates of return has been due to the unsustainable conversion of natural capital to human-made capital. To recall from Chap. 2, income is properly defined as the maximum amount that can be consumed in the present without compromising the ability to consume the same amount in the future (Hicks 1946). Income is of relevance here because real rates of return on capital depend on the income streams generated by the capital stock. Should income streams of the past have been the result of an unsustainable conversion of natural capital into human-made capital, then, from a strong sustainability perspective, income steams have been overstated. Likewise, perceived rates of return have exceeded genuine rates of return.

Indeed, given that low-entropy matter-energy is the only true input of the economic process, it follows that real rates of return on human-made capital are entirely dependent on the rate at which human-made capital can generate real goods and services from a sustainable flow of natural resources. This means that real rates of return are equivalent to the regeneration rates of renewable natural capital. These are typically in the range of 1–2 per cent (Lawn 2007).[26] Thus, from a strong sustainability perspective, it could be argued that a descriptive approach to discounting would still lead to the application of a Stern-like discount rate.

Finally, I believe that Tol of Yohe's (2007b) claim that Stern's recommended climate change strategy is sub-optimal is the result of their misunderstanding of Stern's explanation of the changing relationship between the marginal cost and average cost of mitigation. To be fair to Tol and Yohe, I believe their confusion may have arisen from the less-than-adequate explanation of the relationship between the two classes of costs in Box 9.6 of the *Stern Review* (Stern 2007).

To remind the reader once more, Tol and Yohe have argued that an optimal mitigation strategy requires the marginal costs of mitigation to be rising over time as mitigation measures are taken to reduce greenhouse gas emissions. Whilst Stern (2007) stresses the need for marginal mitigation costs to rise in order to reflect the increase in damages caused by a still-rising atmospheric concentration of greenhouse gases, Stern argues that average mitigation costs should be falling. According to Stern, the gradual decline in average mitigation costs is likely to result from technological innovation, learning-by-doing, and the generation over time of larger mitigation-related co-benefits.

Tol and Yohe's confusion, it seems, stems from Stern's claim that it is possible for average mitigation costs to be falling at the same time marginal mitigation costs are rising when the latter are higher than the former. In line with textbook theory, Tol and Yohe (2007b) correctly point out that when marginal costs exceed average costs, rising marginal costs typically imply rising average costs. Figure 5.3 presents a textbook view of the relationship between marginal cost (MC) and average cost (AC). It shows that the MC curve always cuts a U-shaped AC curve at its minimum point. Consequently, when marginal costs are greater than average costs (i.e., when $Q > Q_0$), average costs must be rising as Q increases—the exact opposite of what Stern believes should be happening to average mitigation costs as mitigation measures are ramped up.

To explain how Stern may have reached his conclusion, Tol and Yohe (2007b) run a number of simple thought-experiments. In the end, the only way they can decipher a coherent story from Stern's reasoning is to presume that the short-run mitigation cost curve must be getting steeper over time. Yet, as Tol and Yohe explain, and as Stern argues in Fig. 2.2 of the *Stern Review* (Stern 2007), the short-run mitigation cost curve is likely to shift downwards or become flatter over time. In all, Tol and Yohe believe that Stern's reasoning is contradictory. Moreover, given Stern's assumptions, Tol and Yohe believe it is impossible for a decline in average mitigation costs to be accompanied by a rise over time in marginal mitigation costs.

I disagree with Tol and Yohe. May I say, I believe there other good reasons why the average cost of mitigation could rise over time, which I shall leave to

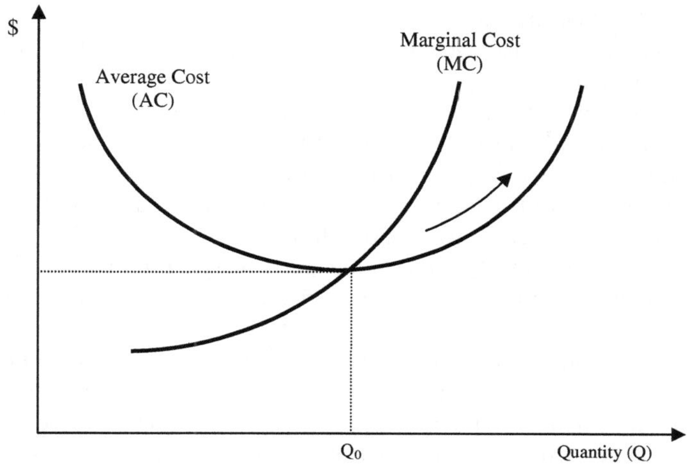

Fig. 5.3 Textbook view of the relationship between average cost (AC) and marginal cost (MC)

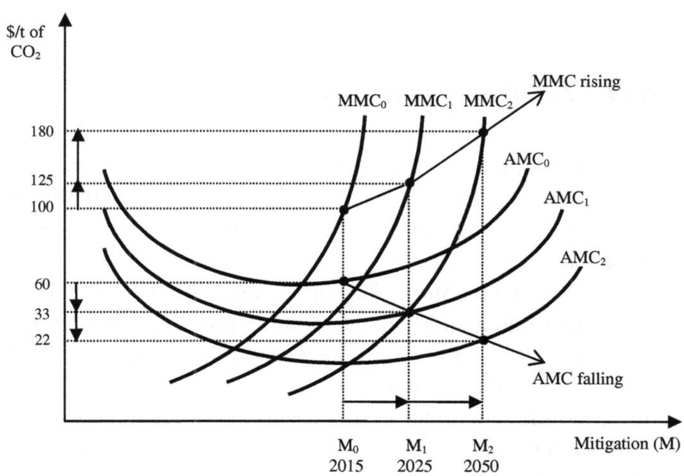

Fig. 5.4 Possible change in marginal mitigation costs (MMC) and average mitigation costs (AMC) as the level of greenhouse gas mitigation rises

Chap. 6. For now, if we assume that technological progress can progressively shift the Marginal Mitigation Cost (MMC) curve and Average Mitigation Cost (AMC) curve downwards, it is entirely possible for marginal costs to rise as average costs decline. Consider a scenario depicted in Fig. 5.4 based on changes in marginal and average mitigation costs anticipated by Anderson (2006). We shall assume that MMC_0 and AMC_0 represent the respective Marginal and Average Mitigation Cost curves in 2015.[27] With the level of mitigation in 2015 at M_0, marginal mitigation costs are higher than average mitigation costs (Note: it has been estimated by Anderson that marginal and average mitigation costs in 2015 will be around \$100 and \$60 per tonne of CO_2 respectively).[28]

Assuming that the Marginal Mitigation Cost curve and the Average Mitigation Cost curve shift downwards to MMC_1 and AMC_1 by 2025, and the level of climate change mitigation rises to M_1, Anderson expects marginal mitigation costs to rise to \$125 per tonne of CO_2 and average mitigation costs to fall to \$33 per tonne of CO_2. Finally, by 2050, as the two curves shift to MMC_2 and AMC_2, and the level of mitigation increases to M_2, Anderson expects marginal mitigation costs to rise to \$180 per tonne of CO_2. Over the same 25-year period, Anderson expects average mitigation costs to fall to \$22 per tonne of CO_2.

Of course, whether these changes in mitigation costs eventuate will depend on the relative shifts of the Marginal Mitigation Cost and Average Mitigation Cost curves plus the extent to which climate change mitigation measures are implemented over time. Provided technological progress can sufficiently shift the two curves downwards, a contemporaneous rise in marginal mitigation costs and decline in average mitigation costs, as depicted in Fig. 5.4, are theoretically possible, thus supporting the viability of Stern's argument.

5.5.2 Further Criticisms of the Stern Review

Mainstream economists are not the only group of economists critical of the *Stern Review*. Economists from other schools of thought have also highlighted what they consider to be its major shortcomings, although these are often unrelated to the criticisms outlined in Sect. 5.3.

In a major attack on the *Stern Review*, Spash (2007) asks how Stern can acknowledge the many intractable climate change problems that render a benefit-cost analysis unsuitable for generating policy recommendations, yet still prescribe a greenhouse mitigation strategy based, to a significant degree, on the results generated from an economic modelling exercise. Instead of dealing with the problems appropriately, Spash believes that Stern has sidelined them by crafting what many perceive is a 'state-of-the-art' benefit-cost approach. Although few people would doubt that Stern's economic analysis includes a number of novel innovations—e.g., endogenous discounting, an ethical basis for risk aversion, and the inclusion of extreme climate change possibilities—Spash argues that Stern has maintained an allegiance with mainstream economic orthodoxy by perpetuating the standard myth that GDP growth is both ecologically sustainable and desirable. Despite what impression Stern gives to the contrary, Spash believes the *Stern Review* can in no way be regarded as an alternative heterodox approach to the climate change crisis.

More specifically, Spash (2007) highlights a number of apparent deficiencies of the *Stern Review*. The first of these is the constant framing of the *Review* in terms of GDP growth. Not only does Stern regard the continued growth of GDP an essential requirement for ongoing increases in human welfare, which Spash hotly disputes (see also Easterlin 1974, 1995; Daly and Cobb 1989; Redefining Progress 1995; Kahneman and Krueger 2006; Lawn and Clarke 2008), Stern also expresses the cost of climate change damages and mitigation in terms of equivalent GDP losses. Yet if GDP growth beyond a certain threshold level no longer increases human well-being (see Figs. 2.7 and 2.8), it is misleading to articulate costs in this manner, since the forgone GDP no longer constitutes a net welfare cost. Furthermore, as Spash reminds us, even in the presence of energy-saving technological progress, GDP growth is highly correlated with energy use and is therefore a significant driver of greenhouse gas emissions. Despite this, Stern, in typical mainstream fashion, assumes that the impact of GDP growth on emissions is unproblematic, claiming instead that global warming is the consequence of an untreated environmental externality.

Secondly, Spash (2007) is critical of the expected-utility analysis used by Stern to conduct his economic modelling of climate change. Although expected-utility analysis is neat, compelling, and tractable, many observers believe it fails to accurately represent typical human behaviour (Perrings 2003; Quiggin 2008). Empirical evidence suggests that individuals apply considerably more weight to low-probability extreme events than that implied by expected-utility analysis (Kahneman and Tversky 1979; Weitzman 2007, 2009).

Thirdly, despite Stern's acknowledgement of the inevitable change in the relative prices of human-made and environmental goods, Stern's analysis is based on comparative statics—in other words, on the shift over time from one equilibrium outcome to another. The problem with comparative statics, according to Spash (2007), is that it conceals complex processes of change, such as the evolution of natural resource and final-goods prices and the allocation of resources that follows. More than this, the use of comparative statics allows Stern to miraculously convert unknown and unknowable future outcomes (ignorance) to events with seemingly known probabilities. In the process, it enables Stern to generate precise climate-change outcomes and cost estimates when it is clear that computer-generated values are seriously incomplete and subject to very wide margins of error (Howarth 2003; McGuire 2006; Spash 2007). I shall return to the issue of ignorance and its implications for the use of benefit-cost analysis in Chap. 6.

Fourthly, Spash (2007) is damning of the way in which Stern 'internalises' catastrophic events when carrying out his economic modelling exercise. In Chap. 3 of the *Stern Review*, Stern emphasises the existence of ecosystem thresholds and the potential for extreme outcomes to emerge should thresholds be exceeded. At the same time, Stern models potentially catastrophic outcomes in Chap. 6 as 'GDP loss events' of known probability with positive and increasing risk. As part of the modelling exercise, Stern assumes that climate change catastrophes are most likely to occur when global temperatures rise by 5 °C above pre-industrial levels.[29] In doing so, Stern assumes that the catastrophic threshold is as much as 3 °C above the critical threshold level assumed by many climatologists (i.e., 3 °C above a 'safe' temperature rise of 2 °C). According to Spash, these assumptions give the impression that 'moderate' global warming of 3–5 °C is acceptable even if very costly in terms of forgone GDP. More critically, it means that Stern is essentially relegating potentially irreversible and surprising disasters to the status of known, bounded, threshold events expressed merely as GDP losses. Yet, as Spash (2007) highlights, catastrophic events, which could be triggered by average temperature rises of 3–5 °C, are inherently surrounded in unknowns and do not fit within normal probability density functions.

Fifthly, Spash believes that most rational and morally-driven individuals, if informed about the potential implications of exceeding a catastrophic temperature threshold, would avoid catastrophic events at almost any cost. Consequently, Spash contends, along with Perrings (2003), Baer (2007), and Ackerman (2007), that an appropriate policy response to the climate change crisis requires the adoption of the 'precautionary principle', which implies a climate change policy explicitly designed to prevent the atmospheric concentration of greenhouse gases exceeding a safe upper limit, such as the CO_2-e level of 450 ppm recommended in Chap. 1. By failing to rule out any greenhouse concentration level, Stern's modelling exercise excludes any consideration of safe upper limits, which is best exemplified by Stern's recommended target of up to 550 ppm—a level many climatologists regard as 'dangerously' high.

Finally, despite Stern's claims to the contrary, Spash (2007) believes that the moral framework employed by Stern is seriously deficient. Spash's criticism begins with Stern's conversion of stated moral theories into a concern for three objects of desire most likely to be affected by climate change—human health, environmental quality, and income/consumption (Stern 2007). Although Stern invites the reader to view this approach with great circumspection, the manner in which Stern treats the three categories of desire assumes they are commensurable. For example, it is implicitly assumed that the benefits of increased consumption can offset the cost of having to relocate people *en masse*. Worse still, the increase in consumption deemed necessary to compensate a person expected to be displaced in an impoverished nation is assumed to be less than the increase in consumption required to compensate a relocated person in a high-GDP country. Rather than explain the ethical basis for this approach, Spash (2007) believes that Stern has done little more than answer these and other ethical dilemmas implicitly, thus failing to adequately outline the moral framework used to select the intergenerational objective function employed in the *Stern Review*.

May I say, I don't entirely agree with Spash's final point. As previously revealed, Stern has spent considerable time and effort discussing the ethical dimensions of the climate change problem (e.g., the need to adopt a James Meade-like policy philosophy) and has warned about the dangers of elevating economic analyses above ethical concerns. Having said this, Spash (2002) is correct to point out that the commensurability assumed in many economic studies of climate change has resulted in such absurdities as equating the value of greater recreational benefits in the USA with the loss of human life in low-GDP countries. Although Stern does not explicitly make this assumption, and should be given credit for the low value used to represent the aversion to economic inequality among different generations (i.e., $\eta = 1$ in the Ramsey equation), Spash (2007) believes this does little to address the commensurability predicament inherent in his analysis.

The intractable problem with incommensurability is that it raises issues that cannot be resolved through the use of economic analyses alone (Martinez-Alier et al. 1998; Aldred 2002). This leads to the conclusion that only non-economic tools of analysis can fully recognise the ethical and ecological values that must be taken into account to determine and ultimately prescribe an emissions trajectory capable of averting a potential ecological catastrophe (Spash 2007; Baer 2007).

Along similar lines to Spash's incommensurability concerns, Neumayer (2007) argues that the *Stern Review* fails to deal with the most pertinent of all climate change issues—specifically, to what extent should greenhouse gas emissions be reduced to prevent global warming exacting an irreversible loss of non-substitutable natural capital. By 'non-substitutable', Neumayer implies instances where the loss of natural capital cannot be offset by increases in the consumption of manufactured goods and services. Seen in this light, Neumayer (1999b, 2003, 2007) believes that debates surrounding appropriate discount rates are misguided given there is no point in comparing the net present value of different mitigation strategies should many involve the irrecoverable loss of *critical* natural capital.

In what essentially amounts to a 'strong sustainability' view of the climate change issue, Neumayer (2007) believes the justification for immediate and decisive mitigation measures requires a convincing case explaining why the current generation should undertake costly action to avoid climate change when the ensuing damages will predominantly affect consumption-richer future generations. According to Neumayer, such a case can be argued on the basis of one of two approaches to the non-substitutability issue. The first is a utility-based perspective that involves embedding the disutility of climate change damages in the utility or welfare function of society. This achieved, it is then necessary to demonstrate that the disutility inflicted by excessive climate change is likely to become so great that no amount of consumption growth can prevent a decline in the net welfare of future generations.

The second line of reasoning involves the abandonment of utilitarianism and an appeal to a deontological or rights-based approach. In this instance, it is necessary to argue that the actions of the current generation should be based on a moral obligation to respect the right of future generations to enjoy a climate system as comfortable and amenable to life as it is at present.[30] According to Neumayer (2007), the successful prosecution of such a case would require a rigorous demonstration that global warming beyond a critical temperature threshold would violate this inalienable right, in which case the current generation would be duty bound to take the measures necessary to prevent the threshold from being exceeded. Exactly where this temperature threshold exists remains a moot point, although if we take it to be a 'safe' upper limit of global warming as defined by Article 2 of the United Nations Framework Convention on Climate Change, then, as argued in Chap. 1, it would constitute a 2 °C rise and an upper limit of atmospheric greenhouse gases of 450 ppm of CO_2-e.

Irrespective of which approach is adopted, Neumayer (2007) believes they both lead to the same conclusion and therefore an identical mitigation strategy.

Whilst I agree with Neumayer with regards to Stern's failure to adequately deal with the potential loss of non-substitutable natural capital, I don't agree with Neumayer's belief that the utilitarian and rights-based approaches justify similar climate change action. In view of our inability to know the preferences of future generations and a similar inability to accurately estimate the cost of future climate change damages (Funtowicz and Ravetz 1994; Howarth 2003; Ackerman 2007), there is no reason why a utilitarian approach would not lead to a gross underestimation of natural capital depletion costs and a resultant conclusion that mitigation measures need only limit the rise in average global temperatures to, say, 3 °C. If so, this would amount to limiting the atmospheric concentration of greenhouse gases to a dangerously high 550 ppm of CO_2-e (IPCC 2007b), which would constitute a violation of the rights-based approach to climate change mitigation.

Continuing on with the non-substitutability theme, Sterner and Persson (2008) believe that Stern's recommended mitigation strategy, or one more stringent, can be supported by a strong conviction that the cost of future climate change damages will be much higher than currently anticipated. The reason why Sterner and Persson believe damage costs will be higher than expected is that climate change will significantly reduce humankind's access to environmental goods and services. In doing so, climate change will increase the relative price of the category of environmental amenities that are crucial to human welfare and are largely non-substitutable. Compounding this problem is that even greater price pressures are likely to emerge in the future given that the goods and services provided by natural capital are particularly vulnerable to non-market damages. Despite this, many of these damages are completely overlooked in cost-projection studies.[31]

To reinforce their point, Sterner and Persson (2008) have incorporated into a benefit-cost analysis the relative price changes they envisage occurring as the scarcity of environmental goods and services gradually increases. What they find is that damage costs are not only higher than anticipated, but the additional cost outweighs almost any cost disparities that arise from the application of different discount rates. Indeed, so much higher do Sterner and Persson expect future damage costs to be, they believe that greenhouse gas targets of 450 ppm of CO_2-e and less can be justified on pure economic grounds.

With this in mind, Sterner and Persson (2008) are critical of the *Stern Review* for two reasons. Firstly, they argue that the *Review* gives insufficient weight to many non-market damages—a weakness readily conceded by Stern.[32] Secondly, although Stern raises the non-substitutability issue and expresses concern about the impact of climate change on relative prices, Sterner and Persson argue that the economic analysis conducted in the *Review* fails to analyse the effect of changing relative prices on the composition of future output and its consequential impact on future welfare. Furthermore, despite the fact that Stern downplays the primacy of his economic analysis, Sterner and Persson believe that Stern's failure to fully grasp the impact of climate change on relative prices has tainted the *Review's* final conclusions and recommendations.

I believe Sterner and Persson are entirely correct in their assessment of relative price changes and their future impact on climate change damage costs. However, it is my contention that the impact of price changes will have more extreme ramifications than Sterner and Persson envisage. In my view, inevitable damages to natural capital caused by climate change together with the likely introduction of policies designed to ensure markets better reflect the true scarcity of the ecosphere's source and sink capacities will for some time dramatically increase the price of all natural resources. Just how much resource prices are likely to rise will, for a given rate of technological progress, depend a lot on the rate of growth in GWP, since a larger growth rate will increase the scarcity of natural resources. Given that natural resources constitute the only true input of the production process, a sharp rise in the prices of natural resources would significantly increase the cost of everything. By 'everything', I mean not only the cost of supplying consumption goods, but also the cost of providing throughput-reducing and low carbon-emitting capital/infrastructure plus the cost of all prospective climate change mitigation and adaptation projects.

The reason why I believe Sterner and Persson underestimate the impact of climate change on future costs is that they view the potential damage to natural capital from a very narrow perspective. They do this in two related ways. Firstly, they assume that climate change will merely reduce our direct consumption of environmental goods and services. Secondly, Sterner and Persson (2008) do little more than question whether the reduced consumption of environmental goods and services can be adequately compensated by increases in the consumption of human-made goods and services (i.e., whether an extra $1 of human-made goods and services can compensate for a $1 loss of goods and services directly provided by the ecosphere). Crucially, Sterner and Persson ask this question on the further assumption that the aggregate quantity of human-made goods and services available for consumption can be increased indefinitely.

On what basis do Sterner and Persson make this latter assumption? Although Sterner and Persson (2008) recognise the limits imposed by non-substitutable natural capital on *physical* production possibilities, they argue that rising consumption levels can continue to be enjoyed so long as the composition of human-made output is shifted from goods to services. This 'dematerialisation' argument employed by Sterner and Persson is based on the widely held view that, unlike physical goods, the supply of services requires the use of few if any natural resources. This is a fallacy. Services, or what essentially constitutes the output of the tertiary sector of a nation's economy, have a considerable physical dimension. Unbeknown to many, the tertiary sector does not exist independently of the primary and secondary sectors. Indeed, the output of the secondary sector (manufacturing), which is made possible by the output of the primary sector (agriculture and resource-extractive industries), effectively constitutes the input of the tertiary sector. Consequently, the resource demands of the primary and secondary sectors are a subset of the total resource demands of the tertiary sector, thus suggesting that the tertiary sector is virtually as resource-intensive as any other (Costanza 1980; Ayres and Ayres 1999). Clearly, any attempt to increase consumption levels, irrespective of its composition, will almost certainly place greater pressure on resource prices and increase the future cost of climate change.

What's more, the biophysical limits imposed on all sectors of the economy casts serious doubt over whether future generations will, in a severe climate change-affected world, be as consumption-rich as many expect. Therefore, one must not only question whether much higher consumption levels can fully compensate for the loss of environmental goods and services, but whether there will be enough human-made goods to provide the bare minimum level of compensation. Based on the simulation exercises conducted in Chap. 4, where it is doubtful that growth-as-usual output levels can coexist with a 'safe' greenhouse gas target of 450 ppm of CO_2-e, it would appear not.

The final criticism of the *Stern Review* I would like to bring to attention is one raised by Baer (2007). Baer believes the *Review* is a highly political document in the sense that its recommendations are largely driven by the need to appease various 'audiences'. For example, given the dominant view that continued GDP growth is a development requirement, Baer argues that Stern's conclusion that GDP growth will not be seriously impaired by climate change mitigation measures was an expedient attempt to avert any widespread rejection of the *Review's* recommendations. Moreover, Baer believes that Stern's dismissal of a greenhouse gas target below 450 ppm is at odds with many of the crucial premises outlined in the *Review* that support an extremely stringent emissions target.

I agree with Baer and believe that Stern has contradicted himself in the *Stern Review* and in many of the climate change papers and books he has since had published. For example, as indicated earlier, Stern believes that the results of economic modelling should not be given precedence when making climate change recommendations. Given Stern's propensity to elevate the importance of risk and his recognition of the enormous risks associated with a greenhouse gas concentration above 450 ppm of CO_2-e, one must ask why Stern ruled out a target at or below this level? In the end, one is left believing that Stern came to his conclusions by rejecting his own advice and giving primacy to his economic analysis. Why, otherwise, would Stern contend that a target below 450 ppm is not 'economically' justified, or, equivalently, that avoiding potentially catastrophic climate change is not 'economically' warranted?

Like Baer, I believe that what matters most is our moral obligation to future generations, not whether efforts to ensure a 'safe' greenhouse gas concentration are too costly for the current generation to bear. As stressed, what is urgently needed is a climate change policy premised on the precautionary principle—a principle enshrined in Article 3.3 of the United Nations Framework Convention on Climate Change. Honouring such a principle would require policy measures aimed, first and foremost, at preventing the global concentration of greenhouse gases from exceeding a safe or ecologically sustainable level. Once guaranteed, it would then be acceptable to think in terms of allocative efficiency, which, to achieve, would require separate policy measures to stabilise greenhouse gases somewhere in the vicinity of the most efficient of *all possible safe levels*.[33]

What does this ultimately imply? Assuming that the safe upper level is 450 ppm of CO_2-e, it would mean stabilising at the most efficient greenhouse gas concentration at or below the 450 ppm level.

5.5.3 An Ecological Economics Perspective of the Stern Review

To my knowledge, no official position on the *Stern Review* has ever been formally released by the International Society for Ecological Economics or by a separate collective of ecological economists. What therefore follows is my own interpretation of an ecological economics perspective of the *Stern Review*.[34]

There are many elements of the *Stern Review* that I believe ecological economists would wholeheartedly support. These include Stern's acceptance of the climate-change science; the ethical arguments underpinning Stern's attitude to risk aversion, as reflected by Stern's use of $\eta = 1$ and $\delta = 0.1$ per cent in the Ramsey equation; Stern's exploration and elucidation of various mitigation and adaptation possibilities; Stern's concern for the extreme vulnerability of the world's poor nations and the need for rich countries to assist them; and, finally, the proposals put forward by Stern to facilitate international collective action on climate change.

There are, however, a number of key aspects of the *Review* that ecological economists would be highly critical of. Above all, in view of the primacy afforded by ecological economics to the goal of ecological sustainability, ecological economists would consider Stern's greenhouse stabilisation target of 550 ppm of CO_2-e as being too high. Equally, they would regard Stern's recommended rate of emissions cuts to be insufficiently stringent, although they would agree with Stern that severe emissions cuts are required immediately.

Secondly, ecological economists would be critical of the manner in which the *Stern Review* failed to position the climate change issue within the broader context of sustainable development. By this I mean the failure of the *Stern Review* to: (i) understand the inextricable relationship between sustainable development, GDP growth, and greenhouse gas emissions; (ii) identify the three major policy goals (outlined in Chap. 3) that require adequate resolution to avert catastrophic climate change in the most cost-effective and equitable way; and (iii) recognise the distinct policy instruments that need to be applied to resolve the three sustainable development goals. Had Stern adopted a broader contextual framework based on an ecological economics perspective of sustainable development, it is doubtful he would have contemplated a stabilisation target above 450 ppm of CO_2-e.

As for the mainstream criticisms of the *Stern Review* outlined earlier in the chapter, ecological economists would broadly disagree with the mainstream contention that Stern overestimated the cost of climate change damages. To the contrary, ecological economists would argue that the *Review* grossly underestimated future damage costs. There are basically three reasons why. Firstly, Stern neglected many non-market impacts of climate change and, by his own admission, overlooked many of its social impacts. Secondly, as already explained, Stern did not adequately account for the likely loss of non-substitutable natural capital, both in terms of its direct impact on the consumption of environmental goods and services, and its indirect impact on production possibilities. Thirdly, in view of the second reason, Stern overlooked the full effect of relative price changes on the future cost of having to use much scarcer and poorer quality natural resources. For the same reasons, ecological economists would also argue that Stern underestimated the cost of climate change mitigation, particularly given his assumed rate of GWP growth.

Ecological economists would also go much further than mainstream economists in their criticism of Stern's estimation of damage costs. In keeping with Spash's (2007) concerns, ecological economists would disapprove of the manner in which Stern calculated damage costs in terms of GDP losses, particularly given that the approach is based on the false assumptions that continued growth is desirable and sustainable, and that the nexus between real GDP and greenhouse gas emissions can in essence be severed. In response, ecological economists would insist on a different means of estimating the cost of climate change that would consequently lead to a different representation of the benefits of climate change mitigation.[35]

Although ecological economists would be less critical of Stern's use of a low discount rate—in particular, Stern's use of an endogenous discount rate—they would be quick to point out that matters concerning the most appropriate discount rate are largely irrelevant in terms of climate change policy. This is because a discount rate is a form of 'price'. A price essentially represents what an individual or society is willing to forego in order to obtain something else. For example, if having 'A' means having to forego 'B', then B constitutes the price of choosing A. In an inter-temporal sense, a discount rate represents how much the current generation is willing to forego a future benefit in order to receive a benefit now. As a means of illustration, a 10 per cent annual discount rate implies that, in a year's time, the current generation would need to receive a benefit that is 10 per cent more than a benefit it could receive today to be indifferent between the two benefits on offer (i.e., the current generation would be indifferent between \$100 offered now and \$110 offered in 12 months' time). Thus, in a very real sense, a discount rate constitutes an 'inter-temporal' price. Moreover, the net present value calculated via the application of a particular discount rate is nothing more than an inter-temporal efficiency quotient.

What is the significance of this? In Chap. 3, it was explained that allocative efficiency does not guarantee ecological sustainability. Since a low discount rate does little more than generate a different efficiency quotient than a high discount rate, the application of a low discount rate cannot guarantee a 'sustainable' outcome. For example, while low discount rates promote environmentally-benign activities, they also encourage a greater volume of activities. Should the 'scale' effect of more projects exceed the 'low impact' benefit of having a greater proportion of environmentally-benign activities, it is conceivable for low discount rates to have a negative net impact on the natural environment—a case of the Jevons' Paradox explained in Chap. 3. It is not surprising, therefore, that almost all benefit-cost studies of the climate change problem yield stabilisation targets that, regardless of the discount rates used, carry unacceptably high probabilities of inducing catastrophic global warming.[36]

In terms of the benefit-cost analysis conducted in the *Stern Review*, ecological economists would concur with the mainstream criticism that Stern's analysis involved no attempt to equate the marginal benefits of climate change mitigation with the marginal damage costs of global warming. That said, ecological economists would draw attention to the fact that Stern's use of an endogenous discount rate (i.e., one that reflects the per capita consumption associated the emissions trajectory under investigation) is a major improvement on most standard approaches. However, once again, ecological economists would stress that the use of an endogenous discount rate is largely irrelevant given that a benefit-cost analysis cannot guarantee a greenhouse gas strategy that is ecologically sustainable.

Worse still, as we shall see in Chap. 6, a benefit-cost analysis is incapable of determining the most efficient emissions trajectory. This is because future climate change outcomes are unknowable—in other words, our understanding of future climate change outcomes is plagued by 'ignorance' rather than by 'risk and uncertainty'. Also unknowable are future rates of technological progress; the co-benefits of the mitigation measures likely to be undertaken to reduce greenhouse gas emissions; future growth rates of real GDP; and eventual changes in resource scarcity and resource prices, all of which are likely to affect future emissions levels and our capacity to respond to the climate change crisis. As a consequence, damage and mitigation costs cannot be estimated by climate change models with any degree of precision. Complicating matters further is the fact that, as explained earlier in the chapter, the positions of the Total Damage Cost, Marginal Damage Cost, and Total Emissions Cost curves depend on future emissions levels as well as past and present emissions. Thus, an absurd situation exists where determining an efficient emissions pathway requires having to make an initial assumption on what the pathway is. I shall return to this illogical circular reasoning in Chap. 6.

Given the above concerns, ecological economists believe that a least-cost greenhouse target generated by a benefit-cost analysis is unlikely to lie within the range of 'safe' atmospheric concentration levels (i.e., one that is at or below 450 ppm of CO_2-e). Ecological economists therefore reject the use of a benefit-cost analysis as a dominant form of climate change investigation. Despite this, ecological economists believe that much can be learned from closely examining the likely source and evolution of climate change damage costs and mitigation costs. For this reason, considerations of costs will be comprehensively explored in the next chapter, including an explanation as to how future changes in costs are likely to affect mitigation responses. Also to be explained in greater detail will be the arguments underpinning the inherent shortcomings of benefit-cost analyses. By the end of Chap. 6, it should be abundantly clear why resolving the climate change crisis will require the atmospheric concentration of greenhouse gases to be restricted to a range of safe levels prior to permitting an adjustment—essentially through the agency of a market mechanism—to a concentration level that is economically efficient (i.e., one that contributes to an optimal scale of economic activity).

Notes

1. By negative impacts, mainstream economists essentially mean the detrimental impact of climate change on future consumption together with the cost of adapting to a warmer climate.
2. The damages include some non-market as well as market damages which, in total, would be the equivalent of losing 5–20 per cent of GWP. Importantly, it has been estimated that the damages would be disproportionately borne across the planet with many of the world's poorest nations suffering the most.
3. The details of PAGE2002 are described in Chap. 6 of the *Stern Review*.
4. The 14.4 per cent reduction in per capita consumption equivalents is surrounded by a 5th and 95th percentile range from 2.7 per cent to 32.6 per cent. It is based on a high climate-change scenario (Stern 2007, Table 6.1, p. 163).
5. This conversion is based on one unit of carbon being equivalent to 3.67 units of CO_2.
6. See Quiggin (2008) for an explanation of the logic behind the Ramsey Equation.
7. This is based on the assumption that the relative difference in the utility enjoyed by persons A and B upon them both receiving an extra dollar for spending purposes is represented by $\Delta U_A/\Delta U_B = 1/(\text{Income}_A/\text{Income}_B)^{\eta}$. As a means of comparison, if $\eta = 0$, it is assumed that an extra dollar received by both persons will increase their utility by the same amount (e.g., $1/2^0 = 1$). If, however, $\eta = 2$, it is assumed that the utility of B will increase four times as much as the utility of A (e.g., $1/2^2 = 1/4$).
8. The 1.3 per cent value is based on different GWP projections than the projections used in the growth-as-usual simulation in Chap. 4 (Table 4.3).
9. As uneasy as Tol and Yohe (2006a) are with the numbers produced by Stern, they admit to being more uneasy with their perceived view that the numbers were unsupported by new science and/or analytical techniques.
10. The average social cost of carbon from a survey of 100 estimates in 2005 was US$43 per tonne of carbon (IPCC 2007a).
11. DICE is an acronym for a Dynamic Integrated model of Climate and the Economy.
12. The discount rate of 6 % = 2 % + [2 × 2 %] and the savings rate of 33 % = [6 %−2 %]/12 %.
13. The discount rate of 4.3 % = 3 % + [1 × 1.3 %] and the savings rate of 30 % = [4.3 %−3 %]/4.3 %.
14. The discount rate of 4 % = 0.1 % + [3 × 1.3 %] and the savings rate of 32.5 % = [4 %−1.3 %]/12 %.
15. Marginal damage costs equal $d\text{TDC}/d\text{E}$, where TDC denotes total damage costs and E denotes tonnes of greenhouse gas emissions. Marginal mitigation costs equal $d\text{TDC}/d\text{M}$, where TDC denotes total damage costs and M denotes tonnes of reduced greenhouse gas emissions.

16. In reality, global temperatures will not stabilise immediately upon the stabilisation of greenhouse gases. The assumption that they do is for diagrammatic simplicity only.
17. To recall, a specific stabilisation target can be achieved in a variety of ways—for example: (i) through deep, early, emissions cuts and then a more gradual reduction in emissions; and (ii) minor, initial, emissions cuts followed by drastic reductions in later years. In the second example, it is technically possible to overshoot a particular stabilisation target and then dramatically reduce greenhouse gas concentrations to the desired target.
18. A large team was assembled to undertake the climate change modelling conducted in the *Stern Review*. For the remainder of this chapter, references to 'Stern' will imply references to any paper defending the *Review* published by a member or members of the Stern team.
19. Because of this, most IAMs are based on the 'weak sustainability' position outlined in Chap. 2.
20. The Monte Carlo procedure involves running a particular scenario many times over (e.g., 1000 times), where a set of uncertain parameters are randomly 'drawn' from a pre-determined range of possible values. In terms of climate change, the range of parameter values are, or at least should be, calibrated in accordance with the latest scientific and economic literature on climate change. The benefit of a Monte Carlo procedure is that it generates a probability distribution of results rather than a single point estimate of damage costs.
21. This is evidenced by the support given to Stern in the *IPCC Fourth Assessment Report* (IPCC 2007c).
22. To recall, Stern's choice of $\delta = 0.1$ per cent rather than zero reflects the small probability that an unavoidable catastrophe could abruptly end humankind's existence.
23. See, also, Robinson (1962), Easterlin (1974, 1995), and Daly (1991).
24. Average mitigation costs equal TMC/M, where TMC denotes total mitigation costs and M denotes the total quantity of emissions reductions.
25. Taken from 'PDF file of comments on the Stern Review by leading economists', http://www.hm-treasury.gov.au/media/9F3/38/20061028_Quotes-7.pdf. These four selected economists are all Nobel Laureates.
26. Of course, human-made capital presumably adds value to the incoming resource flow. Hence, *nominal* rates of return on human-made capital may be much higher than 1–2 per cent. But once the price-increasing effect of any value added is eliminated in the calculation of *real* income, real rates of return to human-made capital more closely resemble the real rates of return on natural capital.
27. These cost changes are based on reducing fossil fuel emissions to 18 $GtCO_2$ by 2050.
28. The costs presented here are based on US dollars.
29. A similar assumption has been made by Nordhaus and Boyer (2000).

30. This would amount to a climate change system that offers future generations with the same opportunities that are currently enjoyed by existing people.
31. The non-market impacts of climate change can take the form of biodiversity loss; ecosystem destruction; natural resource depletion; direct impacts on human well-being (e.g., illness, death, and loss of human amenities); the impact of natural disasters and extreme weather events; and socially contingent consequences, such as forced relocation, social unrest, and wars brought on by the international competition for increasingly scarce resources. Many of the non-market impacts expected to affect different regions of the world are outlined in Table 1.4.
32. Sterner and Persson (2008) believe that Stern has done an excellent job of presenting the probable non-market impacts of climate change (e.g., Stern 2007, Table 3.1, pp. 66–67). However, they argue that Stern has given insufficient weight to non-market damages in his overall analysis. In the *Review* itself, Stern admits that efforts to incorporate non-market damages into the PAGE2002 model were seriously incomplete (Stern 2007, pp. 164–173).
33. This would ensure that the goal of allocative efficiency remained subordinate to the primary goal of ecological sustainability.
34. Many of the viewpoints and criticisms outlined in this sub-section were expressed in the previous (Sect. 5.5.2). This simply indicates that many of the criticisms outlined were expressed by ecological economists.
35. To recall, the benefits of mitigation are equivalent to the value of avoided climate change damages.
36. Many people believe that low discount rates are always better for the natural environment than high discount rates. There are two main factors behind this. Firstly, most environmental damages usually emerge well after an environmentally-damaging activity has taken place or has been initiated. Secondly, because of the first factor, future environmental costs are not heavily discounted when low discount rates are applied. Consequently, low discount rates, by generating net present values that are lower for environmentally-damaging activities than for environmentally-benign activities, discourage the uptake of the former.

    However, what is often overlooked is that low discount rates render more investment projects economically viable. This is because an investment project will generally proceed if the rate of return on the project exceeds the discount rate applied. Thus, low discount rates encourage the uptake of more activities. In the end, the net environmental impact depends on whether the 'scale' effect of more projects is larger or smaller than the benefit of having fewer 'high-impact' activities. Ultimately, achieving ecological sustainability boils down to explicitly limiting the rate of resource throughput—via throughput restrictions—so that the rate of throughput does not exceed the ecosphere's sustainable carrying capacity. If done, the choice of discount rate would only determine the extent to which the sustainable incoming resource flow is consumed or invested.

Chapter 6
An Ecological Economic Perspective of the Climate Change Crisis

6.1 Introduction

The previous chapter concluded with the recognition—frequently highlighted by ecological economists—that the use of a benefit-cost analysis to determine the most desirable emissions pathway is rendered ineffectual by the inherent inability to accurately estimate all climate change-related costs. Notwithstanding this, it was suggested that much can be gained from examining the factors affecting these costs and by considering how they are likely to vary over time. As we shall see, an examination of climate change-related costs can provide valuable insights into probable mitigation responses and, in the event that a global emissions-trading system is introduced, how nations might adjust to a gradual tightening of emissions caps.

Before I proceed, it is worth pointing out that the views expressed in this chapter are not based on any climate change modelling on my part, except for some inferences drawn from the simulation exercises conducted in Chap. 4. As such, I will not be attaching monetary values to my predictions. I will simply be declaring what I anticipate climate change costs will be in relation to the cost estimates generated in the past by climate change modellers (i.e., whether they will be higher or lower) and whether I expect these costs to rise or fall over time, and why. What I ultimately aim to show is that many factors affecting climate change-related costs have been seriously overlooked and that these oversights have grossly distorted cost assessments and recommended emissions trajectories.

P. Lawn, *Resolving the Climate Change Crisis*, DOI 10.1007/978-94-017-7502-1_6

6.2 Measuring Climate Change Damage Costs and Mitigation Costs

In this first section of the chapter, a close examination will be made of climate change damage costs and mitigation costs as well as the manner in which they are estimated. To do this, it is expedient to consider whether it is better to embark on climate change mitigation measures; adapt to climate change; or undertake a mixture of both (Henson 2011, p. 322). We have already seen that avoiding catastrophic climate change will require the implementation of mitigation measures that stabilise the atmospheric concentration of greenhouse gases at no more than 450 ppm of CO_2-e. At the same time, adaptation measures are needed if only because detrimental climate change impacts can be expected regardless of what action is taken to curb human-induced climate change.[1] Clearly, without the implementation of vital adaptation measures, future human development will be gravely compromised, especially in poor nations where development in the broadest sense of the word is most needed. Unquestionably, then, it will be necessary to embark on a desirable mix of mitigation and adaptation strategies—a mix that will depend upon a wide range of factors, some of which have been seriously overlooked in much of the climate change literature.

6.2.1 The Total Cost of Climate Change Damages

The total cost of climate change damages includes all the costs associated with the climate change arising from a human-induced increase in the atmospheric concentration of greenhouse gases.[2] It therefore comprises the costs directly and indirectly pertaining to the detrimental impacts of human-induced climate change plus the costs associated with humankind's response to current and expected changes in the Earth's climate. It does not include the costs associated with the measures needed to limit human-induced changes to the Earth's climate system (i.e., climate change mitigation costs). For this reason, the total cost of climate change damages is essentially the sum of two categories of costs.

The first category comprises the direct and indirect costs of *climate change damages*, such as increasing water shortages; declining agricultural output; ecosystem service losses; dwindling forestry and fishery stocks; coastline losses and seawater inundations; damage to physical infrastructure; increases in the frequency and severity of natural disasters; and a likely escalation in resource-based conflicts and wars.[3] The second category involves *adaptation costs* and includes such things as better flood control; the establishment of drought-resistant crops; improvements in water, forestry, and fisheries management to minimise the climate change diminutions of natural capital stocks; the construction and/or expansion of dykes and levees; improved monitoring and control of disease outbreaks; and the establishment of heat-health warning systems and natural-disaster action plans.[4] In view of anticipated regional variations in the impact of climate change and the different response capacities of nations (see Chap. 1), the total cost of climate change damages is likely to vary enormously from country to country.

Although adaptation costs contribute to the total cost of climate change damages, it is important to recognise that adaptation measures can reduce total damage costs. This can be demonstrated by way of Figs. 6.1 and 6.2. Consider Fig. 6.1, where it is assumed that, at time t_0, the potential exists to emit greenhouse gases up to a level of G_3'. The horizontal curve of $DC_1(t_0)$ represents the climate change damage costs at t_0 should the exact amount of adaptation be undertaken to keep damage costs constant across a greenhouse gas emissions range of E_0 to E_0'. The adaptation costs associated with these various adaptation levels are represented by the $AC_1(t_0)$ curve. The shape of the $AC_1(t_0)$ curve reflects the fact that adaptation measures are subject to diminishing returns, which means that as emissions levels rise, it becomes increasingly more difficult to prevent damage costs from escalating (Stern 2007).

The reason why the $DC_1(t_0)$ curve terminates at E_0' is because E_0' represents, for given human know-how, the maximum emissions level where climate change damage costs can be kept at DC_1 by way of additional adaptation measures. Thus, beyond an emissions level of E_0', climate change damage costs rise irrespective of how much additional adaptation is undertaken. At this point, the sensible course of action requires the cessation of further adaptation efforts and, painfully, the acceptance of higher climate change damage costs. This is because the cost of the former now exceeds the latter. Assuming this is done, an examination of damage costs and adaptation costs shifts to consideration of $DC_2(t_0)$ and $AC_2(t_0)$. For much

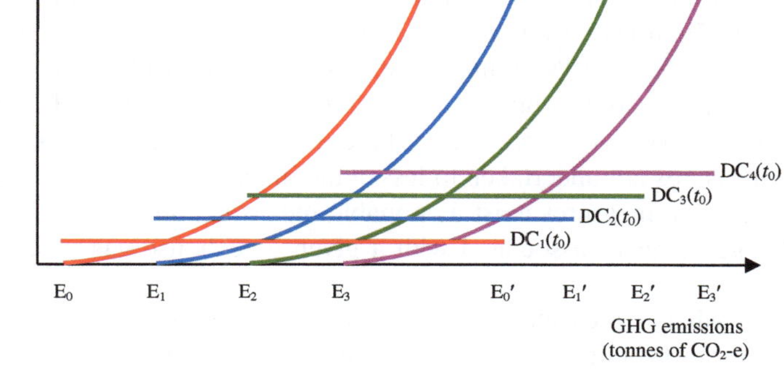

Fig. 6.1 A range of climate change damage costs and adaptation costs for different greenhouse gas emissions levels at time t_0

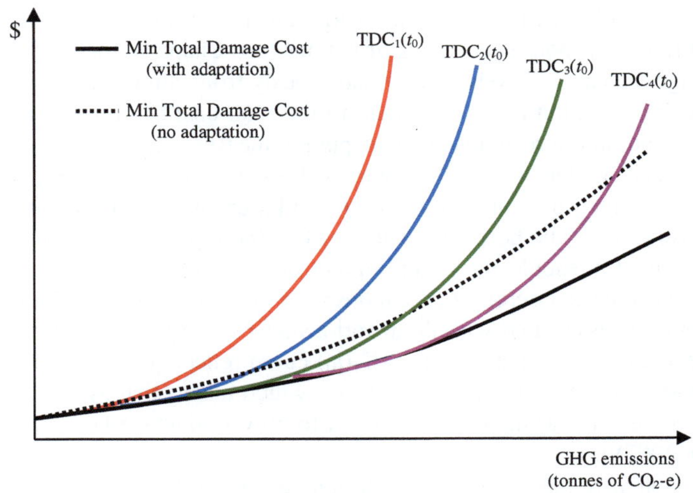

Fig. 6.2 Minimised total damage costs for different emissions levels of greenhouse gases at time t_0 (with and without adaptation)

higher ranges of greenhouse gas emissions, cost considerations move to $DC_3(t_0)$ and $AC_3(t_0)$ and, finally, to $DC_4(t_0)$ and $AC_4(t_0)$.

Because the total cost of climate change damages equals the sum of climate change damage costs and adaptation costs, Fig. 6.2 reveals four Total Damage Cost curves—$TDC_1(t_0)$, $TDC_2(t_0)$, $TDC_3(t_0)$, and $TDC_4(t_0)$. These four curves represent the summation of the four sets of damage cost and adaptation cost curves in Fig. 6.1. If we put climate change mitigation considerations aside for the moment, it is in a nation's best interest to minimise the total cost of climate change damages by engaging in the optimal balance between accepting additional climate change damage costs and attempting to reduce damages costs by way of adaptation. In other words, it is rational for a nation to adopt the adaptation measures necessary at different emissions levels to ensure it operates on the lower envelope of the different Total Damage Cost curves. Operating in this cost-minimising way is represented by the bold 'Minimum Total Damage Cost (with adaptation)' curve in Fig. 6.2. It contrasts sharply with the dotted 'Minimum Total Damage Cost (no adaptation)' curve which represents the cost of climate change damages if no adaptation measures are undertaken to lessen the impact of climate change. As can be seen, adaptation measures exist as a viable means of limiting the total cost of climate change damages associated with a given greenhouse gas emissions level (Stern 2007).

Of course, what is presented in Fig. 6.2 is merely an illustration of how total damage costs can be minimised for different emissions levels given an assumed level of technological progress and human know-how. Improvements over time in adaptation measures through technological advances, learning-by-doing, and economies of scale (i.e., average cost reductions that arise because of the sheer scale of the adaptation measures undertaken) can further limit the damage costs from climate change. However, advances in adaptation can only go so far in lowering damage costs. Indeed, should greenhouse gas emissions continue to rise, no amount of adaptation would be capable of preventing the eventual onset of catastrophic climate change. For this reason, adaptation alone should never be viewed as a means of overcoming the climate change problem. In a roundabout way, this is reflected by the manner in which the 'Minimum Total Damage Cost (with adaptation)' curve in Fig. 6.2 continues to rise as emissions levels increase.

Importantly, the capacity of adaptation measures to reduce the total cost of climate change damages alters what many observers would consider an efficient greenhouse gas emissions level. The mainstream representation of an efficient emissions level was revealed in Chap. 5 (Fig. 5.1). Figure 6.3 illustrates the

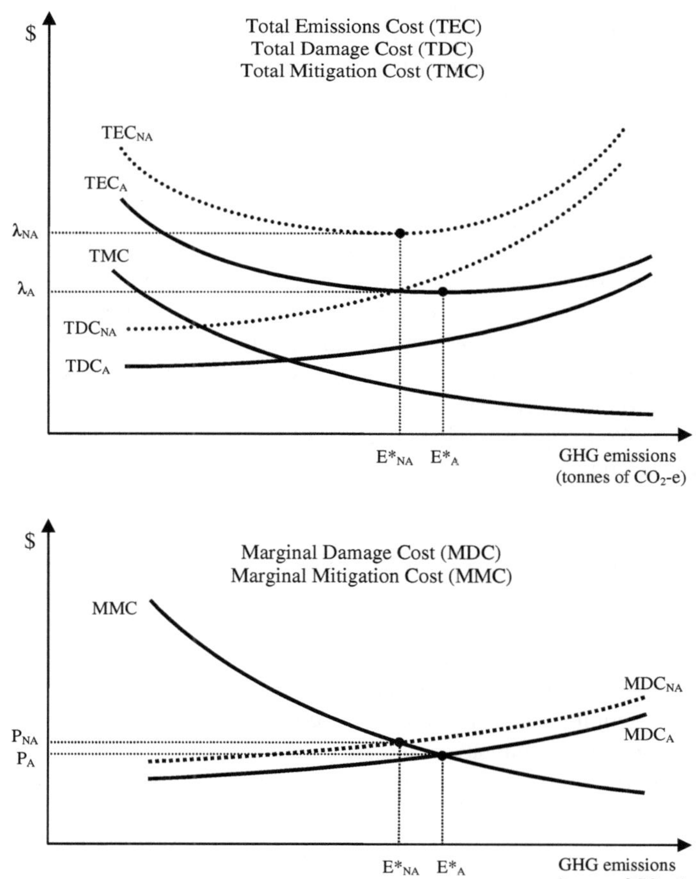

Fig. 6.3 An efficient level of greenhouse gas emissions at time t_0 (with and without adaptation)

difference in an efficient emissions level when adaptation measures have been undertaken and when they have not. The top panel of Fig. 6.3 shows two Total costs where no adaptation measures have been implemented. Conversely, TDC_A represents the total damage costs where adaptation measures have been implemented to keep damage costs to a minimum. Because there are two Total Damage Cost curves, there also happen to be two Total Emissions Cost (TEC) curves. The lower of the two curves (TEC_A) represents the sum total of mitigation and damage costs where adaptation measures have been undertaken. As the top panel shows, the minimum point of the TEC_A curve—which denotes the efficient emissions level—exists to the right of the minimum point of the TEC_{NA} curve. As such, the efficient emissions level in the presence of adaptation (E^*_A) is higher than when there is no adaptation (E^*_{NA}). What's more, the emissions cost associated with this higher efficient emissions level is lower ($\lambda_A < \lambda_{NA}$).

Corresponding to the top panel, the bottom panel reveals that the Marginal Damage Cost curve 'with adaptation' (MDC_A) is lower than the Marginal Damage Cost curve 'without adaptation' (MDC_{NA}). Given that an efficient emissions level is denoted by the intersection of the Marginal Damage Cost and Marginal Mitigation Cost curves, the bottom panel again shows why the efficient emissions level is much higher with adaptation than without ($E^*_A > E^*_{NA}$). The bottom panel also reveals that the efficient price of greenhouse gas emissions—that is, the price per tonne of CO_2-equivalent greenhouse gases—is much lower when adaptation takes place ($P_A < P_{NA}$).

It was pointed out in Chap. 5 that it is nigh impossible to determine the exact positions of the various cost curves from which an efficient emissions level and an efficient price of greenhouse gas emissions can be determined. This raises the question as to why it is worth highlighting the effect of adaptation on an elusive efficient greenhouse gas emissions level. It is worthwhile because, as Fig. 6.3 indicates, adaptation measures can reduce emissions costs and thus reduce the extent to which greenhouse gas emissions need to be cut to achieve a desirable stabilisation target. In saying this, it needs to be stressed that the emissions level must be consistent with a pathway that stabilises the atmospheric concentration of greenhouse gases at no more than 450 ppm of CO_2-e.[5] Thus, in no way am I suggesting that adaptation relieves humankind from the confines of a 450 ppm upper limit. But it may, for example, permit a 430 ppm target that is less costly and more desirable than a 400 ppm target with no adaptation. The fact that the exact positions of the various cost curves cannot be accurately determined, which means that a desirable emissions pathway cannot be ascertained with precision, says something about the shortcomings of conventional benefit-cost analysis, not about a diagrammatic investigation of adaptation and its potential impact on desirable emissions targets. I shall return to this issue later in the chapter.

6.2.2 The Total Cost of Climate Change Mitigation

As already indicated, the total cost of climate change mitigation includes the costs associated with measures undertaken to limit human-induced changes to the Earth's climate system. There are many measures and technologies, both existing and potential, which fall into the mitigation category. They include energy efficiency measures to reduce the rate of fossil fuel use; a general shift from fossil fuels to renewable forms of energy, such as wind, solar, tidal, wave, biomass, and geothermal energy sources; reduced deforestation; reforestation; altered consumption patterns (e.g., the reduced consumption of meat products); increased rates of material recycling; greater product durability; the development and deployment of carbon capture-and-storage technologies; geo-engineering solutions; reductions in real output; and population growth control strategies.

Although some of these mitigation measures and their potential benefits are well understood, others are not. I will therefore briefly outline what some of these measures entail and the benefits they could possibly generate. Carbon capture-and-storage—otherwise known as carbon geosequestration—is designed to prevent fossil fuel-generated CO_2 from entering the Earth's atmosphere. The process involves a number of phases. The first phase entails the capture of CO_2 generated from the burning of fossil fuels. The second phase involves compressing or liquefying the captured CO_2. The final phase involves injecting the compressed or liquefied CO_2 into suitable spaces within the Earth's crust or pumping it directly into deep layers of the world's oceans. Although some observers hold out great hope for carbon capture-and-storage technologies, others are more sceptical.

Geo-engineering technologies come in a variety of forms, each with different aims and purposes. The aim of four of the most commonly proposed technologies is to extract CO_2 from the Earth's atmosphere. The first entails the construction of large devices to facilitate a reaction between CO_2 and lye (a strong alkaline solution) to remove CO_2 from the surrounding air (Keith 2009). The second involves the erection of 'synthetic' trees containing an ion-exchange resin to also remove CO_2 from the atmosphere (Adam 2008; Henson 2011; Markelova 2011). The third approach entails fertilising the ocean with iron filings to promote the rapid growth of CO_2-absorbing phytoplankton. Because dead phytoplankton sinks, it is envisaged that the dead phytoplankton will transport absorbed carbon out of harm's way by sending it to the bottom of the ocean (Buesseler et al. 2004, 2007, 2008; Flannery 2008). The final approach involves the modification of agricultural practices to increase the carbon locked up in soils. As a critical component of the Earth's carbon cycle, soils constitute a huge reservoir of stored carbon. Past agricultural practices have played a significant role in releasing large quantities of carbon into the atmosphere. It is therefore anticipated that a shift towards sustainable agricultural practices can vastly increase the quantity of carbon stored in the Earth's soils (Lehmann 2007; Laird 2008; Fynn et al. 2009).

The remainder of the commonly proposed geo-engineering technologies involves efforts to reduce the solar radiation reaching the Earth. Hence, rather than reduce the global warming potential of rising greenhouse gases, these measures are designed to reduce the warming potential of incoming solar radiation. The most popular measure of this type entails a large-scale release of aerosols (sulphates) into the stratosphere to reflect sunlight (Crutzen 2006; Rasch et al. 2008; Robock et al. 2009). A second proposed measure involves the deployment of strategically positioned mirrors and lenses in space to deflect and refract sunlight before it reaches the Earth's atmosphere (Early 1989; Stern 2009; Henson 2011).

It would be remiss of me to overlook the fact that many of these mitigation technologies remain unproven and/or carry unknown dangers and major uncertainties. For example, experiments involving the ocean fertilisation of iron filings have yielded disappointing results. An ocean fertilisation experiment performed in 2004 in the Southern Ocean revealed that no significant quantity of carbon had been transported into deep ocean layers via the death of carbon-rich phytoplankton. Indeed, only a miniscule quantity of carbon had fallen in 21 days to a depth of just 100 m. It was therefore concluded that the quantity of carbon likely to be sequestered on the sea floor via the ocean fertilisation of iron filings would be inadequate to absorb any meaningful proportion of the quantity of carbon emitted each year by humans (Buessler et al. 2004). More disturbingly, Buessler et al. found that, when fertilised, certain species of phytoplankton thrived at the expense of others, thus raising the possibility that ocean fertilisation could lead to a loss of ocean biodiversity (Note: phytoplankton serves as a base food source for many aquatic species further up the oceanic food chain).

At present, carbon capture-and-storage technology exists at the embryonic stage of development, with only minor advances having been made over the past decade. Even if rapid improvements materialise, it is widely acknowledged that the application of carbon geosequestration measures will be limited by both the high costs involved (IEA 2009) and the enormous volume of space required to store liquefied CO_2. The magnitude of the space required is a key constraint given that very few subterranean reservoirs are stable enough to be safely exploited for geosequestration purposes (Flannery 2008). Moreover, experiments involving the pumping of compressed CO_2 into the ocean's depths have led to high death rates of organisms in the vicinity of CO_2 plumes—a consequence of the pumped CO_2 turning the adjacent seawater acidic (Barry et al. 2004). This again raises the issue of biodiversity loss and ecosystem degradation which would undoubtedly limit geosequestration possibilities.[6]

Although opportunities to fix carbon in soils appear promising, agriculture involves the cultivation and removal of short-lived vegetation (crops and animal-feed). As a consequence, carbon sequestration via modified agricultural practices can only keep additional carbon out of circulation for short periods of time, even if such practices can lock up large quantities of carbon at any point in time. This shortcoming must be considered in light of the fact that much of the carbon already and likely to be released has been safely locked away for eons (Flannery 2008). Thus, questions remain as to how effective carbon sequestration via altered agricultural practices can seriously reduce the atmospheric concentration of greenhouse gases. Overall, it would seem that carbon sequestration technology in all its varieties will be of some assistance in reducing greenhouse gas concentrations, but will not, alone, come close to resolving the climate change problem. What's more, carbon sequestration technologies are unlikely to be deployed to any extensive level before 2050—a time when most emissions cuts must be made to avert a potential climate change catastrophe.

Finally, the use of gigantic mirrors and lenses to deflect and refract sunlight has been called into question given its possible ecological implications. Climatologists warn that a modified sunlight/carbon mix could radically alter regional and global weather patterns. Should we fear such a possibility? This largely depends on whether the exact amount of sunlight can be deflected and refracted to precisely offset the warming effect of an excessive atmospheric concentration of greenhouse gases. Of course, it would be delusional to believe that human beings could master the technology sufficiently to guarantee a benign climate outcome. Thus, attempting to deflect and refract sunlight could be just as detrimental to ecological and natural resource systems as the global warming that these technologies seek to address (Henson 2011). Making matters worse in this regard is the ever-present danger that a nation could unilaterally embark on a risky geo-engineering exercise. This concern is no better exemplified than by calls emanating from the United Nations Convention on Biological Diversity to impose a de facto moratorium on the development of large-scale geo-engineering strategies until the risks and uncertainties of geo-engineering technologies have been thoroughly investigated. Thus, even at this early stage, the potential hazards of employing geo-engineering technologies have been internationally recognised.

I would like to finish with an important point before moving on to how total damage and total mitigation costs are estimated. One cannot overemphasise the importance that early mitigation action can have in terms of future mitigation costs. Given that the physical capital and technology being used at any point time is inextricably linked to the capital and technology used in the past—a consequence of the *path-dependency*[7] of economic systems (David 1985; Arthur 1989)—early mitigation action, whether mandated or induced, would greatly limit the rise in future mitigation costs (Ackerman 2007; Stern 2009; Ekins et al. 2011). In fact, early action could even reduce mitigation costs. Furthermore, waiting in the sense of taking little action initially and drastically bumping up mitigation action later on requires a radical shift in the use of physical capital and technology. Radical and abrupt shifts in the use of physical capital entails the premature retirement of large chunks of capital infrastructure. This is very costly.

The cost advantages of early action are also magnified by the associated benefits of early learning-by-doing, economies of scale, energy-efficiency gains, and a reduced climate-change impact on the natural resources needed to undertake further mitigation action. May I say, the benefits of early action also apply to adaptation measures. Early preparation to defend physical infrastructure from future climate change impacts lessens the need for costly emergency actions. Furthermore, much can be learned and gained from increasing the urgency with which adaptation measures are undertaken, such as accelerated attempts to develop drought-resistant crops and improve the management of natural resource stocks. In all, early action with regard to adaptation measures is likely to vastly reduce the total damage costs of climate change.

6.2.3 *How Are Total Damage Costs and Mitigation Costs Estimated?*

6.2.3.1 Estimating the Total Cost of Climate Change Damages

As explicated in Chap. 5, estimating the total cost of climate change damages involves the use of Integrated Assessment Models (IAMs) to simulate the impact of human-induced climate change. IAM simulation exercises essentially entail the modelling and linking of the various stages of the climate change process and its impact on human welfare. The first major stage modelled is the link between human activities and the greenhouse gas emissions they generate, such as the influence on emissions levels of real output, population growth, technological progress, and the types of resources used to fuel the economic process. The second major stage modelled is the link between emissions levels and their impact on the Earth's climate system. The third stage modelled is the link between climate change and its likely impact on physical infrastructure, ecosystems, and cultivated natural resources (e.g., timber plantations and crops). In the final major stage, IAMs estimate the impact of affected natural and human-made capital on human well-being.[8]

There are various ways in which IAM estimates of total damage costs are presented. As noted in Chap. 5, they are often presented in the form of annual real GDP losses or annual reductions in per capita consumption equivalents.[9] To achieve this, IAMs estimate the difference between expected real GDP levels with and without the impact of human-induced climate change, albeit the expected future losses of real GDP are discounted to calculate the present value of climate change damages.[10] Whilst real GDP levels in the presence of climate change are derived via the process explained above, the real GDP levels applicable to the 'no climate change' scenario are generated by deactivating the elements within the IAMs that simulate the future impacts of climate change.

In some modelling exercises, such as Stern (2007), a subtraction is made to real GDP to account for the adaptation measures undertaken in response to the negative impacts of climate change. These adaptation measures, which are costs, constitute defensive and rehabilitative expenditures. Although they boost real GDP, they are necessary to maintain a nation's productive capacity in the face of climate change. They are not undertaken for beneficial consumption purposes. Hence, unless these expenditures are subtracted from real GDP, the measure of national income generated in the 'climate change' scenario is overstated in the Hicksian (1946) sense.[11] For this reason alone, the total cost of climate change damages tends to be underestimated by most climate change modellers.

To calculate total damage costs, IAMs initially focus on the climate change damages inflicted upon sectors of the economy where the market prices of the impacted goods and services already exist or can be imputed with relative ease. These sectors are referred to by climate change modellers as 'market' sectors. The obvious shortcoming of the market-sector approach is that it omits non-priced and socially contingent factors. In doing so, the market-sector approach fails to capture many of the direct impacts of climate change on environmental quality, social stability, and human health (Stern 2007).

To overcome this deficiency, many climate change modellers have employed an assortment of techniques to calculate the monetary value of some non-market impacts. This has enabled IAMs to account for a greater proportion of all climate change damages. Nevertheless, in a number of cases, modellers have elected not to include non-market impacts because of ethical and practical difficulties that render the incorporation of many such factors highly problematic. A good example of this is the modelling undertaken in the *Stern Review*, where Stern chose not to incorporate many non-market impacts in the PAGE2002 model (Stern 2007). Furthermore, even though Stern included non-market costs, such as the loss of human life and declines in environmental quality, much of the *Review* presented these costs alongside income comparisons to avoid summarising non-market costs in monetary terms.

In Chap. 1, I outlined a range of surprises and feedbacks with the potential to generate abrupt and large-scale changes to the Earth's climate system. Past modelling work (e.g., Nordhaus and Boyer 2000; Stern 2007) has shown that the inclusion in IAMs of possible extreme temperature changes greatly increases damage cost estimates. Unfortunately, very few IAMs account for potentially radical changes to the Earth's climate system. As for moderate rises in average global temperatures (e.g., 2–3 °C), their impact on damage cost estimates depends on the assumptions incorporated into IAMs. These include assumptions relating to: (i) how climate change damages should be aggregated across regions; (ii) how the cost of climate change should be valued in poor nations/regions relative to rich nations/regions; and (iii) how well nations are able to implement adaptation measures to limit the impact of climate change.

The above assumptions are crucial because the prices of marketed goods and the imputed monetary values of human life and environmental quality are invariably higher in rich countries than in poor countries. Hence, a given climate change impact is often valued much more if it is incurred by a high-GDP nation. Many modellers find this approach unacceptable. In response, they often employ 'equity' weights to reduce the cost discrepancies across nations and regions (e.g., Nordhaus and Boyer 2000; Tol 2002; Stern 2007). Where modellers ignore equity weights—that is, where they estimate the global cost of climate change damages by simply adding the real GDP of all nations—substantial climate change impacts on poor nations are frequently overwhelmed by much smaller impacts on rich nations. Since many poor nations are expected to endure the worst impacts of future climate change, a failure on the part of modellers to incorporate equity weights results in a significant underestimation of global damage costs.

It has already been shown that the implementation of adaptation measures can minimise the cost of climate change damages. How adaptation is modelled and what assumptions are made with respect to adaptation measures greatly influences damage cost estimates. The assumptions with the largest influence on damage costs are those pertaining to: (i) likely adaptation levels, with or without policy incentives; (ii) the effectiveness of adaptation measures in limiting climate change damages; and (iii) how much it will cost to implement specific adaptation measures. There is, unfortunately, very little consistency across IAMs when it comes to the modelling of adaptation responses. Notwithstanding other critical factors, in cases where adaptation is modelled optimistically, estimates of total climate change damage costs tend to be very low (e.g., Mendelsohn et al. 2000b).

With most IAMs, adaptation is treated as an implicit, inseparable component of a mathematical equation that describes the damages applicable to different levels of global warming (e.g., Nordhaus and Boyer 2000). This results in many IAMs failing to recognise the role that an appropriate level of adaptation can play in keeping total damage costs to a minimum (as represented in Fig. 6.2). As a consequence, the manner in which most IAMs are constructed renders it difficult to investigate what constitutes a desirable mix of adaptation and mitigation measures. There are, however, many IAMs that make the contribution of adaptation measures explicit. One of the best examples is the PAGE2002 model used in the *Stern Review*. Assumed in this model was the following (Dietz et al. 2007b):

- Adaptation will reduce the climate change impact on the market sectors of industrialised economies (e.g., agriculture) by 90 per cent at all levels of warming.
- Adaptation will reduce the climate change impact on the market sectors of low-GDP economies in Africa, India, Southeast Asia, and Latin America by 50 per cent at all levels of warming.[12]
- Adaptation will reduce the climate change impact on the non-market sectors of *all* economies (e.g., human health and environmental quality) by 25 per cent at all levels of warming.

According to Stern, the lower adaptive capacity assigned to non-market sectors reflects the difficulties associated with shielding ecosystems and renewable natural resource stocks from rapid shifts in local climatic conditions.[13] Critics of Stern have argued that the adaptation assumptions in PAGE2002—which constitute a subset of Stern's overall assumptions regarding total damage costs—are overly pessimistic. Stern disagrees. In fact, Stern believes his assumptions are optimistic. Why? Although Stern recognises that higher GDP levels and more advanced adaptation technologies are likely to increase humankind's capacity to adapt to climate change, Stern points out that a rise in average global temperatures will increase the extent to which adaptation measures must be applied to maintain their protective effect (Dietz et al. 2007b). Moreover, Stern acknowledges that the assumed ability of industrialised nations to keep reducing the impact of rising temperatures on market sectors by 90 per cent is dubious. Indeed, Stern concedes there is good reason to suppose that the cost of adaptation will rise faster than average global temperatures given that what will be required to limit damage costs will, despite technological progress, become much more difficult to accomplish.[14] If so, this implies that adaptation costs are convex. Yet PAGE2002 and similar IAMs treat adaptation costs as if they are concave (i.e., increase at a slower rate than average global temperatures). Stern has consequently urged climate change modellers to reappraise the assumed relationship between temperatures and adaptation costs.

6.2.3.2 Estimating Climate Change Mitigation Costs

In Chap. 5, it was explained that climate change mitigation costs can be estimated via the use of 'bottom-up' or 'top-down' modelling approaches. To recall, the former involves determining the cost of adopting various mitigation measures to achieve a specific greenhouse stabilisation target, whereas the latter involves comparing key macroeconomic variables under a no-mitigation scenario and alternative mitigation scenarios.

If we consider these two approaches in more detail, the top-down method entails the use of macroeconomic models to simulate substitutions from high-emissions to low-emissions techniques in the face of changing relative prices (Dietz and Stern 2008). In order to design and construct top-down models, assumptions must be made regarding mitigation targets; the price of high-carbon fuels; the development and uptake of new technologies; the flexibility of different mitigation technologies; the spillover savings and co-benefits of mitigation; and the extent and nature of government policies to induce the shift to low-carbon technologies.[15] Past studies reveal that the assumptions made by modellers have greatly affected the mitigation cost estimates generated by top-down models.

To estimate total mitigation costs from a global perspective, users of top-down models first determine the probable rise in real Gross World Product (GWP) should no action be taken to reduce greenhouse gas emissions. They then choose a greenhouse gas stabilisation target and insert into the model the emissions reductions required to achieve the target. The model is subsequently run to estimate the likely decline in real GWP from having to make the necessary emissions cuts. It is the projected loss of real GWP—the real GWP gap—which researchers deem as the total cost of mitigation. This is illustrated in Fig. 6.4.[16]

The estimated fall in real GWP depends on many factors. To understand what they are and how they affect mitigation costs, consider the following equation:

$$GHG = \left(\frac{GHG}{R} \times \frac{R}{GWP}\right) \times GWP \tag{6.1}$$

where GHG = greenhouse gas emissions; R = natural resource (energy) inputs; GWP = real Gross World Product (global real output).

The GHG/R ratio is the ratio of greenhouse gas emissions to energy inputs and can be decreased, for example, by substituting from coal-generated to solar-generated electricity. The R/GWP ratio is the ratio of energy inputs to real output and can be reduced by raising the technical efficiency of production (i.e., by reducing the energy

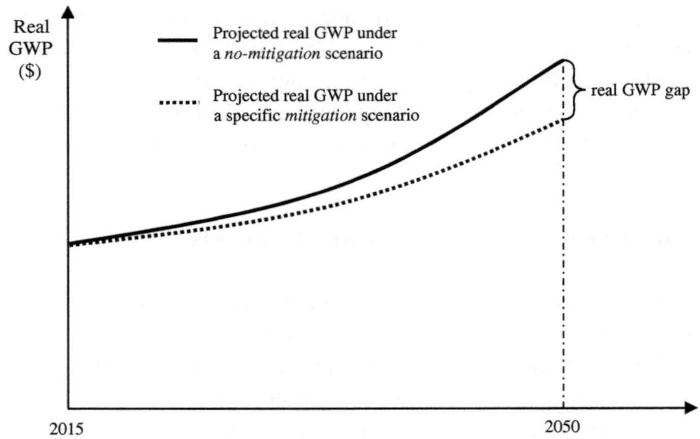

Fig. 6.4 Total mitigation costs (2050) depicted as the loss of real GWP arising from the implementation of mitigation measures to meet a specific stabilisation target. *Source* Adapted from OECD (2008)

resources consumed to produce a given quantity of real output).[17] The product of the two ratios—(*GHG/R* × *R/GWP*)—represents the greenhouse gas emissions-intensity of real GWP and can be lessened by engaging in mitigation measures, such as those described above, which can reduce one or both ratios.

Consider, then, a hypothetical situation where there are no technological advances or shifts over time in the types of resources used. Because there would be no variation in the two ratios in Eq. (6.1), the emissions-intensity of real GWP would remain unchanged. In these circumstances, it would only be possible to reduce greenhouse gas emissions by cutting real GWP. Thus, if a stringent stabilisation target is in place, minimising the decline in real GWP and therefore decreasing what is generally considered the total cost of mitigation, will require structural changes that reduce the *GHG/R* and *R/GWP* ratios.

Given the way in which top-down models are used to estimate total mitigation costs, some of the modelling assumptions have a greater impact on cost estimates than others. Since an assumed high rate of technological progress accelerates the decline in the *GHG/R* and *R/GWP* ratios, it predictably leads to relatively low mitigation cost estimates (i.e., smaller real GWP gaps) (Grubb et al. 2006; Köhler et al. 2006; IARU 2009; Edenhofer et al. 2009). So, too, does an assumed high degree of flexibility between different abatement technologies and various industries within the economy. For example, if it is easy to use a different form of technology or transfer production from one industry to another, the potential for cost-effective emissions reductions increases significantly (Stern 2007). Also having a marked effect on mitigation cost estimates are the assumed spillover savings and the co-benefits arising from the implementation of mitigation measures. If it is assumed that the magnitude and scope of the co-benefits are considerable (e.g., if it is assumed that energy-efficiency savings will be large with the capacity to dramatically reduce production costs across the entire economy), mitigation cost estimates are often substantially lowered. Conversely, more stringent stabilisation targets produce relatively higher mitigation cost estimates, although, as we shall see, it may be possible for total mitigation costs to be significantly reduced if the concomitant need for deeper emissions cuts generates very large scale economies and/or co-benefits.[18] Mitigation cost estimates are also likely to be reduced if emissions cuts are accompanied by lower real output levels arising from nations making the orderly transition to a qualitatively-improving steady-state economy.

As for the bottom-up approach, it does not involve the use of macroeconomic models. Rather, it involves a microeconomic analysis of various mitigation techniques across different sectors of the economy to determine the probable cost of achieving a specific mitigation target (Stern 2009). It is therefore based on much greater detail than the top-down method—an aspect that some observers believe lends it a distinct advantage given that many top-down models are devoid of important information about individual mitigation techniques and abatement technologies.

According to Stern (2009), another advantage of the bottom-up approach is that it facilitates a more practical assessment of different mitigation techniques and provides a do-it-yourself guide for calculating mitigation costs under various assumptions. A good example of how the bottom-up method can simplify the estimation process has been demonstrated by Nauclér and Enkvist (2009). What Nauclér and Enkvist have done is construct a Marginal Mitigation Cost curve by estimating the *net* cost of achieving different mitigation targets by 2030. The curve and the means by which it has been constructed are revealed in Fig. 6.5.[19]

The vertical axis in Fig. 6.5 indicates the additional net cost incurred as annual global emissions are reduced relative to a business-as-usual scenario (measured in Euros per tonne of CO_2-e). The horizontal axis indicates the amount by which annual global emissions can be reduced between now and 2030 by using mitigation techniques currently available or likely to become available in the near future (measured in Gigatonnes of CO_2-e). The various techniques are ranked along the horizontal axis from cheapest to the most expensive.[20]

The net cost reflected on the vertical axis is based on the cost of employing different mitigation techniques minus the spillover savings generated from their use. For instance, if the use of energy-efficient lighting reduces greenhouse gas emissions via reductions in electricity consumption, the net cost of switching to energy-efficient lighting equals the cost of the lighting less the savings from lower electricity expenses.

The net cost 'bar' applicable to each mitigation technique reveals two pieces of information. Firstly, the width of each bar indicates the quantity of greenhouse gases that each technique has the potential to abate. Secondly, the height of each bar shows the marginal cost per tonne of emissions abated. As can be seen from Fig. 6.5, some of the net cost bars lie below the horizontal axis. This indicates that the net cost of employing some mitigation techniques is likely to be negative. In other words, we can expect the spillover savings generated by some mitigation techniques to exceed the cost of employing them. Based on Nauclér and Enkvist's

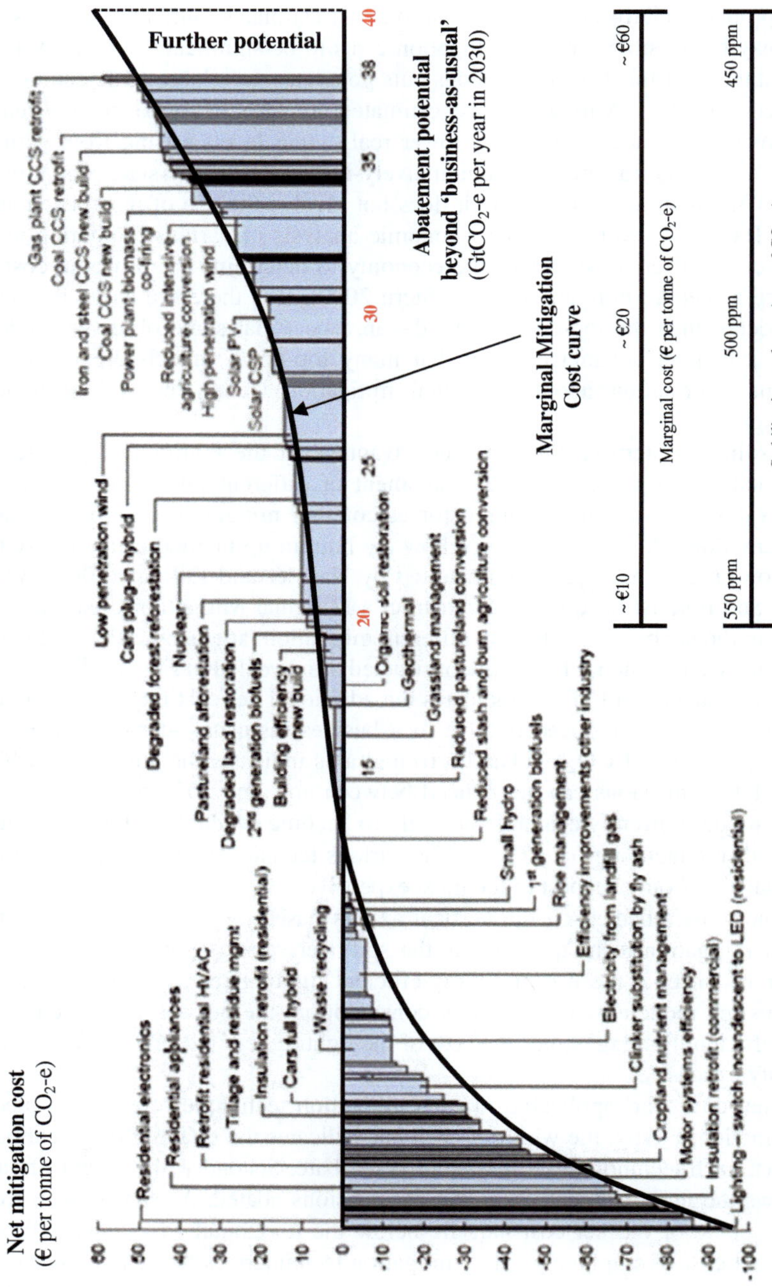

Fig. 6.5 Marginal Mitigation Cost curve—2030. *Source* Adapted from Nauclér and Enkvist (2009)

cost estimates, negative mitigation costs can be obtained by reducing annual emissions by up to 11 $GtCO_2$-e by 2030.

Assume, for a moment, that mitigation measures end up reducing annual global emissions by 2030 by 20 $GtCO_2$-e. If we want to know the net cost of reducing global emissions by an additional tonne, we merely search for '20' on the horizontal axis and look across to the vertical axis to identify the net cost. According to Nauclér and Enkvist's estimates, this is around €10.[21] Returning to the horizontal axis, it can be seen that the most cost-effective way of reducing annual global emissions by an extra tonne beyond 20 $GtCO_2$-e is through the reforestation of degraded land.[22]

Of course, the ultimate reason for reducing global emissions is to stabilise greenhouse gases at a particular concentration level. If one looks closely along the horizontal axis, three mitigation levels are highlighted—20, 30, and 40 $GtCO_2$-e. These three levels approximate the annual emissions reductions required by 2030 to move to stabilisation paths of 550, 500, and 450 ppm of CO_2-e respectively.[23] In order to estimate the total mitigation cost of moving towards one of these stabilisation targets—say, 550 ppm of CO_2-e—one simply aggregates the marginal cost values between 0 and 20 $GtCO_2$-e of abatement.[24] Thus, total mitigation costs are represented by the area between the Marginal Mitigation Cost curve and the horizontal axis. For emissions cuts of up to 20 $GtCO_2$-e per annum, the combined area of negative mitigation costs (0–11 $GtCO_2$-e of cuts) outweighs the combined area of positive mitigation costs (11–20 $GtCO_2$-e of cuts). Hence, according to Nauclér and Enkvist's estimates, the total mitigation cost of moving to a 550 ppm stabilisation target is likely to be negative. Indeed, total mitigation costs only become positive once cuts to annual global emissions exceed around 30 $GtCO_2$-e, which corresponds to stabilisation trajectories below 500 ppm of CO_2-e.

I should emphasise that if there is a high probability of total mitigation costs being negative for low emissions cuts, this would not support calls for limited mitigation action. Minimal mitigation implies the likelihood of high climate change damage costs which must be added to total mitigation costs to determine the total cost of greenhouse gas emissions, as exemplified by the Total Emissions Cost (TEC) curve in Fig. 6.3. Hence, apart from being 'unsafe', minimal mitigation is likely to be very costly in the long-term.

Given that stabilisation at no more than 450 ppm of CO_2-e is necessary to avoid a potentially catastrophic climate change outcome, how might we use Fig. 6.5 to estimate the total mitigation costs of moving to a 450 ppm trajectory by 2030? If we assume that the total cost of reducing annual greenhouse gas emissions by 30 $GtCO_2$-e relative to a business-as-usual scenario is zero, we first need to estimate the marginal cost of having to reduce annual emissions by an extra 10 $GtCO_2$-e to 40 $GtCO_2$-e.[25] With the marginal cost of the additional emissions cuts ranging from €20 to €60 per tonne of CO_2-e, we can reasonably assume that the marginal cost will average around €40 (approximately US$60 per tonne of CO_2-e). If we then multiply the marginal cost of US$60 by the additional 10 $GtCO_2$-e of required emissions cuts, we obtain a total mitigation cost (globally) of US$600 billion.[26] Assuming that US$600 billion is a reasonable estimate of the total cost of moving to a 450 ppm stabilisation trajectory by 2030—which I would refute—I believe that this would be the most appropriate way to present the total cost estimate. For reasons to be given soon, I do not believe that a further step should be taken to present total mitigation costs as real GWP losses or as a percentage of real GWP.[27]

It is important to recognise that Nauclér and Enkvist's (2009) Marginal Mitigation Cost curve is not stationary. Not unlike top-down models, the position and the shape of the curve depend on assumptions with regard to technological progress, learning-by-doing, transition costs, the range of available mitigation options, indirect savings, and the policies implemented by governments to encourage the use of low-emissions techniques. For example, rapid learning-by-doing and greater advances in abatement technologies would increase the abatement potential of mitigation techniques. In doing so, they would shift the Marginal Mitigation Cost curve downward and thus reduce the total mitigation costs applicable to a given mitigation level. The Marginal Mitigation Cost curve would also shift down if the spillover savings generated from the application of mitigation techniques increased (e.g., if mitigation measures led to much lower energy costs).

Although many analysts are aware of the impact that the above factors can have on mitigation costs, most fail to understand the potential effect of other key factors and assumptions. These include the impact of: (i) future real output levels; (ii) the health and integrity of ecological systems; (iii) the scarcity of natural resources and associated natural resource prices; (iv) the value of mitigation co-benefits; (v) information failures; (vi) past investments in mitigation technologies; (vii) technical, deployment, and economic constraints; and (viii) interactions between mitigation techniques. I will return to the impact of these factors on mitigation costs shortly.

6.3 Overlooked Factors Affecting Climate Change Damage Costs and Mitigation Costs

6.3.1 Climate Change Damage Costs—Overlooked Factors

As outlined at length, there are many perspectives on how climate change damages should be estimated and what assumptions should be made when doing so. Although a lot of attention has focused on the fact that IAMs ignore non-market and socially-contingent impacts and inadequately account for the potential effects of extreme global warming, too much attention has centred on discount rates. At the same time, there are justifiable concerns that the assumed commensurability of climate change impacts and a lack of equity considerations have contributed to distorted damage cost estimates. Since these shortcomings lead to the underestimation of damage costs, more must be done to overcome them.

These concerns aside, it is my belief that a number of other factors have been overlooked which, if they were taken into consideration, would vastly alter perceptions of climate change damage costs. In this sub-section, I will discuss three of them: (i) the likely impact of climate change on natural resource scarcity and its flow-on effect on natural resource prices and the real output of the economy; (ii) the increased vulnerability of the ecosphere to climate change arising from higher real output levels; and (iii) the impact on damage costs of temporarily overshooting an upper limit on the safe concentration of greenhouse gases (i.e., exceeding 450 ppm of CO_2-e).

6.3.1.1 The Direct and Indirect Impacts of Climate Change on the Economy

Table 1.4; Fig. 1.10 in Chap. 1 revealed something that cannot be ignored. Should human-induced climate change increase average global temperatures by more than 2 °C above pre-industrial levels, it will have a moderate to significantly detrimental impact on the ecosphere (natural capital).[28] Whilst most IAMs incorporate the cost of these impacts in circumstances where market prices are readily obtainable or easy to impute, the estimated damage costs are usually confined to industries *directly* affected by climate change, such as agriculture, forestry, fisheries, and the energy industry (i.e., the primary sector). Conversely, the damage costs applicable to the industries *indirectly* affected by climate change, such as the manufacturing and service industries (i.e., the secondary and tertiary sectors), are often overlooked or inadequately accounted for.

There are two main reasons for this omission. Firstly, the significance of the impacts on the primary sector is often downplayed on the false belief that because the sector constitutes a relatively small proportion of the GDP of most nations, the damage costs are negligible. This is no better exemplified than by the following claims[29]:

> Agriculture, the part of the economy that is sensitive to climate change, accounts for just 3 % of national output. That means there is no way to get a very large effect on the US economy (Nordhaus 1991).

> Even if the net output of agriculture fell by 50 % by the end of the next century, this is only a 1.5 % cut in GNP (Beckerman 1997).

> In the developed world, [agriculture] is practically the only sector of the economy affected by climate, and it contributes only a small percentage—3 % in the United States—of national income. If agricultural production were drastically reduced by climate change, the cost of living would rise by 1 or 2 %, and at a time when per capita income will likely have doubled (Schelling 1997).

Secondly, there has been a widespread underestimation of the impact that a decline in the primary sector would have on the secondary and tertiary sectors. This lack of recognition is based on an implicit assumption that the primary, secondary, and tertiary sectors are largely independent of each other—the result of a general failure to understand the complementary relationship between natural resources and human-made factors of production. A good example of this is reflected in the following statement:

> Industrial sectors are generally thought to be less vulnerable to the impacts of climate change than other sectors, such as agriculture and water services. This is in part because their sensitivity to climate variability and change is considered to be comparatively lower and, in part, because industry is seen as having a high capacity to adapt in response to changes in climate. The major exceptions are industrial facilities located in climate-sensitive areas (such as coasts and floodplains), industrial sectors dependent on climate-sensitive inputs (such as food processing), and industrial sectors with long-lived capital assets (IPCC 2007c, p. 366).

No-one would dispute that food processing industries are greatly exposed to climate change. However, as stressed in Chap. 5, what is not widely recognised is that material and energy resources—many of which are extremely susceptible to climate change—constitute the only true input of all economic activity. Of course, labour and human-made capital are also required to produce goods and services, but they are made from natural resources and serve only as the resource-transforming agents of the production process. As such, all industries are climate-sensitive, even if some are less sensitive to direct climate change impacts than others. In fact, it is the high susceptibility of the primary sector to the effects of climate change—as the original source of the natural resources used in the production of all up-stream goods and services—that renders the secondary and tertiary sectors extremely vulnerable to climate change.

To better explain how, consider the following hypothetical example represented by Fig. 6.6.[30] The left-hand panel depicts the primary sector of the economy (i.e., agriculture and the resource-extractive industries). The output of the primary sector (Q_P) is a function of the human-made capital (K_P), labour (L_P), and natural resources (R) devoted to the primary sector. The functional relationship between these factors of primary production is described by way of a Bergstrom production function revealed in Chap. 4. In other words, the output of the primary sector is constrained by the quantity of matter-energy embodied in the natural resources devoted to the sector (e.g., land, water, and energy).[31] At time t_0, the output of the primary sector (100 units) and the price of the sector's output ($20) are determined by the intersection of the demand curve (D_P) and the initial supply curve (S_P^0). Because the output of the primary sector comprises essential consumption goods (necessities) as well as the resources needed to produce goods and services in the secondary and tertiary sectors, the demand curve for the primary sector is steeply sloped. Hence, the demand for primary-sector output is relatively insensitive to changes in the price of primary sector goods. The supply curve for the primary sector is also very steep. The reason for this is that many of the natural resources used in the primary sector are either fixed in quantity (e.g., fertile land) or supplied at a naturally regulated rate (e.g., sunlight and rainfall). In addition, any increase in the quantity of labour and human-made capital allocated to the primary sector is only capable of augmenting primary-sector output by advancing the efficiency of primary-sector production.[32] Hence, a very large increase in the quantity of human-made capital and labour is required to make a perceptible impression

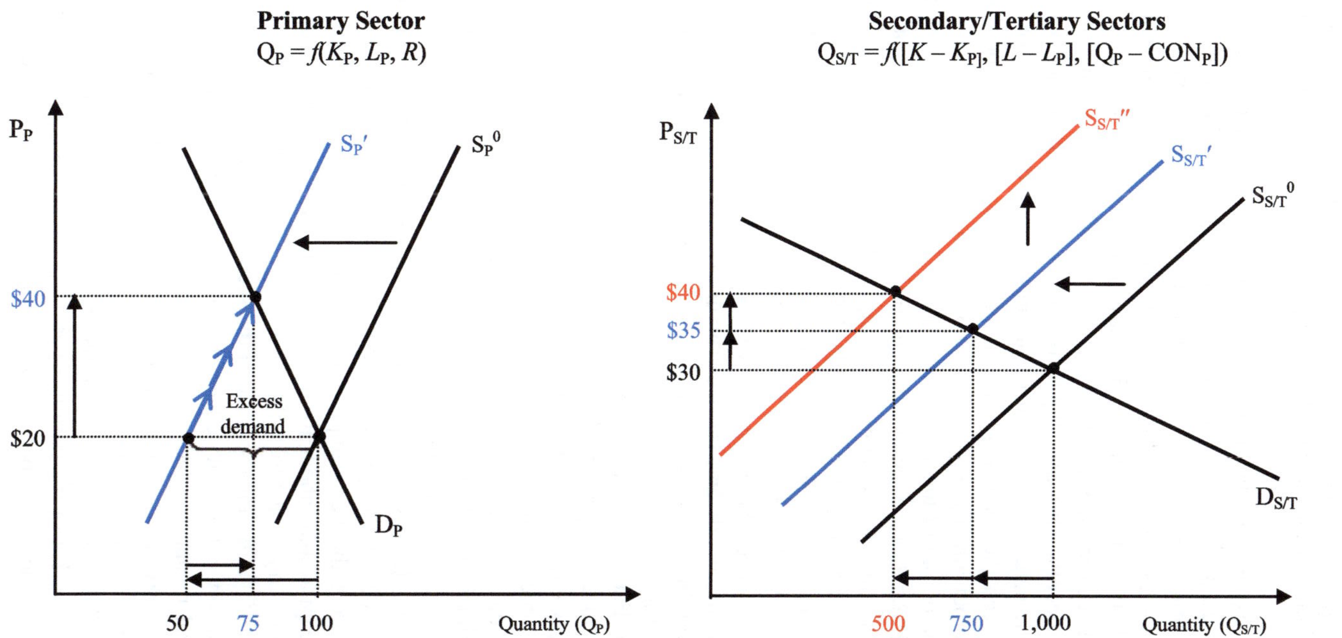

Fig. 6.6 Sectoral interdependencies and the impact of climate change on sectoral prices and output (hypothetical example). *Note* Horizontal axes are not drawn to scale

on primary-sector output. For all these reasons, it is difficult for primary sector producers to significantly increase their output in response to higher commodity prices.

For ease of exposition, the right-hand panel of Fig. 6.6 depicts the combined secondary and tertiary sectors of the economy (i.e., the manufacturing and service industries). In this panel, it is assumed that the human-made capital ($[K - K_P]$) and labour ($[L - L_P]$) employed in the secondary/tertiary sector are what remains after some of both factors of production have been allocated to the primary sector. Because of sectoral interdependencies (i.e., because some of the output of the primary sector becomes resource inputs of the secondary/tertiary sector), it is also assumed that the natural resources used in the secondary/tertiary sector equals the difference between the total output of the primary sector and the portion of primary-sector output already allocated for consumption purposes ($[\mathrm{Q_P} - \mathrm{CON_P}]$). Once again, the output of the secondary/tertiary sector is constrained by the natural resources devoted to the sector, although on this occasion the output of the sector is constrained by the non-consumed output of the primary sector. At time t_0, the output of the secondary/tertiary sector (1000 units) and the price of the sector's output ($30) are determined by the intersection of the demand curve ($\mathrm{D_{S/T}}$) and the initial supply curve ($\mathrm{S_{S/T}}^0$). Unlike the primary sector, the demand curve for the secondary/tertiary sector is relatively flat because the output of the sector contains many non-essential goods and services (luxuries). Consequently, the demand for the secondary/tertiary sector's output is very sensitive to changes in the price of the sector's output.

Imagine the Earth's climate changing so drastically that it precipitates a collapse in agriculture, forestry, and fisheries production between time t_0 and t_1. As a result, the supply curve of the primary sector shifts left to $\mathrm{S_P}'$, thus causing a massive decrease in the output of the primary sector to 50 units at the prevailing price of $20. Given the essential nature of primary-sector output and the now excess demand for primary sector goods, labour and human-made capital from the secondary/tertiary sector are urgently reallocated to the primary sector to boost the latter sector's output. This has two effects. In the primary sector, output increases to 75 units and the price of primary sector goods rises to $40 (as represented by a movement along the new supply curve). Simultaneously, the transference of human-made capital and labour to the primary sector causes the supply curve in the secondary/tertiary sector to shift left to $\mathrm{S_{S/T}}'$. The output of the sector subsequently falls to 750 units and the price of the sector's output increases to $35.

Importantly, since the non-consumed output of the primary sector constitutes the natural resources used in the secondary/tertiary sector, the large rise in the price of primary-sector output increases the cost of secondary/tertiary sector production. This causes the supply curve in the secondary/tertiary sector to shift upwards to $\mathrm{S_{S/T}}''$. Consequently, the output of the secondary/tertiary sector falls to 500 units and the price of the combined sector's output rises to $40. Overall, the climate change impact on the primary sector has the following effects:

- *primary sector*—a moderate decrease in output (100 to 75 units) and a large increase in price ($20–$40);
- *secondary/tertiary sector*—a large decrease in output (1000–500 units) and a moderate increase in price ($30–$40);
- *total economy*—a significant decline in total output (real GDP) and a moderate to large increase in the price/cost of all forms of output.

To better appreciate the possible output effects of climate change, consider the before and after values of nominal and real GDP and the relative contributions of the primary and secondary/tertiary sectors. From Table 6.1, it can be seen that the contribution made by the primary sector towards nominal GDP at time t_0 is $2,000, which is 6.25 per cent of the nation's nominal GDP of $32,000. At the same time, the contribution made by the secondary/tertiary sector towards nominal GDP is $30,000 or 93.75 per cent of the total. It should be noted that I have

Table 6.1 The impact of climate change on total output (real GDP) and sectoral prices and output (hypothetical example)

At time t_0 • *Primary sector* $P_P = \$20$ $Q_P = 100$ units Contribution to nominal GDP (GDP_P) = $20 × 100 = $2,000 • *Secondary/tertiary sector* $P_{S/T} = \$30$ $Q_{S/T} = 1{,}000$ units Contribution to nominal GDP ($GDP_{S/T}$) = $30 × 1,000 = $30,000 • *Economy* Nominal GDP = $GDP_P + GDP_{S/T}$ = $32,000 Real GDP (using t_0 as the base year) = $32,000 Contribution of *primary sector* to nominal GDP = 6.25 % Contribution of *secondary/tertiary sector* to nominal GDP = 93.75 %
At the end of time t_1 • *Primary sector* $P_P = \$40$ $Q_P = 75$ units Contribution to nominal GDP (GDP_P) = $40 × 75 = $3,000 • *Secondary/tertiary sector* $P_{S/T} = \$40$ $Q_{S/T} = 500$ units Contribution to nominal GDP ($GDP_{S/T}$) = $40 × 500 = $20,000 • *Economy* Nominal GDP = $GDP_P + GDP_{S/T}$ = $23,000 Real GDP (using t_0 as the base year) = ($20 × 75) + ($30 × 500) = $1,500 + $15,000 = $16,500 Contribution of *primary sector* to nominal GDP = 13.04 % Contribution of *secondary/tertiary sector* to nominal GDP = 86.96 % Decline in real GDP = 48.44 %

chosen these values so that the representative contribution of the primary sector approximates the relative contribution of climate-sensitive primary-sector output in most industrialised nations.

By the end of time t_1, primary-sector output has fallen by 25 per cent (100–75 units). Based on the views expressed by Nordhaus (1991), Beckerman (1997), and Schelling (1997), the nation's real output (real GDP) should have fallen by around 1.56 per cent (i.e. 1.56 % = 25 % × 6.25 %). Yet not only does real GDP decline appreciably more than the overall fall in primary-sector output (48.44 per cent), the relative contribution of the primary sector rises from 6.25 per cent to 13.04 per cent of nominal GDP.

Could we really expect human-induced climate change to trigger such dramatic changes to prices and output levels as this? The situation just described is only a hypothetical example and it may be that output and price changes would be considerably less than those postulated above, particularly if the concentration of greenhouse gases was stabilised at no more than 450 ppm of CO_2-e. Having said this, should adequate mitigation action not be taken, there is every reason to expect extreme climate changes to bring about substantial changes to prices as well as total and compositional output. For as Daly (2000) has highlighted, even modest climate change would boost the relative GDP contribution of the primary sector given that an inevitable increase in the relative scarcity of primary-sector output would undoubtedly bring about a sizeable increase in natural resource prices. The relative GDP contribution of the primary sector would also be expected to rise because the reallocation of productive factors to recover some primary-sector output losses would lead to a lower-than-expected level of secondary/tertiary-sector output. Moreover, even though a higher price of primary-sector output would flow through to the secondary/tertiary-sector output in the form of higher input costs, any price increase in the secondary/tertiary sector would be relatively smaller than in the primary sector because the former sector's output includes many non-essential goods and services.

It must be said that the accentuated impact of climate change arising from the complementary relationship between natural resources and human-made factors of production has not been entirely overlooked. In a variety of ways, a number of climate change analysts have sought to incorporate the complementary relationship in their assessment of climate change damage costs. However, as mentioned in Chap. 5, they have done so inadequately. For example, Neumayer (2007) and Sterner and Persson (2008) have only sought to address the question of whether an increase in human-made consumption goods can offset a climate change-induced fall in environmental goods/natural resources. By taking this narrow view, they have overlooked the inevitable impact that a large decline in primary-sector output would have on the secondary/tertiary sector and on the total output of the economy as a whole. What's more, Sterner and Persson (2008) have erred further by suggesting that a gradual shift towards 'services' can increasingly dematerialise a nation's total output and overcome any effect that a fall in primary-sector output would have on the secondary/tertiary sector. As explained previously, this view is based on the assumption that services can be supplied with the minimal use of natural resources—a view that has no theoretical or empirical support (Lawn 2009, Chap. 1).

In another effort to address the complementarity issue, Dietz et al. (2007a) claim that some IAMs, such as the PAGE2002 model used in the *Stern Review*, perform admirably when it comes to evaluating the effects of extreme climate change by incorporating a production function that accounts for any deleterious impact that climate change has on total output (see Dietz et al. 2007a, Eq. (3)). Dietz et al. believe the production function serves this useful purpose by embodying a climate change damage equation that incorporates the potential real output effect of abrupt, discontinuous, and large-scale changes to the Earth's climate system (see Dietz et al. 2007a, Eq. (6.1)).[33] According to Dietz et al., this additional specification enables the PAGE2002 model to simulate the effects of high-damage scenarios in circumstances where average global temperatures exceed critical threshold levels.

Despite good intentions, it is my belief that these models fail to properly tackle the complementary issue. As Dietz et al. (2007a) readily concede, the production function employed in the PAGE2002 model excludes natural capital and the environmental goods and services that natural capital generates as separate productive factors. Consequently, the PAGE2002 model suffers from two major weaknesses. Firstly, since human-induced climate change is expected to reduce primary-sector output and natural resource availability, the PAGE2002 model fails to simulate the full impact of rising global temperatures on total output.

Secondly, even if natural resource inputs were acknowledged, the production function employed in the PAGE2002 model is an example of a Cobb-Douglas production function. Unlike the Bergstrom function described in Chap. 4, the Cobb-Douglas production function 'mathematically' permits physically implausible outcomes—more specifically, outcomes that violate the first and second laws of thermodynamics (Georgescu-Roegen 1979). To explain this further, Cobb-Douglas production functions permit increases in the total output of the economy by assuming that sufficiently large expansions in the stock of human-made capital can offset the decline in natural resource inputs caused by diminutions in the stock of natural capital, including diminutions resulting from climate change. As intimated in Chap. 2, there is one major problem with this assumption—the production of a given quantity of physical goods, including human-made capital, requires a minimum input of natural resources (low-entropy matter-energy). Consequently, producing a given quantity of physical goods requires a minimum quantity of resource-providing natural capital. Thus, while Cobb-Douglas production functions assume that human-made capital and natural capital are substitutes, they are unquestionably complementary forms of capital. It therefore follows that the PAGE2002 model incorporates the same erroneous assumption.

Given the failure of climate change analysts to properly account for the complementary relationship between natural and human-made factors of production, can we say that climate change damage costs will be higher than most analysts expect? There is no doubt in my mind that the negative impact of climate change on the real GDP of nations will be much greater than many anticipate. But does this represent a net welfare loss? After all, it has been argued throughout this book that, to increase per capita economic welfare, there is a need for high-GDP nations to immediately begin the transition to a steady-state economy, which will require many of them to reduce their real GDP (de-growth).

Before answering the above questions, it is worth recognising that the extent to which a lower-than-expected level of real GDP amounts to a net welfare loss depends, in part, on the policies pursued by a nation and the institutions it has in place. It also depends on where the nation's economy is with respect to its optimal macroeconomic scale (i.e., with respect to S_* in Fig. 2.4). If a rich nation, which would more than likely be operating beyond its optimal scale, continues with growth-oriented policies and institutions, then, for some time, lower-than-expected levels of real GDP are likely to incur a greater net welfare loss than an equivalent rise in real GDP.[34] Bear in mind, increases in per capita GDP are already reducing per capita economic welfare in most rich countries (see Figs. 2.7 and 2.8) and so what we are comparing are the net welfare losses from 'uneconomic' growth and the potential net welfare losses arising from a growth-oriented economy that is failing to grow at a targeted rate because of climate change.[35]

The reason why, in a growth-oriented setting, a lower-than-expected real GDP is likely to incur a larger net welfare loss than a rise in real GDP is because growth-oriented policies primarily focus on boosting production, throughput-increasing technological progress, and the hope that income will trickle down from the rich to the poor. Hence, welfare gains from growth are heavily dependent on the benefits that emerge from a rise in the consumption of human-made goods and services. This is not the case with a qualitatively-improving steady-state economy where welfare gains are likely to accrue from policies and actions explicitly aimed at producing better rather than more goods; redistributing income from the rich to the poor[36]; increasing leisure time; and reducing environmental losses per unit of economic activity. Consequently, whereas welfare losses from lower-than-expected levels of real GDP would be counter-balanced in a steady-state context by non-consumption-related welfare gains, the same would not occur in a growth-oriented setting. Indeed, a continuation of growth-oriented policies would almost certainly lead to net welfare losses closely approximating the real output losses caused by climate change.

This final point is an important one because it highlights that the real GDP losses emanating from climate change cannot be automatically classed as climate change damage costs—a false conclusion made by virtually all climate change analysts. For example, whether a 2 per cent loss of real GDP constitutes a damage cost of the same monetary value depends on its broader welfare implications. As just explained, a 2 per cent real output loss may be significant in a growth-oriented setting; however, it may be relatively insignificant in a steady-state context. For this and other reasons outlined in Sect. 3.7 on national accounting reform, the cost of climate change should not be assessed in terms of its impact on real GDP. Given the uneven global impact of climate change and the different development stages that rich and poor nations find themselves, there is even less justification for assessing the cost of climate change in terms of its aggregate impact on real GWP. Ultimately, damage costs should be assessed and expressed in terms of the impact that climate change has on the Genuine Progress Indicator (GPI) or some other broad indicator of human welfare.

Naturally, I do not want to give the impression that there would be no climate change damage costs in a steady-state context. Climate change-induced reductions in real output would exert downward pressure on the GPI regardless of a nation's policy-orientation. This is because two of the benefit items that make up the GPI are 'private consumption' and 'non-defensive public consumption' (Lawn 2007; Lawn and Clarke 2008). Both items would be lower than desired as a result of climate change which, to recoup, would require the costly use of more resources. Needless to say, there would be no overwhelming desire to restore the value of these items in a steady-state setting since greater effort would be devoted towards boosting non-consumption-related benefits and reducing social and environmental costs. Nonetheless, lower-than-expected consumption levels in a steady-state setting would constitute a negative component in a full assessment of climate change damage costs. The important point is that total damage costs would be much lower in a steady-state setting than in a growth-oriented setting.

By and large, what I have just discussed applies to a high-GDP country. What about the impact on poor countries? As stressed in Chap. 2, poor countries desperately need a further dose of real GDP growth to enjoy increases in per capita economic welfare. With this in mind, a climate change-induced decline in the real GDP of nations would incur significant welfare losses on the world's impoverished countries. More particularly, severe climate change impacts would have disastrous consequences for the world's most disadvantaged people.

Finally, in answer to an earlier question, I believe it is safe to conclude that climate change damage costs will be much higher than most analysts expect. Although the extent of the damage costs will depend, in part, on prevailing policies and where a nation's economy is in relation to its optimal macroeconomic scale, there is little doubt that the widespread failure of climate change analysts to recognise sectoral interdependencies and the complementary relationship between natural and human-made factors of production has resulted in the gross underestimation of future damage costs.

6.3.1.2 High Real Output Levels Further Increase Damage Costs by Rendering the Ecosphere More Vulnerable to Climate Change

We have just seen how growth-oriented policies, by failing to generate significant non-consumption-related benefits, are likely to increase the net welfare losses arising from climate change. There is, nonetheless, an additional growth-related ramification that would further magnify climate change damage costs. It also happens to be another factor overlooked by climate change analysts when making damage cost assessments.

In Chap. 4, it was revealed that a 2.3 per cent average annual rise in real GWP would increase the ratio of humanity's ecological footprint to global biocapacity to around 3.16 by 2100 (see Table 4.3).[37] Even without climate change, this would undoubtedly lead to the diminution of natural capital stocks and seriously damage or destroy many key ecosystems (Vitousek et al. 1986; Wackernagel et al. 1999;

Rockström et al. 2009). At the same time, it would render the global ecosphere less resilient and more vulnerable to the effects of climate change (Hare 2003; Leemans and Eickhout 2004; Malcom et al. 2006; IPCC 2007c). By doing so, the growth in real GWP would amplify the damage costs associated with a given rise in average global temperatures and any given increase in the intensity and frequency of extreme weather events.

Exactly when might these additional damage costs emerge? Let's assume that, without climate change, a 2.3 per cent annual growth rate of real GWP is achievable up to 2100.[38] Since a growth rate at or near this level would transpire through an emphasis on growth rather than a concerted effort to make the transition to a steady-state economy, any ensuing damage costs from climate change would closely approximate any future climate change-related losses of real GWP. This is because there would be very little increase in non-consumption-related benefits to fill the void.

May I say, this conclusion raises an apparent contradiction. After all, on the one hand, a high growth rate of real GWP would supposedly render the ecosphere more vulnerable to climate change, which would increase the climate change-induced losses of real GWP. On the other hand, any large losses of real GWP would reduce the rate of growth and lessen the vulnerability of the ecosphere to climate change.

Can this apparent contradiction be reconciled? Yes it can, but only by recognising that a more vulnerable ecosphere will not eventuate until there is considerably more growth in real GWP and substantially more damage to ecosystems and natural resource stocks. To put it another way, should growth remain the overriding policy objective, we would expect most of the additional output losses arising from an increasingly vulnerable ecosphere to occur in the latter part of the 21st century and beyond. By this time, much of the targeted increase in real GWP over the first half of this century would have been realised. That said, should growth-related damages to natural capital substantially increase the vulnerability of the ecosphere to climate change, there is every possibility that real GWP could start falling not long after 2050.[39] If so, total damage costs would increase dramatically towards the end of this century.

As for a steady-state setting, climate change-induced losses of real GDP are unlikely to be further amplified because humanity's reduced resource demands would not increase the vulnerability of the ecosphere to climate change. In fact, because the transition to a steady-state economy would return the ecological footprint/biocapacity ratio to a sustainable level, the ecosphere would regain much of its former resilience.[40] This would significantly reduce real GWP losses, although with policies in place to generate most welfare benefits from non-consumption-related activities, damage costs would only be marginally lowered.

Given the aforementioned, what effect would a high rate of growth have on the Total Damage Cost (TDC) curve depicted in Fig. 6.2? Compared to a steady-state economic setting, we would expect the Total Damage Cost curve to rise more rapidly over time as the expanding economy increases the vulnerability of the ecosphere to the damaging impacts of climate change. Moreover, we would

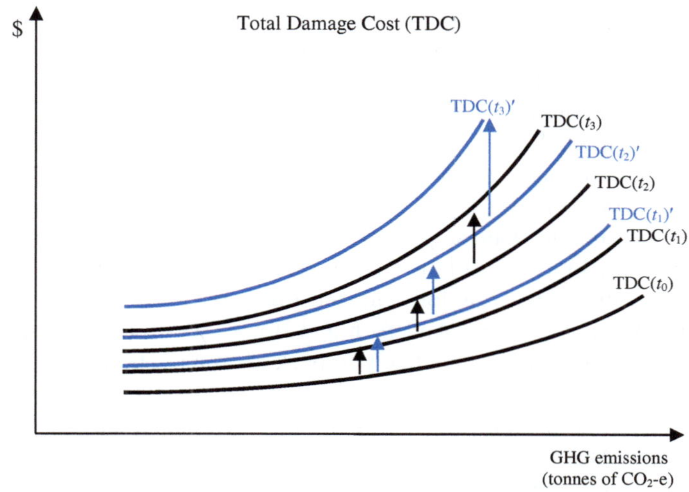

Fig. 6.7 Rising total damage costs (with adaptation) pertaining to two different growth rates of real GWP (real GWP′ > real GWP)

expect the curve to rise more rapidly in the presence of much higher growth rates of real GWP.[41] This is illustrated in Fig. 6.7. At time t_0, total damage costs are represented by the TDC(t_0) curve. As the rise in the atmospheric concentration of greenhouse gases between time t_0 and t_3 increases average global temperatures, the Total Damage Cost curve gradually shifts upwards. In Fig. 6.7, the upward shifts of the Total Damage Cost curve to TDC(t_1)′, TDC(t_2)′, and TDC(t_3)′ correspond to a higher growth rate of real GWP than the shifts to TDC(t_1), TDC(t_2), and TDC(t_3) (i.e., real GWP′ > real GWP).

Although it is impossible to predict the exact position of the shifting Total Damage Cost curves, and hence it is impossible to accurately estimate the impact of growth on damage costs, Fig. 6.7 emphasises that the true cost of climate change damages depends on more than the sheer quantity of greenhouse gas emissions. Furthermore, as we shall later see, the light that Fig. 6.7 sheds on the cost impact of growth can significantly aid our understanding of a desirable greenhouse gas trajectory. Above all, since the impact of real GWP growth on damage costs is regularly ignored, Fig. 6.7 constitutes a further example of how climate change analysts routinely underestimate total damage costs.

6.3.1.3 Overshooting Desired Stabilisation Targets Increases Damage Costs by Inducing Higher Temporary Temperature Rises

Throughout this book, it has been assumed that 450 ppm of CO_2-e constitutes the upper limit on the safe concentration of greenhouse gases. It has also been pointed out that there are many ways to achieve a particular stabilisation target.

For example, it can be achieved by decelerating the rise in the atmospheric concentration of greenhouse gases and stabilising the concentration once the desired target has been reached. Alternatively, it can be achieved by overshooting the target and reducing emissions sufficiently to return the concentration of greenhouse gases to the desired level.

Since the atmospheric concentration of greenhouse gases already stands at approximately 440 ppm of CO_2-e and is rising at a rate of around 2 ppm per year, it will be impossible to stabilise greenhouse gases at 450 ppm without overshooting the upper safe limit.[42] Just how far are we likely to exceed the 450 ppm level? As stringent as the initial emissions cuts must be to replicate the 450 ppm stabilisation trajectory advocated by Anderson and Bows (2008), the atmospheric concentration of greenhouse gases is likely to peak at around 500 ppm of CO_2-e (see den Elzen and Meinshausen 2006). As disconcerting as this seems, bear in mind that it is possible to achieve the 450 ppm target by making smaller initial emissions cuts than Anderson and Bows recommend and much deeper cuts later on. However, exercising this option would simply increase the disparity between the peak and the desired concentration of greenhouse gases. That is, it would lead to the desired stabilisation target being overshot by a greater margin.

From a damage cost perspective, the chosen stabilisation trajectory is crucial. This is because a trajectory taken to achieve a 450 ppm stabilisation target can have a marked effect on near-term concentration and temperature profiles (Schneider and Mastrandrea 2005; Nusbaumer and Matsumoto 2008). Studies have shown that overshooting a stabilisation target is likely to induce higher transient temperature rises. This significantly raises the probability of temporarily or permanently surpassing key climate-system thresholds and vulnerabilities (Hammitt and Shlyakhter 1999; Harvey 2004; O'Neill and Oppenheimer 2004; Hare and Meinshausen 2005; Knutti et al. 2005; Schneider and Mastrandrea 2005; IPCC 2007b). Consequently, overshooting a target increases the climate change damages associated with a particular stabilisation target. Worse still, more extensive overshooting of a desired target further magnifies climate change damage costs.

To represent the effect of overshooting in a way that will be instructive later on, consider Fig. 6.8. Unlike earlier figures, the Total Damage Cost (TDC) curve depicts the total damage costs associated with different atmospheric concentrations of greenhouse gases (stock) as opposed to different greenhouse gas emissions (flow). The distinction is important to understand because it is possible for the atmospheric concentration of greenhouse gases (G) to be rising as emissions (E) are falling. For this to occur, the prevailing emissions level need only be greater than the Earth's greenhouse gas-absorbing capacity (A) (i.e., $E > A$).[43] In Fig. 6.8, a stable concentration of greenhouse gases occurs when the prevailing level of emissions equals the Earth's capacity to assimilate greenhouse gases ($E = A$). As for a situation where the concentration of greenhouse gases is declining, greenhouse gas emissions would be less than the Earth's greenhouse gas-absorbing capacity ($E < A$).

With these dynamics in mind, the TDC curve is upward sloping to represent the rise in total damage costs as the concentration of greenhouse gases increases.

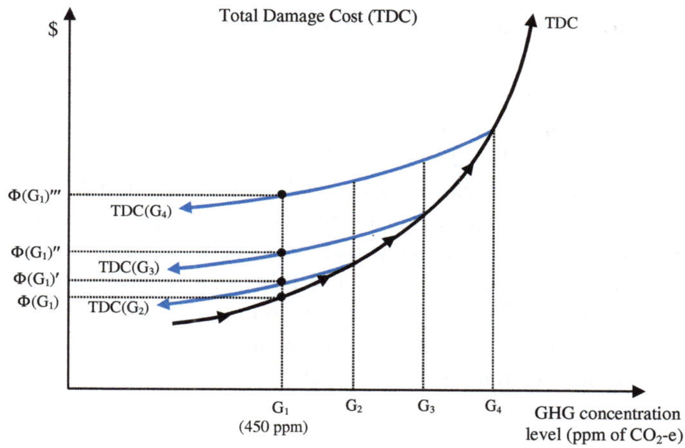

Fig. 6.8 Total damage costs (with adaptation) associated with rising and declining greenhouse gas concentrations

Provided a stabilisation target—say, 450 ppm (G_1)—is approached from a lower concentration level, total damage costs rise in accordance with the TDC curve and settle at $\Phi(G_1)$. However, should the atmospheric concentration of greenhouse gases rise to G_2 before stabilising at G_1 (a minor overshoot scenario), total damage costs do not settle at $\Phi(G_1)$. Because of the harmful effect of a transient boost in global temperatures, they instead settle at $\Phi(G_1)'$. This is represented by the movement back along the TDC(G_2) curve. If, on the other hand, the concentration of greenhouse gases rises to G_3 before stabilising at G_1 (a moderate overshoot scenario), transient temperature increases are much larger and total damage costs settle at a much higher level of $\Phi(G_1)''$. This is represented by the movement along the TDC(G_3) curve. Finally, the fall in the concentration of greenhouse gases from G_4 to G_1 (an extreme overshoot scenario) results in total damage costs settling at an even higher level of $\Phi(G1)'''$. As is clearly evident from Fig. 6.8, the total damage costs associated with a stabilisation target of 450 ppm—or any target for that matter—are greater the more the desired target is exceeded (i.e., the larger is the gap between the peak concentration of greenhouse gases and the desired stabilisation target).

What does this all mean with respect to anticipated damage costs? Although the damage functions used in IAMs to estimate climate change damage costs rely upon global mean temperature as a major cost-influencing variable (see, for example, Stern 2007, p. 660, Eq. (6.1)), the range of temperatures inserted into the models are closely allied to assumed changes in the atmospheric concentration of greenhouse gases. Little regard, however, is given to the potential impact of overshoot scenarios which, as we have seen, can increase the total damage costs associated with a particular concentration level. It is for this reason that climate change analysts further underestimate total damage costs.

6.3.2 *Mitigation Costs—Overlooked Factors*

I mentioned earlier that climate change analysts take account of a number of important factors when making their assessments in connection with climate change mitigation costs—for example, technological progress, learning-by-doing, indirect savings from the mitigation measures likely to be undertaken in the future, and government policy inducements to encourage the uptake of low-emissions techniques. However, just like damage cost assessments, analysts fail to recognise the impact that a high growth rate of real GWP is likely to have on climate change mitigation costs. Along with an explanation as to how the growth in real GWP and higher-than-expected natural resource prices would affect mitigation costs, I will briefly discuss a range of mitigation cost factors frequently overlooked by climate change analysts.

6.3.2.1 The Direct Impact of Real GWP Growth on Mitigation Costs—More Growth Requires More Abatement

In my earlier elucidation of how top-down models are used to estimate climate change mitigation costs, it was explained that global greenhouse gas emissions are closely linked to the real output of the global economy (real GWP). Using Eq. (6.1), it was shown that if real GWP is rising, substantial cuts in greenhouse gas emissions can only be achieved by engaging in measures that reduce the emissions-intensity of global real output. This leads to a self-evident but rather important corollary—the higher is the prevailing growth rate of real GWP, the greater must be the rate of abatement to achieve a specific emissions target. This is reflected in Fig. 6.9. The top curve represents the assumed emissions path pertaining to a 2.3 per cent annual growth rate of real GWP should no mitigation take place. The middle curve represents the emissions path pertaining to the steady-state economic scenario simulated in Chap. 4—again, if no mitigation takes place. The bottom curve represents the emissions path recommended by Anderson and Bows (2008) to achieve a 450 ppm stabilisation target.

As can be seen in Fig. 6.9, to move to a 450 ppm stabilisation trajectory, it is necessary to reduce annual greenhouse gas emissions to 30 $GtCO_2$-e by 2030 (also see Table 4.1). In order to achieve this emissions target under a growth-as-usual scenario, annual emissions in 2030 would have to be 40 $GtCO_2$-e lower than they would without mitigation. Conversely, annual emissions would have to be just 29 $GtCO_2$-e lower under a steady-state scenario.[44]

Since more abatement is required under a growth-as-usual scenario, the cost of abatement *vis-à-vis* the steady-state scenario is much higher. In fact, if we borrow the mitigation costs estimated by Nauclér and Enkvist (2009) to derive the Marginal Mitigation Cost curve in Fig. 6.5, the net mitigation cost associated with moving to a 450 ppm stabilisation pathway by 2030 would be around zero for the steady-state scenario. This compares with the previously calculated US$600

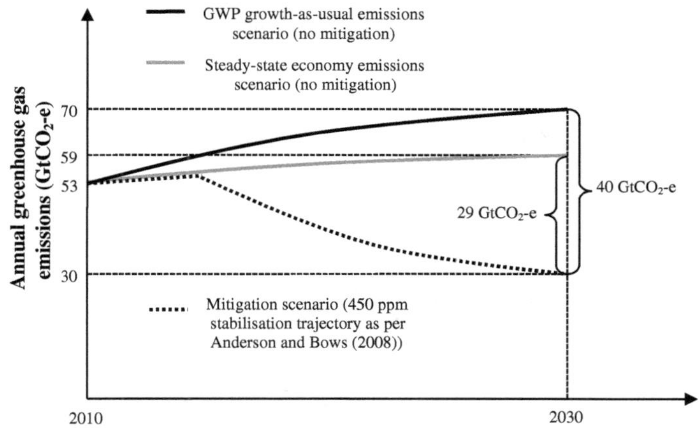

Fig. 6.9 Greenhouse gas emissions abatement required to move to 450 ppm stabilisation trajectory—GWP growth-as-usual versus steady-state economy

billion for the growth-as-usual scenario. Unfortunately, most climate change modellers confine their mitigation cost estimates to calculating the cost of replacing cheaper high-emissions technologies with more expensive low-emissions techniques. In doing so, they take no account of the underlying growth rate of real GWP. Hence, they overlook the cost discrepancy identified above.

There are, it must be said, a number of analysts who recognise that a larger gap between emissions targets and the probable emissions levels under a no-mitigation scenario leads to higher mitigation costs (e.g., Stern 2007, p. 272). However, these analysts assume that a larger-than-expected mitigation gap is solely attributable to the inappropriate choice of both production technologies and energy sources rather than the growth rate of real output. Yet, as Eq. (6.1) highlights, the mitigation gap and any additional mitigation costs associated with it depend on the latter as well as the former combination of factors.

To demonstrate how a rising rate of real GWP growth might affect total mitigation costs, consider Fig. 6.10. For convenience, we shall assume that the rate of resource-saving and pollution-reducing technological progress is less than the rate of growth in real output (Note: this implies that emissions would continue to rise in a 'no-mitigation' scenario). The TMC(t_0) curve represents the total mitigation costs associated with the real GWP generated at time t_0. As real GWP expands between time t_0 and t_3, the emissions cuts required to achieve a particular stabilisation target rises.[45] More particularly, additional mitigation measures are required merely to keep emissions at t_0 levels. Since this increases total mitigation costs, the Total Mitigation Cost curve gradually shifts up. The upward shifts to TMC(t_1)′, TMC(t_2)′, and TMC(t_3)′ correspond to a higher growth rate of real GWP than the shifts to TMC(t_1), TMC(t_2), and TMC(t_3) (i.e., real GWP′ > real GWP).

Although Fig. 6.10 shows the Total Mitigation Cost curve shifting upwards, many factors would be dampening the upward pressure on total mitigation costs. These

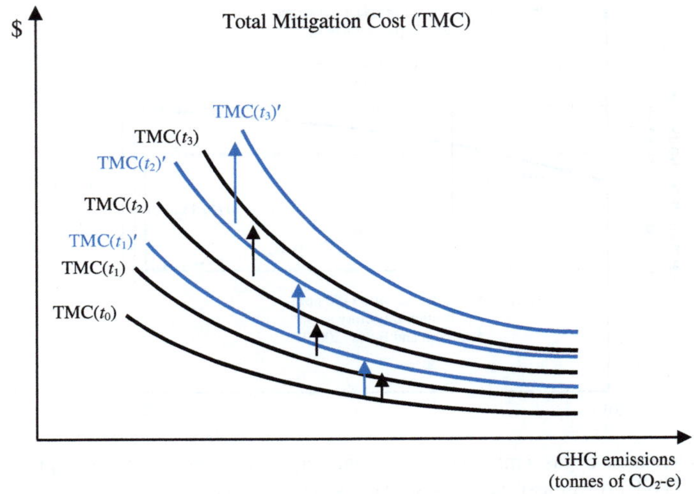

Fig. 6.10 Rising total mitigation costs pertaining to two different growth rates of real GWP (real GWP < real GWP′)

would include improved mitigation techniques, greater resource-use efficiency, learning-by-doing, and any increased co-benefits from mitigation. I shall return to these factors shortly. For now, it should be clear that the failure to acknowledge the impact of real GWP growth on required mitigation levels is likely to contribute to the underestimation of climate change mitigation costs. More crucially, it is likely to result in climate change analysts overlooking the enormous potential that exists to reduce mitigation costs by moving to a lower rate of growth or, better still, by making the transition to a qualitatively-improving steady-state economy.

6.3.2.2 The Indirect Impact of Real GWP Growth on Mitigation Costs—More Growth Increases the Cost of Using All Technologies

As explained, climate change mitigation costs are conventionally estimated by calculating the difference between the cost of using high-emissions methods to generate a unit of real output and the cost of employing low-emissions technologies to generate the same unit of real output but with fewer greenhouse gas emissions. Regardless of whether top-down or bottom-up models are used to perform this exercise, the cost difference is assumed to be a function of the price gap between low-emissions technologies and the high-emissions technologies they would displace—the latter often referred to as 'marker' technologies (Anderson 2006). An example of the price gap is the difference between the price of a kilowatt-hour of wind-generated electricity (low emissions) and a kilowatt-hour of cheaper coal-generated electricity (high emissions).

Using the price gap as an indicator of net cost, bottom-up models involve making a detailed microeconomic assessment of individual mitigation technologies.[46] As we have seen, the net cost of employing the various low-emissions techniques is summed to obtain the total cost of achieving a specific emissions target (see Fig. 6.5). In the case of top-down models, where mitigation costs are viewed from a macroeconomic perspective, the price gap is used as a determining factor in the evolution of consumer preferences, structural modifications to the economy, and the development and uptake of low-emissions techniques and low-carbon fuels. These adjustments are subsequently used to estimate the likely impact on real GWP of having to reduce greenhouse gas emissions to achieve a particular emissions target (see Fig. 6.4).

It is important to understand how a change in the cost of marker technologies and high-carbon fuels affects conventional estimates of mitigation costs. Consider, firstly, bottom-up approaches to mitigation cost estimates. Should the cost of marker technologies and fossil fuels increase—either through market forces or government policies—it is assumed that the price gap between low-emissions and high-emissions technologies automatically narrows, thus making the former technologies economically more viable to employ. Since this would shift the Marginal Mitigation Cost curve in Fig. 6.5 downwards, it is believed that the increase in the cost of marker technologies and fossil fuels would reduce total mitigation costs.

With top-down models, it is also assumed that a rise in the cost of marker technologies and high-carbon fuels would narrow the price gap between low-emissions and high-emissions technologies. Because this would increase the attractiveness of low-emissions goods and services, it is further assumed that the smaller price gap would alter consumer preferences and induce producers to substitute towards low-emissions techniques as well as develop cleaner production methods. Since this lowers the emissions-intensity of production, it decreases the extent to which real output must be reduced to achieve a specific emissions target. It is the lessened need to reduce real GWP which reinforces the belief amongst mainstream commentators that an increase in the cost of marker technologies and fossil fuels would reduce total mitigation costs.

One of the biggest problems with this estimation method is the assumption that an increase in the cost of high-emissions technologies and fossil fuels will have little or no effect on the cost of low-emissions techniques, which implies that an increase in the cost of high-emissions technologies will always reduce the price gap between low-emissions and marker technologies. This ignores the fact that high-emissions technologies and fossil fuels must be used to install low-emissions techniques and establish renewable energy systems on a broad scale. As Georgescu-Roegen (1978–79, p. 1053) once emphasised, "Any use of a presently feasible recipe [...] is a parasite of the current technology". Hence, not unlike the flow-on effect of higher primary-sector prices on the cost of secondary/tertiary-sector production, an increase in the cost of high-emissions technologies and fossil fuels would inflate the cost of employing low-emissions technologies. Consequently, the price gap between low-emissions and high-emissions technologies would not decrease to the extent expected, if at all.

Given this, a number of implications follow. Firstly, in the case of the bottom-up approach to mitigation cost estimates, there will be a tendency to underestimate the net cost of employing low-emissions technologies. Secondly, in the case of the top-down models, the extent of any substitution towards low-emissions technologies will be overestimated. In both instances, total mitigation costs will be undervalued. Thirdly, it is reasonably safe to conclude that a high growth rate of real GWP will seriously inflate the cost of high-emissions technologies and fossil fuels—a consequence of the impact of a high rate of growth on natural resource prices and its flow-on effect on the secondary-tertiary sector.[47] Hence, the higher is the growth rate of real GWP, the more likely it is that total mitigation costs will be undervalued. Furthermore, because a high rate of growth increases the cost of *all* technologies, it would undoubtedly increase the total cost of mitigation relative to a steady-state scenario.

6.3.2.3 Overshooting Desired Stabilisation Targets Increases Mitigation Costs

In much the same way that overshooting a stabilisation target would increase the total damage costs from climate change, so would overshooting a desired target increase total mitigation costs. To begin with, overshooting would bring on the harmful effects of a transient boost in global temperatures, which would increase the cost of employing low-emissions technologies. More than this, overshooting a desired target by a significant margin would lead to irreversible investments in emissions-intensive human-made capital (David 1985; Arthur 1989; Dietz and Stern 2008). This would lock economic systems into high-emissions technologies and foreclose numerous mitigation possibilities.

The impact of overshooting a desired stabilisation target on total mitigation costs is represented by Fig. 6.11. The four curves are not so much mitigation cost curves, but what I have labelled Total Greenhouse Gas Stock-Reducing Cost curves (TGSRC curves). The reason for the name change is because the TGSRC curves depict the total costs associated with reducing the atmospheric concentration of greenhouse gases (G) rather than reducing emissions per se. Like Fig. 6.8, a rightward movement along the horizontal axis represents an increase in the atmospheric concentration of greenhouse gases, which occurs whenever emissions levels (E) exceed the Earth's greenhouse gas-absorbing capacity (A) (i.e., $E > A$). As previously explained, this may or may not coincide with a reduction in greenhouse gas emissions. A decrease in the concentration of greenhouse gases is represented by a leftward movement along the horizontal axis, which can only occur if the cut in emissions is sufficient to reduce emissions levels below the Earth's greenhouse gas-absorbing capacity ($E < A$). No movement either way along the horizontal axis represents a stable atmospheric concentration of greenhouse gases ($E = A$).

The four TGRSC curves in Fig. 6.11 are upward-sloping to reflect the fact that shrinking the stock of greenhouse gases requires ever-more emissions cuts, which

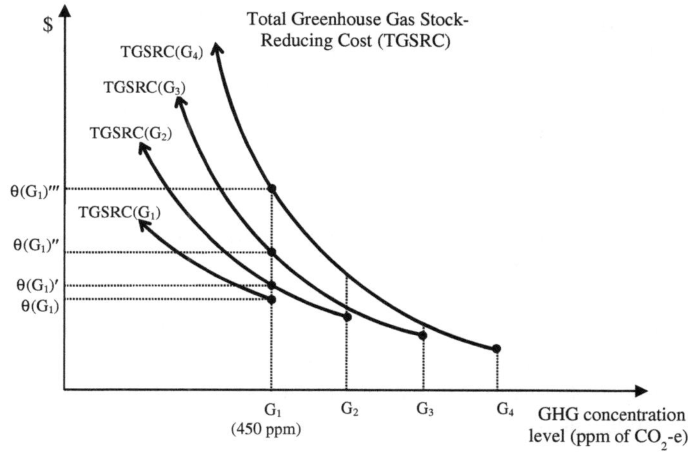

Fig. 6.11 Total mitigation costs associated with reducing the concentration of greenhouse gases from different peak levels to 450 ppm ($G_4 > G_3 > G_2 > G_1$)

become increasingly costly to accomplish.[48] Which of the TGSRC curves we happen to be operating on depends on the prevailing concentration of greenhouse gases and whether the stock of greenhouse gases is rising or falling. For example, if the atmospheric concentration of greenhouse gases has been rising and is currently at G_1 (450 ppm of CO_2-e), the cost associated with reducing the stock of greenhouse gases is depicted by TGSRC(G_1).[49] At the same time, the cost of remaining at 450 ppm of CO_2-e is $\theta(G_1)$. If, however, the concentration of greenhouse gases rises to G_2, the increased mitigation costs associated with exceeding G_1 means that the cost of reducing the stock of greenhouse gases is depicted by TGSRC(G_2). Consequently, the cost of stabilising at 450 ppm of CO_2-e increases to $\theta(G_1)'$. Should the concentration of greenhouse gases swell to G_3 or G_4, the relevant cost curves become TGSRC(G_3) and TGSRC(G_4). The cost of stabilising at 450 ppm of CO_2-e therefore rises to $\theta(G_1)''$ and $\theta(G1)'''$ respectively. As Fig. 6.11 clearly shows, the cost-effective achievement of a desirable stabilisation target requires deep initial emissions cuts, not the reverse as recommended by some commentators.

With this in mind, it is worth considering how well mainstream analysts have accounted for the impact that overshooting will have on mitigation costs. Many mainstream analysts, like Stern (2007, 2009), Ackerman (2007), and Garnaut (2008) have implicitly recognised the importance of limiting the extent of any overshooting by acknowledging the cost-moderating effect of early and sizeable emissions reductions. They have also warned against the dangers of waiting and becoming locked into long-lived, emissions-intensive, human-made capital. Nevertheless, the mainstream argument for deep, initial emissions cuts has centred on the cost-reducing benefits of technological advances, economies of scale, and learning-by-doing—all of which should rapidly emerge if governments institute the necessary policies to kick-start an early and vigorous mitigation programme. Although these developments are crucial in terms of keeping mitigation costs down, the cost-inflating impact of transient temperature rises has largely been ignored. It is because of this oversight that climate change modellers have further underestimated total mitigation costs.

6.3.2.4 Other Key Factors Overlooked in the Estimation of Mitigation Costs

In what is an important critique of bottom-up estimates of mitigation costs, Ekins et al. (2011) have recently outlined a number of factors which also explain why total mitigation costs are generally underestimated by climate change analysts. In their examination of Marginal Mitigation Cost curves, such as those derived by Nauclér and Enkvist's (2009) (see Fig. 6.5), Ekins et al. cast serious doubts over their applicability as a bottom-up tool for estimating mitigation costs and as a practical means of determining the cost-effectiveness of various mitigation techniques.[50]

In the first of their concerns, Ekins et al. (2011) question some of the input assumptions used to derive Marginal Mitigation Cost curves. They argue that because the curves are based on circumstances expected in twenty or more years' time, assumptions regarding the life-time of capital assets, technological progress, and investment costs are highly conjectural. Ekins et al. therefore believe it is incumbent upon practitioners to explain the reasoning behind many of the assumptions used in their cost calculations. Apart from failing to do this, the likes of Nauclér and Enkvist (2009) also fail to examine how sensitive their mitigation cost estimates are with respect to slight variations in their underlying assumptions.[51] According to Ekins et al., this makes it difficult to ascertain whether bottom-up estimates of mitigation costs are plausible.

The second concern of Ekins et al.'s (2011) relates to expected mitigation outcomes in the presence of particular market conditions. It was earlier revealed that Nauclér and Enkvist's Marginal Mitigation Cost curve indicates that annual emissions reductions of up to 11 $GtCO_2$-e by 2030 can be obtained at a negative net cost. This implies that a large proportion of the mitigation measures required by 2030 can more than pay for themselves, even in the absence of a greenhouse gas-emissions price.[52] According to Ekins et al. (2011), this is incompatible with the existence of an efficient market, since, if markets were working effectively, mitigation measures with negative net costs would have already been undertaken. Because many have not, Ekins et al. argue that either Nauclér and Enkvist's negative mitigation costs are illusionary or there are factors obstructing the implementation of what would ordinarily be viable mitigation activities. Assuming that Nauclér and Enkvist have accurately estimated project costs, Ekins et al. believe that the absence of viable mitigation measures must be the upshot of one or more of the following factors: (i) an insufficiently extensive definition of mitigation costs; (ii) non-financial barriers to the implementation of mitigation measures; and

(iii) the asymmetric application of discount rates. I would argue that another reason why some mitigation measures have not been undertaken is that many of them involve the construction of large-scale infrastructure and human-made capital with significant public goods characteristics (e.g., waste recycling and renewable-electricity generation). Consequently, many mitigation projects are unlikely to be undertaken by the private sector. Thus, without sufficient public-sector investment, mitigation levels are likely to be considerably less than what Fig. 6.5 predicts.

With regards to the first of these factors, Ekins et al. discover that Nauclér and Enkvist's definition of mitigation costs is narrowly confined to pure 'project' costs, which simply includes the cost of installing and operating a low-emissions activity. As a result, Nauclér and Enkvist ignore technology, sectoral, and broader macroeconomic costs in their overall estimate of mitigation costs.[53] Ekins et al. consequently believe that Nauclér and Enkvist's Marginal Mitigation Cost curve fails to represent the full extent of the real-world costs of implementing various mitigation projects. Furthermore, since many activities with a low pure project cost have high external costs, Ekins et al. believe that the Marginal Mitigation Cost curve fails to accurately represent the ascending cost order of individual mitigation projects.

As for implementation barriers, Ekins et al. (2011) outline a range of market imperfections that significantly impede or discourage the uptake of mitigation measures.[54] They include:

- *Agency issues* or split incentives. This occurs when the spillover savings from the implementation of mitigation measures are largely enjoyed by external parties (e.g., where the savings benefits from installing solar panels on the roof of rented properties are enjoyed by tenants rather than landlords).[55] From a mitigation perspective, the inability of investors to capture all financial savings reduces their incentive to invest in low-emissions technologies.
- *Information failures*. This is where investment in low-emissions technologies is stifled by uncertainty about future energy prices and a lack of knowledge concerning energy-efficiency options and potential savings opportunities.
- *Financing hurdles*. This situation arises when individual households and businesses have difficulty accessing financial capital markets. The ability to finance low-emissions projects is also rendered problematic by the fact that investments in low-emissions technologies often involve a combination of large upfront costs and meagre short-term pay-offs.
- *Inertia* and satisficing behaviour. This is where individual households and businesses act habitually or according to existing cultural norms. Consequently, they invariably overlook economically viable low-emissions technologies.
- *Path-dependency*. As previously mentioned, this occurs when businesses get locked into emissions-intensive production methods because: (i) many capital assets are long-lived; (ii) the general characteristics of on-coming physical capital are closely related to the features of currently and previously utilised capital, which is often emissions-intensive; and (iii) the use of productive capital often involves large sunk costs.

The final impediment to efficient market outcomes raised by Ekins et al. (2011) concerns the discount rate applied by investors when making decisions regarding low-emissions technologies. To construct their Marginal Mitigation Cost curve, Nauclér and Enkvist (2009) apply a 4 per cent discount rate to future benefits and costs.[56] However, for the Marginal Mitigation Cost curve to represent the potential for annual global emissions reductions in 2030, the 4 per cent discount rate must be applicable to all investors.[57] At best, this discount rate is an average rate that is largely pertinent to industrialised nations. It is not particularly relevant to investors in many low-GDP countries where nominal borrowing costs are much higher than 4 per cent. Yet Nauclér and Enkvist claim that their Marginal Mitigation Cost curve represents global mitigation possibilities.

To complicate matters, the 4 per cent discount rate reflects what might be regarded as a social discount rate, albeit many people would consider a 4 per cent rate to be well above the desirable social rate (see Chap. 5). Putting aside the debate about appropriate discount rates, whether individuals choose to invest in low-emissions technologies depends largely on their private discount rate, not the social discount rate. Investment decisions also depend on project-related risks and interest rates on loans, which are likely to be higher than the rate faced by governments. When all of these factors are taken into account, observed discount rates in the private sector are likely to be much higher than Nauclér and Enkvist's assumed 4 per cent discount rate (DeCanio 1993).

All things considered, private-sector decisions regarding investment in low-emissions technologies are based on the marginal abatement costs faced by individual investors which are inadequately represented by a Marginal Mitigation Cost curve derived from a range of social factors averaged across the global economy. There is, therefore, little doubt that Nauclér and Enkvist's Marginal Mitigation Cost curve cannot be used to predict the level of mitigation expected from the private sector with or without a government-generated price signal for greenhouse gases.

On top of these impediments to efficient market outcomes, Ekins et al. (2011) reveal many other weaknesses with Nauclér and Enkvist's (2009) Marginal Mitigation Cost curve. The most glaring is the fact that the curve only serves as a snapshot for a single year (e.g., 2030). It therefore provides no information about what has happened prior to the year in question or what is likely to occur thereafter. Yet insights into the timing and rate of investments in each mitigation technology are important because, as already illuminated, past investments in low-emissions technologies as well as government policies to encourage the uptake of low-emissions techniques can greatly affect current mitigation potentials and costs. This, in turn, can have a significant impact on the position and shape of the Marginal Mitigation Cost curve.

A less obvious weakness of the Marginal Mitigation Cost curve is the manner in which each mitigation measure is independently assessed in terms of its cost and abatement potential. In reality, changes caused by the implementation of mitigation measures are dynamic in nature. As Ekins et al. (2011) explain, the installation of one particular mitigation technology alters the abatement potential of all other low-emission technologies. For example, a shift to low-emissions electricity generation dramatically reduces the mitigation potential of all electricity-saving technologies. Crucially, this dynamism goes completely unrecognised when mitigation measures are assessed separately. Thus, by ignoring system interdependencies, Nauclér and Enkvist seriously overestimate abatement potentials. In the process, they underestimate many of the costs that collectively comprise the real-world costs of mitigation.

Although Nauclér and Enkvist (2009) can be accused of underestimating many mitigation-related costs, Ekins et al. (2011) also highlight a number of the potential co-benefits of mitigation overlooked by Nauclér and Enkvist that would help reduce net mitigation costs. To recall, the net mitigation costs represented by the Marginal Mitigation Cost curve includes the spillover savings obtained from the use of various mitigation techniques (e.g., lower electricity expenses arising from the use of energy-efficient lighting). However, the net mitigation costs used to derive the Marginal Mitigation Cost curve do not include many wider ancillary benefits that, if accounted for, would lower net mitigation costs. To support their point, Ekins et al. (2011) detail a number of mitigation co-benefits ignored by Nauclér and Enkvist. They include the positive health benefits from lower air-pollution levels (Pearce et al. 1996; OECD et al. 2000; Syri et al. 2001; Woodcock et al. 2009); a reduced reliance on fossil-fuel imports and a subsequent increase in national energy security; and improved indoor air quality in residential buildings located in low-GDP countries (Jakob 2006).[58]

Although taking account of these co-benefits would lower net mitigation costs, it is unlikely they would outweigh the previously-mentioned costs that Nauclér and Enkvist have overlooked. Altogether, it seems safe to assume that the Marginal Mitigation Cost curve would underestimate net mitigation costs by a substantial margin. Having said this, a combination of direct climate change policies and sustainable development policies of the type outlined in Chap. 3 would significantly increase the co-benefits generated by mitigation activities, which would go a long way towards quelling future rises in mitigation costs.

Finally, Ekins et al. (2011) believe that excessive reliance upon a Marginal Mitigation Cost curve to assess mitigation options can create the false impression that abatement potential is exclusively about cost-minimisation. According to Ekins et al., this can lull decision-makers into believing it is sufficient to establish a greenhouse gas-emissions price—either through a tax or an emissions-trading system—and that all the abatement potential rendered economically viable at the new emissions price will be realised. Unfortunately, because of the above-mentioned hidden costs, non-financial barriers, and complex interactions between the various mitigation technologies, government intervention beyond mere emissions pricing is necessary to achieve emissions targets in the most cost-effective manner. The need for further government intervention—in particular, government investment in low-emissions infrastructure—is also required given, as mentioned, a considerable portion of this infrastructure will have public goods characteristics. Some of the additional government policies and economic participation required to supplement an emissions-trading system will be outlined in Chaps. 7 and 10.

6.4 Future Changes in Marginal Damage Costs and Marginal Mitigation Costs and Its Implications for the Social Cost of Carbon

Given the findings of the previous section, we can conclude that climate change analysts have almost certainly underestimated climate change damage costs and mitigation costs. As significant as this conclusion might be, we are still left to consider the following policy-related questions:

- Are marginal damage costs and marginal mitigation costs likely to rise or fall over time and to what extent will the growth rate of real GWP play its part in their change?
- If marginal damage costs and marginal mitigation costs are likely to decline at some point, when might they begin to fall and under what circumstances?
- What effect will the gradual change in marginal damage costs and marginal mitigation costs have on the social cost of carbon and the price of greenhouse gas emissions over time?

Before answering these questions, it is important to understand the distinction between marginal mitigation costs and average mitigation costs—a distinction first raised in Chap. 5. Whereas the marginal cost of mitigation refers to the cost of reducing greenhouse gas emissions by one tonne (the next tonne), the average cost of mitigation refers to the average cost of reducing greenhouse gas emissions by a particular quantity of tonnes. That is, average mitigation costs equal total mitigation costs divided by the total quantity of emissions reductions. Whilst the marginal cost of mitigation provides some valuable insight into likely mitigation responses—and is therefore very important from a policy perspective—average mitigation costs indicate the cost impact across the entire economy of meeting specified mitigation targets. In view of how crucial it is to reduce average mitigations costs, I will have something to say about likely changes in this category of costs as well as probable changes in marginal mitigation costs and marginal damage costs.

6.4.1 Likely Changes in a GWP Growth-as-Usual Context

As alluded to, the future change in marginal damage costs and marginal mitigation costs is of great policy significance. This is because changes in both categories of costs can indicate what is likely to happen to the social cost of carbon. The social cost of carbon is of considerable importance because the price of CO_2-equivalent greenhouse gas emissions, which ought to reflect the social cost of carbon in the broadest sense,[59] plays a crucial role in determining the probable level of greenhouse gas mitigation. Indeed, if an emissions-trading system is in operation, the price of greenhouse gas emissions will go a long way towards determining how cost-effectively we respond to tightening emissions caps.

Because of the impact of real GWP on costs, it is best to consider the likely changes in marginal damage costs and marginal mitigation costs from different macroeconomic contexts. I will therefore outline the probable change in both categories of costs from a GWP growth-as-usual perspective and a steady-state economic perspective. In the process, I will assume our aim is to reduce greenhouse gas emissions in accordance with the emissions trajectory advocated by Anderson and Bows (2008) to achieve a 450 ppm stabilisation target.

6.4.1.1 Marginal Damage Costs and Marginal Mitigation Costs in a GWP Growth-as-Usual Context

In Chap. 4, it was shown that a 450 ppm stabilisation target will be impossible to achieve if real GWP continues to grow at an estimated rate of 2.3 per cent per annum. This is because growth of this magnitude would require an unobtainable 53.9-factor increase in the real GWP/emissions ratio by the end of the 21st century. It was also questioned whether a 2.3 per cent annual growth rate can be sustained over the present century given that the necessary resource demands would raise the global footprint/biocapacity ratio to around 3.16 by 2100—something that would seriously degrade the world's ecosystems and renewable resource stocks. Without wanting to downplay these concerns, to consider what might happen to marginal damage costs and marginal mitigation costs in a GWP growth-as-usual context, it will, for convenience only, be assumed that an annual growth rate of 2.3 per cent is possible, at least until 2100. It will also be assumed that the emissions cuts required to achieve a 450 ppm stabilisation target are technically feasible.

If we begin with marginal damage costs, we can expect this category of costs to increase well into the foreseeable future. There are many reasons why. To start with, even if stabilising the atmospheric concentration of greenhouse gases at 450 ppm of CO_2-e restricts the rise in average global temperatures to 2 °C above pre-industrial levels, a further 1.2 °C increase remains in the pipeline. Furthermore, for a temperature rise of 2 °C, moderate increases in climate change damages can still be expected (see Fig. 1.10). Secondly, as previously mentioned, the severe emissions cuts required to achieve a 450 ppm target will not prevent the concentration of greenhouse gases from peaking at around 500 ppm of CO_2-e. Because this would temporarily boost average global temperatures, it would further increase the damages associated with each additional tonne of greenhouse gas emissions.

Thirdly, with the global footprint/biocapacity ratio likely to rise to 2 by 2060–2070 and over 3 by the end of the century, the ecosphere would become increasingly vulnerable to the detrimental effects of climate change. This would exert additional upward pressure on the Marginal Damage Cost curve. Fourthly, the increased scarcity of renewable resource stocks plus a more rapid rate of non-renewable resource depletion would dramatically inflate natural resource prices. This would increase adaptation costs and reduce the capacity of future generations to limit the damage costs associated with present and future greenhouse gas emissions.

Finally, because growth-as-usual policies are designed to increase the production and consumption of real output and are less focused on boosting non-consumption benefits, any real output losses caused by growth-accentuated climate change damages—which are likely to be significant—would equate very closely to net welfare losses. This would further increase any upward pressure on the Marginal Damage Cost curve.

Of course, at the very same time, there would be numerous factors exerting downward pressure on marginal damage costs. These include technological progress, learning-by-doing, and economies of scale, which would improve the effectiveness of adaptation measures and lower the cost of adaptation strategies. Overall, however, the factors increasing marginal damage costs would far outweigh the factors reducing marginal damage costs. For this reason, we can expect the Marginal Damage Cost curve to keep shifting in an upwards direction in a GWP growth-as-usual context.

As for marginal mitigation costs, a number of factors affecting damage costs would also exert upward pressure on the Marginal Mitigation Cost curve. These include: (i) higher natural resource prices that would increase the cost of employing low-emissions technologies and negate some of the spillover savings and co-benefits generated by mitigation measures; (ii) the lack of policy emphasis on boosting non-consumption benefits, which would increase the net welfare cost of any consumption losses arising from the allocation of resources for mitigation purposes; and (iii) the overshooting of the 450 ppm stabilisation target, which would further inflate resources prices and increase the costs associated with efforts to reduce greenhouse gas emissions.

In terms of factors reducing marginal mitigation costs, there would be many. Once again, technological progress, learning-by-doing, and economies of scale would improve the effectiveness of low-emissions technologies and lower the cost of their use. These advances would similarly increase the spillover savings and co-benefits generated by many mitigation strategies, although, as mentioned, the inflated natural resource prices induced by a consistently high growth rate would negate some of these benefits. Technological progress would also increase the flexibility of mitigation technologies, which would facilitate the cost-effective application of available mitigation techniques.

Another important cost-reducing factor is government policy. Provided governments introduce policies to overcome many of the barriers impeding the effective implementation of mitigation measures, it should be possible to further promote the cost-effective use of mitigation techniques. If successful, this would exert additional downward pressure on the Marginal Mitigation Cost curve.

Taking everything into consideration, what is likely to happen to the Marginal Mitigation Cost curve in a GWP growth-as-usual context? The general consensus amongst climate change analysts is that the Marginal Mitigation Cost curve will shift downwards regardless of the growth rate of real GWP (see, for example, Stern 2007, Box 13.2, p. 343).[60] Because of the cost-increasing factors overlooked by most observers, I do not subscribe to this view. To the contrary, I believe the Marginal Mitigation Cost curve will shift upwards for some considerable time and, at best, eventually stabilise sometime towards the end of the 21st century.

When comparing the two cost curves, I would expect the rise in the Marginal Damage Cost curve to be greater than the rise in the Marginal Mitigation Cost curve. The main reason for this is the lasting damage caused by the prolonged period that greenhouse gases remain in the Earth's atmosphere. In addition, of the two cost categories that make up marginal damage costs—namely, the cost of adaptation and the cost of climate change damages—technological progress can do little to suppress the latter. What's more, the suppression of adaptation costs is unlikely to exceed whatever suppression of costs can be procured on the mitigation cost side.

6.4.1.2 The Social Cost of Carbon in a GWP Growth-as-Usual Context

Now that we have a better understanding of what is likely to happen to the Marginal Damage Cost curve and the Marginal Mitigation Cost curve in a GWP growth-as-usual context, we are well placed to say something about the social cost of carbon. To assist in this regard, refer to Fig. 6.12. The left-hand diagram shows the likely shifts of the Marginal Damage Cost (MDC) curve and the Marginal Mitigation Cost (MMC) curve from time t_0 to t_3. The duration between each point in time is assumed to be 30 years, such that t_0, t_1, t_2, and t_3 approximate the years 2010, 2040, 2070, and 2100 respectively.

Between t_0 and t_3, the Marginal Damage Cost curve consistently shifts upwards from MDC(t_0) to MDC(t_3). Conversely, the Marginal Mitigation Cost curve shifts upwards by a lesser amount and only between t_0 and t_2 (i.e., up from MMC(t_0) to MMC(t_2)). By t_2, or around 2070, the position of the Marginal Mitigation Cost curve effectively stabilises. The vertical 'blue' lines represent the maximum permissible emissions levels at each point in time to ensure eventual stabilisation at 450 ppm of CO_2-e. The narrowing of the gap between the blue lines represents the decline over time in required emissions cuts—a reflection of the need for most emissions cuts to occur in the first half of the 21st century.

The left-hand diagram in Fig. 6.12 reveals two entirely different emissions trajectories. The first trajectory (E_0, E_1, E_2, and E_3) constitutes the variation over time in what mainstream economists refer to as an 'efficient' emissions outcome. The efficient emissions trajectory is represented by the changing intersection of the shifting Marginal Damage Cost and Marginal Mitigation Cost curves.[61] The second trajectory (E_0, E_1', E_2', and E_3') constitutes the change over time in what ecological economists view as an 'optimal' emissions outcome—in essence, an outcome which is both cost-effective and consistent with a 'safe' stabilisation pathway.

Apart from one exceptional circumstance, the optimal emissions trajectory is represented by the changing intersection of the vertical blue lines and the shifting Marginal Mitigation Cost curves. As we shall see, the exception occurs when an efficient outcomes lies to the left of the intersection of a vertical blue line and a time-relevant Marginal Mitigation Cost curve. Crucially, Fig. 6.12 shows that the

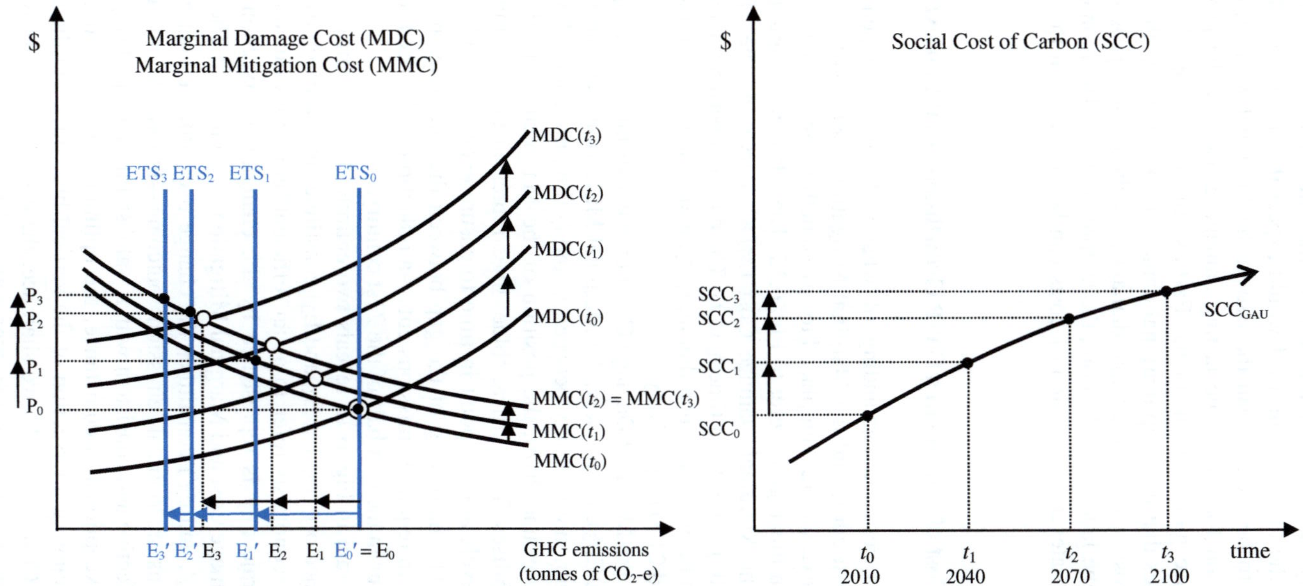

Fig. 6.12 Shifting Marginal Damage Cost (MDC) and Marginal Mitigation Cost (MMC) curves plus the change in the social cost of carbon (SCC) over time—a GWP growth-as-usual context. ○ = mainstream view of an 'efficient' emissions outcome. ● = ecological economics view of an 'optimal' emissions outcome

emissions cuts associated with an optimal trajectory are more severe than the cuts pertaining to an efficient emissions trajectory.

Without government intervention, something approximating the ‘efficient’ and the ‘optimal’ emissions trajectories cannot be realised. At this point, I will not discuss the relative merits of the policy mechanisms currently being promoted to achieve emissions targets. This will be dealt with in some detail in Chap. 7. Suffice to say, it is my belief that an optimal pathway can only be attained via the introduction of an effective global emissions-trading system. I will therefore continue the analysis on the assumption that such a system has been introduced. Given this assumption, the vertical blue lines in Fig. 6.12 represent the emissions caps that would be gradually tightened if a global emissions-trading system (ETS) was operating with the express aim of stabilising the concentration of greenhouse gases at no more than 450 ppm of CO_2-e.

With an emissions-trading system in place, the price of greenhouse gas emissions (P) would be determined by the price of tradeable emissions permits that would initially be auctioned off in a government-created permit market (Note: the key features of an emissions-trading system will be outlined in Chap. 7). Like all things traded in a market, permit prices would be determined by interacting demand and supply forces. The supply of emissions permits would be set by a designated authority and would be reduced over time to comply with the need to lower emissions caps. The changing supply of emissions permits is represented in the left-hand diagram in Fig. 6.12 by the shifting vertical ETS lines. The demand for permits would predominately emanate from greenhouse gas-emitting firms, although some residual demand would be registered by individuals and other organisations with a strong desire to possess permits. In view of the dominant influence of greenhouse gas-emitting firms, the changing demand for permits is represented in Fig. 6.12 by the shifting Marginal Mitigation Cost curves. Overall, the price of greenhouse gas emissions would mildly fluctuate around an equilibrium position located at the intersection of a vertical ETS line and a time-relevant Marginal Mitigation Cost curve.

Why, it might be asked, does the Marginal Mitigation Cost curve constitute a demand curve for emissions permits? In the first instance, it is safe to assume that a greenhouse gas-emitting firm will purchase emissions permits if the price of a permit is less than the cost incurred by the firm to reduce greenhouse gas emissions (i.e., if the cost of emitting an additional tonne of greenhouse gases is less than the cost of avoiding the emission of the same tonne of greenhouse gases). Secondly, it is also safe to assume that once the price of permits is greater than a firm's marginal mitigation costs, it is preferable for the firm to exit the permit market and incur the cost to reduce its greenhouse gas emissions. It is because of this that the Marginal Mitigation Cost curve is likely to provide a reasonable indication of the total number of permits demanded by all greenhouse gas-emitting firms at different permit prices. In other words, the Marginal Mitigation Cost curve is likely to serve as a good approximation of the demand curve for emissions permits.

It is also worth explaining at this point why, in the minds of ecological economists, the changing intersections of the vertical ETS lines and shifting Marginal

Mitigation Cost curves constitute an optimal emissions trajectory. It is optimal firstly because it is an emissions trajectory consistent with a 450 ppm stabilisation pathway. Hence, it is both 'safe' and ecologically sustainable. Secondly, with the price of greenhouse gases mildly fluctuating around the marginal cost of mitigation, it is likely that most of the required emissions reductions will be conducted by firms with the lowest mitigation costs. This means that the emissions levels consistent with the 450 ppm target will be generated in the most cost-effective manner. Unlike mainstream economists, ecological economists would not consider the efficient emissions trajectory depicted in Fig. 6.12 as optimal for the simple reason that it involves unsafe emissions levels—a further demonstration that efficiency need not be consistent with ecological sustainability.

There is one final aspect of the optimal emissions trajectory worthy of note. At t_1, t_2, and t_3, the optimal emissions levels of E_1', E_2', and E_3' exist where marginal mitigation costs exceed marginal damage costs. Although this imbalance would indicate the presence of some sacrificed efficiency, the associated welfare loss is best construed as the cost that must be incurred to ensure the atmospheric concentration of greenhouse gases is stabilised at no more than 450 ppm of CO_2-e. In other words, the imbalance would represent the burden that the next two or three generations would shoulder to ensure the primacy of ecological sustainability.[62] Whilst it is true that attaining the mainstream 'efficient' emissions trajectory would overcome these immediate welfare losses, it would leave in its wake a dangerous atmospheric concentration of greenhouse gases and therefore risk the welfare of future generations. In saying this, by employing an emissions-trading system to cost-effectively achieve the required emissions cuts, the welfare losses pertaining to an optimal emissions trajectory would be kept to a minimum.[63]

If we turn our attention to the right-hand diagram in Fig. 6.12, we can see that the price of greenhouse gas emissions in a growth-as-usual setting would, with an emissions-trading system in place, rise over the course of the 21st century. Whether the price path of greenhouse gas emissions—which constitutes the trend change in the social cost of carbon—would resemble the price path depicted in Fig. 6.12 depends largely on future movements of the Marginal Mitigation Cost curve.[64] Should this curve continue to shift upwards over time, as we would expect if the cost-increasing effect of higher resource prices is sufficient to more than offset the cost-reducing impact of technological progress, the social cost of carbon would almost certainly rise steeply over time.

6.4.1.3 Average Mitigation Costs in a GWP Growth-as-Usual Context

Before examining likely future events in a steady-state context, it is important to say something about average mitigation costs in a GWP growth-as-usual setting. To recall, average mitigation costs constitute the average cost of reducing greenhouse gas emissions by a particular quantity. Average mitigation costs are important because they indicate the cost impact across the economy of meeting mitigation targets. We have already seen that the Marginal Mitigation Cost curve in a GWP growth-as-usual setting will rise over time and possibly stabilise towards the end of the 21st century. The same is also likely to happen to the Average Mitigation Cost curve. This means that average mitigation costs are unlikely to decline in a GWP growth-as-usual setting as many people envisage. Indeed, along with the magnitude of the values themselves, I would seriously call into question Anderson's (2006) estimated decline in average mitigation costs from \$61 per tonne of CO_2 in 2015 to \$22 per tonne of CO_2 by 2050. In all, I believe the cost impact across the global economy of making the transition to a 450 ppm stabilisation target would continue to rise in a GWP growth-as-usual setting, thus rendering any attempt to resolve the climate change crisis in such a context very costly as well as highly problematic (see Chap. 4).

6.4.2 Likely Changes in a Steady-State Economic Context

6.4.2.1 Marginal Damage Costs and Marginal Mitigation Costs in a Steady-State Economic Context

Unlike growth-as-usual, it was shown in Chap. 4 that a 450 ppm stabilisation target is feasible if the world's nations make the collective transition to a steady-state global economy. It was also shown that increases in real GWP in a steady-state setting can be ecologically sustainable provided there is sufficient efficiency-increasing technological progress. Indeed, with realistic rates of technological progress, it was shown that the global footprint/biocapacity ratio would fall to around 0.9 by 2054 and 0.76 by 2100 even as real GWP more or less doubles between 2010 and the end of the 21st century (see Table 4.1).

Despite what would clearly constitute a transition to ecological sustainability, we would again expect the Marginal Damage Cost curve to continue shifting upwards in a steady-state economic setting. One of the main reasons for this is that a further 1.2 °C increase in average global temperatures will transpire regardless of the growth rate of real GWP. In addition, temporary temperature rises will occur due to the overshooting of the 450 ppm stabilisation target. Hence, even in a steady-state context, some moderate rises in climate change damages are inevitable. On the positive side, with the global footprint/biocapacity ratio declining to a sustainable level by the middle of the century, damages would not be amplified by an ecosphere rendered more vulnerable to climate change, as would occur in a growth-as-usual setting. In fact, with the global footprint/biocapacity ratio falling to 0.76 by 2100, it is possible that the ecosphere would become more resilient to exogenous shocks. If so, this would relieve some of the upward pressure on the Marginal Damage Cost curve.

As for adaptation costs, they are likely to increase for some time before falling. The increase would initially occur because natural resource prices will almost certainly keep rising for the next fifty years. For example, with the global ecological footprint likely to exceed global biocapacity until around 2044, the cost of renewable resources will almost certainly rise until the middle of the century. On top of this, non-renewable resource prices are likely to increase at an even faster rate and for longer given that: (i) many non-renewable resources will be in short supply within the next fifty years; and (ii) the non-renewable resource requirements of the global economy will remain high until 2060–2070 (see Table 4.2). Finally, even in a steady-state setting, the ability to maintain the energy inputs required to fuel the global economy will depend heavily on an expanded supply of nuclear energy until 2060. This will undoubtedly be very costly.

Although the concomitant rise in adaptation costs would restrict our capacity to reduce the damage costs associated with present and future greenhouse gas emissions, the impact would not be as severe in a steady-state setting and may recede altogether during the latter half of the 21st century, particularly as renewable resource prices stabilise and nuclear and non-renewable energy demands plummet. If we also take into account the many factors that would lower the cost of adaptation measures, such as technological progress, learning-by-doing, and economies of scale, we would expect the upward shift of the Marginal Damage Cost curve in a steady-state economic context to decelerate over time.

Moving to the marginal cost of mitigation, many of the factors that would exert upward pressure on the Marginal Mitigation Cost curve in a GWP growth-as-usual setting would do likewise in a steady-state context. This particularly applies to the rise in natural resource prices, although the eventual stabilisation of renewable resource prices would dramatically reduce cost-increasing pressures and allow for greater realisation of the spillover savings and co-benefits generated by mitigation measures. Also helping to lower marginal mitigation costs in a steady-state context would be the increasing policy emphasis on non-consumption benefits, since this would significantly reduce the net welfare cost of any consumption losses emanating from the allocation of resources for mitigation purposes. Lastly, it is not unreasonable to assume that all the remaining factors that are likely to reduce marginal mitigation costs in a GWP growth-as-usual context would also exert downward pressure on the Marginal Mitigation Cost curve in a steady-state setting.

All things considered, whilst most of the factors that would increase marginal mitigation costs would have their greatest impact during the first half of the 21st century, many cost-reducing factors would be felt beyond 2050. For this reason, I believe the Marginal Mitigation Cost curve would shift upwards over the next 30 years or so and then start shifting downwards. It is worth making the point that this conclusion vastly contradicts the more optimistically held belief that, with reasonable climate change policies in place, the Marginal Mitigation Cost curve is likely to begin shifting downwards almost immediately.

6.4.2.2 The Social Cost of Carbon in a Steady-State Economic Context

Figure 6.13 reveals the probable change in the social cost of carbon in a steady-state economic setting. Once again, the left-hand diagram illustrates the likely shifts of the Marginal Damage Cost (MDC) curve and the Marginal Mitigation

Cost (MMC) curve from time t_0 to t_3. It shows: (i) the Marginal Damage Cost curve shifting upwards at a decelerating rate between t_0 and t_3 (i.e., up from MDC(t_0) to MDC(t_3)); and (ii) the Marginal Mitigation Cost curve shifting upwards between t_0 to t_1 (i.e., up from MMC(t_0) to MMC(t_1)), and then downwards from t_1 to t_3 (i.e., down from MMC(t_1) to MMC(t_3)).

Not unlike Fig. 6.12, the left-hand diagram in Fig. 6.13 includes vertical ETS lines to represent the reduction in capped emissions should a global emissions-trading system be installed to stabilise the concentration of greenhouse gases at more than 450 ppm of CO_2-e. The left-hand diagram also depicts an 'efficient' and an 'optimal' emissions trajectory.

Because the price of greenhouse gas emissions (P) would be determined by the price of emissions permits, it would again fluctuate around an equilibrium position located at the intersection of a vertical ETS line and a time-relevant Marginal

Mitigation Cost curve. The one exception to this is the situation at time t_3. This is where the efficient outcome lies to the left of the above-mentioned intersection point. In these circumstances, the efficient outcome constitutes the 'true' optimum in the sense that E_3 is both a 'safe' and 'cost-effective' emissions level. It is also an outcome which is likely to prevail in the presence of an emissions-trading system because individuals and environmental groups would have an incentive to reduce greenhouse gas emissions below the maximum permissible level of E_3', which they could achieve by purchasing emissions permits and not using them. As will be explained in Chap. 7, individual citizens and environmental groups would have an incentive to reduce emissions to E_3 because, between E_3 and E_3', the price of emissions permits would be less than prevailing marginal damage costs.[65] Consequently, the total cost of buying and not using the necessary number of permits to reduce emissions from E_3' to E_3 would be less than the present value cost of the damages that the avoided emissions would otherwise inflict. Hence, by further reducing emissions to E_3, it is possible for society to enjoy a net welfare gain.

May I say, anyone interested in purchasing some permits would not know the prevailing marginal damage costs with precision simply because it is impossible to forecast the future impact of climate change. Thus, apart from the strong likelihood that the Marginal Damage Cost curve at t_3 would be positioned elsewhere to that shown in Fig. 6.13, its actual position would be unknowable. Although this would alter the eventual outcome at t_3, it is impossible for greenhouse gas emissions to exceed a safe level because emissions are capped at E_3'. However, assuming that the true Marginal Damage Cost curve is situated near MDC(t_3), we could expect the emissions level to be somewhere near the true optimum of E_3' and for the greenhouse gas price to be very close to P_3.

The probable price path of greenhouse gas emissions in a steady-state economic context is revealed in the right-hand diagram of Fig. 6.13. Even allowing for doubts over the exact price at t_3, it shows that the social cost of carbon is likely to rise at a diminishing rate over most of the present century and begin declining during its latter stages. As for the timing of any directional change in the social cost of carbon, it would depend on the deceleration rate of the shifting Marginal Damage Cost curve and the point at which the Marginal Mitigation Cost

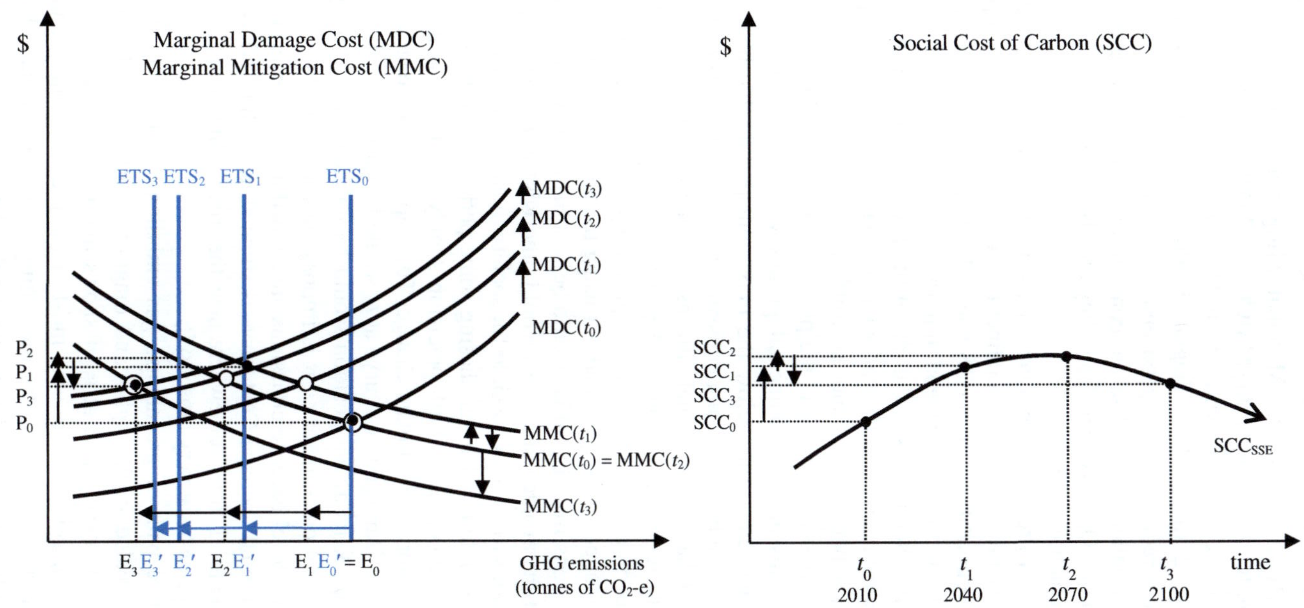

Fig. 6.13 Shifting Marginal Damage Cost (MDC) and Marginal Mitigation Cost (MMC) curves plus the change in the social cost of carbon (SCC) over time—a steady-state context. ○ = mainstream view of an 'efficient' emissions outcome. ● = ecological economics view of an 'optimal' emissions outcome

curve begins to shift downwards. It would also depend on how accurately marginal damage cost can be estimated and how easily and effectively individuals and organisations can reduce greenhouse gas emissions below capped levels via permit acquisitions. As we shall see in Chap. 7, the latter will depend very much on the design features of a fully operational emissions-trading system.

6.4.2.3 Average Mitigation Costs in a Steady-State Economic Context

With the Marginal Mitigation Cost curve likely to rise until around 2050 and shift downwards thereafter, we can expect the Average Mitigation Cost curve to do something similar. As a result, average mitigation costs will probably increase until the middle of the century before eventually declining. This suggests that the overall cost across the economy of meeting stringent mitigation targets will rise in the short-term. This does not mean that making deep initial cuts to greenhouse gas emissions would be undesirable. Nor does it mean that the economic welfare of humanity must be sacrificed in the process of moving to a steady-state economy to achieve a 450 ppm stabilisation target. Nevertheless, it will almost certainly mean that average mitigation costs will be higher than the levels estimated by most climate change analysts, including the average mitigation cost values predicted by Anderson (2006).[66]

It would be wrong, however, to blame the higher average mitigation costs on an emissions-trading system designed to ensure a safe atmospheric concentration of greenhouse gases. Should average mitigation costs be higher in the future, it will essentially be the result of insufficient action so far taken to combat escalating greenhouse gas emissions plus humankind's failure to prevent the global economy growing beyond its maximum sustainable scale.

6.5 The Shortcomings of Benefit-Cost Analysis

It should now be evident that ecological economists and mainstream economists not only differ in terms of what constitutes an optimal emissions trajectory, but also in terms of how the optimum should be ascertained and achieved. For ecological economists, an optimal emissions trajectory is one where the atmospheric concentration of greenhouse gases is stabilised at a safe level and where the necessary reductions in greenhouse gas emissions are achieved in the most cost-effective manner. As explained, ecological economists believe that a trajectory of this nature is best achieved by introducing an emissions-trading system that entails: (i) the sale of transferable emissions permits (emissions rights); (ii) the tightening of emissions caps to stabilise the concentration of greenhouse gases at no more than 450 ppm of CO_2-e; and (iii) a heavy reliance on permit prices to facilitate the cost-effective reduction in greenhouse gas emissions.

In direct contrast, mainstream economists believe it is the most efficient emissions pathway that constitutes the optimal trajectory. They also believe that an efficient pathway should be ascertained by deploying a benefit-cost analysis to determine the emissions levels where marginal damage costs equal marginal mitigation costs, or equivalently, where the present value of total emissions costs are minimised (see Fig. 5.2). Upon estimating the efficient emissions pathway, mainstream economists believe that an appropriate policy instrument should be introduced and fine-tuned over time to hit the relevant emissions targets. Thus, unlike the ecological economic approach of guaranteeing a safe concentration of greenhouse gases and then permitting a market-based adjustment to the optimum (which may lead to emissions cuts beyond the levels necessary to ensure a safe outcome), the mainstream economic approach involves pre-determining the optimal pathway and trying to meet emissions targets with the use of a policy instrument. Invariably the policy instrument recommended by mainstream economists to achieve the targets is an emissions tax.

Ecological economists have three fundamental objections to the mainstream use of a benefit-cost analysis. Firstly, they highlight that the efficient emissions trajectory determined by a benefit-cost analysis need not be optimal. As previously shown, this is because an efficient trajectory does not always coincide with a safe concentration of greenhouse gases. Secondly, as mentioned a number of times, the marginal damage costs and marginal mitigation costs which must be equated to ascertain an efficient emissions trajectory cannot be estimated with any degree of precision. Thirdly, climate change-related costs depend on future emissions levels as well as past and current emissions. Consequently, a benefit-cost analysis involves determining an efficient emissions pathway by using cost values that depend on prior assumptions about future emissions levels. The problem with this is that it leaves analysts with the absurd task of assuming what the future emissions pathway will be prior to conducting an exercise to establish what the future emissions pathway should be!

Given these three objections, I would like to conclude this chapter by explaining why, once and for all, a benefit-cost analysis should not be used to determine the most desirable emissions trajectory. To do this, I will briefly focus on the second objection above before returning to the first. Starting with the second concern, many observers have shown that the validity of employing a benefit-cost analysis as a decision-support tool is severely compromised if the problem at hand is complicated by human *ignorance* rather than risk and uncertainty (e.g., Howarth 2003; McGuire 2006; Spash 2007; Dietz et al. 2007a). Assuming this is so, as I believe it is, this raises the question as to whether our understanding of the climate change issue is confined to risk and uncertainty, or whether it is plagued by our inability to comprehend all climate change-relevant factors and their respective influences (ignorance).

To answer this question properly, we first need to consider what each of these terms mean. Risk is something we face when we know all the possible outcomes of an event as well as the probability of each outcome occurring. For example, risk would be something we would confront if we knew with certainty that: (i) X, Y, and Z are the only possible outcomes of a particular event; and (ii) that the probability of each outcome occurring is 60, 30, and 10 % respectively.

Uncertainty is something we face when we know all the possible outcomes but do not know the probability of each outcome occurring. Thus, we might know that X, Y, and Z are the only possible outcomes of a particular event, but have no knowledge of their respective probabilities. In some cases, it is possible for learning and the application of the scientific method to reveal the respective probabilities of the known possible outcomes. When this occurs, uncertainty is reduced to risk.

Ignorance prevails when all the possible outcomes of an event are unknown. By inference, this means that the probability distribution is also unknown. It is sometimes possible for learning and the application of the scientific method to reduce ignorance to uncertainty or, if a major breakthrough is made, to reduce ignorance to risk. I say "sometimes" because many forms of ignorance are irreducible. These include instances where the possible outcomes of an event remain unknowable due to the complex nature of the attendant circumstances and/or because there is always the potential for surprising and unpredictable outcomes to emerge (novelty).

If one considers all climate change-related factors and their potential cost implications, it is beyond dispute that humankind is operating in a state of ignorance. For example, no precise estimate has yet been made of the impact on average global temperatures of a given rise in the atmospheric concentration of greenhouse gases. As things stand, estimates of likely temperature changes are always presented in the form of a range of probable temperature increases (see, for example, IPCC 2007a, Table TS.5, p. 66). Moreover, the temperature ranges are always reported on the understanding that there is an indeterminable probability that more/less extreme temperature rises are possible.[67] Climatologists have also conceded that a precise estimate of rising global temperatures will never be forthcoming—an exemplary instance of irreducible ignorance.[68]

There are also other cost-related factors that cannot be predicted with certainty, even though it is possible to say something about the direction and speed in which some of them might move in the future. These include: (i) the extent of the damages inflicted on the natural and built environment by a given temperature increase; (ii) rates of technological progress; (iii) adaptation potentials; (iv) mitigation co-benefits and spillover savings; (v) the flexibility of mitigation technologies within and across different sectors of the economy; (vi) future growth rates of real GDP/GWP; (vii) government policies; and (viii) variations in natural resource prices caused by changes in natural resource scarcity. In sum, there is no way that a climate change modeller or policy-maker can possess the cost information required to know what constitutes an 'efficient' emissions trajectory.

Finally, let's assume that it is possible to obtain reasonably accurate estimates of climate change-related costs. Let's also assume that the cost information can be presented in the form depicted in Fig. 6.14. Not unlike Figs. 6.8 and 6.11, the horizontal axis represents the prevailing concentration of greenhouse gases in the Earth's atmosphere. Of the three curves appearing in Fig. 6.14, the Total Damage Cost (TDC) curve represents the damage costs pertaining to different atmospheric concentrations of greenhouse gases plus all the adaptation costs incurred to limit

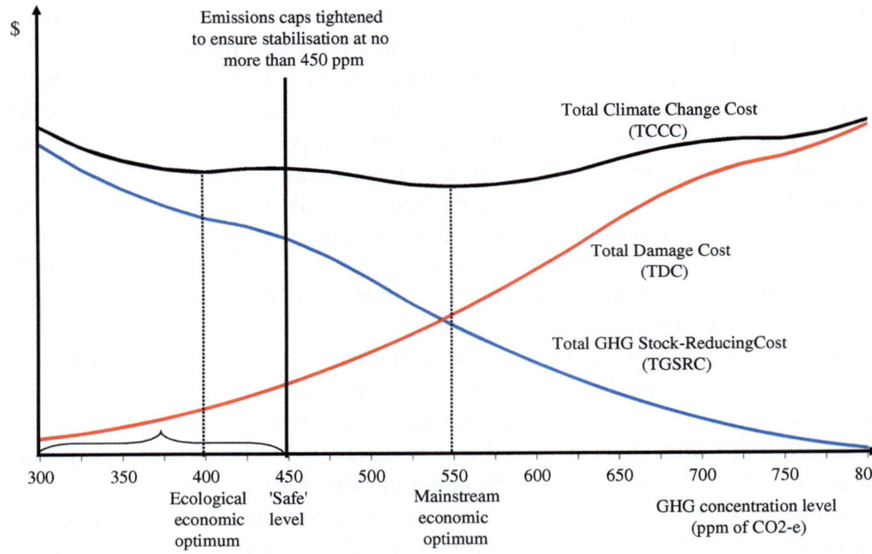

Fig. 6.14 Benefit-cost analysis and the difference between an ecological economic optimum and a mainstream economic optimum

climate change damages. The second major curve—the Total Greenhouse Gas Stock-Reducing Cost (TGSRC) curve—represents the total costs associated with stabilising the concentration of greenhouse gases somewhere between 300 and 800 ppm of CO_2-e. Lastly, the Total Climate Change Cost (TCCC) curve represents the sum total of all climate change-related costs.

The position and shape of the Total Climate Change Cost curve reflects a number of factors, one of which is assumed for explanatory purposes. Firstly, given that the prevailing atmospheric concentration of greenhouse gases is already around 440 ppm of CO_2-e, both ends of the Total Climate Change Cost curve correspond to concentration levels that are unlikely to be reached until some point in the distant future. For example, a concentration level of 300–350 ppm of CO_2-e would only follow decades of significant emissions cuts. Similarly, a high concentration level of 750–800 ppm of CO_2-e would require decades of elevated greenhouse gas emissions. Hence, the kinks at both ends of the curve reflect probable cost savings from learning-by-doing, scale economies, and rising mitigation co-benefits. Secondly, it is assumed that the atmospheric concentration of greenhouse gases will increase to 500 ppm of CO_2-e regardless of where it eventually stabilises. This means that stabilisation at a concentration level below 500 ppm must involve some overshooting. On the other hand, stabilisation at somewhere above 500 ppm would more than likely involve a gradual slowing of the rise in the concentration level prior to its eventual cessation. If so, no overshooting would be involved. Based on these assumptions, the section of the Total Climate Change Cost curve corresponding to concentration levels below 500 ppm of CO_2-e is slightly higher than the section where overshooting is avoided (i.e., above 500 ppm).

Given the hypothetical circumstances depicted in Fig. 6.14, what would constitute the mainstream economist's recommended stabilisation level (i.e., the mainstream economic optimum)? It would constitute the level where total climate change costs are minimised. In other words, it would amount to an atmospheric concentration level of 550 ppm of CO_2-e despite the fact that it exceeds what is broadly considered to be safe—namely, 450 ppm of CO_2-e.

Some people would question how an unsafe stabilisation level could be 'cost-minimising' if it has the potential to generate a catastrophic and therefore very costly outcome. The reason why it can is because the various costs used in a benefit-cost analysis are based on present value estimates. This makes it possible for bourgeoning future costs to be discounted to such an extent as to make the mainstream economists' 'unsafe' recommendation appear optimal.

What, on the other hand, would be the ecological economist's recommended stabilisation level? Since ecological economists would rule out all stabilisation targets above the 'safe' level, the ecological economic optimum would occur where total climate change costs are at their lowest between 300 and 450 ppm of CO_2-e. As per Fig. 6.14, this would constitute an atmospheric concentration of 400 ppm of CO_2-e.

Having said this, ecological economists would never bother to undertake a benefit-cost analysis because they appreciate the fact that climate change-related costs cannot be accurately estimated. What they would do is follow expert climatological advice and recommend the introduction of an emissions-trading system as a means of reducing emissions to the levels required to stabilise greenhouse gases at nor more than 450 ppm of CO_2-e. Should firms, organisations, and individual citizens have reasonable knowledge of their own mitigation and adaptation costs as well as the possible impact of future climate change damages, ecological economists would argue that the buying and selling of emissions permits—and the greenhouse gas price subsequently generated—would facilitate an adjustment to a stabilisation level somewhere near 400 ppm of CO_2-e, should this be the true optimum. In my opinion, no clearer distinction between the ecological economic and mainstream economic approaches to the climate change crisis can be better demonstrated than this.

Notes

1. To recall from Chap. 1, average global surface temperatures are expected to rise by a further 1 °C as a consequence of the greenhouse gases already emitted (IPCC 2007b).
2. By human-induced increase in the atmospheric concentration of greenhouse gases, I'm also including any rise in the concentration of greenhouse gases caused by natural positive feedback processes triggered by human-related actions.

3. For a more comprehensive list of likely climate change damages, see Table 1.4 and Fig. 1.10, and Stern (2007, 2009).
4. For a more comprehensive description of potential adaptation measures and costs, see Stern (2007, 2009).
5. Preferably it would be a pathway that, as much as possible, avoids the overshooting of the 450 ppm target.
6. The high rate of geosequestration assumed in the sustainable emissions scenario in Chap. 4 is due more to the dramatic decline in the rate of fossil fuel use than any massive increase in geosequestration potential.
7. Path-dependency very often leads to technological lock-in—a tendency of economic systems to become locked into human-made capital and technologies that are often sub-optimal (Waddington 1977; Silverberg 1988; Dosi and Metcalfe 1991).
8. See Stern (2007, Table 6.1, p. 165) for an illustration of how IAMs typically model human-induced climate change from greenhouse gas emissions right through to the final impacts on human well-being.
9. As also noted in Chap. 5, these losses can be expressed in terms of a social cost of carbon, which represents the net present value of the extended impact of climate change caused by the emission of one additional tonne of carbon into the atmosphere today.
10. At the global level, total damage costs are represented in terms of lost Gross World Product (GWP).
11. By making a subtraction for adaptation measures, the adjusted measures of real GDP generated by some IAMs come much closer to a Hicksian measure of national income. Unfortunately, the adjusted measures of real GDP almost always fall short of the Hicksian benchmark (see Eq. 3.5 in Chap. 3).
12. The difference in impacts between rich and poor countries reflects their different adaptive capacities rather than any difference in the valuation of impacts of climate change.
13. This includes natural resources that are humanly cultivated and managed.
14. The strong likelihood of this is reflected in the shape of the Adaptation Cost (AC) curves presented in Fig. 6.1.
15. The key assumptions typically embodied in top-down models are outlined and discussed in Stern (2007, Chap. 10).
16. In Fig. 6.4, the real GWP in 2050 would represent the projected mitigation cost for 2050.
17. For more on the technical efficiency of production, see Chap. 4.
18. This is particularly so if deeper emissions cuts reduce the detrimental impact on natural capital and thus keep resource costs down.
19. The Marginal Mitigation Cost curve is upward sloping in Fig. 6.5 rather than downward sloping as in Fig. 6.3 because the horizontal axis in the latter figure represents the emissions being generated. Consequently, in Fig. 6.3, a reduction in emissions involves a movement to the left along the horizontal axis. Because the horizontal axis in Fig. 6.5 represents emissions reductions, an increase in abatement levels involves a movement to the right along the horizontal axis. It therefore results in an inversion of the Marginal Mitigation Cost curve.

20. It is appropriate to rank mitigation techniques from the cheapest to the most expensive because the cheapest technique is likely to be the first employed and the most expensive the last employed.
21. The €10 net cost is the global average. It is likely to be higher in some countries and lower in others.
22. This assumes that mitigation techniques are being deployed from the cheapest technique to the dearest.
23. Cutting emissions by 2030 to these magnitudes does not guarantee stabilisation at the respective target levels. It would merely put us on the pathway to achieving the stabilisation targets. A further reduction in annual emissions cuts beyond 2030 would still be required.
24. To estimate the total mitigation cost of moving towards stabilisation targets of 500 ppm and 450 ppm of CO_2-e, it is necessary to aggregate the marginal cost values between 0 and 31 $GtCO_2$ and 0 and 42 $GtCO_2$ respectively.
25. The 40 $GtCO_2$-e of emissions cuts relative to business-as-usual emissions that are needed by 2030 to move to a 450 ppm stabilisation target is calculated on the basis that business-as-usual emissions are expected to be around 70 $GtCO_2$-e in 2030 (see Fig. 6.9) and that a 450 ppm stabilisation pathway requires emissions in 2030 to be 30 $GtCO_2$-e (see Table 4.1).
26. A more recent estimate of net mitigation costs by Beinhocker et al. (2008) suggests that the total mitigation costs associated with moving to the same stabilisation trajectory by 2030 is around US$1000 billion.
27. In another bottom-up approach undertaken by Anderson (2006), mitigation costs were initially estimated in terms of an average cost of mitigation (i.e., dollars per tonne of CO_2 reduced). To calculate total mitigation costs, the average cost was multiplied by a level of CO_2 abatement consistent with a stabilisation trajectory of 550 ppm of CO_2-e. As a means of demonstration, Anderson estimated the cost of reducing CO_2 emissions to 18 Gigatonnes per year by 2050 at US$930 billion. This figure, which helped inform the *Stern Review*, was the product of an average mitigation cost of US$22 per tonne of CO_2 and a projected emissions abatement level of 42.6 Gigatonnes (note: (i) the 42.6 Gigatonnes of emissions abatement was assumed to be the difference between the 18 Gigatonne emissions target for 2050 and an expected 60.6 Gigatonnes of emissions under a business-as-usual scenario; and (ii) the estimated mitigation cost of US$930 billion is approximately equal to 42.6 billion tonnes of abatement $\times$ US$22/tonne). Upon adding additional costs that are likely to be incurred to reduce other greenhouse gases, Stern (2007) estimated total mitigation costs at around 1 per cent of the projected GWP for 2050.
28. The extent of the detrimental impact would largely increase in line with the rise in average global temperatures.
29. These claims were first revealed in Daly (2000).
30. In addition, the units for output of the primary and secondary/tertiary sectors, as represented in Fig. 6.6, are not commensurate in terms of the matter-energy embodied within them. For example, one unit of primary-sector output could denote one tonne of cotton, whereas one unit of secondary sector output could

denote one cotton shirt. Clearly, in this case, there would be more matter-energy embodied in the one tonne of cotton than in the single shirt. Moreover, the one unit of primary-sector output would be capable of providing the resource inputs needed to produce many shirts and other cotton garments.

31. In all, because of the Entropy Law, the matter-energy embodied in the final output of the primary sector must be less than the matter-energy embodied in the natural resources utilised in the sector (Georgescu-Roegen 1971; 1978–79).
32. The reason why the addition of more labour and human-made capital cannot, without augmenting the efficiency of production, increase primary-sector output is because the marginal product of both labour and human-made capital is effectively zero. This is due to the complementary relationship between natural resources and human-made factors of production, which means that an increase in primary-sector output requires the application of more natural resources as well as more labour and/or human-made capital.
33. Examples given by Dietz et al. include a temperature-induced weakening of the ecosphere's carbon absorption capacity (Friedlingstein et al. 2006) and an increase in the natural methane released from wetlands and thawing permafrost (Gedney et al. 2004).
34. I say "more than likely be operating beyond its optimal scale" given that the economy of every rich country so far subjected to a Genuine Progress Indicator (GPI) study appears to be operating beyond its optimal scale.
35. To recall, 'uneconomic' growth occurs when the additional benefits of growth are outweighed by the additional costs. It occurs when economies grow beyond their optimal scale.
36. As explained in Chap. 2, aggregate welfare can be increased by redistributing income from the rich, whom have a low marginal utility of consumption, to the poor, whom have a high marginal utility of consumption.
37. A 2.3 per cent annual average growth rate of real GWP until 2100 has been forecast by the IPCC (IPCC 2000, Table 2–3).
38. In Chap. 4, the question was raised as to whether it would be possible to achieve the expected growth rate of GWP if the ecological footprint/biocapacity ratio rises to 3.16. The question was raised because a footprint/biocapacity ratio of this magnitude would seriously impact on natural capital stocks and therefore greatly reduce the natural resources required to produce the predicted output levels.
39. Assuming growth remains the over-riding global policy objective, there are three compelling reasons as to why the ecosphere will become ever-more vulnerable to climate change towards the end of the 21st century. Firstly, by around 2070, the ecological footprint/biocapacity ratio will exceed a dangerous threshold value of 2.0 (see Table 4.3). Secondly, the human appropriation of the terrestrial products of photosynthesis will be around 80 per cent. Finally, it is highly probable that all nine 'planetary boundaries' identified by Rockström et al. (2009) will have been exceeded (Note: three of these boundaries have already been transgressed and four are fast being approached).

40. The sustainable scenario simulated in Chap. 4 showed that the ecological footprint/biocapacity ratio could fall to as little as 0.76 by 2100 (see Table 4.3).
41. There would, of course, be factors at play exerting downward pressure on total damage costs. Since total damage costs includes the cost of adaptation, any technological advances, learning-by-doing, and/or economies of scale that lower adaptation costs would restrict the upward shift of the Total Damage Cost curve.
42. This is based on the radiative forcing of the following six Kyoto Protocol gases: carbon dioxide (CO_2), methane (CH_4), nitrous oxide (N_2O), hydrofluorocarbons (HFCs), perfluorocarbons (PFCs), and sulphur hexafluoride (SF_6). See Chap. 1 on what constitutes CO_2-equivalent gases and how the total concentration of CO_2-e gases is estimated.
43. In effect, $\Delta G = E - A$.
44. The 70 $GtCO_2$-e of annual emissions in 2030 under a growth-as-usual scenario (without mitigation) equals 30 $GtCO_2$-e of annual emissions in 2030 (Anderson and Bows (2008) trajectory) plus the 40 $GtCO_2$-e of emissions cuts required by 2030 to meet the 450 ppm stabilisation pathway. The 29 $GtCO_2$-e of required emissions cuts under the steady-state scenario is based on identical changes in the *GHG/R* and *R/GWP* ratios as the growth-as-usual scenario (see Eq. (6.1)). It is therefore calculated on the basis that the difference in emissions reflects the relative difference between 2010 and 2030 in the estimated increase in real GWP under the two scenarios. Thus:

$$\begin{aligned}
\text{Abatement}_{\text{SSE(2030)}} &= \Big[\big[\big(GWP_{\text{SSE(2030)}} - GWP_{(2010)}\big)/\big(GWP_{\text{GAU(2030)}} - GWP_{(2010)}\big) \\
&\quad \times \big(GHG_{\text{GAU(2030)}} - GHG_{(2010)}\big)\big] + GHG_{\text{450ppm(2030)}}\Big] - GHG_{(2010)} \\
\therefore \quad \text{Abatement}_{\text{SSE(2030)}} &= \big[\big[(\$72,874 - \$60,988)/(\$96,107 - \$60,988\big) \\
&\quad \times (70\,\text{Gt} - 53\,\text{Gt})] + 53\,\text{Gt}] - 30\,\text{Gt} \\
\therefore \quad \text{Abatement}_{\text{SSE(2030)}} &= 6\,\text{Gt} + 53\,\text{Gt} - 30\,\text{Gt} = 29\,\text{GtCO}_2\text{-e}
\end{aligned}$$

45. This is reflected by the growing gap between the 'no-mitigation' and 'mitigation' emissions pathways in Fig. 6.9.
46. The price gap would also include any indirect savings associated with the use of low-emissions and high-emissions technologies.
47. Once again, it should be remembered that the ratio of humanity's ecological footprint to global biocapacity is expected to be around 3.16 by 2100 if real GWP increases at an average rate of 2.3 per cent per annum (see Table 4.3). Even if this is feasible, it would involve significant diminutions of natural capital and much higher natural resource prices.
48. As we shall soon see, technological progress, economies of scale, and learning-by-doing can decelerate this increase in cost.
49. This assumes that the least-cost mitigation methods are being used.
50. I believe that many of Ekins et al.'s (2011) criticisms apply equally to top-down estimates of mitigation costs.

51. To reinforce their point, Ekins et al. (2011) provide a simple sensitivity analysis that shows that a shift to hybrid (low-emissions) vehicles is extremely sensitive to changes in vehicle investment cost, fuel efficiency, vehicle life-time, and the applied discount rate.
52. This also implies that the discounted savings of the required mitigation measures exceed the implementation costs (Ekins et al. 2011).
53. See Ekins et al. (2011) for an explanation of each cost category.
54. The ways and means to overcoming some of these market imperfections (market failures) were outlined in Chap. 3.
55. In economics parlance, a split incentive of this nature would be referred to as a 'positive externality'.
56. A 4 per cent rate was chosen by Nauclér and Enkvist for two reasons. Firstly, it closely reflects the nominal interest rate on long-term government bonds. Secondly, governments can readily borrow at the 4 per cent rate should they wish to incentivise capital-intensive mitigation opportunities.
57. Likely emissions reductions would also depend on the price of greenhouse gas emissions, since a mitigation technique is only economically viable if the going greenhouse gas-emissions price—should there be one—is greater than the marginal (net) mitigation cost of employing the technique.
58. A huge range of mitigation co-benefits has also been identified by Stern (2007), most of which are not included in Nauclér and Enkvist's Marginal Mitigation Cost curve.
59. As described in Chap. 5, the social cost of carbon equals the present value of the extended impact of climate change caused by the discharge into the atmosphere of one additional tonne of carbon. Given the radiative forcing potential of all greenhouse gases, the term 'social cost of carbon' is best extended to include all greenhouse gas emissions.
60. Needless to say, as Fig. 5.4 showed, a downward shifting Marginal Mitigation Cost curve does not mean that marginal mitigation costs have to be falling as mitigation levels rise.
61. To recall, the mainstream notion of an efficient outcome occurs where marginal damage costs equal marginal mitigation costs. The mainstream version of an efficient emissions trajectory has previously been illustrated in Figs. 5.2 and 6.3.
62. Despite shouldering this burden, there is no reason why the economic welfare of the next two or three generations would not continue to increase, as demonstrated in Fig. 4.6.
63. This assumes that governments have introduced a range of policies to overcome many of the barriers impeding the effective implementation of mitigation measures.
64. It would also depend on the design of the emissions-trading system, since a defective system would not generate an accurate price signal. The defective nature of the European Union's emissions-trading system—the EU-ETS—is very much to blame for the absurdly low price of emissions permits/greenhouse gas emissions associated with that particular system. I will have more to say about defective emissions-trading systems in Chap. 7.

65. The role of environmental and other groups in buying permits is vital given that collective action tends to overcome some of the 'free riding' that would occur if permit purchases were merely undertaken by individual citizens. For more on the key role of collective action, see Olson (1965) and Ostrom (1990).
66. Average mitigation costs are likely to be higher than Anderson's (2006) estimates, not only because the Average Mitigation Cost curve would be much higher than Anderson expects, but because Anderson's estimates are based on lower mitigation levels than those required to stabilise greenhouse gases at no more than 450 ppm of CO_2-e. Higher mitigations levels involve operating further up the Average Mitigation Cost curve.
67. In Table TS.5 of the IPCC's *Fourth Assessment Report*, the probable range of temperature increases are described as 'likely', which implies a probability of more than 66 per cent that the temperature increases will fall within the reported range. Alternatively, 'likely' implies a probability of 0–34 per cent that they won't. Whichever way one interprets the values, there is no certainty attached to them.
68. The lack of precise estimates and the reasons for them are well explained in the IPCC's *Fourth Assessment Report*. See IPCC (2007a, Frequently Asked Question 8.1, pp. 117–118).

Part III
A Way Forward and the Road to Sustainable Development

Chapter 7
The Case for an Emissions-Trading System to Help Resolve the Climate Change Crisis

7.1 Introduction

Before outlining an emissions-trading system to help resolve the climate change crisis, it is first necessary to argue the case for its introduction. There are a range of mechanisms available to directly regulate or influence greenhouse gas emissions. Most observers rule out the use of a blanket emissions quota on the basis that some form of market or pricing instrument is required to ensure emissions targets are achieved in the most efficient manner (Weitzman 1974; Pearce 1991; Nordhaus 2007b; Stern 2007, 2009).

As already demonstrated, whatever system is instituted, it will be insufficient unless there is a concerted effort by nations to move, when appropriate, towards a qualitatively-improving steady-state economy. The system must therefore be one that can be readily incorporated into a comprehensive policy framework aimed at achieving the broader goal of sustainable development. Given this imperative, an emissions-regulating system is required to help resolve the following policy goals: (i) *ecological sustainability*—which requires, among other things, stabilising the atmospheric concentration of greenhouse gases at no more than 450 ppm of CO_2-e; (ii) *distributional equity*—which involves efforts to narrow the income gap between rich and poor and to ensure the needy have access to basic necessities; and (iii) *allocative efficiency*—which requires a price to be assigned to greenhouse gas emissions to induce a shift away from fossil fuels; maximise the use value generated from the emission of greenhouse gases; and encourage the development and use of greenhouse gas-abatement technologies. The system must also be designed to resolve these policy goals in the above order.

It is my contention that these requirements can only be satisfied by introducing an emissions-trading system—the case for which I began in the latter part of the previous chapter. There are, however, many observers who are of the belief that an emissions-trading system is inferior to an emissions tax (e.g., Kahn and

P. Lawn, *Resolving the Climate Change Crisis*, DOI 10.1007/978-94-017-7502-1_7

Franceschi 2006; Stiglitz 2006; Nordhaus 2007b; Hansen 2009; Shapiro 2009). The preference for an emissions tax is based on what I believe are a number of false assertions. They include that: (i) taxes are much simpler to implement than an emissions-trading system; (ii) the price flexibility associated with an emissions-trading system leads to market uncertainty, which discourages the development and uptake of low-carbon technologies; (iii) emissions-trading systems include allowances for carbon offsets/credits that are vulnerable to abuse; (iv) countries still in the process of rapidly growing their economies (e.g., China) are more likely to agree to an emissions tax than an internationally-negotiated emissions cap; (v) emissions-trading systems fix an emissions 'floor' as well as an emissions cap, which renders voluntary action to further reduce greenhouse gas emissions futile; (vi) emissions-trading systems promote speculation by encouraging the buying and selling of emissions permits for the sole purpose of earning windfall profits; (vii) unlike emissions taxes, emissions-trading systems are highly corruptible and subject to the self-interested motives of political lobby groups; (viii) emissions-trading systems enable big firms to purchase large quantities of permits and control product markets; and (ix) the failure of existing emissions-trading systems to substantially reduce greenhouse gas emissions is a clear indication that emissions-trading systems are inherently deficient.

For much of the remainder of this chapter, I will endeavour to repudiate these assertions as well as highlight many other advantages that emissions-trading systems have over emissions taxes. Before this can be done, it is first necessary to outline the key features of an emissions tax and an effective emissions-trading system. This will allow for a straightforward comparison of the two systems as well as make it easier to explain why the poor design and implementation of many emissions-trading systems has led to their widespread failure.

7.2 Key Features of an Effective Emissions-Trading System

An emissions-trading system can exist in a variety of forms. My aim in this section is to outline the basic features of an effective emissions-trading system at the national level.[1] I will not therefore be describing the fine details of such a system, since these will vary from nation to nation to accommodate a country's unique circumstances. I'll have more to say about the fine details of an emissions-trading system, at both the national and global levels, in Chap. 10.

The most basic feature of an effective emissions-trading system would be the presence of a central-government authority to oversee the system's operation. The authority would have a number of crucial responsibilities. The most important of these would be the task of ensuring the nation's greenhouse gas emissions do not exceed maximum permissible limits. It would do this by restricting the number of emissions permits it issues (sells) so that it corresponds with a greenhouse gas emissions cap in keeping with the nation's emissions targets. It is highly likely that, into the future, the emissions cap would be dictated by a post-Kyoto emissions protocol.

The second major responsibility of the central-government authority would be to organise and conduct regular public auctions of emissions permits. Public auctions would be open to all eligible individuals, government establishments, and privately-owned organisations.[2] The frequency of auctions would depend largely upon the lifespan of the permits. Since drastic annual emissions reductions are required from many countries to meet emissions targets, the life of a permit would ideally be restricted to one year, which would preclude the banking of permits and their use at a later date.[3] To achieve this, a new batch of emissions permits would be auctioned every year. In addition, all existing permits would expire one year following their sale at a previous public auction. Should the central government wish to vary the number of permits between annual auctions, the central-government authority would be required to sell additional permits to increase their number or buy back existing permits to decrease the stock. Besides the issuing of greenhouse-gas offsets, at no stage would the authority distribute permits free of charge. The reason for this is that the free allotment of permits reduces potential competition in the industries where the permits have been distributed and unjustifiably transfers wealth to greenhouse gas-emitting firms (Parry 2003; Goulder 2005; Martinez and Neuhoff 2005; Parry et al. 2005a; Stern 2007).[4]

The third responsibility of the central-government authority would be its administration of a compliance mechanism. To achieve this role successfully, all major greenhouse gases must covered by the emissions-trading system—if not initially, then eventually. Wherever feasible, the monitoring process would involve the direct metering of emissions by the central-government authority. However, in circumstances where direct metering is impracticable, greenhouse-gas emitters would be obliged to regularly report their emissions to the authority. Reporting would encompass a requirement on the part of greenhouse-gas emitters to provide evidence to substantiate their declarations. To ensure integrity of the system, the authority would randomly inspect emitters to ensure entities are operating in accordance with the emissions rights inscribed in the permits they possess.

To improve the effectiveness of the compliance mechanism, there would be circumstances where the authority would vary the so-called 'point of obligation' by regulating a particular stage in the supply chain of a product rather than directly monitor the emissions generated by the production or consumption of the product.[5] The reason for this was raised in Chap. 3—it is often more feasible to regulate the extraction rate of a resource (e.g., oil) than regulate the emissions emanating from the end use of the resource (e.g., vehicle-exhaust emissions). Thus, in the particular case of oil, the central-government authority would require oil companies to purchase emissions permits, not vehicle-users.[6]

Clearly, in order for individuals and entities to know whether they must obtain emissions permits, the central-government authority must stipulate what activities it wishes to regulate. As we shall see in Chap. 10, it is imperative in terms of linking national emissions-trading systems for the authority to be consistent about the types of activities it regulates. It is also important that the monitoring procedures meet global monitoring standards. The latter would ensure consistency of supervision across nations, which would vastly improve greenhouse gas accounting procedures and the overall compliance process.

The fourth major responsibility of the authority would be the need to impose appropriate penalties to deal with breaches of the system. The authority would also extinguish permits commensurate with the actual emission of greenhouse gases and destroy all permits left unused by their expiry date. Of all the possible breaches, the most critical would be instances where entities emit greenhouse gases without possession of the required number of permits. To minimise the number of transgressions, which is fundamental to achieving a specific emissions target, the penalty for emitting greenhouse gases illegally must be ruinously harsh. At the very least, the penalty must involve the confiscation of an offender's emissions permits and a lengthy period where the transgressor is prohibited from possessing permits, which would prevent the offender from emitting greenhouse gases and/or undertaking a stipulated emissions-generating activity. A more appropriate penalty would extend to the imposition of a financially crippling fine, the confiscation of business assets, or, in very serious cases, imprisonment. Given that even the harshest of penalties would not avert all transgressions, the maximum level of greenhouse gas emissions must be reduced by an amount equal to the quantity of illegal emissions. This can be achieved by reducing the number of emission permits sold at subsequent public auctions.

What about emissions permits? In their simplest form, one permit would entitle the possessor the right to emit a specific quantity of carbon dioxide-equivalent gases within a limited time period. Ideally, a single permit would entitle its possessor the right to emit one tonne of carbon dioxide-equivalent gases at any time during a one-year period between a public auction and the permit's expiration date. In cases where a central-government authority chooses to regulate upstream emissions-related activities (e.g., oil extraction) rather than greenhouse gases generated at the end of the supply chain (e.g., petrol consumption), a single permit would enable an entity to engage in an activity that eventually results in the generation of one tonne of carbon dioxide-equivalent gases. To assist entities engaged in such activities, a conversion schedule would reveal the quantity of a particular resource that, when extracted and used, equates to the emission of one tonne of carbon dioxide-equivalent gases. This would enable fossil-fuel extractors (e.g., oil companies) to determine the number of permits they must purchase to undertake their resource-extraction operations. To obtain emissions permits, an individual citizen or entity must purchase permits at a public auction or from individuals or entities already in possession of legally valid permits.

Although the central government could instruct the authority to set and fix a sale price for permits, the permit price is best determined by the permit market. For this reason, the price of emissions permits would be entirely determined by demand-side and constrained supply-side forces.[7] Because demand-side forces would constantly vary between auctions, permit prices would fluctuate over the course of their one year lifespan.

It is my belief that an emissions-trading system would operate most effectively if all emissions permits remained with the central-government authority but were registered in the name of the individual or entity possessing the legal right to emit greenhouse gases. Upon an entity emitting a particular quantity of greenhouse gases, the authority would reduce the entity's right to emit greenhouse gases by an amount equal to the emissions it has generated (i.e., the authority would destroy the appropriate number of permits registered in the entity's name). Under this arrangement, there would be no physical exchange of emissions permits between the buyers and sellers of permits. There would simply be a transfer of emissions rights from the latter to the former on the authority's register. By making it easier for the authority to keep abreast of the owners of emissions permits, this arrangement would facilitate a more effective monitoring of greenhouse-gas emitters and increase the authority's ability to identify transgressors of the system. It would also allow the authority to act as a clearing-house for yet-to-be-used emissions permits, thus ensuring that the going price of permits—or the price of a category of emissions units—represents their true value and is always publicly available. This would significantly reduce transaction costs and lessen the need for permit buyers to hire permit brokers.

To further assist owners of emissions permits, the central-government authority would regularly inform them of the quantity of greenhouse gases they have emitted; the quantity of permits remaining in their possession; and the total quantity of annually-allotted permits that remain unused. The latter is important insofar as it indicates how many permits remain available for sale on the permit market, which would assist greenhouse-gas emitters to better plan their operations. The authority would also maintain a website to enable all citizens to access the above information at any time, albeit information on the ownership of permits would be restricted to the owners themselves (e.g., via password access).

As explained in relation to cap-auction-trade systems in Chap. 3, the number of emissions permits that any one person or entity can possess must be restricted. This is necessary to maintain competitive markets and prevent owners of permits from controlling product markets at the expense of society's welfare. Having emissions permits held by the central-government authority and registered in the name of permit owners would contribute significantly towards the effective implementation of any competition-promoting restriction on permit ownership.[8]

One of the more contentious elements of an effective emissions-trading system is the inclusion of greenhouse-gas offsets. Offsets allow emitting entities to generate and emit greenhouse gases beyond the permitted level in cases where they: (i) have successfully undertaken greenhouse-gas sequestration projects (e.g., have cultivated a greenhouse gas-sequestering timber plantation); or (ii) have altered their operations in such way as to augment the ecosphere's sequestration capacity (e.g., where a farming establishment increases the carbon content of its soils by altering its agricultural practices). Greenhouse-gas offsets would be granted to the relevant entities in the form of 'free' emissions permits, which the central-government authority would issue by increasing the quantity of allowable emissions registered in the relevant entity's name. For obvious reasons, the number of new permits issued must reflect the increase in the quantity of greenhouse gases sequestered as a consequence of the sequestration projects.

Contrary to some opinions, the rise in emission allowances emanating from an offset arrangement would not come free of charge to the relevant entities, since the cost of obtaining the permits would equal the cost of the sequestration initiatives. Presumably, a firm or entity would only invest in a sequestration project if the cost of the project is less than the cost of purchasing additional emissions permits. By offering entities the flexibility to embark on the lowest-cost response to an emissions cap, greenhouse-gas offsets increase the efficiency with which specific greenhouse-gas concentration targets can be achieved.

For all their potential benefits, there are legitimate concerns about greenhouse-gas offsets that must be adequately addressed. Firstly, it is vital that no offsets be granted for the mere prevention of an activity that would have led to the discharge of greenhouse gases. For example, greenhouse-gas offsets would not be granted to an entity that purchases and preserves a patch of forest sequestering greenhouse gases even if the intention of the previous owner was to clear the forest. The reason for this is that forest preservation does not increase the ecosphere's capacity to sequester greenhouse gases. Nor, therefore, does it increase a nation's capacity to safely emit greenhouse gases.[9] In addition, an effective emissions-trading system would require an individual or entity planning to clear vegetation, as an emissions-generating activity, to purchase emissions permits. Hence, although any action taken to preserve a patch of forest earmarked for clearance would not boost the supply of emissions permits, it would beneficially free up existing permits to enable other greenhouse gas-emitting activities to take place.[10]

Secondly, the central-government authority must thoroughly assess offset applications to ensure greenhouse gas sequestration projects are legitimate and additional. Even when approval is granted, projects must be continuously monitored to measure actual sequestration gains. Only when it is demonstrated that increases in sequestration levels have been realised would additional permits be issued to the relevant entities.

Thirdly, a penalty must be imposed in circumstances where sequestration gains are lost due to the failure of an entity to maintain a sequestering asset. An example would be a permit-receiving entity which logged a timber plantation it had previously cultivated for sequestering purposes. In most cases, this will simply require the owner of the asset to acquit the offsets originally received. However, where the offsets have been sold, the owner of the erstwhile asset would be subject to the penalties imposed on entities for illegally emitting greenhouse gases. In such circumstances, the central-government authority must decrease the number of available emissions permits to reflect the ecosphere's reduced sequestration capacity. This can be achieved by confiscating the offsets/permits forfeited by the contravening entity. However, in cases where the offsets have been sold, the central-government authority would need to sell fewer permits than planned at future permit auctions.[11]

7.3 Key Features of a Tax-Based System

Not unlike an emissions-trading system, an emissions tax-based system can exist in a myriad of forms. Many of the key features I will be outlining in this section are typical of the emissions-tax systems proposed by most emissions tax advocates.

Some readers would be surprised to find that many features of an emissions-tax system are similar to those of an effective emissions-trading system. For example, an emissions-tax system at the national level would require a central-government authority to oversee its operation. As for the authority's responsibilities, it would be a major task of the authority to determine the appropriate level of greenhouse gases to be emitted over a specific period, although, again, this would probably be dictated by a new emissions protocol. Crucially, whereas greenhouse gas emissions would be capped under an emissions-trading system, the desired level of emissions under a tax-based system would merely constitute a target that the authority would 'hope' to achieve. It would not be a target that the authority could guarantee because greenhouse gas emissions would be not be capped.

A second major responsibility of the central-government authority would be to estimate and set the emissions tax rate deemed necessary to achieve an emissions target. As much as a tax-based system would do away with public auctions of emissions permits, the need for a central authority to vary the price of greenhouse gas emissions would always remain. For example, should the authority wish to alter the nation's emissions target, it must set a different emissions tax rate to have any chance of achieving it.

As with an emissions-trading system, the third responsibility of the central-government authority would be the administration of a compliance mechanism. Transgressions would be largely confined to such misdemeanours as avoiding emissions detection, sabotaging emissions-measuring equipment, and the deliberate under-reporting of emissions levels. An additional penalty would not be imposed on entities for 'excessively' emitting greenhouse gases, since entities of all kinds would be at liberty to emit whatever quantity of greenhouse gases they desire. They would simply be required to pay the emissions tax for the privilege, which would involve no more than increasing their tax burden.

The emissions tax itself would consist of a charge of $X for every tonne of carbon dioxide-equivalent gases that an individual or entity emits. Not unlike an emissions-trading system, it would often be more feasible, if not preferable, to tax the extraction of a resource that, when used, leads to the emission of greenhouse gases.[12] In these instances, an emissions tax of $X would be charged against the resource-extraction activity (e.g., coal mining) for every tonne of carbon dioxide-equivalent gases that would eventually be generated from the use of the extracted resource.

Although very few advocates of an emissions-tax support the inclusion of greenhouse-gas offsets, there is no reason why offsets could not be incorporated into a tax-based system. Within the system, greenhouse-gas offsets would be conferred in the form of a tax credit. For example, a tax credit would be granted to an entity that has successfully undertaken a greenhouse gas sequestration project. In such cases, the tax credit would be received in the form of a reduction in an entity's tax bill equal to $X for every tonne of carbon dioxide-equivalent gases sequestered by the newly-created sequestration asset (Note: the $X would correspond to the prevailing emissions tax rate). The granting of a tax credit to a particular entity would not constitute a 'free lunch' because the cost incurred to receive the tax credit would equal the cost of the sequestration project.

Finally, if greenhouse-gas offsets were to be included in an emissions-tax system, all concerns with offset arrangements would be dealt with in much the same way as if an emissions-trading system was in place. For example, no tax credit would be granted to an entity that merely prevented an activity that would have resulted in the emission of greenhouse gases. In addition, penalties would be imposed on entities which, having received a tax credit, failed to maintain the sequestering asset. Lastly, to ensure tax credits are only granted where it is appropriate and legitimate, the central-government authority would rigorously monitor sequestration projects to ensure they are generating sequestration gains as promised.

7.4 A Theoretical Backdrop to Examine the Pros and Cons of Different Policy Mechanisms

To explain the superiority of an emissions-trading system, it is useful to provide a theoretical backdrop from which comparisons can be made between the different policy mechanisms. The theoretical backdrop I wish to outline is based on the mainstream economic perspective of the climate change problem, albeit it will be a slight variation of the mainstream framework outlined in Chaps. 5 and 6. This may seem a strange choice. After all, it was evident from these two chapters that there are significant weaknesses with the mainstream approach. This said, it will soon become apparent that, even from a mainstream perspective, an emissions-trading system is far superior to an emissions tax. My ultimate aim is to complete the comparison between the two systems by altering the assumptions underpinning the mainstream framework to take account of concrete realities. Once this is done, the superiority of an emissions-trading system should be unequivocal.

As a starting point, consider Fig. 7.1, which illustrates a mainstream view of the social welfare loss that arises if a government fails to take any climate change action. For simplicity, we shall assume that all the benefits and costs associated with greenhouse gas emissions are known by a relevant government authority. We shall also assume that all the greenhouse gases emitted in a nation over a one year period are generated by three profit-maximising firms—ABC Ltd, JKL Ltd, and XYZ Ltd. The marginal benefit (*mb*) curves in Fig. 7.1 represent the marginal profits that the three firms respectively earn from emitting greenhouse gases (i.e., from selling the goods and services produced from their greenhouse gas-emitting

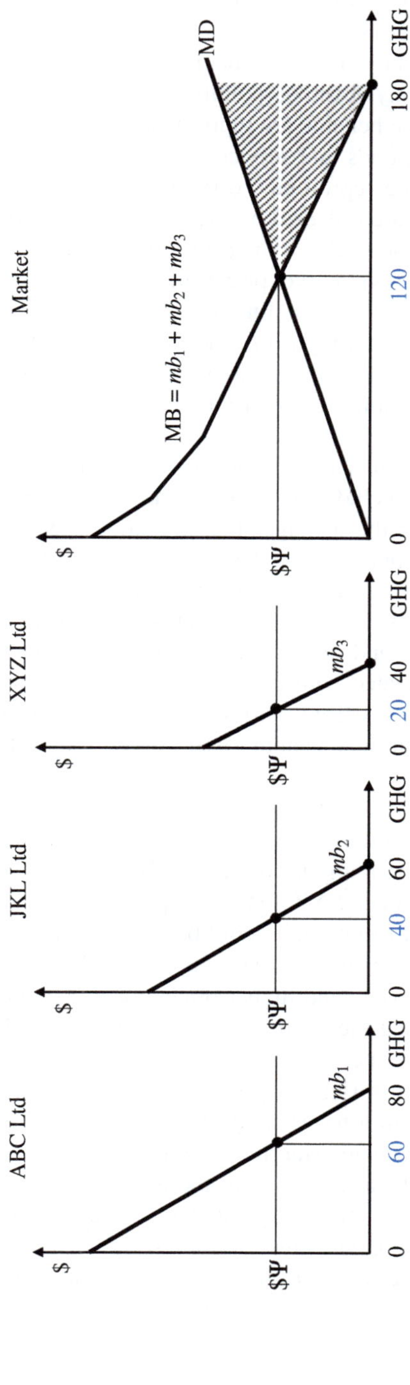

Fig. 7.1 The social welfare loss associated with excessive greenhouse gas emissions (no policy response). *Note* GHG denotes tonnes of greenhouse gas emissions

activities). The downward sloping nature of the marginal benefit curves indicates that marginal or additional profits typically diminish as the three firms increase their greenhouse gas emissions.

How are the marginal benefit curves depicted in Fig. 7.1 derived? Often the benefits associated with any form of pollution are represented as consumption benefits. In other words, the benefits are represented in terms of the utility enjoyed from the consumption of goods and services produced by the polluting activities. These same benefits can be represented differently. This is because, at the margin, consumption benefits are reflected by the prices consumers are willing to pay for various goods and services. Since the prices paid by consumers are the prices received by sellers (producers), market prices effectively represent the revenue that sellers receive from producing and selling an additional unit of output.[13] Thus, in the case of greenhouse gas-emitting firms, the marginal revenue earned from emitting an additional tonne of greenhouse gases effectively equals the marginal consumption benefits that consumers enjoy from consuming the additional output they generate.

At the same time, there are a range of costs associated with greenhouse gas-emitting activities. These include the costs of the natural resources, capital, and labour used in production; the damages that greenhouse gas emissions inflict on the natural and built environment; and the cost of the adaptation measures that a nation must undertake to limit the environmental impact of greenhouse gas emissions. For our purposes, it will be assumed that the three greenhouse gas-emitting firms incur capital, labour, and resource costs but do not incur the environmental and adaptation costs associated with their emissions. It is by subtracting the capital, labour, and resource costs from marginal revenue that we obtain, in Fig. 7.1, the marginal profit or marginal benefit (*mb*) curves for each firm (Pearce and Turner 1990, pp. 68–69).

The different position of the marginal benefit curves indicates that the marginal profits earned from emitting greenhouse gases vary between the three firms. As depicted in Fig. 7.1, the marginal profit of ABC Ltd is greater than JKL Ltd, which, in turn, is greater than XYZ Ltd. A disparity of this nature could be the result of many factors. For example, it could be the consequence of ABC Ltd generating more use value in production (i.e., producing the highest quality goods), which would enable it to command higher prices for its output.[14] It could also be the result of ABC Ltd being the most efficient producer, which would allow it to produce its output at a lower per unit cost.

The right-hand diagram in Fig. 7.1 portrays the market conditions confronting the three greenhouse gas-emitting firms. It includes an aggregate Marginal Benefit (MB) curve and an aggregate Marginal Damage (MD) curve. The Marginal Benefit curve represents the aggregation of the individual marginal benefit curves of the three firms. Conversely, the Marginal Damage curve represents the environmental damage cost of total greenhouse gas emissions.[15]

If there is no policy response from the government to regulate greenhouse gas emissions, total emissions are determined entirely by the profit-maximising desires of the greenhouse gas-emitting firms. The profit-maximising emissions level of each firm occurs where the marginal benefit or marginal profit from emitting greenhouse gases equals zero.[16] As shown in Fig. 7.1, the profit-maximising emissions level of each firm is 80 tonnes of greenhouse gases per year for ABC Ltd; 60 tonnes per year for JKL Ltd; and 40 tonnes per year for XYZ Ltd. Consequently, the total emission of greenhouse gases is 180 tonnes per year, which is also indicated by the point where the aggregate Marginal Benefit curve cuts the horizontal axis.

From a mainstream economic perspective, the welfare-maximising level of greenhouse gas emissions is determined by the intersection of the Marginal Benefit and Marginal Damage curves.[17] In Fig. 7.1, this intersection corresponds to an annual emissions rate of 120 tonnes. The intersection of the Marginal Benefit and Marginal Damage curves also gives rise to the welfare-maximising price of greenhouse gases, which is $\$\Psi$ per tonne of greenhouse gas emissions. By equating this price with the marginal benefits of the three greenhouse gas-emitting firms, one can determine the welfare-maximising emissions level of each firm, which is 60 tonnes of greenhouse gases per year for ABC Ltd; 40 tonnes per year for JKL Ltd; and 20 tonnes per year for XYZ Ltd. In the presence of no government policy action, the total emissions level exceeds the welfare-maximising level by 60 tonnes, with each firm exceeding their welfare-maximising emissions by 20 tonnes. The consequential welfare loss is represented by the shaded area in the right-hand diagram in Fig. 7.1.

7.4.1 The Social Welfare Impact of Introducing a Non-tradeable Emissions Quota

Let's assume that the government wants to reduce annual greenhouse gas emissions to the welfare-maximising level of 120 tonnes. Figures 7.2, 7.3, and 7.4 illustrate the impact of implementing three different policy mechanisms. Figure 7.2 reveals the social welfare effect of instituting a non-tradeable greenhouse gas emissions quota where the annual quota is set at 120 tonnes and is equally divided among the three greenhouse gas-emitting firms (i.e., each firm is subject to a firm-level emissions quota of 40 tonnes).[18] To prevent the total quantity of greenhouse gas emissions exceeding the quota, we shall assume that the government introduces a penalty harsh enough to deter each firm from exceeding their firm-level quota of 40 tonnes (e.g., confiscation of business assets or imprisonment).

Despite the introduction of the emissions quota, much of what appears in Fig. 7.2 is the same as in Fig. 7.1. What differs is the inclusion of a vertical bold line in the right-hand diagram. This line represents the total emissions quota of $\underline{120}$ tonnes. There is also now a vertical bold line in each of the firm diagrams to denote a firm-level emissions quota of $\underline{40}$ tonnes.

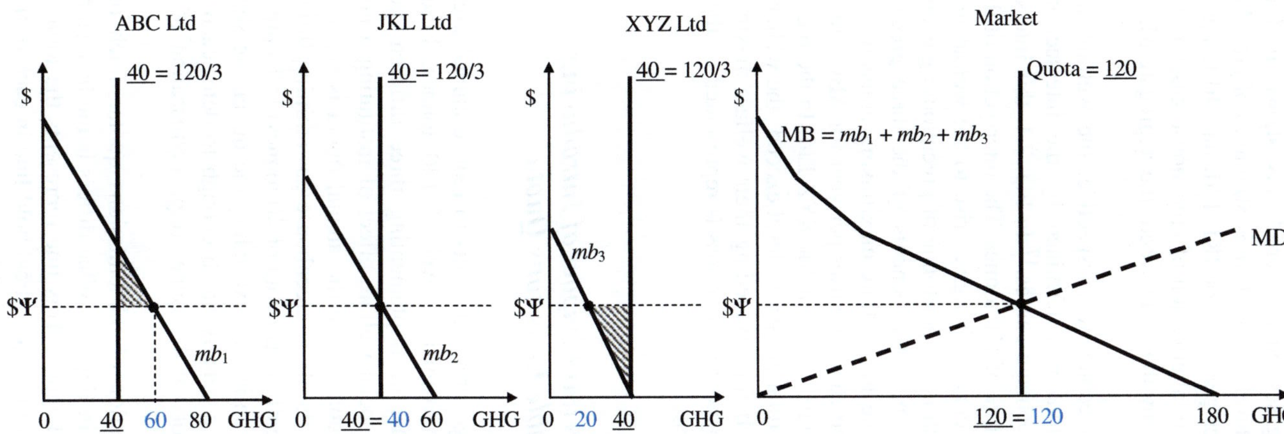

Fig. 7.2 The inefficiency of a non-tradeable greenhouse gas quota). *Note* GHG denotes tonnes of greenhouse gas emissions

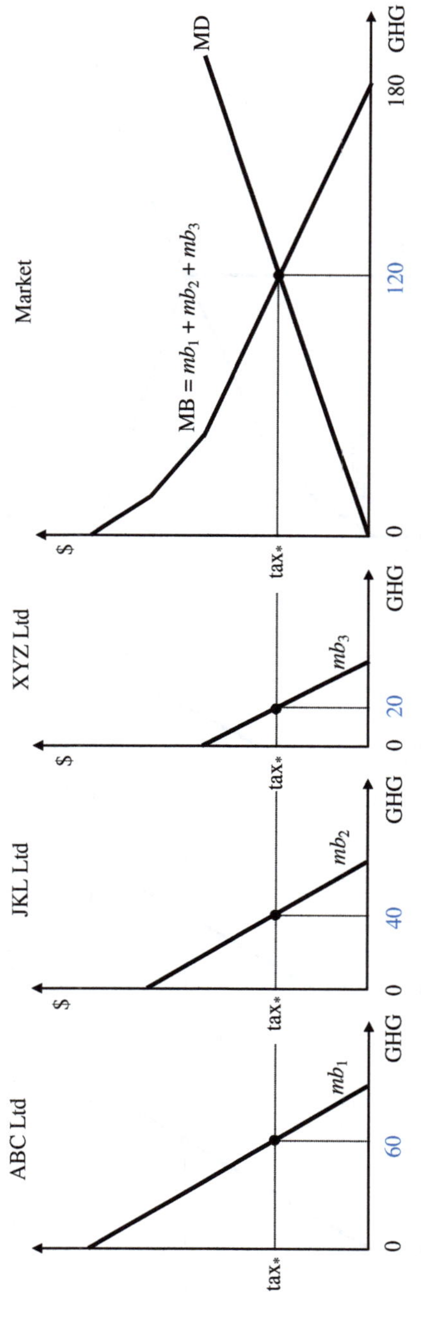

Fig. 7.3 Achieving a welfare-maximising outcome with the use of an emissions tax. *Note* GHG denotes tonnes of greenhouse gas emissions

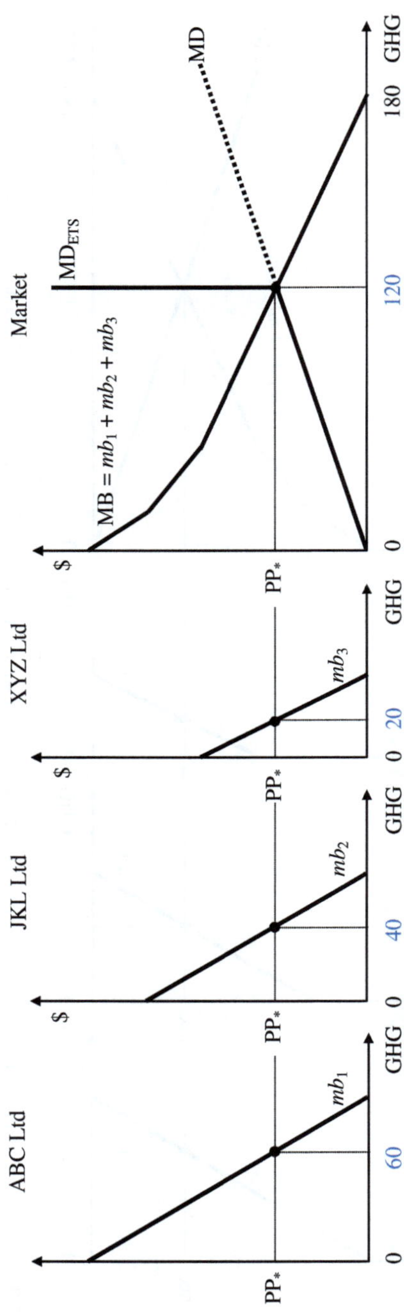

Fig. 7.4 Achieving a welfare-maximising outcome with the use of an emissions-trading system. *Note* GHG denotes tonnes of greenhouse gas emissions

To consider the impact of the quota on the three firms, it first needs to be recognised that an emissions quota does not have the effect of internalising the damage cost of greenhouse gas emissions. In other words, the quota does not require the three firms to bear the spillover cost of their production activities.[19] For this reason, the quota does not alter the firms' profit-maximising emissions levels. Consequently, without a quota, all three firms would continue to operate as indicated in Fig. 7.1. Because the introduction of an emissions quota limits the emissions of all three firms to 40 tonnes per year, XYZ Ltd is unaffected. The same cannot be said of ABC Ltd and JKL Ltd. ABC Ltd is compelled to reduce its annual greenhouse gas emissions by 40 tonnes, whilst JKL Ltd is forced to reduce its annual emissions by 20 tonnes.

Although the non-tradeable emissions quota guarantees a welfare-maximising outcome at the macro level, its inflexible nature—the consequence of prohibiting the firms from trading all or part of their emissions quota—ensures that only JKL Ltd operates in manner consistent with welfare-maximisation at the micro level. That is, with all three firms subject to a firm-level emissions quota of 40 tonnes, only JKL Ltd operates where $\$\Psi = mb_2$. In contrast, ABC Ltd operates where $\$\Psi < mb_1$ (i.e., 20 tonnes short of its welfare-maximising emissions level), whilst XYZ Ltd operates where $\$\Psi > mb_3$ (i.e., 20 tonnes above its welfare-maximising emissions level). Together, the two shaded areas in Fig. 7.2 represent the social welfare losses arising from the inflexibility of a non-tradeable emissions quota. By comparing the sum of these two shaded areas and the shaded area in Fig. 7.1, it is obvious that a non-tradeable emissions quota is preferable to no government policy action. However, it is the continued existence of a social welfare loss that discourages most policy-makers from implementing a non-tradeable emissions quota as a means of climate change mitigation.

7.4.2 The Social Welfare Impact of Introducing an Emissions Tax

Consider, on this occasion, the social welfare effect of instituting a greenhouse gas emissions tax (Fig. 7.3). Unlike a quota, an emissions tax internalises the damage cost of greenhouse gas emissions. In so doing, it forces greenhouse gas-emitting firms to bear the spillover cost of their activities, thus compelling them to take account of the tax when determining emissions levels.

Ideally, from a mainstream economic perspective, the government would set an emissions tax at the welfare-maximising price for greenhouse gases. That is, it would set the tax rate at $\$\Psi$ per tonne of greenhouse gas emissions. Should the government do this, Fig. 7.3 is almost identical to Fig. 7.1. The main difference is that 'tax$*$' now replaces $\$\Psi$ in Fig. 7.3. The other difference is the adjusted emissions levels of the three greenhouse gas-emitting firms. Provided the firms are short-term profit-maximisers, they would all reduce their annual greenhouse gas emissions to the welfare-maximising level.

The reason why the firms would cut their greenhouse gas emissions is straightforward. Consider ABC Ltd, which, if did not have to pay an emissions tax, would maximise its short-term profits by emitting 80 tonnes of greenhouse gases per year. With the tax imposed, it is only profitable for ABC Ltd to emit an additional tonne of greenhouse gases if its marginal profit—that is, the additional profit it earns from emitting an extra tonne of greenhouse gases—is greater than the unit cost of the tax. As can be seen from Fig. 7.3, this occurs up to an annual emissions level of 60 tonnes. Once ABC Ltd's emissions reach 60 tonnes, the tax rate coincides with its marginal profit. At this point, its profits are maximised. Clearly, any increase in emissions above 60 tonnes would reduce ABC Ltd's profits. Assuming that ABC Ltd is a short-term profit-maximiser, we would expect it to emit no more than 60 tonnes of greenhouse gases per year. As for JKL Ltd and XYZ Ltd, we would expect them to emit 40 and 20 tonnes of greenhouse gases per year respectively, thus reducing the total annual quantity of greenhouse gas emissions from 180 to 120 tonnes.

Ultimately, the introduction of an emissions tax maximises the welfare of society insofar as the welfare loss associated with no policy action—that is, the shaded area in Fig. 7.1—is entirely eliminated. In addition, the welfare loss arising from the inflexibility of a non-tradeable emissions quota is avoided because an emissions tax allows the most efficient producers (i.e., the firms that can most afford to pay the emissions tax) to emit the majority of the reduced quantity of emissions. Thus, ABC Ltd emits the most greenhouse gases because it generates larger consumption benefits per tonne of greenhouse gases emitted and/or because it generates fewer greenhouse gas emissions per unit of output produced.[20] Having ABC Ltd emit the most greenhouse gases and XYZ Ltd the least maximises the welfare benefits to society.

There is, nonetheless, a significant weakness associated with an emissions tax. As we shall soon see, this weakness constitutes a major reason why an emissions-trading system is a superior policy mechanism. The major problem with an emissions tax is that it is not severe enough to prevent the welfare-maximising emissions level—indeed, any chosen emissions target—from being exceeded. For example, if set at $\$\Psi$ per tonne of greenhouse gas emissions, a tax would serve as a minor penalty for any firm that chooses to emit beyond its short-term profit-maximising level.

Why, it might be asked, would a firm forego short-term profits? There are many reasons. Firstly, not all firms are profit-maximisers. For various reasons, some firms operate as sales revenue-maximisers, even if it is only for a short period of time. Secondly, should the owners (shareholders) of a firm explicitly desire profit-maximisation, it is well known that the separation of ownership and management can result in a firm failing to operate in a profit-maximising manner (Baumol 1959; Williamson 1964; Alchian 1965; Jensen and Meckling 1976). Thirdly, many firms prefer to maximise their market share and forego short-term profits in the belief that it can reduce future competition and lead to long-term profit-maximisation.

Whatever the case, achieving a particular emissions target with an emissions tax requires the tax-setter to make assumptions about the behaviour of greenhouse gas-emitting firms that may not reflect reality. Whilst there is little doubt that an emissions tax, by internalising the spillover cost of greenhouse gas emissions, can induce greater efficiency, thus leading to the emission of fewer greenhouse gases per unit of output produced, it is also clear that there is nothing inherent about the tax to prevent the total emissions of greenhouse gases exceeding a desired target level—a classic example of the Jevons' Paradox explained in Chap. 3.

7.4.3 The Social Welfare Impact of Introducing an Emissions-Trading System

Consider, lastly, the social welfare impact of introducing an emissions-trading system (Fig. 7.4). We shall assume that a government-authority limits the number of emissions permits it auctions to 120 in order to cap the total emissions of greenhouse gases at the welfare-maximising level of 120 tonnes per year.[21] We shall also assume, firstly, that any organisation emitting greenhouse gases without possession of a permit is severely penalised, and secondly, that the penalty is harsh enough to deter firms from emitting illegally.

With an emissions-trading system of this kind in place, Fig. 7.4 is much like Fig. 7.3. However, in Fig. 7.4, the Marginal Damage curve is vertical at 120 tonnes of greenhouse gases to reflect the emissions cap imposed by the government (MD_{ETS}). In addition, the usual Marginal Damage curve is bold up to the welfare-maximising emissions level and dotted beyond it. The reason for this is that the price that firms must now pay for emissions permits internalises the damage cost of their greenhouse gas emissions. The dotted section of the curve simply indicates that the usual Marginal Damage curve beyond the cap is no longer relevant.

The other difference in Fig. 7.4 is the nature of the Marginal Benefit (MB) curve. Since it indicates the total quantity of greenhouse gases that would be generated at different permit prices, it effectively serves as the aggregate demand curve for emissions permits. Correspondingly, the mb_1, mb_2, and mb_3 curves constitute the demand curves for permits of each of the three firms.

Because the price of greenhouse emissions is now established in the emissions permit market, it is determined by the intersection of the Marginal Benefit (MB) and Marginal Damage (MD_{ETS}) curves. Given the position of the Marginal Benefit and the Marginal Damage curves in Fig. 7.4, the permit price is PP_*. At this price, we would expect ABC Ltd, JKL Ltd, and XYZ Ltd to emit 60, 40, and 20 tonnes of greenhouse gases per year respectively (120 tonnes in total). Not unlike an emissions tax set at tax_*, the social welfare loss depicted by the shaded area in Fig. 7.1 is eliminated. Furthermore, the welfare loss associated with a non-tradeable emissions quota is avoided because an emissions-trading system allows the most efficient producers to purchase the largest quantity of permits and emit the majority of the capped emissions. Thus, once more, the welfare of society is maximised by having ABC Ltd emit the most greenhouse gases and XYZ Ltd the least.[22]

Despite the subtle differences between Figs. 7.3 and 7.4, it can be seen that the outcomes associated with an emissions tax and an emissions-trading system are virtually the same. One might ask in what way an emissions-trading system is superior to a tax? There are numerous reasons why the former is superior—many of which I will explain soon—but a major reason is the difference in the penalty incurred if emissions exceed the welfare-maximising level. As we saw with an emissions tax, a tax rate set at $\$\Psi$ per tonne of greenhouse gas emissions is unlikely to deter firms from exceeding their profit-maximising emissions level, should they choose to do so. The same does not occur with an appropriately designed emission-trading system because the severity of the penalty is sufficient to deter firms from emitting illegally.

It is important to understand that if all greenhouse gas-emitting firms purchase and use the permits they require to maximise their profits, it is impossible for a firm to legally exceed their profit-maximising emissions level. This is because once all firms have reached their profit-maximising emissions level, no unused permits remain.[23] Hence, exceeding the profit-maximising emissions level must entail the emission of additional greenhouse gases without possession of the necessary permits. Of course, prior to the use of all permits, it is still possible for a firm to legally exceed its profit-maximising emissions level. However, it requires the firm to purchase additional permits from other firms. It also requires the firm to pay a sufficiently high permit price to persuade other firms to surrender some of their unused permits. In fact, we would only expect short-term profit-maximising firms to sell some of their permits if the price being offered exceeds the marginal profit they would otherwise earn from their use (i.e., from the production and sale of additional goods).

The crucial point here is that, with an emissions-trading system in place, the desire of firms to emit large quantities of greenhouse gases cannot undermine a government's emissions target. Firms wanting to emit large quantities of greenhouse gases can do little more than outbid competing firms during the initial auctioning process or, if they find themselves short of permits, pay a higher price to obtain permits already possessed by other firms. By functioning in this manner, an emissions-trading system allows the permit price to serve as an adjustment mechanism to buffer against fluctuations in the demand for emissions permits.[24] Conversely, with an emissions tax, it is the quantity of emissions that serves as the adjustment mechanism. Yet, as stressed in Chap. 6, resolving the climate change crisis requires 'quantity certainty' (i.e., meeting an emissions target), not price certainty. In all, an emissions-trading system promotes allocative efficiency whilst ensuring that the targeted emissions level is not exceeded. Thus, unlike a tax, an emission-trading system circumvents the Jevons' Paradox.

It is also worth pointing out that the inability of an emissions tax to prevent a rise in greenhouse gas emissions is not just a theoretical claim. It also has empirical support (Lawn 2006a; IPCC 2007d). A study of four nations—Denmark, Sweden, the Netherlands, and Finland—has revealed that each nation significantly reduced the CO_2 intensity of its real output following the initial imposition of

CO_2 and energy taxes in the early-1990s. That is, the taxes significantly reduced the ratio of CO_2 emissions to real GDP. However, in three of the four countries, the scale effect of the increase in real GDP was more than enough to bring about a rise in total CO_2 emissions over the study period (Lawn 2006a).[25] In the one instance where CO_2 emissions fell (Sweden), the decline over the fourteen-year study period was a meagre 2.4 per cent, which could be attributed to a larger-than-usual reduction in emissions intensity (−35.1 per cent). Whilst the 35.1 per cent decline in Sweden's emissions intensity amounted to an average annual reduction rate of 2.2 per cent, the negligible overall decrease in total emissions fell well short of what would be required of all nations if they are to make a meaningful contribution to a safe atmospheric concentration of greenhouse gases. Notwithstanding the small decrease in CO_2 emissions in Sweden, it is clear from the study that the growth in real GDP could not be slowed to the extent needed to prevent the onset of the Jevons' Paradox.

Of course, advocates of the emissions tax will argue that a tax can sufficiently limit the scale effect of rising real output provided the tax rate is set high enough. Whilst true, this still begs the question: How high is high enough? Furthermore, what if the tax rate is set well above the level needed to reduce greenhouse gas emissions? Although I would argue that ensuring a safe concentration of greenhouse gases is more important than maximising allocative efficiency, I see no good reason to sacrifice some efficiency by imposing an excessive tax rate when an emissions-trading system can guarantee an emissions target as well as generate a greenhouse gas price with the capacity to achieve outcomes every bit as efficient as an emissions tax.

7.5 Refuting the Main Criticisms of Emissions-Trading Systems

7.5.1 Criticism # 1—An Emission-Trading System Leads to Undesirable Market Uncertainty

One of the key arguments against an emissions-trading system is that, by facilitating a flexible emissions price, it would generate market uncertainty that would stifle investment in existing greenhouse gas-abatement technologies as well as discourage the development of more advanced technologies. Critics claim that uncertainty of this kind would not occur with an emissions tax because a government can fix the tax rate for as long as it desires.

There are many weaknesses with this criticism, and I will discuss some of them shortly. The weakness I would like to focus on now is that, contrary to popular opinion, an emissions tax system does not guarantee price stability. Assuming that an emissions tax is designed to achieve a welfare-maximising emissions target, the tax rate must be altered each and every time there is a shift of the Marginal Benefit and Marginal Damage curves.

7.5.1.1 Responding to a Shift of the Marginal Benefit Curve

To illustrate my point, refer to Fig. 7.5. In the three left-hand diagrams, all firms experience a rightward rotation of their marginal benefit curve, which could be due to productivity improvements (i.e., increases in the output generated per tonne of greenhouse gases emitted). This leads to a rightward rotation of the aggregate Marginal Benefit curve and a new welfare-maximising emissions level of 150 tonnes of greenhouse gases per year. Assuming that all three firms are short-term profit-maximisers, the welfare-maximising emissions level can only be attained if the emissions tax rate is raised to tax_*'. In fact, if the government fails to alter the tax rate in the belief that a stable greenhouse gas price is beneficial, 180 tonnes of greenhouse gases would be emitted each year. This would result in a social welfare loss equivalent to the shaded area in the right-hand diagram in Fig. 7.5. Given that it is reasonable to expect regular shifts in the Marginal Benefit curve, the government would be required to continuously alter the emissions tax to avoid social welfare losses.

Advocates of an emissions tax often respond by arguing that governments should abandon futile attempts to manipulate the emissions tax rate to continuously maximise social welfare. They stress that a tax rate should be set with the sole aim of achieving a specific greenhouse gas emissions target. Even so, it is obvious from Fig. 7.5 that a fixed tax rate in the presence of a regularly shifting Marginal Benefit curve would lead to fluctuating emissions levels.[26] Clearly, if a government maintained a fixed tax rate in the belief that price stability is desirable, it would almost certainly fail to achieve an emissions target of any sort.

Compare this situation to one where an emissions-trading system is operating. If the government is determined to maximise social welfare rather than restrict emissions to a particular level, shifts in the Marginal Benefit curve would compel it to vary the number of permits in the emissions permit market. If there is a need to increase the number of permits, the central-government authority must sell additional permits. Conversely, to reduce the number of permits, the government must buy back existing permits.[27]

In the situation depicted in Fig. 7.6, the rightward rotation of the aggregate Marginal Benefit curve requires the government to sell an additional 30 permits. Should the government do this, the vertical section of the Marginal Damage curve shifts rightward from MD_{ETS} to MD_{ETS}'. Although there are more emissions permits in the market, we would expect the permit price to rise to reflect the new welfare-maximising price of PP_*'. The rise in the permit price can be expected because the increase in permits is less than the increase in the demand for permits arising from the rightward rotation of the Marginal Benefit curve.

Interestingly, any failure on the part of the government to increase the number of permits would not lead to excessive greenhouse gas emissions, as would be expected if the government failed to raise the emissions tax rate (see Fig. 7.5). On this occasion, government inaction would result in ‘too few’ permits and a permit price that exceeds the new welfare-maximising price of PP_*' (i.e., $PP' > PP_*'$). It would also result in a social welfare loss represented by a different shaded area in

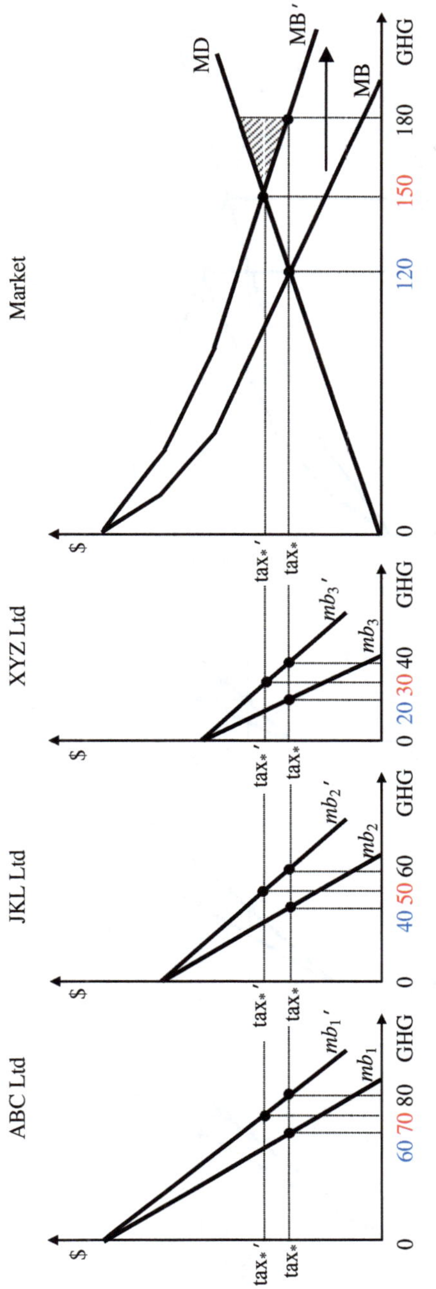

Fig. 7.5 Instituting an emissions tax in the presence of a shifting Marginal Benefit (MB) curve. *Note* GHG denotes tonnes of greenhouse gas emissions

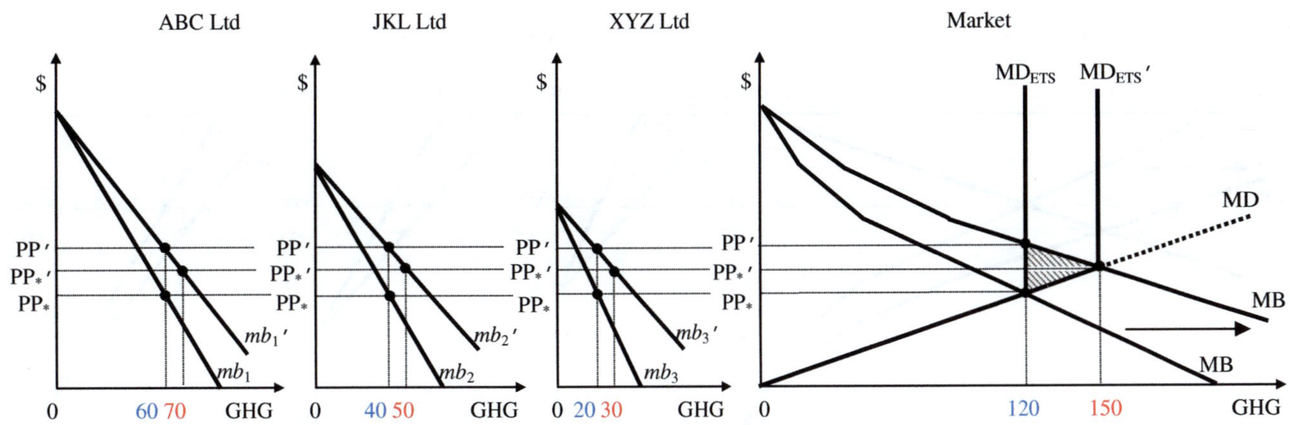

Fig. 7.6 Instituting an emissions-trading system in the presence of a shifting Marginal Benefit (MB) curve. *Note* GHG denotes tonnes of greenhouse gas emissions

the right-hand diagram in Fig. 7.6 (Note: remember, we are presently assuming a mainstream economic perspective of the climate change problem. By referring here to a social welfare loss, we are ignoring the fact that an efficient outcome does not guarantee a 'safe' concentration of greenhouse gases).

7.5.1.2 Responding to a Shift of the Marginal Damage Curve

Matters change considerably if the adjustment to market conditions involves a shift of the Marginal Damage curve. In Fig. 7.7, for example, there is a leftward rotation of the Marginal Damage curve, which might be the result of new evidence indicating that the climate change impact of a given quantity of greenhouse gas emissions is more severe than previously thought.[28] Because of the shift of the Marginal Damage curve, the welfare-maximising emissions level falls to 90 tonnes of greenhouse gases per year. Once again, if firms are short-term profit-maximisers, attaining the welfare-maximising emissions level with an emissions tax requires the tax rate to be raised—this time to tax_*''. Failure to raise the tax rate results in all three firms continuing to emit the same annual quantity of greenhouse gases—specifically, 60, 40, and 20 tonnes by ABC Ltd, JKL Ltd, and XYZ Ltd respectively. Thus, government inaction means that a total of 120 tonnes of greenhouse gases would continue to be emitted each year.

The impact of government inaction differs when there is a shift of the Marginal Damage curve because, as previously explained, the quantity of greenhouse gases emitted by the profit-maximising firms is determined by the point at which the marginal benefit or marginal profit of their operations equals the emissions tax rate. We would expect the three firms to continue to emit the same annual quantity of greenhouse gases because the shift in the Marginal Damage curve does not alter their profit-maximising circumstances. Of course, this doesn't mean that a social welfare loss is avoided. The now 'excessive' quantity of emissions—30 tonnes above the welfare-maximising level—leads to a social welfare loss depicted by the new shaded area in the right-hand diagram in Fig. 7.7.

Should an emissions-trading system be in place, the leftward rotation of the Marginal Damage curve requires the government to buy back 30 existing permits (see Fig. 7.8). The reduction in the number of permits (i.e., the tightening of the emissions cap) causes the vertical section of the Marginal Damage curve to shift leftward from MD_{ETS} to MD_{ETS}''. The permit price now rises to the new welfare-maximising price of PP_*''. This induces ABC Ltd, JKL Ltd, and XYZ Ltd to reduce their annual greenhouse gas emissions to 50, 30, and 10 tonnes respectively.

On this occasion, any failure on the part of the government to buy back 30 permits results in 'too many' permits and an unaltered permit price that is lower than the welfare-maximising price of PP_*'' (i.e., $PP_* < PP_*''$). The resultant social welfare loss is represented by the shaded area in the right-hand diagram in Fig. 7.8.

What, ultimately, does this analysis indicate? Whether the change in market conditions is brought about by a shifting Marginal Benefit curve or a shifting

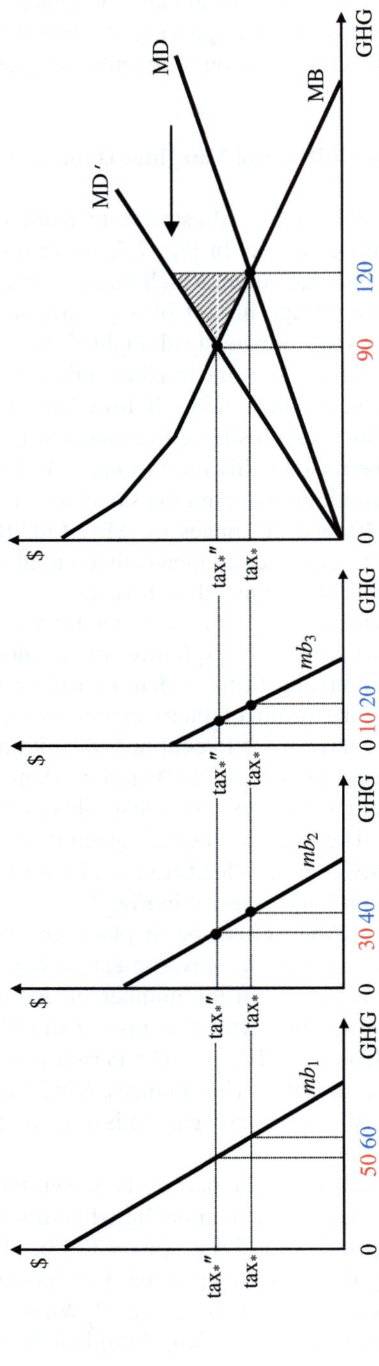

Fig. 7.7 Instituting an emissions tax in the presence of a shifting Marginal Damage (MD) curve. *Note* GHG denotes tonnes of greenhouse gas emissions

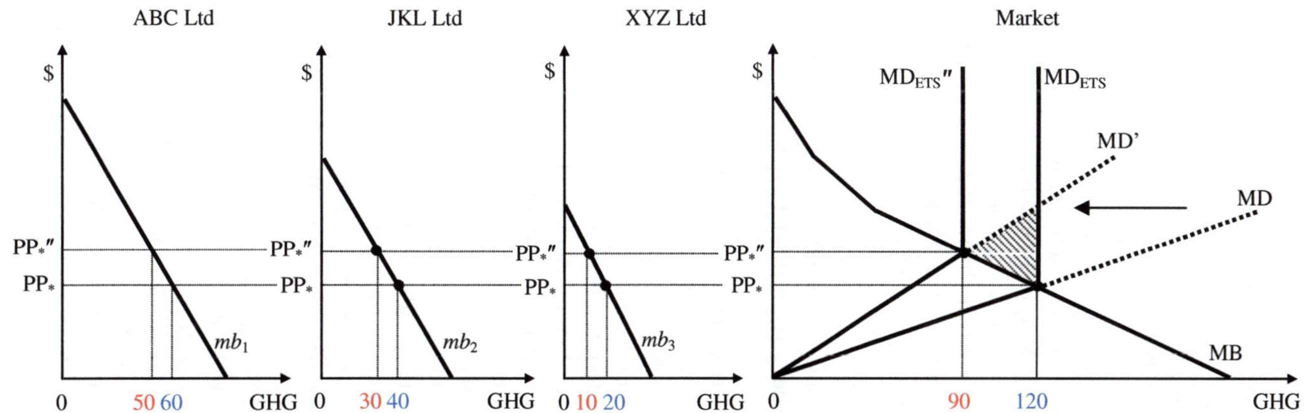

Fig. 7.8 Instituting an emissions-trading system in the presence of a shifting Marginal Damage (MD) curve. *Note* GHG denotes tonnes of greenhouse gas emissions

Marginal Damage curve, the overall conclusion regarding the superiority of an emissions-trading system remains intact. With an emissions-trading system in place, the government can always achieve its emissions targets—including altered targets—in the knowledge that the permit price will adjust to ensure its targets are met in an efficient manner (OECD 2008). The same cannot be said for an emissions tax. In the presence of a fixed emissions tax rate, a shifting Marginal Benefit curve leads to fluctuating emissions levels. If the government wishes to achieve a particular emissions target, it has no option but to vary the emissions tax rate. Whilst it is true that an emissions target can be achieved with a fixed tax rate if no more than a shift of the Marginal Damage curve takes place, a problem arises insofar as a leftward shifting Marginal Damage curve renders the chosen emissions target insufficiently stringent. In response, the government must tighten the emissions target, which it can only do by raising the emissions tax rate. In all, suggestions that an emissions-trading system leads to price flexibility but an effective (sic) emissions tax does not are fallacious.

7.5.1.3 Investment Decisions Are Based on Long-Term Price Projections Not Short-Term Price Fluctuations

A further weakness of the criticism directed towards an emissions-trading system is the perception that the price flexibility it promotes would stifle investment in greenhouse gas-abatement technologies. It is simply untrue that strong investment of any sort requires price stability. In favourable circumstances, strong investment occurs across all industries despite significant short-term fluctuations in input costs and output prices—a reality often overlooked along with the fact that short-term price fluctuations are part-and-parcel of a market-based economy. Surely critics do not believe that investment in abatement technologies would be enhanced via the imposition of Soviet-style price controls?

What, then, affects investment? Along with interest rates and many non-price factors, investment levels are greatly influenced by long-term price projections. Mining companies, for example, do not base their capital investment plans on the day-to-day fluctuations of ore prices. They base their investment decisions on where they envisage ore prices to be in 10, 20, and 30 years' time. With an emissions-trading system in place, we would expect the majority of greenhouse-gas-emitting firms to adopt the same long-term investment approach in relation to greenhouse-gas abatement technologies. In other words, we would expect them to predicate their investment decisions on the estimated long-term price trend of greenhouse gases (Stern 2007; OECD 2008).

This raises the issue of how greenhouse gas prices are likely to vary over the next four to five decades and how this might affect investment in greenhouse gas-abatement technologies. Quite obviously, the long-term price trend of greenhouse gases will depend heavily on the climate change policies that governments implement. As indicated in Chap. 6, it will also depend on the rate of technological progress; the growth rate of real GWP; and many other government policies that affect climate change damage costs and mitigation costs. Let's assume that an emissions-trading system is introduced and the cap on emissions is gradually tightened in line with the emissions trajectory advocated by Anderson and Bows (2008) to achieve a 450 ppm stabilisation target.[29] Let's also assume that there is a fall in the demand for emissions permits that corresponds to a reduction in the demand for non-renewable energy as per the sustainable emissions scenario (see Table 4.2).

Given these assumptions, Fig. 7.9 provides an indication of what we might expect in terms of short-term variations in the price of greenhouse gases (i.e., the price fluctuations between annual permit auctions) and the long-term trend of *real* greenhouse gas prices.[30] The left-hand diagram shows the vertical section of the Marginal Damage curve (MD_{ETS}) shifting to the left as the emissions cap is gradually reduced from E_0 to E_3. The narrowing of the gap between the MD_{ETS} curves reflects the reduction in required emissions cuts over time (i.e., deep initial cuts and smaller cuts later on). At the same time, the upward-sloping section of the Marginal Damage curve is likely to shift up at a decelerating rate (see Fig. 6.13), albeit the high probability that the Marginal Benefit (MB) curve would intersect the vertical section of the Marginal Damage curve means that the upward-sloping section of the Marginal Damage curve is likely to be of little relevance.[31] The same left-hand diagram also shows the Marginal Benefit curve shifting to the left, which reflects the gradual decline in the demand for emissions permits.

The price of emissions permits at the beginning of each time period is determined by the intersection of the Marginal Benefit and Marginal Damage curves for each respective time period (PP_0 to PP_3). Although there is a gradual leftward shift over time of the Marginal Benefit curve, we would expect the Marginal Benefit curve to frequently shift up and down during each time period (i.e., between each permit auction). These shifts, and their range, are denoted by the upward and downward arrows along each MD_{ETS} curve. It is because of these shifts that the short-term price of emissions permits is likely to fluctuate significantly. This is represented by the erratic price movements during each time period in the right-hand diagram of Fig. 7.9.[32] Despite these short-term price fluctuations, the right-hand diagram indicates a more stable long-term price trend.

In what direction the long-term real price of emissions permits is likely to move is a moot point. Given the extent of the emissions cuts initially required to achieve a safe concentration of greenhouse gases, it is highly likely that real permit prices would trend upwards for a considerable period of time (e.g., t_0 to t_2 in Fig. 7.9).[33] We would therefore expect a dramatic shift away from fossil fuel use and a long period of robust investment in the use and development of greenhouse gas-abatement technologies.

At some point, it is very likely that the fall in demand for emissions permits would exceed the reduction in the available number of permits. Once this point is reached (e.g., t_3 in Fig. 7.9), real permit prices would begin to trend downwards, just as we would expect the social cost of carbon to eventually fall in a steady-state economic setting (see Fig. 6.13).[34] The fall in permit prices would not, however, induce a collective shift back to fossil fuels or a widespread uptake of high

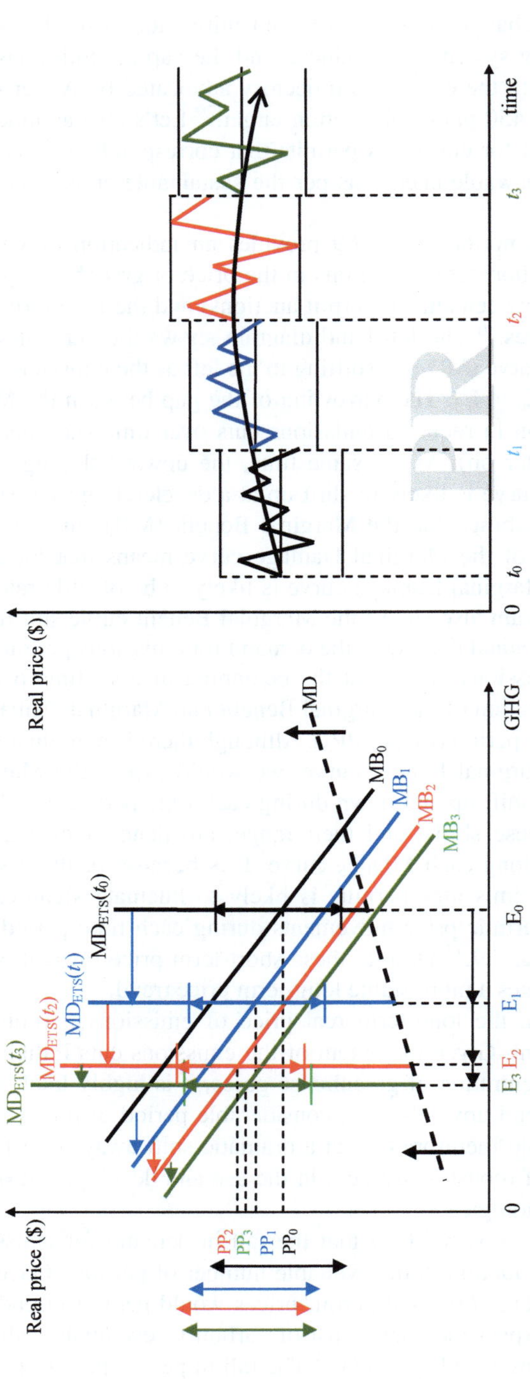

Fig. 7.9 Probable short-run price fluctuations and the long-run trend of real permit prices when using an emissions-trading system to reduce greenhouse gas emissions—drastic initial emissions cuts; smaller emissions cuts later. *Note* GHG denotes tonnes of greenhouse gas emissions

greenhouse gas-emitting practices. There are three main reasons why. Firstly, the very stringent emissions cap would prohibit any collective rise in greenhouse gas emissions. Secondly, although some firms would be tempted to exploit lower real permit prices and adopt high-emissions practices (i.e., by purchasing a large quantity of low-priced permits), any aggregate increase in the demand for emissions permits would precipitate a sharp rise in permit prices. Consequently, any incentive for firms to exploit lower real permit prices would be immediately negated by the price-increasing effect of the very tight emissions cap. Thirdly, by time t_3, the use of low-emissions technology would be the rule rather than the exception. Indeed, much of the energy-supplying infrastructure would be of the low-emissions variety. As a consequence, the adoption of high-emitting practices would become extremely costly in the sense that the national economy would be incapable of readily and cheaply serving any desire that a firm might have to shift back to high greenhouse gas-emitting practices.

7.5.2 Criticism # 2—Taxes Are Simpler to Implement Than an Emissions-Trading System

A large number of people believe that an emissions-tax system is much simpler to introduce and operate than an emissions-trading system. They therefore claim that the tax system's simplicity enhances its chances of success (e.g., Nordhaus 2005; Humphries 2007). Moreover, they believe that the simplicity of a tax-based system increases its political palatability, thus making it a preferred climate change solution within the decision-making fraternity.

There is no doubt that a tax-based system would eliminate some of the logistical tasks required of a government authority charged with administering an emissions-trading system. These include regular public auctions of emissions permits and the need to oversee the operation of an emissions permit market. However, as mentioned, the presence of a tax-based system would require the central-government authority to determine the nation's appropriate emissions level (if not already determined via international negotiation), and install an effective compliance mechanism. More than this, the need to constantly set and charge a new emissions-tax rate to achieve a pre-determined emissions target would entail a layer of bureaucracy not required with an emissions-trading system in place. To make matters worse, linking the national system with a global emissions-tax system would be complicated by the contaminating effect of fluctuating exchange rates on emissions tax rates. The same would not occur with an emissions-trading system in place because a global emissions target would be guaranteed by a global emissions cap. That is, the emissions target would not be undermined by the potential impact that fluctuating exchange rates would have on the cost of purchasing emissions permits.

Another perceived complication of an emissions-trading system is the belief that anyone engaged in an activity that generates greenhouse gases must purchase emissions permits. In addition, there is a belief that all greenhouse gas emissions must be directly monitored by a central-government authority. As alluded to previously, this wouldn't be the case in both instances. Only individuals and entities directly involved in the emission of greenhouse gases and/or stipulated emissions-generating activities would be required to purchase and acquit emissions permits. Thus, very few consumers of a final product would be required to purchase permits. Furthermore, only individuals and entities that must purchase permits would have their emissions monitored or be required to report their emissions levels. Whilst it is true that an emissions-trading system would force everyone to pay higher prices to consume final goods and services, in most cases the price hikes would entail firms transferring some of the cost from having to purchase emissions permits to the next stage of the supply chain.[35] At what point during the supply chain the cost of permit prices would be passed on would be governed by the central-authority's chosen point of obligation. This, in turn, would depend on the point during the supply chain where greenhouse gas emissions and the acquittal of permits are easiest to measure and monitor.

Perhaps the best way to think about the simplicity of the process is to visualise the production and sale of timber furniture. Initially, the raw timber used to manufacture timber furniture is purchased by a timber mill from a logging company. The timber mill then processes the raw logs and sells the processed timber to a furniture manufacturer. The manufacturer, having produced the furniture, sells its output to a furniture retailer. Finally, the retailer sells the timber furniture to consumers. Throughout the process, the value of the service provided by each firm (the logger, miller, manufacturer, and retailer) adds value of the initial service provided by the forest. Ultimately, the consumer, in paying the final price for timber furniture, pays for the cost of the timber used in its manufacture. He or she does not initially purchase the timber. Nor, if there is a cap-auction-trade system in place to regulate the rate of timber extraction, would a consumer of timber furniture be required to purchase timber-buying permits.

A similar situation would occur with respect to a petrol-fuelled car. A driver of the car would have no need to purchase the oil that is used to produce petrol. Assuming that oil extraction is a designated emissions-generating activity requiring the purchase and acquittal of emissions permits, the car-driver would have no obligation to engage in the emissions-permit market. He or she would simply pay a higher price to consume a litre of petrol. In this instance, only the oil companies would be required to purchase and acquit emissions permits. Furthermore, the oil companies, not the car-driver, would have their activities monitored.

Lastly, advocates of an emissions tax point to the presence of greenhouse-gas offsets in most emissions-trading systems as further evidence of the latter's complexity. Although I believe that rigorously assessed offsets should be included in an emissions-trading system, the incorporation of offset arrangements is not mandatory. In any event, offsets can, as already shown, be incorporated into an emissions-tax system. Hence, the potential to include greenhouse-gas offsets does not make emissions-trading systems any more complex than tax-based systems. In sum, there is nothing inherently simpler about an emissions-tax system *vis-à-vis* an emissions-trading system. Whilst a tax-based system does away with some logistical functions and requirements, it includes a number of bureaucratic responsibilities, such as the regular setting and charging a new emissions-tax rate, that are not required to successfully operate an emissions-trading system.

7.5.3 Criticism # 3—Emissions-Trading Systems Often Include Allowances for Carbon Offsets/Credits that Are Vulnerable to Abuse

Critics question the validity of the greenhouse-gas offsets contained within many emissions-trading systems by arguing that offset arrangements are vulnerable to abuse. To support their argument, critics highlight cases where organisations have engaged in dubious offset projects by exploiting the lack of effective oversight typical of most offset arrangements (e.g., the Kyoto Protocol's Clean Development Mechanism).[36] The failure of some offset projects has allowed many firms to obtain carbon offsets/credits that far exceed any sequestration gains secured, thus undermining the principal purpose of imposing emissions caps in the first instance.

Whilst abuse of offset arrangements cannot be denied, there are two points worth making. Firstly, as explained earlier in the chapter, offsets/credits should only be granted to firms if offset projects: (i) have been thoroughly and rigorously assessed before obtaining approval; and (ii) can be continuously monitored to ensure projected gains are realised. Short of this guarantee, greenhouse-gas offsets should not be incorporated into an emissions-trading system.[37] Secondly, since greenhouse-gas offsets can be incorporated into an emissions-tax system (e.g., by way of tax credits/rebates), whatever deficiencies apply to the offset arrangements found in emissions-trading systems apply equally to offset arrangements in tax-based systems.

7.5.4 Criticism # 4—Emissions-Trading Systems Are Highly Corruptible and Subject to Negative Influences by Political Lobby Groups

Even if, in theory, emissions-trading systems can achieve desired greenhouse gas objectives, there is a strong belief amongst some critics that emissions-trading systems will always fail because they are highly corruptible and subject to the self-interested motives of political lobby groups (Grubb and Neuhoff 2005; Lohmann 2006). For example, it is argued that powerful lobbying and corruption can result in: (i) the issuance of free emissions permits to certain industries and greenhouse

Table 7.1 Equivalent concessions from corruption and/or effective lobbying

Emissions-trading system	Emissions-tax system
• Grandfathering (initial issuance of free permits)	• Tax exemptions (tax-free emissions)
• Unsubstantiated greenhouse-gas offsets granted in the form of additional free permits	• Unsubstantiated greenhouse-gas offsets granted in the form of additional tax credits/rebates
• Too many permits = excessive emissions	• Insufficient tax rate = excessive emissions
• Monitoring oversights = excessive emissions	• Monitoring oversights = excessive emissions
• Failure to extinguish permits in accordance with actual emissions	• Failure to charge the emissions tax in accordance with actual emissions

gas-emitting firms—a phenomenon referred to as 'grandfathering'; (ii) the granting of unsubstantiated greenhouse-gas offsets in the form of free additional permits; (iii) the sale/issuance of too many permits, which results in an excessive level of greenhouse gas emissions; (iv) monitoring oversights, which also contribute to excessive emissions levels; and (v) the failure to extinguish emissions permits in accordance with the greenhouse gas emissions of certain entities and organisations.

The fact remains that emissions-regulating systems of all types are vulnerable to corruption and political influence. The relevant issue is whether emissions-trading systems are more susceptible to corruption and political lobbying than tax-based systems. This is a difficult to say. What can be said is that, for every potential concession associated with emissions-trading systems, an equivalent concession can be identified in relation to tax-based systems (see Table 7.1). For instance, in cases where grandfathering can occur in the form of freely issued emissions permits, it can also occur with a tax-based system in the form of tax exemptions (e.g., tax-free emissions for certain industries and organisations). In Australia, where a carbon tax was introduced on 1 July 2012, coverage of the tax was confined to certain industries and specific forms of carbon emissions.[38] Excluded from the tax were household-transport fuels; all forms of light-vehicles used for business purposes; and off-road fuel use by the agriculture, forestry, and fishing industries (Australian Government 2012). It is difficult to believe that the granting of these tax exemptions was the result of anything other than the successful lobbying of industry-based pressure groups.

As for unsubstantiated greenhouse-gas offsets, these can be granted in a tax-based system in the form of tax credits or rebates rather than free emissions permits. Furthermore, while the public sale and/or issuance of too many permits can lead to an excessive level of greenhouse gas emissions, so can a woefully inadequate emissions-tax rate aimed at appeasing powerful industry lobby groups. Finally, the failure to extinguish emissions permits in line with greenhouse gas emissions can be mirrored in a tax-based system by the failure on the part of a government authority to appropriately charge the emissions tax.

Overall, there is no reason to believe that emissions-trading systems are any more susceptible to corruption and political lobbying than tax-based systems. Although there are questions hanging over many of the emissions-trading systems currently in operation, including whether some of the inadequacies are due to political lobbying, there are as just many questions hanging over some emissions-tax systems (e.g., the Australian carbon-tax system during the 2012–2014 period).

7.5.5 *Criticism # 5—Emissions-Trading Systems Impose an Emissions 'Floor' as Well as an Emissions Cap*

It has recently come to the attention of some commentators that emissions-trading systems not only set a cap on greenhouse gas emissions, but potentially impose an emissions 'floor' below which emissions cannot fall (Denniss 2008; Brook 2009). If this observation is correct, there are two impending problems with emissions-trading systems. At the micro level, it means that actions taken by individuals to reduce greenhouse gas emissions below the cap (e.g., the household installation of solar panels) are rendered futile. At the macro level, it becomes virtually impossible for a national economy to adjust over time to an optimal emissions trajectory should the optimum exist at an atmospheric concentration of greenhouse gases which is less than the safe concentration (see Fig. 6.14).[39]

In what way can an emissions-trading system impose an emissions floor? Consider Fig. 7.10. Let's assume that the current price and quantity of greenhouse gas emissions are determined by the intersection of the Marginal Benefit (MB) curve and the Marginal Damage (MD) curve. This means that the price of

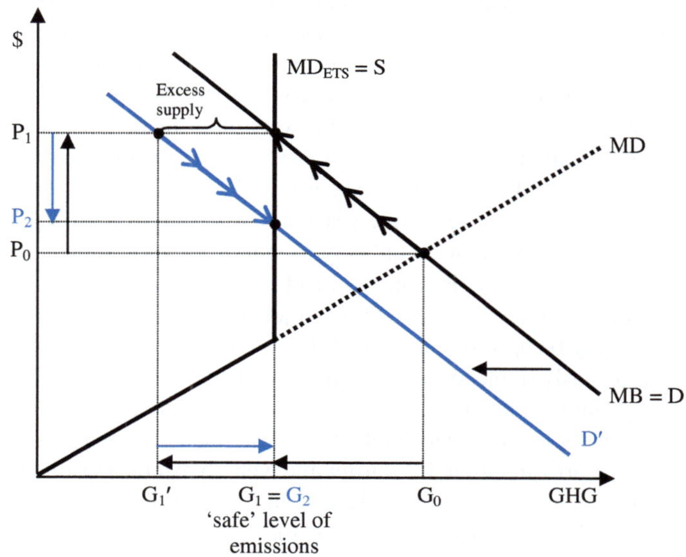

Fig. 7.10 The argument that an emissions-trading system sets a 'floor' as well as a cap on greenhouse gas emissions

greenhouse gas emissions would be P_0 per tonne and the quantity of greenhouse gas emissions would be G_0 tonnes. Because G_0 exceeds the safe emissions level, the government introduces an emissions-trading system to cap greenhouse gas emissions at G_1. The Marginal Damage curve now becomes vertical (MD_{ETS}) at the capped quantity. This leads to a new equilibrium price for greenhouse gas emissions of P_1, which becomes the new market price for emissions permits.

Because of the substantial rise in the price of greenhouse gas emissions, the generators of emissions react in two ways. Firstly, they purchase fewer emissions permits and reduce their greenhouse gas emissions as best they can in the short-run (G_o to G_1). This is represented by a movement along the Marginal Benefit curve. Secondly, they undertake measures to substitute away from greenhouse gas-intensive products and technologies in order to reduce their long-term demand for emissions permits. This is represented by a leftward shift of the Marginal Benefit curve to D' and a further reduction in greenhouse gas emissions to G_1'.

At the prevailing permit price of P_1, there is an excess supply of emissions permits. The permit price subsequently falls to P_2. As it does, the demand for emissions permits rises, as represented by a movement along the D' curve (G_2). Consequently, greenhouse gas emissions return to the original level set by the government (i.e., $G_2 = G_1$). In the end, everything the public does to reduce the emission of greenhouse gases below the cap simply decreases the demand for emissions permits; lowers permit prices; and induces an increase in greenhouse gas emissions back to the capped level. Thus, individual action to reduce greenhouse gas emissions below the cap becomes a pointless exercise.

There is no doubt that the potential exists for an emissions-trading system to impose a floor as well as a ceiling on greenhouse gas emissions. However, whether this eventuates depends on how the emissions-trading system is designed and operated. It also depends on the behaviour and purchasing intentions of permit holders. For instance, if emissions permits have a very long life—which I have argued against—permit prices will continue to fall every time there is substantial action taken to reduce greenhouse gas emissions. Consequently, the demand for emissions permits will rise and total emissions will return to the capped rate. However, if permits have a short, limited life (e.g., one year), any lowering of permit prices arising from actions to reduce emissions is unlikely to induce a return to the capped rate of emissions. The reason for this is that, as already explained, the actions of individuals and firms will be guided by long-term projections of real permit prices, which will almost certainly be upwards for some time as the capped rate of greenhouse gas emissions is vigorously tightened. What's more, with emissions permits expiring one year after every annual auction, there is no way that a firm can accumulate permits (bought at a relatively low price) to shield itself from higher future permit prices.

Even in the short-run—that is, during the year between permit auctions—it is worth pondering whether householders would massively increase their use of carbon-intensive electricity simply because the price of emissions permits and, presumably, the price of electricity has temporarily declined. Would, for example, householders rush off to purchase electricity-hungry appliances to exploit a single month of lower electricity prices? Would they abandon plans to invest in solar panels because, despite believing that the price of greenhouse gas emissions will rise over the next decade or two, electricity prices have fallen one month prior to the next public auction of emissions permits? I think not. As it is, people do not directly benefit from using electricity. They benefit from the appliances powered by electricity. Thus, if the weather is no colder or hotter than normal at the time electricity prices are falling, people do not increase their use of air-conditioning systems just because the cost of powering them has temporarily declined.

Much of the same logic applies to greenhouse gas-emitting firms. In general, they will not boost production levels simply because the cost of emitting greenhouse gases has temporarily fallen. Nor will they shelve plans to invest in low carbon-intensive capital goods because permit prices have dramatically declined just short of an upcoming auction of new permits.

It is true that producers with the ability to hoard their goods may opportunistically increase production levels during periods when the price of emissions permits is relatively low. However, firms have limited productive capacity and are restricted in their ability to boost production levels during times of lower permit prices. Furthermore, should enough firms adopt this operating strategy, the additional demand for permits would increase permit prices, thus negating the benefit that the strategy would otherwise deliver (see the effect of a temporary rightward shift of the Marginal Benefit curve in Fig. 7.6).[40]

Another important factor that critics overlook is that many firms will purchase more permits than they anticipate needing to ensure they are not devoid of permits and/or forced to pay exorbitant permit prices to continue operating just prior to a permit auction. Consequently, nearing an auction of new permits, it is likely that there will be a temporary 'glut' of unused permits. If so, permit prices are likely to drop just prior to their expiration date—although, as explained, if enough firms hold out to exploit these favourable circumstances, the fall in permit prices could be minimal.[41] Assuming a price plunge, a massive increase in greenhouse gas emissions is unlikely to occur for the reasons outlined above. Indeed, it is almost certain that many emissions permits would go unused, thus suggesting that the quantity of greenhouse gas emissions generated during any year is likely to be less than the cap imposed on emissions for that year.

Do the unused permits constitute an unnecessary cost to society? Not at all. Whilst the unused permits qualify as a cost in the sense that greenhouse gases that could have been safely emitted were not generated, the cost serves as a form of 'insurance', which is borne by the firms left with the expired permits. In this sense, the cost does not differ to any other insurance cost which, like all insurance costs, is willingly incurred because the benefits of the insurance—in this case, not running out of permits—outweigh the costs of being left with unused permits.

There is one final reason why the actual quantity of greenhouse gas emissions is likely to be less than the emissions cap. Unlike an emissions tax, an emissions-trading system enables individuals and environmental groups who would prefer very low emissions levels to purchase emissions permits and not use them. With an emissions tax, there is nothing these individuals can do to prevent emissions levels exceeding a safe rate. Admittedly, following the imposition of an emissions tax, the increased cost of emitting a tonne of greenhouse gases remains intact regardless of how firms and individuals respond to the tax (unlike declining permit prices). As such, individual action to help reduce greenhouse gas emissions is not immediately negated. But it maybe negated eventually. As explained a number of times, a tax cannot regulate the aggregate quantity of greenhouse gas emissions. Because the savings generated by the initial reduction in emissions (lower tax bills) are likely to be spent on goods and services, it is highly probable, particularly if there is an over-riding desire to grow the national economy, for the ensuing increase in the volume of economic activity to overwhelm the positive effect of the actions taken to reduce greenhouse gas emissions (the Jevons' Paradox). If this occurs, total emissions would eventually rise above the pre-action level. The same situation would not occur with an emissions-trading system installed, since the pre-action level of greenhouse gas emissions is capped.

Altogether, with an effective emissions-trading system in place, individual and/or collective action that takes the form of permit purchases and the subsequent non-use or destruction of them would unquestionably reduce greenhouse gas emissions. As for individual action designed to reduce one's long-term greenhouse gas emissions, it is also likely to reduce total emissions if it is premised on the expectation that real permit prices will trend upwards for a lengthy period of time. In addition, so long as permits have a short lifespan, many of the emissions permits freed up in the short-run by emissions-reducing actions will expire unused. There is, therefore, every reason to believe that actions taken to reduce greenhouse gas emissions will bring about an emissions level that is below the emissions cap.

7.5.6 Criticism # 6—Emissions-Trading Systems Promote Speculative Trading Behaviour

One of the basic premises underpinning the efficacy of an emissions-trading system is the belief that permit prices can help a nation achieve its emissions targets in the most efficient manner. Although no market price is ever a perfect reflection of marginal benefits and marginal costs, if only because marginal benefits and marginal costs are never fully internalised by markets, market prices are inevitably distorted if individuals engage in markets for the sole purpose of capturing economic rents—that is, if they purchase something in the belief that its price will rise and that, through possession and a later sale, they can earn windfall profits. The distortion arises because the windfall profits do not reflect any genuine creation of real wealth. Hence, economic rents constitute 'unearned income'. Crucially, it is the prevalence of economic rents in some markets that promotes speculative trading behaviour.

In Chap. 3, it was explained that economic rents invariably emerge when there is a powerful scarcity influence on the supply-side of a market that exerts upward pressure on prices. Because an emissions-trading system involves the deliberate imposition of an emissions cap—which serves as an artificial supply-side constraint—many argue that the system lends itself to the speculative buying and selling of emissions permits (e.g., Spash 2010). They therefore believe that a distorted market price for emissions permits is inevitable, which impedes the efficient achievement of emissions targets. Furthermore, they raise concerns that permit price distortions lead to price bubbles (grossly inflated permit prices), which have the potential to destabilise permit markets.

Again, whether an emissions-trading system promotes speculative behaviour depends on how the system is designed and operated. Should emissions permits have a one-year lifespan, as I have recommended, it is almost impossible to profit substantially through speculation, since earning economic rents requires permits to maintain their inflated value. Yet any inflated value that permits might gain over a given year vanishes the moment they expire. Even within the year, there is no guarantee that permit prices would rise. Indeed, as I have pointed out, there is a strong likelihood that permit prices would decline towards the end of a year—the very time speculators would be endeavouring to procure their windfall profits.

Of course, whenever an asset is created that can potentially be bought and sold for profit, there will always be some people willing to engage in speculative trading behaviour. In the case of a well-designed emissions-trading system, the number of speculators is likely to be small. Consequently, I would expect any distortionary effect that speculators might have on permit prices to be negligible.

7.5.7 Criticism # 7—Emissions-Trading Systems Allow Big Firms to Buy a Large Quantity of Permits to Control Markets

Virtually all existing emissions-trading systems allow a single entity or individual to purchase as many emissions permits as they can afford. Given that most economic activities result in the generation of greenhouse gases, the ability to possess a large quantity of emissions permits allows large-scale permit holders to exert an undue influence on many product markets. To prevent this, an effective emissions-trading system would limit the number of emissions permits that any one person or entity can possess. Hence, concerns about emissions-trading systems conferring market power to large corporations apply to poorly-designed emissions-trading systems. They do not apply to the type of system being recommended in this book.

7.5.8 Criticism # 8—Countries Requiring More Growth Are Less Likely to Agree to an Emissions Cap Than an Emissions Tax

In view of the positions taken by many low-GDP nations at recent climate change conferences, there is a widespread belief that they are more likely to agree to an emissions tax than an emissions cap (Kahn and Franceschi 2006). The reason why low-GDP nations would prefer an emissions tax is difficult to fathom. One of the reasons given is that low-GDP nations regard climate change as a problem largely created by the excessive generation of greenhouse gases by high-GDP countries. Thus, low-GDP nations consider it unfair that they should make substantial sacrifices in the form of emissions cuts when they urgently need to grow their economies over coming decades. Consequently, they remain wary of an international agreement—such as a global emissions-trading system—that might unduly limit their greenhouse gas emissions. On the other hand, low-GDP nations appear more receptive to an emissions tax, particularly if the tax is set at a lower rate than it is for high-GDP countries and is accompanied by financial incentives to help low-GDP nations move towards low-emitting production methods.

May I begin my response by saying that I agree with the view that the past emissions of high-GDP countries are essentially to blame for the current climate change crisis. Hence, there are valid reasons why high-GDP countries should bear the greatest burden when it comes to emissions cuts. At the same time, there is no escaping the fact that low-GDP countries must play a role in achieving a safe atmospheric concentration of greenhouse gases. For some low-GDP countries, this may imply acceptance of emissions cuts, albeit smaller cuts initially than those required of high-GDP countries, whereas, for severely impoverished nations, it may imply a period of emissions growth before having to reduce emissions.

Either way, to contribute towards a safe atmospheric concentration of greenhouse gases, low-GDP countries will need to meet specific emissions targets. The fact that low-GDP countries may be more willing to accept an emissions tax when, as shown, a tax does not compel a nation to meet an emissions target raises the concern that low-GDP countries are not committed to resolving the climate change crisis. But, then, why wouldn't a nation accept an emissions-trading system if it is more likely to achieve an emissions target than an emissions tax? My only answer is that an emissions tax leaves open the possibility of a nation exceeding an emissions target without having to incur a harsh penalty, notwithstanding the fact that domestically-located firms would have to pay more tax.

In no way do I want to give the impression that the failure of UNFCCC Parties to establish an effective emissions protocol can be solely attributed to the intransigent behaviour of low-GDP countries. The obstructive actions of many high-GDP countries at UNFCCC conferences is testimony that few of them seem genuinely committed to cutting emissions to the extent required to achieve a safe atmospheric concentration of greenhouse gases. Indeed, some high-GDP countries appear hell-bent on sabotaging efforts to reach a desirable outcome.

This aside, should low-GDP countries (and some high-GDP countries) only be prepared to accept the imposition of an emissions tax, it may be more fruitful to institute an emissions-tax regime in some countries before making the transition to a global emissions-trading system. Whatever the case, the greater likelihood of the world's countries agreeing to an emissions tax should not be seen as justification for maintaining an emissions-tax regime and abandoning efforts to install a global emissions-trading system. The bottom line is that only the latter will achieve a safe concentration of greenhouse gases at lowest cost.

7.5.9 Criticism # 9—Emissions-Trading Systems so Far Instituted Have Failed to Significantly Reduce Emissions

A number of emissions-trading systems have been implemented in different countries and in one region of the world. Some of the better-known systems include the European Union Emissions Trading System (EU-ETS), the United Kingdom Emissions Trading Scheme (UK-ETS), the Japanese Voluntary Emissions Trading Scheme (JVETS), and the New Zealand Emissions Trading Scheme (NZ-ETS).[42] In each case, the system has failed to reduce greenhouse gas emissions much beyond the targets imposed by the Kyoto Protocol. In addition, the price of greenhouse gases has failed to reach the heights needed to induce a widespread shift to low-carbon technologies. On top of this, the above systems contain flawed greenhouse-gas offset arrangements (Block 2008). Critics cite these examples as proof that emissions-trading systems are inherently incapable of resolving the climate change crisis.

I will be the first to admit that the emissions-trading systems so far instituted have not been very successful. However, there is a good reason for this—all such systems have been inadequately designed, implemented, and operated. None of them resemble the type of emissions-trading system recommended in this chapter. For example, Phase I of the EU-ETS (1 January 2005 to 31 December 2007) covered just 11,500 heavy energy-consuming installations in the power generation and manufacturing industries which were responsible at the time for a mere 40 per cent of all greenhouse gas emissions in the European Union (European Commission 2008). Alas, for all its lack of coverage, too many emissions permits were issued to these organisations of which around 95 per cent of the permits were allocated free of charge. Consequently, permit prices were too low to encourage a broad regional shift to low-carbon technologies and production methods (Skjærseth and Wettestad 2008). Worse still, despite a collapse in permit prices from €30 in April 2006 to just €0.10 in September 2007, a German electricity company was able to earn $US6.4 billion during Phase I of the EU-ETS by charging the market value of permits it originally received for free (Spash 2010; Martinez and Neuhoff 2005; Hepburn et al. 2006).

Phase II of the EU-ETS (1 January 2008 to 31 December 2012) included some minor improvements on Phase I, such as the issuing of 6.5 per cent fewer emissions permits and a reduction in the number of permits issued free of charge (European Commission 2008). However, the coverage of greenhouse gas-emitting entities remained inadequate, whilst the number of circulating permits still exceeded the quantity necessary to markedly reduce emissions (Spash 2010).

Can we conclude that emissions-trading systems are inherently defective? I don't believe so. Perhaps it would be more accurate to say that it will always be difficult for policy-makers to implement a truly effective emissions-trading system, particularly at the global level. If so, this does not prove that emissions-trading systems have an inherent weakness. What ultimately matters is that an emissions-trading system must be well designed and operated to be effective. Provided the global community is genuinely committed to achieving a safe atmospheric concentration of greenhouse gases in the most efficient and equitable manner, I see no good reason why are a global emissions-trading system of the type to be outlined in Chap. 10 would not be widely accepted and implemented.

7.6 Further Ecological Economic Reasons for Preferring an Emissions-Trading Systems Over an Emissions Tax

So far, the comparison between an emissions-trading system and an emissions tax has been conducted from a largely mainstream economic perspective. In this section of the chapter, I will continue the case for an emissions-trading system by adopting a strict ecological economic position. To do this, I will alter some of the basic assumptions of the mainstream framework to take account of the concrete realities revealed in Chaps. 5 and 6. At the same time, I will also demonstrate that an emissions-trading system can be incorporated into a comprehensive policy framework aimed at achieving sustainable development. That is, I will show that an emissions-trading system can help resolve the three crucial goals of ecological sustainability, distributional equity, and allocative efficiency.

7.6.1 Ecological Sustainability

As explained many times throughout this book, achieving ecological sustainability requires the entropic rate of throughput (i.e., the input of low-entropy resources and the output of high-entropy wastes) to remain within ecosphere's regenerative and waste-assimilative capacities.[43] From a climate change perspective, this means stabilising the atmospheric concentration of greenhouse gases at no more than 450 ppm of CO_2-e. To achieve this, emissions targets must be met. Moreover, the policy mechanism instituted to meet such targets must be capable of achieving 'quantity certainty'. From what we have seen so far, an emissions tax cannot do this. Conversely, an emissions-trading system can.

There were, however, two weaknesses associated with the analysis undertaken earlier in this chapter. Firstly, the analysis was conducted on the assumption that a central-government authority can accurately estimate the costs relating to climate change damages and the mitigation measures needed to reduce greenhouse gas emissions. Furthermore, it was assumed that the authority can calculate the marginal benefits of emitting greenhouse gases. Yet, as explained in Chaps. 5 and 6, a central-government authority is incapable of predicting future rates of technological progress, the co-benefits of mitigation measures, and eventual changes in resource scarcity and resource prices. Hence, it cannot know the exact whereabouts of the Marginal Benefit and Marginal Damage curves.

Secondly, the earlier analysis was predicated on a mainstream economic perspective. Hence, it was assumed that it is socially optimal to set an emissions cap or emissions tax at the point where the Marginal Benefit and Marginal Damage curves intersect. Importantly, this pre-supposed that the so-called 'efficient' rate of greenhouse gas emissions coincided with a safe emissions trajectory. However, as explained numerous times, allocative efficiency does not guarantee ecological sustainability. Consequently, it is possible for the efficient rate of greenhouse gas emissions to exceed the rate needed to stabilise the atmospheric concentration of greenhouse gases at no more than 450 ppm of CO_2-e.

If we now assume ignorance on the part of a central-government authority and the need to limit greenhouse gas emissions to a safe rate, it makes a significant difference to our earlier analysis. To show how, consider Fig. 7.11, where a comparison is made between an emissions-trading system and an emissions tax on the assumption that: (i) the position of the Marginal Benefit curve is unknown; and (ii) in order for a nation to make a contribution towards a safe concentration of greenhouse gases, it must limit its greenhouse gas emissions in the current year to 90 tonnes. As can be seen from Fig. 7.11, 90 tonnes is 70 tonnes less than the efficient level of 160 tonnes of greenhouse gas emissions.[44]

If the central-government authority employs an emissions-trading system, the approach it adopts is straightforward—it sets an emissions cap by auctioning 90 emissions permits.[45] Because the Marginal Damage curve is vertical at the emissions cap (MD_{ETS}), the permit price is determined by the intersection of the MD_{ETS} curve and the 'true' Marginal Benefit curve (MB_{True}).[46] Hence, the going price for a permit is PP_*, which is much higher than the efficient price of $\$\Psi$.

Why is the permit price determined by the intersection of the MD_{ETS} and MB_{True} curves? As explained in Chap. 6, even if the central-government authority does not know the position of the Marginal Benefit curve, individual greenhouse gas-emitting firms will be acutely aware of the marginal benefits (marginal profits) they enjoy from emitting greenhouse gases. Being cognisant of their own marginal benefits and the number of permits available for sale, individual firms will reveal their permit-buying intentions at and between each permit auction. These intentions determine the permit price and, thus, the quantity of greenhouse gases each firm generates. Assuming that all three firms are profit-maximisers, we would expect ABC Ltd, JKL Ltd, and XYZ Ltd to respectively emit 50, 30, and 10 tonnes of greenhouse gases during the current year. I should point out that if one or more of the three firms have objectives that do not involve the maximisation of profits, the share of the greenhouse gas emissions generated by the three firms would more than likely differ. However, the total would not exceed 90 tonnes.

Should an emissions tax be employed to achieve a target of 90 tonnes of greenhouse gas emissions, the eventual emissions levels would remain at the discretion of the greenhouse gas-emitting firms, since they are free to generate whatever quantity of greenhouse gases they desire so long as they are willing to pay the emissions tax. Assuming that all greenhouse gas-emitting firms are profit-maximisers, each firm will emit up to the level where the tax rate equals the marginal profit of the last tonne of greenhouse gases generated. In other words, each firm will determine their emissions level based on their mb_{True} curves.

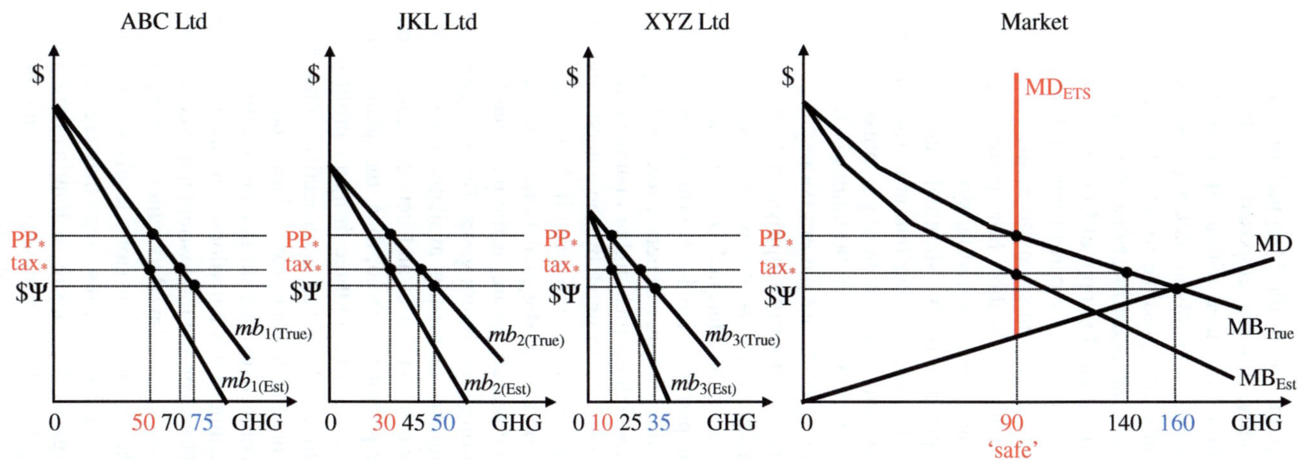

Fig. 7.11 Policy choice when a 'safe' GHG concentration requires GHG emissions to be less than the perceived 'efficient' level (position of the Marginal Benefit curve unknown to the central-government authority). *Note* GHG denotes tonnes of greenhouse gas emissions

Recognising this, but not knowing the actual location of the mb_{True} curves, a tax-setting government authority must estimate the position of the marginal benefit curves of each firm and the Marginal Benefit curve prevailing in the market. It must then set the tax rate based on the estimated position of the curve (MB_{Est}). If we assume, as in Fig. 7.11, that the authority has underestimated the position of the Marginal Benefit curve (i.e., the MB_{Est} curve is to the left of the MB_{True} curve), it will set a tax rate of $tax*$ to achieve the safe level of greenhouse gas emissions. However, at this tax rate, ABC Ltd, JKL Ltd, and XYZ Ltd will emit 70, 45, and 25 tonnes of greenhouse gases respectively. In total, 140 tonnes of greenhouse gases will be emitted, which is 50 tonnes above the safe level.

Clearly, in these circumstances, the tax rate will be insufficient to achieve the target level of emissions despite the authority knowing what the safe level is. Indeed, even if the authority could correctly estimate the position of the Marginal Benefit curve, it would still have to rely on each greenhouse gas-emitting firm being a profit-maximiser. This is because firms with alternative objectives might emit more greenhouse gases, pay the extra tax, and forego some short-term profits. In the process, the total quantity of emissions would exceed the safe level. Crucially, should firms have the same alternative objectives but be operating under an emissions-trading system, the authority's emissions target would not be threatened.[47] Instead, the increased desire to emit greenhouse gases would lead to a heightened demand for emissions permits and a permit price above $PP*$.[48]

This, then, begs the question: Why would a suitably resourced authority rely on a tax to achieve an emissions target that can be realised with absolute certainty with an emissions-trading system? Employing an emissions tax is like trying to hit a bull's-eye whilst being blind-folded when you have the option of seeing the dart-board and placing the dart wherever you like. Given the primacy of ecological sustainability, there is no doubt that the biggest problem with an emissions tax is that it cannot achieve quantity certainty—something that an effective emissions-trading system can always guarantee and something that is paramount if stabilisation at a safe concentration of greenhouse gases is ever to be realised.

7.6.2 Distributional Equity

In Chap. 3, a number of policies were outlined to achieve distributional equity. One of the policies involved a maximum income limit on the rich to limit the range of income inequality between a nation's richest and poorest citizens. To recall, achieving this requires a 100 per cent marginal tax rate on incomes above a certain income threshold.

Although the price paid for emissions permits effectively constitutes a tax, an emissions-trading system could never be relied on to set and impose a maximum limit on personal incomes. However, because of two factors, it could be harnessed to help reduce the income gap between rich and poor. Firstly, a government's ability to limit the range of income inequality depends, in part, on its ability to tax the economic rents generated by economic rent-earning assets, which are disproportionately owned by the rich. Secondly, one of the important features of an emissions-trading system is its capacity to capture the economic rents generated by the restriction it places on the rate of greenhouse gas emissions.

How do these two factors help a government bridge the income gap between rich and poor? A key feature of an effective emissions-trading system is that it artificially creates economic rents (scarcity rents) by internalising an ecologically-determined constraint into the price of emissions permits. It does this by capping the number of permits to ensure the rate of greenhouse gas emissions is consistent with a safe atmospheric concentration of greenhouse gases. At the same time, an emissions-trading system captures the very economic rents it creates by forcing individuals and entities to purchase the limited number of permits on offer. Furthermore, provided the emissions permits have a short lifespan, as I have recommended, an effective emissions-trading system eliminates any economic rents that a permit owner might earn from selling the permits later on at inflated prices.

Importantly, by ensnaring the economic rents it creates, an emissions-trading system is able to generate the revenue or spending room that a government can exploit to fully or part-finance a number of equity-improving initiatives.[49] These include: (i) reductions in the marginal tax rates on low incomes; (ii) the establishment of a Job Guarantee to achieve full employment—a vital policy given that low-income people would be most affected by the employment implications of an emissions-trading system; (iii) the construction and low-cost provision of a wide range of high-quality public goods and critical infrastructure; and (iv) in the case of high-GDP countries, the transfer of aid money to impoverished nations. Hence, rather than provide opportunities for the rich to get richer, an emissions-trading system can help reduce the income gap between rich and poor, both within and across nations. Overall, since the government sale of emissions permits has the same revenue-earning potential as an emissions tax, an emissions-trading system can contribute equally well towards achieving the goal of distributional equity.

7.6.3 Allocative Efficiency

As we have seen, by auctioning a limited number of emissions permits and subsequently allowing the permits to be bought and sold in a permit market, an emissions-trading system generates a market price for greenhouse gas emissions. Significantly, an emissions-trading system provides not one, but two key price signals that an emissions tax does not. The first is a short-term price signal that is represented by the day-to-day fluctuation in the price of circulating emissions permits. It is the short-term variation in permit prices that allows individuals and firms to make immediate and effective adjustments in response to changing demand-side and supply-side forces in the permit market.

The second is a long-term price signal that exists in the form of the probable trend change in permit prices. Since the central-government authority would publicise its intention to gradually tighten the emissions cap, we would, as Fig. 7.9 showed, expect real permit prices to trend upwards for a considerable period of time. This would encourage individuals and firms to substitute away from fossil fuels and invest in the development and uptake of greenhouse gas-abatement technologies. Not only would this help a nation efficiently meet its greenhouse gas-emissions targets, it would generate additional efficiency benefits, such as reductions in the energy-intensity of economic activity.

Conversely, should the central-government authority set a fixed emissions-tax rate, the short-term variation in the price of greenhouse gas emissions would be precluded. Such inflexibility would impede short-term effective responses to changing market conditions. Of course, the central-government authority could overcome this inflexibility by varying the emissions-tax rate on a regular basis. However, because the authority would not know the marginal benefits and marginal damage costs associated with greenhouse gas emissions, the adjusted tax rate would fail to maximise the effectiveness with which individuals and firms respond to changing market circumstances.

As for the long-term price signal, a central-government authority could increase the emissions-tax rate over time to induce the development and uptake of greenhouse gas-abatement technologies. The problem the authority would face on this occasion is that it would not know how quickly and to what level the tax rate should be increased. Although the authority could set the tax at a very high rate in an effort to prevent a particular emissions level from being exceeded, if the tax is set too high, some allocative efficiency would be unnecessarily sacrificed. This is something that an emissions-trading system would not do because permit prices automatically adjust in line with changing demand-side and supply-side forces in the permit market—no more, no less.[50]

In the end, an effective emissions-trading system would achieve emissions targets much more efficiently than an emissions tax. This should scarcely come as a surprise, even to mainstream economists. After all, it is widely recognised that markets have unique efficiency-facilitating qualities that stem from their unrivalled capacity to generate price signals that rapidly respond to changes in demand-side and supply-side market forces, including, in the case of an emissions-trading system, a regulated supply-side force (i.e., an emissions cap).

7.7 The Need for Additional Government Intervention

In this chapter, I have endeavoured to put forward the case for an emissions-trading system to help resolve the climate change crisis. We have seen that one of the main advantages of an emissions-trading system over an emissions tax is that it can achieve quantity certainty and thus guarantee a safe atmospheric concentration of greenhouse gases. At the same time, it can generate a greenhouse gas price to facilitate the cost-effective realisation of greenhouse gas emissions targets.

An important aspect worth recognising is that the greenhouse gas emissions price generated by the initial sale and the subsequent trade in emissions permits merely 'facilitates' efficient market outcomes. It does not guarantee the most cost-effective mitigation solutions. Mind you, if it were possible to set an emissions tax rate equal to the going price of emissions permits, it would also fail to guarantee efficient market outcomes. The reason for this was explained at some length in Chap. 6—simply assigning a price to greenhouse gas emissions does not guarantee the realisation of all economically viable abatement potential.

There are various reasons for this. One of them, as previously highlighted, is that a range of financial capital constraints, hidden costs, and complex interactions between various mitigation technologies can obstruct the implementation of economically viable mitigation strategies (Ekins et al. 2011). Other reasons include: (i) a lack of financial incentives over and above a price on greenhouse gas emissions to promote the development and uptake of low-emissions technologies (Johnstone 2002; Menanteau et al. 2003; IPCC 2007d); (ii) the public goods feature of many forms of knowledge, which can lead to insufficient research and development; the under-provision of low-emissions infrastructure; and a dearth of information on viable, energy-efficient options (Popp 2004; Stern 2007, 2009; Garnaut, 2008); and (iii) 'carbon leakage', which can arise when emissions-intensive forms of production are relocated to countries with an inadequate price on greenhouse gas emissions and/or countries completely devoid of emissions targets (Saddler et al. 2006; OECD 2008; Ekins et al. 2011; Henson 2011).[51]

Although an emissions-trading system out-performs a non-market-based mechanism when it comes to cost-effectively reducing greenhouse gas emissions, this does not mean that other non-market-based policies should be overlooked. In some cases, non-market-based instruments can improve the efficiency of trading mechanisms by addressing market imperfections and non-market barriers (Sorrell and Sijm 2003; Sijm 2005).[52] This is particularly so in low-GDP countries and former-communist nations which lack effective market-based institutions (Blackman and Harrington 2000; Sterner 2003; Harrington et al. 2004b). In other instances, firms not particularly responsive to market-generated price signals can be spurred on by carefully crafted regulations and energy-efficiency standards to dramatically reduce the emissions intensity of their operations. These include publicly-owned enterprises and firms operating in relatively non-competitive market settings (Wätzold 2004; Montero 2005; Freeman and Kolstad 2006).

Because of these realities, achieving emissions targets in the most cost-effective manner will require additional government intervention to overcome the market imperfections that currently discourage and/or impede the development and uptake of low-emissions technologies. Consequently, any such government intervention should be seen as a means of assisting an emissions-trading system to realise the efficient outcomes it has been specifically designed to achieve. It should not be viewed as an anti-market policy response (Stern 2009).

With this in mind, what complementary forms of intervention might a government engage into cost-effectively achieve emissions targets? There are many, but given the urgent need to drastically reduce greenhouse gas emissions, a number of them must be aimed at boosting research and development in low-emissions technologies. Government intervention is needed in this area because many of the new technologies/knowledge generated by research and development have public goods characteristics (Popp 2004). To recall from Sect. 3.5.1, the use of a public good by one person does not preclude someone else from using it (non-rivalry). In addition, without government intervention, it is impossible to prevent someone from using a public good should they not pay for its use (non-excludability). In the case of knowledge, creators of new knowledge incur a charge equal to the cost of the resources expended to create it. However, the users of the new knowledge incur little or no cost at all. Hence, the additional or marginal cost of supplying new knowledge is effectively zero.

Given its public goods nature, economic logic dictates that all new knowledge should be freely available for everyone to exploit. The dilemma, however, is that whilst some people have sufficient resources to create new knowledge and to generously make it available to the rest of society, most people or organisations do not. Thus, in order to render the creation of new technologies economically viable, the creators of the technologies need to secure an adequate financial return on their investment. Alas, no financial returns are possible if the technologies can be freely used by others.

In Chap. 3, I pointed out that incentives to create new technologies are generally offered by governments in the form of patents and copyright laws which, for a period of time, allow the creators of new technologies to profit from their exclusive use. I also highlighted two negative implications of patents—firstly, they deprive individuals and organisations from the use of cost-effective technologies; secondly, they restrict the ability of organisations to use new technologies for the social good and/or to make further technological breakthroughs. Neither of these negative side-effects is desirable at a time when greenhouse gas emissions must be markedly reduced. I would therefore reiterate my call for the introduction of 'intellectual royalty rights', which would ensure complete access to new technologies whilst rewarding the creators of new technologies with a royalty payment each time the technologies are used. In many cases, the royalty payment would involve individuals and organisations having to purchase a licence to secure unlimited access to a new technology for a specified period. In all, intellectual royalty rights would provide the incentive for firms to generate new technologies without placing undue restrictions on their application.

Because of uncertainties about costs and eventual returns on investment, intellectual royalty rights would not, however, be sufficient to ensure adequate research and development in low-emissions technologies. For this reason, governments must increase their direct funding of research and development activities (e.g., public research centres, public-private research partnerships, and science and engineering education and training), which is currently at less than acceptable levels. Governments should also stimulate greater private-sector research via the introduction of research subsidies, capital grants, tax credits/rebates, and research seed-funding (Popp 2004; Anderson 2006; IPCC 2007d; OECD 2008).[53]

It is one thing to increase research and development into energy-efficient and low-emissions technologies. It is another to induce households, firms, and organisations to take up new technologies. Although a price on greenhouse gas emissions can help enormously in this regard, the presence of market failures and of institutional and financial capital barriers referred to above call for governments to provide greater deployment support to the private sector, especially with regards to the uptake of fledgling technologies with proven benefits, and even more so in cases where benefits emerge from learning-by-doing (Stern 2009). Some of the measures capable of increasing the uptake of low-emissions technologies include investment tax credits/rebates (Stern 2007; IPCC 2007d; IARU 2009); low-interest loans; feed-in tariffs to reward the generation of low-emissions electricity (Ackermann et al. 2001; Menanteau et al. 2003; Anderson 2006); and procurement policies to guarantee a sufficient market for goods and services generated from the use of low-carbon production methods (Anderson 2006; Stern 2007).

Unfortunately, even if governments offered these inducements, the potential uptake of low-emissions technologies is unlikely to be fully realised if households and private and public organisations remain ignorant of emerging new technologies and the costs and benefits associated with their use. To overcome this, governments need to provide easy and free access to relevant information. They also need to establish demonstration centres and websites that exhibit new technologies and offer accurate and up-to-date technical advice (Garnaut 2008). Access to useful information can also be increased through product-labelling laws requiring energy-efficiency and emissions ratings to be displayed on selected goods (IPCC 2007d).

Finally, as explained a number of times, much of the infrastructure and capital goods required to make the transition to a low-emissions economy will have public goods characteristics. It is therefore unlikely that the private sector will provide these goods in sufficient quantities. This is especially the case with respect to the infrastructure and related transmission and supply networks of the energy, electricity, and transportation industries (Stern 2007, 2009). Hence, much of the future investment in low-emissions infrastructure will need to be undertaken by governments, just as governments were prominent in establishing the fossil fuel-dominant infrastructure in the industrialised world during the early and middle parts of the 20th century.

Given the growing scarcity of natural resources and limits on national productive capacity, crucial political choices will have to be made regarding the allocation of the incoming flow of natural resources to the private and public sectors of the economy—sometimes referred to as the macro-allocation of natural resources (Daly and Farley 2004).[54] In my opinion, resolving the climate change crisis will require greater involvement of governments, not less. Consequently, the citizens of all nations will need to consider how many and what types of private goods they are willing to forego to produce the public goods required to enjoy a high quality, low-emissions (ecologically sustainable) life-style. Bear in mind, this choice has greater ramifications for the high-GDP nations facing the prospect of having to reduce their real output when making the transition to a qualitatively-improving steady-state economy (Lawn 2011). Furthermore, increases in government infrastructural investment and many of the above-listed financial inducements will also be necessary to boost climate change adaptation activities, which are required to minimise the total damage costs of climate change (see Fig. 6.2). It is also worth adding that the inability of the governments of low-GDP nations to provide public goods in sufficient quantities means that high-GDP countries will need to commit funds and newly-developed technologies to assist poor nations in their mitigation and adaptation endeavours—a theme I will revisit many times throughout the remainder of the book.

As for other international factors, such as the shift of emissions-intensive forms of production to low-cost nations, there are, as I argued in Chap. 3, good reasons to allow governments to impose 'compensating' tariffs on imported products to avert the problem of carbon leakage and industrial flight. Since my aim is to outline a global emissions-trading system in the context of a new global climate change protocol, I will deal with these international factors in upcoming chapters.

Notes

1. Since an emissions-trading system is effectively a specifically designed cap-auction-trade system, many of its basic features are similar to those outlined in relation to cap-auction-trade systems in Chap. 3.
2. Eligibility would be restricted to individuals and organisations that have not been barred from participating in public auctions or in possession of the maximum allowable number of permits.
3. There are many observers who believe there should be no time restriction on the use of emissions permits (e.g., Garnaut 2008). The basis for this is two-fold: (i) the hoarding of permits does not breach emissions budgets; and (ii) it reduces the cost of achieving emissions targets (i.e., increases the efficiency of an emissions-trading system) by allowing permit holders to hoard or bank permits and use them when they have their greatest value. It is certainly true that having longer life permits increases the flexibility and efficiency of an emissions-trading system. However, as stipulated at the beginning of the chapter and earlier in the book, an emissions-trading system needs to be based on the principles of ecological sustainability, distributional equity, and allocative efficiency. As we shall see, allowing permit holders to hoard their permits promotes speculative behaviour, which, in turn, allows windfall profits or economic rents to be earned. This is not only inequitable, but price-distorting—the latter of which can dramatically reduce economic efficiency. Even allowing for a small net improvement in efficiency by issuing long-life permits, such a small efficiency gain is worth sacrificing in order to ensure distributional equity.
4. The free issuance of permits can act as a market-entry barrier for new firms.

5. The point of obligation defines the party liable for surrendering emissions permits under an emissions-trading scheme (Garnaut 2008). The point of obligation may exist anywhere in the supply chain of a final good or service—from the extraction of the resources used to produce the goods that eventually lead to the release of greenhouse gases; to production processes that directly generate greenhouse gas emissions; and to the consumption of emissions-related final goods. The point of obligation is best imposed at the stage in the supply chain where monitoring and reporting of emissions is easiest and most accurately and cost-effectively achieved.
6. Other examples would include: (i) the owners of coal-powered electricity-generators having to purchase emissions permits rather than electricity-users; (ii) farmers clearing land to grow crops having to purchase permits rather than the consumers of agricultural products; and (iii) fertiliser-producing companies having to purchase permits rather than farmers. Of course, the buyers of emissions permits would be free to pass on some of the cost of having to purchase permits to the consumers of the relevant products. However, this should encourage end-users to purchase products made from renewable-energy sources and low emissions-intensive production methods.
7. Supply-side forces are constrained by the emissions cap set by the central-government authority.
8. It may well be that a nation's anti-trust legislation could already deal successfully with this requirement. For example, in Australia, the Competition and Consumer Act enables the Australian Competition and Consumer Commission (ACCC) to prohibit any action it construes would 'significantly reduce competition'. There are various sections of the Act that would allow the ACCC to limit the ownership of greenhouse gas-emissions permits.
9. Although some people would claim that the failure to grant additional permits would deter individuals and entities from taking action to prevent the release of greenhouse gases, it should be remembered that appropriate sustainable development policies would already be preventing a decline in natural capital that, in turn, would be reducing the release of greenhouse gases.
10. Presumably, most if not all the greenhouse gas-emitting activities would be undertaken by the entity that has invested to preserve the forest.
11. Further emissions reductions would be required if the ecosphere's diminished sequestration capacity exceeded the reduction in allowable greenhouse gas emissions arising from the confiscation of the transgressor's permits.
12. This would depend, again, on the ease with which the source of greenhouse gas emissions can be monitored.
13. This assumes that the three firms operate in a perfectly competitive market.
14. High quality goods generally command higher market prices than similar goods of lesser quality.
15. Similar to Chap. 6, the Marginal Damage curve in Fig. 7.1 includes both direct damage costs plus adaptation costs.

16. The reason for this is that if the marginal profit of a particular emissions level is positive, an increase in emissions boosts profits. A profit-maximising firm would therefore increase its emissions. Assuming a downward sloping marginal benefit (*mb*) curve, the increase in profits is exhausted once marginal profit equals zero. At this point, profit is maximised. Any further increase in emissions reduces a firm's profit.
17. This is just a different perspective of the welfare-maximising outcome illustrated in Fig. 5.1. I should also add that the welfare-maximising outcome depicted in Fig. 7.1 assumes that the social benefits from emitting greenhouse gases are confined to the private consumption benefits generated from consuming the goods and services produced from greenhouse gas-emitting activities. In other words, it is assumed that there are no spillover benefits from the greenhouse gas-emitting activities.
18. By non-tradeable, I mean that a firm is prohibited from trading part or all of its emissions quota with another firm, organisation, or individual.
19. To reflect this, the Marginal Damage curve is a dotted instead of being an unbroken line.
20. It is because of this that ABC Ltd's marginal profit exceeds the emissions tax rate over a much larger output level than JKL Ltd and XYZ Ltd. Thus, ABC Ltd can afford to emit more greenhouse gases than the other greenhouse gas-emitting firms.
21. This assumes that one emissions permit equates to one tonne of greenhouse gas emissions.
22. In this analysis, it has been assumed that the marginal benefits (marginal profits) of each firm are different and that the marginal abatement costs of each firm are the same. In some analyses, it is assumed that the marginal benefits of each firm are identical and that it is the marginal abatement costs of each firm that differ. In this alternative situation, the firms with the lowest abatement costs purchase fewer permits and undertake considerably high levels of emissions abatement. Conversely, the firms with high abatement costs find it cheaper to undertake minimal abatement and purchase a large quantity of emissions permits. A good textbook example of this latter type of analysis can be found in Pearce and Turner (1990).
23. This assumes that the number of permits initially auctioned by the government is limited to the welfare-maximising level.
24. This also important given that permit prices can automatically adjust to take account of price inflation. The impact of an emissions-tax is eroded by inflation.
25. The rises in total CO_2 emissions for the other three countries were: (i) Denmark (2.6 per cent, 1990–2002); (ii) The Netherlands (13.1 per cent, 1990–2002); and (iii) Finland (9.6 per cent, 1990–2002) (Lawn 2006a).
26. Referring back to Fig. 7.5 does not imply recourse to the mainstream economic framework. Regardless of what a firm's economic motives are, if market conditions change but the emissions tax rate does not, the firm will almost certainly vary its emissions levels.

27. If permits have a short lifespan, it will also be necessary for the government to reduce the number of permits it sells in subsequent auctions.
28. This, for example, could take the form of evidence showing that the ecosphere's greenhouse gas-absorbing capacity has been overestimated; that greater-than-expected deforestation has reduced the rate of carbon sequestration; or that positive feedback processes (e.g., the 'ice-albedo feedback') are accelerating.
29. As we shall see in Chap. 10, the extent and the speed of emissions cuts in the world's high-GDP countries are likely to be greater than what is applied to the rest of the world.
30. The *real* greenhouse gas price takes account of the general rate of price inflation (i.e., the real GHG price = nominal GHG price ÷ general price level).
31. This also means that permit prices will be above the conventional 'efficient' level (i.e., where MB = MD).
32. Although these are hypothetical short-term price fluctuations, they would not be dissimilar to the short-term price variations being generated by the emissions-trading systems currently operating around the world.
33. It is worth noting that economic modelling exercises consistently lead to the conclusion that the social cost of greenhouse gas emissions—of which the price of emissions permits should reflect—is likely to increase for many decades (Watkiss et al. 2005). See also OECD (2008, Fig. 6).
34. The eventual decline in the price of permits (social cost of carbon) goes very much against popular opinion. Garnaut (2008), for example, believes the price of permits should continue to increase at the market rate of interest, much like the Hotelling (1931) model predicts in relation to the price of a non-renewable resource. Garnaut considers there to be a parallel between the two markets because, although greenhouse gas emissions do not resemble a non-renewable or exhaustible resource, the remaining allowable emissions budget does. Assuming that the Hotelling model is an applicable one, it should be remembered that an upward-trending price for a non-renewable resource is based on a number of parameters remaining constant, such as the demand for the resource, the marginal cost of extraction, and the choke price for the resource (i.e., the price at which would-be users of the resource switch to a cheaper substitute).

    In the case of greenhouse gas-emissions permits, the equivalent parameters would be the demand for permits, the cost of low-carbon technologies, and the permit price that would choke off demand for high-carbon technologies. It is not unreasonable to expect the demand for permits to decline as firms shift to low-carbon technologies and for the cost of employing low-carbon technologies to eventually fall because of technological progress and economies-of-scale effects (i.e., as the deployment of low-carbon technologies increases). In addition, we would expect the falling cost of low-carbon technologies to lower the choke price of emissions permits. Should these changes occur, the Hotelling model suggests that these forces would exert downward pressure on the price of emissions permits. Since these forces would eventually be significant in magnitude and on-going, it is highly likely that they would over-ride any price-increasing influence of market interest rates, thus resulting in a declining permit price.

35. In almost all instances, firms will not be able to pass the entire cost of purchasing emissions permits onto the next stage of the supply chain. The reason for this is that the market demand and supply curves for a particular product will invariably be downward-sloping and upward-sloping respectively. Because of this, firms will be forced to bear some of the cost of purchasing emissions permits and therefore some of the increased cost of supplying goods and services. Consumers will be required to bear the remainder. In all, the price of final goods and services will almost certainly rise by something less than the increase in the cost of producing goods and services caused by greenhouse gas-emitting firms having to purchase emissions permits.
36. The Clean Development Mechanism is an offset arrangement under the Kyoto Protocol that allows greenhouse-gas emitters in the world's high-GDP (Annex I) countries to invest of projects that reduce greenhouse gas emissions in low-GDP nations. The Clean Development Mechanism exists as a means by which greenhouse-gas emitters in high-GDP nations can avoid more expensive emissions reductions in their own countries. Although the Clean Development Mechanism operates under the auspices of the United Nations Framework Convention on Climate Change (UNFCCC), there are concerns that spurious emissions credits have been granted due to inaccurate estimates of reduced greenhouse gas reductions (overstated baselines); inadequate monitoring of offset projects; and an incentive for firms located in low-GDP countries to initiate high-emissions operations in order to be paid to stop polluting (Rosenthal and Lehren 2012). I'll have more to say about the Clean Development Mechanism in Chaps. 8 and 9.
37. In the case of the Clean Development Mechanism, which currently exists as an international offset arrangement, it is critical that the Mechanism be better designed and resourced to minimise abuse of the system.
38. The carbon tax in Australia was repealed in 2014.
39. It is an emissions trajectory that stabilises the atmospheric concentration of greenhouse gases at or below 450 ppm of CO_2-e.
40. As anyone who has studied first-year undergraduate economics will know, the current position of the Marginal Benefit (demand) curve for a particular good is influenced by future expectations of the price of the good. Should people believe that the price of the good will rise in the future (or is temporarily low), they will demand more of the good now and less in the future. Consequently, there will be a rightward shift of the demand curve for the good.
41. It is worth considering what happened to the price of emissions permits at the end of Phase I of the European Union's Emissions Trading System (EU-ETS), where holders of unused permits could not bank the permits and use them in Phase II of the EU-ETS. The price of these permits almost fell to zero. I would not expect a similar price plunge if there was a more effective emissions-trading system in place.

42. For more information on each of these emissions-trading systems, see: (i) EU-ETS (Ellerman and Buchner 2007; http://ec.europa.eu/clima/policies/ets/index_en.htm); (ii) UK-ETS (Smith and Swierzbinski 2007); (iii) JVETS (Sudo 2006); and (iv) NZ-ETS (Bertram and Simon 2010).
43. This condition was the basis for the four sustainability precepts outlined in Chap. 2.
44. Given that a greenhouse-gas offset arrangement can be incorporated into an emissions-trading system and an emissions-tax system, we shall ignore the ability of offsets to allow for a higher 'safe' rate of greenhouse gas emissions.
45. This again assumes that one permit equates to one tonne of greenhouse gas emissions.
46. Because permit prices would be determined by the intersection of the MD_{ETS} and MB_{True} curves, the lack of knowledge about the exact whereabouts of the upward-sloping Marginal Damage (MD) curve is irrelevant. It is only relevant if the aim is to achieve an efficient outcome, which is itself irrelevant if, firstly, the efficient rate of greenhouse gas emissions is greater than the safe rate, and secondly, achieving a safe rate is the primary objective of the central-government authority.
47. This assumes that the penalties for emitting greenhouse gases without an emissions permit are severe enough to deter illegal emissions-related activities.
48. In Fig. 7.11, the higher demand for emissions permits would be represented by a further rotation of the mb_{True} and MB_{True} curves.
49. As explained in endnote 8 in Chap. 3, taxes do not technically generate the spending power that currency-issuing central governments require to spend. See Mitchell and Muysken (2008).
50. No market operates perfectly. Hence, permit prices would closely approximate rather than precisely reflect the demand and constrained supply-side forces in the permit market.
51. In essence, carbon leakage occurs when reductions in greenhouse gas emissions are negated by increases in emissions caused by the relocation of emissions-intensive forms of production to other parts of the world.
52. As Stern (2009) stresses, as much as markets are the most effective efficiency-promoting mechanisms, they rarely function to their full potential without some form of government assistance.
53. The importance of this is exemplified by the fact that public and private spending on research and development into new and more efficient forms of energy has been declining globally (IEA 2007; Stern 2007, 2009).
54. Of course, the inflow of natural resources should never be greater than the maximum sustainable rate.

Chapter 8
International Climate Change Institutions and the Greenhouse-Gas Emitting Performance of Nations

8.1 Introduction

It stands to reason that a proposal to establish a new global protocol will only be successful if it is realistically achievable. This infers two things. Firstly, it must be institutionally and logistically feasible to implement and operate. Secondly, there must be a high probability that the proposal will be agreed upon by most if not all the world's nations.[1] The latter is essential given the difficulties encountered in trying to reach a consensus at past United Nations climate change conferences. I believe the chances of success are increased if the new global protocol is built on the basic features of the existing Kyoto architecture. Nevertheless, I agree with Stern (2009) that successfully taking a new global protocol forward will require the Kyoto architecture to be modified. I'll have more to say about a modified climate change architecture in Chap. 9.

In the meantime, my intention is to lay the foundations to support the global protocol I plan to reveal in Chap. 10. I will begin this exercise by illuminating the features of some key climate change institutions and related mechanisms. This will be followed by an examination of the greenhouse gas-emitting performance of individual countries. Together, these should support the case for modifying the Kyoto architecture and the need to adopt a 'common but differentiated convergence' approach when setting and embodying national emissions targets in a post-2020 protocol.

P. Lawn, *Resolving the Climate Change Crisis*, DOI 10.1007/978-94-017-7502-1_8

8.2 Climate Change Institutions and Related Mechanisms

8.2.1 The United Nations Framework Convention on Climate Change

To recall from Chap. 1, the first international institution established to deal with the climate change crisis was the United Nations Framework Convention on Climate Change (UNFCCC) in 1992 (UNFCCC 1992). Initially signed by 154 countries, the UNFCCC provides the legal framework to foster the international collaborative action deemed necessary to prevent any dangerous anthropogenic interference with the Earth's climate system (Rübbelke 2011).

As of early-2014, there were 195 UNFCCC signatory nations belonging to one of the following four UNFCCC-defined groups (see Fig. 8.1):

- *Annex I*: The Annex I group includes 43 Parties, many belonging to the Organisation for Economic Co-operation and Development (OECD), of which a number are 'economies in transition' (EITs). The EITs constitute the former centrally-planned economies of the now-defunct USSR and Eastern Europe. Included in this group is a single Annex I Party referred to as the European Union-15 (EU-15).
- *Annex II*: Existing as an elite assembly of 24 Parties within Annex I, the Annex II group comprises high-GDP countries and the EU-15. Whilst all the Parties

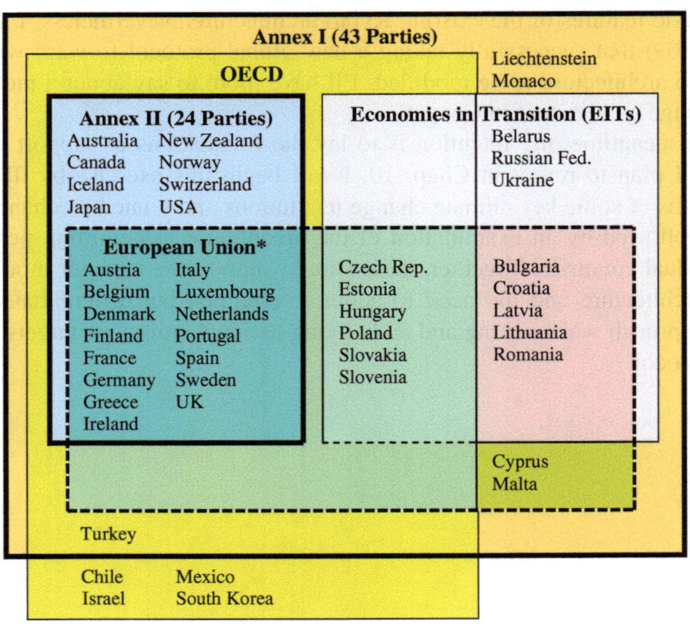

Fig. 8.1 Current Annex I and Annex II Parties of the UNFCCC. *Source* Adapted from Höhne et al. (2005). *Note* * The European Union exists as an individual UNFCCC Party

in Annex II are members of the OECD, the group does not include EITs or the OECD countries of Turkey, Israel, South Korea, Chile, and Mexico.
- *Non-Annex I*: This residual group includes all UNFCCC Parties not listed as Annex I nations. The group largely consists of the world's low-GDP countries.
- *Least-Developed Countries* (LDCs): The final group comprises 48 Parties within the group of non-Annex I nations.[2] As impoverished countries, LDCs are afforded special status in view of their more urgent development needs and their limited capacity to both reduce greenhouse gas emissions and adapt to the undesirable effects of climate change.

The above four groups were created to facilitate effective climate change negotiations and to reflect the key principle that industrialised countries should take the lead in cutting greenhouse gas emissions and provide climate change-related assistance to low-GDP countries (Winkler et al. 2002). In recognition of this, Annex I nations initially agreed to reduce their greenhouse gas emissions to 1990 levels by the year 2000 (Article 4.2). In addition, the 'common but differentiated responsibilities' enshrined in Article 3.1 of the UNFCCC emphasised the need for low-GDP countries to take up emissions targets once they reached a reasonable level of wealth or, if reached sooner, a high per capita rate of greenhouse gas emissions (Depledge 2002; Michaelowa et al. 2005a).

Along with the main objective of preventing dangerous human interference with the climate system (Article 2), it is a requirement of the UNFCCC to operate by three guiding principles specified in Article 3 of the Convention. The first is the 'precautionary principle', which requires all UNFCCC Parties to act in a cost-effective way to avoid serious or irreversible threats from climate change (Article 3.3). The second concerns the right of all UNFCCC Parties to implement measures to achieve sustainable development, including a commitment to promote such measures (Article 3.4). The third principle amounts to an obligation on the part of all Parties to co-operate in terms of: (i) sharing information about climate change matters and concerns; (ii) exchanging new technologies; and (iii) co-ordinating national action plans (Article 3.7).

Since the UNFCCC became effective in 1994, signatory nations have met annually at what are referred to as the Conference of the Parties (COP). The aim of these meetings has been to advance the process towards achieving the objectives of the UNFCCC—that is, to assess the progress being made towards dealing with the climate change crisis and establish emissions targets and the measures required to achieve them. At the first meeting in Berlin in 1995 (COP-1), a major decision was made to select the Global Environment Facility (GEF) as the interim financing mechanism of the UNFCCC.[3] From a climate change perspective, the main function of the Global Environment Facility has been to facilitate the transfer of funds and new technologies from high-GDP to low-GDP countries as per Articles 4.3 and 4.4 of the Convention (Rübbelke 2011).

During the Berlin conference, the Parties to the UNFCCC concluded that efforts to encourage Annex I nations to reduce their greenhouse gas emissions were grossly inadequate and that more stringent emissions targets were necessary to avert a climate change crisis (IPCC 2007d). This conclusion would be reinforced by evidence contained within the *IPCC Second Assessment Report* showing that global warming was a reality and that the contribution made towards it by human activities was irrefutable (IPCC 1996a, b, c).

8.2.2 The Kyoto Protocol—Obligations During the First Commitment Period of 2008–2012

In response to the IPCC conclusions, the first mandatory emissions treaty was successfully negotiated at the third UNFCCC meeting in 1997 (COP-3). Known as the Kyoto Protocol, the ground-breaking treaty was designed on the understanding that industrialised countries were largely responsible for the elevated concentration of greenhouse gases in the Earth's atmosphere and that, under the principle of 'common but differentiated responsibilities', industrialised countries should bear the major brunt of any first-up attempt to reduce global greenhouse gas emissions.[4] Thus, in the spirit of the UNFCCC, the Kyoto Protocol aimed to limit the global generation of greenhouse gases in a manner reflecting national disparities in per capita income; past and present greenhouse gas emissions; and the capacity of countries to achieve emissions cuts (Grubb 2003).

Under the Kyoto Protocol, Annex I Parties, which were designated under the Protocol as Annex B Parties[5], agreed to legally-binding emissions targets pertaining to six major greenhouse gases. Non-Annex I nations, on the other hand, would be spared emissions targets, albeit they would be urged to take whatever measures they could to limit their greenhouse gas emissions. The six greenhouse gases included in the Kyoto Protocol were carbon dioxide (CO_2), methane (CH_4), nitrous oxide (N_2O), hydrofluorocarbons (HFCs), perfluorocarbons (PFCs), and sulphur hexafluoride (SF_6), where the latter five gases would be expressed in terms of their CO_2-equivalent warming effect (see Chap. 1 and Table 1.1).[6] Expression in this form would allow the six greenhouse gases to be measured and tracked as a uniform bundle of gases, thus making it easier to formalise and achieve emissions targets (Henson 2011).[7]

The emissions quotas that applied to each Annex I Party were referred to as Assigned Amount Units (AAUs). Collectively, Annex I Parties agreed to reduce their greenhouse gas emissions by 5.2 per cent relative to 1990 emissions levels during what would turn out to be the first Kyoto commitment period of 2008–2012 (see Table 8.1). Although the final emissions targets were weaker than those proposed by some UNFCCC Parties (e.g., the Alliance of Small-Island States, Peru, the Philippines, and G-77/China), they were more stringent than targets proposed by a number of other Parties (e.g., Canada, USA, and the Russian Federation) (Depledge 2000).[8] To meet the 5.2 per cent reduction target, most Annex I Parties were called upon to reduce their greenhouse gas emissions by between five and eight per cent (see Table 8.1, column *a*). Exceptions were the Russian Federation,

Table 8.1 Greenhouse gas emissions targets of the Annex I Parties to the Kyoto Protocol

Region/Party	Kyoto target (2008–2012) (relative to 1990 levels)	Kyoto target following EU redistribution of targets (2008–2012) (relative to 1990 levels)	Kyoto target (2013–2020) (relative to 1990 levels)
	a	*b*	*c*
Nth America			
Canada*	−6 %	N/A	N/A
Europe			
EU-15	−8 %	N/A	−20 %
Austria	−8 %	−13 %	−20 %
Belgium	−8 %	−7.5 %	−20 %
Cyprus	N/A	N/A	−20 %
Denmark	−8 %	−19.6 %	−20 %
Finland	−8 %	0 %	−20 %
France	−8 %	0 %	−20 %
Germany	−8 %	−21 %	−20 %
Greece	−8 %	+25 %	−20 %
Iceland	+10 %	N/A	−20 %
Ireland	−8 %	+13 %	−20 %
Italy	−8 %	−6.5 %	−20 %
Liechtenstein	−8 %	N/A	−16 %
Luxembourg	−8 %	−28 %	−20 %
Malta	N/A	N/A	−20 %
Monaco	−8 %	N/A	−22 %
Netherlands	−8 %	−6 %	−20 %
Norway	+1 %	N/A	−16 %
Portugal	−8 %	+27 %	−20 %
Spain	−8 %	+15 %	−20 %
Sweden	−8 %	+4 %	−20 %
Switzerland	−8 %	N/A	−15.8 %
UK	−8 %	−12.5 %	−20 %
EITs			
Belarus	N/A	N/A	−12 %
Bulgaria	−8 %	N/A	−20 %
Croatia	−5 %	N/A	−20 %
Czech Republic	−8 %	N/A	−20 %
Estonia	−8 %	N/A	−20 %
Hungary	−6 %	N/A	−20 %
Latvia	−8 %	N/A	−20 %
Lithuania	−8 %	N/A	−20 %
Poland	−6 %	N/A	−20 %
Romania	−8 %	N/A	−20 %
Russian Fed.†	0 %	N/A	N/A
Slovakia	−8 %	N/A	−20 %
Slovenia	−8 %	N/A	−20 %
Ukraine	0 %	N/A	−24 %

(continued)

Table 8.1 (continued)

Region/Party	Kyoto target (2008–2012) (relative to 1990 levels)	Kyoto target following EU redistribution of targets (2008–2012) (relative to 1990 levels)	Kyoto target (2013–2020) (relative to 1990 levels)
	a	*b*	*c*
Asia/Oceania			
Australia	+8 %	N/A	−0.5 %
Japan†	−6 %	N/A	N/A
New Zealand†	0 %	N/A	N/A
Other			
Turkey†	N/A	N/A	N/A

Notes
- *Canada officially withdrew from the Kyoto Protocol in 2012
- †The Russian Federation, Japan, Turkey, and New Zealand are not subject to emissions targets during the second Kyoto commitment period (2013–2020). Japan and the Russian Federation indicated their intentions in 2010 not to participate in the second commitment period; New Zealand chose to set an economy-wide reduction target under UNFCCC supervision for the period 2013–2020; and Turkey was not given a greenhouse gas emissions target for the second Kyoto commitment period
- N/A denotes Not Applicable

Sources UNFCCC (1998, 2013b), EEA (2009/2012)

Ukraine, and New Zealand, which were required to maintain emissions at 1990 levels; and Norway, Australia, and Iceland, which were permitted to increase emissions by one, eight, and ten per cent respectively (see Table 8.1, column *a*).[9]

Despite the European Union (EU) being allotted a target to reduce greenhouse gas emissions by 8 per cent, it was granted the flexibility to redistribute the targets of each country within the overall EU constraint. The rationale for redistribution was based on the belief that some member nations would find it easier and cheaper to achieve emissions reductions than other members (Liverman 2009; Henson 2011). Following redistribution, the greenhouse gas emissions targets varied from reductions as large as 28 per cent for Luxembourg and 21 per cent for both Denmark and Germany through to increases of 27 per cent for Portugal, 25 per cent for Greece, 15 per cent for Spain, and 13 per cent for Ireland (see Table 8.1, column *b*).

8.2.3 The Kyoto Protocol—Obligations During the Second Commitment Period of 2013–2020

As explained in Chap. 1, with the end of the Kyoto commitment period in sight, a roadmap was created at the 2007 UNFCCC meeting in Bali (COP-13) with the aim of establishing a new, legally-binding protocol to take effect at the end of 2012. The desire at the time was for the successor treaty to be fully negotiated at the COP-15 conference in Copenhagen in 2009. During the Bali conference, European Union nations stressed the need for global emissions to be severely reduced beyond 2020. The strategy, however, was strongly opposed by a group of countries led by the USA. Indeed, whilst the USA remained a signatory to the UNFCCC, it continued with its policy not to ratify the Kyoto Protocol. Conversely, on the back of a change of government in Australia in late-2007, the new Prime Minister, Kevin Rudd, ratified the Protocol.

By late-2008, the attention of most of the world's leaders had shifted to the deepening global 'GDP' recession. Given the perceived need to stimulate the global economy, concerns were raised that the Copenhagen conference (COP-15) would fail to produce the effective new Protocol that many had been anticipating. By the end of the Copenhagen conference, the widespread concerns were confirmed with the conference producing little more than a new undertaking—the Copenhagen Accord—that, while recognising the need to restrict temperature rises to no more than 2 °C above pre-industrial levels, included no legally-binding commitments to reduce greenhouse gas emissions. On a positive note, the Copenhagen Accord included a proposal to establish a Green Climate Fund to help finance mitigation and adaptation measures in non-Annex I nations and facilitate the transfer of green technologies from the world's richest to poorest countries.

Despite considerable opposition to the Copenhagen Accord, most countries eventually endorsed it during 2010. However, in what constituted a further blow to the process, Japan and the Russian Federation indicated their desire in December 2010 to be relieved of emissions obligations at the completion of the first Kyoto commitment period (UNFCCC 2013b).

The lack of a genuine commitment to climate change mitigation continued at the COP-16 conference in Cancún, Mexico (2010), where the world's governments again failed to broker a new, legally-binding protocol. At the 2011 conference in Durban (COP-17), a decision was made to extend the Kyoto Protocol by way of a second commitment period (2013–2020) and to establish a new protocol in 2015 to take effect at the end of 2020. The so-called Durban Platform was reified at the 2012 conference in Doha (COP-18). Some of the Doha amendments to the Kyoto Protocol included: (i) an expansion in the number of Annex I (Annex B) Parties; (ii) the inclusion of a seventh greenhouse gas—Nitrogen trifluoride (NF_3)—for accounting and performance evaluation purposes; (iii) a commitment by most Annex I nations to reduce their collective greenhouse gas emissions by at least 18 per cent below 1990 levels during the period of 2013–2020; and (iv) a pledge to distribute a share of the value of Assigned Amount Units exchanged under the Protocol's International Emissions Trading system to help cover administrative expenses and contribute to the cost of adaptation strategies in the world's most vulnerable nations (UNFCCC 2013b). 2012 ended with Canada being officially expunged from the Kyoto Protocol one year after notifying its intention to withdraw from the Protocol on 15 December 2011 (UNFCCC 2013b).[10]

To achieve the new 18 per cent emissions reduction target, most Annex I nations agreed to cut their greenhouse gas emissions by 20 per cent below 1990 levels during the second Kyoto commitment period (see Table 8.1, column *e*). Following the negotiations, two countries were handed more severe reduction targets than the remainder of the group—Monaco (22 per cent) and the Ukraine (24 per cent). In direct contrast, Switzerland, Belarus, and Australia were granted more lenient reduction targets of 15.8, 12, and 0.5 per cent respectively (see Table 8.1, column *e*).[11]

8.2.4 Mitigation—The Kyoto Protocol's Flexibility Mechanisms

To help Annex I countries meet their emissions targets and achieve them at the lowest cost, the Kyoto Protocol embodies a range of sophisticated flexibility mechanisms.[12] Under these arrangements, an Annex I nation can exceed its annual emissions allowance by offsetting its excess emissions through the use of a flexibility mechanism recognised by the UNFCCC. The three flexibility mechanisms initially registered under the Kyoto Protocol were: (i) an International Emissions Trading system (Article 17); (ii) the Clean Development Mechanism (Article 12); and (iii) a Joint Implementation facility (Article 6). These three mechanisms and the manner in which they can reduce the cost of achieving emissions targets will now be outlined and explained.

8.2.4.1 International Emissions Trading (IET)

The Kyoto Protocol's International Emissions Trading is a system currently confined to Annex I (Annex B) nations, which allows such countries, should they better their Kyoto targets, to sell their surplus Assigned Amount Units to Annex I nations that have exceeded their emissions quotas. The main purpose of the system is to promote the cost-effective achievement of emissions targets. Cost reductions are attained by allowing the emissions units to be reallocated, via exchange, from low-cost to high-cost mitigation countries.[13]

To explain in simple terms how the Kyoto Protocol's International Emissions Trading system works, consider Table 8.2. Imagine that only three Annex I nations exist—countries *A*, *B*, and *C*—each with an emissions allowance (AAUs) of 100 units per year.[14] Imagine, also, that the projected annual emissions of *A*, *B*, and *C* for the upcoming year are 110, 90, and 100 units respectively (300 units in total). Based on these projections, country *C* would exactly meet its emissions target; country *A* would exceed its annual allowance by 10 units (+10); and country *B* would possess 10 unused AAUs at the end of each year (−10). To enable county *A* to meet its emissions target, country *B* could sell its 10 unused or surplus AAUs

Table 8.2 Using the International Emissions Trading system to achieve greenhouse gas (GHG) emissions targets

Group	Country	GHG allowance (AAUs)	Projected GHG emissions	Projected GHGs relative to target	Adjusted allowance (AAUs)	Eventual GHG emissions	GHGs relative to adjusted target
Annex I	*A*	100	110	+10	110	110	0
Annex I	*B*	100	90	−10	90	90	0
Annex I	*C*	100	100	0	100	100	0
Annex I	Total	300	300	0	300	300	0

every year to country *A*. The sale of AAUs would increase *A*'s annual missions allowance to 110 units and reduce *B*'s annual allowance to 90 units, thus allowing each Annex I Party to meet its adjusted emissions target.

How does International Emissions Trading reduce mitigation costs? The cost-reducing feature of an emissions-trading system was explained in some detail in Chap. 7. Hence, what is presented here is a crude example. Recognising that an International Emissions Trading system allows a nation to purchase surplus AAUs to cover any excess emissions, it is conceivable that country *A* might exceed its initial emissions allowance in the knowledge that its marginal cost of mitigation is higher than the cost of buying AAUs. At the same time, it is conceivable that country *B* could be operating below its emissions target because it is a low-cost mitigation nation. Of course, there is also the possibility that the surplus AAUs in country *B*'s possession could be the upshot of its emissions allowance far exceeding its prevailing greenhouse gas emissions.[15]

Assuming the former reason, country *B* would be willing to sell its surplus AAUs if the annual revenue it receives (e.g., $50 million) is more than the cost of limiting its greenhouse gas emissions each year to 90 units (e.g., $20 million).[16] Similarly, country *A* would be willing to purchase the AAUs it requires to legally emit an extra 10 units of greenhouse gases if the cost of acquiring the AAUs (e.g., $50 million) is less than the cost of reducing its annual emissions from 110 to 100 units (e.g., $90 million). In these circumstances, the annual benefit of emissions trading to countries *A* and *B* would be $40 million and $30 million respectively.[17] All up, the combined benefit of $70 million would represent the total mitigation costs saved each year by the Annex I group of nations.

It is important to recognise that the Kyoto Protocol does not confine emissions trading to the buying and selling of AAUs. The International Emissions Trading system also allows countries to trade emissions units acquired through the Clean Development Mechanism and the Joint Implementation facility (Garnaut 2008). As we shall see, this not only promotes these two Kyoto flexibility mechanisms, it facilitates the international transfer of funds from high-GDP to low-GDP nations.

8.2.4.2 Clean Development Mechanism (CDM)

The Clean Development Mechanism is a project-based offset system that awards emissions credits—Certified Emissions Reduction units (CERs)[18]—to Annex I nations which invest in approved emissions-reduction projects in non-Annex I countries (Garnaut 2008). The Mechanism is supervised by a CDM Executive Board under the guidance of the Conference of the Parties serving as the meeting of the Parties to the Kyoto Protocol. Although the CDM Executive Board is the point of contact for any nation or entity seeking CDM-project registration, Designated Operational Entities (DOEs)[19], which many regard as the 'extended arm' of the Board, are responsible for the validation of proposed CDM-projects and the verification and certification of emissions reductions (Schneider 2007; UNFCCC 2008a).[20]

In order to receive CERs, projects must satisfy an 'additionality' test, which means that the reduced emissions in the recipient country must be additional to reductions that would have transpired if the Clean Development Mechanism did not exist (Carbon Trust 2009).[21] The quantity of CERs received is calculated by determining the difference between the emissions levels generated in the recipient nation following the implementation of a CDM-project and the emissions that would have been generated in the absence of the project—the latter referred to as the emissions 'baseline' (IPCC 2007d). Because CDM-projects are undertaken in countries without Kyoto targets, the additionality test plays an important function in preventing Annex I nations from claiming fictitious emissions credits through business-as-usual activities (IPCC 2007d).[22]

Besides enabling investor countries to earn CERs, it is a requirement under the Kyoto Protocol for CDM-projects to assist host countries to achieve the goal of sustainable development.[23] It is also necessary for the CDM registration process to involve adequate consultation with all major stakeholders to ensure approved projects do not adversely affect third parties (Schneider 2007).

Upon a successful project receiving CERs, the acquired emissions units increase the investor nation's permissible greenhouse gas emissions. This allows Annex I nations which would have exceeded their emissions allowances to meet their emissions targets. It also allows Annex I nations to meet their targets at lowest cost.

To better understand how the Clean Development Mechanism works, consider Table 8.3. Imagine, again, that there are only three Annex I nations, each with an emissions allowance of 100 units per year.[24] On this occasion, imagine that the projected annual emissions of countries *A*, *B*, and *C* are 110, 100, and 100 units respectively. This would imply that country *A*, with an expected emissions level 10 units above its emissions allowance (+10), would fail to meet its annual emissions target. It would also mean that, at 310 units, the annual emissions of the group of Annex I nations would exceed its collective target of 300 units (+10). At the same time, it has been projected that the emissions of country *X*—a non-Annex I nation with no emissions target—will be 40 units in the upcoming year.

Table 8.3 Using the Clean Development Mechanism to achieve greenhouse gas (GHG) emissions targets

Group	Country	GHG allowance (AAUs)	Projected GHG emissions	Projected GHGs relative to target	Adjusted allowance (AAUs + CERs)	Eventual GHG emissions	GHGs relative to adjusted target
Annex I	*A*	100	110	+10	110	110	0
Annex I	*B*	100	100	0	100	100	0
Annex I	*C*	100	100	0	100	100	0
Annex I	Total	300	310	+10	310	310	0
Non-AI	*X*	–	40	–	–	30	–

Despite the existence of the International Emissions Trading system, let's assume that there are no surplus AAUs for country *A* to purchase to meet its emissions target. Although a number of options would be available to country *A*, we shall assume that it chooses to fund a CDM-project in country *X*. This is expected to reduce *X*'s annual emissions by 10 units to 30 units per year. As a consequence, 10 CERs are created each year and credited to country *A*, thus increasing *A*'s annual emissions allowance to 110 units. This allows country *A* and the Annex I group as a whole to meet their adjusted emissions targets. Furthermore, since it only makes sense for country *A* to undertake a CDM-project if the annual cost of the project (e.g., $30 million) is less than the cost of limiting its greenhouse gas emissions each year to 100 units (e.g., $90 million), the Clean Development Mechanism assists *A* to reduce its mitigation costs—in this case, by $60 million per year.[25]

There are two additional points worth highlighting with respect to the Clean Development Mechanism. Firstly, the desire of an Annex I nation to undertake a CDM-project need not be induced by the inability to purchase surplus AAUs from other Annex I nations. An Annex I nation would still prefer the CDM option if the cost of a CDM-project is less than the cost of purchasing surplus AAUs. This is precisely the situation depicted above where the annual cost to country *A* of investing in and maintaining the CDM-project is $30 million, whilst the cost of purchasing 10 AAUs is $50 million per year (see the International Emissions Trading example). Thus, even if 10 AAUs were available for purchase, country *A* would still be $20 million better off each year by investing in the CDM-project.

Secondly, a significant weakness of the Clean Development Mechanism is that it only assists the Annex I group of nations to meet its emissions target—it does not prevent global greenhouse gas emissions from rising (Garnaut 2008). For example, in the hypothetical situation above, global emissions would be higher in the upcoming year if country *X*'s annual emissions were anything less than 40 units in the current year.[26] In addition, even though the CDM-project reduces country *X*'s annual greenhouse gas emissions to 30 units in the upcoming year, it does nothing to stop *X*'s annual emissions from increasing to 40 units and beyond in future years.[27] Nor does it prevent the emissions of other non-Annex I nations from rising over time. Whilst it is true that additional CDM-projects undertaken in non-Annex I nations could restrict the group's emissions to current levels, continuation of this process would require an ongoing desire of Annex I nations to ramp up investment in CDM-projects. Such a desire would largely depend on the cost of new CDM-projects remaining below the cost of employing additional mitigation strategies at home.[28] Unfortunately, this is unlikely since one would expect the cost of new CDM-projects to rise as Annex I nations undertake the easiest and cheapest project options first. Thus, at some point, further investment in CDM-projects would become economically unviable.

Moreover, as much as any additional mitigation in Annex I nations would reduce the group's greenhouse gas emissions, there is no guarantee that the reduction in emissions would exactly match the upsurge in emissions in non-Annex I nations. Even if it did, the Annex I nations in question would be free to sell the surplus AAUs generated by the additional mitigation measures.[29] This would allow buyer-nations to legally increase their greenhouse gas emissions, thus resulting in an overall rise in global emissions.

Turning to some real numbers, a total of 7,516 CDM-projects had been registered or were in the process of registration as of 1 June 2014. At the same point in time, 1,183 CDM-projects were undergoing validation, which is the qualifying stage prior to submission to the CDM Executive Board for formal registration (http://www.cdmpipeline.org/overview.htm). A worrying trend has been the recent fall in both the number of projects submitted for validation and the number of new project registrations. Compared to 1,891 submissions in 2012, submissions fell to 226 in 2013. As for new registrations, they declined from 3,428 to 338 during the same year (World Bank 2014a). The concern is heightened once it is recognised that some projects took until 2013 to be registered because the CDM Executive Board was unable to process the huge number of projects submitted in 2011 and 2012. The backlog of submissions was due to the hasty submission of many projects hoping to beat the 31 December 2012 deadline for acceptance of CERs in Phase III of the EU-ETS (more on this later). Given the reduced number of submissions in 2013 and the fact that the backlog of projects has essentially been cleared, the number of new registrations in 2014 was expected to fall further. The situation is unlikely to be helped by the probable decline in new submissions arising from the recent collapse of CER prices and the expectation that no price recovery is likely in the foreseeable future (World Bank 2014a).

With respect to emissions units, 1.46 billion CERs—equal to 1.46 Gigatonnes of CO_2-equivalent greenhouse gases—had been issued by the CDM Executive Board as of 1 June 2014 (http://cdm.unfccc.int).[30] In keeping with the recent fall in newly-registered CDM-projects, there was a significant decline in the number of CERs issued in 2013—265 million as compared to 339 million in 2012 (World Bank 2014a). This decline has continued into 2014, with March 2014 bearing the lowest monthly CER issuance since 2010 (World Bank 2014a). Notwithstanding the fall in CER issuance, it has been estimated that the Clean Development Mechanism has saved Annex I nations with Kyoto commitments approximately US$4 billion in mitigation and compliance costs (UNFCCC 2013c).

8.2.4.3 Joint Implementation (JI)

The Joint Implementation facility is another project-based offset mechanism that offers emissions units to Annex I nations, this time for subsidising or investing in emissions-reduction projects in other Annex I countries (UNFCCC 1998). Under the Joint Implementation scheme, emissions credits—referred to as Emission Reduction Units (ERUs)[31]—are transferred to the investor nation from the project-recipient's pool of Assigned Amount Units (AAUs). Not unlike the Clean Development Mechanism, the calculation of Emission Reduction Units is based on the difference between the emission levels generated by a host nation following implementation of a JI-project and the emissions that would have been generated in the project's absence (IPCC 2001d). In a similar vein, the granting of emissions units is conditional upon a JI-project satisfying an 'additionality' test.[32] However, the issues of additionality and the related problem of fictitious emissions units are less problematic compared to the Clean Development Mechanism (Carbon Trust 2009). This is because host nations must forfeit AAUs each time they accept a JI-project. Hence, in most cases, they are likely to be more vigilant to ensure the ensuing greenhouse gas emissions reductions are genuinely additional to what would have eventuated in the absence of the JI-project.[33]

Notwithstanding this, the issuance of ERUs still requires verification, which involves an eligible host nation having to verify that emissions reductions or the enhancement of greenhouse gas removals from a JI-project are additional to what would have occurred.[34] Upon verification, the relevant quantity of ERUs can be issued by the host country and transferred to the investor nation (referred to as a 'Track 1' procedure). If a host nation does not meet all Track 1 eligibility requirements, verification must be undertaken via the Joint Implementation Supervisory Committee (JISC). Under this so-called 'Track 2' procedure, an independent auditor accredited by the JISC must ascertain whether the necessary project conditions have been met before the host nation can issue and transfer ERUs to the investor nation (Carbon Trust 2009; UNFCCC 2013d).[35]

Putting aside the technicalities, the general manner in which a nation can harness the Joint Implementation facility to accomplish its emissions targets is revealed in Table 8.4. Imagine, once more, that there are three Annex I nations, each with an emissions allowance of 100 units per year.[36] Much like the previous example, imagine that the projected annual emissions of countries *A*, *B*, and *C* are 110, 100, and 100 units respectively, which implies that country *A* and the Annex I group as a whole will exceed their annual emissions allowance by 10 units (+10).

Assuming there are no surplus AAUs for country *A* to purchase, country *A* funds a JI-project in country *B* which reduces *B*'s annual greenhouse gas emissions by 10 units. Consequently, 10 of country *B*'s AAUs are converted each year to ERUs and credited to country *A*. This increases *A*'s annual emissions allowance to 110 units and reduces country *B*'s allowance to 90 units. In the process, country *A* is able to meet its adjusted emissions target. Crucially, the total emissions allowance of the Annex I group remains unchanged at 300 units per year.

Table 8.4 Using the Joint Implementation facility to achieve greenhouse gas (GHG) emissions targets

Group	Country	GHG allowance (AAUs)	Projected GHG emissions	Projected GHGs relative to target	Adjusted allowance (AAUs + ERUs)	Eventual GHG emissions	GHGs relative to adjusted target
Annex I	*A*	100	110	+10	110	110	0
Annex I	*B*	100	100	0	90	90	0
Annex I	*C*	100	100	0	100	100	0
Annex I	Total	300	310	+10	300	300	0

Not unlike the Clean Development Mechanism, it is only viable for country *A* to invest in the JI-project if the annual cost of the project (e.g., $20 million) is less than cost of limiting its own annual emissions to 100 units (e.g., $90 million). In this instance, the use of the Joint Implementation facility would reduce *A*'s mitigation costs by $70 million per year.

There is, nonetheless, an additional cost that country *A* may wish to consider before embarking on a JI-project. It is the annual cost of undertaking a CDM-project, which was assumed to be $30 million per year (see the Clean Development Mechanism example). Since the annual cost of the JI-project is only $20 million, we would expect country *A* to invest in the JI-project. However, should country *A* need to increase its emissions allowance by a further 10 units—for example, if its rising real output demands were to increase its annual emissions to 120 units per year—the preference for a CDM-project or a second JI-project would depend on the cost of the latter option. Should the annual cost of undertaking a second JI-project be $40 million per year, it would make sense for *A* to undertake a CDM-project.

I earlier referred to the Joint Implementation Supervisory Committee (JISC), which is the body responsible for supervising the Joint Implementation facility and overseeing the Track 2 verification procedure. The JISC is a 10-member Committee drawn from the Parties to the Kyoto Protocol.[37] In accordance with the Joint Implementation guidelines, the JISC is required, *inter alia*, to undertake any task assigned to it by the Conference of the Parties serving as the meeting of the Parties to the Kyoto Protocol (https://ji.unfccc.int/Ref/Documents/Rules.pdf).

Turning once again to some concrete numbers, 604 JI-projects had been registered under the Joint Implementation facility as of 1 June 2014. Of this total, 555 projects had been registered via the Track 1 process and 49 through the Track 2 procedure. Significantly, 373 of the registered JI-projects had taken place in the Russian Federation and the Ukraine (more on this later). At the same point in time, 849.5 million ERUs—equal to 849.5 Megatonnes of CO_2-equivalent greenhouse gases—had been issued to investor nations, which, on top of the various benefits generated by the Clean Development Mechanism, had brought hundreds of millions of dollars of savings to Annex I nations in the form of avoided mitigation and compliance costs (http://www.cdmpipeline.org/ji-projects.htm).[38]

Disconcertingly, the number of new JI-projects entering the Joint Implementation pipeline decreased markedly in 2013. Compared to 229 new projects in 2012, only 26 new projects entered the Joint Implementation pipeline in 2013. Unsurprisingly, the issuance of ERUs also fell steeply—down from 526 million ERUs in 2012 to 184 million ERUs in 2013. Worse still, analysts were predicting that the number of ERUs issued in 2014 would fall by a further 40 per cent, with the downward trend likely to continue as fewer new projects enter the Joint Implementation pipeline (World Bank 2014a).

8.2.5 Mitigation—The Kyoto Protocol's Treatment of Land Use, Land-Use Change, and Forestry (LULUCF)

8.2.5.1 Afforestation, Reforestation, and Avoided Deforestation

Although not strictly acknowledged as a Kyoto flexibility mechanism, Articles 3.3, 3.4, and 3.7 of the Kyoto Protocol provide opportunities for Annex I Parties to implement 'land use, land-use change, and forestry' (LULUCF) activities to lower the cost of meeting their emissions targets.[39] In terms of forestry, the opportunity is made possible by two realities. The first is the recognition that forests, as carbon sinks, can remove carbon dioxide from the atmosphere (carbon sequestration). The second is the fact that *net* emissions are taken into account when measuring a nation's greenhouse gas performance (UNFCCC 1998). Thus, upon demonstrating that a domestically-implemented afforestation or reforestation project will remove greenhouse gases from the atmosphere, an Annex I nation can be issued emissions credits (offsets) referred to as Removal Units (RMUs) (UNFCCC 2013e).[40] Table 8.5 shows how an Annex I nation can receive RMUs from afforestation and reforestation projects to meet its emissions targets.

Imagine an Annex I nation—country *A*—with an emissions allowance of 100 units per year. It has been projected that *A*'s annual emissions will rise to 110 units. Hence, country *A* is expected to exceed its emissions allowance by 10 units (+10). To meet its emissions target, country *A* chooses not to reduce its greenhouse gas emissions but to undertake a reforestation project. Because the project will remove 10 units of carbon each year from the atmosphere for a ten-year period, country *A* receives 10 RMUs on an annual basis for ten years.[41] This increases *A*'s annual emissions allowance to 110 units. In doing so, it allows *A* to meet its emissions target for the next decade.

Table 8.5 Using afforestation and reforestation to achieve greenhouse gas (GHG) emissions targets

Group	Country	GHG allowance (AAUs)	Projected GHG emissions	Projected GHGs relative to target	Adjusted allowance (AAUs + RMUs)	Eventual GHG emissions	GHGs relative to adjusted target
Annex I	*A*	100	110	+10	110	110	0

Whether an Annex I nation invests in forestry projects for the purposes just described depends largely on the cost of undertaking them relative to the cost of alternative mitigation strategies. Provided the marginal cost of forestry projects is less than the marginal cost of the alternatives, it makes sense for an Annex I Party to undertake afforestation and reforestation projects rather than implement additional mitigation measures and/or undertake additional CDM-projects and JI-projects.[42] Moreover, it enables an Annex I nation to reduce the total cost of achieving its Kyoto targets.

Of course, in the same way that afforestation and reforestation projects can reduce a nation's net emissions, deforestation can increase net emissions. For this reason, Annex I nations must, as per Article 3.3, report all greenhouse gas emissions from deforestation. From an operational perspective, deforestation can result in Annex I nations having to forfeit AAUs or some or all of the RMUs acquired from the forestry activities expected to remove greenhouse gases.

The ability of Annex I nations to undertake forestry projects to achieve their emissions targets is not confined to the domestic domain. The Kyoto Protocol's flexibility mechanisms contain provisions which allow Annex I nations to achieve their targets by implementing LULUCF activities in other Annex I countries (Joint Implementation) and in non-Annex I countries (Clean Development Mechanism). However, in both cases, eligible LULUCF activities are restricted to projects which increase greenhouse gas removals by sinks (i.e., via sequestration). They do not extend to LULUCF projects which reduce emissions by sources (UNFCCC 2013e). Furthermore, in the case of the Clean Development Mechanism, eligible LULUCF activities are confined to afforestation and reforestation activities. Projects resulting in 'avoided deforestation' in non-Annex I countries do not qualify as a source of Certified Emissions Reduction units (CERs).

8.2.5.2 Agriculture

Land-based activities do not simply involve forestry-related projects. They also include agricultural activities and the revegetation of disused and/or marginal agricultural lands. Along with increased levels of carbon sequestration arising from revegetation and better forestry management, sustainable forms of agriculture can dramatically boost the carbon stored in soils and the biomass they support (IPCC 2007d; Lehmann 2007; Laird 2008; Fynn et al. 2009). For this reason, improved agricultural practices can play a valuable role in removing carbon from the Earth's atmosphere. Indeed, with agriculture accounting for nearly 15 per cent of global greenhouse gas emissions, it has been estimated that sustainable land management practices could remove around 3.3 Gigatonnes of carbon dioxide per year up to 2050 ($\pm$1.1 Gigatonnes) (Lal 2004).[43]

As previously highlighted (see Table 1.1), nitrous oxide (N_2O) emissions have contributed significantly to the rise in the atmospheric concentration of greenhouse gases. Nitrous oxide is generated by a host of human agricultural activities, including nitrogenous fertiliser use, the burning of biomass, and the discharge and management of livestock manure (Garnaut 2008).[44] It has been shown that nitrous oxide emissions could be dramatically reduced through improved fertiliser use, better soil and water management, the application of organic fertiliser additives, and a reduction in the burning of vegetation (de Klein and Eckard 2008).

Because land-based activities, such as sustainable forms of agriculture, can reduce emissions and increase sequestration rates, Article 3.3 of the Kyoto Protocol requires Annex I nations to include their full impact when accounting for anthropogenic emissions and greenhouse gas removals. In addition, the Protocol allows Annex I nations to earn Removal Units (RMUs) in circumstances where domestically-located agricultural activities result in the net removal of greenhouse gases. In the same manner described in Table 8.5, this provides Annex I nations with a further opportunity to lower the cost of meeting their emissions targets.

There are, however, restrictions in terms of how nations can acquire emissions units from LULUCF activities undertaken via engagement with the Clean Development Mechanism and the Joint Implementation facility. As already stressed, eligible LULUCF activities are limited to CDM-projects and JI-projects that increase greenhouse gas removals by sinks, meaning that LULUCF-related activities which reduce emissions by sources cannot generate emissions units for investor nations. In the case of the Clean Development Mechanism, a further restriction exists because eligible LULUCF activities are confined to afforestation and reforestation projects. Consequently, Annex I countries cannot acquire CERs by investing in land-based projects which improve agricultural and other land-use practices in non-Annex I nations. The same restriction does not apply in relation to the Joint Implementation facility. In this instance, eligible activities extend to investments in sequestration projects involving improvements in agricultural and other land-use practices.

8.2.6 Mitigation—The Kyoto Protocol's Treatment of Aviation and Shipping

Of all the greenhouse gas emissions generated by human activities, emissions from international aviation and maritime transport—commonly referred to as 'bunker-fuel' emissions—constitute a small share of the global total (3.6 per cent of global CO_2 emissions in 2010) (IEA 2012a).[45] However, emissions generated from these two transport-related sources are rising steadily and at a rate much faster than most emissions-generating activities. Between 1990 and 2010, CO_2 emissions from international aviation and international maritime transport rose by 78.3 and 77.6 per cent respectively, whereas total CO_2 emissions grew by a much lower 44.4 per cent over the same period (IEA 2012a).[46]

Despite the rapid increase, the greenhouse gas emissions generated by international aviation and shipping activities are not subject to UNFCCC regulations. Consequently, they are not included in the emissions-reduction commitments of Annex I Parties under the Kyoto Protocol (UNFCCC 1998, 2013f). Furthermore, although UNFCCC Parties are obliged to calculate bunker-fuel emissions as part of their national greenhouse gas inventories, they are required to exclude this category of emissions from their national emissions totals and disclose them via a separate reporting means (UNFCCC 2013f).

Given the growing relevance of bunker-fuel emissions, efforts have intensified to address the greenhouse gas emissions from international aviation and maritime activities. At the COP-13 meeting in Bali (2007), the UNFCCC-aligned Ad Hoc Working Group on Long-Term Cooperative Action (AWG-LCA) was instructed to develop ways to limit bunker-fuel emissions. In 2011, six options were tabled by the AWG-LCA. This was later extended to nine. Around the same time, UNFCCC Parties recognised the potential role that the International Civil Aviation Organization (ICAO) and the International Maritime Organization (IMO) could play in limiting bunker-fuel emissions.[47] Accordingly, the UNFCCC has worked very closely with the secretariats of both agencies in recent years.

To date, UNFCCC Parties have had enormous difficulty bridging their divergent views on the various options proposed by the AWG-LCA—a problem magnified by difficulties of attribution and concerns about the impact that regulatory measures could have on the international competitiveness of high-export countries (UNFCCC 2013f). In addition, while the ICAO has long been working on an emissions-trading system for international civil aviation, no such system has materialised. In sum, little progress has been made by the UNFCCC to regulate the greenhouse gas emissions emanating from international aviation and maritime activities.

Despite the lack of progress at the UNFCCC level, the European Union has had greater success in reducing aviation-related emissions within Europe. As of 2012, the EU-ETS began regulating the CO_2 emissions from flights beginning and ending at airports located in all EU member nations plus airports in Norway, Iceland, and Liechtenstein (EEA 2013). In order to set its limits on aviation emissions, the EU has used the average emissions generated by the aviation industry in Europe between 2004 and 2006 as its emissions baseline.[48] In the most recent phase of the EU-ETS (Phase III), which began on 1 January 2013, aviation-related emissions were capped at 95 per cent of the baseline emissions level (EEA 2013).

As expected, the emissions reductions achieved by the EU have been vastly overwhelmed by the growth of aviation-related emissions outside of Europe. Hence, they have had little effect on global emissions levels. Nonetheless, the EU-ETS example has offered a potential template for the UNFCCC to follow when engaging in future efforts to regulate global aviation and maritime activities.

8.2.7 *Mitigation—The Kyoto Protocol's Enforcement Mechanisms*

The enforcement mechanisms embodied in the Kyoto Protocol are designed to reinforce its environmental integrity, guarantee transparent accounting of greenhouse gas emissions, and support the credibility of global carbon markets

(UNFCCC 2013g). To ensure the enforcement mechanisms are strong and effective, they are overseen by a Compliance Committee consisting of a Facilitative Branch and an Enforcement Branch. Both sub branches comprise ten members—one from each of the five regional United Nations groups[49]; one from the Alliance of Small-Island States (AOSIS); and two each from the Annex I and non-Annex I group of nations (Dessai 2001).

The aim of the Facilitative Branch is to provide advice to assist the Parties with the implementation of the Kyoto Protocol and to promote compliance with Kyoto commitments. The Facilitative Branch is also responsible for addressing issues related to the mitigation measures undertaken by Annex I Parties to meet their Kyoto targets, including the use of the Protocol's flexibility mechanisms and ensuring the adverse side-effects of their measures on poor countries are kept to a minimum (UNFCCC 2013g). By giving prior notice of potential non-compliance, the Facilitative Branch also serves as an early-warning system for all UNFCCC Parties with Kyoto obligations (Dessai 2001; UNFCCC 2013g).

The Enforcement Branch is responsible for determining whether Annex I Parties are complying with: (i) emissions targets; (ii) methodological and reporting requirements for greenhouse gas inventories; and (iii) eligibility requirements pertaining to the Protocol's flexibility mechanisms (UNFCCC 2013g).[50] In cases where the Enforcement Branch identifies instances of non-compliance, a different course of action is undertaken depending on the nature of the breach. Where the Branch believes a Party has exceeded its emissions allowance, the Party has 100 days to make up the shortfall through acquisitions of emissions units.[51] Should it fail to do so, the Enforcement Branch initially declares the Party as 'Kyoto non-compliant'. The transgressing Party is thereupon required to submit a compliance action plan and make up for the excessive emissions plus a 30 per cent penalty during a subsequent commitment period (UNFCCC 2013g).[52] Until the Party is reinstated as 'Kyoto compliant', it is suspended from any international trading in emissions units (Stern 2007).

Through its two branches, the Compliance Committee considers questions related to the implementation of the Kyoto Protocol, which may be raised by: (i) an expert review team operating under Article 8 of the Protocol; (ii) a Party concerned about its own affairs and actions; or (iii) a Party concerned about the actions of another Party. Upon a question of implementation being raised, the Compliance Committee allocates the matter to the most appropriate of its two branches. In making its deliberations, the Compliance Committee must take into account the flexibility afforded to the Annex I nations attempting to make the transition to a market economy (i.e., the EITs in the group). The Facilitative Branch must also consider the 'common but differentiated responsibilities' of the various Parties and the circumstances pertaining to questions of implementation tabled before it (UNFCCC 2013g).

Where a disagreement exists between a Party and an expert review team assessing a Party's emissions inventory, the Enforcement Branch must determine whether to: (i) apply adjustments to greenhouse gas inventories (Article 5, Paragraph 2); or (ii) correct the compilation and accounting database of emissions units (Article 7, Paragraph 4). The Enforcement Branch is required to make a final decision within 12 weeks of being informed in writing of a disagreement. No fixed deadlines for action are required of the Facilitative Branch. Generally speaking, Parties are unable to appeal the decisions taken by the Compliance Committee. Exceptions include decisions made by the Enforcement Branch concerning emissions targets, although appeals are confined to circumstances where a Party claims to have been denied due process (UNFCCC 2013g).

8.2.8 Climate Change Financing Mechanisms and Institutions

The primary purpose of climate change financing mechanisms is to direct financial capital, and ultimately real resources, to activities designed to mitigate and adapt to climate change. Although the private sector is expected to provide the bulk of the financial resources, additional financing mechanisms are required for a number of reasons. Firstly, the UNFCCC contains key principles to promote an equitable sharing of climate change costs—in particular, principles aimed at transferring funds and technology from high-GDP to vulnerable low-GDP countries.[53] Hence, financing mechanisms are required to serve as international redistribution instruments.

Secondly, although the aim of having a price on greenhouse gas emissions is to stimulate investment in greenhouse gas-reducing measures, it was revealed in Chap. 7 that a number of financial capital constraints, hidden costs, and complex interactions between various mitigation technologies can impede the implementation of viable mitigation strategies. Financing mechanisms specific to climate change can play a crucial role in overcoming these financial impediments.

Thirdly, it has been explained at various stages that a large slice of low-emissions technologies and infrastructure have public goods characteristics and require significant government-funded investments to ensure their adequate provision and widespread availability. Once again, financing mechanisms, particularly those created and sourced by national governments, will be crucial in establishing the low-emissions infrastructure of the future.[54]

It is because of these needs that a multitude of financing mechanisms and institutions exist to support climate change-related programmes and activities. A number of them have already been mentioned in this chapter—for example, the Global Environment Facility, the Green Climate Fund, and the three main Kyoto flexibility mechanisms. Figure 8.2 puts into perspective the most prominent climate change financing mechanisms and the manner in which climate change funds flow from upstream sources to downstream recipients.

As can be seen from Fig. 8.2, climate change funds are initially provided and/or generated by governments and capital markets. They are subsequently channelled through bilateral and multilateral finance institutions, development agencies, the UNFCCC, and private-sector organisations (Atteridge et al. 2009).[55] In most instances, the funds identified in Fig. 8.2 assist the climate change endeavours of non-Annex I nations. Residual flows include: (i) funds provided by the national and sub-national governments of Annex I nations for domestic climate change purposes; (ii) financial capital generated by Annex I countries from the sale of emissions units in international carbon markets; and (iii) funds invested by Annex I nations undertaking JI-projects in other Annex I countries.

Whilst Fig. 8.2 provides a useful, simplified view of financial flows, it highlights a number complex challenges associated with tracking the finance available for climate change mitigation and adaptation purposes (Atteridge et al. 2009). The first of these is the need to determine a clear distinction between the funds available for climate change activities and those pertaining to official development assistance (ODA) (denoted in Fig. 8.2 by the two dashed text-boxes). Making this distinction has proven to be problematic in climate change negotiations and will require further clarification to ensure an effective 'finance' architecture emerges within any newly-created climate change protocol.

The second challenge concerns the arbitrary way in which the finance earmarked to support climate change activities in non-Annex I nations is assessed as being 'new and additional'. To recall, the additionality test plays an important function in preventing Annex I nations from claiming fictitious emissions units through business-as-usual activities. The additionality requirement also thwarts attempts by non-Annex I nations to access climate change funds for non-legitimate purposes. At present, assessments of finance are subjectively made by participant institutions, especially with regards to adaptation-related activities. To bolster the integrity of the system and ensure greater harmonisation between the relevant institutions, there is an urgent need to better define and codify the parameters used in this important assessment process (Atteridge et al. 2009).

Finally, data on private financial flows is very difficult to interpret given the multitude of private-sector actors involved and the lack of a centralised reporting forum from which to access aggregated information. A central data repository is desperately required to help analysts and climate change negotiators make better sense of the private-sector efforts being undertaken to combat climate change and minimise its harmful effects.[56] Bearing these challenges in mind, as well as the flow of funds represented by Fig. 8.2, a brief description will now be given of the prominent climate change financing institutions. These institutions and their governing bodies, aims, and sources of funds are summarised in Table 8.6.

8.2.8.1 The Global Environment Facility and GEF Trust Fund

As mentioned earlier in the chapter, the Global Environment Facility was chosen as the UNFCCC's interim financing mechanism at the first Conference of Parties in 1995. At the time, the Global Environment Facility existed as an independent organisation responsible for providing new and additional funding to meet the incremental costs of projects related to five environmental areas of concern: (i) biodiversity; (ii) international waters; (iii) land degradation; (iv) ozone depletion;

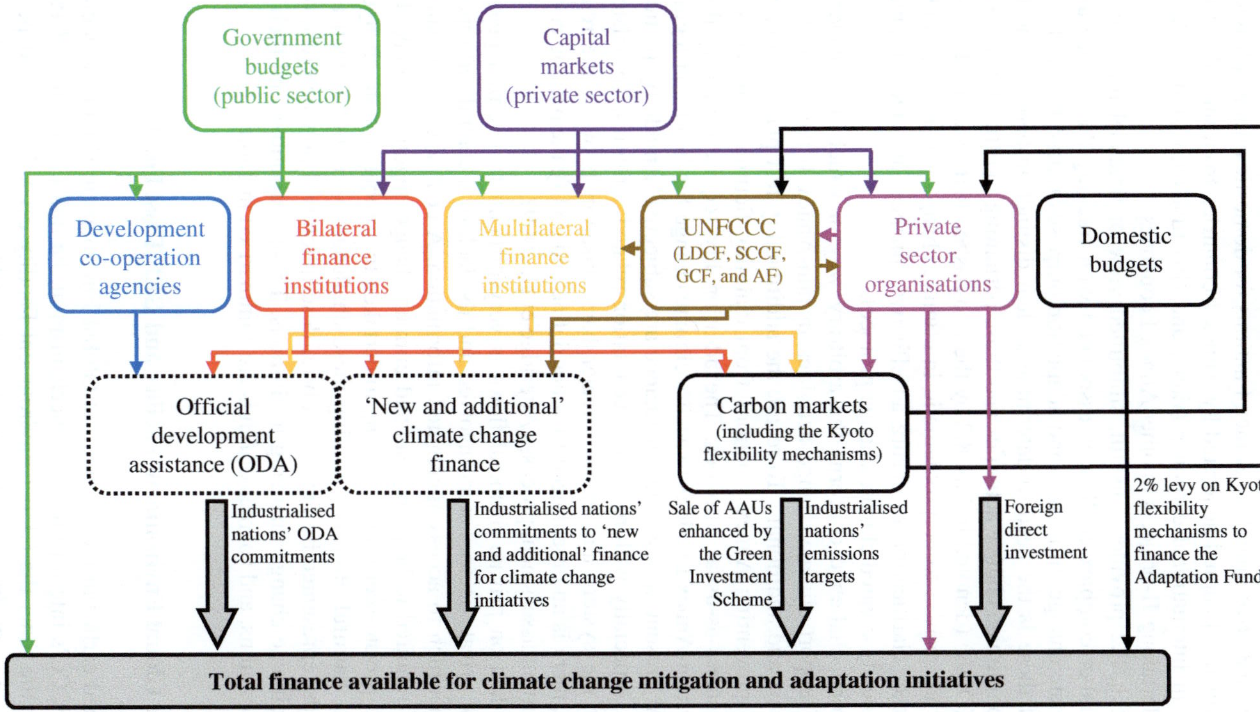

Fig. 8.2 Financial flows to support climate change mitigation and adaptation activities. *Source* Adapted from Atteridge et al. (2009)

Table 8.6 Major sources of climate change finance

UNFCCC climate change financing institutions and mechanisms			**Other funding sources**
Global Environment Facility (GEF)	**UNFCCC (Article 11)**	**Kyoto Protocol (KP)**	**Multilateral Finance Institutions**
GEF Trust Fund (ended in 2009) • Aim: To support adaptation projects in developing countries that generate global environmental benefits • Method: Provide financial support to projects in non-Annex I nations that reduce vulnerability to climate change and build adaptive capacity • Trustee: World Bank (since 1994) • Source: Voluntary contributions from UNFCCC Parties every four years *Least Dev. Countries Fund (LDCF)* • Aim: To support adaptation programmes in LDCs • Method: Provide financial support to National Adaptation Programmes of Action (NAPAs) in LDCs • Trustee: World Bank • Source: Funded by Annex II nations and other capable Annex I countries *Special Climate Change Fund (SCCF)* • Aim: To provide climate change finance to non-Annex I countries • Method: Provide financial support for adaptation activities and technology transfer; catalyst to leverage funds from additional sources • Trustee: World Bank • Source: As per the LDCF	*Green Climate Fund (GCF)* • Aim: ➢ Short-term: To provide mitigation and adaptation finance to non-Annex I countries ➢ Long-term: To become the main multilateral financing mechanism to support climate change action in developing countries • Method: A country-driven approach involving relevant institutions and stakeholders to support projects, programmes, and policies in non-Annex I countries, taking into account the needs of nations particularly vulnerable to climate change • Governance: GCF Board with the World Bank as the interim Trustee (for a period of three years). A permanent Trustee will eventually replace the World Bank as Trustee • Source: Voluntary contributions from Annex I nations and alternative public-sector and private-sector sources	*Adaptation Fund (AF)* • Aim: To support adaptation projects in non-Annex I Parties to KP • Method: Support concrete adaptation projects and programmes in vulnerable developing countries • Governance: Adaptation Fund Board with the World Bank as the Trustee • Source: Funded by a 2 % levy on: ➢ CERs generated by CDM-projects ➢ first international transfers of AAUs ➢ ERUs generated by JI-projects ➢ RMUs generated by domestic afforestation/reforestation projects ➢ Adaptation Fund also resourced by voluntary contributions from governments and the private sector *International Emissions Trading (IET)* • Proceeds obtained from the sale of emissions units—enhanced by *Green Investment Scheme (GIS)*[57] *Clean Development Mechanism (CDM)* • Transfer of funds from Annex I to non-Annex I nations via CDM-projects; value of the CERs created *Joint Implementation (JI)* • Transfer of funds from one Annex I nation to another via JI-projects; value of ERUs created	• World Bank • Inter-American Dev. Bank (IDB) • Asian Development Bank (ADB) • African Development Bank (AfDB) • Europ. Bank for Rec.& Dev. (EBRD) • Carbon Fund of the Forest Carbon Partnership Facility (FCPF) **Bilateral Finance Institutions** • French Development Agency (AFD) • German Development Bank (KfW) • Japan Int. Co-op. Agency (JICA) **Other institutions and agencies** • United Nations Development Programme (UNDP) • United Nations Environment Programme (UNEP) • Food and Agriculture Organization of the United Nations (FAO) • International Fund for Agricultural Development (IFAD) • Organisation for Economic Co-op. and Development (OECD) • Bilateral Dev. Co-operation Agencies • National and sub-national governments • Private sector (various entities, institutions, and private-sector carbon markets)

and (v) persistent organic pollutants (IPCC 2007d). Upon the UNFCCC's decision to utilise the Global Environment Facility as an interim financing mechanism, climate change became the sixth area of the Facility's concern. The Facility was eventually installed as the UNFCCC's formal financing mechanism following a decision at the COP-4 meeting in 1998 (Möhner and Klein 2007).

It has already been stressed that Article 4.3 of the UNFCCC stipulates that climate change funds earmarked for distribution to developing nations must be *additional* to official development assistance. In keeping with this mandate, the Global Environment Facility only funds activities that are incapable of proceeding without the Facility's support (Rübbelke 2011). The operating principles of the Global Environment Facility also require funded activities to generate broader global benefits and adhere to the concept of sustainable development (GEF 2011).

One of the appealing features of requiring financial transfers to be conditional upon the generation of additional climate change-related benefits is that the funds can serve as a subsidy to reduce the net cost of mitigation activities in recipient nations. This has the desirable effect of boosting the demand by non-Annex I nations for mitigation projects. Furthermore, by providing a financial incentive to increase mitigation efforts in countries where mitigation measures can be achieved at lowest cost, the funding policies of the Global Environment Facility have helped to minimise the global cost of climate change mitigation (Rübbelke 2011).

As Table 8.6 shows, there are three Global Environment Facility Funds that non-Annex I nations have been able to draw upon for climate change assistance. Until 2009, the GEF Trust Fund supported projects which aimed to reduce the vulnerability of developing countries to climate change and/or build adaptive capacity. To assist in this regard, the Global Environment Facility created the 'Strategic Priority on Adaptation' or SPA in 2004 by way of a US$50 million allocation inside the GEF Trust Fund.[58] The SPA used the funds to provide financial support to kick-start pilot adaptation projects in essentially non-Annex I nations (GEF 2005; Rübbelke 2011).[59]

By September 2009, the entire US$50 million of allocated funds had been exhausted (Rübbelke 2011). Despite the fifth replenishment of the GEF Trust Fund covering the 2010–2014 period, the Global Environment Facility decided not to allocate new funds to the SPA. This has meant that, since 2009, all adaptation-related activities supported by the Global Environment Facility have been financed through the Facility's newly established Least Developed Countries Fund and Special Climate Change Fund (Rübbelke 2011; GEF 2012). In all, the SPA has supported 26 pilot projects involving the allocation of US$658 million of direct and indirect funding (GEF 2012).[60]

8.2.8.2 The Least Developed Countries Fund (LDCF)

Although significant gains can be had from requiring financial transfers to be conditional upon the generation of additional climate change-related benefits, ascertaining a clear distinction between development assistance and adaptation support is invariably complex and difficult. There are also instances where separating development and adaptation policies can result in the loss of policy synergies (Rübbelke 2011). With these complexities in mind, the Parties to the UNFCCC issued new guidance to the Global Environment Facility at the COP-7 meeting in 2001 which led to the establishment of the above-mentioned Least Developed Countries Fund and Special Climate Change Fund (see Table 8.6). Launched at the end of 2001, the two Funds were created to simultaneously address the UNFCCC's development and adaptation objectives in the belief it would reduce the loss of policy synergies. Both Funds were made operational in 2002 (www.climatefundsupdate.org).

The Least Developed Countries Fund and the Special Climate Change Fund are very different in terms of their roles and the UNFCCC Parties they support. The Least Developed Countries Fund supports a Work Programme to assist the group of 48 least-developed countries (LDCs) to carry out, among other things, the preparation and implementation of 'National Adaptation Programmes of Action' (NAPAs) (UNFCCC 2013h).[61] To underpin the Least Developed Countries Fund, Annex II Parties and other capable UNFCCC Parties are regularly invited to contribute financial resources to support the implementation and eventual completion of the LDC Work Programme. Although the Fund is administered by the Global Environment Facility, the World Bank serves as the Fund's Trustee (Möhner and Klein 2007).

The NAPAs embodied in the LDC Work Programme provide LDCs with the opportunity to identify priority activities that address their urgent and immediate adaptation needs, especially activities where commencement delays are likely to increase the vulnerability of LDCs to climate change and/or magnify future adaptation costs (UNFCCC 2013i).[62] To gain approval, NAPAs must be action-oriented and country-driven. They must also be flexible and based on national circumstances. Since a fast-track assessment process is necessary to address urgent and immediate adaptation needs, the Global Environment Facility requires NAPA documents to be presented in a format that can be easily understood by stakeholders and policy-makers (UNFCCC 2013i).

Initially, the Least Developed Countries Fund was used solely to meet the agreed cost of preparing NAPAs. At the COP-9 meeting in 2003, the UNFCC Parties requested the Global Environment Facility to take account of the criteria established to support the implementation of NAPAs on an agreed full-cost basis. In 2005 (COP-11), guidance from the UNFCCC Parties was extended in the form of a request to the Global Environment Facility to provide full-cost funding to meet the immediate adaptation needs of LDCs as identified and prioritised in the NAPAs (Möhner and Klein 2007; UNFCCC 2013h).

Following guidance at the COP-18 meeting in 2012, the Global Environment Facility was further advised to (UNFCCC 2013h):

- continue its mobilisation of resources to ensure full implementation of the LDC Work Programme, including implementation of programme elements other than NAPAs;
- facilitate greater access to the Least Developed Countries Fund;

- enhance the country-driven nature of NAPA projects and the broader LDC Work Programme;
- raise awareness of the need to maintain adequate and predictable resources in the Least Developed Countries Fund to ensure the full implementation of the LDC Work Programme;
- enhance communication with implementing agencies on any updated operational guidelines pertaining to the Least Developed Countries Fund.[63]

In terms of the disbursement of funds, the Global Environment Facility adopts the position that there is no need for projects funded by the Least Developed Countries Fund to generate global environmental benefits—a direct contrast to the Facility's previous approach with respect to the GEF Trust Fund. Furthermore, the financial resources allocated from the Fund are not subject to the incremental cost provisions formerly applied to SPAs. Instead, the Global Environment Facility has developed the concept of 'additional costs', which constitute the extra costs incurred from having to ensure development-promoting activities are climate change-resilient (GEF 2006).[64] A similar funding methodology is also adopted with respect to the Special Climate Change Fund (Möhner and Klein 2007).

As at the end of 2013, cumulative pledges to the Least Developed Countries Fund amounted to US$879.8 million, of which US$831.5 million had been deposited into the Fund. Of this, US$726.3 million had been allocated to finance NAPAs and other adaptation projects.[65] In addition, projects supported by the Fund had attracted around US$2 billion in co-financing (GEF 2013a). Despite dealing with the world's poorest nations, many with minimal capacity to adapt to climate change, the Least Developed Countries Fund has supported the preparation of 50 NAPAs—one for each current and former LDC—of which all but one NAPA have been successfully finalised.[66] Except for Angola and Eritrea, the Fund has approved at least one implementation project in all the LDCs which have submitted a NAPA. Evidence suggests that completed projects or those in the process of implementation have generated considerable adaptation benefits (GEF 2012).

8.2.8.3 Special Climate Change Fund (SCCF)

Unlike the Least Developed Countries Fund, which is confined to LDCs, the aim of the Special Climate Change Fund is to financially support adaptation activities and the transfer of technologies to *all* vulnerable non-Annex I countries. The Special Climate Change Fund primarily finances long-term and short-term adaptation activities in the areas of water resources management, agriculture, natural resources management, health, infrastructure, fragile ecosystems, and integrated coastal zone management (GEF 2012). To a lesser extent, the Fund supports the monitoring of disease vectors, capacity building for disaster-risk management, and the establishment of information networks to facilitate rapid responses to extreme weather events (Möhner and Klein 2007). Given its central objective, the Special Climate Change Fund embodies two active funding windows or programmes—an

Adaptation Programme (SCCF-A) and a more enabling Technology Transfer Programme (SCCF-B) (UNFCCC 2013j).

The Special Climate Change Fund is operated by the Global Environment Facility with the World Bank serving as the Fund's Trustee (Möhner and Klein 2007). Although the Special Climate Change Fund was made operational in 2002, the current operational basis for financing activities through the Fund did not receive GEF Council approval until 2004 (UNFCCC 2013j).

As the designated administrator of the Special Climate Change Fund, the Global Environment Facility has been instructed to implement innovative adaptation and technology-transfer projects by: (i) adhering to guidance from the UNFCCC Parties (relevance); (ii) addressing adaptation needs through innovative schemes that emphasise project sustainability (effectiveness); and (iii) ensuring its operations are cost-effective (efficiency). Projects funded by the Special Climate Change Fund must be demand-driven and consistent with a recipient country's national plans and climate change strategies (GEF 2013b).

It was mentioned above that the financing methodology used to fund projects under the Special Climate Change Fund is much the same as it is for the Least Developed Countries Fund. Where the methodology differs is in terms of the sliding scale used to estimate the 'additional costs' of a proposed project. Under guidance from the UNFCC Parties, a project of the same monetary value receives proportionally less funding from the Special Climate Change Fund than the Least Developed Countries Fund (see Möhner and Klein 2007, Table 1).

As explicit as this variation in funding methodology is, at the COP-17 meeting in 2011, the Global Environment Facility was requested to clarify the concept of additional costs as it applies to the funding of different adaptation projects under the Special Climate Change Fund.[67] Furthermore, the Global Environment Facility was requested to allocate financial resources to help non-Annex I nations establish observation and monitoring networks and strengthen those already in place (UNFCCC 2013j). In a similar manner to the Least Developed Countries Fund, the Global Environment Facility was urged to raise awareness of the need for donor countries to maintain sufficient and predictable resources to ensure the Special Climate Change Fund can adequately support country-driven adaptation activities in non-Annex I nations.

As at the end of 2013, US$333.1 million had been pledged to the Special Climate Change Fund, of which US$299.1 million had been paid into the Fund by donor nations. To this date, 66 countries had accessed US$242.3 million from the Special Climate Change Fund to support 58 individual projects. Of this, US$201.8 million had been allocated towards 50 projects under the Adaptation Programme (SCCF-A), with the remaining US$40.5 million granted to eight projects under the Technology Transfer Programme (SCCF-B) (GEF 2013b). Through the agency of both funding windows, the Special Climate Change Fund has mobilised more than US$1.5 billion in co-financing.

Of the US$242.3 million so far allocated from the Fund, the majority has been distributed to non-Annex I Parties belonging to the African and the Asian groups of countries (29 per cent and 28 per cent respectively) (GEF 2013b). In terms of project types, the largest share of funding has been directed towards projects designed to enhance the resilience of water resources and agriculture, with 27 per cent of funds having been allocated to both industries (54 per cent in total). Apart from generating promised adaptation benefits, the Evaluation Office of the Global Environment Facility has provided evidence showing that the Special Climate Change Fund has significantly advanced the development agendas of beneficiary countries and the innovative nature of the implementation measures financed by the Fund. The Evaluation Office has also noted that the Fund's management costs are the lowest of all comparable funds (GEF 2012).

8.2.8.4 Green Climate Fund (GCF)

Created more recently than the above-outlined Funds, the Green Climate Fund was established by the UNFCCC Parties in response to the growing urgency and seriousness of climate change. The concept of the Green Climate Fund was originally conceived at the COP-15 meeting in 2009 and was adopted as a UNFCCC financing mechanism in accordance with Article 11 of the Convention at the COP-17 meeting in 2011 (GCF 2011; UNFCCC 2013k). The Fund finally became operational in early-2014. By 2020, it is anticipated that the Green Climate Fund will be raising US$100 billion per year to help finance climate change-related activities in non-Annex I nations. It is also expected that the Fund will eventually overtake the Least Developed Countries Fund and Special Climate Change Fund and become the UNFCCC's main multilateral financing mechanism in support of climate change action in developing countries.

Given the motivation for its establishment, the aim of the Green Climate Fund is to make a significant and more ambitious contribution towards accomplishing the goals set by the international community to reduce greenhouse gas emissions and minimise the impact of climate change on the world's most vulnerable nations. To achieve this aim, the Green Climate Fund will be used to promote a paradigm shift in the promotion and establishment of low-emissions and climate change-resilient pathways in developing countries (GCF 2011). The Green Climate Fund will attempt to realise this goal by financing the full and incremental costs of projects, programmes, and policies in non-Annex I nations that seek to: (i) address climate change mitigation ambitions (including efforts to reduce emissions emanating from deforestation and land degradation); (ii) meet climate change adaptation needs; (iii) facilitate the development and transfer of green technologies (including geosequestration technologies); and (iv) build institutional capacity and self-reliance. The Fund also exists to help non-Annex I nations prepare national climate change reports.

Because the Least Developed Countries Fund and the Special Climate Change Fund have essentially been used to finance adaptation activities, the Green Climate Fund will aim for a 50:50 balance between its support for mitigation and adaptation measures (UNFCCC 2014a).[68] The Fund will pursue a country-driven approach and promote and strengthen engagement at the country level through effective involvement of relevant institutions and stakeholders.[69] The Fund will endeavour to operate as an evolving institution guided by transparent and accountable monitoring and evaluation systems (GCF 2011).

In terms of attracting and disbursing funds, the Green Climate Fund will seek voluntary contributions from the governments of Annex I nations and alternative institutional sources (see Fig. 8.2). It is also anticipated that the Fund will play a key role in channelling new and additional financial resources to developing countries and catalyse public-sector and private-sector sources of climate change finance at both the national and international levels. Through a process of 'direct access', it also hoped that the Green Climate Fund will provide a simpler and improved means by which developing countries can acquire climate change funds. To assist in this regard, a facility has been created within the Fund to make it possible for developing countries to access financial resources via multilateral implementing entities, such as accredited multilateral development banks (e.g., the Inter-American Development Bank and the Asian Development Bank) and United Nations agencies (e.g., the United Nations Development Programme). In addition, a private-sector facility has been established to enable the Fund to directly and indirectly finance private-sector mitigation and adaptation activities at the national, regional, and international levels (GCF 2011).[70]

Despite being guided by the collective decisions of UNFCCC Parties, the Green Climate Fund is governed and supervised by a GCF Board, which is fully responsible for all funding decisions (see Table 8.6). The GCF Board is comprised of 24 members with an equal number of members drawn from the Annex I and non-Annex I groups of nations. Membership from the non-Annex I group includes representatives from the various United Nations regional constituencies as well as one representative from the group of small-island developing states (SIDS) and least-developed countries (LDCs) (GCF 2011).

For the first three years, and until a permanent Trustee is selected, the World Bank will serve as the interim Trustee of the Green Climate Fund. The Trustee will be charged with the responsibility of managing the financial assets of the Fund in accordance with the relevant decisions of the GCF Board. The Trustee will be accountable to the Board for its performance as trustee of the Fund.

Being a relatively new financing mechanism, a meagre US$35.7 million had been pledged towards the Green Climate Fund as of February 2014. This amount had increased to US$10.2 billion by the end of 2014 following the first major pledges to the Fund during the COP-21 meeting in Lima. To date, disbursements have been small and used primarily to cover administrative costs, such as the convening of Board meetings, hiring consultants, and establishing the Fund's headquarters in Incheon City, South Korea (GCF 2014; UNFCCC 2014a).

8.2.8.5 The Kyoto Protocol's Adaptation Fund (AF)

The Adaptation Fund is a supplementary adaptation-related mechanism that was established by the UNFCCC in 2001 within the framework of the Kyoto Protocol.

The Fund was officially launched in 2007 and was eventually operationalised in 2009 (www.climatefundsupdate.org). Because the Adaptation Fund operates under the Kyoto Protocol, its existence is not guaranteed beyond the conclusion of the second Kyoto commitment period in 2020.

There are two main reasons why the Adaptation Fund was created. First and foremost, the Fund was designed to ramp up efforts to reduce the adverse effects of climate change on vulnerable nations, communities, and climate-sensitive sectors of national economies.[71] Secondly, by making access to the Adaptation Fund conditional upon being a Kyoto signatory, it was hoped that the Fund's creation would encourage non-Annex I Parties to ratify the Kyoto Protocol (UNFCCC 2013l).[72] At present, all non-Annex I nations, except South Sudan, are Kyoto signatories and therefore qualify for financial assistance from the Adaptation Fund.

In an effort to support the most vulnerable non-Annex I Parties, priority access to the Adaptation Fund is given to eligible LDCs unable to secure assistance from the Least Developed Countries Fund (www.climatefundsupdate.org). The various activities supported by the Adaptation Fund are similar to those supported by the Special Climate Change Fund—namely, activities pertaining to natural resources management, human health, fragile ecosystems, the monitoring of disease vectors, and the erection of disaster-risk management systems (UNFCCC 2013l).

Like most climate change Funds, the Adaptation Fund is financed through voluntary contributions from donor Annex I nations and private-sector organisations and institutions (see Fig. 8.2). However, as Table 8.6 highlights, the Adaptation Fund is heavily capitalised by injections of funds from a share of the proceeds generated by the various Kyoto flexibility mechanisms. Since the inception of the Adaptation Fund, financial resources have been raised through a 2 per cent levy on the Certified Emissions Reductions (CERs) issued through the Clean Development Mechanism, albeit CDM-projects undertaken in least-developed countries are exempt from the levy (Garnaut 2008).[73] In an effort to augment the Adaptation Fund, the 2 per cent levy was recently imposed on the Assigned Amount Units (AAUs) exchanged for the first time under the Protocol's International Emissions Trading framework; Emission Reduction Units (ERUs) issued through the Joint Implementation facility; and Removal Units (RMUs) generated by domestically-implemented afforestation and reforestation projects.[74] The extension of the 2 per cent levy took effect at the commencement of the second Kyoto commitment period in 2013 (UNFCCC 2013l).

Although, at present, the World Bank serves as the Trustee of the Adaptation Fund, the Fund is supervised and managed by a 16-member Board comprised of: (i) two representatives from the five United Nations regional groups; (ii) one representative from the group of small-island states (SIDS); (iii) one representative from the group of least-developed countries (LDCs); (iv) two additional representatives from the Annex I group; and (v) two additional representatives from the non-Annex I group of nations. Where possible, decisions made at Adaptation Fund Board meetings are arrived at via consensus. In circumstances where a consensus cannot be reached, decisions are determined by a two-thirds majority of members present at the Board meeting (www.climatefundsupdate.org).

As at the end of 2013, the Adaptation Fund had accrued US189.8 million in proceeds from the 2 per cent levy on the Kyoto flexibility mechanisms. In addition, cumulative donations to the end of 2013 had grown to US$205.5 million, which was just short of the US$223.6 million pledged by donor nations (World Bank 2014b).[75] Over the same period of time, the Adaptation Fund Board had approved the transfer of US$224.2 million in support of adaptation projects and programmes in 29 recipient countries. Taking account of cumulative revenues, donations, and cash transfers, US$170.9 million remained in the Fund at the end of 2013 to support the future funding decisions of the Adaptation Fund Board (World Bank 2014b).

Of considerable concern is the fact that revenues generated during 2013 by the 2 per cent levy on Kyoto emissions units was just US1.8 million. This was well down on the US$100.2 million raised in 2010. The massive drop in proceeds was the result of two factors. The first was the huge decline in the number of CERs and ERUs issued in 2013. The second was the dramatic fall in the value of emissions units, most notably between early-2011 and 2013 when the price of CERs fell from around US$20 per tonne of emissions units to just five US cents in February 2013. Although the market price of CERs rose to around 75 US cents by the end of 2013, many observers believe this is well below the value required for the Kyoto offset mechanisms to drive the transition to low-emissions technologies (World Bank 2013).[76]

Notwithstanding the meagre proceeds generated during 2013, the total funds available in the Adaptation Fund at the end of 2013 were US$35.4 million higher than at the close of 2012. This increase over 2013 was due to large donations from the governments of Germany, Norway, and Switzerland—the size of which may not be received again (World Bank 2014b). Should the current price and estimated issuances of the various Kyoto emissions credits remain close to 2013 levels, as many predict, it is estimated that the Adaptation Fund will receive revenues of around US$15–30 million between 2014 and 2020. When added to outstanding pledges and the current level of available funds, the Fund should provide approximately US$210 million through to 2020 for new adaptation-related projects and programmes (approximately US$30 million per year) (World Bank 2014b).[77]

8.2.8.6 The Kyoto Flexibility Mechanisms

Given the importance of global carbon markets, it can be seen from Fig. 8.2 and Table 8.6 that the Kyoto flexibility mechanisms constitute major generators and distributors of climate change funds. The manner and extent to which these mechanisms function as financing instruments differ markedly. With regards to the International Emissions Trading system, climate change funds are generated from the sale of emissions units that, in turn, can be used to finance new and additional mitigation and adaptation projects. However, the importance of the International Emissions Trading system extends beyond the mere generator of climate change funds. The system also serves a valuable redistribution role—a role that is likely to increase in importance over coming decades. To date, a large portion of the emissions units sold in carbon markets have been the surplus AAUs possessed by countries with transition economies (EITs). Hence, even as we speak, the opportunity to sell these units has facilitated the redistribution of funds from wealthier to poorer Annex I countries.

Looking into the future, non-Annex I nations will almost certainly be required to meet emissions targets in upcoming international protocols. Not unlike the Annex I nations currently subject to Kyoto obligations, non-Annex I countries are likely to be issued with emissions allowances similar to AAUs.[78] This will inevitably draw non-Annex I nations into some sort of international emissions-trading arrangement, although not necessarily one that immediately involves integration with Annex I nations. Regardless of how these arrangements evolve, as a growing number of low-GDP countries become subject to emissions obligations, there is little doubt that international emissions trading will serve as a more prominent means of transferring climate change funds from wealthy to poor nations.

At this point, it is worth mentioning something about the Green Investment Scheme. This Scheme, which has existed for over a decade, has become an important element of the International Emissions Trading system. The aim of the Scheme is to increase the marketability of AAUs. The need to increase the attractiveness of AAUs has arisen because the expanded application of the Clean Development Mechanism and Joint Implementation facility has increased the quantity of emissions units possessed by Annex I nations. This has forced many Annex I nations—in particular, countries with transition economies (EITs)—to offer an additional incentive for nations to purchase their surplus AAUs (Korppoo 2003; Carbon Trust 2009). The Green Investment Scheme has accomplished this by directing the proceeds from the sale of AAUs into projects and programmes that reduce greenhouse gas emissions in AAU-selling nations (Blyth and Baron 2003).[79] As an added bonus, the Scheme's flexibility and ability to generate upfront finance appears to have increased investment in emissions-reducing activities beyond the levels fostered by the Clean Development Mechanism and the Joint Implementation facility alone (Carbon Trust 2009).[80] Overall, by raising extra funds and generating additional benefits for both the buyers and sellers of emissions units, the Green Investment Scheme has significantly boosted the cost-effectiveness and attractiveness of the International Emissions Trading system.

Having said this, there are considerable restrictions on the ability of Annex I nations to buy and sell AAUs for compliance purposes. For example, the European Union Emissions Trading System—the EU-ETS—limits the extent to which European Union nations can rely upon the trade in emissions units to achieve emissions targets.[81] Specifically, European Union countries must meet at least 50 per cent of their emissions reductions via domestic mitigation activities. However, in an attempt to boost mitigation efforts, some European Union nations have agreed to much stricter limits on their capacity to purchase emissions units to meet their emissions obligations. The restrictions range from 8 per cent in the Netherlands (which means 92 per cent of emissions reductions must be achieved via domestic mitigation measures) to 50 per cent in Spain and Ireland (no additional restrictions) (Larson et al. 2013).

Beyond the EU-ETS, decisions made at the COP-18 meeting at Doha in 2012 also imposed considerable restrictions on the use and sale of the AAUs carried over from the first Kyoto commitment period (2008–2012) into the second commitment period (2013–2020). The restrictions include (World Bank 2013):

- All Annex I countries not participating in the second Kyoto commitment period are prohibited from selling carry-over AAUs. This will have a significant impact on the Russian Federation and the market for emissions units generally given the Russian Federation's non-participation in the second commitment period and its large possession of surplus AAUs at the end of 2012.[82]
- AAUs acquired by an Annex I country during the first Kyoto commitment period can no longer be sold if they were first purchased from Annex I nations not subject to emissions obligations during the second commitment period.[83]
- The quantity of carry-over AAUs that an Annex I nation can purchase is limited to a maximum of 2 per cent of the total AAUs granted to it over the second Kyoto commitment period.[84]

On top of these restrictions, most of the AAUs carried over from the first Kyoto commitment period have effectively been eliminated by a political declaration made by the European Union and a large number of Annex I countries not to purchase carry-over AAUs for compliance purposes during the second commitment period. In addition, the 2012 Doha amendments to the Kyoto Protocol mean the automatic cancellation of any positive difference between the AAUs allotted to an Annex I country for the second Kyoto commitment period and eight-times its average 2008–2010 emission levels (Carbon Market Watch 2013).[85] These decisions are likely to significantly reduce the quantity of emissions units available for trading in international carbon markets.

Because of these limitations and the fact that current international trading arrangements under the UNFCCC are restricted to the Kyoto Protocol, Parties to the UNFCCC have been investigating new market instruments to form part of a future climate change protocol. Two potential approaches are presently being explored. The first is a New Market-Based Mechanism (NMM), which would provide incentives to increase mitigation activities in non-Annex I nations beyond the levels occurring under existing Kyoto market-based arrangements. To be effective, all UNFCCC Parties have agreed that the NMM must stimulate emissions reductions across broad segments of the economy to ensure a net decrease in global greenhouse gas emissions. Moreover, UNFCCC Parties share the position that the governance of the NMM and the responsibility to develop the Mechanisms' rules and modalities must, for the time being at least, rest with the UNFCCC (World Bank 2012, 2013, 2014a; UNFCCC 2013m).

The second market mechanism under consideration is a so-called Framework for Various Approaches (FVA). The FVA would allow individual nations to establish and implement market-based schemes based on their own standards and methodologies subject to UNFCCC approval (World Bank 2013; UNFCCC 2013n).

To gain UNFCCC recognition, schemes would have to "meet standards that deliver real, permanent, additional, and verified mitigation outcomes; avoid double-counting of effort; and achieve a net decrease and/or avoidance of greenhouse gas emissions" (UNFCCC 2014b). Under the FVA, emissions units issued by approved domestic schemes would be recognised by the UNFCCC. The emissions units would, as a consequence, be available for trading purposes in international carbon markets (World Bank 2013).

Turning now to the remaining Kyoto flexibility mechanisms, there are three main ways in which they provide a source of climate change finance. Firstly, there is the value of the CERs and ERUs generated from CDM-projects and JI-projects undertaken by Annex I nations. Where permissible, recipients of these emissions units can sell them and use the proceeds to fund, either at home or abroad, additional mitigation and/or adaptation projects. Secondly, as we have seen, the 2 per cent levy on the value of the emissions units generated by projects induced by the Kyoto Protocol's flexibility mechanisms provides revenue to assist in the capitalisation of the Protocol's Adaptation Fund.[86] Finally, there is the value of the CDM-projects and JI-projects themselves. Besides drawing funds directly from the governments of investor countries, these projects provide an important channel for private-sector participation in the financing of low-emissions technologies in host-nations (Stern 2007).[87] Moreover, the projects often generate significant multiplier benefits in the regions where the projects take place.

There are, however, a number of additional points worth highlighting with respect to the fund-raising capacity of the Clean Development Mechanism and the Joint Implementation facility. Let me begin with the Clean Development Mechanism. Under current Kyoto regulations, nuclear-energy activities with the capacity to reduce greenhouse gas emissions in non-Annex I nations do not qualify as eligible CDM-projects (Larson et al. 2013). Nor, as previously mentioned, do land-use activities that reduce emissions by sources, rather than sinks. Even then, the eligibility of sink-related land-use activities is confined to afforestation and reforestation projects, which means that activities that prevent deforestation or reduce emissions via improved agricultural practices are ineligible for registration as CDM-projects. Furthermore, the Kyoto Protocol precludes Annex I nations not subject to emissions targets during the second Kyoto commitment period (2013–2020) from trading in CERs (Dinar et al. 2013). This latter condition effectively closes off the Clean Development Mechanism at present to the USA, Canada, the Russian Federation, and Japan.[88]

At the same time, additional restrictions apply to EU nations. As explained above, the EU-ETS requires member countries to achieve a minimum percentage of emissions reductions via domestic mitigation activities. Complicating matters further, the CERs generated from CDM-projects registered after 31 December 2012 are only eligible for trading under Phase III of the EU-ETS if the projects are hosted by Least-Developed Countries (LDCs) or non-Annex I nations with bilateral agreements with the European Union. Moreover, post-April 2013, all CERs generated from CDM-projects involving the destruction of trifluoromethane hydrofluorocarbon-23 (HFC-23) and nitrous oxide (N_2O) emissions from adipic acid production were rendered ineligible for trading under the EU-ETS (World Bank 2013). The same also applies to the CERs generated through forestry-related CDM-projects. Because these rulings restrict the ability of EU nations to generate CERs, they dampen their incentive to invest in CDM-projects. All things considered, the Kyoto regulations and the EU-ETS restrictions reduce the pool of funds that can be generated by the Clean Development Mechanism.

On the upside, the concept of a Programme of Activities (PoA) was approved at the COP-11 meeting in 2005 to increase the attractiveness of the Clean Development Mechanism. Under a PoA, an unlimited number of projects across an industry, region, or country can be registered under a single administrative umbrella. This allows large-scale emissions reductions to be achieved from aggregating smaller project activities that, if judged on an individual basis, would be economically unviable (UNFCCC 2013c).

In addition, the revised rules covering the modalities and procedures of the Clean Development Mechanism—which were passed at the COP-19 meeting in Warsaw—have improved the Mechanism's efficiency and effectiveness. There are also efforts underway to streamline the CDM-project cycle and simplify the regulatory framework relating to the 'additionality' test and the project validation process. Reform in this area has already led to certain project types receiving automatic additional classification (Kachi et al. 2014).[89] Taken together, these reforms have reduced the transaction costs that have hitherto served as a disincentive for Annex I nations to undertake CDM-projects (World Bank 2013).

Overall, since 2001, the Clean Development Mechanism has leveraged over US$300 billion in investment to support mitigation projects and activities in non-Annex I nations (UNFCCC 2013c). As well as contributing towards the establishment of over 100 Gigawatts of additional renewable energy capacity, the Clean Development Mechanism has underpinned a number of environmentally-beneficial programmes that have generated a range of co-benefits in non-Annex I nations, such as the transfer of new technologies; the creation of additional employment and income-generating activities (a beneficial multiplier effect); increased educational opportunities; enhanced access to electricity in rural areas; and improved air quality (UNFCCC 2013c).

As for the Joint Implementation facility, the restrictions imposed by the EU-ETS on the ability of European Union nations to trade in emissions units also limit the capacity of member countries to achieve emissions targets through investment in JI-projects (Larson et al. 2013). This capacity has been further constrained by new European Union rules governing the registration infrastructure underpinning the EU-ETS. In an effort to avoid double-counting, the issuance of ERUs is not permitted in cases where projects are: (i) hosted in European Union nations directly or indirectly related to activities covered by the EU-ETS Phase II after 31 December 2012; and (ii) related to activities newly covered by Phase III of the EU-ETS after 30 April 2013. In addition, ERUs transferred to the EU-ETS registry after 1 May 2013 by countries not participating in the second Kyoto commitment period (2013–2020) can only remain in the EU-ETS registry if they represent emissions reductions achieved before 31 December 2012.[90] Crucially, ERUs which do not qualify for EU-ETS registration cannot be used for compliance purposes (World Bank 2013).

At a broader international level, Kyoto regulations impose two main restrictions on the use of the Joint Implementation facility. The first, as previously highlighted, involves the ineligibility of land-use activities which reduce emissions by sources. The second restriction arises because of the combined effect of three decisions taken at the COP-18 meeting in Doha—one of which prohibited Annex I nations from issuing ERUs during the second Kyoto commitment period until the new AAUs for the period were allotted. Whilst this decision did not directly limit the use of the Joint Implementation facility, a second decision stipulated that the issuance of new AAUs was dependent upon a nation's participation in the second Kyoto commitment period. Together, these two decisions effectively prohibited non-participating countries from engaging in the Joint Implementation facility (e.g., the Russian Federation) (World Bank 2013).

On top of this, there is the impact of a third decision to cancel any AAUs covered by the second Kyoto commitment period that exceed eight-times a nation's average 2008–2010 emissions levels. Besides reducing the quantity of AAUs that can be converted to ERUs, this decision weakened the position of the Annex I nations with transition economies (EITs) to such an extent that the Ukraine, Belarus, and Kazakhstan have threatened to withdraw their participation in the second Kyoto commitment period (Carbon Market Watch 2013). Given this threat and the absence of the Russian Federation, there is great uncertainty as to whether Annex I nations in possession of large quantities of emissions units will be able to supply ERUs in the same quantities as they did during the first Kyoto commitment period. If they cannot, this will limit the fund-raising capacity of the Joint Implementation facility.

To alleviate the impact of these legitimate restrictions, the UNFCCC Parties have been investigating ways to reform the Joint Implementation facility to stimulate future investment in JI-projects. As a means of simplifying the process, a proposal has been forwarded to merge the Track 1 and Track 2 verification procedures. Should the proposal be adopted, it is expected that the ensuing single-track procedure would involve a stronger role for host Parties and verifiers accredited by the Joint Implementation Supervisory Committee (JISC). It is also likely that mandatory standards for accreditation would be specified by a newly-established governing body (World Bank 2013). Above all, the aim of the reform process would be to widen the scope of participation in the Joint Implementation facility without compromising its environmental integrity. If accomplished, the reforms would undoubtedly increase its effectiveness and fund-raising potential.

In terms of actual finance, it is more difficult to ascertain the total value of the funds that the Joint Implementation facility has leveraged to support emission-reducing projects compared to the Clean Development Mechanism. Nevertheless, given the nature of JI-projects and the combined value of ERUs traded in international carbon markets, it is safe to say that it would be in the tens of billions of US dollars (World Bank 2012, 2013).[91]

8.2.8.7 Multilateral Finance Institutions

To complete the climate change financing picture, it is important to have a good understanding of the vast array of non-UNFCCC organisations and agencies that serve as both conduits and generators of climate change funds. As Fig. 8.2 shows, one of the prominent collectives in this regard is the assemblage of multilateral finance institutions. Multilateral finance institutions are organisations with a core banking basis into which multiple nations contribute funds and share in their ownership (Atteridge et al. 2009). Some of the multilateral institutions heavily engaged in the area of climate change finance include the World Bank, the Inter-American Bank (IDB), the Asian Development Bank (ADB), the African Development Bank (AfDB), and the European Bank for Reconstruction and Development (EBRD). To date, the World Bank has been the most influential, particularly in view of its auxiliary role as the Trustee of a number of UNFCCC climate change funds.

There are various ways in which multilateral finance institutions provide or garner financial resources to support climate change initiatives. The most obvious way is through their lending activities. The second is via dedicated trust funds, which the various institutions have established to directly finance climate change projects and programmes. The third, which supports the Kyoto Protocol, entails the purchase of emissions reduction units generated by Annex I nations from the use of the Clean Development Mechanism and Joint Implementation facility (Atteridge et al. 2009). The final method involves the leveraging of additional funds, which multilateral finance institutions have achieved by: (i) establishing new financing instruments to support private-sector investment in climate change activities; (ii) improving the financial viability of emissions-reducing projects through policy dialogue, regulatory reform, and capacity building; and (iii) addressing barriers to the adoption of energy-efficiency projects by providing technical advice and investment grants (www.ebrd.co/sei; IDB 2013; AfDB 2013; ADB 2014).

According to the Climate Policy Initiative (2013), multilateral finance institutions contributed around US$38 billion in climate change funding in 2012, although much of this took the form of low-cost loans rather than direct grants or subsidies. At approximately US$11 billion, the largest proportion of the total funds provided by multilateral finance institutions in 2012 was used to support sustainable transport projects. Multilateral finance institutions also made a significant contribution to adaptation measures—a noteworthy aspect given that global support for adaptation in 2012 constituted just 6.1 per cent of the total funds allocated for climate change purposes (Climate Policy Initiative 2013).[92]

Although not endowed with a large amount of financial capital at present, a dedicated fund supported by multilateral finance institutions with the potential to expand enormously in terms of financial resources and influence is the Carbon Fund of the Forest Carbon Partnership Facility (FCPF). The FCPF Carbon Fund is a multilateral funding mechanism created to support projects which reduce greenhouse gas emissions from deforestation. The Fund also promotes the sustainable management of forests in developing countries—a collection of activities referred to as REDD+ (World Bank 2013).[93] The overall aim of the Carbon Fund is to provide result-based payments to nations possessing large areas of tropical and/or sub-tropical forests as a reward for emissions reductions achieved through REDD+ activities.[94] With a capitalised value of US$466.5 million as at 1 June 2014, the FCPF Carbon Fund is well positioned to finance emissions reductions delivered by five selected jurisdictional and national scale programmes, should the reductions eventually be achieved.[95] The average result-payment for each programme is expected to be around US$70 million (http://www.forestcarbonpartnership.org/sites/fcp/files/2013/june2013/CF%20Origination-web_0.pdf). That said, the Carbon Fund requires considerably more resources to adequately fulfil its function. Governments of high-GDP nations need to play a greater role in this regard, which they could do by making explicit commitments to the Fund.

8.2.8.8 Bilateral Finance Institutions

Another important cluster of organisations engaged in attracting and disbursing climate change funds is the group of bilateral finance institutions (see Fig. 8.2). Bilateral finance institutions have a core banking basis but differ to multilateral finance institutions in that the contributions used to capitalise them are provided by the governments and capital markets of a single nation. Some of the key bilateral finance institutions involved in climate change finance include the French Development Agency[96] (AFD), the German Development Bank (KfW), and the Japan International Co-operation Agency (JICA).

The various means by which bilateral finance institutions support climate change activities are similar to those employed by multilateral finance institutions.[97] Nevertheless, they differ greatly across institutions. The reason for this is that the organisational structures and mandates of bilateral finance institutions vary according to the relationship they have with other institutions in their country of origin. This has the effect of influencing the type of projects and programmes that individual institutions support and the manner in which they support them (Atteridge et al. 2009).[98] For example, whereas the French Development Agency is able to operate independently when assessing and financing projects in line with the French Government's official development assistance (ODA) policies, the German Development Bank is forced to operate in the knowledge that Germany's international development operations are shared between different government agencies. Moreover, despite the recent merging in Japan of its international development operations into an all-encompassing Japan International Co-operation Agency, the Japanese Ministry of Foreign Affairs continues to play an active role in administering ODA loans and grants, including many that are directly relevant to Japan's contribution to global climate change action (Atteridge et al. 2009).

Not surprisingly, as a portion of overall institutional activities, levels of climate change finance vary considerably from one bilateral finance institution to the next. Also differing between the institutions is the ratio of public-sector to private-sector entities in receipt of allocated climate change funds, although public-sector entities tend to receive the largest slice of funding regardless of the disbursing institution (Atteridge et al. 2009; Climate Policy Initiative 2013).

A further feature of the climate change funding of bilateral finance institutions is the strong leaning towards mitigation projects (Climate Policy Initiative 2013). This is partly due to the fact that adaptation measures require significant grant funding which bilateral finance institutions have difficulty delivering given their heavy reliance on debt and equity instruments (Atteridge et al. 2009). It is also believed that support for adaptation has been hampered by the project-based mindset of many financial institutions and an associated lack of support for programmatic approaches to adaptation funding.

In 2012, bilateral finance institutions contributed approximately US$15 billion in climate change finance, although, when combined with the funds provided by national development banks, it was much closer to US$84 billion.[99] Because of a strong emphasis on mitigation support, around 65 per cent of the funds provided by bilateral finance institutions in 2012 were allocated to renewable energy and energy-efficiency projects (Climate Policy Initiative 2013).

8.2.8.9 Development Co-operation Agencies

As Fig. 8.2 shows, development co-operation agencies are yet another important group of establishments involved in climate change finance. Generally speaking, the principal climate change function of development co-operation agencies is not to raise funds for climate change purposes. It is to channel climate change funds to needy recipients by filling knowledge gaps and bringing together key financiers and project facilitators (Atteridge et al. 2009).[100] A good example of the conveyancing role of these agencies can be found in the form of the United Nations Environment Programme (UNEP). The UNEP promotes the evolution of clean-energy markets in developing countries whilst actively mobilising finance to create new economic opportunities and enhance access to sustainable-energy systems. Working closely with governments and private-sector institutions, the UNEP also assists in the development of appropriate systems and cross-cutting institutional arrangements to facilitate effective decision-making and disbursement of climate change funds (http://www.unep.org/climatechange/finance).

The UNEP is not the only organisation with environmental and/or development mandates actively engaged in the area of climate change finance. Many other United Nations agencies have been drawn into the climate finance arena if only because climate change has the potential to undermine their ability to achieve their mandated objectives. The risk that climate change poses to human welfare, water supplies, and food security (see Table 1.10 and Fig. 1.4) has meant that the United Nations Development Programme (UNDP), the Food and Agriculture Organization of the United Nations (FAO), the United Nations Industrial Development Organization (UNIDO), and the International Fund for Agricultural Development (IFAD) have become heavily involved in redirecting climate change funds. The approaches used by each of these organisations to promote climate change finance are similar to those described with respect to the UNEP.[101]

Given its sheer size and influence, another development co-operation agency worthy of note is the Organisation for Economic Co-operation and Development (OECD). The OECD has been instrumental in guiding its member countries to integrate climate change finance into their broader development strategies (Atteridge et al. 2009). Furthermore, the OECD has successfully persuaded its members to collaborate on a range of climate change finance matters. This has significantly reduced unnecessary duplication and waste. At the same time, the OECD has the honour of being the only international institution to which bilateral finance institutions report their development funding with respect to climate change action. This privilege allows the OECD, which issues guidelines instructing bilateral finance institutions on how to report their financing activities, to influence the way in which bilateral finance institutions design, monitor, and assess the climate change activities they support (Atteridge et al. 2009).[102]

8.2.8.10 National and Sub-national Governments

We have already seen the pivotal role that governments play in terms of capitalising the national development banks and financial institutions engaged in climate change finance. It has also been pointed out that governments actively fund domestic climate change activities and, in the case of Annex I nations, support climate change projects and programmes in non-Annex I countries. Excluding the funds provided to capitalise development banks and finance institutions, national and sub-national governments contributed approximately US$13.6 billion in climate change funds in 2012. Around US$6 billion of this total was channelled through bilateral co-operation agencies (US$5.2 billion) and United Nations institutions (US$0.8 billion). Of the remaining US$7.6 billion, approximately US$6 billion involved government spending by both Annex I and non-Annex I nations on domestic climate change interventions, of which US$5 billion was directly allocated to renewable-energy projects (Climate Policy Initiative 2013).[103] A further US$0.7 billion was spent by sub-national governments in support of local climate change activities, and approximately US$0.3 billion was allocated, primarily by national governments, to boost domestic exports of low-emissions technologies (Climate Policy Initiative 2013).

In addition, national governments contributed a further US$1.6 billion towards various nation-based and internationally-administered climate change funds (Climate Policy Initiative 2013). The international climate change funds include the Least Developed Countries Fund (LDCF), the Special Climate Change Fund (SCCF), the Clean Technology Fund[104], and the soon to be expanded Green Climate Fund (GCF). As for the nation-based funds, most have been established by Annex I countries. They include, for example, the Danish Carbon Fund, the Netherlands European Carbon Facility, the Italian Carbon Fund, and the Spanish Carbon Fund (World Bank 2008). In general, these nation-based funds are used to:

(i) develop CDM-projects and/or JI-projects; (ii) purchase emissions units, such as CERs and ERUs, to assist home countries to meet their Kyoto obligations; and (iii) support capacity building and the transfer of low-carbon technologies to non-Annex I nations.

8.2.8.11 Private-Sector Sources of Funds

As mentioned earlier, the private sector is the major provider of climate change finance, although its availability is heavily reliant on government incentives and the enabling effect of public-sector investments. Evidence of the private-sector's ascendancy over the public sector is the US$224 billion it contributed towards climate change programmes and activities during 2012—almost two-thirds of the total funds made available by all agents during the year. Within this total, project facilitators, such as privately-owned energy utilities, independent electricity producers, and renewable-energy suppliers contributed US$102 billion, of which around 61 per cent was used to finance climate change investments in developing countries (Climate Policy Initiative 2013).[105] A further US$66 billion was invested by manufacturers and corporate end-users as a means of reducing production costs, whilst households spent US$33 billion on domestic solar hot-water and solar electricity-generating systems.[106] Of the remaining US$23 billion of private-sector funding in 2012, just over US$21 billion was forwarded by private financial institutions; US$1.2 billion was contributed by venture-capital, private-equity, and privately-owned infrastructure funds; and US$0.4 billion was provided by institutional investors (Climate Policy Initiative 2013).

Before finishing this sub-section on financing mechanisms and institutions, it is worth mentioning something about voluntary carbon markets as a means of promoting the private-sector flow of climate change funds. Voluntary carbon markets cater to organisations wanting to voluntarily reduce their greenhouse gas emissions via the use of carbon offsets. In most instances, the use of voluntary carbon markets has been driven by regulatory vacuums in many countries and/or the anticipation of forthcoming compliance legislation (World Bank 2013). At present, the volume of transactions in voluntary carbon markets constitutes a tiny fraction of the various emissions units transacted in global carbon markets (0.1 per cent in 2011) (World Bank 2013). Hence, voluntary carbon markets have yet to raise large quantities of climate change funds. However, as long as the quality of emissions reductions in voluntary markets can be guaranteed—which depends on the existence of adequate voluntary offset standards—there is no reason why voluntary carbon markets could not generate significantly larger private-sector flows of climate change funds between now and when a new climate change protocol comes into existence.

8.2.9 The Kyoto Architecture—A Summary

In Chap. 9, I will argue that a new global protocol should be built on a modified version of the existing Kyoto architecture. This, of course, raises the question as to what are its main features. Given what has been outlined so far in this chapter, we are now in a position to do this. Keeping things simple, I would summarise the Kyoto architecture in terms of five basic elements.[107] They are:

1. The grouping and separate treatment of countries according to their per capita GDP and/or their current and historical greenhouse gas emissions. This distinction has been institutionalised through the categorisation of countries as either Annex I (Annex B) nations or non-Annex I nations.
2. The exclusive imposition of greenhouse gas emissions targets on the world's wealthiest and/or heavily industrialised nations (i.e., Annex I nations).
3. The existence of three flexibility mechanisms—the system of International Emissions Trading, the Clean Development Mechanism, and the Joint Implementation facility—to assist countries subject to obligations to cost-effectively achieve their emissions targets.
4. A (non-binding) penalty for failing to comply with emissions targets comprising of: (i) the need for a non-compliant nation to make up for any excessive emissions plus an additional 30 per cent during a subsequent commitment period; and (ii) suspension from international trading in emissions units.
5. The transfer of funds and technology from the world's richest to poorest countries to enable the latter to undertake mitigation activities and adapt to the damaging impacts of climate change.

8.3 The Greenhouse Gas-Emitting Performance of Nations

8.3.1 The Greenhouse Gas-Emitting Performance of Annex I Nations

In Chap. 4, it was explained that stabilising the atmospheric concentration of greenhouse gases at no more than 450 pp of CO_2-e will require a 4 per cent annual reduction in global CO_2-e emissions beyond 2015.[108] It has also been argued that, in view of the historical emissions of Annex I countries and their significantly higher per capita emissions levels, much larger emissions cuts will be demanded of them in coming decades. Of course, when considering the contribution that all countries must make towards resolving the climate change crisis, the greenhouse gas emissions of non-Annex I nations cannot be ignored, particularly given the rapid rise in their per capita emissions in recent decades and the fact that, in aggregate terms, China and India have respectively become the world's largest and third-largest emitters of greenhouse gases.

In view of the required reductions in global emissions and the relative disparities in emissions between countries, it is worth revealing the greenhouse gas-emitting performance of individual countries and the four UNFCCC-defined groups of nations referred to earlier in the chapter. Doing this will put into perspective the performance of Annex I nations already subject to emissions targets; the performance of countries that are currently target-free but likely to have emissions targets in the near future; emissions trends, both at the national and the UNFCCC-defined group levels; and the extent to which individual nations will need to overturn their current performance to achieve future emissions cuts. It will also paint a clearer picture of the practical and diplomatic challenges that are destined to lie ahead as individual countries and alliances negotiate a new global climate change protocol at the scheduled COP-21 meeting in Paris in 2015.

Starting with Annex I Parties, columns *a* and *b* of Table 8.7 reveal, in slightly different forms, the Kyoto target of each Annex I Party during the first Kyoto commitment period. In column *a*, the Kyoto target is presented as the maximum aggregate quantity of greenhouse gases that each Party was permitted to emit between 2008 and 2012. Column *b* presents the Kyoto target as the maximum (average) annual quantity of greenhouse gases that each Party was allowed to emit over the same five-year period.[109] As is evident, three Parties—Cyprus, Malta, and Turkey—were not given emissions targets, whilst the USA was target-free as a consequence of not ratifying the Kyoto Protocol. Although Canada received an initial Kyoto target, its withdrawal from the Kyoto Protocol in 2011 absolved it from its Kyoto obligations. As for EU-15 nations, columns *a* and *b* account for the redistribution of Kyoto targets by the European Union (see Table 8.1, column *b*).

Taking account of process-related emissions[110] plus the emissions and greenhouse gas removals from land-use, land-use change, and forestry (LULUCF) activities, column *c* of Table 8.7 reveals the percentage difference between each Annex I Party's 1990 and 2012 *net* emissions levels. Column *c* shows that the net emissions of most of the Annex II Parties decreased over the 1990–2012 period. The exceptions were Australia (+2.4 %), Austria (+11.7 %), Canada (+42.2 %), Greece (+5.3 %), Iceland (+9.8 %), Ireland (+4.6 %), Japan (+8.6 %), New Zealand (+111.4 %), Spain (+18.0 %), and the USA (+2.7 %). Because of the huge decline in emissions that immediately followed the collapse of communism in Eastern Europe and the former-USSR in the early-1990s (Olivier et al. 2011), the net emissions of all transition economies (EITs) in the Annex I (non-Annex II) group of nations decreased between 1990 and 2012. However, the net emissions of the three Parties with no Kyoto targets—Cyprus, Malta, and Turkey—were much higher in 2012 than in 1990 (Cyprus +52.1 %; Malta +57.7 %; and Turkey +133.4 %).

Column *d* of Table 8.7 presents the emissions targets of Annex I Parties during the first Kyoto commitment period that applied to greenhouse gas emissions not covered by an emissions-trading system—commonly referred to as 'non-ETS targets'. Since the notion of non-ETS targets is not immediately clear, let me say something about them and the manner in which they were calculated.[111]

Table 8.7 Greenhouse gas emissions targets and performances of Annex I Parties to the UNFCCC during the first Kyoto commitment period (2008–2012)

Group/Party	Kyoto target (2008–2012) (max. total GHGs) (Kilotonnes of CO_2-e)	Kyoto target (2008–2012) (max. ave. annual GHGs) (Kilotonnes of CO_2-e) (*a*/5)	Percentage difference in GHGs (2012 compared to 1990) (including LULUCF)	Non-ETS target (2008–2012) (max. ave. annual GHGs) (Kilotonnes of CO_2-e)	Ave. annual non-ETS GHGs (2008–2012) (excluding LULUCF) (Kilotonnes of CO_2-e)	Percentage gap between ave. annual GHGs and target (non-ETS) (2008–2012) (excluding LULUCF) $(e - d)/d \times 100$ %	Ave. annual non-ETS GHGs (2008–2012) (including LULUCF) (Kilotonnes of CO_2-e)	Percentage gap between ave. annual GHGs and target (non-ETS) (2008–2012) (including LULUCF) $(g - d)/d \times 100$ %	Percentage gap between ave. annual GHGs and adjusted target (non-ETS) (2008–2012) (inc.LULUCF and Kyoto flexibility mechanisms)
	a	*b*	*c*	*d*	*e*	*f*	*g*	*h*	*i*
Annex II									
Australia	2,957,579	591,516	2.4 %	591,516	542,231	−8.3 %	540,799	−8.6 %	d.n.a.
Austria†§	343,866	68,773	11.7 %	36,343	52,818	45.3 %	49,744	36.9 %	−7.2 %
Belgium†§	673,996	134,779	−19.0 %	77,524	76,643	−1.1 %	75,398	−2.7 %	−10.4 %
Canada*	2,791,793	558,359	42.2 %	558,359	703,907	N/A	733,652	N/A	N/A
Denmark†§	278,827	55,765	−31.3 %	31,590	35,265	11.6 %	34,774	10.1 %	2.5 %
EU-15	19,621,382	3,924,276	−16.8 %	2,403,521	2,266,123	−5.7 %	2,066,565	−14.0 %	−17.2 %
Finland†§	355,018	71,004	−38.0 %	33,046	32,154	−2.7 %	3,792	−88.5 %	−87.9 %
France†§	2,819,627	563,925	−15.7 %	424,679	393,795	−7.3 %	352,861	−16.9 %	−16.6 %
Germany†§	4,868,097	973,619	−23.5 %	563,818	486,425	−13.7 %	481,384	−14.6 %	−14.2 %
Greece†§	668,670	133,734	5.3 %	69,031	56,787	−17.7 %	53,876	−22.0 %	−22 %
Iceland	18,524	3,705	9.8 %	3,705	4,671	26.1 %	5,458	47.3 %	47.3 %
Ireland†§	314,184	62,837	4.6 %	40,576	42,310	4.3 %	38,456	−5.2 %	−9.9 %
Italy†§	2,416,278	483,256	−14.3 %	280,430	301,433	7.5 %	276,970	−1.2 %	−2.0 %
Japan§	5,928,258	1,186,592	8.60 %	1,186,592	1,278,461	7.70 %	1,203,858	1.50 %	d.n.a.
Lux'bourg†§	47,403	9,481	−13.9 %	6,531	9,578	46.6 %	9,156	40.20 %	−2.7 %
N'lands†§	1,001,262	200,252	−9.1 %	113,526	117,457	3.5 %	120,591	6.20 %	−1.9 %
NZ§	309,565	61,913	111.4 %	61,913	74,560	20.4 %	43,628	−29.5 %	d.n.a.

(continued)

Table 8.7 (continued)

Group/Party	Kyoto target (2008–2012) (max. total GHGs) (Kilotonnes of CO_2-e)	Kyoto target (2008–2012) (max. ave. annual GHGs) (Kilotonnes of CO_2-e) (*a*/5)	Percentage difference in GHGs (2012 compared to 1990) (including LULUCF)	Non-ETS target (2008–2012) (max. ave. annual GHGs) (Kilotonnes of CO_2-e)	Ave. annual non-ETS GHGs (2008–2012) (excluding LULUCF) (Kilotonnes of CO_2-e)	Percentage gap between ave. annual GHGs and target (non-ETS) (2008–2012) (excluding LULUCF) $(e - d)/d \times 100$ %	Ave. annual non-ETS GHGs (2008–2012) (including LULUCF) (Kilotonnes of CO_2-e)	Percentage gap between ave. annual GHGs and target (non-ETS) (2008–2012) (including LULUCF) $(g - d)/d \times 100$ %	Percentage gap between ave. annual GHGs and adjusted target (non-ETS) (2008–2012) (inc.LULUCF and Kyoto flexibility mechanisms)
	a	*b*	*c*	*d*	*e*	*f*	*g*	*h*	*i*
Norway[§]	250,577	50,115	−35.3 %	41,831	33,786	−19.2 %	6,332	−84.9 %	−94.9 %
Portugal[†§]	381,938	76,388	−5.4 %	44,071	45,495	3.2 %	29,803	−32.4 %	−36.0 %
Spain[†§]	1,666,196	333,239	18.0 %	179,100	219,459	22.5 %	185,829	3.80 %	−17.9 %
Sweden[†§]	375,189	75,038	−34.8 %	52,233	40,589	−22.3 %	5,126	−90.2 %	−90.2 %
Switzerland[§]	242,838	48,568	−1.3 %	45,268	49,607	9.6 %	48,498	7.10 %	0.5 %
UK[†§]	3,396,475	679,295	−26.2 %	448,152	355,917	−20.6 %	348,805	−22.2 %	−22.2 %
USA[*]	N/A	N/A	2.7 %	N/A	6,758,528	N/A	5,812,810	N/A	N/A
Non-Annex II									
Belarus[*]	N/A	N/A	−42.3 %	N/A	88,934	N/A	60,538	N/A	N/A
Bulgaria[§]	610,046	122,009	−44.7 %	82,252	26,535	−67.7 %	18,243	−77.8 %	−74.2 %
Croatia	148,780	29,756	−18.6 %	29,756	28,875	−3.0 %	22,185	−25.4 %	−25.4 %
Cyprus[*]	N/A	N/A	55.30 %	N/A	9,958	N/A	9,895	N/A	N/A
Czech Rep.[§]	893,542	178,708	−35.5 %	92,470	61,192	−33.8 %	55,110	−40.4 %	−12.4 %
Estonia[§]	196,063	39,213	−45.8 %	26,085	5,684	78.2 %	1,163	−95.5 %	−38.4 %
Hungary[§]	542,367	108,473	−39.8 %	83,361	43,687	−47.6 %	39,583	−52.5 %	−45.9 %
Latvia[§]	119,182	23,836	−120.8 %	19,181	8,402	−56.2 %	−4,791	−125.0 %	−82.8 %
L'stein[§]	1,056	211	−0.1 %	193	228	17.9 %	220	14.2 %	−11.7 %
Lithuania[§]	227,306	45,461	−69.5 %	37,521	16,015	−57.3 %	6,347	−83.1 %	−45.5 %

(continued)

Table 8.7 (continued)

Group/Party	Kyoto target (2008–2012) (max. total GHGs) (Kilotonnes of CO_2-e)	Kyoto target (2008–2012) (max. ave. annual GHGs) (Kilotonnes of CO_2-e) (*a*/5)	Percentage difference in GHGs (2012 compared to 1990) (including LULUCF)	Non-ETS target (2008–2012) (max. ave. annual GHGs) (Kilotonnes of CO_2-e)	Ave. annual non-ETS GHGs (2008–2012) (excluding LULUCF) (Kilotonnes of CO_2-e)	Percentage gap between ave. annual GHGs and target (non-ETS) (2008–2012) (excluding LULUCF) $(e-d)/d \times 100\,\%$	Ave. annual non-ETS GHGs (2008–2012) (including LULUCF) (Kilotonnes of CO_2-e)	Percentage gap between ave. annual GHGs and target (non-ETS) (2008–2012) (including LULUCF) $(g-d)/d \times 100\,\%$	Percentage gap between ave. annual GHGs and adjusted target (non-ETS) (2008–2012) (inc.LULUCF and Kyoto flexibility mechanisms)
	a	*b*	*c*	*d*	*e*	*f*	*g*	*h*	*i*
Malta*	N/A	N/A	57.70 %	N/A	3,042	N/A	3,035	N/A	N/A
Monaco	495	99	−14.7 %	99	95	−4.8 %	94	−4.9 %	d.n.a.
Poland§	2,648,181	529,636	−16.7 %	323,767	202,191	−37.6 %	170,761	−47.3 %	−46.4 %
Romania§	1,279,835	255,967	−56.0 %	181,741	71,195	−60.8 %	47,975	−73.6 %	−71.7 %
Russian Fed.	16,617,095	3,323,419	−50.3 %	3,323,419	2,235,371	−32.7 %	1,676,189	−49.6 %	d.n.a.
Slovakia§	331,434	66,287	−46.1 %	33,734	22,928	−32.0 %	16,525	−51.0 %	−26.1 %
Slovenia§	93,629	18,726	−14.2 %	10,488	11,562	10.2 %	7,155	−31.8 %	−26.1 %
Turkey*	N/A	N/A	163.3 %	N/A	401,469	N/A	343,181	N/A	N/A
Ukraine	4,604,185	920,837	−57.1 %	920,837	398,296	−56.7 %	378,242	−58.9 %	d.n.a.

Notes

- *Canada officially withdrew from the Kyoto Protocol in 2012 and therefore was no longer subject to its initial Kyoto target. The USA did not ratify the Kyoto Protocol and was not subject to an initial Kyoto target. Cyprus, Malta, Belarus, and Turkey were not subject to emissions targets despite being Annex I nations to the Kyoto Protocol
- †The Kyoto emissions targets of the EU-15 nations during the first commitment period reflect the redistribution of targets by the European Union
- §Refers to the Parties that introduced a domestic emissions-trading system or opted to participate in a regional emissions-trading system. Although emissions-trading systems were introduced in New Zealand and Japan during the Kyoto first commitment period, for various reasons, neither country designated a cap on the emissions to be covered by the system. Therefore, their non-ETS target was the same as their Kyoto target
- N/A denotes Not Applicable; d.n.a. denotes data not available. A 'negative' sign on the values in columns *f*, *h*, and *i* indicates emissions levels below the Kyoto target

Sources UNFCCC (2008b, 2014c); EEA (2012, 2013)

Prior to the first Kyoto commitment period, many Annex I Parties introduced a domestic emissions-trading system or opted to participate in a regional emissions-trading system to cost-effectively achieve their Kyoto targets. Upon implementing or partaking in an emissions-trading system, countries were required to designate the emissions caps to be imposed on the greenhouse gases generated by the industries covered by the system. As a result, these countries made an explicit decision to share their target-achieving efforts between the industries covered by the emissions-trading system and all remaining industries of the economy. A good example of this sharing of effort occurred in the form of the EU-ETS.[112] To recall from Chap. 7, the EU-ETS initially encompassed the emissions of around 11,500 installations that were responsible in 2005 for approximately 40 per cent of all greenhouse gas emissions in the European Union.

Upon a national government setting its own ETS emissions cap, a quantity (equal to the cap) of the Assigned Amount Units (AAUs) allocated to it under the Kyoto Protocol was subsequently converted to ETS emissions allowances.[113] By setting an ETS emissions cap, a nation effectively did two things. Firstly, it fixed the contribution that the emissions-trading system made towards achieving its Kyoto target. Secondly, it determined the remaining AAUs that applied to the greenhouse gases generated by the industries not covered by the emissions-trading system. That is, it indirectly assigned itself a 'non-ETS target'. It is this non-ETS target which appears in column *d* of Table 8.7. The target itself is determined by the following equation:

$$\text{Non-ETS target} = \text{Kyoto target} - \text{ETS emissions allowances} \tag{8.1}$$

Because some Annex I countries did not have a domestic emissions-trading system or did not participate in a regional emissions-trading system during the first Kyoto commitment period, it was not necessary for them to set an ETS emissions cap. Hence, they had no need to convert some of their allocated AAUs to ETS emissions allowances. Consequently, the non-ETS targets of these countries were the same as their Kyoto targets.

I should point out that by splitting the effort required to achieve their Kyoto targets, the Annex I Parties that chose to adopt or participate in an emissions-trading system in no way absolved themselves of their Kyoto obligations. They merely opted to have the greenhouse gas emissions of some industries constrained by the regulations of a UNFCCC-approved emissions-trading system and the emissions of all remaining industries constrained by the conventions of the Kyoto Protocol. As mentioned, this was a policy that many Annex I parties undertook in the belief that it would reduce the cost of achieving their Kyoto targets.

Column *e* of Table 8.7 reveals the average annual quantity of non-ETS emissions generated by Annex I countries during the first Kyoto commitment period (excluding LULUCF activities). By non-ETS emissions, I mean process-related emissions generated by industries not covered by an emissions-trading system (Note: the greenhouse gases generated by the industries covered by an emissions-trading system are referred to as 'verified ETS emissions'). In the case of Annex

I Parties governed by a national or regional emissions-trading system, their non-ETS emissions were ascertained by applying the following formula:

$$\begin{array}{c}\text{Non-ETS emissions}\\ \text{(exc. LULUCF)}\end{array} = \begin{array}{c}\text{all process-related}\\ \text{emissions}\end{array} - \begin{array}{c}\text{verified ETS}\\ \text{emissions}\end{array} \tag{8.2}$$

In a similar vein to the values in column *d*, because some Annex I countries were not governed by an emissions-trading system during the first Kyoto commitment period, none of their greenhouse gas emissions fell into the category of verified ETS emissions. As a result, their non-ETS emissions equated to their entire quantity of process-related emissions.[114]

Using the values in columns *d* and *e*, column *f* indicates the percentage gap between each Party's non-ETS target and the average annual quantity of non-ETS emissions that each Party generated during the first Kyoto commitment period (excluding LULUCF activities). Column *f* shows that the process-related emissions of the following Annex II Parties exceeded their first-round targets: Austria (+45.3 %), Denmark (+11.6 %), Iceland (+26.1 %), Ireland (+4.3 %), Italy (+7.5 %), Japan (+7.7 %), Luxembourg (+46.6 %), the Netherlands (+3.5 %), New Zealand (+20.4 %), Portugal (+3.2 %), Spain (+22.5 %), and Switzerland (+9.6 %). With EITs making up the majority of the Parties in the Annex I (non-Annex II) group, only two Parties within this group exceeded their first-round targets—Liechtenstein (+17.9 %) and Slovenia (+10.2 %).

Column *g* of Table 8.7 goes further than column *e* to include the net emissions from LULUCF activities in the calculation of non-ETS emissions generated by Annex I Parties during the first Kyoto commitment period. Whereas the values of the Parties not subject to an emissions-trading system encompass all their greenhouse gas emissions—including the emissions and removals from LULUCF activities—the values of the Parties governed by a national or regional emissions-trading system were determined by the following equation:

$$\begin{array}{c}\text{Non-ETS emissions}\\ \text{(inc. LULUCF)}\end{array} = \begin{array}{c}\text{total emissions}\\ \text{(inc. LULUCF)}\end{array} - \begin{array}{c}\text{ETS emissions}\\ \text{allowances}\end{array} \tag{8.3}$$

Not unlike column *f*, column *h* indicates the percentage gap between the values in column *g* and the non-ETS target of each Annex I Party (column *d*). Column *h* shows that almost all Annex I nations improved their greenhouse gas-emitting performance through net removals from LULUCF activities—the exceptions being Canada, Iceland, and the Netherlands. In some cases, the net removals from LULUCF activities were sufficient to reduce a Party's average non-ETS emissions below its first-round target having initially exceeded it through process-related emissions alone. The relevant Parties were: Ireland (+4.3 % to −5.2 %); Italy (+7.5 % to −1.2 %); New Zealand (+20.4 % to −29.5 %); Portugal (+3.2 % to −32.4 %); and Slovenia (+10.2 % to −31.8 %).

Column *i* is included in Table 8.7 to demonstrate how the government use of Kyoto flexibility mechanisms and the Emission Reduction Units (ERUs) generated through Joint Implementation projects assisted some Annex I Parties to meet their Kyoto obligations. The column achieves this by revealing the percentage gap between the average annual quantity of non-ETS emissions generated during the 2008–2012 period and any adjusted Kyoto targets arising from the use of the International Emissions Trading system, the Clean Development Mechanism, and/or the Joint Implementation facility. A comparison of columns *h* and *i* shows that the use of the Kyoto flexibility mechanisms enabled the following Parties to achieve their first-round Kyoto targets: Austria (+36.9 % to −7.2 %); Luxembourg (+40.2 % to −2.7 %); the Netherlands (+6.2 % to −1.9 %); Spain (+3.8 % to −17.9 %); and Liechtenstein (+14.2 % to −11.7 %).

It is worth recognising that the sale of emissions units by the governments of some Annex I Parties reduced their overall emissions allowance and therefore worsened their final emissions performance (e.g., Bulgaria, the Czech Republic, Estonia, Hungary, Latvia, Lithuania, Romania, the Russian Federation, Slovakia, and Ukraine).[115] However, in all instances, the Parties in question remained well within their Kyoto targets.

Overall, except for Denmark (+2.5 %), Iceland (+47.3 %), Japan (+1.5 %), and Switzerland (+0.5 %) all Annex I Parties with a Kyoto target were able to meet their first-round Kyoto obligations. Despite the non-compliant status of these four Parties, there are a number of proceedings and actions that could bring their emissions into line with their Kyoto targets. To begin with, the first-round Kyoto compliance assessment will not be finalised until 2015 (EEA 2013). Hence, further adjustment to the recorded emissions of all Annex I Parties is still possible. Secondly, should a final assessment of each Party's emissions still indicate that Denmark, Iceland, Japan, and Switzerland have exceeded their first-round Kyoto targets, all four transgressors will have 100 days to undertake the action necessary to meet their commitments (see Sect. 8.2.7). For Denmark, Japan, and Switzerland, taking the necessary action should not be difficult or burdensome, since compliance will require a small purchase of the many surplus emissions units currently in existence.[116]

As for Iceland, it appears to be an exceptional case. Previous assessments of the emissions of Annex I nations indicated that Iceland was well on track to achieve its first-round Kyoto target (EEA 2013). However, recent recalculations of the emissions of all Annex I nations (released in May 2014) reveal a dramatic rise in the recorded emissions of Iceland—both in terms of process-related emissions and net emissions from LULUCF activities. The re-estimation of Iceland's emissions runs counter to the newly recorded emissions of most Annex I nations, which tend be lower than previous estimates, especially with regards to net emissions/removals from LULUCF activities. Unless the final compliance assessment reveals an error in the recalculation of Iceland's emissions, Iceland will need to embark on a substantial purchase of emissions units to achieve its first-round target.[117] As of mid-2014, the Icelandic Government had made no plans to engage the Kyoto flexibility mechanisms to meet its Kyoto obligations (EEA 2013). Presumably, this may change as a result of the recent recalculation of emissions.[118]

8.3.2 *The Performance of Non-annex I Nations*

Table 8.8 reveals the greenhouse gas emissions (including LULUCF activities) of the non-Annex I group of nations over the 1990–2000, 2000–2010, and 1990–2010 periods. The table also presents the greenhouse gas emissions of some major greenhouse gas-emitting non-Annex I Parties, including a small number of least-developed countries (LDCs).

As can be seen from column *c*, the emissions of the non-Annex I group of nations increased by a considerable 69.1 per cent over the 1990–2010 period. Disconcertingly, most of the rise took place between 2000 and 2010 (+48.8 %)—a consequence of the substantial increase in the emissions of non-LDCs during the 2000–2010 period (+52.6 %) and a decline in the emissions of the group of LDCs during the 1990–2000 period (−18.8 %). Also of note is the increase in the greenhouse gases emitted by non-LDCs between 1990 and 2010 compared to the increase in greenhouse gases generated by LDCs over the same period (+85.3 % compared to +2.0 %). In large part, this enormous disparity in emissions was the result of the former group's much higher growth rate of real GDP over the 1990–2010 period.

In terms of individual Parties, the greenhouse gas emissions of virtually all the major emitters in the non-LDC group increased substantially between 1990 and 2010. What's more, the emissions of China, India, and Indonesia—currently the world's largest, third-largest, and fifth-largest generators of greenhouse gases—rose dramatically over the 1990–2010 period (China +189.0 %; India +95.6 %; and Indonesia +67.5 %). On top of this, most of the increase occurred in the second half of the period. Not surprisingly, the emissions of four major oil-producing non-LDCs increased markedly between 1990 and 2010 (Indonesia +67.5 %; Saudi Arabia +142.7 %; the United Arab Emirates +186.7 %; and Venezuela +47.9 %).

Unlike the group of non-LDCs, the greenhouse gas-emitting performances of the major LDCs were mixed, although the percentage changes in their emissions over the 1990–2010 period were quite substantial (e.g., Central African Republic +103.4 %; Sudan +110.1 %; Myanmar −58.7 %; and Zambia −63.3 %). In general, the outcomes at the national level were heavily influenced by the LULUCF activities occurring within individual LDCs, especially the greenhouse gas emissions/removals resulting from deforestation/afforestation activities.

8.3.3 *Comparing the Greenhouse Gas-Emitting Performances of the UNFCCC-Defined Groups of Nations*

To put into clearer perspective the greenhouse gas-emitting status of the various UNFCCC-defined groups and their emissions relative to global emissions levels, consider Table 8.9.[119] Using comparable data, columns *a*, *b*, and *c* show that the

Table 8.8 Greenhouse gas emissions of the non-Annex I group and some major greenhouse gas-emitting non-Annex I Parties and LDCs (1990–2010)

Group/Party	Percentage change in GHG emissions (1990–2000) (including LULUCF)	Percentage change in GHG emissions (2000–2010) (including LULUCF)	Percentage change in GHG emissions (1990–2010) (including LULUCF)
	a	*b*	*c*
Non-Annex I	+13.7 %	+48.8 %	+69.1 %
Non-Annex I (non-LDCs)	+21.5 %	+52.6 %	+85.3 %
Non-Annex I (LDCs)	−18.8 %	+25.5 %	+2.0 %
Non-Annex I (non-LDCs)			
Algeria	+25.1 %	+27.3 %	+59.2 %
Argentina	+12.2 %	+5.4 %	+18.2 %
Brazil	−8.8 %	+10.8 %	+1.0 %
China	+31.1 %	+120.4 %	+189.0 %
Egypt	+39.1 %	+50.6 %	+109.4 %
India	+36.1 %	+43.7 %	+95.6 %
Indonesia	+24.4 %	+34.6 %	+67.5 %
Iran	+58.1 %	+17.9 %	+86.4 %
Ivory Coast	+10.1 %	−1.5 %	+8.4 %
Kazakhstan	−47.9 %	+64.0 %	−14.7 %
Malaysia	+28.3 %	+29.9 %	+66.6 %
Mexico	+16.2 %	+16.2 %	+35.0 %
Nigeria	+24.3 %	+6.4 %	+32.2 %
Pakistan	+42.2 %	+38.7 %	+97.2 %
Saudi Arabia	+51.9 %	+59.7 %	+142.7 %
South Africa	+13.5 %	+6.9 %	+21.3 %
South Korea	+70.8 %	+26.4 %	+115.9 %
Taiwan	+81.5 %	+18.4 %	+114.9 %
Tanzania	+3.4 %	−28.6 %	−26.2 %
Thailand	+36.1 %	+45.8 %	+98.5 %
United Arab Emirates	+59.5 %	+79.8 %	+186.7 %
Uzbekistan	−0.5 %	+4.4 %	+3.8 %
Venezuela	+20.8 %	+22.5 %	+47.9 %
Vietnam	+57.5 %	+96.3 %	+209.2 %
Non-Annex I (LDCs)			
Bangladesh	+12.6 %	+29.8 %	+46.1 %
Central African Rep.	−24.1 %	+168.0 %	+103.4 %
Dem. Rep. of Congo	−24.7 %	+7.4 %	−19.2 %
Myanmar	−35.8 %	−35.7 %	−58.7 %
Sudan	+27.0 %	+65.5 %	+110.1 %
Uganda	+9.5 %	+42.3 %	+55.8 %
Zambia	−27.3 %	−49.6 %	−63.3 %

Source http://edgar.jr.ec.europa.eu/overview.php?v=GHGts1990-2010

Table 8.9 Greenhouse gas emissions status and performances of the UNFCCC-defined groups of nations

UNFCCC-defined group	Percentage change in GHG emissions (1990–2000) (including LULUCF)	Percentage change in GHG emissions (2000–2010) (including LULUCF)	Percentage change in GHG emissions (1990–2010) (including LULUCF)	Per capita GHG emissions (2010) (including LULUCF) (tonnes of CO_2-e)
	a	*b*	*c*	*d*
World	+5.2 %	+24.5 %	+31.0 %	7.69
Annex I	−3.1 %	−3.2 %	−6.2 %	16.19
Annex II	+8.8 %	−3.8 %	+4.6 %	18.02
Annex I (non-Annex II)	−27.6 %	−1.3 %	−28.5 %	12.71
Non-Annex I	+13.7 %	+48.8 %	+69.1 %	5.56
Non-Annex I (non-LDCs)	+21.5 %	+52.6 %	+85.3 %	5.93
Non-Annex I (LDCs)	−18.8 %	+25.5 %	+2.0 %	3.55

Source http://edgar.jr.ec.europa.eu/overview.php?v=GHGts1990-2010

percentage changes in greenhouse gases emitted by the various groups differed enormously over the 1990–2000, 2000–2010, and 1990–2010 periods.[120] The three columns also show major differences in: (i) the percentage changes in the emissions of each group; and (ii) the percentage change in each group's emissions *vis-à-vis* the percentage change in global emissions.

More specifically, Table 8.9 reveals that global greenhouse gas emissions rose by an alarming 31.0 per cent over the 1990–2010 period, with a much larger increase occurring between 2000 and 2010 than in the preceding decade (+24.5 % compared to +5.2 %). Of particular significance is the fact that the 31.0 per cent increase in global emissions took place despite a 6.2 per cent fall in the greenhouse gases emitted by the Annex I group of Parties over the same period. This appears to support the widespread prediction that the Kyoto Protocol's confinement of emissions targets to Annex I nations would do little to prevent global emissions from rising above their 1990 levels.

As for the 6.2 per cent decrease in the Annex I group's emissions, it is clear from Table 8.9 that it was only made possible by the enormous decline in the greenhouse gases emitted by the Annex I (non-Annex II) group of Parties over the 1990–2010 period (−28.5 %). In direct contrast, and notwithstanding the 3.8 per cent decrease in emissions between 2000 and 2010, the Annex II group's emissions increased by 4.6 per cent between 1990 and 2010. What is also noteworthy is that the greenhouse gases emitted by the Annex I (non-Annex II) group fell by just 1.3 per cent between 2000 and 2010. As mentioned previously, the decline in this group's emissions after 1990 was principally due to the sharp fall in the real GDP of EITs in the early-1990s. Barring a severe global or regional GDP depression, it is highly unlikely that a decrease in emissions similar to that experienced by EITs over the 1990–2000 decade (−27.6 %) will occur again soon. This suggests that the cuts to greenhouse gas emissions required of the Annex I group in coming decades will necessitate a much greater reliance on policy measures aimed at reducing the emissions-intensity of economic activity.[121]

If one compares the emissions of the non-Annex I group of nations with that of the Annex I group, there is little doubt that stabilising the atmospheric concentration of greenhouse gases at no more than 450 ppm of CO_2-e will require a new climate change protocol that imposes emissions targets on all the world's nations. Furthermore, in view of the large rise in greenhouse gases generated by the non-Annex I group of nations between 2000 and 2010 (+48.8 %), a future protocol must include binding emissions targets on non-LDCs and, at least to begin with, the imposition of emissions-intensity targets on LDCs.

Last but not least, column *d* of Table 8.9 presents the per capita greenhouse gas emissions of the world as a whole and of each UNFCCC-defined group of Parties. Table 8.9 shows that global per capita emissions were in the order of 7.69 tonnes per person in 2010 (including LULUCF activities). It is worth remembering from Chap. 4 that achieving a 450 ppm stabilisation target will necessitate reductions in global per capita emissions to around 1.61 tonnes per person by 2050 and 0.96 tonnes per person by 2100 (see Table 4.1).[122] With this in mind, it goes without saying that the 2010 per capita emissions of the Annex I group (16.19 tonnes per person) far exceeded the current global average as well as the long-run sustainable average. This exceptionally large figure also puts into perspective the disproportionate quantity of per capita emissions generated by the world's wealthiest countries and the extent to which many of them will need to dramatically reduce their greenhouse gas emissions in coming decades to help achieve the 450 ppm stabilisation target—something that will no doubt continue to be an important negotiating issue at future COP meetings.

Within the Annex I group of Parties, Table 8.9 shows that the per capita emissions of the Annex I (non-Annex II) group (12.71 tonnes per person) was much lower than the Annex II group (18.02 tonnes per person). Although this suggests it will be easier for the Annex I (non-Annex II) group to reduce its greenhouse gas emissions, some countries in this group have a per capita GDP less than the optimum and therefore require some additional GDP growth.[123] The challenge for these countries will be how they can reduce their greenhouse gas emissions whilst increasing their per capita GDP to the optimal level. This challenge further underlines the urgency with which all countries must reduce the emissions-intensity of their economic activities.[124]

Turning now to the non-Annex I group of nations, of great concern is the group's 2010 per capita emissions, which, at 5.56 tonnes per person, was still considerably higher than the long-run sustainable average despite being lower than the global average. This concern is magnified by the fact that many countries in the non-Annex I group need to substantially augment their real output to raise their per capita GDP to the optimal level. Given this need, it will be necessary to allow the per capita emissions of many non-Annex I countries to remain near current levels for the remainder of the decade, and let the per capita emissions of LDCs creep higher for a decade or so. I say 'many non-Annex I countries' because it is highly likely that wealthy non-Annex I nations, such as Brunei, Israel, Kuwait, Singapore, and South Korea, will be promoted to the Annex I group of Parties and therefore be subject to strict emissions targets. Irrespective of the exact outcome, most of the emissions cuts over the next two decades will need to be made by the Annex I group—in particular, the countries currently in the Annex II group of Parties.

Notes

1. Even if some countries do not agree to a final protocol, it can still be effective if the total emissions of the nations willing to abide by the protocol constitute the great majority of the world's future greenhouse gas emissions.
2. Until recently, the group of least-developed countries (LDCs) consisted of 49 nations. In 2013, both Cape Verde and The Maldives graduated out of the group of LDCs, whereas South Sudan, as a new nation, was added. Equatorial Guinea and Vanuatu will also graduate out of the LDC group in the next few years.

3. The Global Environment Facility was established in 1991 by the World Bank as a $1 billion pilot programme to help protect the global environment and promote sustainable development. Projects funded by the Global Environment Facility were initially implemented by the United Nations Development Programme (UNDP), the United Nations Environment Programme (UNEP), and the World Bank (GEF 2011).
4. Raupach et al. (2007) have estimated that, between 1750 and 2004, UNFCCC-defined Annex I countries had contributed around 77 per cent of all energy-related CO_2 emissions. A more recent study indicates that Annex I nations had contributed 71 per cent of the world's cumulative energy-related CO_2 emissions between 1850 and 2010 (29 per cent by non-Annex I nations) (den Elzen et al. 2013). However, the same study suggests that the contribution made by Annex I nations is just 52 per cent when non-CO_2 and land-use CO_2 emissions are included (48 per cent by non-Annex I nations).
5. Annex I Parties of the UNFCCC are referred to as Annex B Parties in the Kyoto Protocol because they are listed in Annex B of the Protocol.
6. For obvious reasons, carbon dioxide is automatically expressed in terms of its CO_2-equivalent warming effect.
7. National emissions targets specified in the Kyoto Protocol exclude emissions generated by international aviation and shipping activities (UNFCCC 1998).
8. The Alliance of Small Island States includes a group of island nations highly vulnerable to the impact of warming-induced sea-level rise. During the Kyoto Protocol negotiations, the G-77 represented 133 low-GDP nations of which China, at the time, was an associate rather than a member (Depledge 2000; Dessai et al. 2001).
9. Five UNFCCC Parties had alternative base years to 1990. They were: (i) Bulgaria (1988); (ii) Hungary (average of 1985–1987); (iii) Poland (1988); (iv) Romania (1989); and (v) Slovenia (1986).
10. Initially, Canada committed to cutting its greenhouse gas emissions to 6 per cent below 1990 levels by 2012. However, in 2009, Canada's emissions were 17 per cent higher that in 1990. With Canada unlikely to achieve its Kyoto obligations and therefore facing significant penalties, it opted to withdraw from the Kyoto Protocol before the first commitment period concluded at the end of 2012 (Toronto Star 2011).
11. Table 8.1 (column *e*) indicates that Canada, Japan, the Russian Federation, New Zealand, and Turkey have no emissions target for the second Kyoto commitment period. As mentioned, Japan and the Russian Federation chose not to be subject to a new round of emissions obligations. New Zealand elected to set an economy-wide reduction under the UNFCCC, whilst Turkey was spared a greenhouse gas emissions target for the 2013–2020 period. As for Canada, it withdrew from the Kyoto Protocol in 2012. The USA (not included in Table 8.1) also has no emissions target for 2013–2020 because it is not a signatory to the Kyoto Protocol.
12. The economic basis behind the creation of a number of flexibility mechanisms is that the marginal cost of mitigation differs between countries (IPCC

2001d). Hence, by allowing Annex I countries to trade in emissions allowances (Assigned Amount Units) and to invest in emissions-reduction projects in other Annex I nations and/or non-Annex I nations (i.e., where the marginal cost of emissions reduction is much lower), it is possible to reduce the overall cost of reducing greenhouse gas emissions. This philosophy is similar to that explained in Chap. 7 with regard to different greenhouse gas-emitting firms (see Fig. 7.4).

13. Upon a Party transferring some of its Assigned Amount Units in accordance with the provisions of Article 17 relating to International Emissions Trading, the Assigned Amount Units shall be added to the Units allotted to the acquiring Party (Article 3.10). Conversely, the Assigned Amount Units shall be deducted from the Units allotted to the transferring Party (Article 3.11).
14. One AAU is equal to one tonne of CO_2-equivalent greenhouse gases. In any one year, an Annex I nation can typically emit millions of tonnes of CO_2-equivalent greenhouse gases. Hence, the assumption of annual emissions of 100 units or 100 tonnes of CO_2-equivalent greenhouse gases is made here for illustrative purposes only.
15. This latter situation has occurred in the case of Annex I nations with 'economies in transition' (EITs). In the aftermath of communism, the productive capacity of the former USSR and most Eastern European countries declined significantly. When Annex I nations were assigned AAUs under the Kyoto Protocol in the late-1990s, the AAUs were determined relative to their greenhouse gas emissions in 1990 (with a few minor exceptions; see Endnote # 9). 1990 coincided with the collapse of communism. Hence, the AAUs of many EITs far exceeded their capacity to generate greenhouse gas emissions. This left many EITs with a windfall of unused AAUs which they could sell to non-EITs within the Annex I group of nations (Carbon Trust 2009).
16. The $20 million cost to country *B* could exist in the form of the cost of reducing the emissions-intensity of its real output or, should it not do this, the value of the goods and services it must forego to limit its greenhouse gas emissions to 90 units.
17. The $40 million benefit to country *A* equals $90 million *less* $50 million; the $30 million benefit to country *B* equals $50 million *less* $20 million.
18. Just like one AAU, one CER is equal to one tonne of CO_2-equivalent greenhouse gases.
19. A Designated Operational Entity (DOE) is an independent auditor accredited by the CDM Executive Board to validate project proposals and verify whether approved CDM-projects have achieved greenhouse gas emissions reductions, as promised. Verification by a DOE is used to determine the quantity of CERs that should be issued to a project (https://cdm.unfccc.int/DOE/index.html).
20. To assist in the management of the Clean Development Mechanism, the Executive Board is also supported by various other panels and working groups. For more information on them, see (http://cdm.unfccc.iny/EB/governance.html).
21. To qualify as a CDM-project under Article 12 of the Kyoto Protocol the project must deliver "real, measurable, and long-term benefits related to the mitigation

of climate change" including "reductions in emissions that are additional to any that would occur in the absence of the certified project activity".

22. In other words, the additionality test prevents an Annex I nation from acquiring Certified Emissions Reduction units (CERs) from projects that would have taken place in non-Annex I countries regardless of the Clean Development Mechanism.
23. This is a requirement under Article 2 of the Kyoto Protocol.
24. Once again, the assumption of annual emissions of 100 units or 100 tonnes of CO_2-equivalent greenhouse gases is made for illustrative purposes only.
25. The annual cost of the CDM-project would constitute the initial and ongoing costs amortised over the lifetime of the project.
26. Let's assume that, in the current year, country *X*'s annual emissions are 35 units and the annual emissions of countries *A*, *B*, and *C* are 100 units each. The combined emissions of the four nations would be 335 units ($100 + 100 + 100 + 35$). With the emissions of countries *B* and *C* remaining at 100 units each, and the emissions of countries *A* and *X* expected to rise to 110 and 40 units respectively, the combined emissions for the upcoming year, should no CDM-project be undertaken, would be 350 units ($110 + 100 + 100 + 40$). By undertaking the CDM-project to reduce *X*'s upcoming emissions to 30 units in order to allow *A* to legally emit 110 units, the project does not prevent combined emissions rising over the upcoming year by 5 units to 340 units ($110 + 100 + 100 + 30$).
27. That is, the Clean Development Mechanism does not prevent 'carbon leakage' to countries not subject to emissions targets.
28. In the case of European Union countries, it would also depend on whether they had reached the limits imposed by the EU-ETS on their capacity to meet their emissions targets via investments in CDM-projects.
29. Restrictions on the ability to sell surplus AAUs could arise if the Annex I nations are European Union countries which have reached their permissible limits under the EU-ETS.
30. This was the number of CERs issued as of 1 June 2014. It has been estimated that around 5.3 billion CERs are likely to be issued between the beginning of 2014 and the end of 2020 (http://www.cdmpipeline.org/overview.htm).
31. One ERU is equal to one tonne of CO_2-equivalent greenhouse gases.
32. To qualify as a JI-project under Article 6 of the Kyoto Protocol the project must deliver "a reduction in emissions by sources, or an enhancement of removals by sinks, that is additional to any that would otherwise occur". Given the large technology gap between transition economies (EITs) and Annex I nations with established market economies, the Joint Implementation facility was created on the expectation that most JI-projects would be undertaken in EITs. As of late-2010, more than 200 JI projects had been initiated in fourteen host countries, the great majority of which had taken place in the Russian Federation and the Ukraine (Henson 2011).
33. If the country hosting the JI-project has a comfortable AAU surplus (i.e., its AAUs far exceed its likely emissions), it will have less concern over whether

the emissions reductions arising from the project are additional. This is because it is unlikely that the host country will need to undertake any action to make up for non-additional emissions. I'll have more to say about this in Chap. 9.

34. By eligible, I mean that the host nation meets all Joint Implementation eligibility requirements. In this situation, the host nation supervises the JI-project.
35. The independent auditors are referred to as Accredited Independent Entities (AIEs).
36. The assumption of annual emissions of 100 units or 100 tonnes of CO_2-equivalent greenhouse gases is again made for illustrative purposes.
37. The JISC consists of: (i) six Annex I countries—three of which must be countries undergoing the transition to a market economy (EITs); (ii) three non-Annex I countries; and (iii) one member from the group of small-island developing states (SIDS). To become a member of the JISC, a country must be nominated by a relevant constituency and then be elected by the Conference of the Parties serving as the meeting of the Parties to the Kyoto Protocol.
38. 824,098 ERUs had been issued to Track 1 projects; 25,386 ERUs to Track 2 projects; and 46 ERUs to 'programmes of activities' (PoAs).
39. Articles 3.3, 3.4, and 3.7 of the Kyoto Protocol also discourage activities in Annex I nations that lead to deforestation, which is a form of carbon removal.
40. One RMU equals one tonne of CO_2-equivalent greenhouse gases. RMUs can be traded under the Kyoto Protocol's International Emissions Trading (IET) system (Article 17).
41. A forestry project is only a net-sequester of greenhouse gases during the forest's 'growth' phase. Upon full maturity of a forest or timber plantation, it ceases to be a net-sequester of greenhouse gases. This is because, at full maturity, the greenhouse gases that a forest sequesters via regeneration are offset by the release of greenhouse gases caused by natural tree death. The forest thus becomes a steady-state store of carbon. In the hypothetical example used for Table 8.5, a ten-year growth period prior to full maturity is assumed. In reality, this may be many decades, although the annual rate of net sequestration is likely to decline as the forest nears full maturity, meaning that the quantity of RMUs generated by the project will eventually taper off.
42. Of course, as more afforestation and reforestation projects are undertaken, the marginal cost of remaining projects increases. At some point, the marginal cost of afforestation and reforestation must exceed the marginal cost of mitigation, in which case it makes more sense to abandon the implementation of new forestry-based sequestration projects and undertake additional strategies to reduce greenhouse gas emissions.
43. Eventually, soil carbon reaches a new equilibrium, upon which time additional carbon removals in soils cease. In the meantime, the estimated 3.3 Gigatonnes of annual carbon dioxide removals should be considered in light of: (i) the 53.2 Gigatonnes of CO_2-equivalent greenhouse gases that were emitted worldwide in 2010; and (ii) the 13.7 Gigatonnes of CO_2-equivalent greenhouse gases that would be emitted in 2050 if emissions cuts were in line with those

recommended by Anderson and Bows (2008) to stabilise the atmospheric concentration of greenhouse gases at 450 ppm of CO_2-e. In the second case, 3.3 Gigatonnes would constitute around one-quarter of total greenhouse gas emissions in 2050.

44. Nitrous oxide is also generated by fossil fuel combustion and industrial activities, such as the manufacture of nylon (Garnaut 2008).
45. These shares are based on estimates of global CO_2 emissions using the 'sectoral' approach.
46. In terms of the average annual rate of growth, the difference between the rate of increase in bunker-fuel emissions and total CO_2 emissions between 1990 and 2010 was 2.9 per cent and 1.9 per cent respectively.
47. The ICAO is a specialised agency of the United Nations assigned with the task of promoting the safe and orderly development of international civil aviation throughout the world. Addressing climate change forms a key element of the work of the ICAO's Environment Branch (UNFCCC 2013f). The ICAO has long been working on an emissions-trading system for international civil aviation (Garnaut 2008). The IMO is another specialised agency of the United Nations responsible for the safety and security of international shipping and the prevention of marine pollution by ships. The limitation of greenhouse gas emissions from international shipping forms a vital component of the work performed by the IMO's Marine Environment Division (UNFCCC 2013f).
48. For all participating countries, the EU baseline for aviation activities was set at 221.4 Megatonnes of CO_2 emissions.
49. The 193 member states of the United Nations are divided up into five regional groups. They are: (i) the African Group (54 member states); (ii) the Asia-Pacific Group (54 members); (iii) the Eastern European Group (23 member states); (iv) the Latin American and Caribbean Group (33 member states); and (v) the Western European and Others Group (28 member states plus one state (USA) which attends, as an observer, the meetings of this regional Group).
50. To participate in the Kyoto Protocol's flexibility mechanisms, an Annex I Party must meet, among other things, the following eligibility requirements (UNFCCC 2013o):

    - ratification of the Kyoto Protocol;
    - calculation of Assigned Amount Units (AAUs) in terms of tonnes of CO_2-equivalent greenhouse gas emissions;
    - installation of a national system for estimating emissions and removals of greenhouse gases within its own territory;
    - installation of a national registry to record and track the creation and movement of Emissions Reduction Units (ERUs), Certified Emission Reduction units (CERs), Assigned Amount Units (AAUs), and Removal Units (RMUs);
    - the annual reporting of information on all emissions and removals to the Kyoto secretariat.

51. The emissions credits can be any one or a combination of Assigned Amount Units (AAUs), Emissions Reduction Units (ERUs), Certified Emission Reduction units (CERs), or Removal Units (RMUs).
52. This means that a non-compliant Party will be allocated fewer emissions units than it otherwise would have received in a subsequent commitment period.
53. The equitable sharing of costs is based on the recognition of common but differentiated responsibilities as reflected by the respective capabilities of different nations.
54. It has been estimated that, in the energy sector alone, the additional investment required to stabilise the concentration of greenhouse gases at the level needed to prevent average global temperatures rising 2 °C above pre-industrial levels is US$910 billion per year until 2050 (IEA 2012b). Since much of the energy-sector infrastructure has public goods characteristics, governments will have to provide most of this funding. To put the funding requirements into perspective, the US$910 billion per year represents nearly three times the estimated US$337 billion of climate change finance used for mitigation purposes in 2012 (Climate Policy Initiative 2013).
55. Of these agents, bilateral and multilateral finance institutions play a central and unique role in directing funds from both public and private sources (Atteridge et al. 2009). See, also, Climate Policy Initiative (2013).
56. As things stand, the UNFCCC is probably the best place to locate a repository of this nature, although the repository would probably shift to a new international climate change agency, should one emerge in the future.
57. The Green Investment Scheme (GIS) is a means of promoting the environmental efficacy of transactions involving the buying and selling of surplus Assigned Amount Units (AAUs). The GIS operates by earmarking some of the funds generated by international emissions trading for use in environmentally-related projects. Each GIS project is set up by the seller of surplus AAUs and operates as a domestic scheme within their climate change policy framework, albeit the full operational details must ultimately be agreed upon on a bilateral basis between buyer and seller nations (Blyth and Baron 2003; Carbon Trust 2009; World Bank 2011).
58. The GEF Trust Fund is supported by resources committed every four years by donor nations through a formal replenishment process (Möhner and Klein 2007). Since 1994, the World Bank has served as the Trustee of the GEF Trust Fund, which requires the Bank to perform a fiduciary role and provide administrative services on behalf of the Global Environment Facility (GEF 2011).
59. The one exception was an allocation of funds to support a pilot adaptation project in Hungary—an Annex I nation.
60. A full list of the 26 projects funded under the SPA is provided in Annex I of GEF (2012). Of the US$658 million allocated to the 26 projects, US$49.3 million was provided directly from the GEF Trust Fund and a further US$608.7 million was made available from co-financed sources. US$0.7 million of the US$50 million allocated to the SPA was used for administrative and logistic purposes.

61. The Least Developed Countries Fund also supports other elements of the work programme designed to assist LDCs, such as the provision of training and the strengthening of the capacity of meteorological and hydrological services (Möhner and Klein 2007). The Fund is not accessible to other non-Annex I Parties.
62. The steps for the preparation of a NAPA include a synthesis of available information and: (i) a participatory assessment of the vulnerability of a nation to current climate variability; (ii) identification of key adaptation measures as well as the criteria for prioritising activities; and (iii) a selection of a prioritised short list of adaptation-related activities. The development of a NAPA also includes short profiles of projects and/or activities intended to address the urgent and immediate adaptation needs of a LDC Party (UNFCCC 2013i).
63. In partnership with the Global Environment Facility are the following ten implementing agencies: the African Development Bank (AfDB), the Asian Development Bank (ADB), the European Bank for Reconstruction and Development (EBRD), the Inter-American Development Bank (IDB), the International Fund for Agricultural Development (IFAD), the United Nations Development Programme (UNDP), the United Nations Environment Programme (UNEP), the Food and Agriculture Organization of the United Nations (FAO), the United Nations Industrial Development Organization (UNIDO), and the World Bank.
64. The additional costs are the difference between a 'baseline' scenario (i.e., one where development activities would be pursued in the absence of climate change) and an alternative GEF adaptation scenario (i.e., one where activities respond to the adverse impacts of climate change) (GEF 2006). Given the difficulties associated with making an *ex ante* estimation of the additional costs of adaptation, the Global Environment Facility applies a sliding scale for funding under the Least Developed Countries Fund. The scale serves as a proxy for estimating additional costs (see Möhner and Klein 2007, Table 1). Under the sliding scale formula, small projects receive proportionally more funding than large projects on the assumption that the former have a higher adaptation component. (GEF 2006).
65. To date, the Least Developed Countries Fund has financed 138 country-level projects, all with the aim of addressing the urgent and immediate adaptation needs of LDCs. See http://www.thegef.org/gef/LDCF (accessed 12 February 2014).
66. The last completed NAPA was submitted by Equatorial Guinea in November 2013. The only remaining NAPA to be finalised is that by South Sudan—a recent addition to the group of LDCs. The details of the completed NAPAs are available from https://unfccc.int/adaptation/workstreams/national_adaptation_programmes_of_action/items/4585.php.
67. The same clarification of additional costs was also sought with respect to the Least Developed Countries Fund given that a similar funding methodology is used.
68. A further aim of the Green Climate Fund is to set a 50 per cent floor on the adaptation allocation to highly vulnerable countries, in particular,

least-developed countries (LDCs), small-island developing states (SIDS), and African nations (GCF 2014).

69. Stakeholders are defined in the Green Climate Fund Governing instrument as private-sector actors, civil-society organisations, vulnerable groups, women, and Indigenous Peoples (GCF 2011).
70. In keeping with the country-driven nature of the Green Climate Fund, National Designated Authorities are to ensure that private-sector interests are aligned with national climate change policies (www.climatefundsupdate.org).
71. To recall from Chap. 6, since all sectors of the economy depend on the input of natural resources, all sectors are in some way climate-sensitive. In this context, 'climate-sensitive' implies the sectors of the economy that are directly rather than indirectly affected by climate change, such as the agricultural and resource-extractive industries that make up the primary sector of a national economy.
72. Since the great majority of non-Annex I nations ratified the Kyoto Protocol after the Adaptation Fund was established, it could be argued that the Fund has been successful in facilitating ratification of the Protocol.
73. Also exempt from the levy are small-scale afforestation and reforestation CDM-projects (UNFCC 2008a).
74. The 2 per cent levy does not apply to any subsequent transfer of the same AAUs.
75. This amounts to a cumulative total of US$394.8 million in the Adaptation Fund since its inception or US$412.9 million if outstanding pledges are eventually received.
76. The CER price averaged US$0.51 in 2013 (World Bank 2014a).
77. This figure does not include any future donations that would obviously boost the Adaptation Fund.
78. Of course, whether non-Annex I nations are issued with emissions allowances will depend entirely on the design of future protocols.
79. Although projects generated by the Green Investment Scheme can take various forms, they are generally divided into the category of 'hard greening' and 'soft greening'. The former involves activities that directly reduce emissions in ways that can be easily monitored and quantified (e.g., renewable energy projects and improvements in the energy efficiency of new and existing buildings). The latter involves capacity building to help increase the potential for future emissions reductions (Carbon, Trust 2009). Classifying the Green Investment Scheme in this way has also overcome the additional hurdle posed by EU State Aid legislation that, by prohibiting government funds from being used to fund many commercially-related activities, has the potential to restrict the nature of some emissions-reducing projects. It has been estimated that, between 2008 and 2012, around 458 million AAUs valued at approximately €1.6 billion (US$2.3 billion) were traded through the application of Green Investment Schemes (Tuerk et al. 2013).
80. It has been shown that the Green Investment Scheme can promote activities that the Clean Development Mechanism and the Joint Implementation

programme have great difficulty facilitating, such as investments in land use/bioenergy projects and increased energy efficiency of buildings and transport (Carbon Trust 2009).

81. To recall from Fig. 8.1, all European Union countries belong to the Annex I group of nations.
82. At the end of 2012, the Russian Federation possessed AAUs equivalent to around 5,800 MtCO_2-e of greenhouse gas emissions. This constituted 45 per cent of the estimated of surplus AAUs in existence at the end of the first Kyoto commitment period (equivalent to around 13,000 MtCO_2-e of greenhouse gas emissions).
83. This edict was designed to maintain the integrity of the emissions targets applied during the first Kyoto commitment period.
84. There is no equivalent restriction on the sale of surplus AAUs from the second Kyoto commitment period.
85. The amendments compel a country wishing to partake in the second Kyoto commitment period (2013–2020) to submit a greenhouse gas emissions target that is no greater than its average 2008–2010 emissions levels. The reason for doing this is to avoid the huge surpluses of AAUs that EITs enjoyed during the first commitment period—a policy aimed at preventing countries from accumulating so-called 'hot air' (Carbon Market Watch 2013).
    The formula used to cancel AAUs is as follows (Note: the average emissions levels of the 2008–2010 period are multiplied by eight because the second Kyoto commitment period is eight years long; AAU_{CP2} denotes the AAUs allotted for the second Kyoto commitment period; GHG denotes greenhouse gas emissions):

$$\text{AAUs cancelled} = [AAU_{\mathrm{CP2}} - (\text{ave. } 2008 - 2010\, GHG \times 8)]$$

    If, for example, the following applied to an Annex I nation ($AAU_{\mathrm{CP2}} = 800$ units and ave. 2008–2010 $GHG = 80$ units), then:

$$\begin{aligned}\text{AAUs cancelled} &= [800 - (80 \times 8)] \\ &= [800 - 640] = 160\end{aligned}$$

86. To recall, the CDM-projects undertaken in least-developed countries (LDCs) are exempt from the levy.
87. It has been estimated that a CDM-project can leverage new private-sector and public investment in the order of 6–8 times the amount of the initial funds used to finance the project (Ellis et al. 2004).
88. Canada and the USA are not eligible in any case because they are not signatories to the Kyoto Protocol.
89. Modification of the rules regarding automatic additionality was first approved at COP-16 in Cancún in 2010.
90. The amendment to the Regulation governing the registry infrastructure underpinning the EU-ETS was approved on 23 January 2013 and took effect later that year.

91. The value of the primary and secondary ERU markets in 2011 was US$339 million and US$780 million respectively (World Bank 2012).
92. Global adaptation support in 2012 was estimated at US$22 billion—much less than the US$337 billion allocated in support of mitigation projects (US$359 billion in total).
93. The concept of REDD+ was first proposed at the COP-11 meeting in 2005 as a policy mechanism to decelerate, stabilise, and reverse the loss the global forests. The concept began as an initiative to reduce greenhouse gas emissions from deforestation and forest degradation (REDD). It was later expanded to incorporate the enhancement of forest carbon stocks and the conservation and sustainable management of forests in developing countries (REDD+).
94. Result-based payments differ from conventional forms of up-front finance insofar as financial support is provided *ex post* pending the verified accomplishment of pre-defined outcomes (World Bank 2013).
95. The US$466.5 million includes pledges as well as committed funds (https://www.forestcarbonpartnership.org/sites/fcp/files/2014/June/FMT%20Note%20CFM-2014-1%20FCPF%20CF%20Budget%20FY15%20final.pdf).
96. Referred to in French as the Agence Française de Developpement.
97. Having said this, bilateral institutions seldom contribute finance to carbon funds through purchases of emissions units.
98. Despite the differences between the institutions, lending constitutes the greatest portion of participating institution's activities (Atteridge et al. 2009).
99. Although not strictly bilateral finance institutions, national development banks share similar characteristics. The main differences are the mandates of the two categories of institutions, their objectives, and the types of activities they support.
100. United Nations agencies and programmes come under the grouping of 'Development co-operation agencies' in Fig. 8.2.
101. It is also worth pointing out that the UNEP and UNDP perform the function of multilateral implementing agencies of the Global Environment Facility. Since 2002, both organisations have served as implementing entities of the Least Developed Countries Fund (LDCF) and Special Climate Change Fund (SCCF) (Atteridge et al. 2009). Moreover, as indicated earlier, the UNEP and UNDP serve as implementing agencies associated with the Green Climate Fund's 'direct access' mechanism (GCF 2011).
102. The OECD reporting system measures the climate change funding of bilateral finance institutions in two distinct ways. Firstly, it measures funding activities with respect to three OECD policy objectives, including 'aid to environment', which helps bilateral finance institutions assess their progress in achieving Millennium Development Goals. Secondly, it measures funding activities with respect to certain United Nations conventions—such as the UNFCCC—thus allowing bilateral finance institutions to judge their performance against so-called 'Rio Markers' (Atteridge et al. 2009).
103. Domestic government spending on other mitigation initiatives is not included in these figures (Climate Policy Initiative 2013).

104. The Clean Technology Fund is not specifically designed to deal with climate change, but has been set up by the US, UK, and Japanese Governments to accelerate the deployment of clean, energy-efficient technologies. Administered by the World Bank, it is hoped that the Fund will indirectly assist non-Annex I nations to reduce the greenhouse gas emissions-intensity of its economic activities (Garnaut 2008; Paulson et al. 2008).
105. A significant amount of this investment has been made possible by the Clean Development Mechanism (Stern 2007).
106. 83 per cent of household mitigation spending took place in industrialised nations (Climate Policy Initiative 2013).
107. These five elements are a variation and extension on the four elements presented in Aldy et al. (2003).
108. This assumes that global greenhouse gas emissions peak in 2015.
109. To obtain average values in column *b*, the values in column *a* are divided by 5.
110. Process-related emissions include greenhouse gas emissions from energy sources, industrial processes, solvent and other product use, waste, and sundry activities. In sum, they include net greenhouse gas emissions from all sources except land-use, land-use change, and forestry activities (LULUCF).
111. A more detailed explanation can be found in EEA (2013).
112. The weaknesses of the EU-ETS were outlined in Chap. 7.
113. The conversion takes place on the basis that one ETS-related emissions allowance equates to one AAU.
114. This simply means that, for countries not governed by an emissions-trading system, the quantity of 'verified ETS emissions' in Eq. (8.2) equals zero.
115. Although column *i* of Table 8.7 does not provide data on the use of the Kyoto flexibility mechanisms by the Russian Federation and the Ukraine, both Parties have been active sellers of excess AAUs and willing Joint Implementation participants.
116. To make up for the shortcomings, Denmark, Japan and Switzerland respectively require emissions units equal to 0.8 million, 17.3 million, and 0.2 million tonnes of CO_2-e per year. The surplus AAUs alone from the first Kyoto commitment period have been estimated at 13 Gigatonnes or 13,000 million tonnes of CO_2-e (World Bank 2013). Hence, there is no lack of emissions units available for purchase. To date, there is no indication that the Danish and Swiss Governments intend to purchase additional emissions units to achieve their first-round Kyoto targets. However, Japan has shown a keen interest to acquire a substantial quantity of emissions units from Ukraine, Hungary, Poland, the Russian Federation, and the Czech Republic in conjunction with the Green Investment Scheme (Stern 2007; Government of Japan 2008).
117. The purchase of emissions units would be large in terms of the percentage of Iceland's net emissions (47.3 per cent). In absolute terms, it would require permits equivalent to around 1.8 million tonnes of CO_2-e per year, which is well within the available surplus of emissions units and much less than what Japan needs to purchase to meet its first-round target of 17.3 million tonnes of CO_2-e per year.

118. At the time the manuscript for this book was submitted, there was no news on whether Iceland's emissions had significantly exceeded its targets or whether the Icelandic Government had taken action to address the problem.
119. The values in Table 8.9 include emissions/removals from LULUCF activities. Hence, they represent net greenhouse gas emissions.
120. To ensure comparable data on each of the UNFCCC-defined groups over the 1990–2010 period, the data on Annex I Parties used in Table 8.9 differs slightly to the data used to compile Table 8.7.
121. Having said this, it is worth noting that reducing greenhouse gas emissions would be made easier in future if the wealthiest Annex I countries reduced their real GDP in order to transition to an optimal macroeconomic scale. I'll have more to say on this in Chap. 10.
122. These values depend on the world's population peaking at 8.55 billion in 2060 and falling to 7.95 billion by 2100 (see Table 4.1). Should the world's population be higher than this, a more severe cut in per capita global emissions will be required to achieve the 450 ppm target.
123. In Chap. 4, it was assumed, based on the GPI studies of many countries, that the optimal scale of a national economy exists at a per capita GDP of Int$15,000 (2004 prices).
124. Reducing the emissions-intensity of economic activities would be akin to a nation increasing its real GDP/emissions ratio.

Chapter 9
Laying the Foundations to Support a New Global Climate Change Protocol

9.1 Introduction

Since the Kyoto Protocol was first established, considerable criticism has been directed towards its lack of explicit long-term goals, its inadequate coverage, its weak first-round emissions targets, and what many believe are its defective enforcement and flexibility mechanisms (Gupta and Bhandari 1999; Aldy et al. 2003; Corfee-Morlot and Höhne 2003; Grubb 2003; Atteridge et al. 2009; World Bank 2013). However, most national governments and commentators are less critical of the Protocol's architecture having recognised that, as a product of years of political, diplomatic, and intellectual endeavour, it would take enormous time and effort to supplant with no guarantee of a superior replacement (Michaelowa et al. 2005b).[1] For this reason, most observers believe that the existing Kyoto architecture should serve as the foundation upon which a new global protocol is established (Hamilton et al. 2005; Stern 2007, 2008).

As we saw in the previous chapter, the present Kyoto architecture makes a clear institutional distinction between Annex I (Annex B) and non-Annex I Parties in recognition of the different responsibilities and obligations that the world's nations have in resolving the climate change crisis. Despite this, I believe the urgency and extent to which greenhouse gas emissions must be reduced and the fact that a number of non-Annex I nations are now significant greenhouse gas emitters, demands a major modification to the existing architecture.[2] Although I'll have more to say about the modified architecture later in the chapter and again in Chap. 10, I will be upfront and contend that a more explicit institutional distinction needs to be made between the wealthy and poor members of the Annex I and non-Annex I groups of nations. I also believe that a number of wealthy non-Annex I nations should be included in the Annex I group of Parties once the second Kyoto commitment period concludes in 2020. All told, the existing architecture should be modified to create four new groups—they being: (i) an Annex II group

P. Lawn, *Resolving the Climate Change Crisis*, DOI 10.1007/978-94-017-7502-1_9

of Parties (much like the existing Annex II group); (ii) an Annex I (non-Annex II) group of Parties; (iii) a group consisting exclusively of LDCs; and (iv) an Annex III group of Parties, which would be similar to the existing non-Annex I group minus the LDCs.

Although, initially, Annex III countries would be subject to fewer and less stringent obligations than Annex I and Annex II nations, they would be required to make more demanding commitments than the group of LDCs. In this way, the modified architecture would accommodate the need for LDCs to rapidly grow their real GDP in order to experience much-needed increases in per capita economic welfare. It would also recognise the minimal contribution that LDCs have made to the elevated concentration of greenhouse gases and, thus, their greater claim to future allowable emissions.

For obvious reasons, it is crucial that all future growth in the per capita GDP of LDCs is as equitable and efficient as possible. This would be facilitated by the domestic implementation of the many policies outlined in Chap. 3 plus the transfer of funds and technology from the Annex II and Annex I (non-Annex II) groups of nations. It would also be promoted by requirements under a new protocol for LDCs to gradually reduce their reliance on fossil fuels and energy-intensive production methods along with further inducements tied to the transfer of funds and emissions-reducing technologies from richer nations. Eventually, all LDCs would be subject to binding emissions targets—ideally, as soon as 2031.

As for Annex III countries, they would be subject to binding emissions targets as of 2021, albeit the required emissions cuts would be far less stringent than those required of Annex I and Annex II nations. Indeed, whereas Annex I and Annex II countries would be required to all but de-carbonise their industrial production during the decades of 2041–2050 and 2051–2060 respectively, there would be no need for Annex III nations to do this until around 2070.[3] Coupled with the transfer of funds and new technologies from Annex I and Annex II nations, these lesser demands would allow Annex III countries to grow their per capita GDP, as required, before making the eventual transition to a steady-state economy. I should point out that Annex III nations would not receive the same level of funding and transfer of new technologies as the group of LDCs. Furthermore, the transfer of funds and new technologies to Annex III nations would cease much sooner.

In sum, despite all five elements of the Kyoto architecture continuing to exist in some form (see Sect. 8.2.9), the first two elements would be dramatically altered, whilst the remaining three would undergo minor changes. As we shall see, a new climate change architecture would include two additional elements—one to better account for the common but differentiated responsibilities of each nation; another as a means of equitably distributing national emissions rights within the safe global emissions budget (i.e., one that stabilises the atmospheric concentration of greenhouse gases at no more than 450 ppm of CO_2-e).

In order to continue laying the foundations to support a new global protocol (Chap. 10), I will spend most of this chapter outlining some of the views expressed by various commentators on the Kyoto Protocol and how they (and I) believe international climate change institutions and mechanisms can be greatly improved. Finally, I will assess some of the proposed approaches to the distribution of national emissions rights before arguing the case for a 'common but differentiated convergence' approach when setting national emissions targets.

9.2 Views on the Kyoto Institutions and Related Mechanisms

9.2.1 Mitigation—The Kyoto Protocol's Flexibility Mechanisms

9.2.1.1 International Emissions Trading (IET)

In principle, the Kyoto Protocol's system of International Emissions Trading is an extension of the emissions-trading concept described in Chap. 7. It can exist in the form of a single expanded system covering two or more countries or the linking of established national systems. For this reason, many of the recognised benefits of the Protocol's International Emissions Trading system are similar to those outlined in Chap. 7. To reiterate, the main benefits include the potential to achieve emissions targets without having to unduly sacrifice allocative efficiency, which increases the likelihood of achieving emissions targets at relatively low cost (Stern 2007; UNFCCC 2013o).

There are, however, many observers who have been very critical of the Kyoto Protocol's International Emissions Trading system. Some of the criticisms are based on objections to the concept of emissions trading, while others are based on their inappropriate design, not the least being the EU-ETS. Since I have already dealt with the concerns regarding the concept of emissions trading in Chap. 7, most of my attention here will focus on the latter apprehensions.

Perhaps the greatest criticism of the Kyoto Protocol's International Emissions Trading system relates to its present lack of coverage, both in terms of greenhouse gas-emitting activities and the number of participating countries (Torvanger and Ringius 2000; Corfee-Morlot and Höhne 2003; Stewart and Weiner 2003; Schmidt et al. 2006; Böhringer and Welsch 2006; Spash 2010). As pointed out in Chap. 8, the Protocol's International Emissions Trading system is confined to Annex I nations, which means it excludes a number of large greenhouse gas-emitting nations, such as China, India, and Indonesia.[4] Furthermore, aviation and shipping emissions (bunker-fuel emissions) plus various agricultural and LULUCF activities with the capacity to reduce greenhouse gas emissions and/or sequester greenhouse gases are not covered by the system. Apart from contributing to the failure of the Kyoto Protocol to prevent the rise in global greenhouse gas emissions, the system's lack of coverage diminishes the scope for achieving cost-effective emissions cuts—paradoxically, the very goal that the International Emissions Trading system seeks to accomplish.

A second major criticism of the Protocol's International Emissions Trading system is its inability to ensure the quantity of emissions permits issued reflects the emissions trajectories that must be met to achieve a safe atmospheric concentration of greenhouse gases (Garnaut 2008; Block 2008). Because of misgivings regarding the stringency of Kyoto emissions targets, some observers believe that the Protocol's trading system fails the 'environmental integrity' test. These same critics also point to the fact that the system's environmental integrity is dependent upon transparent monitoring and reporting that should strictly comply with standards and procedures accepted by the international community, especially by major permit buyers (Stern 2007; Garnaut 2008).[5] Given the disparity in the rules governing national emissions-trading systems and the consequent lack of system compatibility, critics question whether the Kyoto Protocol's International Emissions Trading system meets universal standards and co-ordinated emissions caps (Bramley et al. 2009; Heitzig 2012; Haita 2013). Moreover, they question whether the UNFCCC, with its enormous range of functions and responsibilities, is capable of ensuring that minimum standards, where they exist, are being adequately fulfilled. According to the Carbon Trust (2009), these concerns are supported by the near absence of intergovernmental emissions-trading systems and/or the linking of existing national systems, with the EU-ETS the only major exception.

Even if national emissions-trading systems were sufficiently harmonised to guarantee environmental integrity, a third major criticism of the International Emissions Trading system is the uncertainty surrounding the setting of emissions targets. For the trading system to be truly effective, investors in low-emissions technology must have reasonable knowledge of likely future emissions targets and be confident that the targets will be maintained into the future (Stern 2007). Yet the international process of setting emissions targets is at best ad hoc. Hence, there are no assurances over the long-term emissions targets that are likely to be embodied in future global protocols, nor any certainties regarding the extent and timing of future emissions cuts.

A fourth widespread criticism of the Kyoto Protocol's system of International Emissions Trading is that it does not protect vulnerable nations from the possibility of highly inequitable outcomes. There are four main ways that inequities can occur. Firstly, the initial distribution of tradeable emissions permits to participant nations can be unfair, either because it does not take account of disparities in per capita GDP (ability to pay) or because it overlooks historical emissions. Secondly, irrespective of whether a country becomes part of a large multilateral system or links its own national emissions-trading system with the systems of other nations, engaging in the international trading of emissions units reduces the control that a national government has over the permit price paid by domestically-located firms (Garnaut 2008). Although price variability, by reflecting changing demand and supply-side forces, can have positive efficiency benefits at the broader system level, a persistently high permit price can render emissions permits unaffordable to many entities operating in poorer participating nations. Without the ability to rapidly shift towards low-emissions production, this could make it difficult for poor nations to achieve their real GDP targets.

It is true that an international emissions-trading system can be designed to redistribute permit revenue in order to compensate poor nations having difficulty purchasing emission permits (Barnes et al. 2008). However, there is the danger that poor nations could become 'welfare-reliant' in the sense that a considerable portion of their consumption could depend on the redistribution of permit revenue rather than on domestic production. Over time, this could result in the loss of the productive capacity required by many poor nations to reduce the emissions-intensity of their production activities and raise their per capita GDP to the optimum level (Int$15,000 at 2004 prices).

Thirdly, the distribution of tradeable emissions rights in excess of the emissions requirements of some nations, such as the circumstances enjoyed by transition economies (EITs) during the first Kyoto commitment period, provides an opportunity for some countries to influence permit prices to the disadvantage of others (Maeda 2003). It also leaves open the possibility of cartels forming which can further disadvantage vulnerable nations.

Fourthly, as explained in Chap. 7, if tradeable emissions units can be accumulated (banked) for future use, as is possible with many types of units under the Kyoto Protocol, it can lead to speculative trading behaviour and the potential for wealthy corporations and entities to earn windfall profits at the expense of other agents engaged in the trading system. Furthermore, by distorting permit prices, speculation can reduce investment in low-emissions technologies. This, in turn, can undermine the goal of achieving emissions targets at lowest cost.

The final criticism of the Kyoto Protocol's International Emissions Trading system centres on its questionable ability to achieve least-cost outcomes. The criticism has arisen because some commentators see no evidence of the system supporting the government policies needed to overcome the many factors that impede the implementation of low-cost mitigation strategies, such as financial capital constraints, hidden costs, and the public goods feature of low-emissions technologies (Johnstone 2002; Menanteau et al. 2003; IPCC 2007d; Stern 2007, 2009). In fact, some commentators are concerned that the system is being employed as a substitute for additional regulatory measures. If so, there is the danger that the International Emissions Trading system will be viewed as a 'silver bullet' that eliminates the need for complementary forms of government intervention.

Whilst the Protocol's International Emissions Trading system has many other weaknesses, I believe the above criticisms are most in need of addressing if a new system is to effectively achieve emissions targets in an equitable and cost-effective manner. Notwithstanding this, it is worth stressing that an international emissions-trading system cannot threaten global emissions targets if the emissions-generating activities and the nations not covered by the system are bound by regulatory and enforcement mechanisms that guarantee emissions targets are met. What an inadequately designed emissions-trading system can still threaten is the ability to reduce the emissions-intensity of economic activity sufficiently to increase the real GWP/emissions ratio by a factor of 13.6 between 2010 and 2100. To recall from Chap. 4, a 13.6-factor increase in the real GWP/emissions ratio is necessary in a steady-state economic setting to reach an optimal per capita GWP of Int$15,000 without compromising the 450 ppm stabilisation target. Apart from my concern that no country is contemplating the transition to a steady-state economy, I fear that unless the above weaknesses are satisfactorily addressed, a future international emissions-trading framework will fail to drive the necessary reductions in the emissions intensity of global economic activity. It is also unlikely to achieve equitable outcomes both across and within nations.

One final point. Since, all nations and all activities must be subject to emissions caps to ensure an international emissions-trading system does not threaten global emissions targets, maximising the coverage of emissions targets should be a priority negotiating issue at the COP-21 meeting in Paris in 2015. It should remain one until full coverage is attained.

9.2.1.2 Clean Development Mechanism (CDM)

Despite the sizeable benefits generated by the Clean Development Mechanism, it is the most complex and controversial of the three Kyoto flexibility mechanisms (Carbon Trust 2009). Because of this, it has been subject to considerable and ongoing criticism. Perhaps the greatest criticism of the Clean Development Mechanism concerns the issue of 'additionality'—the need for a CDM-project to deliver reductions in greenhouse gas emissions over and above any that would have occurred in the project's absence (Lohman 2006; IPCC 2007d; Schneider 2007; Stern 2007, 2009; Garnaut 2008; Wara and Victor 2008; Carbon Trust 2009; Gillenwater and Seres 2011). As stressed earlier, proof of additionality is required to avoid the generation of fictitious CERs and ensure environmental integrity of the Clean Development Mechanism (Schneider 2007).

Although efforts have been made to ensure additionalty is at all times present, a number of critics have revealed deficiencies in the way in which the additionality of projects is assessed, both in terms of the methodologies and the eligibility criteria used in the evaluation process (Gillenwater and Seres 2011). In many of the cases identified, critics have shown that registered projects—especially renewable-energy projects—would have proceeded without CDM-accreditation (Carbon Trust 2009).[6] To make matters worse, the additionality problem has been exacerbated by: (i) concerns over the adequacy of staff numbers and resources to enable the CDM Executive Board to properly review and monitor projects; (ii) potential conflicts of interest between the Executive Board's different roles; and (iii) doubts over the integrity, expertise, and impartiality of some third-party auditors (Carbon Trust 2009; Gillenwater and Seres 2011).

In an effort to overcome these shortcomings, the resources allocated to the CDM Executive Board to administer the Clean Development Mechanism steadily increased from 40 staff members with an annual budget of US$11.2 million in 2006 to 177 staff members with an annual budget of US$39.7 million in 2011 (Gillenwater and Seres 2011; UNFCCC 2013c). More recently, the quantity of allocated resources has declined slightly. As of 2014, it stood at 154 staff members working with an annual budget of US32.9 million. Notwithstanding the recent falls, the overall increase in resources since 2006 has enabled the Executive Board to introduce more detailed project requirements and establish a separate registration team to carefully assess each project and highlight potential issues for the Board's review. This has markedly improved the Board's ability to evaluate the quality of submitted projects. In addition, the Board is now better placed to scrutinise third-party auditors—indeed, so much so, it has been able to endorse several auditing companies in recent years (Gillenwater and Seres 2011).

Despite these improvements, some critics still believe the quantity and quality of the resources allocated to administer the Clean Development Mechanism are inadequate and that too many roles remain concentrated in the hands of the CDM Executive Board. Moreover, they believe that not enough has been done to separate the governance of the Mechanism from executive project decisions and to establish an independent appeals procedure (Carbon Trust 2009).[7] In all, there is little doubt that more reform of the Executive Board and its crediting rules is needed to ensure the CERs generated by CDM-projects represent the realisation of emissions reductions beyond those arising from business-as-usual activities.

A second major criticism of the Clean Development Mechanism is one that has already been highlighted in Sect. 8.2.4.2—namely, the Mechanism does not prevent the rise in global greenhouse gas emissions. As previously explained, Annex I countries ostensibly acquire CERs to enable them to increase their emissions, especially when the cost of acquiring CERs is less than the cost of domestic mitigation.[8] Whilst the CERs generated by CDM-projects are able to offset the rising emissions of Annex I nations, they do nothing to put a lid on the aggregate emissions of non-Annex I nations, which, unlike Annex I countries, are not subject to emissions targets.[9] This predicament is aggravated by a number of additional shortcomings, of which the above-mentioned additionality problem is one of them. Other shortcomings include: (i) the poor monitoring and verification of emissions reductions (Carbon Trust 2009; Gillenwater and Seres 2011; Nature Code 2014); (ii) the leakage of emissions arising from the inadequate boundary setting of projects and the shift in emissions-intensive forms of production from Annex I to non-Annex I nations (IPCC 2007d; Gillenwater and Seres 2011; Henson 2011); and (iii) the winding down of some verification activities caused by the recent decline in CER prices and the subsequent reduction in the revenue needed to cover project-monitoring costs (World Bank 2013, 2014a).

The boundary-setting problem, which hasn't been discussed as yet, refers to greenhouse gas emissions (leakages) occurring outside a CDM-project that can be indirectly linked to a project's broader impacts and influences (UNFCCC 2006; IPCC 2007d). Examples of emissions leakages include the emissions caused by land clearance undertaken to offset the agricultural land lost following inundation by hydro-based CDM-projects. Others include renewable energy projects that require the emissions-intensive manufacture of solar panels, wind-turbines, and energy-storage batteries (Working Group on Baseline for CDM/JI Project 2001; Geres and Michaelowa 2002; Kartha et al. 2002; Gillenwater and Seres 2011). To ensure the CERs earned from CDM-projects accurately reflect net emissions savings, more needs to be done to include emissions leakages when assessing

CDM-projects. To achieve this, estimates need to be made of project boundaries to better account for emissions savings and any emissions indirectly brought about by individual CDM-projects (Gillenwater and Seres 2011).

How then, should the rising emissions of non-Annex I nations be quelled if they cannot be contained by the Clean Development Mechanism? In the end, containment can only be guaranteed by subjecting non-Annex I nations at some point to emissions targets. For this reason, a one-sided offset mechanism, such as the Clean Development Mechanism, should only operate whilst there are countries free of quantitative emissions obligations. Indeed, the Clean Development Mechanism should be gradually phased out as all countries become subject to emissions targets (Garnaut 2008; Stern 2009).[10] As we shall see in Chap. 10, the Clean Development Mechanism should eventually be subsumed by the expansion of the Joint Implementation facility, albeit the use of this latter facility is also likely to decline over time.

A third common criticism of the Clean Development Mechanism is that the quantity of CDM-projects and the international flow of funds generated by the Mechanism is insufficient to induce a radical shift towards renewable energy and low-emissions production methods in non-Annex I countries. The reason for this is that the need to demonstrate additionality on a largely case-by-case basis imposes significant *transaction costs* in terms of validating, verifying, and scrutinising projects. For example, when the Clean Development Mechanism was first introduced, the registration process took around ten months to complete (UNFCCC 2013p). Although the length of this process has fallen to around two months, transaction costs for individual projects regularly exceed US$0.5 million. In many cases, the sheer magnitude of these transaction costs is prohibitive, especially when combined with other cost factors, such as CDM-registration fees; long and/or delayed pay-back periods; learning costs associated with new technologies; and the 2 per cent levy charged on CERs to help capitalise the Kyoto Protocol's Adaptation Fund (Stern 2007; Carbon Trust 2009).[11] The projects most disadvantaged by large transaction costs tend to be small in scale or specific to the building, transport, and forestry industries. Given the potential for projects in these industries to deliver substantial low-cost emissions reductions, the lack of CDM-investments in these areas is of great concern to the critics of the Clean Development Mechanism (Carbon Trust 2009).

Of course, the often prohibitive nature of transaction costs has not gone unnoticed by the CDM Executive Board. The Board has endeavoured to lower transaction costs by streamlining the rules regarding eligibility and proof of additionality. To do this, the Board has maintained the existing baseline-and-credit methodology, but has moved from a strictly project-based system to a more wholesale approach. The Board has achieved this via three main reforms. Firstly, the Board has introduced a system of 'programmatic' qualification. As outlined in Chap. 8, the concept of a Programme of Activities (PoA) was introduced in 2005 as a means of bundling discrete projects into a single programme in order to reduce transaction costs and overcome other cost barriers. PoAs can also hasten the registration and crediting of projects that might be developed in the wake of newly-introduced emissions-reducing policies (Carbon Trust 2009).[12]

Secondly, the Board has instituted baseline benchmarks to standardise the criteria used to assess projects and programmes—an approach that many hope will reduce the risk of carbon leakage. Thirdly, as mentioned, the Board has established a register of project types that are automatically deemed as additional (Kachi et al. 2014). The register includes the following micro-scale projects: (i) renewable-energy projects (solar, wind, and wave) involving less than five Megawatts of installed capacity[13]; (ii) energy-efficiency projects delivering energy savings of no more than 20 Gigawatt-hours per year[14]; (iii) eligible project types reducing greenhouse gas emissions by less than 20 Kilotonnes per year[15]; and (iv) many 'first-of-a-kind' projects (Shishlov and Bellassen 2012; Buen 2013).

Also available to a non-Annex I nation is the opportunity for its Designated National Authority (DNA)[16] to submit for approval a 'positive' list of micro-scale renewable energy projects that it believes are additional—in particular, activities that would not proceed without Annex I nation investment via the Clean Development Mechanism (Kachi et al. 2014). To date, the CDM Executive Board has approved the positive lists submitted by nine countries (UNFCCC 2013c).[17]

As important as these reforms are, they are unlikely to facilitate any massive expansion in the number of sanctioned CDM-projects, as evidenced by the relatively small number of PoAs registered[18] since programmatic qualification was introduced plus the pitiful approval of just six standardised baselines (UNFCCC 2013c).[19] Not that this should be a cause for concern. The regulatory complexity and high level of scrutiny needed to maintain performance standards virtually guarantee moderately high transaction costs. What's more, since it is impossible to apply standardised baselines everywhere, project-by-project assessment is likely to remain dominant in many industries/sectors.

In view of this stark reality, should persistently high transaction costs matter? I don't believe so. In fact, I believe it would be a mistake to streamline eligibility rules to increase the attractiveness of the Clean Development Mechanism if it undermines the Mechanism's environmental integrity. Given the inalienable need to achieve a safe atmospheric concentration of greenhouse gases, environmental integrity must remain paramount. Moreover, while a moderately successful Clean Development Mechanism may not have the resource-transferring impact that many of its supporters would like, there are numerous ways in which rich countries can help non-Annex I nations begin the transition to renewable-energy and low-emissions technologies. In sum, sacrificing some cost-effectiveness to ensure the Clean Development Mechanism is environmentally sound is a small price to pay to achieve ecologically sustainable outcomes.

The importance of ecological sustainability leads to us the fourth major criticism of the Clean Development Mechanism. Contrary to requirements under Article 2 of the Kyoto Protocol, critics question whether the Mechanism fulfils its objective of assisting host-nations to achieve sustainable development. Very few critics deny the Clean Development Mechanism's record of delivering significant economic and social benefits to non-Annex I countries. However, they point to a lack of procedures within the Mechanism to ensure the benefits of CDM-projects are equitably distributed and to prevent project activities from depleting natural capital stocks (Schneider 2007; Gillenwater and Seres 2011; Shishlov and Bellassen 2012).

Heightening the concern of critics is the fact that the criteria used by Designated National Authorities (DNAs) to assess the sustainable development contribution of CDM-projects vary greatly from nation to nation. In addition, whilst the criteria often include a wide variety of economic, social, and environmental principles, the social and environmental benchmarks tend be weaker than the benchmarks applied to economic factors (Shishlov and Bellassen 2012). Furthermore, CDM-projects need only advance some of the sustainable development principles, and, in many cases, just one (Schneider 2007). On top of this, the assessment of the environmental and social impacts of projects has yet to become common practice in many non-Annex I countries (Nature Code 2014). Thus, it is not unusual for a CDM-project to be approved because it satisfies employment and poverty-alleviation concerns even though it may undermine environmental principles and/or impose spillover costs on adjacent communities.

In an effort to improve the sustainable development outcomes of CDM-projects, the CDM-registration process requires all relevant stakeholders to be adequately consulted. Unfortunately, project-initiators and DNAs lack clear guidance on how to conduct and validate stakeholder consultations, which can be blamed on poorly defined, regulated, and documented consultation requirements. For example, it is commonplace for communities that might be adversely affected by a CDM-project to be ill-informed of the project and its potential impacts (Nature Code 2014). Where they are informed, comments are often invited from a select group of stakeholders (Schneider 2007). Furthermore, there are few avenues available for stakeholders to raise concerns once a project is registered and operating. In other words, the Clean Development Mechanism includes no effective appeals procedure or grievance mechanism (Nature Code 2014).

The evidence regarding the sustainable development contribution of CDM-projects is scant. A number of researchers have closely reviewed hundreds of projects and discovered many examples where harm has been inflicted on external parties and/or local and adjacent ecosystems. They have also found that very few projects address the immediate needs of the poor and that a large number of renewable-energy projects in rural areas disproportionately benefit rich farmers and urban populations (Michaelowa and Michaelowa 2007; Olsen 2007; Sutter and Parreño 2007; Olsen and Fenhann 2008; Boyd et al. 2009; Shishlov and Bellassen 2012; Nature Code 2014). Even an assessment by the UNFCCC could only find 151 of 2,250 registered projects that had improved and/or protected natural resources and 10 projects that had passed the UNFCCC's poverty alleviation test (UNFCCC 2011a).

To what extent the Clean Development Mechanism meets the sustainable development goal is likely to remain a moot point if only because a final answer depends on the chosen assessment criteria. This aside, there is no doubt that the modalities and procedures of the Clean Development Mechanism need to be reformed to improve the Mechanism's consultation process, including the DNA's response to stakeholder feedback. This can be achieved by: (i) establishing a formal communication channel between local stakeholders and DNAs; (ii) providing project-initiators and DNAs with best-practice guidelines on how to conduct stakeholder consultation; (iii) creating a framework to modify projects and redirect project benefits to alleviate stakeholder concerns; and (iv) establishing an effective grievance mechanism to enable stakeholders to appeal the decisions made by the CDM Executive Board.

Also urgently required is an appropriate and comprehensive assessment criteria to assist DNAs to better evaluate the sustainable development contribution of CDM-projects. By appropriate, I mean a standardised assessment criteria based on objective and quantifiable indicators, much like the sustainability precepts outlined in Chap. 2. Establishing a framework of this nature is a task that could be achieved at annual COP meetings. It is also important that safeguards are introduced to avoid projects that violate some of the sustainability principles embodied in the assessment criteria (CDM Policy Dialogue 2012). Where violation occurs, a project should only proceed if measures are taken to overcome the project's deficiencies, such as redistributing some of the income generated by the project to compensate adversely affected third parties or the establishment of new assets to replace the assets lost as a result of the project's activities (e.g., restoration of natural capital stocks).

In terms of the economic assessment of CDM-projects, there is a desperate need to incorporate the 'user cost' principle referred to in Chap. 3. Inclusion of this principle ought not to be controversial. Economists have recommended its inclusion—in effect, the cost of capital depreciation—for some time (Keynes 1936; Daly 1996). However, traditionally, only human-made capital has been deemed relevant in the calculation of a project's user cost. Yet, as explained in Chap. 3, to satisfy the condition of 'strong sustainability', the user cost also needs to account for the depletion of natural capital.[20] This is not being done with respect to CDM-projects. If accounting for the cost of natural capital depletion were to become a standard CDM-assessment practice, it would automatically expunge many environmentally-destructive projects from the list of projects requiring further appraisal. Moreover, it would bring into contention many environmentally-sound projects that investors would otherwise overlook. In sum, full user cost accounting would improve the quality of CDM-projects and increase the overall contribution they make towards sustainable development in the host countries.

Irrespective of the above reforms, there is always the possibility that emissions reductions or carbon sequestration arising from CDM-projects may eventually be reversed. This is particularly so in the case of afforestation and reforestation projects which became eligible CDM-projects subsequent to the Marrakesh Accords in 2001. For example, CO_2 sequestered by a cultivated tree plantation or restored native forest can be released into the atmosphere following its harvesting or destruction by fire, disease, or insect infestation (Gillenwater and Seres 2011). It is because of the potential impermanence of a project that a fifth major criticism of the Clean Development Mechanism has surfaced—specifically, the Mechanism cannot prevent the reversal of emissions reductions that are designed to offset excessive emissions in Annex I nations.

To date, the UNFCCC has adopted two approaches to deal with the potential non-permanence of forestry-sequestered greenhouse gases. Firstly, it has issued two classes of non-permanent CERs—'temporary' CERs (*t*CER) and 'long' CERs (*l*CER) (IPCC 2007d). Temporary CERs are emissions units which expire at the end of the commitment period that immediately follows the period during which the emissions units were issued. This means that a *t*CER issued during the first Kyoto commitment period (2008–2012) would expire at the conclusion of the second Kyoto commitment period (December 2020). Somewhat differently, long CERs expire at the end of a project's crediting period (Fair Climate Network 2014).[21] Whether a forestry-related project earns *t*CERs or *l*CERs depends on the nature of the project and the likely permanence of the sequestered greenhouse gases. Despite the fact that temporary CERs can only be issued following verified sequestration from approved CDM-projects, the European Union (EU) has prohibited the exchange of *t*CERs and *l*CERs under the EU-ETS.

It is worth noting that the nation in possession of temporary CERs, not the host-nation, is liable to replace the units when they expire or when they are annulled due to the loss of associated carbon-storing forests from logging, clearance, or natural destruction (IPCC 2007d). In order to replace the expired or annulled temporary CERs, the relevant Annex I nation must cancel an equivalent number of emissions allowances, which may be any one or a combination of AAUs, RMUs, CERs, ERUs, and temporary CERs (Fair Climate Network 2014).[22] Notwithstanding this requirement, if, through re-verification, it can be demonstrated that the carbon-storing forests/plantations have remained intact, temporary CERs are re-issued. By offsetting the cancelled emissions allowances, the re-issued units are able to restore the Annex I nation's emissions quota.

The second approach employed by the UNFCCC to deal with potential non-permanence involves the imposition of strict qualifying rules regarding the land on which a forestry-related project can take place. CDM-projects only qualify as eligible *afforestation* projects if the area in question has been devoid of tree-cover for a minimum of 50 years. To qualify as an eligible *reforestation* project, the area must have been devoid of tree-cover since at least 31 December 1989 (http://www.fairclimate.com/about_cdm/tutorial/lcers.aspx). These restrictions are designed to discourage land-owners from clearing land for the purpose of undertaking afforestation or reforestation projects in order to earn valuable *t*CERs and/or *l*CERs.

Because of these qualification restrictions, temporary CERs trade at much lower prices than their permanent counterparts (Carbon Trust 2009).[23] Whilst many accept that the UNFCCC's two approaches plus the EU's refusal to accept temporary CERs have adequately addressed the non-permanence issue, they also believe that the low prices for temporary CERs are responsible for the extremely poor take-up of forestry-related CDM-projects—a major concern given the low-cost nature of many forestry activities and their long-term capacity to sequester carbon. Furthermore, since deforestation accounts for nearly 20 per cent of global greenhouse gas emissions, the Clean Development Mechanism has been further criticised because it does not issue CERs as a means of rewarding and therefore incentivising 'avoided deforestation'.[24]

Given the potential positive contribution of forestry-related activities, a number of proposals have emerged to deal with and prevent reversals. These include: (i) directing a portion of the *t*CERs and *l*CERs generated by projects at risk of reversals into a collective insurance pool; (ii) having project-investors purchase assurance bonds sufficient in value to replace any *t*CERs and/or *l*CERs that might be lost as a consequence of reversals; (iii) introducing compulsory project-insurance, whereby insurance-providers must replace confiscated *t*CERs and/or *l*CERs arising from reversals on behalf of insurance policy-holders (i.e., project-investors)[25]; and (iv) compelling project-investors to establish a replacement reserve—sometimes referred to as a 'shadow-project'—to cover potential project failures (Pearce and Turner 1990; Baalman and Schlamadinger 2008; Carbon Trust 2009; Gillenwater and Seres 2011).

Great care needs to be taken when designing ways to harness the potential benefits of forestry-related projects and avoided deforestation. I believe the current practice of issuing non-permanent CERs should be retained, as should the policy of not issuing CERs for avoided deforestation. I also see no reason to impose an outright ban on the trade in forestry-related CERs, although there is considerable merit in limiting the quantity of CERs that can be traded in order to encourage greater domestic mitigation.

Of the alternative proposals put forward to deal with sequestration reversals, I believe the concept of project-insurance is best. However, insurance-providers should only be required to replace confiscated *t*CERs and/or *l*CERs if it can be demonstrated that the possessors of annulled emissions units *are not responsible* for the project's failure (e.g., if reversals result from unavoidable forest destruction by natural causes).[26] This qualification is needed to ensure the generators or possessors of *t*CERs and/or *l*CERs have an incentive to prevent reversals. Whilst the establishment of shadow-projects is worthy of consideration, this approach is best confined to CDM-projects with the potential to damage or destroy existing forests. It should not apply to forests or plantations cultivated through the Clean Development Mechanism (more on this soon).

As for the claim that the low prices of non-permanent CERs are the principle cause for the low take-up of forestry-related CDM-projects, there are three points worth making. Firstly, should this claim be true, it is a further example of the price being paid to ensure the Clean Development Mechanism remains environmentally sound. Secondly, upon more stringent global emissions targets being introduced, I would expect the prices of all types of emissions units to be considerably higher than they are at present. I would also expect prices to rise for some considerable time (see Figs. 6.13 and 7.9). Higher future prices would undoubtedly increase the attractiveness of all offset projects, including forestry-related CDM-projects. Thirdly, if the framework used to assess the sustainable development contribution of CDM-projects was improved—in particular, if the user cost principle was included in the economic evaluation process—it would dramatically increase the viability of forestry-related projects relative to alternative projects. This would further boost the attractiveness of forestry-related CDM-projects.

Finally, deforestation in non-Annex I nations should not be addressed through the Clean Development Mechanism. The problem should be resolved through the establishment of conservation areas (e.g., National Parks); legislation to control land clearance; and payments for ecosystem services. As explained in Chap. 3, offering payments to maintain the ecosystem services provided by forests and other forms of native vegetation is a just means of compensating parties no longer able to exploit the resources contained within newly-established conservation zones (Stern 2007). Such payments also serve as a way of rewarding/compensating land-owners who, through choice or through regulation, are unable to use portions of their land for agricultural and other purposes. Thus, the latter would involve paying land-owners to incorporate carbon sequestration (carbon farming) in their portfolio of income-generating activities.[27]

It is also worth noting that encouraging land-owners to preserve native vegetation without penalising deforestation could lead to the perverse situation where land-owners intentionally clear vegetation and subsequently establish a timber plantation or allow the vegetation to regrow to gain valuable emissions units (Stern 2007).[28] In other cases, the failure to penalise deforestation and compensate land-owners for having vegetated land taken out of potential economic production has already led to the conversion of forests and associated ecosystems into mono-cultured crops. A prime example of this is the practice of clearing huge areas of rainforest to produce bio-fuels and the establishment of commercially lucrative palm-oil plantations in countries like Indonesia and Malaysia (Fitzherbert et al. 2008; Butler and Laurance 2009; Swain 2014).

I will have more to say about how to finance the payments for maintaining ecosystem services soon. Suffice to say now, the use of the FCPF Carbon Fund to achieve emissions reductions through REDD+ activities is a good example of how payments for ecosystem services can be employed at an international level to help reduce deforestation. Expansion of the Carbon Fund or a similar fund is urgently required to further slow the global rate of deforestation, as is the need to adopt a similar funding approach within nations to discourage native vegetation clearance.

The final major criticism I wish to focus on is the limited application of the Clean Development Mechanism in the world's poorest nations. This has raised serious concerns about its distributional ramifications (Jung 2006; Atteridge et al. 2009; Shishlov and Bellassen 2012). As of late-2014, 49.6 per cent of all registered CDM-projects had taken place in China; 20.3 per cent in India; 4.4 per cent in Brazil; 3.3 per cent in Vietnam; 2.5 per cent in Mexico; but only 2.4 per cent throughout the whole of Africa, and just 1.9 per cent in Sub-Saharan Africa (1.2 per cent if South Africa is excluded) (UNFCCC 2014d). Hence, whilst the use of the Clean Development Mechanism has resulted in a healthy transfer of resources—including low-emissions technologies—to a small number of non-Annex I nations, the majority of needy countries have missed out, especially those in Africa. The main reason for this imbalance is the greater mitigation potential in the over-represented countries and their more favourable investment climate—the latter due to their superior political and institutional stability (Jung 2006; Shishlov and Bellassen 2012).

Concerted efforts to overturn this imbalance have long been underway. The UNFCCC together with the CDM Executive Board have attempted to increase the number of CDM-projects in poor countries by establishing CDM Regional Collaboration Centres in four strategic locations: (i) Lomé, Togo, in partnership with BanqeUest Africaine de Devéloppment; (ii) Kampala, Uganda, in conjunction with the East African Development Bank (EADB); (iii) St George's, Grenada, in partnership with the Windward Islands Research and Education Foundation; and (iv) Bogota, Colombia, supported by the Banco de desarollo de Amèrica Latina. Establishment of a fifth Regional Collaboration Centre is planned for Asia (UNFCCC 2013c).

The dominant aim of Regional Collaboration Centres is to increase the project attractiveness of under-represented countries by building capabilities within their borders—that is, by overcoming the barriers caused by a lack of resources and technical expertise—and by reducing risks for potential investors. The Centres endeavour to achieve this by assisting poor countries and investors to initially develop CDM-projects (UNFCCC 2013c). Having completed this first step, the Centres usher the projects through the full CDM-project cycle. Other important functions of the Regional Collaboration Centres include the setting of standardised baselines to streamline the registration process; supporting local Designated National Authorities (DNAs); creating investor-community partnerships; and establishing a pipeline of future CDM-projects (UNFCCC 2013c).

Given the meagre increase to date in the number of CDM-projects in many poor countries, the UNFCCC has tried to stimulate the process further by implementing a CDM Loan Scheme. Established in 2012, the Scheme provides interest-free loans for projects undertaken in least-developed countries (LDCs) and in non-Annex I countries hosting fewer than ten registered CDM-projects.[29] The Scheme is jointly operated by the UNFCCC, the United Nations Environment Programme (UNEP), and the United Nations Office for Project Services (UNOPS) (World Bank 2013; UNFCCC 2013c). It will be some time before a judgement can be made of the success or otherwise of the Regional Collaboration Centres and the CDM Loan Scheme.

There are, nonetheless, some observers who believe that the under-representation problem will, in due course, be naturally resolved by existing PoAs (UNFCCC 2013c; Buen 2013). The reasoning for this is twofold. Firstly, the PoAs registered in Africa constitute 32.4 per cent of the current total—a figure much larger than that pertaining to single projects registered in Africa (2.4 per cent). Secondly, assuming that the 32.4 per cent figure can be maintained or improved, any increase in the scalability of the Clean Development Mechanism facilitated by PoAs, which many expect in coming years, should automatically boost the percentage of all projects taking place in under-represented countries.

As plausible as this seems, I don't share the same level of optimism, largely because, for reasons given earlier, I don't believe there will be sufficient growth in the number of registered PoAs. Then again, since one of the aims of the Clean Development Mechanism is to limit the growth in greenhouse gas emissions in non-Annex I nations, this shouldn't significantly matter provided alternative approaches can reduce the greenhouse gas emissions of the world's poorest countries without impeding their capacity to grow and later stabilise their per capita GDP at the optimum of Int$15,000 per person (at 2004 prices). I am much more optimistic about the likely success of the alternative measures despite acknowledging that their success will depend on the world's high-GDP countries beginning the transition to a steady-state economy in the very near future.

9.2.1.3 Joint Implementation (JI)

As alluded to above, expansion of the Joint Implementation facility will be necessary to fill the gap left by the cessation of the Clean Development Mechanism once all nations become subject to greenhouse gas emissions targets. To promote the Joint Implementation facility and ensure maximum gains from its application, it will be necessary to overcome its shortcomings whilst strengthening its benefit-yielding features. Apart from fostering the transfer of emissions-reducing technologies and facilitating the cost-effective realisation of emissions targets, many additional benefits of the Joint Implementation facility have been identified, including some that were never envisaged at the time of its conception.

One of the unexpected benefits of the Joint Implementation facility has been the manner in which it has operated as a 'frontier mechanism'. By revealing and, in some instances, generating new information on greenhouse gas emissions, the facility has activated the search function of the private sector by providing firms with novel opportunities to test new technologies and estimate mitigation costs. This has enabled many firms to cost-effectively improve their greenhouse gas-emitting performance. In addition, the facility has led to advances in the methods used to compile national greenhouse gas inventories and has improved the setting of emissions benchmarks (Shishlov et al. 2012).

On the other side of the ledger, the Joint Implementation facility contains many weaknesses that urgently need addressing. For starters, the Joint Implementation process is cumbersome and beset with institutional obstacles (Korppoo 2005; IPCC 2007d). These hurdles have deterred investment in projects with the potential to both dramatically reduce greenhouse gas emissions and yield substantial economic benefits (Carbon Trust 2009). Secondly, where verification of JI-projects must be undertaken via the Track 2 procedure, the obligatory oversight by the Joint Implementation Supervisory Committee (JISC) has proven to be time-consuming and costly. Thirdly, the freedom granted to countries to stipulate their own assessment guidelines under the Track 1 option has resulted in large discrepancies in the stringency with which JI-projects have been approved and ERUs have been issued (Karousakis 2006). Finally, there is a severe shortage of accredited auditors—so-called Accredited Independent Entities (AIEs)—to properly assess JI-projects via the Track 2 procedure (Shishlov et al. 2012).[30]

In recent years, the JISC has sought to overcome these complications by simplifying the accreditation process and aligning it with the Clean Development Mechanism. It has done this by streamlining the assessment rules and allowing project applications to be submitted in the form of a JI-programme of activities (JPAs). Much like PoAs with respect to the Clean Development Mechanism, JPAs involve the aggregation of individual JI-projects into a single programme in order to reduce transaction costs and harness the benefits of economies of scale (World Bank 2013).[31] As for concerns about the Track 1 and Track 2 verification procedures, they have been allayed to some degree by the increasing willingness of Annex I nations to align their assessment approaches with the JISC's Track 2 procedure and to employ AIEs to assess projects and verify emissions reductions.

Notwithstanding these developments, efforts have long been underway to merge the two procedural tracks in order to create a simpler and more effective assessment and verification process. As indicated previously, included in this reform proposal is a plan to strengthen the accreditation role of AIEs and establish a new governing body to: (i) oversee the Joint Implementation facility; (ii) set out mandatory JI-accreditation standards; and (iii) ensure the rules applying to the facility are correctly applied (World Bank 2013).

There are three additional shortcomings of the Joint Implementation facility worth mentioning. The first of these is the risk of double-counting greenhouse gas emissions reductions—a problem that can emerge when the Joint Implementation facility operates in tandem with a domestic or regional emissions-trading system, such as the EU-ETS. Double-counting can occur either directly or indirectly. *Direct* double-counting transpires when a JI-project is undertaken by a firm covered by an emissions-trading system that frees up some of the emissions allowances it possesses within the system. *Indirect* double-counting occurs when a JI-project generating ERUs reduces the emissions of another firm covered by an emissions-trading system (Jung et al. 2008). An example of the latter is where a renewable-energy project reduces the demand for electricity from an electricity supplier that is already operating within the confines of an emissions-trading system (Shishlov et al. 2012). In this situation, the JI-project simultaneously earns the investor firm ERUs and, by reducing the emissions generated by the existing electricity supplier, liberates some of the emissions allowances that the latter possesses within the emissions-trading system cap.

To help prevent direct double-counting, the European Union has mandated that all ERUs issued by firms covered by the EU-ETS must be matched by the cancellation of an equal quantity of EU emissions allowances—referred to as EUAs. As for indirect double-counting, the problem was averted for a short time through the creation by some EU-ETS countries of a reserve of EUAs, which were cancelled on a *quid pro quo* basis as ERUs were issued to JI-investors. This policy also played an important role in successfully facilitating investment in renewable energy-related JI-projects, particularly in countries lacking a system of financial incentives (e.g. feed-in tariffs) to promote such investment (Mizerny 2011).[32] However, recent modifications to the rules governing the EU-ETS have brought an end to this practice. This is because the rule changes meant that, as of the beginning of Phase III of the EU-ETS (2013–2020), projects which indirectly reduced the greenhouse gas emissions covered by the EU-ETS could no longer receive ERUs, which obviated the need for countries to create EUA reserves (Shishlov et al. 2012). Whilst these rule changes have eliminated the risk of indirect double-counting, they have stifled investment in renewable energy projects in countries relying heavily on the Joint Implementation facility to increase the deployment of renewable-energy technologies.[33]

It is my belief that the proposal to merge the Track 1 and Track 2 procedures, which includes the plan to create a new body to oversee the Joint Implementation facility, can solve the double-counting problem without deterring appropriate forms of investment. Nevertheless, success will require the rules governing the facility's operation to include a directive requiring ERU issuance to be matched by the forfeiting of AAUs or equivalent allowances covered by an emissions-trading system.

The next shortcoming of the Joint Implementation facility worthy of mention concerns the discontinuity of the JI-process that arises when the emissions allowances for upcoming commitment periods have yet to be determined and issued, or when Annex I nations refrain from participating in a new commitment period (e.g., the second Kyoto commitment period). As should be clear from the above, avoiding direct double-counting requires AAUs or equivalent allowances to be cancelled each time ERUs are issued. The problem with this is that if emissions allowances for future commitment periods have yet to be determined, which is what happened towards the end of the first Kyoto commitment period, it is impossible for Annex I nations to issue ERUs.[34] The same incapacitation also applies if an Annex I nation does not commit itself to greenhouse gas emissions targets. This is because a commitment is necessary for a country to receive AAUs.

During the early stages of 2013, the delayed allotment of AAUs for the second Kyoto commitment period interrupted the issuance of ERUs, thus creating an issuance 'gap'. With the Russian Federation opting out of the second commitment period and other ERU providers, such as the Ukraine, Belarus, and Kazakhstan threatening to do likewise, significant investor uncertainty emerged regarding the cash-flows from ERU sales needed to part-finance new JI-projects and support existing JI-projects. This situation, and the potential for it to occur again, raised serious questions about the future role of the Joint Implementation facility. Indeed, such was the decline in the use of the facility, the JISC expressed concern that the knowledge and institutional capacity it had built up over many years was in danger of being lost (UNFCCC 2014e).

To help restore investor confidence, the JISC has refocused its efforts on promoting the long-term development of the Joint Implementation facility. In particular, the JISC has encouraged Parties to expedite the issuance, transfer, and acquisition of ERUs for the second Kyoto commitment period in order to address the current 'gap' problem. It has also urged Parties to allow JI-host countries with targets for the second commitment-period to receive advanced allotments of AAUs, which would allow them to issue ERUs (World Bank 2013). As of October 2014, approval of advanced allotments had yet to be granted by the UNFCCC Parties (UNFCCC 2014e).

Given the need to expand the Joint Implementation facility as the Clean Development Mechanism is phased out (i.e., as all countries become subject to emissions targets), it is critical for future climate change protocols to include a timetable, even if it is provisional, with future commitment periods and emissions targets explicated well before their commencement. This would facilitate the advanced allotment of AAUs, which would enable ERUs to be seamlessly issued and the Joint Implementation facility to function without interruption. As we shall see, the global climate change protocol to be revealed in Chap. 10 will spell out a timetable that includes emissions targets and other obligations that, among other things, would prevent the undermining of the Joint Implementation facility.

The final shortcoming of the Joint Implementation facility worth raising concerns the issue of additionality. It was pointed out in Chap. 8 that the potential for bogus additionality claims is less of a problem with the Joint Implementation facility than the Clean Development Mechanism because host-nations, obliged to forfeit AAUs when issuing ERUs, are generally more vigilant to ensure emissions reductions are additional to what would have occurred in the absence of a JI-project.[35] Although vigilance is commonly practiced, there are three ways in which non-additional projects can illegitimately pass the additionality test. Firstly, as explained above, the assessment process under the Track 1 verification procedure differs from country to country. Until there is a standardised assessment approach, or until the two tracks are merged, there is always the possibility that some non-additional JI-projects will gain assessment approval. Secondly, much like the CDM Executive Board, the JISC is resource poor. It therefore has difficulty performing its oversight role—in particular, ensuring countries meet the eligibility criteria to employ the Track 1 verification procedure.

Thirdly, the incentive for host countries using the Track 1 procedure to ensure projects satisfy the additionality test depends largely on whether their AAUs are fully committed. To explain why, consider the situation depicted in Table 9.1. Imagine there are two Annex I nations—country *A* and country *B*—both with an emissions allowance (AAUs) of 100 units per year.[36] Imagine, also, that the projected annual emissions of *A* and *B* are 120 and 70 units respectively. Hence, while country *A* is expected to exceed its emissions allowance by 20 units, country *B* is expected to be left with 30 surplus AAUs.

In order to meet its emissions target, country *A* invests in a series of JI-projects in country *B*, which reduces *B*'s annual greenhouse gas emissions by 20 units.[37] Accordingly, 20 of country *B*'s AAUs are converted to ERUs and subsequently

Table 9.1 Track 1 verification (joint implementation) and the stringency of the additionality test—surplus AAUs

Group	Country	GHG allowance (AAUs)	Projected GHG emissions	Projected GHGs relative to target	Adjusted allowance (AAUs + ERUs)	Eventual GHG emissions	GHGs relative to adjusted target
Annex I	*A*	100	120	+20	120	120	0
Annex I	*B*	100	70	−30	80	55	−25

issued to country *A*. This increases *A*'s annual emissions allowance to 120 units and reduces *B*'s allowance to 80 units per year.

Imagine that country *B* has been negligent in its assessment of the JI-projects it plans to host and only 15 units of the projected emissions reductions are genuinely additional (i.e., 5 units of emissions cuts would have occurred anyway). In these circumstances, country *B*'s annual emissions would not decrease from 70 to 50 units as they would if all emissions reductions generated by the JI-projects were additional. They would only decline to 55 units per year. Nonetheless, *B*'s annual emissions would remain 25 units below its adjusted emissions allowance. Since country *B* would be at no risk of Kyoto non-compliance, it would have no need to make up the AAU deficit of 5 units in order to meet its emissions target.

Mind you, country *B*'s leniency would still come at a small cost. The cost would equate to the value of the 5 fewer AAUs that country *B* can sell each year to Annex I nations by engaging in the International Emissions Trading system. Having said this, for the arrangement with country *A* to be worthwhile, the cost to country *B* would almost certainly be exceeded by the benefits associated with hosting the JI-projects, such as the receipt of emissions-reducing technologies and the creation of additional employment and income-generating activities. Hence, the cost of country *B*'s negligence would do little to encourage it to be more vigilant when subjecting JI-projects to the additionality test.

Compare, now, the above situation with one where the AAUs of a host-nation are fully committed (see Table 9.2). Let's assume that the circumstances are the same as above except that country *B*'s projected annual emissions are 100 units per year—exactly equal to its maximum emissions allowance. In order to meet its emissions target, country *A* again invests in a series of JI-projects in country *B*, which results in 20 of country *B*'s AAUs being converted to ERUs and transferred to country *A*.

Just like the previous situation, we shall assume that country *B* is negligent in its assessment of the JI-projects and only 15 units of the projected emissions reductions are genuinely additional. As a consequence, *B*'s annual emissions fall to 85 units, not 80 units per year. The problem confronting country *B* on this occasion is that, barring further action, its greenhouse gas emissions will exceed its maximum emissions allowance by 5 units. Country *B* must therefore undertake additional mitigation measures to limit its annual emissions to 80 units per year (i.e., to remain Kyoto compliant). This is likely to be very costly. Presumably, the

Table 9.2 Track 1 verification (joint implementation) and the stringency of the additionality test—fully committed AAUs

Group	Country	GHG allowance (AAUs)	Projected GHG emissions	Projected GHGs relative to target	Adjusted allowance (AAUs + ERUs)	Eventual GHG emissions	GHGs relative to adjusted target
Annex I	*A*	100	120	+20	120	120	0
Annex I	*B*	100	100	0	80	85 (−5)	0

cost would be significant enough to induce country *B* to strictly apply the additionality test in the first instance.

All told, where a host-nation's AAUs are fully committed, it is unlikely to be lenient when applying the additionality test to JI-projects.[38] Conversely, if a host-nation has a comfortable AAU surplus, it has little incentive to strictly apply the additionality test. Worse still, as Shishlov et al. (2012) have highlighted, a country in the latter situation is also likely to be less ambitious when determining benchmarks for baseline setting under the Track 1 verification procedure. The ramifications of this are significant because negligence or leniency on the part of host-nations leads to higher than anticipated aggregate emissions by the group of Annex I nations. Whilst this need not result in Kyoto non-compliance, it calls into question the environmental integrity of the Joint Implementation facility insofar as it makes it difficult to know how much Annex I nations are collectively reducing their greenhouse gas emissions. This has the potential to distort the setting of future emissions targets.

Ultimately, I believe this last weakness heightens the need and urgency to merge the Track 1 and Track 2 procedures and align the JI-accreditation process with that of Clean Development Mechanism. In fact, I see no reason why the institutional capacity and experience built up within the Clean Development Mechanism could not be transferred to the Joint Implementation facility as the former institution is phased out and replaced by the latter. Although greater stringency could increase project transaction costs, it would no doubt improve the environmental integrity of the Joint Implementation facility, in which case the elevated cost would constitute a further but worthy price to pay to achieve a safe atmospheric concentration of greenhouse gases. However, rather than impede the Joint Implementation facility, a more stringent accreditation process would free up more AAUs for potential conversion to ERUs. Combined with the greater certainty that a reformed facility would deliver, I believe the number of legitimate JI-projects would increase, which would promote the cost-effective attainment of greenhouse gas emissions targets.

Before moving on, it is important to highlight the difference in the kind of emissions allowances currently issued for forestry projects under the Joint Implementation facility compared to the Clean Development Mechanism. To recall, the potential for sequestration reversals from forest losses has prompted the UNFCCC to issue temporary CERs (*t*CERs and *l*CERs) for forestry-related CDM-projects. Moreover, upon the expiration or cancellation of temporary CERs, the UNFCCC has made it the responsibility of the nation in possession of the temporary CERs to cancel an equivalent number of emissions units. The reason for doing this is because the host-nation of a CDM-project will be a non-Annex I country free of emissions targets. Hence, there is no requirement on the part of the host-nation to acquit emissions units, which is necessary to offset the increase in emissions allowances initially brought about by the issuance of the temporary CERs. By holding the possessor of the temporary CERs responsible for cancelling the emissions allowances, this accomplishes the compensating role not performed by the non-Annex I nation.

Conversely, with forestry-related JI-projects, there is no need to issue temporary ERUs because the host-nation is an Annex I nation subject to emissions targets. It is therefore responsible for emissions within its own borders and must cancel an equivalent number of emissions units to offset any reversals resulting from the failure of a JI-project.

By the way, the lack of any requirement to issue temporary emissions allowances applies equally to the RMUs generated by domestic forestry-related projects. Since projects of this type can only generate RMUs if they occur within Annex I nations, the responsibility to cancel emission units following a project failure again lies with the country where the project took place. Once more, this cancellation of emissions units has the effect of offsetting sequestration reversals. The current policy of issuing temporary CERs, but not temporary ERUs and RMUs, should be retained in a new global climate change protocol.

9.2.2 Mitigation—The Kyoto Protocol's Treatment of Land Use, Land-Use Change, and Forestry (LULUCF)

Much has been said about the Kyoto Protocol's treatment of land use, land-use change, and forestry (LULUCF) activities and how these activities should be treated in future global protocols (see IPCC 2007d; Benndorf et al. 2007; Schlamadinger et al. 2007; Stern 2007, 2009; Garnaut 2008; Ellison et al. 2014). The close attention paid to these activities can be attributed to three key factors highlighted in Chap. 8: (i) emissions reductions and greenhouse gas removals from improved agricultural and forestry-related activities can be realised at relatively low cost; (ii) deforestation and agriculture respectively account for around 20 and 15 per cent of global greenhouse gas emissions; and (iii) sustainable forestry and agricultural practices generate many co-benefits, such as watershed preservation, soil erosion control, biodiversity conservation, and the maintenance of critical ecosystem services.

On the positive side, most observers acknowledge the importance of requiring Annex I nations to include the emissions and removals associated with LULUCF activities when reporting greenhouse gas inventories for compliance purposes. However, considerable criticism has been directed at the inadequate coverage of LULUCF emissions by the Kyoto flexibility mechanisms and its potential to undermine efforts to achieve cost-effective forms of mitigation. Additional criticism has also centred on the lack of commitment to build the institutional and technical capacity needed to adequately monitor the greenhouse gas emissions and removals pertaining to *individual* activities, which is an essential precondition for any workable incorporation of agricultural and forestry activities into emissions-trading and emissions-offset systems.

Provided the monitoring issue can be dealt with, many believe that agriculture, forestry, and other land-use activities should be fully incorporated into the entire range of flexibility mechanisms to drive the net sequestration of greenhouse gases within a comprehensive greenhouse gas accounting and monitoring framework (Stern 2007, 2009; Garnaut 2008; Ellison et al. 2014). I wholeheartedly agree with these sentiments. However, the different impacts of agriculture and forestry on greenhouse gas emissions and removals, together with their important ecological ramifications, demand that both categories of land-use activities be governed by a distinct set of rules and regulations both within and external to the flexibility mechanisms. Some of these regulations have already been outlined in earlier chapters and previous sub-sections in this chapter. By taking account of the reforms proposed by a number of climate change analysts, I will endeavour to summarise the regulations that should form part of a future global protocol plus the national policies required to complement them.

9.2.2.1 Afforestation, Reforestation, and Avoided Deforestation

Let me begin by reiterating some of the Kyoto rules regarding the treatment of forestry-related activities:

- Domestic forestry-related projects that remove greenhouse gases *by sinks* can earn Removal Units (RMUs), which can be used for compliance purposes and be traded on international carbon markets.
- Forestry-related projects undertaken through the Clean Development Mechanism and the Joint Implementation facility can respectively earn Certified Emission Reduction units (CERs) and Emission Reduction Units (ERUs), but only if they involve greenhouse gas removals *by sinks*. Like RMUs, the acquired CERs and ERUs can be used for compliance purposes and be traded on international carbon markets.
- Forestry-related projects that reduce emissions *by sources* do not qualify as eligible CDM/JI-activities and cannot, therefore, earn CERs or ERUs.
- Where forestry-related removals of greenhouse gases are potentially reversible, two types of non-permanent CERs may be issued—'temporary' CERs (*t*CER) and 'long' CERs (*l*CER).
- Afforestation activities undertaken in non-Annex I countries can only qualify as eligible CDM-projects if the land earmarked for use has been devoid of tree-cover for a minimum of 50 years. Reforestation activities undertaken in non-Annex I nations can only qualify as eligible CDM-projects if the land in question has been devoid of tree-cover since at least 31 December 1989.
- Annex I nations not subject to Kyoto targets are precluded from engaging in the Kyoto flexibility mechanisms. Hence, these countries are unable to earn CERs, ERUs, and RMUs from forestry-related activities.
- Avoided deforestation does not attract CERs, ERUs, and RMUs.

On top of these regulations, the European Union has imposed further restrictions on its member countries. These restrictions, which affect the number of forestry-related activities occurring worldwide, include:

- CERs generated from forestry-related CDM-projects registered after 31 December 2012 are only eligible for trading under Phase III of the EU-ETS if the projects are hosted by Least-Developed Countries (LDCs) or non-Annex I nations with bilateral agreements with the European Union.
- ERUs cannot be issued in cases where forestry-based projects are hosted in European Union nations directly or indirectly related to activities covered by the EU-ETS Phase II after 31 December 2012.
- ERUs cannot be issued in cases where forestry-based projects are related to activities covered by Phase III of the EU-ETS after 30 April 2013.
- ERUs generated from forestry-related JI-projects which do not qualify for EU-ETS registration (i.e., ERUs held by EU-member countries not participating in the second Kyoto commitment period that were generated prior to 31 December 2012) cannot be used for compliance purposes.
- Non-permanent CERs generated from forestry-related CDM-projects (i.e., *t*CERs and *l*CERs) cannot be used for compliance purposes or be exchanged under the EU-ETS.

There are many observers who are critical of these rules and regulations, although there are different opinions on how they should be reformed. The suggested changes vary from dilution to outright deletion. Whilst I believe the Kyoto rules should be retained in a future global climate change protocol, various reforms should be implemented to increase the coverage of the Kyoto flexibility mechanisms and improve their environmental integrity and cost-effectiveness. As for the European Union, it should endeavour to link its EU-ETS with the emissions-trading systems of other Annex I nations, as they emerge. It should also rid the EU-ETS of some of the above-described restrictions on the issuance and trading of emissions units provided the changes do not compromise the environmental integrity of the system. Finally, it should overturn the outright banning of non-permanent CERs.

The first of the reforms to the Kyoto rules would allow a specific category of forestry activities that reduce emissions by sources to qualify as CDM/JI-projects. The activities in question would be instances where a project-investor assumes control of an existing forestry operation and employs more sensitive logging practices and/or management techniques. The CERs and ERUs generated by the project would equate to the reduction in the carbon-storing vegetation lost from harvesting a given quantity of timber. Upon issuance, the CERs and ERUs would be eligible for compliance purposes and be tradeable on international carbon markets.

To prevent investors claiming CERs and ERUs from logging operations that would not have otherwise occurred, the activities must be confined to forests already being exploited for timber or forests earmarked for (sustainable) exploitation prior to a specific date.[39] The need for such a restriction is also necessary to ensure CERs and ERUs are not issued for ‘avoided’ deforestation. Because of the manner in which the CERs and ERUs would be generated, this reform would improve the management of forests in host-nations, which would increase the *growth efficiency* and *exploitative efficiency* of natural capital in these countries (Ratios 3 and 4 in Chap. 2). This is of particular importance in non-Annex I nations where forestry practices are invariably sub-standard.

A second reform would involve removing the requirements regarding the eligibility of the land used for afforestation and reforestation projects under the Clean Development Mechanism—i.e., that the land must be devoid of tree-cover for a minimum of 50 years to qualify as an afforestation project, and no later than 31 December 1989 to qualify as a reforestation project. Having said this, because the restriction is commendably aimed at deterring land-owners from clearing native vegetation for forestry purposes, its non-application in a host-nation should be contingent upon the nation in question introducing other key reforms and national policies governing forestry activities—in particular, land-clearance controls (more on these reforms soon).

Thirdly, to encourage afforestation, reforestation, and help eliminate deforestation, the Kyoto Protocol's International Emissions Trading system should be modified via the introduction of a couple of major reforms. Where feasible, there should be full coverage of all types of activities within the system, including afforestation and reforestation activities. Likewise, all national and regional/group emissions-trading systems should be expanded to encompass forestry-related activities. This means that emissions credits earned from forestry projects, such as RMUs, CERs, ERUs, and temporary units should be eligible for trading. Once again, certain conditions should be applied to safeguard the integrity of the system as well as promote equitable and cost-effective outcomes. For example, to encourage greater levels of domestic mitigation and to limit the quantity of emissions credits generated from forestry-related CDM-projects in countries with no emissions targets, the number of CERs and *t*CERs that can be traded and used for compliance purposes should be restricted—a qualification that could be easily agreed upon by participating Parties and varied in line with changing mitigation responses.

Of course, before forestry can be incorporated into an emissions-trading system, it is necessary to resolve a number of issues regarding the measurement and monitoring of greenhouse gas emissions and removals pertaining to specific forestry activities (Garnaut 2008). This constitutes a great challenge. For not only does the carbon content and sequestration potential of forests vary according to the density, age, and species of the standing trees, the detection of forest loss and forest degradation often requires the application of sophisticated ground-based and remote-sensing techniques (Stern 2007). Fortunately, the cost of advanced sensing methods is falling rapidly. This is likely to make the detection of forest loss for compliance purposes increasingly viable for wealthy nations already in possession of strong and stable compliance-focused institutions. The same, however, cannot be said of a large number of non-Annex I nations, many with the greatest potential to reduce net emissions via forestry-related activities. As these poorer countries become subject to emissions targets and obligations, it will be incumbent upon high-GDP nations to help them establish the institutional and technical capacity to estimate and monitor emissions and removals from individual forestry-related activities. More specifically, future global protocols will need to embody greater commitments from high-GDP nations to transfer funds to poor countries, including the establishment of transfer mechanisms that go well beyond the present reliance on voluntary contributions from Annex I nations (e.g., the Green Climate Fund) and the 2 per cent levy on the full range of Kyoto emissions units.

At the same time, all new forestry-related initiatives incorporated in a new global climate change protocol should be aligned and integrated into national or regional development programmes, with nation-based policies developed in collaboration with the governments of major forest-possessing countries (Stern 2009). This implies that, on top of the funds currently provided by UNFCCC-financing mechanisms and multilateral and bilateral finance institutions, the success of forestry-related initiatives in low-GDP nations will largely depend on the adequate funding of development programmes. For this to happen, high-GDP countries should immediately deliver on past foreign-aid commitments and promised overseas development assistance (Stern 2009).[40]

Funding in support of forestry-related initiatives should not, I believe, stop here. It would be particularly beneficial if the world's richest nations funded the creation of an international body to randomly scrutinise forestry-related activities. This would pressure national authorities to accurately monitor and assess activities as well as identify institutional and capacity shortcomings at the national level. Given the importance of forestry in reducing greenhouse gas emissions, the body would ideally operate under the auspices of the UNFCCC, much like the CDM Executive Board operates under the guidance of UNFCCC-affiliated Parties.[41]

I have already referred to possible measures to address deforestation in my discussion on reforms to the Clean Development Mechanism. Most of these measures, which should be supplemental to emissions-trading and emissions-offset systems, need to be implemented at the national level (e.g., land-clearance controls and the establishment of conservation reserves). Nonetheless, they should also be integrated into the development programmes of low-GDP countries, not only because deforestation is driven by many development-related forces, but because an overarching global strategy is necessary to prevent deforestation shifting from one location to another (Stern 2009).[42]

For obvious reasons, addressing deforestation will require the close monitoring and detection of lost vegetation. To meet this end, the institutional and technical capacity that is needed to promote afforestation and reforestation activities should also be employed to reduce and ultimately eliminate deforestation. Once again, efforts to prevent deforestation in low-GDP countries should be supported by financial transfers from the world's richest nations.[43] In particular, a portion of the funds raised through new transfer mechanisms embodied in a future global protocol should be directed into national Environmental Trust Funds (see Chap. 3) and subsequently used to provide adequate compensation payments to land-owners deprived of bringing land into economic production (e.g., payments to maintain ecosystem services). Some of the funds should also be used to help establish replacement economic activities (e.g., tourism) or retrain affected land-owners to enable them, if required, to obtain alternative forms of employment.

What if a country does not have legislation in place to control vegetation clearance? There are two further ways to address deforestation. Both entail creating an economic disincentive to clear land. The first approach involves requiring land-owners to acquire and acquit emissions units upon releasing greenhouse gases through the clearance of native vegetation—a requirement that should also extend to the harvesting of native forests and cultivated timber plantations (Garnaut 2008). Although, with emissions caps in place, this approach would prevent a nation's total emissions exceeding its targets, it would not halt deforestation altogether because it would still be possible to generate emissions from vegetation clearance whilst winding back other emissions-generating activities.

The second approach involves nullifying the negative impact of economic activities on native vegetation by requiring culpable business-operators to establish replacement forests and/or cultivated plantations (shadow-projects).[44] There are, however, a couple of things worth noting in relation to shadow-projects. Firstly, such a requirement need not apply to projects or activities that initially sequester greenhouse gases because, upon any future vegetation losses caused by the same projects or activities, the emissions units originally created from the successful sequestration would be forfeited. To demand a shadow-project on top of cancelling the emissions units would amount to a double-penalty.[45] Secondly, shadow-projects only succeed in a truly holistic sense if the replacement forest or plantation is able to restore the ecosystem services previously provided by the defunct vegetation. If this cannot be achieved, as is common when an old-growth forest is logged or damaged, the shadow-project will fail to avert the loss of natural capital services. Indeed, while the amount of greenhouse gas sequestering vegetation would remain unchanged, the area of 'forest'—in the sense of mature, native, biodiverse ecosystems—would decline. From a strictly ecological perspective, this would amount to deforestation.

All up, since none of these approaches guarantee avoided deforestation, they should be viewed as last-resort measures in cases where there are no legislative controls on vegetation clearance. The bottom-line remains that the equitable elimination of deforestation requires a prohibition on vegetation clearance reinforced by adequate institutional capacity and complemented by appropriate compensation for land-owners.

9.2.2.2 Agriculture

The Kyoto rules governing agriculture are similar but critically different to those pertaining to forestry-related activities. The rules so far mentioned can be summarised as the following:

- Domestic agricultural activities that remove greenhouse gases *by sinks* can earn Removal Units (RMUs), which can be traded on international carbon markets.

- Agricultural projects that reduce emissions *by sources* or remove greenhouse gases *by sinks* do not qualify as eligible CDM-activities. They cannot, therefore, earn CERs.
- Agricultural projects undertaken through the Joint Implementation facility can earn Emission Reduction Units (ERUs), but only if they involve greenhouse gas removals *by sinks*. The acquired ERUs can be traded on international carbon markets.
- Agricultural projects that reduce emissions *by sources* do not qualify as eligible JI-activities and cannot earn ERUs.
- Annex I nations not subject to Kyoto targets are unable to earn ERUs and RMUs from agricultural activities.

As with forestry, the European Union has imposed further restrictions on its member countries in relation to agricultural activities. These include:

- In the case of the Joint Implementation facility, ERUs cannot be issued where agricultural projects involving greenhouse gas removals *by sinks* are covered by Phase III of the EU-ETS after 30 April 2013 (Note: this restriction is irrelevant with regards to agricultural projects that reduce emissions *by sources* given that the Kyoto rules already preclude the issuance of ERUs to such projects).
- ERUs generated from agricultural JI-projects which do not qualify for EU-ETS registration (i.e., ERUs held by EU-member countries not participating in the second Kyoto commitment period that were generated prior to 31 December 2012) cannot be used for compliance purposes.

As with forestry, there are divergent views on how these rules and regulations should be reformed. I believe that all but the second and the fourth Kyoto rules should be retained in a future global protocol. Once again, reforms should be introduced to fully incorporate agriculture into the Kyoto flexibility mechanisms and enhance the environmental integrity and cost-effectiveness of each mechanism. Likewise, the European Union should remove the above restrictions in instances where it would not undermine the environmental integrity of the EU-ETS.

Many of the necessary reforms to the Kyoto rules in relation to agriculture follow on from the reforms with regards to forestry-related activities. One of the most important reforms would involve: (i) allowing agricultural activities that reduce emissions by sources or remove greenhouse gases by sinks to qualify as eligible CDM-projects; and (ii) allowing agricultural activities that reduce emissions by sources to qualify as eligible JI-projects. The CERs and ERUs generated in both cases should, subject to previously described restrictions, be tradeable on international markets.

Because agricultural removals of greenhouse gases are potentially reversible, the second reform would involve the issuance of temporary CERs or *t*CERs for all removals resulting from agricultural activities undertaken through the Clean Development Mechanism. New *t*CERs should be re-issued to cover a new crediting period provided the accumulated soil-carbon has not been lost through soil erosion, unsustainable land-management practices, or natural factors.

Of course, like forestry projects, reversals related to agricultural activities can occur because of factors beyond a land-owner's control, such as droughts and floods. To deal with these circumstances and encourage land-owners to undertake carbon-sequestering farming practices, insurance schemes are needed to compensate those whose emissions allowances are confiscated because of unforeseen reversals, much like the insurance scheme concept recommended earlier for forestry-related reversals.

Thirdly, the Kyoto Protocol's International Emissions Trading system should be amended to include all agricultural activities. Wherever feasible, agriculture should be covered by national and regional/group emissions-trading systems. Furthermore, all emissions units generated from agricultural activities should be eligible for trading, with restrictions only in place to safeguard the environmental integrity of the system and/or promote greater levels of domestic mitigation—the latter catered for by the same limitation recommended in relation to forestry.

For obvious reasons, once agriculture is included in an emissions-trading system, it will be necessary for farmers to acquire and subsequently acquit emissions units upon releasing greenhouse gases through their agricultural activities. There are, however, two major impediments that need to be overcome before agriculture can be realistically included in an emissions-trading system. Before delving into them, it is worth reconsidering how carbon can be stored in soils and how improved agricultural practices can reduce the nitrous oxide (N_2O) released into the atmosphere.

Much of the carbon found in soils results from the removal of carbon dioxide by active plant roots and its storage in soil humus. Essential to the carbon-sequestration process is the level of microbe activity in soils and the nature and extent of the vegetation covering them (Post and Kwon 2000; Parr and Sullivan 2005; Chan 2008; Jones 2008). In terms of agriculture, soil-carbon can be augmented through conservation tillage (Valzano et al. 2005; Chan 2008; Mangalassery 2014); improved cropping practices (Chan 2008); the application of mulch, compost, and calcium-bearing silicates (Ryals et al. 2014); and by altering vegetation/crop coverage (Jones 2008). Besides increasing soil-carbon, these measures can improve soil structure and the plant availability of soil minerals and other nutrients. They can also increase the ability of soils to retain moisture (Grace et al. 2004; Jones 2007; Lal 2007; Chan 2008). Crucially, these advances can boost the productivity of the agricultural sector, thereby contributing to increases in the productivity of natural capital (Ratio 3 in Chap. 2).

As for reducing the nitrous oxide released through agriculture into the atmosphere, I explained in Chap. 8 that this can be achieved through better soil and water management methods; the use of organic fertiliser additives; the application of nitrification inhibitors, and a reduction in the burning of vegetation (de Klein and Eckhard 2008). Ironically, it has been shown that building up soil-carbon, especially in arable soils, has the potential to increase nitrous oxide emissions (Changsheng et al. 2005). It will therefore be necessary to strike an appropriate balance between soil-carbon accumulation and limiting the effect that soil-carbon can have on nitrous oxide emissions—a balance that would be best achieved by incorporating agriculture into an emissions-trading system.

Why would an emissions-trading system be the best mechanism to attain this balance? Because if the measures adopted by farmers to build up soil-carbon contributed more to the radiative forcing of nitrous oxide emissions than they reduced the radiative forcing of atmospheric carbon dioxide, the farmers in question would be required to purchase emissions units. This is because the RMUs received for achieving the former would be insufficient to cover the emissions units needed to offset the latter. With an emissions-trading system in place, the need to purchase emissions units (emissions permits) would make economic sense to farmers if the combined cost of the sequestration measures and required emissions units was less than the value of the productivity benefits gained from improved soils. Importantly, since greenhouse gas emissions would be capped with the cap diminishing over time, emissions targets would not be exceeded. There would simply be a relative rise in agricultural emissions and a relative reduction in the greenhouse gas emissions generated by non-agricultural activities.

Mind you, there could just as easily be a relative decline in the radiative forcing of agricultural emissions (i.e., a net sequestration of agriculture-related greenhouse gases), whereupon farmers would be entitled to receive emissions units (RMUs), which they could then sell in order to earn additional revenue. In this situation, the farmers' additional revenue would represent the amount that operators engaged in other emissions-generating activities would be willing to pay to purchase the RMUs earned by farmers rather than incur a much larger cost to reduce their own emissions. In all, emissions targets would not be compromised, but a cost-effective outcome would ensue. For reasons outlined in Chap. 7, a similar outcome would not occur if all emissions-generating activities were subject to an emissions tax.

The aforementioned brings us to the first impediment requiring attention before agriculture can be confidently included in an emissions-trading system—that is, the need to reliably measure the net greenhouse gas emissions of the agricultural sector.[46] Unless greenhouse gas emissions and removals can be attributed to specific agricultural activities and, more crucially, to individual agricultural establishments, it is impossible to determine the number of emissions units that farmers must acquit or are entitled to receive if they net-sequester greenhouse gases. Unfortunately, the different emissions and sequestration profiles of each farming establishment plus the fact that soil-carbon is prone to spatial, seasonal, and annual variations means that greenhouse gas emissions and removals by soils are inordinately difficult to estimate.[47] Furthermore, and this raises the second impediment, the high cost of estimating agricultural emissions and removals is rendering measurement and attribution prohibitively expensive.

Because of these impediments, resources urgently need to be directed, especially by governments and representative farming organisations, to discover low-cost methods of attributing emissions and removals to individual agricultural establishments (Conant 2010).[48] To accomplish this, technical know-how and the expertise to monitor and verify emissions must be developed. Also required are training courses, information, and demonstration centres to help farmers adopt greenhouse gas-reducing practices and measure their own emissions and removals. Provided farmers can become proficient in monitoring their activities, the most cost-effective outcome is likely to involve farmers self-reporting their emissions and greenhouse gas removals and having officials from a relevant authority randomly inspect farming establishments to ensure compliance.

For farmers, however, the cost of engaging in an emissions-trading system does not end here. With the point of obligation likely to be installed at the farm or establishment level, it will be very costly and time-consuming for farmers to regularly engage in an emissions-trading system to buy and sell emissions units for compliance purposes. Consequently, there will be a large role for collective action among farmers, with private broking firms likely to play an important function in reducing the transaction costs of each farmer's engagement in the system (Olson 1965; Ostrom 1990; Stern 2007; Garnaut 2008). Governments will also be important in this regard, since the quality of the national institutions overseeing the operation of the emissions-trading system will greatly affect the nature and magnitude of the transaction costs involved.

In the meantime, should it be impracticable to incorporate agriculture in an emissions-trading system, or should a nation remain free of emissions targets for some time, policies of the type outlined in Sect. 3.3.3 should be relied upon to begin the mitigation process until agriculture can be feasibly covered by the system (Garnaut 2008). I say, "begin the mitigation process", because containment of agricultural emissions cannot be guaranteed until agriculture is included in an emissions-trading system or alternative 'capping' arrangement.

Given the high cost involved and the technical and institutional capacity required to effectively incorporate agriculture into an emissions-trading system, the funding transfers from high-GDP to low-GDP nations recommended with respect to forestry-related activities are just as vital in relation to agriculture. Funds from high-GDP countries will also be needed to help farmers in low-GDP nations shift to sustainable and low carbon-emitting practices plus provide compensation payments for any initial reduction in agricultural output caused by the transition. Once again, these initiatives should be integrated into the national or regional development programmes of low-GDP countries, which, as I emphasised earlier, also need to be adequately funded.

9.2.3 Mitigation—Incorporating Funding Commitments in a New Global Protocol to Boost Mitigation Efforts

In Chap. 7, it was explained why the establishment of a price on greenhouse facilitates but does not guarantee the most cost-effective mitigation solutions. Some of the factors included financial capital market constraints; a lack of economic incentives to promote the uptake of low-emissions technologies; the public goods feature of many forms of knowledge, which can lead to insufficient research and development and an under-supply of low-emissions infrastructure; and 'carbon leakage', which can arise when firms shift their emissions-intensive forms of production to countries either completely devoid of emissions targets or harbouring an inadequate greenhouse gas price. In Chap. 8, we saw that a range of financial mechanisms and institutions can alleviate capital market constraints and other financial impediments. However, it was also made clear that a much larger injection of financial capital will be required to promote a low-emissions future, with a large percentage of the funds likely to come from the public sector, especially the governments of high-GDP nations.

Given the important enabling role of governments, I will put forward some arrangements that I believe should be incorporated in a new global climate change protocol to scale up and improve the public-sector's contribution to mitigation efforts. Obviously, there are limits to what can be included in a global protocol to compel national governments to take appropriate action. As a consequence, there will always be considerable reliance on the policies outlined in Chap. 3 to lighten the policy load, albeit they would be more effective and better funded if they were incorporated into national policy frameworks and development plans. For this reason, I will refrain from saying anything in this sub-section about national policies other than to reiterate the importance of adequate public investment in low-emissions infrastructure and the need for government regulations where price signals fail to induce appropriate mitigation responses from the private sector.

9.2.3.1 Technology Transfer, R&D, and Promoting Low-Emissions Technologies

As explained in Sect. 8.2.8, one of the weaknesses of the global response to the climate change crisis has been the inadequate public funding of mitigation action. In particular, a number of analysts have been at pains to point out that levels of research and development spending on renewable energy, which is an important element in the transition to a low-emissions future, have declined in recent years (Nemet and Kammen 2007; IPCC 2007d; Garnaut 2008; Stern 2009; IEA 2014).[49] Significantly, it has also come to the attention of critics that the UNFCCC has had very little to say about research and development, which in part explains the lack of emphasis on research and development in the Kyoto Protocol (Garnaut 2008). Furthermore, while the need for technology transfers from high-GPD to low-GDP countries features prominently in the Kyoto Protocol, most observers agree that Annex I nations have failed to deliver on their pledges.

Admittedly, various public-sector and private-sector funds plus the Clean Development Mechanism have helped facilitate the transfer of low-emissions technologies, but it is nowhere near the scale required to activate a widespread mitigation response in the world's poorest nations, let alone in Annex I nations where there has been a moderate increase in renewable energy use over the last decade (Garnaut 2008). Notwithstanding the establishment of the Green Climate Fund, which has prompted large injections of financial resources towards global climate change solutions, a number of critics have stressed the need for governments of wealthy nations to do much more to finance the development and uptake of low-emissions technologies in their own countries as well as trigger and sustain a shift to low-emissions energy sources and production methods elsewhere in the world (Carmody and Ritchie 2007; Garnaut 2008; Stern 2009).

Approximately how much public funding will be required to stabilise the atmospheric concentration of greenhouse gases? This is a difficult question to answer, for two reasons. Firstly, it will depend on the chosen stabilisation target, with a more ambitious target requiring more funding. Secondly, the extent to which the emissions-intensity of economic activity must be reduced to stabilise at a chosen concentration level will depend, as demonstrated in Chap. 4, on the growth rate of real GWP. The higher the growth rate, the more difficult it will be to achieve a given target, which again implies a higher funding requirement.[50] As we shall see in Chap. 10, the need for Annex II countries to reduce their real GDP to operate at an optimal per capita GDP of Int$15,000 (2004 prices) will render the shift a lot less demanding than most observers believe. What's more, it could be easier than the task confronting some of the non-Annex I nations still requiring further GDP growth, thus re-emphasising the urgent need for funds to be redistributed to the world's poorest nations.

Returning to the above question, one early study by Popp (2004) estimated the global research and development spending needed to restrict annual greenhouse gas emissions to 1995 levels at around US$20 billion per year in 2015; US$30 billion per year in 2050; and US$50 billion per year in 2100.[51] In another study, Kammen and Nemet (2005) put the figure to achieve a 550 ppm stabilisation target at US$15–30 billion per year until stabilisation is reached. However, a more recent study by Bosetti et al. (2009), which is also premised on a 550 ppm stabilisation target, estimated the annual figures for present and 2050-required spending at US$30 and US$50 billion respectively.

Although a significant portion of the research and development funding will emerge from the private sector, these studies greatly understate the public funding requirement. This is because they fail to include the funds needed to: (i) transfer technologies from rich to poor nations; (ii) establish adequate institutional and technical capacity in low-GDP countries, which is needed to effectively direct funds, provide stable incentives, and promote the commercialisation of new technologies; (iii) overcome financial capital market constraints, which are most acute in needy countries; and (iv) construct some of the physical infrastructure needed to support research and development activities. The underestimation can also be attributed to the assumed emissions targets in these studies being far higher than the 450 ppm safety level. It is also worth noting that the development and early uptake of low-emissions technologies, which is needed to prevent countries locking themselves into high-emissions pathways, will require a large slice of any additional funds to be delivered over the next two decades.[52]

Taking account of the extra funding requirements of low-GDP nations, the UNFCCC has estimated that, by 2030, additional global financial flows of US$200 billion per year will be needed to help constrain greenhouse gas emissions to mid-2000 levels (UNFCCC 2007d). Of this, the UNFCCC believes that US$100 billion needs to be directed each year to low-GDP countries. Once more, a large chunk of these funds can be expected to emanate from private-sector sources. However, until international carbon markets are able to generate large financial flows, which may not occur until a global emissions-trading system is firmly established, a significant reliance on public-sector funding can be expected for at least the next 20–30 years (Garnaut 2008). Nevertheless, the magnitude of the funds will be much greater than what the UNFCCC has estimated simply because, by 2030, global greenhouse gas emissions will need to be around 40 per cent lower than the mid-2000 levels assumed in the UNFCCC study.

9.2.3.2 Embedding Funding/Transfer Commitments in Future Protocols

Whatever the true cost of promoting low-emissions technologies is, the importance of adequate financial flows and appropriate international burden-sharing highlights the need to embed funding commitments in a future climate change protocol.[53] To be consistent with the UNFCCC principle of 'common but differentiated responsibilities', the funding commitments should apply exclusively to Annex I nations (Garnaut 2008). This said, Annex I countries are only likely to agree to funding commitments if a reciprocal relationship between Annex I and non-Annex I nations can be established that provides the incentives needed to render a global protocol effective plus the 'glue' to maintain its structural integrity (Stern 2009).

Ideally, this reciprocal arrangement would involve Annex I countries agreeing to meet most of the mitigation costs that non-Annex I nations face in coming decades on the condition that non-Annex I nations are willing to take on emissions obligations and increasingly fund their own mitigation efforts. At the same time, non-Annex I nations would be expected to reduce their greenhouse gas emissions on the expectation that they would receive adequate funding from Annex I countries for mitigation and adaptation purposes (more on adaptation soon).

To guarantee a minimum transfer of funds from Annex I nations, a compulsory 'mitigation' commitment should be embodied in a future global protocol.[54] The commitment, which would be used to inject additional financial resources into the Green Climate Fund[55], would require Annex I countries to allocate a small proportion of their real GDP into the Fund. The funds should be supplementary to existing funding commitments, such as the Monterrey (2002) and Gleneagles (2005) promises in relation to overseas development assistance.[56]

Given the magnitude of the funds required, the contribution rate should be set at approximately 0.25 per cent of GDP and should apply to all Annex I nations with a per capita real GDP of Int$10,000 or more (at 2004 prices). Over time, this commitment would decline as market-based mechanisms become more effective; as technological breakthroughs emerge and the cost of low-emissions technologies falls (the latter more likely in a steady-state setting); and as low-GDP countries become increasingly self-reliant in terms of supporting their mitigation actions.

Importantly, the funds provided by this and other commitments would cover the development and uptake of new technologies across a wide variety of fields. Besides the transport and stationary-energy sectors, the funds would support

LULUCF activities and the development of carbon capture-and-storage technologies (geosequestration). In the case of LULUCF activities, support would extend to research and development into agricultural practices that increase the carbon content of soils and reduce nitrous oxides emissions. Support would also encompass research into improved forestry management.

Of course, to qualify as recipients of the additional funds, low-GDP countries would be required to uphold their half of the reciprocal relationship by complying with commitments embodied in a new climate change protocol. These would include, initially, non-binding commitments to meet renewable and energy-efficiency targets, and eventually binding commitments to reduce greenhouse gas emissions. Not unlike the mitigation initiatives I referred to earlier, the use of these funds and the technology-promoting activities they support should be incorporated into the national policy frameworks and development programmes of low-GDP countries, thus ensuring mitigation efforts are well co-ordinated with minimal duplication and waste.

9.2.4 Mitigation—The Kyoto Protocol's Enforcement Mechanisms

Without an effective enforcement system, a global climate change protocol will not achieve designated emissions targets. As previously mentioned, there are many who believe that the Kyoto Protocol's enforcement mechanisms are weak or outright defective. Although a number of these critics acknowledge the Annex I group of Parties' successful achievement of its first-round Kyoto targets, they argue that it had little to do with the strength of the Protocol's enforcement mechanisms and more to do with the bounty created by the decline in the greenhouse gas emissions of former-communist countries in the early-1990s—a claim well supported by the evidence presented in Table 8.9.[57]

Of all the criticisms levelled at the Kyoto Protocol's enforcement mechanisms, the greatest has been directed at the primary penalty meted out to nations which fail to meet their emissions targets—specifically, the need to make up for their excessive emissions plus 30 per cent in a subsequent commitment period. There are various reasons why this penalty has been so heavily criticised. Firstly, the penalty serves as a very weak enforcement mechanism if the emissions targets pertaining to future commitment periods have not been stipulated beforehand (Garnaut 2008). This is because non-specified future targets provide an incentive for countries to exceed their current target and negotiate weaker upcoming targets in order to absorb the penalty (Barrett 2003; IPCC 2007d).

Secondly, in a world of sovereign states, nations cannot be compelled to sign agreements or, should they do so, meet their obligations (Victor 1999; Schelling 2002; Garnaut 2008). Since this implies that a global climate change protocol can only be 'notionally' binding, the obligation to make up for excess emissions in subsequent commitment periods cannot prevent a nation from postponing restoration indefinitely (Barrett 2003; Nentjes and Klaassen 2004).

Thirdly, it is widely agreed that getting countries to set and adhere to appropriate emissions targets requires enforcement mechanisms that restrain the collective and individual desire to transgress the system (Hovi and Areklett 2004; Stern 2009). In other words, there is a strong conviction that a compliance system must fulfil the self-interest of each Party whilst promoting a collegial approach to target-setting and compliance—often referred to as *soft* compliance management (Murase 2002). In direct contrast, critics believe the primary Kyoto penalty amounts to a formalised means of 'international punishment' that belongs to the less effective category of *hard* compliance management (Stern 2009).

It should be said that although the second of the two Kyoto non-compliance penalties—namely, suspension from international trade in emissions units—is punitive rather than facilitative, it is considered incentive-based in the sense that it renders it more difficult for a nation to cost-effectively adhere to a future emissions target. It is also legally enforceable. Hence, it serves as a powerful incentive not to exceed an emissions target in the first instance.

As explained in Sect. 8.2.7, an enforcement system cannot succeed without adequate monitoring. Through the agency of the Enforcement Branch of the Kyoto Protocol's Compliance Committee, most observers believe that the Protocol has established a strong institutional basis for monitoring, reporting, and verifying greenhouse gas emissions (e.g., Stern 2007). Nevertheless, many see enforcement as an ongoing international challenge, not only because of the shortcomings associated with the primary Kyoto penalty, but because the system is extremely vulnerable to corruption. Moreover, the system relies heavily on participating countries having the institutional capacity to: (i) reliably measure and report greenhouse gas emissions and removals; (ii) competently administer a national registry to keep track of emissions units (e.g., AAUs, CERs, ERUs, and RMUs); and (iii) transparently and scrupulously operate a domestic emissions-trading system compatible with a group or global emissions-trading system. Although the issue of institutional capacity is of little concern to Annex I nations, it is very problematic in the case of non-Annex countries. It therefore constitutes a major mitigation hurdle given the strong likelihood that non-Annex I countries will be subject to various commitments and emissions obligations in the near future.

I have already talked about what high-GDP countries can do from a financing perspective to help equip low-GDP countries with the institutional capacity to facilitate the development and uptake of low-emissions technologies. To reinforce the global response to the climate change crisis, the same financing sources should be used to help low-GDP nations monitor and report their greenhouse gas emissions as well as administer a sound emissions-trading system. Since the global trade in emissions units must be open and transparent, there is good reason to establish a major international authority to oversee the operation of international carbon markets (Stern 2009). An authority of this type should function under the auspices of the UNFCCC. The authority could be a newly-created organisation or, should existing compliance institutions survive beyond the Kyoto Protocol, be part of an expanded Enforcement Branch.

9.2.4.1 Incentives and Compliance

It is my belief that there are two key issues at stake with respect to emissions targets and enforcement. They are: (i) how to induce or, if need be, compel countries to set commitments and obligations consistent with Articles 2–4 of the UNFCCC; and (ii) how to ensure nations abide by their commitments. Although the general consensus is that a hard compliance approach is highly undesirable, I believe punitive penalties are essential. Whilst I recognise the important role that a facilitative approach can play in encouraging and incentivising countries to negotiate an effective climate change protocol, a soft compliance approach is unlikely to guarantee compliance, especially with respect to greenhouse gas emissions targets. Indeed, sole reliance on soft compliance management could hinder the emergence of an effective climate change agreement.

One of the main reasons why a hard approach is spurned is because it is believed that the fear of incurring harsh penalties for failing to meet emissions obligations can deter nations from agreeing to stringent emissions targets. It is my contention that if a country fears stringent targets and is not genuine about tackling climate change, it will consistently push for lenient targets regardless of whether a soft or hard approach is adopted.[58] Conversely, if a government is serious about participating in a global quest to resolve the climate change crisis and is confident that transgressors *will be harshly penalised*, it will welcome stringent targets and tough penalties. After all, it is the fear of obeying stringent targets whilst recalcitrant countries go un-penalised that constitutes the greatest deterrent to setting tough targets. Why agree to harsh emissions cuts and play your part in meeting them if other countries can flout their obligations with impunity?[59] The key is to have penalties that are sufficiently severe and to ensure non-compliant nations are compelled to incur them—something that cannot be said of the present Kyoto enforcement mechanisms.

To support what I've said, consider Fig. 9.1. Figure 9.1 is a pay-off matrix depicting a hypothetical world involving two participating parties—country *A* and country *B*. Let's assume that both countries have been compelled by an international authority to make the large emissions cuts necessary to achieve a safe atmospheric concentration of greenhouse gases. Let's also assume that the penalties for failing to comply with emissions targets are unenforceable. Despite their obligations, both countries can choose to meet their emissions targets or exceed them by undertaking smaller than required cuts to their greenhouse gas emissions. If they both choose the former option, both countries enjoy moderate short-term gains as well as significant long-term benefits in the form of avoided climate change damages. In Fig. 9.1, this is represented by an overall pay-off of US$120 billion to both *A* and *B*.

If, on the other hand, both nations make small emissions cuts, it leads to catastrophic global warming. Although this option generates large short-term benefits, they are greatly offset by the huge cost of future climate change damages. For both countries, the overall pay-off is a much smaller US$70 billion each.

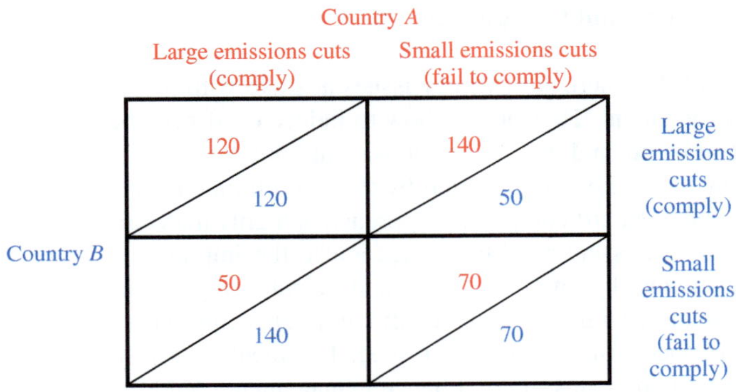

Fig. 9.1 Pay-off matrix for countries *A* and *B* involving large and small emissions cuts (penalties for non-compliance are unenforceable). *Note* Assumed pay-offs in billions of US dollars

In between, there are two hypothetical cases—the first involving country *A* making large emissions cuts and country *B* violating its emissions target; the second involving the opposite. In both instances, it is assumed that the eventual climate change outcome is not catastrophic but is devastating enough to carry with it large long-term damage costs. It is also assumed that the nation which drastically decreases its greenhouse gas emissions suffers a competitive disadvantage which reduces its short-term benefits and whatever long-term benefits remain once the long-term costs of devastating global warming are subtracted. This is represented in Fig. 9.1 by a pay-off of US$50 billion. Conversely, the country which only marginally reduces its emissions enjoys significant short-term benefits—indeed, so much so, it receives an overall pay-off worth US$140 billion despite incurring large long-term damage costs.[60]

Given these alternatives, it is safe to assume that neither country would take the chance of being rendered significantly worse off by meeting its emissions target whilst its counterpart did not. In the end, both countries have an incentive to exceed their emissions target in the off-chance that its counterpart will behave in a similar manner. In other words, they would rather receive US$70 billion than US$50 billion. Thus, if both nations were given the opportunity to set emissions targets rather than have them imposed externally, neither would agree to stringent emissions targets.

Certainly, relative to a US$70 billion pay-off, it would be mutually beneficial for countries *A* and *B* to adhere to their targets (US$120 billion each). However, both countries will always have an incentive to renege on their commitments—if not to enjoy a US$140 billion pay-off, then to avoid receiving the lowest of all possible pay-offs (US$50 billion each). Given the circumstances depicted in Fig. 9.1, it is clear that a soft or facilitative approach to target-setting would not lead to a safe atmospheric concentration of greenhouse gases.

Consider the same initial situation except there is now an enforceable non-compliance penalty equivalent to US$50 billion. As Fig. 9.2 shows, the pre-penalty

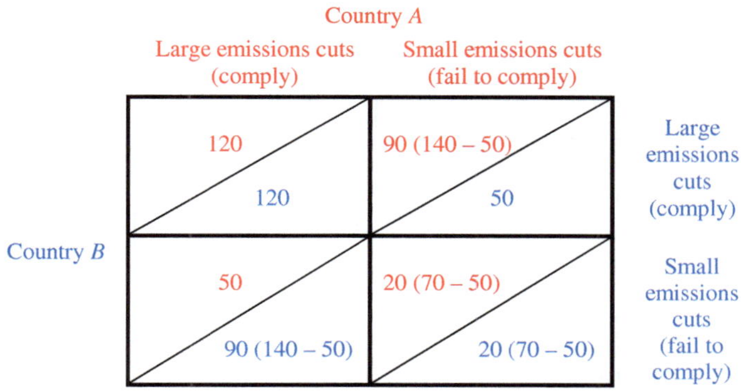

Fig. 9.2 Pay-off matrix for countries *A* and *B* involving large and small emissions cuts (enforceable penalty worth US$50 billion for non-compliance). *Note* Assumed pay-offs in billions of US dollars

pay-offs associated with the different scenarios are exactly the same. However, the post-penalty pay-offs are considerably different, as are the corresponding incentives. For example, if both nations make small emissions cuts, both will be non-compliant and incur a US$50 billion penalty. For countries *A* and *B*, the overall post-penalty pay-off is just US$20 billion each. This is much lower than the pay-off enjoyed if a nation meets its target and its counterpart does not (US$50 billion). Furthermore, the post-penalty pay-off enjoyed by a recalcitrant nation (US$90 billion) is less than the pay-off it would receive if it adhered to its emissions target (US$120 billion).

It is worth recognising that if countries *A* and *B* were given the option to set emissions targets and they agreed and adhered to weak targets, their pay-offs of US$70 billion would be much less than if they took the more austere route. As a consequence, there is now an incentive for both countries to set and meet stringent emissions targets. Just as importantly, since the US$50 billion penalty is necessary to prevent a recalcitrant nation from benefiting through free-riding, both countries have an incentive to establish a tough, enforceable penalty.

Of course, the outcomes depicted in Figs. 9.1 and 9.2 depend on the extent of the pay-offs pertaining to each scenario plus the magnitude of any non-compliance penalty, should it be enforceable. Consider Fig. 9.3, where all but one of the pay-offs are the same as in Fig. 9.1. The only exception is the pay-off received by a nation that exceeds its emissions target while its counterpart remains compliant. On this occasion, the pay-off is just US$100 billion—the result of the short-term benefits from free-riding being less than the previous case.

Unlike the situation depicted in Fig. 9.1, the incentive for a nation to free-ride has now completely evaporated. This is because the pay-off for a country that cheats while its counterpart adheres to its emissions target is US$100 billion, yet only US$70 billion when its counterpart also opts to cheat. Both pay-offs are less than the US$120 billion gained if countries *A* and *B* adhere to their emissions

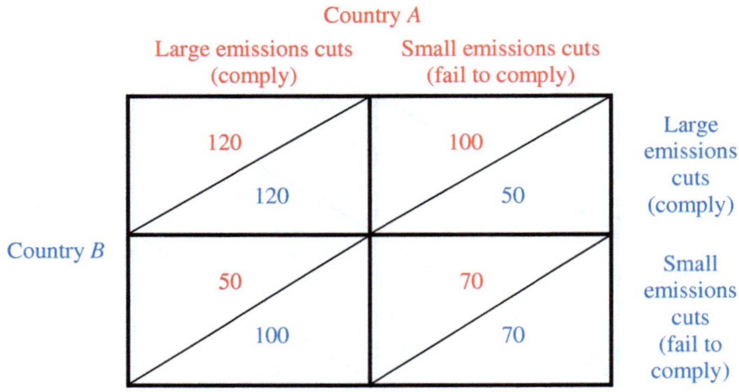

Fig. 9.3 Pay-off matrix for countries *A* and *B* involving large and small emissions cuts (penalties for non-compliance are unenforceable). *Note* Assumed pay-offs in billions of US dollars

targets. Even without an enforceable penalty, both countries have an incentive to set and meet stringent emissions targets. Hence, in these circumstances, it would appear that a soft or facilitative approach to target-setting could generate an agreement capable of achieving a safe concentration of greenhouse gases.

There are, however, three reasons why, even in the situation depicted in Fig. 9.3, a facilitative approach might not lead to a desirable climate change outcome. Firstly, it is impossible to know what the true pay-offs are.[61] Consequently, the fear of being rendered worse off by adhering to stringent targets whilst other nations free-ride is likely to result in all countries adopting a maximum loss-aversion strategy (assuming the penalties are unenforceable).[62] If so, a facilitative approach is likely to produce weak targets. Only if there is high probability of a strict and enforceable penalty emerging from the negotiations between the two countries would we expect them to agree to stringent targets. However, we then enter into the realm of hard compliance management.

Secondly, even if the pay-offs can be estimated with some degree of accuracy, it is possible that the government of a participating party may discount the long-term damage costs of climate change and only concern itself with short-term gains. To understand the possible implications of this, the pay-offs in Fig. 9.1 are split into hypothetical short-term and long-term benefits and presented in Fig. 9.4. The greatest long-term gains are enjoyed if both countries agree to and subsequently meet stringent emissions targets (US$80 billion each). The smallest long-terms gains occur if both nations choose to violate their emissions targets (US$5 billion each). Where one country adheres to its emissions target and the other does not, the respective long-terms gains are US$20 and US$25 billion. We would expect a free-rider to enjoy slightly larger long-term benefits than a compliant nation because it would continue to gain a competitive advantage from emitting a higher quantity of greenhouse gases.

Let's assume that country *B* has chosen to disregard the long-term damage costs of climate change and is only concerned with its short-term pay-offs.

Country B \ Country A	Large emissions cuts (comply)	Small emissions cuts (fail to comply)
Large emissions cuts (comply)	A: 120 = 40 + 80; B: 120 = 40 + 80	A: 100 = 75 + 25; B: 50 = 30 + 20
Small emissions cuts (fail to comply)	A: 50 = 30 + 20; B: 100 = 75 + 25	A: 70 = 65 + 5; B: 70 = 65 + 5

Fig. 9.4 Pay-off matrix for countries *A* and *B* involving large and small emissions cuts (short-term and long-term pay-offs; penalties for non-compliance are unenforceable). *Note* Assumed pay-offs in billions of US dollars

Country *B*'s short-term benefits are underlined in Fig. 9.4. From *B*'s perspective, the ideal scenario is one where: (i) it agrees to stringent emissions targets; (ii) it violates its own target; and (iii) country *A* remains compliant. This would generate a pay-off to country *B* worth US$75 billion. Since, in all likelihood, country *A* would also defy its emissions target, the pay-off for country *B* would be US$65 billion. Either way, country *B* would view the alternative pay-offs from non-compliance as superior to the pay-offs enjoyed from adhering to its target (US$40 billion or US$30 billion).

Conversely, because country *A* would be concerned with long-term as well as short-term gains, it would regard compliance by both countries as an ideal outcome. Nevertheless, it would recognise that, without an enforceable penalty, it would be left with a meagre US$50 billion pay-off should country *B* opt to free-ride. Although country *A* is unlikely to know whether or why country *B* has a desire to cheat, uncertainty about *B*'s emissions intentions would almost certainly result in country *A*, like country *B*, not agreeing to stringent emissions targets.

A third factor—where a government rejects the findings of climate change science and the notion of long-term damage costs—would have much the same impact on incentives as the second factor. In this case, the government in climate change denial is likely to have aired its views on climate change science and its unwillingness to drastically reduce its greenhouse gas emissions.[63] Presumably, a counterpart nation would be sufficiently cautious to avoid agreeing and adhering to stringent emissions targets in fear of receiving the smallest of all possible pay-offs.

All up, even when various pay-off scenarios suggest that a soft compliance approach has the potential to produce a desirable climate change protocol, issues surrounding the ignorance of true pay-offs and the uncertainty about the emissions intentions of other nations point to the strong compliance approach being the most likely means of reaching a suitable climate change agreement.

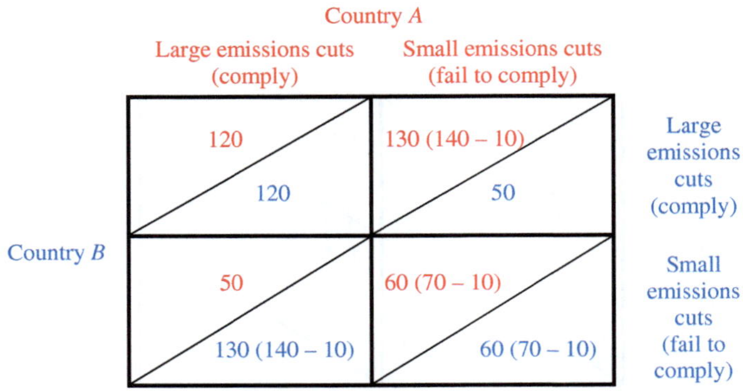

Fig. 9.5 Pay-off matrix for countries *A* and *B* involving large and small emissions cuts (enforceable penalty worth US$10 billion for non-compliance). *Note* Assumed pay-offs in billions of US dollars

The final example I wish to present is designed to indicate the importance of installing a sufficiently harsh non-compliance penalty. To demonstrate my point, consider Fig. 9.5. On this occasion, the pay-offs are the same as in Fig. 9.1, except there is a penalty for non-compliance worth US$10 billion. Without going through all the pay-off scenarios, Fig. 9.5 shows that the post-penalty pay-off from cheating on a compliant counterpart is higher than the pay-off received if both countries meet their emissions targets (US$130 billion *vis-à-vis* US$120 billion). It also shows that the post-penalty pay-off when both nations are violating their emissions target is larger than the pay-off received when a country suffers from the free-riding behaviour of its counterpart (US$60 billion *vis-à-vis* US$50 billion). Although the US$10 billion non-compliance penalty reduces the pay-offs obtained from violating a stringent emissions target, it is insufficient to alter the incentives confronting countries *A* and *B*. Consequently, both nations are unlikely to agree to stringent emissions targets.

How big must the penalty be in the above situation to alter the incentives and behaviour of both nations? It must be large enough to reduce the post-penalty pay-off from cheating on a compliant counterpart below that of the pay-off received if both countries adhere to their targets (i.e., less than US$120 billion). Assuming that both countries are able to estimate the pay-offs of each scenario with reasonable accuracy, the non-compliance penalty must be worth at least US$21 billion to have any positive effect (Note: a US$21 billion penalty would reduce the pay-off from cheating on a compliant counterpart to US$119 billion). Since, as highlighted, it is impossible to know what the true pay-offs are, large and enforceable non-compliance penalties will be required under all circumstances to ensure nations have the incentive to agree to stringent emissions targets and to quell any desire to violate them.

9.2.4.2 Establishing Appropriate Penalties

Given the need for hefty and enforceable penalties, in what form should they take? Because meeting global emissions targets requires the excessive emissions of nations to be offset by a larger than planned decrease in future emissions, the current penalty of compelling countries to make up for their shortfall plus 30 per cent should remain in a future global protocol. To ensure the penalty has greater potency, the current period-by-period approach to target-setting should be replaced by an emissions-reducing schedule that indicates the timing of future emissions cuts and when non-Annex I nations will be subject to binding emissions targets.

For reasons given above, the suspension from international trading in emissions units plus a requirement to submit a compliance action plan should also be retained in a new global protocol. However, these penalties and current compliance procedures should be supplemented by trade sanctions and border-tax adjustments (i.e., tariff penalties). By tariff penalties, I mean tariffs sanctioned by an international institution to offset any competitive advantage that a nation might enjoy from not reducing its greenhouse gas emissions. Having said this, I believe trade sanctions and tariff penalties should only be applied when a nation: (i) commits a second and subsequent transgression; (ii) refuses to discharge a presently imposed penalty; or (iii) refuses to participate in climate change negotiations and wantonly generates greenhouse gas emissions at levels which undermine the achievement of global emissions targets. Confining trade penalties to a second or subsequent transgression makes an important allowance for first-up violations which may occur because of an obscure or unavoidable circumstance.

Although the idea of imposing trade sanctions and tariff penalties has its detractors, they are needed to provoke intransigent nations into playing their part in achieving a safe atmospheric concentration of greenhouse gases.[64] Of course, many would regard such action as a violation of national sovereignty and therefore unlawful. However, there are a couple of factors worth recognising. Firstly, trade sanctions and tariff penalties would not compel a nation to reduce its greenhouse gas emissions. A country could continue to operate as it pleases, but would be proportionately penalised for doing so. Secondly, trade sanctions and tariff penalties can already be accommodated by international law. Under Article 39 of the United Nations Charter, the United Nations Security Council is well within its rights to class the climate change crisis as a threat to international security, as many believe it is (Campbell 2008; Art and Waltz 2009; Matthew 2011; Moran 2012; also see Table 1.4; Fig. 1.10). Should the Security Council take this step, it could then invoke Article 41 of the Charter and impose economic and other non-military sanctions to give effect to a United Nations-approved climate change protocol (Depledge and Feaken 2012; Gilley and Kinsella 2013).

Whilst the use of trade sanctions and tariff penalties may not be considered necessary at present, sentiment amongst committed nations could shift very quickly once the detrimental impacts of climate change intensify (Campbell 2008; Gilley and Kinsella 2013). I believe we are fast approaching the point where nations would regard the international enforcement of emissions targets as legally justified. All things considered, the use of trade sanctions and tariff penalties as an additional enforcement mechanism is entirely feasible (Ismer and Neuhoff 2007) and probably not far from becoming a reality.

Would trade sanctions and tariff penalties work? There is no guarantee they would, although there are many instances where they have, such as the trade sanctions which helped overturn the apartheid policy in South Africa. If there is one further complication worth noting it concerns whether climate change-related trade penalties would conflict with the World Trade Organisation (WTO) rules on international trade (Sindico 2008). This essentially depends on the nature of the penalties and how they are imposed. As highlighted in Chap. 3, Article XX of the General Agreement on Tariffs and Trade (GATT) allows for trade-related measures to be imposed for environmental reasons, but only under very strict and narrowly-defined circumstances (Biermann and Brohm 2003; Frankel 2005; Ismer and Neuhoff 2007).[65] Given the continuing uncertainty surrounding what type of trade-related penalties are legal under the WTO's international trade rules, it would be prudent for the WTO to clarify what action can be taken and to work more co-operatively on these matters with the UNFCCC. There is also good reason to believe that the WTO rules should be modified to better accommodate legitimate trade penalties. I'll have more to say about compensating tariffs and the relationship between the WTO and the UNFCCC in the next sub-section.

9.2.5 Mitigation—Border-Tax Adjustments/Green Tariffs to Limit Carbon Leakage and Promote Genuine Efficiency

In Chap. 3, it was explained how, in a global economy dominated by the free mobility of international capital, transnational corporations can avoid nationally instituted non-price rules and cost-internalisation policies by shifting their operations to countries with low wages, poor working conditions, and feeble environmental standards. It was also explained how this can have a degenerative impact on the rate of natural resource use and its subsequent allocation for economic purposes. For instance, it was shown that globalisation can lead to resources being allocated to nations where the cost of production is lowest, but not where production activities are genuinely the most efficient. Given the threat this poses, governments in wealthy countries often limit the loss of industries by diluting environmental standards and the regulations protecting wages and conditions of employment. In other cases, governments deal with nascent social or environmental problems by introducing regulations that are too weak to be fully effective (Garnaut 2008, p. 342).

To overcome the degenerative effects of globalisation, it was recommended that countries with similar wages, tax regimes, and environmental standards should trade freely with each other but be permitted to impose 'compensating' tariffs on countries with lower standards.[66] To discourage governments from protecting inefficient domestic industries, it was stressed that compensating tariffs must reflect the cost advantage arising from disparities in standards, not from genuine differences in the efficiency of production. Because ensuring tariffs are imposed in this way would require strict international oversight, it was suggested that the WTO should be responsible for assessing tariff applications and the sanctioning of compensating tariffs and other border-tax adjustments.

9.2.5.1 The Need for Border-Tax Adjustments in Relation to Climate Change

With respect to climate change, globalisation has probably had as much of a detrimental impact on efforts to combat it as any other environmental problem. Evidence of this is best reflected in the growing pressure being exerted on non-Annex I nations to accept greater responsibility for their greenhouse gas emissions. This pressure, which has been heavily applied at recent COP meetings, has not been solely aimed at urging non-Annex I nations to make a greater contribution in the fight against rising global emissions. It has also been aimed at easing the concerns about the economic impact of emissions-intensive industries shifting from Annex I to non-Annex I countries (Elliott et al. 2010a). Many observers believe that the threat posed by industrial flight was a major factor behind the refusal of the US Government to ratify the Kyoto Protocol (Saddler et al. 2006; Henson 2011). Others believe it has plagued the setting of much harsher greenhouse gas emissions targets (e.g., Garnaut 2008).

Crucially, the stultifying impact of industrial flight on efforts to combat climate change does not stop with its potential to impede the emergence of an effective emissions protocol. As alluded to earlier, the relocation of emissions-intensive forms of production to countries with no emissions targets can result in carbon leakage. The problem of carbon leakage occurs when firms operating within a country subject to emissions targets transfer their emissions-intensive production activities to unregulated countries and sell (export) the goods back to the country from which the firms departed. While this practice allows the importing-nation to meet its emissions targets, some or all of the spared emissions take place in the new country of operation. Hence, it is possible for there to be little or no reduction in aggregate emissions. In fact, if the displacement involves the use of more emissions-intensive production methods—which often occurs following the shift in production from a high-GDP to a low-GDP nation—it is possible for aggregate emissions to rise.

Just as importantly, it has been shown that the relocation of production activities can reduce the economic incentive for low-cost nations to develop and employ low-emissions production methods (Saddler et al. 2006). This not only increases the probability of aggregate emissions rising following displacement, it locks low-GDP countries into high-emissions pathways. Given that non-Annex I nations will inevitably be subject to emissions targets, this will make it very difficult for low-GDP countries to achieve their targets in a cost-effective manner.

There are, however, three additional things worth recognising with respect to border-tax adjustments. Firstly, the need for Annex I countries to impose adjustments for climate change-related purposes is likely to diminish over time. As we shall see in Chap. 10, the emissions targets of the four recommended groups of nations should eventually converge along with the greenhouse gas price they are confronted with. Since this would bring about a level playing field for trade-exposed industries, the need for border-tax adjustments would vanish (Saddler et al. 2006; Garnaut 2008).[67]

Secondly, there are many people who believe that the rate of carbon leakage is very low and that the argument supporting border-tax adjustments is very weak. Although some estimates of carbon leakage rates support this conclusion (e.g., Paltsev 2001; Kuik and Gerlagh 2003; Baylis et al. 2014), one study by Babiker (2005) has shown that carbon leakage rates have been high enough to counteract the emissions reductions taking place in Annex I nations. Another study found that many EU nations have been off-loading more than 30 per cent of their emissions-intensive activities to countries outside the region (Davis and Caldeira 2010).

Who should we believe? Whatever the answer, it needs to be acknowledged that estimating the rate of carbon leakage is problematic given the impossibility of identifying all cases where increases in the emissions of non-Annex I nations can be attributed to industrial flight. Notwithstanding this, the following should be borne in mind. There are clearly instances where Annex I countries have obtained emissions reductions through the relocation of production activities to nations devoid of emissions targets (IEA 2004; IPCC 2007; Davis and Caldeira 2010; Elliott et al. 2010b). There is also hard evidence indicating that the emissions reductions secured by Annex I countries have been dwarfed by the enormous increase in greenhouse gases generated by non-Annex I nations (see Table 8.9). Despite the technological gap, it is difficult to believe that the difference in the greenhouse gas-emitting performances of the Annex I and non-Annex I group of nations can be principally attributed to the former group having massively reduced the emissions-intensity of its production whilst the latter group has hopelessly failed. The disparity can only be explained by accepting that, to a significant degree, the imbalance has been caused by the relocation of emissions-intensive activities from the former to the latter group.

Thirdly, even when carbon leakage is considered a large enough problem to justify border-tax adjustments, some commentators believe that only a few industries are being adversely affected by the competiveness impact of disparate greenhouse gas price signals (Saddler et al. 2006; Stern 2007; Garnaut 2008, 2011). Hence, they rule out the need for the across-the-board tax adjustments being advanced by many border-adjustment advocates.

There are two points worth making here. Firstly, if these commentators are correct, the misguided callings for strong corrective action should be exposed via a thorough assessment of compensating tariff applications by the WTO or other institution entrusted with the responsibility. Secondly, it is my contention that the number of industries whose competitiveness has been affected by disparate greenhouse gas price signals is far greater than what many have identified through empirical analyses and modelling exercises. Because of sectoral interdependencies, I believe some trade-exposed industries in the services sector of countries subject to a greenhouse gas price have been adversely affected by output price rises in the emissions-intensive industries of their own primary and secondary sectors. To recall from Chap. 6, outputs from these two latter sectors serve as inputs to the services (tertiary) sector. Hence, any increases in output prices in the primary and secondary sectors caused by assigning a price to greenhouse gas emissions flow on as increased input costs for the service-sector. Yet virtually all the empirical studies undertaken to examine this issue ignore the potential competitiveness impact that a greenhouse gas price can have on the services sector.[68] More than this, these studies invariably assume an emissions price that is well below what I would consider adequate (i.e., one reflecting the true social cost of carbon).[69] Should a more appropriate price be assumed, it would magnify the competitiveness impact on emissions-intensive industries and increase the number of adversely affected tertiary-sector industries.

Overall, there is more than enough evidence to suggest that something needs to be done to reduce or negate the detrimental impact that globalisation is having on efforts to resolve the climate change crisis. Since the mere threat of industrial flight has served as a powerful obstacle in the setting of necessary emissions cuts, and the industrial flight which has occurred has resulted in clear examples of carbon leakage and international resource allocation inefficiencies, there is an urgent need to grant countries subject to high greenhouse gas price signals the right to apply border-tax adjustments. Nevertheless, as mentioned, these adjustments should be sanctioned by an appropriate international authority to ensure they do not unduly protect inefficient industries nor impede the genuine benefits that international trade can generate.

9.2.5.2 Border-Tax-Adjustments and the Kyoto Protocol

There is nothing within the Kyoto Protocol that explicitly facilitates the imposition of border-tax adjustments to extinguish the cost gap between countries exposed to a greenhouse gas price signal and nations which are not. In part, this is the result of the UNFCCC Parties eschewing the opportunity at the COP-3 meeting in Kyoto to include multilateral sanctions of any sort in the Kyoto Protocol (Frankel 2005). Despite this, the Protocol's Articles do leave open the opportunity for countries to adopt trade-related measures to overcome the competitive disadvantage brought on by differences in national mitigation action. For example, Article 2.1(a) encourages Annex I nations to develop and implement policies which: (i) enhance energy efficiency; (ii) phase out tax exemptions and subsidies at variance with the UNFCCC's objective of preventing dangerous interference with the climate system; and (iii) reduce the emission of greenhouse gases not regulated by the Montreal Protocol.[70] However, to comply with Article 2.3 of the Protocol, the measures must be undertaken in such a way as to limit any adverse effects on international trade as well as minimise social, economic, and environmental impacts on other Parties, especially non-Annex I nations.

This raises the question as to whether border-tax adjustments would satisfy Articles 2.1(a) and 2.3 of the Kyoto Protocol. Given that the aim of trade-related adjustments is to: (i) internalise the spillover costs of climate change; (ii) protect genuinely efficient domestic industries; and (iii) eradicate carbon leakage, it would seem they are well suited to Article 2.1(a). As for satisfying Article 2.3, this depends on what is meant by "adverse effects on international trade" and "impacts on other Parties". If an adverse effect on international trade means reducing the volume of internationally traded goods and services, then border-tax adjustments could be considered in violation of Article 2.3. However, WTO rules do not class all policies which reduce the volume of international trade in this way, since the WTO forbids the use of child labour in the production of internationally-traded goods even though this practice augments international trade. Given that environmental concerns constitute legitimate grounds to impose trade-related measures under Article XX of the GATT, there is no reason why such measures, even if they reduced the volume of international trade, would not be treated in the same way as child labour. After all, trade is a potential means to a higher level of economic welfare, not an end in itself. There is, therefore, every reason to believe that in the context of the Kyoto Protocol, border-tax adjustments could be viewed just as favourably as they are by the WTO's rules on international trade.[71]

Perhaps more contentious is whether border-tax adjustments, which competitively disadvantage many low-GDP countries, would unduly affect some of the world's poorest nations. In Chap. 3, it was explained how compensating tariffs need not have a negative impact on the welfare of low-GDP nations. To reiterate, whilst compensating tariffs make it more difficult for low-GDP nations to export their goods to high-GDP countries, they force policy-makers in the world's low-GDP nations to focus their attention on boosting domestic spending on domestically-produced goods. Achieving this not only requires policies which bolster local purchasing power and its equitable distribution, but policies that increase the relative allocation of a nation's scarce resources to the production of goods which meet the needs and desires of its own citizens. Furthermore, should producers in low-GDP nations wish to compete on international markets, they are compelled to genuinely improve efficiency and reduce the emissions-intensity of their production activities, since these advances would be required to negate the cost-raising impact of a border-tax adjustment.[72] Thus, if introduced, compensating tariffs would encourage poor nations to pull themselves up to the standards set by high-GDP countries rather than preserve their competitiveness by maintaining a welfare-reducing standards gap between themselves and the world's richest nations. In sum, there is no reason why border-tax adjustments should be deemed in violation of Article 2.3 of the Kyoto Protocol.

There is, however, one issue concerning border-tax adjustments that needs urgent clarification. When making its rulings on international trade matters, the WTO employs the principle that nations must treat imported goods no less favourably than 'like' goods produced domestically. Thus, a tariff can only be applied to an imported good if a tax or duty of the same magnitude has been imposed on a similar domestically-produced good.

For some time, it was the belief that the definition of a 'like' product was confined to the physical characteristics and performance of the good in question. Consequently, it was argued that differences in the way in which goods are produced—commonly referred to as 'process and production methods' (PPMs)—could not be used to justify a border-tax adjustment even though the differences could have social and ecological implications as well as significant impacts on the relative cost of production (Saddler et al. 2006). However, recent WTO rulings suggest that differences in PPMs can serve as sufficient grounds to justify border-tax adjustments. In what became a highly publicised case, the WTO allowed the US Government to sanction the importation of fish products involving fish-harvesting practices that contravened US regulations designed to protect endangered sea turtles (e.g., the US regulation of shrimp imports) (Deal 2002; Frankel 2005). Importantly, when making this decision, the WTO adopted the policy that restrictive trade measures are legitimate provided they are not "arbitrary or unnecessarily discriminatory" (Saddler et al. 2006). Since this case, the WTO has made similar rulings on precisely the same grounds. Presumably, then, as long as border-tax adjustments comply with the 'non-discriminatory' principle, there is no reason why compensating tariffs would not be permitted in instances where the cost disadvantages confronting trade-exposed producers arise because of clear differences in the emissions-intensities of the PPMs employed by domestic and foreign producers.

Despite what would appear as legitimate ground for imposing compensating tariffs, the current situation regarding border-tax adjustments and PPMs is vague and not at all conducive to long-term mitigation decisions. For this reason, WTO rules should be modified to clarify where and how differences in PPMs can be used to justify border-tax adjustments. In doing this, UNFCC Parties should work with the WTO to conduct a thorough analysis of relevant WTO rules prior to designing a border-tax adjustment system that would allow nations to make legitimate adjustments based on cost differences caused by discrepancies in greenhouse gas price signals, not disparities in the emissions-intensities of production methods.

9.2.5.3 What Should Border-Tax Adjustments Entail and How Should They Be Implemented?

The aim of a border-tax adjustment is to preserve the international competitiveness of emissions-sensitive producers whilst leaving intact any greenhouse gas price signal established within the domestic economy (Saddler et al. 2006). The adjustment should exist in two forms. The first is a *tax rebate* paid to emissions-sensitive 'exporters' to offset the increase in production costs arising from the introduction of an emissions tax or emissions-trading system. The second is a *compensating tariff* applied to imported goods to offset any cost disadvantage that an emissions tax or emissions-trading system would exact on trade-exposed domestic producers. Although the tariff would be applied to all relevant imported goods, the tax rebate would only be paid to domestically-located firms to cover their exported product. This means that the remainder of their emissions-intensive output would not attract a tax rebate. By confining the rebate in this way, all domestically-produced output consumed locally would continue to be subject to a greenhouse gas price signal to induce a general a shift towards less emissions-intensive forms of production (Saddler et al. 2006; Garnaut 2008, 2011).

There is, however, a weakness and associated dilemma pertaining to the payment of a tax rebate to domestically-located exporters. In terms of preventing carbon leakage, there is no reason to grant exporters a rebate if the destination (importing) countries are subject to emissions targets. The reason for this is that, even with disparate greenhouse gas price signals, there is no way that excessive aggregate emissions can result from industrial flight, since the emissions of both the origin and destination nations would be capped. On the other hand, if a tax rebate is not granted to domestically-located exporters, destination countries would be rewarded in the short-term for employing emissions-intensive production methods—i.e., by allowing them to reduce competition from exports—but would be locked into high-emissions pathways. Complicating the issue further, if a national government grants a tax rebate to exporters, the production decisions made by firms with a predominant export-focus would be influenced by an inadequate greenhouse gas price signal. This would result in the excessive allocation of resources to export-based industries, which would draw resources away from less emissions-intensive industries and increase the cost of achieving domestic emissions targets.

Thus, policy-makers face the seemingly intractable problem of having domestically-located exporters subject to an inadequate price signal should they grant them a tax rebate or, if they do not, would-be destination countries being locked into high-emissions production methods. As much as both situations pose a problem, I believe the former is less so because, firstly, the exporting of product is a residual activity for many firms, and secondly, any reduction in exports and associated job losses would be counteracted to some extent by an expansion of job opportunities in emerging low-emissions industries. Should the loss of existing jobs exceed the number of new jobs created, this too would not be a cause for concern if the central government instituted a Job Guarantee as described in Chap. 3. In this case, the resources freed up by the decline in emissions-intensive export industries would be acquired by the central government and largely provided in the form of public goods.

Given the aforementioned, I believe a tax rebate should be granted by a nation to eligible exporters until such time as any greenhouse gas price gap between it and other nations vanishes. This aside, the trade-related dilemma faced by policy-makers demonstrates that border-tax adjustments are a stop-gap measure only and that the ultimate solution is greenhouse gas price convergence and ensuring all nations are subject to emissions targets (Garnaut 2008, p. 231).

Contrary to what many people believe, border-tax adjustments are not rare. They have been applied in the past to ozone-depleting chemicals in the US (see Hoerner 1998) and are a common feature of energy tax systems, such as fuel excises. In Australia, for example, petrol produced domestically is subject to a fuel excise if sold within Australia, but is excise-exempt if exported. To ensure parity, imported petrol is subject to an import duty equal to the fuel excise imposed by the Australian Government on domestically produced and consumed petrol—a practice entirely in keeping with the WTO rules on international trade (Saddler et al. 2006). If border-tax adjustments were introduced for climate change purposes, one would expect a similar approach to be adopted to bridge the cost gap between foreign producers and trade-exposed domestic producers.

A desirable feature of border-tax adjustments is that they can be administered relatively simply and transparently. Nonetheless, should border-tax adjustments be broadened to shield industries indirectly affected by a greenhouse gas price signal, as I believe they should, the system would be administratively more complex to operate. Although this would increase administrative costs, I believe they would be much less than the cost of inaction.

One of the major advantages of a border-tax adjustment over other correction methods is that it offers the greatest potential to achieve a multilateral or international solution to the dual problems of carbon leakage and international resource misallocation. As we have seen, it is possible for a nation to unilaterally introduce border-tax adjustments in keeping with the WTO's rules on international trade. However, should many countries wish to take such action, a much broader solution could be achieved through multilateral negotiations. Since the ensuing agreement would provide explicit acknowledgement of the right of nations to impose border-tax adjustments, the solution would be substantially more robust and legitimate than a collection of unilateral courses of action (Saddler et al. 2006).

In light of existing Kyoto arrangements, Saddler et al. (2006) have referred to the possibility of a coalition of Annex I nations establishing a border-tax adjustment scheme. Whilst this is worthy of consideration, I believe the scheme would be stronger, fairer, and more effective if it was negotiated by all UNFCCC Parties in co-operation with the WTO and subsequently embodied in a new global climate change protocol. This would not only encourage non-Annex I nations to take on emissions obligations, since it would lessen and ultimately abrogate the need for border-tax adjustments, it would increase the likelihood of non-Annex I countries accepting the scheme, particularly if the border-tax arrangements were accompanied by a package offering funds and new technologies to assist them with their mitigation and climate change adaptation endeavours.

9.2.5.4 Designing a Feasible Border-Tax Adjustment Scheme

A border-tax adjustment scheme should be designed on the basis that adjustments are best applied by adopting the 'destination' principle as opposed to the 'origin' principle.[73] In the case of emissions-intensive imported goods, application of the destination principle would involve the imposition of a tariff to compensate for the cost impact of a gap in the greenhouse gas price at home (the destination country) and that of the exporting country (Saddler et al. 2006). As for emissions-intensive exports, the adjustment would be much the same as described earlier—namely, a tax rebate equal to the increase in production costs arising from having to absorb a price on greenhouse gas emissions when foreign producers of 'like' goods do not.

Since border-tax adjustments should only be applied where a disparity in the domestic and foreign greenhouse gas price results in a clear competitive disadvantage to domestic producers, there is a need to determine which products are eligible for a tax adjustment. There is also a need to determine the size of the adjustment on eligible products. Because greenhouse gas prices differ between countries, it would be impractical to apply a border-tax adjustment to all traded products. An adjustment is only required where the greenhouse gas price gap significantly raises the cost of domestically-produced goods *vis-à-vis* foreign-produced goods, and where the goods in question are trade-exposed (Saddler et al. 2006). To ensure the scheme is administratively manageable, it is necessary to install an objective-eligibility criteria.

Garnaut (2008) has defined an eligibility threshold—in effect, a 'cost-burden threshold test'—in terms of a three-step assessment process. In the case of emissions-intensive exports, it would entail: (i) estimating the expected uplift in the world price of an exported good on the assumption that foreign producers are subject to the same greenhouse gas price as the domestic manufacturers of the same good; (ii) granting eligibility for the domestically-produced good if the uplift in the unit price exceeds 3 per cent of the world price; and (iii) setting the tax rebate equal to the difference in the expected uplift price and the 3 per cent threshold price.[74] To demonstrate how this process would work, Garnaut envisages a world price for a particular good equal to $1,000. Garnaut then posits a $90 increase in the world price if foreign manufacturers were required to absorb the same greenhouse gas price as domestic producers. In this situation, a tax rebate of $60 would be granted to domestic producers of the exported product equal to the difference in the expected world price of $1,090 and the threshold price of $1030 (Note: $1,030 = 1.03 × $1,000).

The same cost-burden threshold test would also apply to domestic producers exposed to foreign imports. Thus, if the world price and the expected uplift price of the good under review were, for argument sake, the same as the example above, a $60 tariff would be applied to the imported product.

Because of the interdependency phenomenon, eligibility for a border-tax adjustment should also extend to trade-exposed goods indirectly affected by input cost increases arising from an introduced greenhouse gas price. Since the entire burden of a greenhouse gas price would rarely be passed on in full by sellers upstream in the supply chain (as input-suppliers to downstream producers), there is less likelihood of a gap appearing between the expected uplift price and the 3 per cent threshold price as one moves down the supply chain. Thus, the number of eligible goods at the end of the supply chain—especially the goods produced in the services sector—would be smaller than those upstream, as would be the size of any border-tax adjustment.

It is also worth recognising that accounting for the indirect burden of a greenhouse gas price would be a more complicated task than accounting for the direct burden. This is because the estimation of the indirect burden would require knowledge of the greenhouse gas price paid upstream in the supply chain as well how much of the price has been passed on to downstream producers. Whilst Saddler et al. (2006) recognise the increased complexity that this presents, they believe the task is eminently manageable given the eligibility test is likely to be confined to industries whose energy and emissions characteristics are well understood. Moreover, like Sinner (2002), Saddler et al. (2006) believe there is no need to ensure the accounting system is perfect, since the objective is not to achieve optimal outcomes, but to set border-tax adjustments with the capacity to alleviate competitiveness problems, minimise carbon leakage, and improve the international allocation of resources without violating international trade rules.

There are two further complications requiring amelioration. The first, which is necessary to calculate a compensating tariff, involves determining the 'greenhouse gas content' of imported goods. This, in turn, requires information on the energy use and production methods employed in exporting nations—information not readily available to authorities in the destination country (Saddler et al. 2006). In the 1980s, an equivalent problem confronted the US Government when it contemplated border-tax adjustments on imported products made with chemicals taxed domestically under the Superfund Chemical Excises legislation. Following negotiations with a number of trading partners and approval from a Conciliation Panel under the GATT (now WTO), a two-tiered system was introduced with the first tier designed to encourage foreign exporters to provide detailed information on the quantities of taxable chemicals used in the production of their goods. The US Government then used the information to calculate and impose import duties based on prevailing US excise rates. Upon a foreign exporter failing to provide the necessary information, the second tier of the system was invoked. This entailed the US Government imposing an import duty using a 'predominant method of production' approach, which involved calculating the import duty based on the quantity of taxed inputs 'used' if the same product was manufactured in the US (Hoerner 1998; Saddler et al. 2006).[75] Because of the success of the system, a similar system was installed by the US Government to make border-tax adjustments in relation to ozone-depleting chemicals under the Ozone Depleting Chemicals Tax (Saddler et al. 2006).

Saddler et al. (2006) believe that a similar two-tiered system could be devised to enable nations to make border-tax adjustments in response to disparate greenhouse gas prices. They also argue that the system would be relatively easy to manage in view of the small number of products likely to attract a border-tax adjustment. Although I agree with Saddler et al. with respect to the applicability and viability of the two-tiered system, as mentioned earlier, I believe there would be a larger number of goods captured by the system than Saddler et al. envisage. This would render the system a little more convoluted than expected. However,

I believe a better and more streamlined process—which would arise if a scheme was negotiated by all UNFCCC Parties in collaboration with the WTO—would help establish uniform guidelines and standardised eligibility tests that could alleviate many potential operational complexities.

A second complication exists where border-tax adjustments are made in the presence of an emissions-trading system. This is because, unlike an emissions tax, the greenhouse gas price is constantly fluctuating. Consequently, eligibility and the size of a border-tax adjustment, which depends on the expected uplift in the world price of competing goods, would be forever changing. To overcome this problem, Saddler et al. (2006) suggest that a greenhouse gas price could be stipulated for tax adjustment purposes on the basis of an initial emissions permit price and then be reset (perhaps, annually) as the movement in the permit price exhibits a relatively predictable trend (see Fig. 7.9).

9.2.5.5 A Standardised Procedure for Assessing and Imposing Border-Tax Adjustments

Given the above, a border-tax adjustment scheme would operate on the basis of the following standardised procedure. Firstly, upon a firm or industry seeking compensatory action, a government authority would employ a UNFCCC-sanctioned 'cost-burden threshold test' to determine whether the domestically-produced goods in question are eligible for protection via a border-tax adjustment. Assuming they are, the government authority would calculate the size of the border-tax adjustments to be imposed—whether it be a compensating tariff in the case of imported products or a tax rebate in the case of emissions-intensive exports. Having done this, the authority would submit a border-tax adjustment application with supporting evidence on behalf of the claimant to the WTO.

Secondly, a WTO panel, working in collaboration with the UNFCCC, would assess the border-tax adjustment application. Having completed their review, the panel would approve or reject the claim for a border-tax adjustment. In some instances, the panel might approve an individual claim but reduce the size of the allowable tax adjustment. All border-tax adjustments would be periodically reviewed and reset as circumstances changed over time—for example, as the gap in the greenhouse gas price between nations narrowed; as inputs varied in line with technological advances; and as the trade-exposure of particular goods waned. To assist the WTO with the ongoing review process, governments would be called upon on to provide updated information not readily available to the WTO panel.

Thirdly, assuming approval for a border-tax adjustment has been granted, countries whose cost advantage would be negated by the sanctioned adjustments would be entitled to appeal the WTO's decision. Countries lodging border-tax adjustment applications would also be permitted to appeal the WTO's rejection of any tax adjustment claims. An appeals tribunal, which might best rest with the UNFCCC to avoid a conflict of interest, would make a final judgement on border-tax adjustments. Lastly, where border-tax adjustments are approved, they would be imposed by the national authority in collaboration with domestic taxation and/or customs officials.

9.2.6 Adaptation—Incorporating Commitments in a New Global Protocol to Boost Adaptation Efforts

As revealed in Chap. 8, a great deal of work at the international level has gone into financing adaptation activities, especially adaptation measures in non-Annex I nations. However, less has been done in a more concrete sense to establish adaptation plans of action. Indeed, one of the criticisms of the Kyoto Protocol has been the lack of explicit adaptation measures embodied within it as well as little if any detail regarding obligations of individual nations (IPCC 2007d).

Early on in the assessment of adaptation measures, Parry et al. (2005b) recognised that effective adaptation would not be realised unless adaptation action was incorporated into broader development policies and practices, including decision-making process at the local, sectoral, and national levels. Stern (2007, 2009) has echoed these sentiments by urging national governments to integrate adaptation into their budget plans and development programmes and by encouraging development banks to promote a coherent adaptation response to climate change. In particular, Stern has stressed the important role that the global community can play in assisting nations, especially low-GDP nations, to develop national development strategies that take account of adaptation requirements across all levels of government (Stern 2007). In the process, Stern believes the national development strategies of Least Developed Countries (LDCs)—as the countries most vulnerable to climate change—should be based on the National Adaptation Programmes of Action (NAPAs) already generated from financial resources allocated from the Least Developed Countries Fund. To recall, NAPAs are designed to identify short-term and long-term adaptation priorities specific to national circumstances. According to Stern, formulating adaptation strategies in this way would assist in the efficient and low-cost allocation of resources for adaptation purposes, which would increase the effectiveness of adaptation programmes.

As important as it is to invest in adaptation infrastructure and institutional capacity at the local, sectoral, and national levels, it is also widely agreed that an effective adaptation response will require adequate investment in adaptation-facilitating public goods at the global level (Stern 2007; IPCC 2007d, 2014a; Garnaut 2008; Atteridge et al. 2009; GEF 2012). As alluded to in Chap. 6, areas where international investment in global public goods is urgently needed include: (i) climate change research, monitoring, and forecasting networks; (ii) information-sharing systems; (iii) international response plans and logistical response capacity[76]; (iv) research and development into climate change-resilient crops and improved irrigation methods; (v) heat-health warning systems; and (vi) the detection and subjugation of climate change-induced disease outbreaks.

Although more must be done both nationally and internationally to put plans into practice, adaptation action has progressed in recent years in response to observable climate change impacts (IPCC 2014a). Most of these actions have been initiated to address local and sectoral concerns of which a large proportion has been reactive rather than proactive in nature (Amundsen et al. 2010; Mullan et al. 2013).[77] Some progress has also been made with regard to the development of nation-level plans and adaptation strategies; however, there is little evidence of these plans being implemented on a broad scale (IPCC 2014a). The reason for this is that the transition from planning to implementation continues to be beset with institutional, resource, and capacity constraints (Patt and Schröter 2008; Bryan et al. 2009; Wolf et al. 2009; Amundsen et al. 2010; Nelson et al. 2010; Olesen et al. 2011; Pfister et al. 2011; Sorte et al. 2011; Runhaar et al. 2012; Carlsson-Kanyama et al. 2013; Lesnikowski et al. 2013). In part, these constraints are the consequence of a lack of global public goods, which indicates that international investment in this area has lagged behind the planning process. Such impediments reinforce the need for the international community to scale up its investment in these increasingly critical public goods.

What about the cost of adaptation? Estimates vary greatly, largely because it is difficult to forecast the exact impact of future climate changes and the action required to minimise the consequential damages (Fankhauser and Burton 2011; Christiansen et al. 2012). Initial estimates put the global cost at between US$50–100 billion for 2015, with the annual cost rising substantially into the future (World Bank 2006; UNDP 2007; Stern 2009). As for low-GDP nations, the annual cost by 2030 is expected to be in the range of US$25–70 billion (UNFCCC 2007), but many times more by 2060 (Parry et al. 2009; Smith et al. 2011).

Unfortunately, the quantity of funds available to support adaptation programmes and activities, especially the adaptation needs of low-GDP nations, falls well short of these estimates. There is, to recall, just US$10.2 billion at present in the UNFCCC's Green Climate Fund (around half of which is earmarked for mitigation) and a miniscule US$0.2 billion in the Kyoto Protocol's Adaptation Fund.[78] Furthermore, only a small fraction of the substantial climate change funds being raised by the private sector and bilateral finance institutions is being directed towards adaptation.[79] Whilst the aim of the Green Climate Fund is to raise US$100 billion a year by 2020, great uncertainty remains as to where these funds will come from and whether such an ambitious target will be achieved without explicit commitments from Annex I nations and various finance institutions. There is also a concern that a heavy reliance on the private sector to capitalise the Green Climate Fund would reduce the financial resources available for adaptation purposes (Abbot and Gartner 2011).

If, as suggested, adaptation needs were to be incorporated into broader development plans and high-GDP nations delivered on their promise to provide 0.7 per cent of their GDP for overseas development assistance, it follows that considerably more funds would automatically be channelled into development programmes generating significant adaptation benefits. As much of a boost that this would be, it would not guarantee sufficient funds to support adaptation activities in non-Annex

I nations. To raise adaptation funding to the required level, a further compulsory funding commitment should be integrated into a future global protocol. This so-called 'adaptation' commitment would exist in the form of a percentage contribution of the real GDP of Annex I nations, or as an injection of a specific quantity of funds based on the estimated future cost of adaptation.[80]

Once again, the funds should be deposited into the Green Climate Fund. Although, strictly speaking, the funds would constitute official development assistance, safeguards would need to be installed to ensure they do not displace funds already committed for other assistance purposes (Garnaut 2008).

Not unlike mitigation, an adaptation funding commitment would eventually taper off as early adaptation measures reduced the growth in damage costs; as technological advances improved the quality of adaptation responses; as mitigation efforts slowed the rate of climate change; and as low-GDP countries became increasingly capable of supporting their own adaptation activities.

In the meantime, and since the adaptation requirements of countries differ considerably, funds allocated to support adaptation action in non-Annex I nations should be determined by assessing their respective needs and capabilities. This should be done through the development of vulnerability metrics which measure: (i) a nation's vulnerability to climate change; (ii) the extent of the action required by a nation to minimise damage costs; and (iii) the capacity of a nation to respond to the impacts of climate change (Srinivasan and Prabhakar 2009; Moss et al. 2012). Having done this, disbursing agencies should employ a rigorous process to select the metrics most applicable to particular adaptation contexts (Preston et al. 2009). As for the metrics themselves, they along with the assessment process should be standardised by the UNFCCC Parties at future COP meetings.

In Table 1.4, it was revealed that the frequency and severity of extreme weather events, such as floods, droughts, heat-waves, and storms, are expected to rise throughout the current century and beyond. Because of this, there will continue to be climate change-induced losses that no amount of adaptation can avert (Verheyen 2013; see, also, the explanation of damage costs in Fig. 6.1). To help pay the cost of emergency responses and restoration programmes in the wake of weather-related disasters, it will become increasingly important to develop private (and public) insurance schemes aimed specifically at dealing with the extreme impacts of climate change (Warner et al. 2009; IPCC 2007d, 2014a). Insurance can reduce the damage cost of climate change relative to ad hoc disaster-relief by spreading climate change risks away from the most vulnerable households, communities, and nations (Stern 2007; Warner et al. 2009).

Unfortunately, the capacity to utilise insurance varies across the world, with insurance markets in many low-GDP countries still in the embryonic stage of development (Churchill 2007; Warner et al. 2009). Given this and the fact that many of the world's poorest nations are extremely vulnerable to severe weather-related events, the issue of 'loss and damage' has become a hotly debated concern since low-GDP countries raised the stakes on the issue at the COP-18 meeting in Doha in 2012. Despite opposition from the USA to the creation of a new funding mechanism to deal with loss and damage cases, the UNFCCC Parties pressed ahead and established the Warsaw International Mechanism for Loss and Damage at the COP-19 meeting in 2013.[81] The principal function of the Mechanism is to promote the implementation of approaches to address losses and damages from climate change in a comprehensive, integrated, and coherent manner (Roberts et al. 2013). In order to cohesively guide the implementation of the Mechanism, the UNFCCC Parties adopted a 2-year work plan developed by the Mechanism's Executive Committee at the COP-20 meeting held in Lima in 2014.

At present, the aim of the Warsaw Mechanism is to deal with risk management, insurance matters, and rehabilitation issues under the guidance of the Conference of the Parties. Roberts et al. (2013, p. 11) believe the Mechanism or any new oversight institution should have autonomous decision-making authority but be confined to "residual loss and damage only", thus leaving adaptation action to other already-existing institutions and funding mechanisms.

It is my belief that a Global Climate Change Emergency Fund should be established to address the loss and damage issue in the manner loosely prescribed by Roberts et al. (2013). The Fund would be used to cover costs caused by a weather-related event that results from, or is accentuated by, climate change. The latter proviso is important in that, for example, an inherently cyclone-prone nation would not be entitled to compensation to cover damage costs arising from a typical cyclonic event.[82] To ensure the Emergency Fund is confined to climate change-induced disasters, indexes should be calculated to determine the typical frequency of a particular weather-related event (e.g., once a decade) and the mean damage impact.[83] Should an event exceed either or both, a nation would be entitled to compensation from the Fund. Where an impoverished nation does not qualify for compensation—i.e., where a destructive event is not deemed to be 'atypical'—assistance would come in the form of foreign aid, as is the present approach when responding to natural disasters. However, to assist low-GDP nations with the cost of 'normal' events, the body overseeing the Emergency Fund should provide an array of insurance products that impoverished countries have had great difficulty accessing in the past. At the same time, the body should play a constructive role in developing insurance markets in countries where they are presently inadequate.

The capitalisation of the Emergency Fund would be expedited by the creation of three separate funding instruments. The first instrument would be a funding commitment from Annex I countries equal to 0.1 per cent of their real GDP. Like other commitments, this would only apply to Annex I nations with a per capita real GDP of Int$10,000 or more (at 2004 prices). The second funding instrument would involve siphoning some of the revenue generated from the emissions permits auctioned under the Kyoto Protocol's International Emissions Trading system and any permits auctioned post-2020 under a new global protocol. I would recommend a siphoning rate of one-quarter of the auction revenue raised.

Because a new global protocol will eventually require non-Annex I nations to meet emissions targets, which will almost certainly draw them into an emissions-trading system, the second funding instrument would result in all countries making at least some financial contribution to the Global Climate Change Emergency Fund. In this sense, the contribution pertaining to the second funding instrument would serve as an 'insurance premium' that all countries would pay towards a larger international insurance pool. The third and final funding mechanism would involve international levies on air passenger travel and bunker fuels.

No doubt, some people would consider the above-recommended 'adaptation', 'mitigation', and 'loss and damage' commitments as excessive financial burdens on Annex I nations, particularly when some of them (e.g., Greece) are dealing with severe economic problems at home. Domestic difficulties aside, these additional commitments should not be viewed as donations but as compensation from the countries most responsible for, and the major beneficiaries of, past greenhouse gas emissions (Stern 2009; Rübbelke 2011). These commitments should therefore be seen as a crucial equity component of a new global protocol and a means of inducing non-Annex I nations to accept greenhouse gas emissions targets.

9.2.7 Finance—Streamlining and Consolidating Climate Change Funding Arrangements

Having extensively outlined existing climate change financing mechanisms and institutions in Sect. 8.2.8, I have no intention of repeating myself here. What I would like to say in this sub-section is something about how climate change funding arrangements could be better co-ordinated to improve their efficiency and effectiveness. I will also take the opportunity to summarise the funding instruments I believe should be embodied in a new climate change regime and how and where the funds they raise should be disbursed.

One of the criticisms of the present climate change architecture, of which the Kyoto Protocol is a major feature, is that climate change funding arrangements are excessively convoluted and poorly co-ordinated, thus leading to unnecessary duplication and waste.[84] As a consequence, maximum benefit is not being obtained from the funds being raised for mitigation and adaptation purposes. To improve the situation, climate change funds should, where possible, be unified and the process of accessing them simplified to facilitate the easier and more effective uptake of financial resources by low-GDP nations (Stern 2007). Achieving this will require many existing financing mechanisms and institutions to be streamlined and consolidated, which would lead to the creation of new funding mechanisms, some of which I have previously outlined.

Given the need for various funding sources to serve different climate change functions, the effective co-ordination of financing mechanisms and institutions is crucial (Gigli and Agrawala 2007; Smith et al. 2011). The importance of co-ordination, as Stern (2009) reminds us, has been recognised by the Paris Declaration on Aid Effectiveness and the Accra Agenda for Action (OECD 2005/2008), which provides constructive guidelines on how to ensure funds from a wide variety of sources are delivered more efficiently and effectively. These include, *inter alia*: (i) the establishment of common arrangements at the national level to improve the planning and evaluation of donor activities; (ii) simplification of donor policies to align them with partner countries' priorities, systems, and procedures; and (iii) efforts to harmonise financial assistance with the broader development agendas of partner nations. Notwithstanding these recommendations, co-ordination would also be improved by mainstreaming mitigation and adaptation across development programmes and encouraging multilateral and bilateral finance institutions to integrate climate change action into their budgets and project evaluation processes (Stern 2007).

On the positive side, one of the major recent developments has been the establishment of the Green Climate Fund to consolidate the UNFCCC's funding mechanisms. As mentioned in Chap. 8, the aim of the Green Climate Fund is to replace the Least Developed Countries Fund and Special Climate Change Fund in the process of becoming the UNFCCC's main multilateral financing mechanism in support of climate change action in low-GDP nations. The Green Climate Fund is also likely to subsume the role of the Adaptation Fund at the conclusion of Kyoto Protocol's second commitment period in 2020. While the termination of the Least Developed Countries Fund would result in the loss of a Fund specifically aimed at supporting the world's most impoverished nations, favourable treatment of these countries would be maintained by devoting a certain percentage of the resources within the Green Climate Fund to LDCs or, as recommended earlier, by applying vulnerability metrics to assess the funding needs of eligible nations, which would naturally favour LDCs.

Given the need to streamline and consolidate the funding arrangements and institutions pertaining to climate change, I believe the following should exist in a new global climate change regime to help fund and facilitate climate change action (see Table 9.3). The first is the already-mentioned Green Climate Fund, which would remain under the control of the UNFCCC. The Green Climate Fund would continue to be used to finance climate change mitigation and adaptation activities in non-Annex I nations—essentially on a 50:50 basis—where the former activities would include the transfer of technology and capacity building. The Fund would be capitalised by five mechanisms/commitments plus a one-off injection of resources from whatever remained in the institutional Funds scheduled to terminate in the near future (e.g., the Adaptation Fund). Two of these five mechanisms already exist—the voluntary pledges made by Annex I nations (approximately Int$100 billion per year by 2020); and the 2 per cent levy on emissions units exchanged for the first time under the existing or a future global protocol (e.g., AAUs, CERs, ERUs, and RMUs). With the rise in emissions-trading and a likely proliferation in the use of the various Kyoto flexibility mechanisms (e.g., the Clean Development Mechanism and the Joint Implementation facility), the levy should raise approximately Int$20 billion per year by 2020.

The remaining three funding mechanisms are new and should be incorporated into the global protocol that supersedes the Kyoto Protocol. The first two are the 'mitigation' and 'adaptation' commitments of Annex I nations that I recommended earlier. With the commitments respectively set at 0.25 per cent of the GDP of Annex I nations, both would raise approximately Int$125 billion. The final mechanism would be a pledge to allocate one-quarter of the revenue raised from emissions permits auctioned under the International Emissions Trading system. Assuming the

Table 9.3 Funding institutions and mechanisms in a new climate change regime—annual funds by 2020

Funding institution	Funding mechanism/source	Int$ (approx.)	Purpose of disbursed funds	Recipients of funds
Green Climate Fund (*GCF*) • Governance: UNFCCC	• Voluntary contributions from Annex I nations (not including private-sector sources)	100 bn	• Climate change mitigation (including technology transfer and capacity building) • Climate change adaptation	• Non-Annex I nations
	• Carry over resources from defunct Funds	5 bn		
	• Mitigation commitment (0.25 % of GDP of AI nations*)	125 bn		
	• Adaptation commitment (0.25 % of GDP of AI nations*)	125 bn		
	• Revenue from IET (25 % of auction revenue)	250 bn		
	• 2 % levy on emissions units	20 bn		
	• Total	(625 bn)		
Global Climate Change Emergency Fund (*GCCEF*) • Governance: UNFCCC	• Loss and damage commitment (0.1 % of GDP of AI nations*)	50 bn	• Emergency relief following a climate change-induced disaster	• Low-GDP countries
	• Revenue from IET (25 % of auction revenue)	250 bn		
	• Levies on air fares/bunker fuels	25 bn		
	• Total	(325 bn)		
Carbon Fund of the Forest Carbon Partnership Facility (*FCPF*) • Governance: FCPF under auspices of the UNFCCC	• Annex I commitment (0.1 % of GDP of AI nations*)		• Sustainable management of forests • Minimise deforestation	• Low-GDP countries
	• Total	(50 bn)		

(continued)

Table 9.3 (continued)

Funding institution	Funding mechanism/source	Int$ (approx.)	Purpose of disbursed funds	Recipients of funds
Overseas Development Assistance (ODA)	• Monterrey and Gleneagles commitments (one-half of 0.7 % of GDP of high-GDP nations) • Total	(175 bn)	• Mitigation and adaptation concerns incorporated into broader development plans funded by ODA	• Low-GDP countries
Central, State/Provincial, and Local Governments	• Public finances • Total	(500 bn)	• Climate change mitigation • Climate change adaptation • Capacity building	• Domestic recipients and beneficiaries: ➢ Individuals, organisations, businesses accessing low-emissions and renewable-energy infrastructure ➢ Research institutes (R&D) ➢ Education and training institutions ➢ Private organisations ➢ Households
Multilateral and Bilateral Finance Institutions; Development Agencies	• Public finance (excluding government spending on domestic climate change activities and funds leveraged from the private sector) • Total	(200 bn)	• Climate change mitigation • Climate change adaptation • Capacity building	• Low-GDP countries
Private Sector	• Capital markets • Private sector income • Total	(1000 bn)	• Climate change mitigation • Climate change adaptation • Capacity building	• All countries
Insurance	• Insurance premiums • Total	(125 bn)	• Climate change adaptation • Compensation for climate change-induced damages	• All countries
Total funds		**3000 bn**		

Note *AI nations denotes Annex I nations

auctioning in 2020 of 20 Gigatonnes of emissions permits at Int$50 per permit, this would raise around Int$250 billion. In all, by 2020, approximately Int$625 billion would be available for climate change funding via the Green Climate Fund.

The second major institution would be a Global Climate Change Emergency Fund which, again, would be overseen by the UNFCCC. Designed to provide emergency relief for non-Annex I countries deleteriously affected by a climate change-induced event, the Emergency Fund would be resourced by: (i) a ‘loss and damage’ commitment from Annex I countries (0.1 per cent of the GDP of Annex I nations) (Int$50 billion); (ii) one-quarter of the revenue raised from emissions permits auctioned under the International Emissions Trading system (Int$250 billion); and (iii) levies on air fares and bunker fuels (Int$25 billion). Altogether, these three mechanisms would generate approximately Int$325 billion of financial resources per year for the Emergency Fund by 2020.

The third major funding institution—the Carbon Fund of the Forest Carbon Partnership Facility (FCPF)—would continue to serve its function of promoting the sustainable management of the world’s forests and minimising the global rate of deforestation. To ensure its adequate capitalisation, the Carbon Fund should be supported by an additional commitment equal to 0.1 per cent of the GDP of Annex I nations (approximately Int$50 billion per year by 2020). Given the nature of the Carbon Fund, which involves collaboration with REDD+ countries in a number of specialised areas, the Fund is probably best left in the hands of the FCPF, although the FCPF would need to work closely with the UNFCCC to limit overlap and wasteful duplication.

The fourth major source of funds would be the overseas development assistance provided by high-GDP nations to impoverished countries. This source of climate change funds would differ to the aforementioned in that it would serve as an indirect means of climate change assistance. This is because the funds would not directly finance climate change activities but be used to finance development programmes with mitigation and adaptation concerns built into them. Assuming that high-GDP countries meet their Monterrey (2002) and Gleneagles (2005) commitments by 2020 (0.7 per cent of GDP), and assuming one-half of all development activities funded by overseas development assistance can indirectly contribute to climate change endeavours, as much as Int$175 billion would be available for climate change purposes.

The fifth major source of climate change funds would be the spending by all levels of government on domestic climate change action. This spending, which would be financed by the usual public financial mechanisms, would be aimed at mitigation and adaptation activities, with a particular emphasis, where required, on institutional and capacity building. Beneficiaries of such spending would include all individuals, organisations, and business entities able to access newly-provided low-emissions and renewable-energy infrastructure that, in turn, would reduce the cost of using ‘clean’ energy/electricity (Note: this assumes there is a price on greenhouse gas emissions). Other recipients of publicly-provided funds would include: (i) research institutes engaged in research and development into renewable-energy and low-emissions/carbon sequestering practices; (ii) education and training institutions involved in creating expertise in renewable-energy and low-emissions technology and a workforce able to utilise such technologies; (iii) private organisations in receipt of subsidies or tax credits/rebates to encourage the development and uptake of low-emissions technology; (iv) households which enjoy the benefits of rebates and feed-in tariffs to encourage the use of energy-efficient appliances; and (v) low-income citizens who benefit from lower income tax rates designed to assist them with the higher cost of fossil fuel-derived electricity (ecological tax reform). At around 1 per cent of Gross World Product (GWP), the spending by governments on domestic climate change activities and infrastructure—excluding government contributions to institutional Funds and development banks/agencies—would amount to approximately Int$500 billion per year by 2020.

The sixth source of climate change funds would be the financial resources provided or indirectly leveraged by multilateral and bilateral finance institutions and development agencies on behalf of low-GDP nations. In view of past funding practices and the need for these institutions to scale up their climate change-related contributions, approximately Int$200 billion could be expected from this sixth source of climate change funds per year by 2020.

Far and away the largest slice of climate change finance will continue to be generated by the private sector, whether it be in the form of business investment in low-emissions technologies, climate change-induced defensive and rehabilitative spending, or household spending on domestic solar hot-water and solar electricity-generating systems in response to higher fossil fuel-generated energy costs. With significant greenhouse gas emission prices and a strong transition towards energy-efficient and low-emissions technologies/capital/appliances likely by 2020, one would expect the private-sector contribution to climate change activities to intensify and be twice that provided by governments for domestic climate change purposes. This would amount to around Int$1 trillion per year by 2020.

Finally, with the development of insurance markets to spread the risks associated with additional adaptation needs and compensation for climate change-induced damages, a further Int$125 billion of climate change finance could be expected per year by 2020—equivalent to the adaptation commitment cost of Annex I nations. This would bring the total funds available by 2020 for climate change-related needs and activities to approximately Int$3 trillion, which is something approaching 5 per cent of GWP by 2020. Incidentally, this is in line with the upper estimates of future climate change costs.[85]

Beyond 2020, some of the funding sources in Table 9.3 would diminish, whilst others would rise. In real terms, the total funding would initially increase, but, should mitigation measures be successful along with efforts to make the transition to a steady-state global economy, funding requirements would eventually fall, especially for mitigation requirements. However, depending on eventual changes in the Earth's climate, it is possible for adaptation funding together with compensation payouts for climate change damages to rise for a considerably long period of time. Whatever the case, there is no reason why climate change funds could not be provided in sufficient quantities and disbursed in a cost-effective manner whilst recognising the different responsibilities and obligations that each nation has in resolving the climate change crisis.

9.3 The Distribution of Emissions Rights

I mentioned at the very beginning of the chapter that a proposal to replace the Kyoto Protocol will only succeed if it is accepted by most if not all the world's nations. I also indicated that acceptance of a new global agreement will largely depend on the establishment of a reciprocal relationship between the existing Annex I and non-Annex I countries that will require concessions from them both. In view of the stances taken at past COP meetings and the 'common but differentiated responsibilities' enshrined in Article 3.1 of the UNFCCC, the following can be expected to remain as key negotiating demands:

- Demands of the existing *Annex I group*:
  - All countries must participate in the global effort to stabilise greenhouse gases at no more than 450 ppm of CO_2-e. This will require non-Annex I nations to be subject to binding emissions targets—some as early as 2021; others later, but all non-Annex I nations by 2031.
  - There is a need for all countries to converge to an equal per capita share of the world's global greenhouse gas emissions. In the meantime, a new emissions-reducing schedule/timetable should be calibrated with respect to per capita emissions as they exist at the end of the second Kyoto commitment period (i.e., the end of 2020).
  - In general, the per capita GDP of non-Annex I nations is rising. As this trend continues, non-Annex I countries should fund an increasing proportion of their own mitigation and adaptation activities. Eventually, all non-Annex I nations should become 'mitigation' and 'adaptation' self-reliant.
- Demands of the existing *non-Annex I group*
  - Non-Annex I nations must emit the majority of the greenhouse gases that remain within the 'safe' (450 ppm) emissions budget. Apart from being fair and equitable, this will allow non-Annex I countries to grow their per capita GDP to the optimum value of Int$15,000 (2004 prices).
  - Annex I nations must increase their commitment towards the transfer of renewable and low-emissions technologies and the funding of mitigation, adaptation, and emergency relief activities in non-Annex I countries.
  - Compared to Annex I nations, non-Annex I countries have a lesser capacity to undertake mitigation action and pay for the rights to emit greenhouse gases. They therefore need the opportunity to: (i) initially increase their greenhouse gas emissions; (ii) reduce their emissions at a slower rate than Annex I nations once subject to emissions targets; and (iii) pay a lower cost to emit greenhouse gases, which will require non-Annex I nations to engage in a separate emissions-trading system to the system covering Annex I countries.

Some of these demands are already accommodated by the existing Kyoto architecture, while others would be met if previously-recommended 'mitigation', 'adaptation', and 'loss and damage' commitments were incorporated into a new global protocol. Because I will be revealing my proposed new protocol in Chap. 10, I will refrain from saying anything precise in the remainder of this chapter about mitigation commitments, the timing of emissions cuts, and the basic framework of an effective global emissions-trading framework. For now, I want to outline some of the methodological approaches that have been proposed to distribute emissions rights beyond the Kyoto Protocol and reveal which approach I support. It will be my favoured approach that will form the basis of the emissions-reducing timetable in my proposed protocol.

Essentially six methodological approaches have been put forward to distribute emissions rights beyond 2020. They include:

1. *Status quo/acquired rights*—This so-called acquired rights approach involves distributing national emissions rights on the basis of current greenhouse gas emissions (Grubler and Nakicenovic 1994; Gupta and Bhandari 1999; Michaelowa et al. 2005b; Böhringer and Welsch 2006; Vivekanandan et al. 2008; Ekholm et al. 2010). Assumed in this approach is that the emissions level at the conclusion of an existing commitment period should constitute a 'status quo' right at the beginning of a new commitment period (den Elzen and Lucas 2005).[86]
2. *Per capita/egalitarian distribution*—The per capita distribution approach is premised on the egalitarian right of every global citizen to an equal share of the greenhouse gases that remain within the safe emissions budget (Grubler and Nakicenovic 1994; Gupta and Bhandari 1999; Baer et al. 2000; Winkler et al. 2002; den Elzen and Lucas 2005; Michaelowa et al. 2005b; Böhringer and Welsch 2006; Höhne et al. 2006; Persson et al. 2006; Garnaut 2008; den Elzen and Höhne 2010; see IPCC 2014b). Unlike the acquired rights approach, the per capita proposal places international justice perspectives on historical responsibility front-and-centre in emissions rights negotiations (Vivekanandan et al. 2008).
3. *Cumulative emissions (Brazilian proposal)*—The cumulate emissions approach is sometimes referred to as the 'Brazilian' approach following promotion of it by the Brazilian Ministry of Science and Technology at the COP-3 meeting in 1997 (Bodansky et al. 2004). The proposal involves sharing the global emissions-reducing burden on the basis of each nation's relative contribution to climate change (Filho and Miguez 2000; La Rovere et al. 2002; den Elzen et al. 2005). More specifically, it involves calculating the cumulative warming effect of each nation's greenhouse gas emissions and assigning the most stringent emissions targets to the countries with the highest historical emissions (La Rovere et al. 2002).
4. *Contraction and convergence*—The contraction and convergence approach recognises the need for global greenhouse gas emissions to contract in order to achieve the safe atmospheric concentration level of 450 ppm of CO_2-e or less. In achieving the desired target, each nation's emissions begin at current emissions levels and converge to an equal per capita share of global emissions by an

agreed year (Meyer 2000; Bodansky et al. 2004; Böhringer and Welsch 2006; Höhne et al. 2006; Persson et al. 2006; Stern 2007; Garnaut 2008; GCI 2008).

5. *Common but differentiated convergence*—The common but differentiated convergence approach is similar to the contraction and convergence proposal in that the emissions of each nation converge to an equal per capita share at some point in the future (Höhne et al. 2006). The variation in the approaches lies with the burden-sharing responsibilities of each nation. With the common but differentiated convergence approach, Annex I countries are not only required to endure more rapid emissions cuts, but, for some time, non-Annex I nations are permitted to increase their emissions and thus enjoy a period where their per capita emissions exceed the global average (Gupta and Bhandari 1999; Bodansky et al. 2004; Höhne et al. 2006; Stern 2007). The difference between this and the contraction and convergence approach is best illustrated by Fig. 9.6 (Note: the different groups depicted in Fig. 9.6 correspond to the four groups recommended as part of a modified Kyoto architecture).
6. *Triptych*—The triptych approach involves the application of a sectoral-based methodology to determine national emissions targets (Groenenberg et al. 2002, 2004; Höhne et al. 2003; Bodansky et al. 2004; den Elzen and Berk 2004; Michaelowa et al. 2005b; Stern 2007). With this approach, variables are assigned to reflect the greenhouse gas-intensity and emissions-reducing potentials of various economic sub-sectors. When the triptych model was first proposed, three sub-sectors were singled out—electricity generation; energy-intensive industries; and households (including transportation). The methodology has since been extended to include methane (CH_4), nitrous oxide (N_2O), and CO_2 emissions from forestry. Once estimated, the variables are assumed to linearly converge to a uniform global value at a future date. It is upon this convergence that future emissions targets are set, thus making the triptych proposal equivalent to a 'sectoral' contraction and convergence approach (Michaelowa et al. 2005b). The triptych approach has been used by the European Union to distribute the emissions rights of its member nations within its overall emissions budget under the Kyoto Protocol (Stern 2007).

To assess each of these burden-sharing approaches, I will borrow an assessment criteria posited by Bodansky et al. (2004) that has been widely applied by a number of analysts. The first part of the criteria is the policy element, where the following needs to be considered:

- *Environmental effectiveness*—Will the distribution approach achieve the goal of stabilising the atmospheric concentration of greenhouse gases at no more than 450 ppm of CO_2-e?
- *Equity*—Is the proposal equitable?
- *Flexibility*—Is the proposal flexible enough to allow for the adjustment of emissions targets in response to changing scientific information or unforeseen economic circumstances?

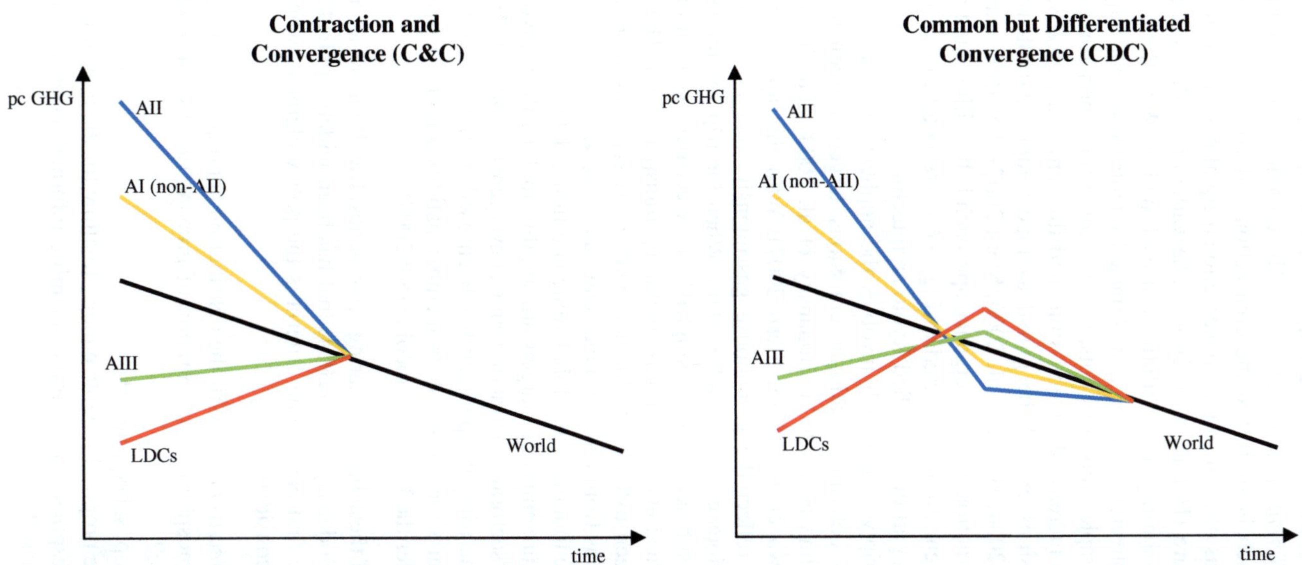

Fig. 9.6 Schematic representation of per capita GHGs under 'contraction and convergence' (C&C) and 'common but differentiated convergence' (CDC) approaches to the distribution of emissions rights. *Source* Modified version of Höhne et al. (2006). *Notes* pc GHG denotes per capita greenhouse gas emissions. AII denotes the Annex II group of nations. AI (non-AII) denotes the Annex I (non-Annex II) group of nations. AIII denotes the Annex III group of nations. LDCs denotes Least-Developed Countries

The second half of the criteria is the political element. On this occasion, the following requires consideration:

- *Negotiable commitments*—Is the proposal sufficiently compatible with the national development goals and demands of Annex I and non-Annex I Parties to be broadly accepted by the world's nations?
- *Implementable commitments*—Can the distribution approach be successfully implemented? In other words, is the approach compatible with the capabilities of the national and international institutions upon which the distribution of emissions rights and associated compliance mechanisms depend?

Starting with the acquired rights approach, it would be possible to achieve the 450 ppm target if, having started from a status quo position, the collective emissions of the world's nations were reduced in line with the cuts recommended by Anderson and Bows (2008). Hence, this approach has the potential to pass the 'environmental effectiveness' test. Nonetheless, without convergence to an equal per capita share of emissions, it fails the 'equity' test. Indeed, while the acquired rights approach has found support among some high-GDP countries, it is strongly opposed by many non-Annex I nations, especially those demanding the right to increase their greenhouse gas emissions to achieve national development goals (Michaelowa et al. 2005b). Given this latter aspect, the acquired rights approach fails the 'negotiable commitments' test.

There is also some doubt as to whether the acquired rights approach could be successfully implemented, even if it could be agreed upon. The reason for this is that the approach would require low-GDP nations to acquire additional emissions rights—most probably through international emissions trading—in order to grow their real GDP. However, it is unlikely that these countries would have sufficient income to purchase the required number of emissions permits, nor, for the time-being, the institutional capacity to participate alongside high-GDP nations in an emissions-trading environment.

Similar to the acquired rights approach, the per capita distribution proposal passes the 'environmental effectiveness' test if per capita emissions are adequately reduced to ensure a safe atmospheric concentration of greenhouse gases. What's more, it is a very equitable approach, although it has been argued that it still favours Annex I nations insofar as their historical emissions lie above the global average and hence they would emit more than their fair share of greenhouse gases. It is for this reason that non-Annex I nations have insisted on non-Annex I countries compensating them in the form of mitigation and adaptation funding.

One weakness of the per capita distribution approach is that it would fail the 'flexibility' test if most countries insisted on maintaining equal per capita emissions rights when, because of a slower-than-expected rate of development in some low-GDP nations, there was a need for impoverished countries to enjoy a period of above-average emissions. A more glaring weakness of the approach is that it would lead to enormous shortfalls in Annex I emissions budgets and correspondingly huge surpluses in non-Annex I budgets (Michaelowa et al. 2005b; Vivekanandan et al. 2008). Consequently, the per capita approach has yet to be accepted by Annex I nations and, furthermore, has not been seriously proposed at COP meetings. Thus, the per capita distribution approach fails the 'negotiable commitments' test.

Moving onto the cumulative emissions proposal, it too passes the 'environmental effectiveness' test provided the negotiated outcome reduces combined emissions sufficiently to achieve the 450 ppm concentration target. Whilst the proposal would also appear to satisfy the 'equity' test in that the toughest emissions targets would be applied to the nations with the highest historical emissions, it doesn't account for the current per capita GDP of countries and their capacity to undertake mitigation activities (Bodansky et al. 2004). There are, for example, some countries with very high historical emissions but, for whatever reasons, a relatively low per capita GDP. This begs the question: Should current and future generations of people living in these countries suffer simply because the high per capita emissions of past generations have been squandered? Clearly not, thus suggesting that the mitigation capacity of a nation must be taken into consideration when determining emissions targets. On top of this, the cumulative emissions proposal is inflexible and unlikely to pass the 'negotiable commitments' test, as evidenced by its failure to receive support at the COP-3 meeting in Kyoto.

Because the aim of the fourth proposal—the contraction and convergence approach—is to cut global emissions sufficiently to achieve a safe concentration of greenhouse gases, it naturally passes the 'environmental effectiveness' test. It is also an ethically appealing approach given that its other main goal is to reduce the greenhouse gas emissions of all nations from existing levels to an equal per capita share of the global aggregate (Michaelowa et al. 2005b). However, because it allows Annex I nations to emit more greenhouse gases in historical terms, it needs to be counterbalanced by compensation for non-Annex I nations in the form of technology transfers and mitigation and adaptation funding.

As for the 'flexibility' test, the contraction and convergence approach is able to pass it in two ways. Firstly, the rate of contraction of global emissions can be slowed or speeded up to accommodate improved scientific knowledge on climate change. Secondly, since future greenhouse gas targets are likely to be set at the group and/or regional level, there is room for negotiations within a particular group to vary national emissions budgets. This approach has already been adopted in the case of the European Union (Bodansky et al. 2004).

Finally, given that UNFCCC-imposed emissions obligations have so far been confined to Annex I nations, which implies that a process of 'convergence' has already been in operation, the contraction and convergence approach clearly passes the 'negotiable commitments' test. Whether it also passes the 'implementable commitments' test will depend on the capacity of non-Annex I nations to meet their greenhouse gas emissions targets once they are subject to emissions obligations. Provided there is sufficient funding and other support from Annex I nations to ensure adequate capacity in non-Annex I countries, this test should also be passed with relative ease.

As explained above, the triptych approach is essentially a sectoral contraction and convergence approach. It therefore passes the assessment criteria in the same way as the contraction and convergence approach does in its generic form. Perhaps the only difference is that the triptych approach relies heavily upon sectoral-based data that is currently absent or not available in a form sufficiently comprehensive to determine the greenhouse gas intensity of key sub-sectors in some non-Annex I economies. Hence, for the time-being at least, the triptych approach would struggle to pass the 'implementable commitments' test should there be some agitation to employ it as the basis for future target-setting at the global level.

Last but not least, the common but differentiated convergence approach appears to pass the full assessment criteria. Better still, it is more equitable than the contraction and convergence approach because, for a limited period, it would allow non-Annex I nations to increase their per capita greenhouse gas emissions above the global average (Gupta and Bhandari 1999; Höhne et al. 2006; Stern 2008). This would enable non-Annex I countries to make up for having historically lagged behind Annex I nations. It would also provide them with the opportunity to rapidly increase their real GDP to the optimal level, which, in turn, would allow them to promptly meet their national development goals.

Perhaps the only query regarding the common but differentiated convergence approach is that its broad acceptance has yet to be fully tested at a COP meeting. It remains to be seen whether Annex I countries would allow non-Annex I nations to enjoy the benefits of emitting greenhouse gases for a period of time above the global per capita level. I believe they would because the additional cost to Annex I nations would be negligible. More importantly, allowing the per capita emissions of non-Annex I nations to exceed the global average would almost certainly persuade them to take on emissions obligations sooner rather than later. Overall, the fact that the common but differentiated convergence approach: (i) satisfies the full assessment criteria; (ii) is the most equitable of all the distribution methodologies proposed; and (iii) would more than likely induce Annex I nations to take on emissions targets in the very near future, makes it the ideal distribution approach upon which to base a new global climate change protocol.

9.4 A Modified Architecture for a New (Post-2020) Climate Change Protocol

We are now in a position to reveal the basic elements of a new climate change architecture for the UNFCCC to take beyond Kyoto and form the basis of a post-2020 protocol. As mentioned earlier in the chapter, the five main elements of the Kyoto Protocol would continue to exist in some form. However, two new elements would be added—the first being a graduation process by which some nations would advance over time and be subject to ever more stringent obligations; the second being the adoption of a 'common but differentiated convergence' approach as a means of setting of national emissions targets. The new international climate change architecture would therefore consist of the following:

1. The continued grouping and separate treatment of countries according to their per capita GDP and/or current and historical greenhouse gas emissions. The existing structure would be modified via the creation of four new groups of nations. The groups, which would come into existence at the beginning of 2016, would be:
   - an Annex II group consisting of currently-existing Annex II Parties;
   - an Annex I (non-Annex II) group consisting of currently-existing Annex I Parties minus the Annex II group;
   - an Annex III group consisting of currently-existing non-Annex I Parties minus LDCs;
   - a group of least-developed countries (LDCs).
2. (New element) Through a graduation process, the Annex I (non-Annex II), Annex III, and LDC groups would advance through time to higher groups until all nations were subsumed by the Annex II group. The graduation process would proceed as follows:
   - the Annex I (non-Annex II) group would graduate to the Annex II group as of 2041;
   - the Annex III group would graduate to the Annex I (non-Annex II) group as of 2041 and to the Annex II group as of 2051;
   - the group of LDCs would graduate to the Annex III group as of 2041; to the Annex I (non-Annex II) group as of 2051; and to the Annex II group as of 2061.

   Because of their high per capita GDP and per capita emissions, the Annex III Parties of Brunei, Israel, Kuwait, Qatar, Saudi Arabia, Singapore, South Korea, and the United Arab Emirates would graduate to the Annex I (non-Annex II) group as of 2021.
3. Except for LDCs, greenhouse gas emissions targets would be imposed on all nations as of 2021. LDCs would be subject to quantitative emissions targets as of 2031.
4. (New element) A unique schedule for emissions cuts would apply to each UNFCCC-defined group. The emissions cuts would be based on a 'common but differentiated convergence' approach to target setting.
5. The Kyoto flexibility mechanisms—the system of International Emissions Trading, the Clean Development Mechanism, and the Joint Implementation facility—would continue to operate to assist countries subject to obligations to cost-effectively achieve their emissions targets. Modifications would include:
   - the phasing out of the Clean Development Mechanism by 2030;
   - the replacement of the Clean Development Mechanism by the Joint Implementation facility (2031) as all nations become subject to emissions targets;
   - a separate emissions-trading system for each of the four groups of Parties.

6. The existing penalties for failing to comply with UNFCCC-imposed emissions targets would be maintained. However, trade sanctions and tariff penalties would also be applied in cases where a nation: (i) commits a second and subsequent transgression; (ii) refuses to discharge a presently imposed penalty; or (iii) refuses to participate in climate change negotiations and generates greenhouse gas emissions at levels which undermine the achievement of global emissions targets.
7. The world's richest countries would continue to transfer funds and technology to enable the world's poorest countries to undertake mitigation activities and adapt to the damaging impacts of climate change. Modifications would include:
   - consolidation of existing financing mechanisms and institutions;
   - new funding commitments from the Annex II and Annex I (non-Annex II) groups of nations, with the former shouldering most of the funding responsibility;
   - the gradual phasing out of funding commitments as Annex III nations and LDCs become increasingly capable of funding their own mitigation and adaptation activities (i.e., as they graduate to the higher groups of nations).

Notes

1. As Michaelowa et al. (2005a) have pointed out, the UNFCCC upon which the Kyoto Protocol has been built has made remarkable progress in a relatively short space of time (as compared to other international regimes); it holds annual meetings of the Parties; and has two subsidiary bodies and a permanent secretariat. In addition, quantified emissions budgets are the basis of the regime's existence. It therefore has the features of a stable and effective regime operating within a well-established institutional framework.
2. Stern (2009) has also argued that existing international institutions will need to evolve in order to deal adequately with the climate change crisis, even going so far as to suggest that new institutions may need to be established.
3. Something approaching the de-carbonisation of industrial production would not be required of all LDCs until 2080.
4. As of 2014, the greenhouse gas emissions of the group of Annex I (Annex B) nations covered by the Kyoto Protocol's system of International Emissions Trading amounted to little more than 40 per cent of total global emissions.
5. As some critics emphasise, in an international emissions-trading environment, the incentive to under-report greenhouse gas emissions is heightened.
6. Many of these critics concede that the scale of such projects would not have been as great without CDM-accreditation. Whilst this means that the true reduction in baseline emissions is greater than zero, critics believe it is, at best, 70–80 per cent of credited emissions reductions (Carbon Trust 2009).

7. A former Board Chairperson once conceded that the regulatory, executive, and quasi-judicial functions of the Clean Development Mechanism are effectively monopolised by the CDM Executive Board (Stehr 2008).
8. To recall, the cost of acquiring CERs is equivalent to the cost of investing in and maintaining CDM-projects.
9. That's not to say that the Clean Development Mechanism has no effect on emissions. Assuming most CDM-projects are appropriately designed and assessed prior to registration, the emissions levels of non-Annex I nations are lower than they would have been without the Clean Development Mechanism. Nonetheless, this merely reduces the growth rate of emissions in non-Annex I countries. It does not lead to lower aggregate emissions.
10. The phasing out of the Clean Development Mechanism would not mean the issuance of CERs would immediately cease. Existing CDM-projects would continue to be credited with CERs until they come to the end of their project life. There would be no registration of new CDM-projects.
11. The CDM Executive Board charges registration fees typically in the order of US$0.20 per CER for projects generating more than 30 Kilotonnes of annual emissions savings (capped at US$350,000) (Carbon Trust 2009). The fee does not, however, apply to projects undertaken in LDCs (UNFCCC 2013p).
12. The policies referred to here could include the introduction of minimum energy-efficiency standards or a national/regional emissions-trading system.
13. Projects qualify if they: (i) are located in a least-developed country (LDC) or small-island developing state (SIDS); (ii) are off-grid projects serving households/small communities; (iii) are off-grid projects serving a small or medium-sized enterprise, provided the installation generates less than 1.5 Megawatts of power; and (iv) the technology used is recommended by the host country's Designated National Authority, is approved by the CDM Executive Board, and its penetration does not exceed 3 per cent of total installed capacity.
14. Projects qualify if they are located in a least-developed country (LDC) or small-island developing state (SIDS) and consist of individual installations each saving less than 0.6 Gigawatt-hours per year, and are exclusively serving households/small communities or small or medium-sized enterprises.
15. This implies projects earning no more than 20,000 CERs per year. Projects qualify if they are located in a least-developed country (LDC) or small-island developing state (SIDS) and consist of individual installations each reducing greenhouse gas emissions by less than 600 tonnes per year, and are exclusively serving households/small communities or small or medium-sized enterprises.
16. Not to be confused with a Designated Operational Entity (DOE), a Designated National Authority (DNA) is an organisation which has been granted responsibility by a Kyoto Party to authorise and approve the

Party's participation in CDM-projects. The establishment of a DNA is one of the pre-requisites for a Party to participate in the Clean Development Mechanism (http://cdm.unfccc.int/DNA/index.html).

17. The nine countries are South Korea, Chile, Uruguay, Brazil, Thailand, Peru, Sri Lanka, Macedonia, and Iran.
18. As of November 2014, 266 PoAs had been registered covering 1,762 projects.
19. The six standardised baselines include emission factors pertaining to the power sectors of Uzbekistan (1), Belize (2), Uganda (3), and the Southern African Power Pool (4); and standardised baselines covering charcoal production in Uganda (5) and the rice mill sector in Cambodia (6) (https://cdm.unfccc.int/methodologies/standard_base/new/sb7_index.html).
20. See Sect. 3.3.2 on how to calculate user cost.
21. An Annex I Party can use *t*CERs and *l*CERs to meet its Kyoto targets for the commitment period during which they were issued. Neither class of non-permanent CERs can be carried over to a subsequent commitment period (UNFCCC 2006).
22. To expedite the replacement process, each Annex I nation is required to establish a replacement account at its national registry for each Kyoto commitment period. For every temporary CER that expires or is cancelled during a particular commitment period, the relevant Annex I nation must transfer one AAU, RMU, CER, ERU, *t*CER, or *l*CER to the replacement account that pertains to that commitment period (Fair Climate Network 2014).
23. Markets often discount non-permanent CERs by up to 75 per cent compared to permanent CERs.
24. Nicholas Stern (2009) has been a prominent person suggesting that the Clean Development Mechanism should be reformed to reward avoided deforestation.
25. Insurance premiums would be included in CDM-registration fees.
26. The responsibility to maintain a forest/plantation and ensure it retains the greenhouse gases it has sequestered remains with the owners of the forest/plantation.
27. Ideally, the payments to farmers to manage native vegetation would reflect the full range of services they would be providing, not simply carbon storage (e.g., biodiversity conservation, watershed management, erosion control, and existence values).
28. Having said this, the restrictions on the land-type that can be used to generate *t*CERs or *l*CERs through afforestation-related or reforestation-related CDM-projects already reduces this risk to some degree.
29. The loans are to be paid back following the project's first issuance of CERs.
30. AIEs determine whether a submitted JI-project meets the requirements and the guidelines for implementation set out in Article 6 of the Kyoto Protocol (UNFCCC 2006).

31. As of January 2012, 12 JPAs had been registered, all in Germany (Shishlov et al. 2012).
32. This was particularly the case in Poland, Estonia, and the Russian Federation.
33. This occurs most notably in countries with transition economies (EITs).
34. To recall from Chap. 8, this was the upshot of the Doha amendments to the Kyoto Protocol (COP-18).
35. I say "generally obliged" because it is possible for emissions allowances not to be cancelled if they are covered by a domestic or regional emissions-trading system with different rules and directives to the Kyoto Protocol.
36. Like the examples in Chap. 8, it is assumed that one AAU is equal to one tonne of CO_2-equivalent greenhouse gases.
37. The series of JI-projects might be undertaken in the form of a JPA.
38. This doesn't mean that cases of non-additionality would be entirely averted.
39. The date would be determined by UNFCCC Parties at a future COP meeting. Also, by 'sustainable', I mean a rate of timber extraction from forests that have already been earmarked for exploitation that is no greater than the forests' natural rate of regeneration.
40. These pledges include the Monterrey (2002 United Nations), the European Union (2005), and Gleneagles (2005 G8) commitments on overseas development assistance in the context of extra development costs arising from the impacts of climate change (Stern 2009).
41. The body best positioned to perform this function is the Forest Carbon Partnership Facility (FCPF). See Sect. 8.2.8.7 on the FCPF.
42. The shifting of deforestation can occur when measures that effectively thwart deforestation in one location drive deforestation to locations where measures to address deforestation are weak or non-existent.
43. It has been estimated that the annual cost to the world's richest countries of helping to halve the rate of global deforestation to 2030 is around US$25 billion (US$17–33 billion) (Eliasch 2008). This works out to a relatively low cost of US$5 per tonne of CO_2-equivalent greenhouse gases (Stern 2009).
44. In order to guarantee that a business can finance a shadow-project, it may be necessary to require all businesses with the potential to adversely affect native vegetation to pay an environmental assurance bond. In the event that damage to native vegetation does take place, the bond would be confiscated and used to finance the shadow-project. Should an economic activity not lead to deforestation, the bond would be returned along with any interest earned. For more on environmental assurance bonds, see Costanza and Perrings (1990) and Lawn (2007).
45. This assumes that that the lost vegetation is equivalent to that originally cultivated (afforestation) or allowed to regrow (reforestation). Should the potential exist for the project or activity to impact on more vegetation than what was originally grown, a shadow-project would still be warranted.

46. Should attribution be too difficult to determine, the point of obligation could be applied elsewhere in the supply chain of agricultural goods. For example, the requirement to acquit emissions units could be imposed on the direct procurers of agricultural output (e.g., food processing firms and textile producers). In this case, the acquittal of emissions units or emissions permits might be based on the quantity of the various forms of agricultural output purchased as production inputs. This approach would be far from perfect, since it would potentially over-penalise low-emissions agricultural practices and under-penalise high-emissions practices. Moreover, it may not effectively limit agricultural emissions. However, it may be adopted if the outcome is considered more acceptable than completely excluding agriculture altogether. Then again, a better outcome could perhaps be achieved by omitting agriculture from an emissions-trading system and introducing policies to promote sustainable agricultural practices, such as those outlined in Chap. 3.
47. Complicating matters is the fact that greenhouse gas emissions from the agricultural sector are also the product of natural processes. Although non-anthropogenic emissions need to be included when measuring total emissions, the emissions attributed to natural processes and land-users need to be separately accounted for given that an emissions-trading system must only make land-owners liable for all anthropogenic emissions.
48. Methods already exist to quantify soil-carbon. The challenge lies in designing a system to cost-effectively sample and estimate soil-carbon stocks (Conant et al. 2010).
49. Governments can support research and development through direct grants, contracts, tax credits, and private/public partnerships (IPCC 2007d). There is evidence that research and development spending, both public and price-sector, is one the rise again (REN21 2014).
50. To recall, in the growth-as-usual scenario, a 53.9-factor increase in the GWP/emissions ratio is needed between 2010 and 2100 to remain on a 450 ppm stabilisation pathway, whereas, for the sustainable scenario, only a 13.6-factor increase is required.
51. The emissions level in 1995 amounted to around 40 Gigatonnes of CO_2-equivalent greenhouse gas emissions.
52. To recall from previous chapters, technological lock-in stems from the path-dependent nature of economic systems (David 1985; Arthur 1989).
53. This would include commitments to deter well-endowed countries from free-riding (i.e., shirking their funding responsibilities).
54. The commitment would operate along similar lines to an International Low-Emissions Technology Commitment proposed by Garnaut (2008), except that the funds would be only used to kick-start and support mitigation action in low-GDP countries.
55. The Green Climate Fund should be the repository of the financial resources given the UNFCCC's desire to have the Fund become its principal financing mechanism in support of mitigation action in developing countries.

56. The Monterrey (2002) and Gleneagles (2005) agreements require high-GDP nations to allocate 0.7 per cent of their GDP for overseas development assistance.
57. As Table 8.9 showed, there is little doubt that the fall in the Annex I group's emissions can be attributed to the large decrease in the emissions of the Annex I (non-Annex II) group of Parties (Note: the emissions of the Annex II group of Parties increased slightly). However, it cannot be claimed that the Kyoto Protocol's enforcement mechanisms had no influence on the total emissions of the Annex I group. After all, it could be argued that, without an effective enforcement system, the emissions of the Annex II group could have risen sufficiently to have resulted in an overall increase in the Annex I group's emissions.
58. A country of this nature has been labelled a 'resistant state' by Mitchell (2005).
59. Mitchell (2005) refers to such a nation as a 'contingent state'—one not prepared to take appropriate action if other nations fail to do likewise, especially if the inaction of other countries renders it worse off. There are other countries that Mitchell refers to as 'committed states', which are countries committed to taking necessary climate change action regardless of what other nations do. Nonetheless, since inaction by some countries reduces the overall effectiveness of mitigation measures and can greatly disadvantage nations which have reduced their greenhouse gas emissions, committed states are still likely to demand enforcement mechanisms that coerce 'intransigent states' into engaging in the global mitigation process.
60. In the situation depicted in Fig. 9.1, there are just two participating nations. In reality, there will be many more. Ultimately, all the world's countries are likely to be involved in the setting and meeting of emissions targets. The larger the number of nations involved, the smaller is the reduction in the long-term pay-off arising from one nation free-riding. This is because the impact of one nation exceeding its emission target on global greenhouse gas emissions is greatly diminished. The magnitude of a nation's emissions relative to global emissions is also a significant factor. A large aggregate emitter, such as China, the USA, or India, would impose greater long-term damage costs on all nations if it exceeded its emissions target by a certain percentage.
61. As explained in Chap. 6, when delving into the field of climate change damage costs, humankind operates in a state of ignorance.
62. Evidence suggests that human behaviour is typically affected more by an aversion to large losses than the prospect of future gains (Kahneman and Tversky 1979; Tversky and Kahneman 1992).
63. Despite this, a government in this position might still agree to small emissions cuts to maintain some semblance of international credibility.
64. Present Kyoto arrangements do not deal with nations that fail to co-operate with Kyoto rules or withdraw from the Kyoto Protocol altogether (e.g., Canada).

65. For example, trade-related regulations must qualify as measures: (i) which aim to protect public morals, human health, cultural artifacts, or environmental resources; and (ii) do not constitute a "means of arbitrary or unjustifiable discrimination" or a "disguised restriction on international trade" (Frankel 2005, p. 13).
66. It was also suggested that measures be imposed to restore comparative advantage as the principle governing international trade (see Chap. 3).
67. Compensating tariffs can still be justified if foreign producers enjoy a cost advantage based on differences in wage rates and other environmental standards.
68. In an exercise conducted by Saddler et al. (2006), the potential impact on the services sector was not considered because the criteria used for the analysis assumed that the services sector is not 'emissions-intensive'. More appropriate criteria would have extended the analysis to many industries in the services sector.
69. Saddler et al. (2006) adopted an assumed price of greenhouse gas emissions of $35 per tonne of CO_2-e. In another major study, Elliott et al. (2010a) adopted an assumed a price of $29 per tonne of CO_2. To put these prices into perspective, Stern's (2007) estimated social cost of carbon was $85 per tonne of CO_2, which was already an underestimate given it was based on a dangerously high stabilisation target of 550 ppm of CO_2-e (Note: the price of a tonne of CO_2-e is higher than the price of a tonne of CO_2).
70. To recall from Chap. 1, the Montreal Protocol is a treaty designed to restrict the generation of ozone-depleting gases.
71. Having said this, the WTO's attitude towards border-tax adjustments, such as compensating tariffs, is less accommodating than its own international trade rules suggests it should it be.
72. In other words, any other cost-saving measure would simply attract a higher compensating tariff.
73. Under the origin principle, traded goods are subject to the taxes, charges, and price levies (e.g., a greenhouse gas price) imposed by the origin or exporting nation.
74. Saddler et al. (2006) recommend a threshold equal to 2 per cent of the world price.
75. There are concerns that the first tier encourages foreign exporters to underreport their emissions, while the second tier encourages inefficient foreign exporters to exploit the 'predominant method of production' approach. The latter is possible because the default quantities of the tax-attracting inputs used to calculate the border-tax adjustment usually underestimate the real quantities used. Saddler et al. (2006) believe that, with expert review, it should be possible to detect cheating and deception, especially in light of the availability of national emissions inventories through the UNFCCC; the improved nature of company-level reporting; and the well-understood nature of production process applicable to most greenhouse gas-intensive goods.

76. This would include the capacity to rebuild affected communities, rehabilitate damaged environments, and, if need be, resettle climate change refugees.
77. These actions include the introduction of improved agricultural practices to combat the impact of climate change on agricultural production (Thomas et al. 2007; Geerts and Raes 2009; Lasco et al. 2011; Marongwe et al. 2011; Olesen et al. 2011); public health measures to reduce heat-related risks (Ebi et al. 2004; Tan et al. 2007; Fouillet et al. 2008); the establishment of early-warning systems to minimise weather-related disaster risks and reduce damage costs (Næss et al. 2005; Wall and Marzall 2006; Storbjörk 2007; Measham et al. 2011; Runhaar et al. 2012); and altered water management practices to cope with changes in rainfall and hydrological regimes (Thomas et al. 2007; Marongwe et al. 2011).
78. As revealed in Chap. 8, the 2 per cent levy on Kyoto flexibility mechanisms, which is used to help capitalise the Adaptation Fund, had raised just US$189.8 million in proceeds by the end of 2013—a tiny fraction of the funds required for adaptation purposes.
79. To recall from Chap. 8, only 6.1 per cent of the total funds allocated for climate change purposes were allocated in 2012 to support adaptation activities.
80. Like the mitigation funding commitment, a funding contribution should only apply to nations with a per capita GDP of Int$10,000 or more (at 2004 prices).
81. The Mechanism can be thought of as an international insurance scheme.
82. In some parts of the world, a cyclone is referred to as a typhoon; in others, a hurricane.
83. Indexes could be constructed by national meteorological bureaus working in conjunction with the World Meteorological Organization.
84. This is clearly evident from Fig. 8.2 and Table 8.6.
85. The upper estimate of mitigation costs associated with stabilising the atmospheric concentration of greenhouse gases at around 450 ppm is around 3 per cent of GWP. For damage costs, which can be a proxy for adaptation costs, it is around 2 per cent of GWP. Together, these costs amount to approximately 5 per cent of GWP. As demonstrated in Chaps. 5 and 6, I believe these figures underestimate the full cost. Hence, I believe the 5 per cent figure is not at all excessive.
86. In other words, it is presumed that past emissions imply no guilt on the part of heavy greenhouse gas emitters for the climate change crisis (Vivekanandan et al. 2008).

Chapter 10
A Post-2020 Protocol and Emissions-Trading Framework to Resolve the Climate Change Crisis

10.1 Introduction

As I have stressed a number of times throughout this book, successfully resolving the climate change crisis will necessitate a legally-binding protocol involving all the world's nations. Achieving this will require international collaboration on an unprecedented scale (Stern 2009).

Most of the fundamental elements of an effective global protocol have been presented in previous chapters, including the national policies required to achieve sustainable development and to therefore overcome the root cause of the climate change crisis. My aim in this chapter is to outline the various elements of a global climate change protocol that would take effect in 2021 without specifying its full details. The details would be formulated in due course via international negotiations that would eventually lead to a lengthy and intricate document—a task well beyond the confines of this book. The important thing to bear in mind is that desirable negotiations are unlikely to transpire unless there is a clear set of principles upon which to base a global protocol as well as a broad understanding of its policy and institutional ramifications. It is an effective set of principles plus the basic features of a global emissions-trading system—including a number of built-in flexibility mechanisms—which I aim to spell out in this chapter. As highlighted in previous chapters, the principles upon which a global protocol must be established are: (i) ecological sustainability; (ii) distributional equity (both within and across nations); and (iii) allocative efficiency.

To facilitate international agreement at future UNFCCC conferences, it may be necessary to compromise somewhat on the principle of allocative efficiency. In my opinion, the sustainability goal is of paramount importance. Hence, the principle of ecological sustainability must never be violated. In addition, all efforts should be made to realise the goal of distributional equity, not only because of its moral dimensions, but because a failure to honour equity considerations would almost

P. Lawn, *Resolving the Climate Change Crisis*, DOI 10.1007/978-94-017-7502-1_10

certainly undermine international negotiations—in particular, the co-operation of the world's poorest and most disadvantaged nations. As we shall see, compromising on efficiency may require some restrictions on the arrangements established to facilitate the international trade in emissions rights (emissions permits). In saying this, I believe that sacrificing some efficiency to achieve the sustainability and equity goals will generate significant long-term welfare benefits for all nations, much like the envisaged increase in per capita economic welfare described in Chap. 4 (see Fig. 4.6).

Before moving on, it is important to bear in mind that the climate change protocol outlined in this chapter will be premised on three key assumptions consistent with what I've demonstrated and advocated in previous chapters. In the first instance, it will be assumed that any new climate change architecture will consist of the four groups of nations recommended in Chap. 9. Secondly, for equity reasons, it will be assumed that the emissions reductions required from each nation will be based on a 'common but differentiated convergence' approach to target setting. Finally, it will be assumed that all countries will, when appropriate, make the transition to a qualitatively-improving steady-state economy. This latter objective will be assumed, not only because the shift to a steady-state economy is necessary for all nations to achieve sustainable development, but because it will be impossible otherwise to stabilise the atmospheric concentration of greenhouse gases at a 'safe' level.

10.2 Fundamental Elements of a New Climate Change Protocol

As important as it is to propose a realistically achievable climate change protocol, it is equally important that it be instituted in full to ensure the principles upon which it is based are adequately met. Any worthwhile proposal should not be seen as a menu of options from which policy-makers can cherry-pick the measures most convenient or politically expedient to implement (Stern 2009). Hence, should negotiations proceed on the basis of abandoning some crucial measures in order to reach agreement on others, any ensuing protocol would be half-baked and ineffective. With this in mind, the following proposal should be viewed as an integrated package comprised of various reinforcing elements—that is, a package of measures considerably more potent than the mere sum of its parts.

In many of the sub-sections that follow, I will not explain the reasons behind the various elements that warrant inclusion in a new climate change protocol, since this was accomplished in Chap. 9. Thus, in some sub-sections, I will merely reiterate and summarise the main elements for the purpose of neatness and completeness.

10.2.1 Ecological Sustainability Requirements—Global Greenhouse Gas Emissions Targets

10.2.1.1 A 'Safe' Global Emissions Trajectory

A global climate change protocol must aim to minimise the likelihood of average global temperatures rising by more than 2 °C above pre-industrial levels. As stressed many times, this will require global emissions cuts that are sufficient in magnitude to stabilise the atmospheric concentration of greenhouse gases at no more than 450 ppm of CO_2-e. It will also require measures that dramatically reduce the rate of deforestation given that native vegetation loss accounts for nearly 20 per cent of all greenhouse gas emissions.

Since overshooting the safe concentration of greenhouse gases is best kept to a minimum (see Sect. 6.3), cuts to global emissions should be based on an emissions trajectory similar to that recommended by Anderson and Bows (2008) (see Fig. 4.1). To adhere to this trajectory, global greenhouse gas emissions must be reduced at an average annual rate of 4 per cent from the beginning of 2016, which implies reducing global emissions between 2015 and 2050 by 75 per cent and by 86 per cent between 2015 and 2100.[1] That said, process-related CO_2 emissions must be reduced to negligible levels by 2100, thus rendering non-CO_2 and land use-related emissions the only appreciable greenhouse gas emissions by the end of the century (see Table 4.1).[2] Accordingly, global industrial production activities must be all but de-carbonised prior to 2100.[3]

Box 10.1 Emissions Cuts to Achieve a Safe Concentration of Greenhouse Gases

- For the remainder of this century, global greenhouse gas emissions must be reduced at an average rate of 4 per cent per annum. This means:
  - global emissions must be reduced by 75 per cent over the 2016–2050 period
  - global emissions must be reduced by 86 per cent over the 2016–2100 period
- Global process-related CO_2 emissions must be reduced to negligible levels by 2100
- Global industrial production must effectively be de-carbonised prior to 2100

10.2.1.2 Compliance Institutions and the Penalties for Non-compliance

Achieving the emissions cuts required to move the global economy to a safe emissions trajectory demands effective compliance institutions coupled with adequate penalties for non-compliance. To meet these demands, the existing Kyoto penalties should be retained in a post-2020 protocol, as should the Kyoto Protocol's Compliance Committee and its two branches—specifically, the Facilitative Branch, which would advise Parties on the implementation of a new global protocol; and the Enforcement Branch, which would remain responsible for discerning whether Parties are complying with: (i) their emissions targets; (ii) greenhouse gas-reporting requirements; and (iii) the rules governing the use of the new protocol's flexibility mechanisms.

To improve the effectiveness of the protocol's compliance arrangements, the existing penalties for non-compliance should be bolstered by trade sanctions and border-tax adjustments (see Sect. 9.2.4.1).

Box 10.2 Facilitating Compliance

- In order to induce all Parties to set and comply with sufficiently stringent emissions targets, non-compliance should be dealt with via the imposition of the following penalties:
  - the making up of excessive emissions plus an additional 30 per cent during a subsequent commitment period
  - submission of a 'compliance action plan' to the Enforcement Branch[4] of the UNFCCC
  - suspension from international trade in emissions allowances (e.g., AAUs, CERs, ERUs, and RMUs) until the transgressing Party is again deemed
  - under particular circumstances,[5] the imposition of trade sanctions and tariff penalties, with tariff penalties severe enough to nullify the incentive for Parties to exceed their emissions targets[6]
- The existing 'Kyoto' Enforcement Branch should be continued but expanded to oversee the operation of international carbon markets—in particular, national, regional, and group-based emissions-trading systems to ensure they are soundly administered
- The World Trade Organisation (WTO) should work closely with the UNFCCC to clarify what action can be taken under current WTO rules to impose trade-related penalties for non-compliance
- Where necessary, the WTO rules should be modified to permit the imposition of legitimate trade-related penalties for non-compliance[7]

10.2.1.3 Border-Tax Adjustments

A new climate change protocol should include a mechanism to permit national governments to make border-tax adjustments to negate potential cost disadvantages caused by the introduction of a greenhouse gas price signal. Border-tax adjustments are required for three main reasons: (i) to prevent carbon leakage; (ii) to help facilitate the efficient international allocation of resources; and (iii) to alleviate competitiveness problems (industrial flight) caused by assigning a price to domestic greenhouse gas emissions. To recall, the fear of the latter is often blamed for the lack of effort on the part of governments to price greenhouse gas emissions and to set stringent emissions targets. Hence, the third reason, along with the first, are especially relevant in terms of achieving a safe atmospheric concentration of greenhouse gases.[8]

Border-tax adjustments should take two forms. The first is a tax rebate paid to emissions-sensitive exporters to offset the increase in productions costs caused by the introduction of a domestic price on greenhouse gas emissions. The second is a compensating tariff applied to imported goods to offset any cost disadvantage that an emissions tax or emissions-trading system would exact on trade-exposed domestic producers. A cost-burden threshold test should be employed to determine product eligibility and the magnitude of the tax rebates and compensating tariffs applied to trade-exposed goods (see Sect. 9.2.5.4). To preserve the incentive for domestically-located firms to reduce the emissions-intensity of production, all goods produced for domestic consumption purposes must remain subject to a domestic greenhouse gas price signal.

In order to prevent governments from abusing a border-tax adjustment system on 'protectionist' grounds, the system should be created via negotiation at UNFCCC meetings and in co-operation with the World Trade Organisation (WTO). The WTO should be granted the responsibility of assessing border-tax adjustment applications and the sanctioning of tax rebates and compensating tariffs. An appeals procedure should be established to enable Parties to appeal the decisions made by the WTO.

Box 10.3 A Border-Tax Adjustment System

- A border-tax adjustment system should be embodied in a new climate change protocol to allow national governments to preserve the international competitiveness of emissions-sensitive producers whilst leaving intact the greenhouse gas price signal established within the domestic economy
- The adjustments should exist in two forms:
  - a *tax rebate* paid to emissions-sensitive exporters to offset the increase in production costs arising from the introduction of an emissions tax or emissions-trading system

  - a *compensating tariff* applied to imported goods to offset any cost disadvantage that an emissions tax or emissions-trading system would exact on trade-exposed domestic producers

- Border-tax adjustments must be confined to situations where a disparity in the domestic and foreign greenhouse gas price results in a clear competitive disadvantage to domestic producers
- A cost-burden threshold test should be used to determine: (i) which products are eligible for a border-tax adjustment; and (ii) the size of the adjustment to be applied to eligible products
- Where a tax rebate/compensating tariff is applied, it should equal the difference in the expected uplift price—i.e., the estimated increase in the world price of an exported/imported good if, hypothetically, foreign producers were subject to the same greenhouse gas price as domestic manufacturers of the same good—and a threshold price set at 3 per cent above the world price
- The full border-tax adjustment process would operate as follows:
  - Upon a firm or industry seeking compensatory action, a government authority would employ a UNFCCC-sanctioned cost-burden threshold test to determine which domestically-produced goods are eligible for protection via a border-tax adjustment
  - Assuming qualification, the government authority would calculate the size of the border-tax adjustments to be imposed—whether it be a compensating tariff in the case of imported products or a tax rebate in the case of emissions-intensive exports
  - The authority would submit a border-tax adjustment application with supporting evidence to the WTO
  - A WTO panel, working in collaboration with the UNFCCC, would assess the border-tax adjustment application
  - The panel would approve/reject the claim for a border-tax adjustment or approve an individual claim but alter the size of the tax adjustment
  - All border-tax adjustments would be periodically reviewed and reset as circumstances changed over time
  - To assist the WTO with the ongoing review process, national governments would be required to provide updated information not readily available to the WTO panel
  - Countries whose cost advantage would be negated by the sanctioned adjustments would be entitled to appeal the WTO's decision
  - Countries lodging border-tax adjustment applications would also be permitted to appeal the WTO's rejection of any tax adjustment claims
  - An appeals tribunal, which should rest with the UNFCCC, would make a final judgement on border-tax adjustments
  - Where border-tax adjustments are approved, they would be imposed by a national authority in collaboration with domestic taxation and/or customs officials

10.2.1.4 Land Use, Land-Use Change, and Forestry (LULUCF)—Forestry

Given that deforestation and agriculture respectively account for around 20 and 15 per cent of global greenhouse gas emissions, a great deal more needs to be done through a new climate change protocol to reduce the emissions generated by land use, land-use change, and forestry activities. As revealed in Chap. 9, there are various Kyoto rules governing the treatment of forestry-related activities (see Sect. 9.2.2.1).[9] Whilst some are designed to advance the ecological integrity of the Kyoto Protocol, others pertain to the effective operation of the Kyoto flexibility mechanisms. I'll focus here on the former and leave the latter to the sub-section on the allocative efficiency requirements of a new protocol.

Under present Kyoto arrangements, all forestry-related activities taking place in currently-existing Annex I nations must be included in the measurement of greenhouse gas emissions by sources and removals by sinks. However, not all forestry-related activities undertaken by Annex I nations can be used to meet their emissions targets. As explained in Chaps. 8 and 9, forestry activities undertaken through the Clean Development Mechanism and Joint Implementation facility can only earn Certified Emission Reduction units (CERs) and Emission Reduction Units (ERUs) if the activities are able to increase greenhouse gas removals by sinks. Eligibility for CERs and ERUs under the Clean Development Mechanism and Joint Implementation facility does not extend to forestry activities which reduce emissions by sources. What's more, in the case of the Clean Development Mechanism, eligible forestry activities are confined to afforestation and reforestation activities. Thus, projects which prevent deforestation in currently-existing non-Annex I countries are unable to generate CERs.[10] In addition, 'avoided deforestation' cannot be used to meet national emissions targets.

Whilst the current policy of not issuing emissions allowances for avoided deforestation should remain, reforms to a number of Kyoto rules are urgently required to better reflect the greenhouse-gas impacts of forestry-related activities. Three reforms stand out. Firstly, where a nation is subject to emissions targets, all individuals and entities operating within it should be required to acquit emissions allowances upon clearing native vegetation or harvesting timber from native forests and/or cultivated timber plantations. Secondly, emissions allowances should be issued whenever a forestry activity occurring in a country subject to emissions targets results in the removal of greenhouse gases. As we shall see, should Annex III nations become subject to greenhouse gas emissions targets as of 2021, these conditions would 'potentially' apply to forestry-related activities undertaken in all nations except LDCs in the immediate post-2020 period.[11] With the prospect of LDCs being subject to emissions targets beyond 2030, acquittals for carbon discharges and the generation of offsets for carbon removals would potentially apply to all forestry-related activities world-wide as of 2031.

An important feature of these first two reforms is that the sustainable harvesting of timber—which keeps forest stocks intact—would generate as many emissions allowances from forest regeneration as the number of permits required for acquittal from timber extraction. In a *net* sense, there would be no need for loggers engaged in sustainable forestry practices to acquire emissions allowances, thus reflecting the stable sequestration of carbon within non-declining forests. Forest loss, however, would require the net acquittal of emissions allowances, which would serve as a penalty for deforestation. Conversely, forest expansion would lead to the net acquisition of allowances—a reward for afforestation/reforestation.

As vital as it is for a new global protocol to cover all forestry-related activities in the above-described manner, it was stressed in earlier chapters that requiring individuals and entities to acquit emissions allowances upon logging/clearing vegetation does not guarantee the preservation of ecosystem services and wildlife habitats.[12] To further advance the forestry-related contribution to climate change mitigation, there is a need for policy measures of the type outlined in Chaps. 3 and 9. Unfortunately, many of these policies lie outside the bounds of what a global climate change protocol can achieve. Nonetheless, through its financing mechanisms, the UNFCCC has the opportunity to work with development agencies and multilateral/bilateral finance institutions to encourage national governments to introduce the policies and institutions required to prevent deforestation and the loss of forest-generated ecosystem services.

Besides integrating forestry-related activities and forestry management into national and regional development programmes, the UNFCCC needs to urge national governments to introduce legislative controls on land clearance. In particular, the UNFCCC needs to encourage nations to establish Environmental Trust Funds to create conservation areas (e.g., National Parks) and provide payments for the maintenance of ecosystem services. To recall, the latter involves compensating land-owners unable to exploit protected portions of their land for forestry and other purposes. Offering payments for ecosystem services also removes most if not all the economic incentive for land-owners to convert forests and associated ecosystems into mono-cultured crops and cultivated timber plantations.

Given that, for some time, Annex III countries and LDCs are unlikely to be in a position to establish their own Environmental Trust Funds, they should be created and capitalised through the transfer of financial resources from Annex II and Annex I (non-Annex II) nations—ideally beginning with the resources made available through the Green Climate Fund and the Carbon Fund of the FCPF. Crucially, with the Green Climate Fund expected to remain an integral component of a post-2020 protocol, there exists considerable scope for the UNFCCC to play a constructive policy-influencing role in the area of avoided deforestation.

In terms of project assessment, the UNFCCC also needs to encourage national planning authorities and development banks and agencies to incorporate the 'user cost' principle when evaluating forestry projects.[13] The UNFCCC should also stress the need for authorities to require project-investors to establish, in specific circumstances, a shadow-project in the form of a replacement timber reserve to guard against the future loss of forest/vegetation cover (see Sect. 9.2.2.1).

Because, as the last point suggests, forestry-related removals of greenhouse gases are potentially reversible, emissions allowances generated by forestry activities through the Clean Development Mechanism should continue to be issued in the form of temporary emissions units or *t*CERs. Upon the expiration of *t*CERs, temporary units should be re-issued to cover a new crediting period provided the timber/vegetation in question remains standing. Where greenhouse gas removals are reversed through land clearance, logging, or natural destruction (e.g., fire, disease, or insect infestation), a suitable quantity of emissions allowances must, as is presently the case, be cancelled to replace the retired temporary units. The responsibility to acquire the emissions allowances for cancellation should continue to lie with the possessors of *t*CERs.

To encourage investors to engage in greenhouse gas-reducing forestry activities, there is a need for insurance schemes to compensate those who have their emissions allowances annulled due to factors beyond their control. Of these schemes, those aimed at compensating investors engaged in domestic forestry-related activities should be established by national/provincial governments or private insurance companies. At the same time, the UNFCCC should create insurance schemes to cover the unforeseen loss of emissions allowances generated through the Clean Development Mechanism and the Joint Implementation facility. Insurance premiums for these schemes should be included in CDM/JI-registration fees. In terms of nation-based insurance schemes designed to cover domestic forestry activities, the UNFCC should play a constructive role in developing insurance markets in low-GDP countries where they are presently inadequate or non-existent.

The third of the reforms to the Kyoto rules would allow a limited category of forestry activities that reduce greenhouse gas emissions by sources to qualify as eligible CDM/JI-projects. The eligible forestry activities should be those that, through the adoption of improved forestry practices, reduce the carbon-storing vegetation lost from timber harvesting. To discourage an acceleration of logging operations, these activities should be confined to forests already being exploited for timber or forests previously earmarked for sustainable exploitation. This restriction would prevent CERs and ERUs from being issued for 'avoided' deforestation.

To adequately incorporate forestry-related activities in a new global protocol, it will be necessary for countries to have the institutional and technical capacity to monitor and detect forest/vegetation loss—in particular, the ability to attribute greenhouse gas emissions and removals to individual forestry-related activities. Institutional capacity must also extend to administering the various mechanisms that must be installed to prevent deforestation. Although the establishment of adequate institutional and technical capacity is likely to remain an acute problem for the Annex III and LDC groups of nations, the need to build this capacity is no less relevant for the many Annex II and non-Annex I (non-Annex II) countries where deforestation, land clearance, and ecosystem decline continues (e.g., Australia). Whereas the richer Annex II and non-Annex I (non-Annex II) countries should have little difficulty in accessing the financial and real resources needed to establish adequate institutional capacity, the Annex III group and LDCs will require further assistance through the agency of the Green Climate Fund and/or the FCPF Carbon Fund.

Box 10.4 Land Use, Land-Use Change, and Forestry—Forestry

- Reforms to the present Kyoto rules regarding the treatment of forestry-related activities are required to accurately estimate greenhouse gas emissions by sources and removals by sinks. These reforms are also needed to ensure emissions targets are being properly met
- The reforms that should be embedded in a new climate change protocol include the following:
    - Where a nation is subject to emissions targets, an individual or an entity must acquit emissions allowances upon clearing native vegetation or harvesting timber from native forests and/or cultivated timber plantations
    - Where a nation is subject to emissions targets, emissions allowances should be issued in instances where a forestry-related activity results in the removal of greenhouse gases
    - Forestry activities involving the use of improved forestry practices that subsequently reduce the carbon-storing vegetation lost from harvesting a given quantity of timber should qualify as eligible CDM/JI-projects. The activities should be confined to forests already being exploited for timber or forests previously earmarked for sustainable exploitation
- Where greenhouse gas removals are reversed through land clearance, logging, or natural destruction, emissions allowances must be forfeited. Insurance schemes should be established by the UNFCCC to compensate the possessors of confiscated *t*CERs and ERUs, albeit compensation should only be provided if it can be shown that the possessors of the annulled emissions units *are not responsible* for reversals. The UNFCCC should encourage the establishment of nation-based insurance schemes to deal with domestic forestry activities and associated losses of RMUs
- The current policy of not issuing emissions allowances for avoided deforestation should be retained in a new global protocol
- A new protocol should attempt to eliminate deforestation by:
    - integrating forestry activities into national and regional development programmes
    - encouraging governments to introduce legislative controls on land clearance
    - encouraging governments to create conservation areas (e.g., National Parks) and provide payments for maintaining ecosystem services to compensate land-owners unable to exploit protected portions of their land for forestry purposes. Assistance to Annex III countries and LDCs should be provided in the form of financial transfers via the Green Climate Fund and/or the Carbon Fund of the FCPF

- encouraging national planning authorities and development agencies/banks to incorporate the 'user cost' principle when evaluating forestry projects
- where necessary, requiring project-investors to establish a replacement reserve to cover for potential project failures (e.g., the eventual loss of forest cover/vegetation)
- building global institutional and technical capacity to: (i) improve the monitoring and detection of forest losses, especially the ability to attribute greenhouse gas emissions and removals to individual activities; and (ii) oversee and administer the various mechanisms installed to prevent deforestation. Once again, assistance to establish institutional and technical capacity should be provided to the Annex III group and LDCs by way of the Green Climate Fund and/or the FCPF Carbon Fund

10.2.1.5 Land Use, Land-Use Change, and Forestry (LULUCF)—Agriculture

Under the Kyoto Protocol, all agricultural emissions by sources and removals by sinks that take place within Annex I nations must be included in the measurement of aggregate greenhouse gas emissions. However, much like forestry, the Kyoto Protocol embodies a number of rules that restrict Annex I nations from counting some emissions reductions and greenhouse gas sequestrations from agricultural activities towards their emissions targets (see Sect. 9.2.2.2).[14] These rules, which aim to enhance the ecological integrity of the Kyoto Protocol, prevent all agricultural activities from qualifying as eligible CDM-projects. Hence, at present, the Kyoto Protocol forbids agricultural activities from earning CERs. Secondly, the Kyoto rules render ineligible all agricultural activities undertaken through the Joint Implementation facility that reduce emissions by sources. Hence, only agricultural projects that remove greenhouse gases by sinks can earn ERUs for compliance purposes.[15]

To appropriately penalise and reward the greenhouse-gas impacts of agricultural activities and encourage farmers to adopt agricultural practices that reduce emissions by sources and remove greenhouse gases by sinks, a number of reforms to the Kyoto rules are required. The first and most important reform is the need to compel farmers operating within nations subject to emissions targets to acquit emissions allowances upon generating greenhouse gas emissions through their agricultural activities. Secondly, there is a need to issue emissions allowances to farmers when their agricultural activities remove greenhouse gases from the atmosphere. In the same way as forestry activities, the likelihood of Annex III nations becoming subject to greenhouse gas emissions targets post-2020 means these first two reforms would apply to agricultural activities in all nations other than LDCs as of 2021.[16] However, if LDCs were to become subject to emissions targets as of 2031, these reforms would apply to agricultural activities world-wide beyond 2030.

A third major reform to the Kyoto rules would involve the incorporation of agricultural activities in the flexibility mechanisms of a new global protocol. This would require two rule changes—firstly, the need to allow agricultural activities that reduce emissions by sources or remove greenhouse gases by sinks to qualify as eligible CDM-projects; and secondly, the need to allow agricultural activities that reduce emissions by sources to qualify as eligible JI-projects.

Assuming this third reform is implemented, a fourth reform would involve the issuance of temporary CERs or *t*CERs in cases where removals of greenhouse gases by agricultural activities undertaken through the Clean Development Mechanism are potentially reversible. Much like forestry, new *t*CERs would be re-issued to cover a new crediting period provided the soil-carbon built up by the CDM-project has been retained.

Finally, the Kyoto Protocol's International Emissions Trading system should be modified to permit the trade in all types of emissions allowances generated by agricultural activities (i.e., RMUs, ERUs, and *t*CERs). To enable this, a central authority must have the capacity to attribute emissions and removals to specific agricultural activities. As explained in Chap. 9, attribution is currently a difficult and costly exercise. To overcome this problem, resources should be directed to develop the technical know-how and expertise to monitor and verify emissions with some degree of ease and affordability—a role that all levels of government, along with the private-sector, must be increasingly willing to play. Since developing this capacity would be costly, funding transfers will be necessary to support Annex III nations and LDCs. Funds should be provided directly by the Green Climate Fund and indirectly through overseas development assistance. The latter would be best facilitated by integrating sustainable and low carbon-emitting agricultural practices into the national or regional development programmes of low-GDP countries.

There is, nonetheless, every reason to believe that building the capacity to attribute emissions and removals to specific agricultural activities will be time-consuming. Should it be some time before countries subject to emissions targets are capable of determining attribution, or should attribution not be immediately possible for many countries as they become subject to emissions targets (e.g., Annex III nations in 2021 and LDCs in 2031), mitigation should be promoted by encouraging nations to implement policies that support the adoption of sustainable agricultural practices (see Sect. 3.3.3). The UNFCCC should play its part by exerting the policy influence afforded to it through its financing mechanisms—in particular, the Green Climate Fund.

Box 10.5 Land Use, Land-Use Change, and Forestry—Agriculture

- To encourage agricultural practices that reduce greenhouse gas emissions by sources and remove greenhouse gases by sinks, a number of the Kyoto rules governing the treatment of agricultural activities require reform. They include:
    - Where a nation is subject to emissions targets, the owner of an agricultural establishment must acquit emissions allowances upon generating greenhouse gas emissions through agricultural activities
    - Where a nation is subject to emissions targets, emissions allowances should be issued in instances where agricultural activities result in the removal of greenhouse gases
    - Provided emissions and removals can be attributed to individual agricultural activities, all agricultural activities should be incorporated into the flexibility mechanisms of a new climate change protocol
    - Agricultural activities that reduce emissions by sources or remove greenhouse gases by sinks should qualify as eligible CDM-projects
    - Agricultural activities that reduce emissions by sources should qualify as eligible JI-projects
- Because agricultural removals are potentially reversible, temporary CERs (*t*CERs) should be issued for all removals resulting from agricultural activities undertaken through the Clean Development Mechanism. New *t*CERs should be re-issued to cover a new crediting period provided the soil-carbon built up by the CDM-project has not been lost
- Where greenhouse gas removals are reversed through soil erosion, unsustainable land-management practices, or drought, emissions allowances must be forfeited. Insurance schemes established by the UNFCCC to compensate those who have their *t*CERs and ERUs confiscated because of forestry-related reversals should be extended to include agriculture-related reversals. Compensation should only be provided if it can be shown that the possessors of the annulled emissions units *are not responsible* for the loss of soil-carbon. The UNFCCC should also encourage the establishment of nation-based insurance schemes to deal with domestic agricultural activities and associated losses of RMUs
- The International Emissions Trading system of a new global protocol should permit the trade in all types of emissions allowances generated by agricultural activities (i.e., RMUs, ERUs, and *t*CERs)
- To enable central authorities of Annex III nations and LDCs to attribute emissions and removals to specific agricultural activities, funds should be provided via the Green Climate Fund and indirectly through the receipt of overseas development assistance to build adequate institutional and technical capacity

10.2.1.6 Aviation and Maritime (Bunker-Fuel) Emissions

It was pointed out in Chap. 8 that UNFCCC Parties have had enormous difficulty agreeing on the methods to regulate greenhouse gas emissions from international aviation and maritime activities. The main reasons for this are the difficulties surrounding attribution and concerns that regulatory measures could harm the international competitiveness of high-export countries. It was also highlighted that the European Union, through its EU-ETS, has successfully limited aviation-related emissions within Europe and that the approach offers a potential solution for the UNFCCC in its battle to regulate both aviation and bunker-fuel emissions.

There are two aspects worth mentioning here. Firstly, it is clear from the EU experience that if the fuels used by the aviation and shipping industries could be regulated within an emissions-trading system by resting the point of obligation with the petroleum and oil-refining industries, concerns about attribution would quickly become irrelevant. Secondly, apprehension surrounding the international competitiveness of trade-exposed products could be overcome if the border-tax adjustments recommended in this book were introduced (see Sect. 9.2.5). As a means of illustration, an exporter of a product shipped by a firm subject to a high greenhouse gas price would be entitled to a tax rebate. In other instances, a compensating tariff would be imposed on an imported product shipped by a firm not subject to the same greenhouse gas price. In all, there is nothing unique about aviation and bunker-fuel emissions that should render them problematic. The problem lies with the deficiencies of some of the measures embodied in the Kyoto Protocol, which could be easily addressed by a new global protocol.

10.2.2 Distributional Equity Requirements[17]

10.2.2.1 The Distribution of Emissions Rights—Factors to Consider

Given the desire to distribute emissions rights in line with a common but differentiated convergence methodology, it is necessary for UNFCCC Parties to determine and agree upon greenhouse gas emissions targets at both the group and national levels. At the group level, targets should be based on per capita and historical greenhouse gas emissions along with the per capita GDP of the countries within them. Per capita GDP is a critical metric for three reasons. Firstly, as we saw in Chaps. 2 and 4, per capita GDP is a major factor driving a nation's emissions. Secondly, per capita GDP closely reflects a nation's technological and institutional capacity to reduce greenhouse gas emissions. Thirdly, the per capita GDP of most Annex II nations is well beyond the optimum of Int$15,000 (2004 prices). The opposite is the case for Annex III nations and LDCs. Consequently, the distribution of emissions rights between the four groups must take account of the GDP growth requirements of the world's poorest nations and the comparative ease that many high-GDP nations should have in absorbing stringent emissions cuts as they reduce their per capita GDP in the process of making the transition to a qualitatively-improving steady-state economy.

As for setting emissions targets at the national level, there should be considerable leeway for the UNFCCC to set slightly different targets for individual nations within each group, just as different targets were applied to some Annex I nations during the first and second Kyoto commitment periods. Flexibility of target-setting should also be extended to any association or union of nations that exists as an individual UNFCCC Party—a policy that allowed the European Union to redistribute targets within its overall emissions constraint. Not unlike the distribution of emissions rights at the group level, intra-group variations should be based on the difference in the real GDP of the nations within each group (albeit the real GDP gap would be narrower than the gap between the groups); the disparity in historical emissions; and the unique needs and circumstances of some countries.

For obvious reasons, national-level target setting would be considerably more complex than the setting of targets at the group level. Given this and the fact that special nation-level considerations are probably best left for UNFCCC Parties to determine, I will focus my attention on emissions targets at the group level for the remainder of this chapter.

Once emissions targets have been agreed upon and embodied in a new climate change protocol, the emissions rights of each group and the Parties within them should be issued in the same form as they have during the operation of the Kyoto Protocol—that is, as Assigned Amount Units (AAUs). Upon issuing AAUs, the permissible emissions levels of a nation would depend on whether it is a net buyer or seller of emissions rights on international permit markets and/or on the CERs, ERUs, or RMUs it might obtain from investing in sequestration and emissions-reduction projects at home or abroad. Thus, a net buyer of emissions permits or a net generator of CERs, ERUs, and RMUs would be entitled to exceed its baseline emissions targets by an amount equal to the emissions rights inscribed in the emissions units it has acquired (see Sects. 8.2.4 and 8.2.5).

10.2.2.2 The Distribution of Emissions Rights—An Emissions-Reducing Timetable

Table 10.1 presents an emissions-reducing timetable for the four groups of nations for the 2016–2100 period. The manner in which the commitments are scheduled for each group is consistent with achieving the 450 ppm stabilisation target. Hence, the timetable meets the common desire of all nations to limit the rise in average global temperatures to no more than 2 °C above pre-industrial levels.

In addition to indicating the rate at which the permissible emissions of each group should be reduced over time, Table 10.1 reveals the decade in which each group must all but de-carbonise its industrial production activities. Table 10.1 also reveals the various stages during the century when the nations belonging to the Annex III and LDC groups would graduate to a higher group before ultimately graduating to the Annex II group level.

Table 10.1 Quantitative mitigation commitments of the four assumed UNFCCC groups of Parties/nations[18]

Quantitative mitigation commitments	2016–2020 equivalent status and obligations	2021–2030 equivalent status and obligations	2031–2040 equivalent status and obligations	2041–2050 equivalent status and obligations	2051–2100 equivalent status and obligations
Annex II (2016 status)	*Annex II* • Non-binding commitment to reduce GHGs by 3 % p.a.	*Annex II* • Binding commitment to reduce GHGs by 6 % p.a.	*Annex II* • Binding commitment to reduce GHGs by 9 % p.a.	*Annex II* • Binding commitment to reduce CO_2 from industrial production to negligible levels	*Annex II* • Binding commitment to keep CO_2 from industrial production to negligible levels
Annex I (*non-Annex II*) (2016 status)	*Annex I* (*non-Annex II*) • Non-binding commitment to reduce GHGs by 3 % p.a.	*Annex I* (*non-Annex II*) • Binding commitment to reduce GHGs by 4.5 % p.a.	*Annex I* (*non-Annex II*) • Binding commitment to reduce GHGs by 6 % p.a.	*Annex II* • Binding commitment to reduce GHGs by 9 % p.a.	*Annex II* • Binding commitment to reduce CO_2 from industrial production to negligible levels
*Annex III** (2016 status)	*Annex III* • Non-binding commitment to achieve GHG-intensity targets • Assume GHG increase of 1 % p.a.	*Annex III* • Binding commitment to reduce GHGs by 1 % p.a.	*Annex III* • Binding commitment to reduce GHGs by 3 % p.a.	*Annex I* (*non-Annex II*) • Binding commitment to reduce GHGs by 6 % p.a.	*Annex II* • Binding commitment to reduce GHGs by 9 % p.a. (2051–2060) • Eventually reduce CO_2 from industrial production to negligible levels (2061–2070)
LDCs (2016 status)	*LDCs* • Non-binding commitment to achieve GHG-intensity targets • Assume GHG increase of 2 % p.a.	*LDCs* • Non-binding commitment to achieve GHG-intensity targets • Assume GHG increase of 1 % p.a.	*LDCs* • Binding commitment to reduce GHGs by 1 % p.a.	*Annex III* • Binding commitment to reduce GHGs by 3 % p.a.	*Annex I → Annex II* • Binding commitment to reduce GHGs by: ➢ 6 % p.a. (2051–2060) ➢ 9 % p.a. (2061–2070) • Eventually reduce CO_2 from industrial production to negligible levels (2071–2080)

*Annex III (2016) includes Brunei, Israel, Kuwait, Qatar, Saudi Arabia, Singapore, SKorea, and UAE which would graduate to the Annex I group in 2021

If we begin with the Annex II group, it can be seen that it would continue to meet its second-round Kyoto commitment to reduce its greenhouse gas emissions by around 3 per cent per annum until 2020. Beyond 2020, the Annex II group would be required to reduce its emissions by 6 per cent per annum over the 2021–2030 decade and by 9 per cent over the 2031–2040 decade. At some stage during the 2041–2050 decade, the group would need to reduce the carbon emissions from its industrial production to negligible levels.

The situation for the Annex I (non-Annex II) group is much the same as the Annex II group except that the required rate of emissions reductions is less severe. For the 2021–2030 decade, the group would need to reduce its greenhouse gas emissions by 4.5 per cent per annum; by 6 per cent per annum over the 2031–2040 decade; and by 9 per cent over the 2041–2050 decade. Of note, the Annex I (non-Annex II) group would graduate to the Annex II level in 2041, which is the point when the two groups would be bound by the same emissions obligations. Despite the eventual merger, there would be no need for the Annex I (non-Annex II) group to all but de-carbonise its industrial production until the 2051–2060 decade—a decade later than the Annex II group.

As for the Annex III group, it would remain free from quantitative emissions obligations until the beginning of 2021. In the meantime (i.e., 2016–2020), and to help shift global emissions onto a 450 ppm stabilisation trajectory, Annex III countries would need to take on greenhouse gas-intensity targets—in effect, renewable-resource and energy-efficiency targets—to limit the collective rise in their greenhouse gas emissions to around 1 per cent per annum up to 2020. In order to facilitate this process, Annex III countries would need to make a currently non-required commitment to meet greenhouse gas-intensity targets at the upcoming COP meeting in Paris (late-2015).

Given that a 1 per cent annual increase in greenhouse gas emissions is less than what the Annex III group has enjoyed over the past decade, one might question whether it would make such a commitment if requested to do so. I believe it would in view of the eagerness that non-Annex I nations have shown towards raising global mitigation ambitions up to 2020 together with their willingness to contribute towards their realisation (Kirby 2013; UNFCCC 2014a). This recent change of mindset has come about not only because non-Annex I nations recognise its global importance, but because they recognise that making a minor concession prior to 2020 would help them seamlessly transition from being target-free to taking on quantitative emissions obligations in the near future. May I say that getting non-Annex I nations to make this concession is likely to depend on Annex I countries boosting their post-2020 funding commitments at the Paris conference.

Beyond 2020, and with a new global protocol in place, the Annex III group would be required to reduce its greenhouse gas emissions by 1 per cent per annum over the 2021–2030 decade; by 3 per cent per annum over the 2031–2040 decade; by 6 per cent per annum over the 2041–2050 decade; and by 9 per cent over the 2051–2060 decade. The Annex III group would graduate to the Annex I (non-Annex II) level as of the beginning of 2041 and to the Annex II level in 2051.

During the 2061–2070 decade, the Annex III group would be required to reduce the carbon emissions from its industrial production to negligible levels.

Finally, Table 10.1 shows that LDCs would graduate in a similar fashion to the Annex III group, albeit the group would not be subject to quantitative emissions obligations until the beginning of 2031. Moreover, the stringency of the LDCs' emissions obligations would be delayed by one decade as would the decade in which the group needs to all but de-carbonise its industrial production activities (2071–2080).

Much like the Annex III group, LDCs would be required to make a commitment to meet greenhouse gas-intensity targets prior to being subject to quantitative emissions obligations. For the 2016–2020 period, the targets would be less demanding than those set for the Annex III group. This would allow LDCs to collectively increase their annual greenhouse gas emissions at a rate of around 2 per cent. The greenhouse gas-intensity targets of LDCs would be tightened for the 2021–2030 decade in order to limit the rise in the group's greenhouse gas emissions to around 1 per cent per annum. Starting from 2031, LDCs would be required to reduce its emissions by 1 per cent per annum; by 3 per cent per annum over the 2041–2050 decade; by 6 per cent per annum over the 2051–2060 decade; and by 9 per cent over the 2061–2070 decade. The virtual de-carbonisation of industrial production activities would not be required of LDCs until the 2070–2080 decade.

Box 10.6 A Timetable for Reducing Greenhouse Gas Emissions

Annex II group

- The Annex II group would be subject to quantitative emissions targets throughout the 21st century. The group would be required to cut its emissions as follows:
  - 2016–2020: 3 % per annum
  - 2021–2030: 6 % per annum
  - 2031–2040: 9 % per annum
  - 2041–2050: With its process-related CO_2 emissions falling to negligible levels, the Annex II group would complete the process of de-carbonising its industrial production activities
  - 2051–2100: The group would commit to keeping CO_2 emissions from industrial production activities to negligible levels

Annex I (non-Annex II) group

- The Annex I (non-Annex II) group would be subject to quantitative emissions targets throughout the 21st century. The group would be required to cut its emissions as follows:
  - 2016–2020: 3 % per annum
  - 2021–2030: 4.5 % per annum

- 2031–2040: 6 % per annum
- 2041–2050: 9 % per annum
- 2051–2060: With its process-related CO_2 emissions falling to negligible levels, the Annex I (non-Annex II) group would complete the process of de-carbonising its industrial production activities
- 2061–2100: The group would commit to keeping CO_2 emissions from industrial production activities to negligible levels

Annex III group

- The Annex III group would only be subject to greenhouse gas-intensity targets during the 2016–2020 period. However, it would be subject to quantitative emissions targets thereafter. Thus, pre-2021:
  - 2016–2020: The group would be subject to greenhouse gas-intensity targets to help limit the increase in the group's emissions to 1 % per annum
- Post-2020, the group would be required to cut its emissions as follows:
  - 2021–2030: 1 % per annum
  - 2031–2040: 3 % per annum
  - 2041–2050: 6 % per annum
  - 2051–2060: 9 % per annum
  - 2061–2070: With its process-related CO_2 emissions falling to negligible levels, the Annex III group would complete the process of de-carbonising its industrial production activities
  - 2071–2100: The group would commit to keeping CO_2 emissions from industrial production activities to negligible levels

LDCs

- The group of LDCs would only be subject to greenhouse gas-intensity targets during the 2016–2030 period. However, it would be subject to quantitative emissions targets thereafter. Thus, pre-2031:
  - 2016–2020: The group would be subject to greenhouse gas-intensity targets to help limit the increase in the group's emissions to 2 % per annum
  - 2021–2030: The group would be subject to greenhouse gas-intensity targets to help limit the increase in the group's emissions to 1 % per annum
- Post-2030, the group would be required to cut its emissions as follows:
  - 2031–2040: 1 % per annum
  - 2041–2050: 3 % per annum
  - 2051–2060: 6 % per annum
  - 2061–2070: 9 % per annum

- 2071–2080: With its process-related CO_2 emissions falling to negligible levels, LDCs would complete the process of de-carbonising its industrial production activities
- 2081–2100: The group would commit to keeping CO_2 emissions from industrial production activities to negligible levels

Should a new global protocol involve the imposition of emissions targets as just described, the aggregate emissions of each group would be markedly different between now (2015) and the end of the 21st century. Figure 10.1 reveals the changing emissions of each group at ten year intervals over the rest of the century. Figure 10.1 shows that the aggregate emissions of the Annex II group would rapidly decline to a very low level by 2050, where they would remain through to 2100. The same would occur to the emissions of the Annex I (non-Annex II) group, except that the rate of decline would be less pronounced. The aggregate greenhouse gas emissions of the Annex III group, which currently constitutes a little over 50 per cent of total global emissions, would increase slightly between 2015 and 2020 before falling quite dramatically thereafter. As for LDCs, their aggregate emissions would rise between 2015 and 2030 before declining to relatively low levels by 2070–2080.

To better compare the aggregate greenhouse gas emissions of each group and put them into global perspective, consider Table 10.2 and Fig. 10.2. The table shows that the respective annual emissions of each group in 2015 as a percentage of global emissions are: Annex II group (30.2 %); Annex I (non-Annex II) group) (11.3 %); Annex III group (52.4 %); and LDCs (5.7 %). By 2050, this distribution will have changed to 8.9, 5.0, 70.2, and 15.9 % respectively. The large fall in the percentages experienced by the Annex II and Annex I (non-Annex II) groups and the increases enjoyed by the Annex III and LDC groups would be the product of extensive up-front cuts to the emissions of the former two groups and the delayed imposition of strict emissions cuts on the latter two groups.

Interestingly, except for LDCs, this trend would be overturned to some extent by the end of the century, with the percentage of global emissions generated in 2100 by each group expected to be 16.4, 9.2, 57.3, and 17.2 % respectively. However, such a reversal would not indicate any leniency towards the Annex II and Annex I (non-Annex II) groups. It would reflect that: (i) from around 2060 to 2100, the annual greenhouse gas emissions of the Annex II and Annex I (non-Annex II) groups would barely change given that both groups would have completed or been close to completing the de-carbonisation of their industrial production by 2050; (ii) that the emissions of the Annex III group would still be in decline during the early part of the second half of the century; and (iii) that any rapid decline in the LDC group's emissions would not take place until the 2061–2070 decade.

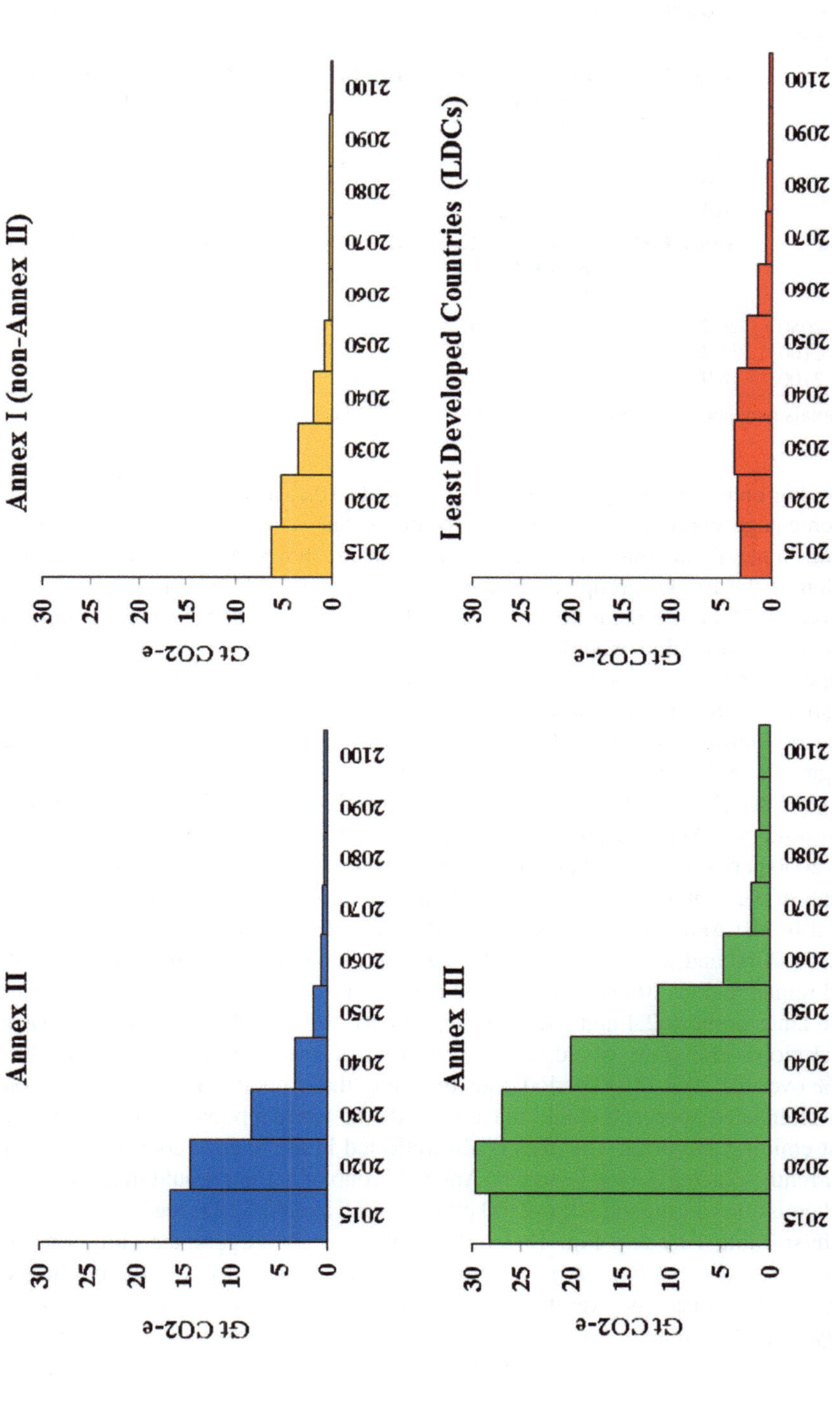

Fig. 10.1 Emissions of each group of Parties consistent with a common but differentiated approach to achieving a 450 ppm stabilisation target

Table 10.2 Projected emissions of each group of Parties based on 'common but differentiated convergence' approach (% of global emissions)

Year	Annex II (%)	Annex I (non-AII) (%)	Annex III (%)	LDCs (%)	Annex I (%)	Non-Annex I (%)
2015	30.6	11.3	52.4	5.7	41.9	58.1
2050	8.9	5.0	70.2	15.9	13.9	86.1
2100	16.4	9.2	57.3	17.2	25.6	74.4
Period	**Annex II (%)**	**Annex I (non-AII) (%)**	**Annex III (%)**	**LDCs (%)**	**Annex I (%)**	**Non-Annex I (%)**
2016–2050	20.2	8.5	62.4	9.0	28.6	71.4
2051–2100	11.2	6.3	64.5	18.0	17.5	82.5
2016–2100	18.9	8.1	62.7	10.3	27.0	73.0

Note Totals may not add up precisely due to rounding of values

Moving onto the top right-hand corner of Table 10.2, it highlights the change in the percentage contribution of global greenhouse gas emissions by the currently-existing Annex I and non-Annex I group of nations. It shows that the percentage contribution of the former group would fall dramatically from 41.9 to 13.9 % between 2015 and 2050 before rising to 25.6 % by 2100. As for the non-Annex I group, its percentage contribution would rise from 58.1 to 86.1 % between 2015 and 2050 before dropping back to 74.4 % by 2100. The reasons for these variations and fluctuations are much the same as those given above in relation to the four groups of nations.

The bottom half of Table 10.2 is different to the top half in that it reveals the *total* greenhouse gases emitted by each group as a percentage of total global emissions over the 2016–2050, 2051–2100, and 2016–2100 periods. It shows that, under my proposal, the Annex II group would emit one-fifth of global emissions during the 2016–2050 period (20.2 %), but only around one-ninth of global emissions in the second half of the century (11.2 %). Although the percentage of global greenhouse gases emitted by the Annex I (non-Annex II) and Annex III groups would vary little over the 2016–2050 and 2051–2100 periods, the contribution made by the former group would remain below 10 per cent (8.5 and 6.3 %) but continue to exceed 60 per cent for the latter group (62.4 and 64.5 %). Importantly, the contribution made by LDCs to total global emissions would increase from 9.0 % over the 2016–2050 period to 18.0 % over the 2051–2100 period, thus reflecting the explicit bias that the common but differentiated approach would make towards the world's poorest and historically lowest-emitting countries. This bias is also reflected in the relative contributions that the currently-existing Annex I and non-Annex I group of nations would make towards global emissions during the 2051–2100 period—17.5 and 82.5 % respectively.

Whilst Table 10.2 and Fig. 10.2 indicate the probable aggregate emissions of the four groups of nations, a more important equity concern is the likely change in per capita emissions over the century. This is revealed in Fig. 10.3, which closely resembles the common but differentiated convergence diagram in Fig. 9.6. Beginning with the Annex II group, Fig. 10.3 shows that the group's per capita

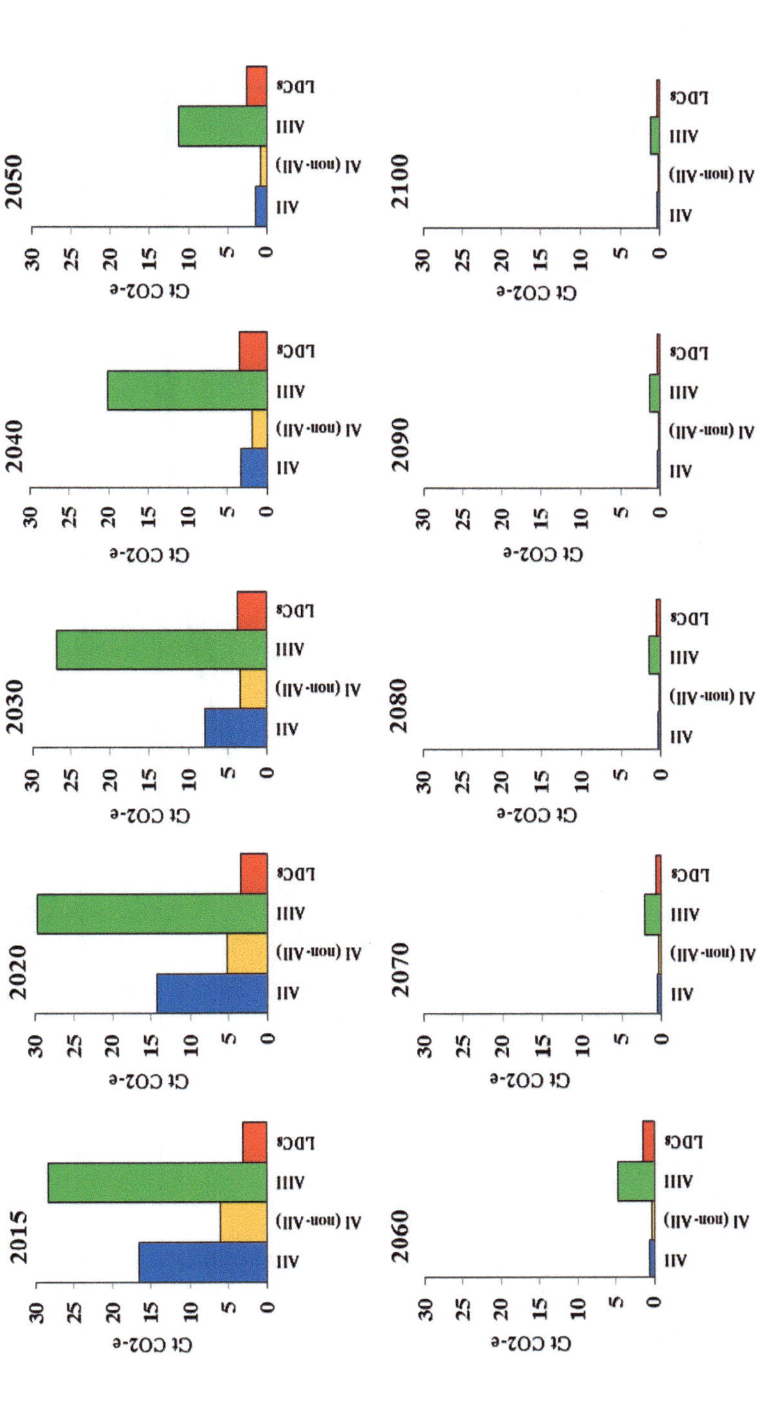

Fig. 10.2 Comparison of the emissions of each group of Parties consistent with a common but differentiated approach to achieving a 450 ppm stabilisation target. *Notes* AII denotes the Annex II group of nations. AI (non-AII) denotes the Annex I (non-Annex II) group of nations. AIII denotes the Annex III group of nations. *LDCs* denotes Least-Developed Countries

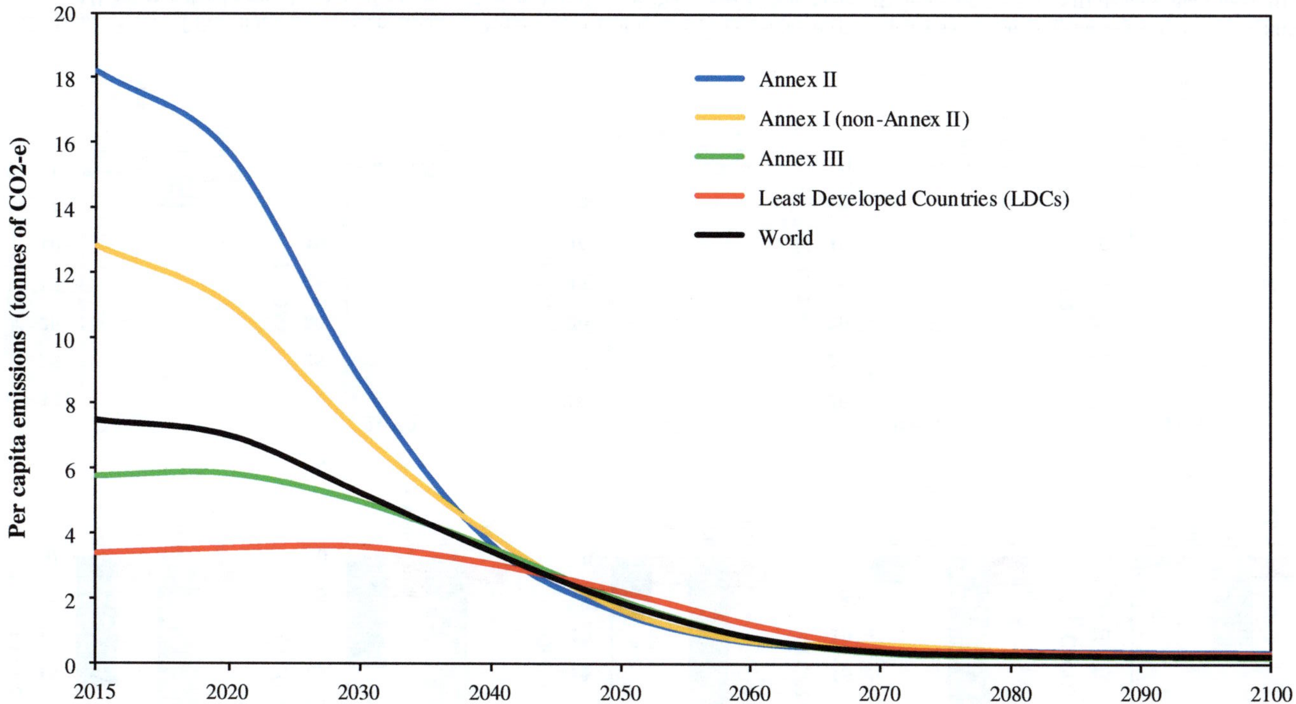

Fig. 10.3 Per capita greenhouse gas emissions of each group of Parties (based on a recommended 'common but differentiated convergence' approach)

emissions would decline precipitously from a starting value of around 18 tonnes of CO_2-equivalent greenhouse gases in 2015 (18.2 tonnes of CO_2-e). As intended, the group's per capita emissions would fall below the global per capita value during the 2041–2050 decade (1.6 tonnes by 2050 compared to the global average of 1.8 tonnes) and eventually converge to the global average of 0.9 tonnes by 2060. A similar change in per capita greenhouse gas emissions would be experienced by the Annex I (non-Annex II) group of nations. The difference is that this latter group's per capita emissions would be 12.8 tonnes in 2015 and 1.7 tonnes by 2050.

Turning to the Annex III group, its per capita emissions would increase slightly during the 2016–2020 period and only start falling in a significant way after 2030. At 2.0 tonnes of CO_2-equivalent greenhouse gases, the Annex III group's per capita emissions would be above the global per capita value of 1.8 tonnes in 2050. The group's per capita emissions would finally converge to the global per capita value somewhere around 2060.

In a similar manner, the per capita emissions of LDCs would be permitted to rise during the 2016–2030 period—a decade longer than the Annex III group of nations—before declining in a marked way post-2040. Because of the delayed requirement for LDCs to meet quantitative emissions targets, the group's per capita emissions would, at 2.2 tonnes of CO_2-e, be 23 per cent higher than global per capita emissions in 2050. Convergence to the global per capita value would not occur for the group until around 2070.

10.2.2.3 Qualitative Mitigation Commitments

Besides imposing quantitative emissions targets to stabilise the atmospheric concentration of greenhouse gases at 450 ppm of CO_2-e, a global protocol must include a range of qualitative mitigation commitments. Their nature and the commitments that each group must make would be appreciably shaped by the equity considerations raised in Chap. 9.

Before specifying what these qualitative commitments would be, it is worth noting that any failure to meet them would not necessarily undermine global attempts to achieve the 450 ppm stabilisation target. Nevertheless, it would reduce the likelihood of increasing the real GWP/emissions ratio sufficiently to realise the 450 ppm target at lowest cost, which would severely undermine humankind's ability to achieve an optimal per capita GWP of Int$15,000 (2004 prices).[19] In particular, it would make it extremely difficult for low-GDP nations to attain high levels of economic welfare whilst also contributing, through emissions cuts, to the global effort to achieve a safe atmospheric concentration of greenhouse gases.

Table 10.3 presents the qualitative mitigation obligations of the four groups of nations over the course of the 21st century. By and large, the qualitative obligations of the Annex II and Annex I (non-Annex II) groups would involve a commitment to undertake self-funded mitigation measures and accept co-funded Joint Implementation (JI) projects from other Parties. The only difference between the

Table 10.3 Qualitative mitigation commitments of the four assumed UNFCCC groups of Parties/nations

Qualitative mitigation commitments	**2016–2020 equivalent status and obligations**	**2021–2030 equivalent status and obligations**	**2031–2040 equivalent status and obligations**	**2041–2050 equivalent status and obligations**	**2051–2100 equivalent status and obligations**
Annex II (2016 status)	*Annex II* • Self-funded mitigation measures and co-funded JI investments	*Annex II* • Self-funded mitigation measures and co-funded JI investments	*Annex II* • Self-funded mitigation measures and co-funded JI investments	*Annex II* • Self-funded mitigation measures and co-funded JI investments	*Annex II* • Self-funded mitigation measures and co-funded JI investments
Annex I (*non-Annex II*) (2016 status)	*Annex I* (*non-Annex II*) • Self-funded mitigation measures and co-funded JI investments	*Annex I* (*non-Annex II*) • Self-funded mitigation measures and co-funded JI investments	*Annex I* (*non-Annex II*) • Self-funded mitigation measures and co-funded JI investments	*Annex II* • Self-funded mitigation measures and co-funded JI investments	*Annex II* • Self-funded mitigation measures and co-funded JI investments
*Annex III** (2016 status)	*Annex III* • AII and AI funded mitigation measures and CDM investments • Non-binding commitment to meet renewable and energy-efficiency targets	*Annex III* • AII and AI funded mitigation measures and JI investments	*Annex III* • Co-funded mitigation measures and JI investments	*Annex I* (*non-Annex II*) • Co-funded mitigation measures and JI investments	*Annex II* • Self-funded mitigation measures and co-funded JI investments
LDCs (2016 status)	*LDCs* • AII and AI funded mitigation measures and CDM investments • Non-binding commitment to meet renewable and energy-efficiency targets	*LDCs* • AII and AI funded mitigation measures and CDM investments • Non-binding commitment to meet renewable and energy-efficiency targets	*LDCs* • AII and AI funded mitigation measures and JI investments	*Annex III* • Co-funded mitigation measures and JI investments	*Annex I → Annex II* • Co-funded mitigation measures and JI investments (2051–2060) • Self-funded mitigation measures and co-funded JI investments

*Annex III (2016) includes Brunei, Israel, Kuwait, Qatar, Saudi Arabia, Singapore, SKorea, and UAE which would graduate to the Annex I group in 2021

two groups is that, until around 2040, one would expect most JI-projects to take place in Annex I (non-Annex II) countries and be instigated by Annex II nations.

Initially (2016–2020), the qualitative mitigation obligations of the Annex III and LDC groups of nations would be very different to those of the two wealthier groups. Although both groups would be expected to maintain their current commitment to pursue mitigation measures and accept Clean Development Mechanism (CDM) projects, many of the former and all of the latter would be supported by the funding commitments of Annex II and Annex I (non-Annex II) countries. Since the Annex III and LDC groups would free from quantitative emissions targets during the 2016–2020 period, they would be expected to make a non-binding commitment to meet renewable-resource and energy-efficiency targets in order to limit the growth in their greenhouse gas emissions (as previously indicated).

With the new climate change protocol coming into effect in 2021, the Annex III group's qualitative mitigation obligations would dramatically alter. Because the group would now be subject to quantitative emissions targets, its previous commitment to satisfy greenhouse gas-intensity targets would be superseded, albeit the need to reduce the emissions-intensity of its economic activities would remain. In addition, the group would be committed to accept JI-projects rather than CDM-projects—the switch being required because the Clean Development Mechanism should only apply to countries free of quantitative emissions obligations (see Sect. 9.2.1.2).

During the 2021–2030 decade, many of the mitigation measures undertaken by the Annex III group would continue to be funded by transfers from the Annex II and Annex I (non-Annex II) groups of nations. The same would also apply to the JI-projects undertaken in Annex III countries. However, as Table 10.3 shows, the mitigation measures and JI-projects occurring in Annex III countries during the 2031–2050 period would be co-funded. This means that Annex III countries would part-fund most of their mitigation measures with the remaining costs supported by a variety of financial resources provided by the Annex II and Annex I (non-Annex II) groups (e.g., the Green Climate Fund). As for JI-projects, they would be jointly funded by the investor-nation and the host Annex III country. Beyond 2050, the mitigation measures undertaken in Annex III countries would be self-funded, although JI-projects would continue to be co-funded. In the second half of the century, and as low-GDP nations become wealthier, an increasing proportion of JI-projects would be undertaken by Annex III nations in Annex II and Annex I (non-Annex II) countries as the Annex III group broadens its search for low-cost ways to meet its emissions targets.[20]

A similar situation would also occur with the qualitative mitigation obligations of LDCs, with the difference being that the process of change would be delayed by a decade. Since the group of LDCs would not be subject to emissions targets until 2031, its commitment to satisfy renewable resource and energy-efficiency targets would continue beyond the 2016–2020 period. It would also be the recipient of CDM-projects during the 2021–2030 decade. However, as of 2031, the group would commence its acceptance of JI-projects. At this point, the registration of new CDM-projects would cease.

Throughout the 2031–2040 decade, the Annex II and Annex I (non-Annex II) groups of nations would fund the JI-projects undertaken in LDCs as well as many mitigation measures embarked upon by the LDC group. In the following two decades (2041–2060), the mitigation measures and JI-projects occurring in LDCs would be co-funded in the manner described above. Starting from 2061, LDCs would be required to self-fund their mitigation endeavours.[21] Conversely, the JI-projects occurring in LDCs would continue to be co-funded. During the 2061–2070 decade, some LDCs would begin to undertake JI-projects in countries belonging to the other three groups, especially those in the Annex III group.

Box 10.7 Qualitative Mitigation Commitments

Annex II and Annex I (non-Annex II) groups

- The Annex II and Annex I (non-Annex II) groups of nations would be subject to the same qualitative mitigation commitments throughout the century. They would involve:
  - the undertaking of self-funded mitigation measures
  - acceptance of JI-projects from other qualifying Parties

Annex III group

- The Annex III group would be subject to the following qualitative mitigation commitments:
  - 2016–2020:
    - the undertaking of Annex II and Annex I (non-Annex II)-funded mitigation measures
    - a non-binding commitment to meet renewable resource and energy-efficiency targets
    - acceptance of Annex II and Annex I (non-Annex II)-funded CDM-projects
  - 2021–2030:
    - the undertaking of Annex II and Annex I (non-Annex II)-funded mitigation measures
    - acceptance of JI-projects from other qualifying Parties
  - 2031–2050:
    - the undertaking of co-funded mitigation measures
    - acceptance of JI-projects from other qualifying Parties
  - 2051–2100:
    - the undertaking of self-funded mitigation measures
    - acceptance of JI-projects from other qualifying Parties

LDCs

- The group of LDCs would be subject to the following qualitative mitigation commitments:

- 2016–2030:
  - the undertaking of Annex II and Annex I (non-Annex II)-funded mitigation measures
  - a non-binding commitment to meet renewable resource and energy-efficiency targets
  - acceptance of Annex II and Annex I (non-Annex II)-funded mitigation CDM-projects
- 2031–2040:
  - the undertaking of Annex II and Annex I (non-Annex II)-funded mitigation measures
  - acceptance of JI-projects from other qualifying Parties
- 2041–2050:
  - the undertaking of co-funded mitigation measures
  - acceptance of JI-projects from other qualifying Parties
- 2051–2100:
  - the undertaking of self-funded mitigation measures
  - acceptance of JI-projects from other qualifying Parties

10.2.2.4 Transfers for Mitigation Purposes

As elucidated throughout this book, mitigation action can be performed in a variety of ways. For example, it can take the form of advances in energy efficiency; the development and uptake of low-emissions technology; increased rates of material recycling; greater product durability; reductions in real output (de-growth); population stabilisation; a switch from fossil fuels to renewable forms of energy; establishment of energy-efficient and low-emissions infrastructure; land clearance controls; afforestation; reforestation; the creation of conservation zones (e.g., National Parks); payments to maintain ecosystem services; improved forestry and agricultural practices; carbon geosequestration; and, despite some uncertainty, the development and application of geo-engineering technologies.

Not unlike efforts to reduce greenhouse gas emissions from LULUCF activities, to embark on these mitigation measures, a nation must have sufficient institutional and technical/productive capacity to undertake them. Although Annex II and Annex I (non-Annex II) countries possess this capacity, most Annex III nations and virtually all LDCs do not. Unfortunately, it takes considerable time for adequate capacity to be established. In the meantime, Annex II and Annex I (non-Annex II) nations must provide funds and technology to enable Annex III nations and LDCs to engage in beneficial mitigation action and establish the institutional and technical capacity required to become 'mitigation' self-reliant.

As argued in Chap. 9, some of these funding requirements must exist as binding funding commitments embodied in a new global protocol. On top of the overseas development assistance that should serve as an indirect source of mitigation funds, a new protocol would include three main funding commitments. The first is a mitigation commitment set at 0.25 per cent of the GDP of Annex II and Annex I (non-Annex II) countries.[22] The funds would be deposited in the Green Climate Fund and be distributed to help fund (or co-fund) the mitigation measures undertaken by Annex III nations and LDCs.[23] The second funding commitment would entail the distribution of some of the revenue raised from the emissions permits auctioned by the relevant central authorities in Annex II and Annex I (non-Annex II) nations (set at 25 per cent of the auction revenue). These funds would also be used to boost the Green Climate Fund. The third commitment, which would be solely aimed at promoting forestry-related mitigation action in low-GDP countries, would be an obligatory contribution to the Carbon Fund of the Forest Carbon Partnership Facility (FCPF) equal to 0.1 per cent of the GDP of Annex II and Annex I (non-Annex II) countries.[24]

Given the large quantity of funds that these commitments represent (see Table 9.3), an obvious question arises as to how they should be distributed. In keeping with the qualitative mitigation obligations above, Table 10.4 presents a timetable for the transfer of financial resources and technology between the four groups of nations for strictly mitigation purposes. Table 10.4 indicates that the Annex II group would provide large transfers for the benefit of currently-existing non-Annex I nations over the 2016–2040 period. The large transfers would be necessary given that, up to 2040, many of the mitigation measures undertaken by LDCs would be entirely funded by currently-existing Annex I nations.

With the Annex III and LDC groups co-funding their own mitigation measures by the 2014–2050 decade, the demands placed on the Annex II group would correspondingly diminish. Annex II countries would therefore provide smaller transfers of financial resources and technology during this decade. The demands on the Annex II group would fall further in the 2051–2060 decade—the upshot of Annex III nations having to self-finance their own mitigation measures—where upon very small transfers would be required of the Annex II group of nations. Beyond 2060, and with the Annex III and LDC groups 'mitigation' self-reliant, transfers from the Annex II group would be phased out.

For good equity reasons, the funding demands of Annex I (non-Annex II) nations would not be as great as those of the Annex II group. This is reflected in Table 10.4 by the fact that large transfers of financial resources and technology would only be required from the Annex I (non-Annex II) group over the 2016–2030 period—a decade shorter than what would be required of the Annex II group. Beyond 2030, the funding demands of the Annex I (non-Annex II) group would gradually diminish and be completely phased out during the 2051–2060 decade.

Table 10.4 shows that the Annex III and LDC groups would be the recipients of financial resources and technology for mitigation purposes. Both groups would enjoy the benefits of large transfers over the 2016–2030 period. In the post-2030 period, the transfers received by the Annex III group would begin to diminish and

Table 10.4 Transfers (technology and funding) for mitigation purposes of the four assumed UNFCCC-defined groups

Transfers for *mitigation* purposes	**2016–2020 equivalent status and obligations**	**2021–2030 equivalent status and obligations**	**2031–2040 equivalent status and obligations**	**2041–2050 equivalent status and obligations**	**2051–2100 equivalent status and obligations**
Annex II (2016 status)	*Annex II* • Large transfers of financial resources and technology to non-Annex I nations	*Annex II* • Large transfers of financial resources and technology to non-Annex I nations	*Annex II* • Large transfers of financial resources and technology to non-Annex I nations	*Annex II* • Small transfers of financial resources and technology to non-Annex I nations	*Annex II* • Very small transfers of funds and technology (2051–2060) • Phasing out of transfer provision (post-2060)
Annex I (*non-Annex II*) (2016 status)	*Annex I* (*non-Annex II*) • Large transfers of financial resources and technology to non-Annex I nations	*Annex I* (*non-Annex II*) • Large transfers of financial resources and technology to non-Annex I nations	*Annex I* (*non-Annex II*) • Small transfers of financial resources and technology to non-Annex I nations	*Annex II* • Very small direct transfers of financial resources and technology to non-Annex I nations	*Annex II* • Phasing out of transfer provision (post-2050)
*Annex III** (2016 status)	*Annex III* • Recipients of large investment flows and technology transfers from Annex I nations	*Annex III* • Recipients of large investment flows and technology transfers from Annex I nations	*Annex III* • Recipients of small investment flows and technology transfers from Annex I nations	*Annex I* (*non-Annex II*) • Recipients of very small investment flows and technology transfers from Annex I nations	*Annex II* • Phasing out of transfer receipt (post-2050)
LDCs (2016 status)	*LDCs* • Recipients of large investment flows and technology transfers from Annex I nations	*LDCs* • Recipients of large investment flows and technology transfers from Annex I nations	*LDCs* • Recipients of large investment flows and technology transfers from Annex I nations	*Annex III* • Recipients of small investment flows and technology transfers from Annex I nations	*Annex I* → *Annex II* • Recipients of very small transfers of funds and technology (2051–2060) • Phasing out of transfer receipt (post-2060)

*Annex III (2016) includes Brunei, Israel, Kuwait, Qatar, Saudi Arabia, Singapore, SKorea, and UAE which would graduate to the Annex I group in 2021. These countries would not be the recipients of investment and technology transfers from Annex I nations during the 2016–2020 period

be phased out altogether during the 2051–2060 decade. On the other hand, the group of LDCs would continue to receive large transfers for mitigation purposes for a further decade (2031–2040). Not until the 2061–2070 decade, and only after a twenty-year period of decline, would mitigation-targeted transfers to LDCs cease.

In view of the reduction over time in mitigation transfers, the three main funding commitments of the Annex II and Annex I (non-Annex II) groups would gradually taper off. At a rate best determined by UNFCCC Parties at future COP meetings, the diminution of transfers should occur through measured reductions in the percentage contribution from: (i) the GDP of Annex II and Annex I (non-Annex II) nations; and (ii) the revenue generated by the auctioning of emissions permits. In making its deliberations, the UNFCCC would need to bear in mind that many Annex II countries would, under ideal circumstances, be undergoing a process of de-growth whilst making the transition to an optimal per capita GDP (i.e., an optimal macroeconomic scale).[25] The associated decline in the real GDP of these nations would automatically serve as a further means of reducing the funds being raised for international redistribution purposes.

Box 10.8 Transfers of Funds and Technology—Mitigation

Annex II group

- For *mitigation* purposes, the Annex II group would be required to provide transfers of financial resources and technology to low-GDP nations until at least 2060. The scheduled transfers are as follows:
  - 2016–2040: Provide large transfers of financial resources and technology
  - 2041–2050: Provide small transfers of financial resources and technology
  - 2051–2060: Provide very small transfers of financial resources and technology
  - 2061 onwards: The provision of transfers would be phased out
- Mitigation-related funding commitments would diminish by way of reductions in: (i) the percentage contribution from the GDP of Annex II nations (declining from 0.25 per cent of GDP); (ii) the real GDP of some Annex II nations as they transition to a steady-state economy (optimal macroeconomic scale); and (iii) the percentage contribution from the revenue generated by the auctioning of emissions permits (declining from 25 per cent of the total revenue raised)

Annex I (non-Annex II) group

- For *mitigation* purposes, the Annex I (non-Annex II) group would be required to provide transfers of financial resources and technology to low-GDP nations until at least 2050
  - 2016–2030: Provide large transfers of financial resources and technology
  - 2031–2040: Provide small transfers of financial resources and technology

  - 2041–2050: Provide very small transfers of financial resources and technology
  - 2051 onwards: The provision of transfers would be phased out

- Mitigation-related funding commitments would diminish by way of reductions in the percentage contribution from: (i) the GDP of Annex II nations (declining from 0.25 per cent of GDP); and (ii) the revenue generated by the auctioning of emissions permits (declining from 25 per cent of the total revenue raised)

Annex III group

- For *mitigation* purposes, the Annex III group would receive transfers of financial resources and technology until at least 2050
  - 2016–2030: Receive large transfers of financial resources and technology
  - 2031–2040: Receive small transfers of financial resources and technology
  - 2041–2050: Receive very small transfers of financial resources and technology
  - 2051 onwards: The receipt of transfers would be phased out

LDCs

- For *mitigation* purposes, the Annex III group would receive transfers of financial resources and technology until at least 2060
  - 2016–2040: Receive large transfers of financial resources and technology
  - 2041–2050: Receive small transfers of financial resources and technology
  - 2051–2060: Receive very small transfers of financial resources and technology
  - 2061 onwards: The receipt of transfers would be phased out

10.2.2.5 Transfers for Adaptation Purposes

In Chap. 1, it was revealed that a host of deleterious impacts can be expected even if the rise in average global temperatures is restricted to 2 °C above pre-industrial levels (see Table 1.4 and Fig. 1.10). Some the expected climate change impacts include: (i) a decline in the water supplies for a large percentage of the world's population; (ii) increased rates of species extinction; (iii) extensive bleaching of coral reefs and rising rates of coral mortality; (iv) a reduced capacity of the terrestrial biosphere to sequester carbon; (v) declining aggregate output of most cereals; (vi) an increased number of people at risk of coastal flooding; (vii) several metres of sea-level rise; and (viii) increasing rates of morbidity and mortality from heatwaves, floods, and droughts.

There are, of course, countless ways that a nation can minimise the cost of the impacts that climate change incurs upon it. Some of the measures suggested throughout this book include better flood control; the establishment of drought-resistant crops; improvements in water, forestry, and fisheries management to minimise the impact on natural capital stocks; the construction and/or expansion of dykes and levees; improved monitoring and control of disease outbreaks; and the establishment of heat-health warning systems and natural-disaster action plans.

Clearly, some of these measures will have a greater moderating effect on the cost of climate change damages than others. However, as argued in Chap. 6, all measures are limited in their capacity to prevent climate change damage costs from escalating as average global temperatures increase (see Figs. 6.1 and 6.2). Indeed, regardless of how effectively mitigation efforts restrict the rise in greenhouse gas emissions, the fact that average global temperatures are likely to increase by at least 2 °C above pre-industrial levels means that adaptation measures will be required for many decades to come.

Not unlike mitigation, Annex III nations and LDCs do not have the institutional and technical/productive capacity to undertake many or, in some cases, any of the above-listed adaptation measures. Hence, again, Annex II and Annex I (non-Annex II) nations will need to provide Annex III countries and LDCs with the funds and technology to limit the rise in their climate change damage costs as they gradually accumulate sufficient institutional and technical capacity to become 'adaptation' self-reliant.

As stressed in Chap. 9, a number of funding obligations must be embedded in a new global protocol to enable Annex III countries and LDCs to meet their adaptation requirements.[26] The first is an adaptation commitment that, like the mitigation commitment outlined above, would constitute 0.25 per cent of the GDP of Annex II and Annex I (non-Annex II) countries.[27] The second is an additional 25 per cent of the revenue raised from the emissions permits auctioned within the emissions-trading systems of high-GDP nations. Because adaptation measures will be required beyond the need to undertake mitigation measures, a third source of adaptation funds would be the continuation of the 2 per cent levy on emissions units generated through the Clean Development Mechanism (CERs), the Joint Implementation facility (ERUs), and greenhouse gas-removing domestic activities (RMUs). All three sources of funds would be deposited in the Green Climate Fund for later redistribution to Annex III nations and LDCs.[28]

It was also explained in the previous chapter why there is a need for a separate Global Climate Change Emergency Fund to cover the damage costs caused by climate change-induced weather events. It was suggested that the Fund should be capitalised by: (i) a further funding commitment equal to 0.1 per cent of the GDP of Annex II and Annex I (non-Annex II) countries; (ii) an additional distribution of 25 per cent of the revenue raised from the auctioning of emissions permits, which would eventually apply to all four groups of Parties[29]; and (iii) international levies on air passenger travel and bunker fuels.

The manner in which adaptation funds and technologies would be transferred over the 21st century is revealed in Table 10.5. Analogous to mitigation, Table 10.5 provides a timetable for the transfer of financial resources and technology for strictly adaptation purposes. Because many of the adaptation measures

Table 10.5 Transfers (technology and funding) for adaptation purposes of the four assumed UNFCCC-defined groups

Transfers for *adaptation* purposes	**2016–2020 equivalent status and obligations**	**2021–2030 equivalent status and obligations**	**2031–2040 equivalent status and obligations**	**2041–2050 equivalent status and obligations**	**2051–2100 equivalent status and obligations**
Annex II (2016 status)	*Annex II* • Large transfers of financial resources and technology to non-Annex I nations	*Annex II* • Large transfers of financial resources and technology to non-Annex I nations	*Annex II* • Large transfers of financial resources and technology to non-Annex I nations	*Annex II* • Large transfers of financial resources and technology to non-Annex I nations	*Annex II* • Small transfers (2051–2060) • Very small transfers (2061–2070) • Phasing out of transfer provision (post-2070)#
Annex I (*non-Annex II*) (2016 status)	*Annex I* (*non-Annex II*) • Large transfers of financial resources and technology to non-Annex I nations	*Annex I* (*non-Annex II*) • Large transfers of financial resources and technology to non-Annex I nations	*Annex I* (*non-Annex II*) • Large transfers of financial resources and technology to non-Annex I nations	*Annex II* • Small transfers of financial resources and technology to non-Annex I nations	*Annex II* • Very small transfers (2051–2060) • Phasing out of transfer provision (post-2060)#
*Annex III** (2016 status)	*Annex III* • Recipients of large investment flows and technology transfers from Annex I nations	*Annex III* • Recipients of large investment flows and technology transfers from Annex I nations	*Annex III* • Recipients of large investment flows and technology transfers from Annex I nations	*Annex I* (*non-Annex II*) • Recipients of small investment flows and technology transfers from Annex I nations	*Annex II* • Recipients of very small transfers (2051–2060) • Phasing out receipt of transfers (post-2060)#
LDCs (2016 status)	*LDCs* • Recipients of large investment flows and technology transfers from Annex I nations	*LDCs* • Recipients of large investment flows and technology transfers from Annex I nations	*LDCs* • Recipients of large investment flows and technology transfers from Annex I nations	*LDCs* • Recipients of large investment flows and technology transfers from Annex I nations	*Annex I → Annex II* • Recipients of small transfers (2051–2060) • Recipients of very small transfers (2061–2070) • Phasing out receipt of transfer (post-2070)#

*Annex III (2016) includes Brunei, Israel, Kuwait, Qatar, Saudi Arabia, Singapore, SKorea, and UAE which would graduate to the Annex I group in 2021. These countries would not be the recipients of investment and technology transfers from Annex I nations during the 2016–2020 period

#The phasing out of transfers does not include funds that would continue to be disbursed through the Global Climate Change Emergency Fund

undertaken by LDCs would be fully funded by currently-existing Annex I nations, Table 10.5 shows that the Annex II group would provide large transfers for the benefit of currently-existing non-Annex I nations up to the middle of the century. Beyond 2050, the Annex II countries would provide small transfers during the 2051–2060 decade and smaller transfers again throughout the 2061–2070 decade. Post-2070, Annex III nations and LDCs would become largely 'adaptation' self-reliant. Hence, transfers from Annex II countries for adaptation purposes would be phased out, with the exception being the funds that would remain available through the Global Climate Change Emergency Fund.

As for the Annex I (non-Annex II) group of nations, Table 10.5 indicates that its funding demands would be substantial over the 2016–2040 period. Post-2040, and with the Annex III group co-funding its own adaptation responses, these demands would dwindle. The demands would further decline during the 2051–2060 decade and be entirely phased out during the 2061–2070 decade.

Compared to mitigation transfers, Annex III nations would enjoy the benefits of large adaptation-related transfers for an additional decade—the consequence of adaptation measures being required beyond the need for mitigation action. With the Annex III group increasingly self-reliant beyond 2040, its receipt of adaptation-related transfers would taper off and eventually cease during the 2061–2070 decade. More favourably, LDCs would receive large transfers for adaptation purposes until 2050. However, following a gradual rate of decline over the 2051–2070 period, adaptation-related transfers to LDCs would be discontinued during the 2071–2080 decade.

Much like mitigation transfers, the adaptation-related funding commitments of the Annex II and Annex I (non-Annex II) groups would diminish to reflect the gradual decline in adaptation transfers. The means by which the commitments would be reduced is much the same as for mitigation—namely, via reductions in the percentage contribution from the GDP of Annex II and Annex I (non-Annex II) nations and a dwindling cut from the revenue generated by the auctioning of emissions permits. Somewhat differently, the need to maintain the Global Climate Change Emergency Fund means that the two commitments to the Fund—namely, the 0.1 per cent share of the GDP of Annex II and Annex I (non-Annex II) nations and the 25 per cent share of auctioned permit revenue—would remain intact.

Box 10.9 Transfers of Funds and Technology—Adaptation

Annex II group

- For *adaptation* purposes, the Annex II group would be required to provide transfers of financial resources and technology to low-GDP nations until at least 2060
    - 2016–2050: Provide large transfers of financial resources and technology
    - 2051–2060: Provide small transfers of financial resources and technology

- 2061–2070: Provide very small transfers of financial resources and technology
- 2071 onwards: The provision of transfers would be phased out

- Adaptation-related funding commitments would diminish by way of reductions in: (i) the percentage contribution from the GDP of Annex II nations (declining from 0.25 per cent of GDP); (ii) the real GDP of some Annex II nations as they transition to a steady-state economy (optimal macroeconomic scale); and (iii) the percentage contribution from the revenue generated by the auctioning of emissions permits (declining from 25 per cent of the total revenue raised)
- Funding commitments to the Global Climate Change Emergency Fund would remain intact throughout the 21st century

Annex I (non-Annex II) group

- For *adaptation* purposes, the Annex I (non-Annex II) group would be required to provide transfers of financial resources and technology to low-GDP nations until at least 2050
  - 2016–2040: Provide large transfers of financial resources and technology
  - 2041–2050: Provide small transfers of financial resources and technology
  - 2051–2060: Provide very small transfers of financial resources and technology
  - 2061 onwards: The phasing out of transfer provision
- Adaptation-related funding commitments would diminish by way of reductions in the percentage contribution from: (i) the GDP of Annex II nations (declining from 0.25 per cent of GDP); and (ii) the revenue generated by the auctioning of emissions permits (declining from 25 per cent of the total revenue raised)
- Funding commitments to the Global Climate Change Emergency Fund would remain intact throughout the 21st century

Annex III group

- For *adaptation* purposes, the Annex III group would receive transfers of financial resources and technology until at least 2050
  - 2016–2040: Receipt of large transfers of financial resources and technology
  - 2041–2050: Receipt of small transfers of financial resources and technology
  - 2051–2060: Receipt of very small transfers of financial resources and technology
  - 2061 onwards: The receipt of transfers would be phased out

- Upon introducing emissions-trading systems to cost-effectively achieve quantitative emissions targets (2021), the Annex III group would begin contributing to the Global Climate Change Emergency Fund. This funding commitment would continue for the remainder of the 21st century

LDCs

- For *adaptation* purposes, the Annex III group would receive transfers of financial resources and technology until at least 2050
  - 2016–2050: Receipt of large transfers of financial resources and technology
  - 2051–2060: Receipt of small transfers of financial resources and technology
  - 2061–2070: Receipt of very small transfers of financial resources and technology
  - 2071 onwards: The receipt of transfers would be phased out
- Upon introducing emissions-trading systems to cost-effectively achieve quantitative emissions targets (2031), LDCs would begin contributing to the Global Climate Change Emergency Fund. This funding commitment would continue for the remainder of the 21st century

10.2.2.6 Intranational Equity Considerations

The revenue generated from the auctioning of greenhouse gas emissions permits cannot only be used to help low-GDP nations finance much needed mitigation and adaptation measures. It can also be used by governments to assist the poor within their own nation to cope with a greenhouse gas price, which, by increasing the price of emissions-intensive goods and services, can have regressive distributional implications.

As part of an ecological tax reform package (see Chap. 3), a government can exploit the opportunity that a national emissions-trading system offers to achieve domestic equity goals by increasing the tax-free income threshold or granting tax cuts to people on low incomes. In addition, some of the revenue can be used to part-finance a Job Guarantee programme to achieve and maintain full employment in a potentially GDP-suppressed but economic welfare-augmented economy.

10.2.2.7 A Global Emissions-Trading System that Sacrifices Some Allocative Efficiency to Ensure Distributional Equity

The final equity matter requiring attention concerns the fear that a global emissions-trading system would adversely affect many low-GDP nations. It has already been explained that an emissions-trading system which allows the price of greenhouse gas emissions to freely fluctuate reduces a government's control over permit prices. This lack of control is magnified in cases where a country links its own national emissions-trading system with the systems of other countries, and would reach a maximum if a single emissions-trading system covered every nation.

Given this loss of government control, the main concern for low-GDP nations is the potential for a global emissions-trading system to generate a persistently high permit price. A high permit price could render permits unaffordable to many entities operating in the poorest participating nations, which would make it difficult for poor countries to achieve their real GDP targets. Consequently, it is unlikely that low-GDP nations would agree to a system that involves paying the same price to emit a tonne of CO_2-equivalent greenhouse gases as that paid by wealthy countries.

To ensure equity across all nations and facilitate effective negotiations at future UNFCCC conferences, a separate emissions-trading system should apply to each of the four groups of nations with restrictions placed on the ability of countries to participate in external systems. Because each group would have a different capacity to pay for emissions permits and a different quantity of permits available for auction/sale, the greenhouse gas price generated by each of the four emissions-trading systems would be vastly different. Presumably, the price pertaining to the system covering the Annex II group of nations would be considerably higher than the price generated by the system covering LDCs.[30] There would, therefore, be some loss of allocative efficiency. This is because the lack of a uniform global price would result in less-than-effective mitigation action in some parts of the world and an elevated greenhouse gas price burden in other parts.[31] For example, in LDCs, the low greenhouse gas price would encourage the use of more emissions-intensive production methods than if LDCs were confronted with a higher global price. Conversely, Annex II nations would be subject to an inflated greenhouse gas price that would force many firms to adopt high-cost mitigation technologies.

Mind you, this undesirable feature of a ‘split-system’ would be overcome to a considerable extent by funding transfers from high-GDP to low-GDP nations that would help subsidise higher-cost mitigation options in the world’s poorest nations. This would encourage the adoption of less emissions-intensive technologies and production methods. At the same time, the inflated greenhouse gas price faced by high-GDP nations would cull many of the firms/industries with high mitigation costs. This would make it easier for many high-GDP nations to de-grow their economies as they transition towards an optimal per capita GDP.

It should also be borne in mind that the emissions-trading systems of the four groups of nations would converge over time. In accordance with Table 10.1, the Annex I (non-Annex II) group would join the emissions-trading system of the Annex II group of nations in 2041. The Annex III group and LDCs would both do likewise in 2051 and 2061 respectively. Consequently, by 2061, there would be full convergence of the greenhouse gas price signal faced by each country as well as convergence of each group’s per capita emissions. Thus, any efficiency sacrificed in order to achieve equity across nations would be temporary.

10.2.3 Allocative Efficiency Requirements

10.2.3.1 National Emissions-Trading Systems

As mentioned a number of times, an emissions-trading framework that encompasses all the world's nations should form the centrepiece of a new climate change protocol. Despite good equity reasons for initially having a separate emissions-trading system for each of the four groups of nations, it is ultimately in the best interests of countries to participate in the largest and broadest possible system given that full participation in such a system would increase the options available to minimise mitigation costs. This, in turn, would promote the efficient allocation of international resources. Nevertheless, countries should always be free to establish their own emissions-trading system and operate independently of a larger group system.[32]

Importantly, the establishment of a domestic-based emissions-trading system need not preclude participation in a designated group system. The Framework for Various Approaches (FVA)—the market-based mechanism being developed by UNFCCC Parties (see Sect. 8.2.8.6)—aims to allow a nation to integrate its own emissions-trading system into a larger system. However, integration would require a domestic system to meet minimum standards regarding the conduct of emissions-trading and ensure permanent, additional, and verifiable emissions reductions (UNFCCC 2014b). In view of the likely embodiment of the FVA in a new global protocol, I shall outline what an acceptable emissions-trading system should look like and how it would be integrated into a designated group system.

The key features of a national emissions-trading system were revealed in Chap. 7. Most of them would prevail in a system that is likely to meet UNFCCC-determined standards. For instance, an independent central-government authority would exist to administer and operate the system. The authority would have a range of functions and responsibilities. They would include, firstly, the issuance of emissions permits in accordance with the AAUs assigned to it by the UNFCCC. Although the number of permits issued would not exceed maximum allowable emissions levels, it could be a great deal less if technical and/or institutional shortcomings prevent the authority from attributing greenhouse gas emissions to particular activities. In these circumstances, the industries where attribution is possible would be covered by the emissions-trading system, but remaining industries would not.[33] In the interim, reductions in the greenhouse gas intensity of activities not covered by the system would be promoted through domestic policies of the type outlined in Chap. 3. At all times, a nation should seek to address attribution problems to maximise the coverage of its emissions-trading system.

Secondly, because all emissions permits would be sold by the government authority (i.e., none would be freely allocated except for permits created through the generation of domestic offsets), the authority would be responsible for conducting regular permit auctions. The auctions would be open to all eligible individuals, government establishments, and privately-owned organisations. Where a nation's emissions-trading system is linked to the systems of other nations, the auctions would be open to overseas entities except those in countries belonging to a higher-ranked group (more on this soon).

Contrary to the preference of many observers, the life of a permit would be restricted to one year, which would preclude the hoarding of permits and their use at a later date. Hence, public auctions would need to be conducted at least once a year.[34] Given the limited lifespan of emissions permits, a single permit would entitle its possessor the legal right to emit one tonne of CO_2-equivalent greenhouse gases during the period between the permit's auction date and its expiration date. Until such time as the permits expire, they would be tradeable. The price a permit would not be set by the government authority, but would be determined by demand and supply-side forces in the permit market.

An important third function of the central-government authority would be its treatment of the CERs, *t*CERs, and ERUs generated by domestic entities engaging in the Clean Development Mechanism and Joint Implementation facility; the RMUs generated by domestic greenhouse gas removals; and emissions reductions and removals achieved in industries not covered by the national emissions-trading system. Given the need to promote domestic mitigation and concerns regarding temporary emissions units, the authority would limit the number of CERs, *t*CERs, and ERUs that can be traded within the national emissions-trading system, which it could do by imposing a restriction based as a fixed proportion of the AAUs allocated to it. Ideally, the limit would be set at around 10 per cent of allocated AAUs, which, in view of the total quantity of CERs, *t*CERs, and ERUs issued to date, might be split as follows: (i) 6 per cent of CERs; (ii) 1 per cent of *t*CERs; and (iii) 3 per cent of ERUs. This would mean that if a country was limited to 100,000 AAUs in 2021, a maximum of 6,000 CERs, 1,000 *t*CERs, and 3,000 ERUs would be available for purchase in the same year within its emissions-trading system.

Assuming a shift from CDM-project investments in the Annex III group to JI-project investments as of 2021, and the same for LDCs as of 2031, the percentage of CERs and *t*CERs available for purchase would be reduced each year after 2021. For example, the percentage of CERs and *t*CERs available for purchase might be reduced by 0.4 per cent and 0.1 per cent respectively (2021, 6.0 %/1.0 %; 2022, 5.6 %/0.9 %; 2023, 5.2 %/0.8 %; etc.). To balance the ledger, the percentage of ERUs might be annually increased by 0.5 per cent (2021, 3.0 %; 2022, 3.5 %; 2023, 4.0 %; etc.). This relative adjustment would continue until the breakup of CERs, *t*CERs, and ERUs in 2030 is 2.4 per cent, 0.1 per cent, and 7.5 per cent respectively. Beyond 2030, the number of tradeable CERs and *t*CERs would be further reduced until it declines to zero (Note: the quantity of CERs and *t*CERs wouldn't be reduced to zero by the end of 2030, even though the registration of new CDM-projects would cease at this time, because registered CDM-projects would continue to be credited with CERs and *t*CERs until they came to an end).

The best way for the central-authority to restrict the number of CERs, *t*CERs, and ERUs that can be exchanged in the national emissions-trading system is to auction 'supplementary permits' equal to the limits it wishes to impose (Garnaut 2008). Based on the above example, the authority would auction 6,000 supplementary-CER permits, 1,000 supplementary-*t*CER permits, and 3,000 supplementary-ERU permits in 2021. For an entity to purchase a supplementary-CER permit, it would need to earn a CER from a CDM-investment and then surrender it upon purchasing the supplementary-CER permit. By sharing the same features as an AAU-derived permit, the supplementary permit could then be used or traded like other permits in the national emissions-trading system. The same approach would also apply to *t*CERs and to ERUs acquired via JI-investments.

Incidentally, the market price of a supplementary-CER permit would equal the expected difference between the price of an AAU and the price of a CER (Garnaut 2008). Hence, if the prices of AAUs and CERs are $15 and $5 respectively, one would expect the price of a supplementary-CER permit to be around $10.[35] The market prices of supplementary-*t*CER permits and supplementary-ERU permits would be determined in a similar fashion.

As for RMUs, they would be granted to the entities that achieve greenhouse gas removals in the form of domestic offsets.[36] Issued in this way, domestic offsets would constitute newly-created emissions permits that reflect the increase in a nation's permissible greenhouse gas emissions (see Table 8.5). The authority would also grant offsets to entities that achieve verifiable emissions reductions and greenhouse gas sequestrations in industries not covered by the national emissions-trading system.[37] All offsets would be available for trade. However, as I will soon explain, some countries would be forbidden from purchasing certain types of offsets within a particular nation's emissions-trading system.

Fourthly, the central-government authority would administer a compliance mechanism which, where feasible, would entail the direct metering of greenhouse gas emissions to ensure the emissions generated by permit holders comply with the emissions rights inscribed in the permits they possess. Where the direct metering of emissions is impracticable, greenhouse-gas emitters would be required to report their emissions to the authority. To promote acquiescence, the authority would randomly inspect self-reporting emitters to ensure they are operating legally.

Fifthly, the government authority would publicise the point of obligation to alert entities of the activities being regulated and provide an information service to enable the public to know if they are required to acquit emissions permits. In cases where the point of obligation lies with emissions-related activities (e.g., oil extraction) as opposed to end-of-line activities that discharge greenhouse gases (e.g., automobile use), the authority would provide a conversion schedule to determine the quantity of a particular resource that, when used, equates to the emission of one tonne of carbon dioxide-equivalent gases. This would enable entities engaged in emissions-related activities to determine the number of permits they must acquire.

Sixthly, the authority would impose penalties in response to breaches of the emissions-trading system. To deter non-compliance, the penalties for emitting greenhouse gases illegally would be severe. At the very minimum, they would include large fines; the confiscation of an offender's unused emissions permits; and a lengthy period where the transgressor would be prohibited from trading in permits.[38] In extreme cases, penalties would include the confiscation of business assets or imprisonment. Upon detection of a transgression, the central-government authority would reduce the maximum level of greenhouse gas emissions to offset the additional greenhouse gases emitted illegally. The authority would do this by reducing the number of emission permits sold at future public auctions.

Finally, the central-government authority would maintain a register of all individuals and entities in possession of emissions permits. As an entity engages in activities being regulated by the emissions-trading system, the authority would reduce the entity's right to emit greenhouse gases by an amount equal to the emissions it has generated. As indicated in Chap. 7, this would involve the authority destroying the appropriate number of permits registered in the entity's name. Hence, there would be no physical exchange of emissions permits between the authority and the initial buyers of permits, nor between the subsequent buyers and sellers of permits. Instead, there would be a transfer of emissions rights from the latter to the former on the authority's register, where the latter would be the authority at the time of a public auction. This arrangement would: (i) facilitate a more effective monitoring of greenhouse-gas emitters; (ii) increase the authority's ability to identify transgressors of the system; and (iii) allow the authority to act as a clearing-house for yet-to-be-used emissions permits. It would also make it easier for a country to links its national emissions-trading system with the systems of other nations in its designated group.

Box 10.10 Key Features of a National Emissions-Trading System

- With the aim of engaging in a group-level system initially and participation in a fully-integrated global system by the middle of the century, countries should establish a national emissions-trading system. The national emissions-trading system would incorporate the following key features:

  - Administration of the system would be the responsibility of an independent central-government authority
  - Emissions permits would be issued with a cap set to ensure emissions do not exceed the AAUs assigned to it as per a new (post-Kyoto) protocol
  - Where technical and/or institutional shortcomings prevent the authority from attributing greenhouse gas emissions to particular activities, the issuance of emissions permits would be reduced so that coverage is limited to the industries where attribution is possible
  - Where attribution is not yet possible, the relevant industries would not be covered by the emissions-trading system. Emissions reductions in these industries would be promoted through alternative sustainable development policies

- There would be regular public auctions of emissions permits (at least one auction per year)
- One emissions permit would entitle its possessor the right to emit one tonne of CO_2-equivalent greenhouse gases during the life of the permit
- The life of an emissions permit would be restricted to one year
- Barring specific restrictions, all greenhouse gas emissions permits would be tradeable—hence, permit prices would be determined by demand and supply forces in the emissions-permit market, not by the authority Limits would be placed on the number of CERs, *t*CERs, and ERUs that can be traded within the national emissions-trading system. The limits would be:
  - set as a fixed proportion of the AAUs allocated to the nation
  - imposed by auctioning 'supplementary permits' that, upon being purchased, would: (i) be exchanged for CERs, *t*CERs, and ERUs; and (ii) be tradeable within the national emissions-trading system
- RMUs would be granted to entities that achieve greenhouse gas removals by sinks in the form of domestic offsets (emissions permits).
- Domestic offsets (emissions permits) would be issued to entities that achieve verifiable emissions reductions and greenhouse gas sequestrations in industries not covered by the national emissions-trading system
- The point of obligation would be publicised to enable entities engaged in emissions-related activities to determine whether they must acquit emissions permits and, if so, their acquittal obligations
- Penalties for breaches of the system, which would be designed to deter non-compliance, would include:
  - large fines
  - the confiscation of an offender's unused emissions permits
  - a lengthy period where transgressors would be prohibited from trading in permits
  - in extreme cases, confiscation of business assets or imprisonment
- The central authority would maintain a register of all individuals and entities in possession of emissions permits. As an entity engages in activities being regulated by the emissions-trading system, the authority would reduce its right to emit greenhouse gases by an amount equal to the emissions it has generated
- Sales of emissions permits would not involve the physical exchange of permits, but the transfer of emissions rights from the seller to the buyer on the authority's register

10.2.3.2 The International Linking of Emissions-Trading Systems—Meeting UNFCCC Standards

As I have stressed, to engage with other Parties in a larger group system, a nation will require a domestic emissions-trading system that meets international standards. Bearing in mind the need to satisfy the sustainable development goals of ecological sustainability, distributional equity, and allocative efficiency, these standards should be based on the following guiding principles (Garnaut 2008)[39]:

1. *Aligning permits with a nation's emissions target*. The number of permits issued by a national government must reflect the emissions targets determined by the UNFCCC. Post-2020, the pertinent emissions targets would be those embodied in a new global protocol, which should resemble the targets recommended in Table 10.1.
2. *Credibility of national emissions-trading institutions*. The institutions that support a nation's emissions-trading system must be reliable, resilient, and credible.
3. *Simplicity of rules*. The rules governing an emissions-trading system must be uncomplicated and straightforward to implement. They must also be applied consistently, which would require the system to be devoid of special rules, concessions, and exemptions. A possible exception would involve a restriction on the number of CERs, *t*CERs, and ERUs that can be traded and used for compliance purposes to encourage more domestic mitigation.
4. *Tradability of permits*. To ensure appropriate permit prices and facilitate the cost-effective achievement of emissions targets, the characteristic features of emissions permits (i.e., lifespan, maximum allowable emissions) must be unambiguous. In addition, there is a need for: (i) commonly understood terms and conditions of exchange; (ii) easy access to permit markets; (iii) minimal transaction costs; and (iv) transparent and readily available offer and bid prices for permits.
5. *Compatibility with other markets*. A national emissions-trading system must be able to coexist and integrate with international markets covering the full range of emissions entitlements (i.e., CERs, *t*CERs, ERUs, and RMUs) as well as financial, natural resource, and goods markets in the international economy. This requires the full transmission of information within and between markets.

If international standards were based on these five principles, the national emissions-trading system recommended in the previous sub-section would have few problems meeting them. For starters, the issuance of emissions permits in line with a nation's allocated AAUs would satisfy the first guiding principle. Secondly, the central-government authority would more than likely be an extension of the authority responsible for overseeing the cap-auction-trade systems covering major renewable resources and different forms of waste. To meet these responsibilities, the authority would need to be well resourced. Assuming this to be the case, the authority would be a credible and resilient institution. Thus, the national emissions-trading system would satisfy the second principle.

Thirdly, in the form just described, the rules applicable to the national emissions-trading system are relatively uncomplicated. They would, therefore, be easy to administer. Furthermore, a major feature of the recommended emissions-trading system would be the consistent application of its rules to all market participants. Hence, the third guiding principle would be satisfied.

The fourth guiding principle—the tradability of permits—would also be met given that: (i) the features of emissions permits would be clearly specified; (ii) all permit types would be tradeable with clearly defined terms and conditions of exchange; (iii) permit prices would be determined by demand and supply-side forces within the emissions-permit market; (iv) apart from possible inter-group restrictions, there would be unfettered access to the emissions-permit market; and (v) the central-government authority's role as a clearing-house for unused permits would ensure offer and bid prices are readily available to market participants.

Last but not least, the tradability of CERs, *t*CERs, and ERUs plus the system's compatibility with a new protocol's flexibility mechanisms would minimise permit-price distortions. Combined with the limited life-span of permits to thwart rent-seeking behaviour, the system would promote the cost-effective realisation of national emissions targets. It would, therefore, satisfy the fifth guiding principle.

I should add that, in order to meet the five guiding principles, it is important for a central-government authority to adequately inform market participants of any impending modifications to the national emissions-trading system, including any significant alteration to the number of permits it plans to issue. All modifications should be kept to a minimum. There are many reasons why changes to the system would be required. They include: (i) increasing the coverage of industries following a technological or institutional advance that augments the capacity of a central-government authority to attribute greenhouse gas emissions to individual activities; (ii) selecting a more cost-effective point of obligation; (iii) regulating additional greenhouse-gas types, which, if it occurred, would probably result from a collective decision by UNFCCC Parties; (iv) a government purchase and release of emissions permits obtained from the emissions-trading system of another nation; and (v) changes in the rules regarding acceptance of certain types of emissions units for compliance purposes (e.g., CERs, *t*CERs, and ERUs).

10.2.3.3 The Coverage of a National Emissions-Trading System

As intimated, the coverage of a national emissions-trading system should be as broad as possible. Within practical limits, the system would require all emitters of greenhouse gases and/or entities engaged in emissions-related activities to acquit emissions permits. Similarly, the system would aim to issue domestic offsets to all entities that remove greenhouse gases through their sequestration activities. For most nations, the industries likely to remain devoid of coverage for some time longer are forestry and agriculture. As important as it is to eventually incorporate these two industries into a national emissions-trading system—particularly given the quantity of emissions generated by agricultural and forestry activities and their considerable sequestration potential—it is also crucial that they not be included until attribution is possible.

Assuming that some Parties belonging to a particular group of nations are in a position to include forestry and agriculture or a previously non-covered industry in their domestic emissions-trading system, the shortcomings of lesser-abled nations should not preclude their participation in a larger group system. This is because a disparity in the coverage of greenhouse gas-generating activities across Parties need not undermine the integrity or effectiveness of a group-level emissions-trading system (Stern 2007; Garnaut 2008). Having said this, achieving emissions targets at the group level will require all countries to implement policies that limit greenhouse gas emissions in non-covered industries.

Should a lesser-abled nation want to include forestry and agriculture within the ambit of its emissions-trading system, its continued participation in the group-level system would depend on its ability to meet the international monitoring and attribution standards pertaining to forestry and agricultural activities. The same proviso would also apply to any other non-covered industry that a national government incorporates into its domestic emissions-trading system. As for the international standards used to assess system compatibility, they would be determined by UNFCCC Parties and regularly re-examined at UNFCCC meetings.

10.2.3.4 Point of Obligation

For a country's emissions-trading system to be linked with the domestic systems of other Parties, it is important that the point of obligation—that is, the point where emissions permits must be acquitted—is applied in a consistent manner. This doesn't mean that the point of obligation must be set at the same stage of the supply chain for all greenhouse gas-emitting activities. But it does mean that the point of obligation should be set at the same stage for a particular category of activities. This latter requirement is important given that the desirable point of obligation will differ across economic sub-sectors.

When determining the most appropriate point of obligation, a natural reference point is the location where greenhouse gases are generated (i.e., the exact source of emissions) (Garnuat 2008). Nonetheless, a more appropriate point of obligation may exist if: (i) measuring the direct source of emissions is difficult or impractical; (ii) transaction costs can be significantly reduced by shifting the point of obligation elsewhere in the supply chain (provided the shift does not affect the ability of a central authority to regulate emissions); or (iii) altering the point of obligation can increase an emissions-trading system's coverage of greenhouse gas-emitting activities.

There are essentially three stages in the supply chain where the point of obligation can be set. The first and most obvious is with the direct source of greenhouse gas emissions. The second is with the final outputs associated with the generation of greenhouse gases. The third is with the resource inputs that, when used, lead to the immediate or eventual release of greenhouse gases.

To put these points of obligation into context, consider automobile use. Should the point of obligation be set at the direct source of emissions, permit acquittal would be required and calculated in terms of the greenhouse gases contained in vehicle exhaust fumes. If, instead, the point of obligation is set with respect to final outputs, permit acquittal would be determined in terms of car mileage. As for resource inputs, this would depend on whether the point of obligation is set at an upstream or downstream stage of the supply chain. If it is the former, permit acquittal would be set at the oil-extraction or petrol-production stage; if it is the latter, permit acquittal would be set at petrol consumption (retail petrol sales).

Clearly, in this automobile example, it would be impractical to monitor the greenhouse gases emanating from car exhaust-pipes or the mileage of every vehicle. It would also be ineffective given there is a loose connection between vehicle mileage and the greenhouse gases generated from vehicle use.[40] It would also be cumbersome and costly to require all vehicle-owners to acquit emissions permits when purchasing petrol. Hence, the point of obligation would be best set at oil extraction or petrol production, which would require permit acquittal to be the responsibility of oil companies, not vehicle-users. Incidentally, setting the point of obligation at either oil extraction or across the production of all petroleum-based fuels would ensure full coverage of aviation and bunker-fuel emissions—two categories of emissions which have largely escaped regulation in the past.

There are a number of complications associated with transport fuels that are also shared by greenhouse gas-generating activities in other industries. To begin with, it is not uncommon for most if not all the transport fuels consumed within a country to be produced in another jurisdiction. Where they are produced domestically, they are often generated from imported oil. This is likely to influence the chosen point of obligation in the sense that a policy to set the obligation at oil extraction could result in the inadequate coverage of emissions from fuel consumption. Insufficient coverage could also occur if the point of obligation is set at transport-fuel production and a large portion of all domestically-consumed fuels is produced externally.

Ultimately, sufficient coverage will depend on whether the imported oil/transport fuels are adequately captured in the exporting country by a national emissions-trading system that meets UNFCCC standards. Provided these products are adequately covered, the oil-extraction and transport-fuel production activities in the importing and exporting countries would not threaten aggregate emissions targets. Nor would a greenhouse gas price disparity exist if the emissions-trading system of the foreign country is linked to the national system of the importing country, since, presumably, the permits covering imported oil/petrol would have much the same price as similar permits for sale in the group-level system.[41] The same could not be said if the oil-exporting country is operating within a different group-level emissions-trading system. In this case, the permits covering the imported oil/petrol would have a vastly different price, with a much higher permit price likely in the Annex II group system compared to the Annex III group system. Fortunately, permit price disparities need not be of great concern if the nations facing higher permit prices can close the price gap by imposing border-tax adjustments in the manner described in Sect. 9.2.5 (also see Box 10.3).

What would be the effect on aggregate emissions if the imported oil/transport fuels are not covered by an emissions-trading system in the exporting country? There would be no major problem if the exporting country has policies in place that successfully limit the greenhouse gases generated by its oil-extraction and transport-fuel production activities.[42] However, if this is not the case, the importing country may choose to capture the emissions from external activities by setting the point of obligation at the importation stage, which would require the domestic importers of oil/transport fuels to acquit emissions permits. Similar factors would also need to be considered by a central government when determining the appropriate point of obligation for other greenhouse gas-related imports.

With all these factors in mind, let us consider how the appropriate point of obligation might be chosen for various sub-sectors of a national economy.

1. *Stationary energy*. An example of stationary energy is the electricity generated by large power-stations. Greenhouse gases generated by this sub-sector can be reliably measured at the facility level. This suggests that the point of obligation for electricity generation should be set at the exact source of emissions. Whilst applying the point of obligation at the power-generation stage avoids the complications of imposing it on the millions of end-users of electricity, it can lead to the problem of double-capture. Double-capture is likely to occur if the point of obligation is, for example, set at coal mining and also at the emissions emanating from coal-generated electricity.
   Given the double-capture possibility, a central-government authority needs to stipulate the emissions it intends to capture at different stages in the electricity supply chain so that power-station operators are aware of the number of emissions permits they need to acquit (Note: the number of permits requiring acquittal would be determined by subtracting the emissions covered by the permits acquitted by resource-extractors from the emissions generated by power-stations).[43] A less desirable alternative would involve setting the point of obligation at the emissions generated by power-stations and establishing a credit system to enable operators to claim back the permit price incurred from purchasing permit-covered inputs (Garnaut 2008).
2. *Transport*. As highlighted with respect to automobile use, greenhouse gas emissions from the transport sub-sector are generated on a very small scale at the individual level, but on a very large scale at the sub-sector level. Although it is technically possible to measure emissions from vehicle exhaust, it would be impractical and costly to set the point of obligation at the level of individual activities. It is therefore best to set the point of obligation at either the oil-extraction or petrol-production stages of the supply chain.
   As indicated above, if some of the oil/petroleum consumed within a country is imported from a nation with insufficient coverage of its oil-extraction and transport-fuel production activities, it may be necessary to set the point of obligation for imported oil/petroleum at the international customs stage of the transport-fuel supply chain.

3. *Industrial processes*. By and large, greenhouse gas emissions from industrial processes can be measured or estimated at the production stage at low cost. Like stationary energy, this implies that the point of obligation for industrial processes should be imposed at the exact source of emissions. However, this can also lead to the problem of double-capture if the point of obligation is set at resource-extraction activities as well as the emissions from industrial processes involving the use of permit-captured resources. There is, therefore, a need for a central-government authority to provide the information required by managers of industrial plants to determine the number of permits they must acquit or, if a credit system exists, to extend it so that operators can claim back permit prices incurred from having to purchase permit-covered inputs.
4. *Fugitive emissions from energy resources*. Fugitive greenhouse gas emissions are the greenhouse gases intentionally or unintentionally released during the extraction, processing, storage, and transportation of energy resources (IPCC 2006). This category of emissions does not include the greenhouse gases that eventually emanate from the end-use or consumption of energy resources. In general, fugitive emissions released through pressure-release pipes, valves, and vents can be easily measured at low cost. There is good reason, therefore, to place the point of obligation for these fugitive emissions at the facility level. Unfortunately, there are other types of fugitive emissions where there are measurement difficulties and site-specific variations that render their estimation problematic (DCC 2008). Complicating matters with respect to these emissions is that it is impossible to capture them simply by setting the point of obligation at an early stage of the energy-resource supply chain (i.e., at the resource-extraction stage). If fugitive emissions were a negligible percentage of the total greenhouse gas emissions from the energy sub-sector,[44] there would be a good case for leaving difficult-to-measure emissions uncovered by a national emissions-trading system and to rely on other policy initiatives to reduce them. Given this is not a desirable strategy, I believe, like Garnaut (2008), that the point of obligation should be imposed at the source of these emissions by utilising existing proxy measures to determine the required acquittal of emissions permits. That said, I also agree with Garnaut that priority should be given to establishing a robust methodology to estimate and attribute the full range of fugitive emissions with greater precision.[45]
5. *Waste*. Greenhouse gas emissions from waste occur mainly in the form of methane releases from organic waste at landfill sites and wastewater treatment plants. At present, there are difficulties associated with covering emissions from waste due to variations in the quantity of emissions and the timing of their release across different site types. Given that the early coverage of waste by an emissions-trading system is highly desirable, establishing a means of accurately measuring and attributing waste-based emissions should, much like fugitive emissions from energy sources, be given high priority (Garnaut 2008). Once established, the point of obligation should be set at the site level. In the meantime, alternative policies should be implemented to promote mitigation in the waste sub-sector.

6. *Land use, land-use change, and forestry* (*LULUCF*). As mentioned earlier, in almost all nations, LULUCF activities remain largely uncovered by an emissions-trading system. In addition, few countries are likely to be in a position to include LULUCF activities in the very near future. It is, nonetheless, critical that these activities be incorporated into an emissions-trading system as early as possible. This demands that issues be resolved regarding the measurement and attribution of greenhouse gas emissions/removals and for considerations to be made with respect to changes in accounting procedures and flexibility mechanisms under the present Kyoto Protocol and in any future protocols (Garnaut 2008).
   In terms of forestry, once issues regarding measurement and attribution have been resolved, reforestation and afforestation activities should be incorporated into a national emissions-trading system, with domestic offsets granted to reward greenhouse gas removals (forest expansion) and acquittal of permits/offsets required to reflect greenhouse gas releases (deforestation). The latter would be most effectively achieved by setting the point of obligation at the establishment level.
   As for agriculture, a similar approach should be adopted with respect to permit acquittals and the generation of domestic offsets. Once more, the point of obligation should be imposed at the establishment/farm level. There are, nonetheless, greater difficulties associated with the measurement and attribution of agricultural emissions and removals which are likely to delay the inclusion of agriculture in a national emissions-trading system. For this reason, there is likely to be much greater reliance on alternative policies to reduce greenhouse gas emissions and to increase greenhouse sequestrations in the agricultural subsector (see Sect. 3.3.3).
   Another option, which has been employed by the New Zealand Government, is to impose the point of obligation on agricultural output processors, such as dairies and abattoirs (www.climatechange.govt.nz/ets). Although this approach has overcome some measurement difficulties and simplified the process by reducing the number of entities that must acquit emissions permits, its success has depended on a strong association between agricultural outputs and greenhouse gas emissions. A concern that this association does not exist appears to have been expressed of late by the New Zealand Government in view of its recent desire to shift the point of obligation to the farm level. A cautious attitude should therefore be adopted before imposing the point of obligation on agricultural outputs.

Box 10.11 Point of Obligation

- The point of obligation—the point in the supply chain where emissions permits must be acquitted by a relevant entity—should be set at the same stage of the supply chain for a particular category of activities, although it may vary from category to category and across economic sub-sectors
- Whilst it is best to set the point of obligation at the exact source of greenhouse gas emissions, it should be set elsewhere in the supply chain if:

- measuring the direct source of emissions is difficult or impractical
- transactions costs can be reduced by shifting the point of obligation elsewhere in the supply chain and provided the shift does not affect the ability to regulate emissions
- altering the point of obligation can increase an emissions-trading system's coverage of greenhouse gas-emitting activities

10.2.3.5 Institutional Structures and the Permit-Auctioning Process

As explained many times throughout this book, achieving allocatively efficient outcomes with respect to the emission of greenhouse gases—i.e., obtaining the highest use value from each tonne of greenhouse gases generated—requires an appropriate greenhouse gas price signal; the accurate measurement, monitoring, and attribution of greenhouse gas emissions; enforcement of the conditions attached to emissions rights; and minimisation of the speculative (price-distorting) buying and selling of emissions permits. To ensure these conditions are met, there is a strong need for effective legal, regulatory, and administrative structures and the presence of a sound permit-auctioning and trading process. As indicated in Chap. 9, these structures are prevalent in high-GDP nations but not so in many Annex III nations and LDCs.

Should a nation with inadequate institutional structures be permitted to operate in a group-level emissions-trading system, associated flaws in its emissions-trading system would contaminate the larger group system. Not only would this threaten a group's aggregate emissions targets, it would lead to a distorted greenhouse gas price signal and quite possibly carbon leakage. The potential for these problems to emerge reinforces the need for the institutional structures pertaining to a nation's emissions-trading system to meet minimum UNFCCC standards before a green-light is given to a country to engage in a larger group system.

10.2.3.6 Distributing the Revenue from Permit Auctions to Promote Allocative Efficiency

Earlier in the chapter, it was recommended that some of the revenue raised by Annex II and Annex I (non-Annex II) nations from the auctioning of emissions permits should be redistributed to satisfy international and intranational equity objectives. The former objectives included the funding of mitigation and adaptation measures in Annex III nations and LDCs and the capitalisation of the Global Climate Change Emergency Fund to cover the damage costs caused by climate change-induced weather events. The latter objectives included reductions in the marginal tax rate on low-income citizens and the part-financing of a Job Guarantee to achieve and maintain full employment.

From an allocative efficiency perspective, the remaining revenue should be used to overcome glaring market failures with respect to information deficiencies, financial capital constraints, and the inadequate provision of low-emissions and energy-efficient forms of infrastructure that exhibit public goods characteristics (see Sect. 7.7). To recall, this would require governments to use the residual revenue generated by the auctioning of emissions permits to: (i) provide the private sector with information on efficiency-increasing and low-emissions technologies; (ii) grant intellectual royalty rights to reward the creation of, and maximise community access to, new technologies; (iii) boost public-sector funding of research and development activities; (iv) increase private-sector research via the introduction of research subsidies, capital grants, tax credits/rebates, and research seed-funding; and (v) increase government investment in the infrastructure needed to directly assist in mitigation and adaptation efforts and to further facilitate the climate change mitigation and adaptation action of the private sector.

Box 10.12 The Distribution of Auction Revenue to Promote Allocative Efficiency

- To promote allocative efficiency, some of the revenue raised from the auctioning of emissions permits should be distributed by governments to:
  - provide information on efficiency-increasing and low-emissions technologies
  - grant intellectual royalty rights to reward the creation of, and maximise community access to, new technologies
  - boost public-sector funding of research and development activities
  - increase private-sector research via the introduction of research subsidies, capital grants, tax credits/rebates, and research seed-funding
  - increase investment in mitigation and adaptation-related public goods

10.2.3.7 Border-Tax Adjustments—Facilitating Allocative Efficiency at the International Level

I do not intend to repeat myself on the need for a mechanism to allow national governments to introduce border-tax adjustments or explain what the system would entail. I only wish to reiterate the fact that, by eliminating production cost gaps caused by international disparities in wages, conditions of employment, and environmental standards (including greenhouse gas prices), border-tax adjustments would increase the likelihood of natural resources and human-made capital being allocated to the genuinely most efficient production locations within the global economy. By doing so, border-tax adjustments—especially if complemented by an IMPEX system of exchange rate management (see Sect. 3.8.2)—would help generate the highest use value from each tonne of globally-emitted greenhouse gases. This would deliver the highest per capita economic welfare for a given per capita Gross World Product (GWP).

10.2.3.8 Flexibility Mechanisms—The Clean Development Mechanism, the Joint Implementation Facility, and Removals by Sinks

Once again, I have no desire to repeat what I have said regarding the workings of the Clean Development Mechanism (CERs), the Joint Implementation facility (ERUs), and the rules applying to greenhouse gas removals by sinks (RMUs). However, it is worth restating that in spite of the criticisms directed towards the Kyoto flexibility mechanisms—many of them warranted—these mechanisms would, if appropriately reformed and embodied in a new climate change protocol, further improve the efficiency of international resource allocation without jeopardising global emissions targets. In particular, they would enable targets to be achieved at lowest cost, which would go a long way towards keeping global mitigation costs to a minimum.

10.3 Governance of a National Emissions-Trading System[46]

In earlier chapters, it was explained that the climate change crisis is the symptom of the larger problem of excessive GDP growth that will not be resolved unless there are policies and institutions in place at the national level to help all nations begin the transition—some much sooner than later—to a qualitatively-improving steady-state economy. There is no doubt that new governance institutions will need to be established at the national level to deal specifically with the climate change crisis. However, should the policies and institutions recommended in Chap. 3 be introduced to achieve the broader goal of sustainable development, they would make it considerably easier for a nation to undertake the measures needed to meet its climate change obligations. In particular, they would enable a nation's institutions to neatly dovetail with a global emissions-trading framework that, in turn, would improve the effectiveness of its national emissions-trading system.

Another crucial influence on the effectiveness of a national emissions-trading system is the division between the policy-making and administrative elements of the scheme. Many of the policy decisions pertaining to the system should always remain the prerogative of a central government (Stern 2007; Garnaut 2008). Consequently, the government should at all times be in control of the policy and legislative dimensions of the national system. However, the hands-on administrative function should be left to an independent authority, whose executive powers should be defined via legislation.

The separate functions of the government and the government authority are summarised in Table 10.6. Beginning with the rules governing the national emissions-trading system, the government should be responsible for determining: (i) the coverage of the system; (ii) the point of obligation; (iii) the acquittal of emissions permits; (iv) the rules pertaining to domestic offsets; and (v) compliance requirements. The government would regularly review these rules and argue its case for rules changes at UNFCCC meetings.

Although the central-government authority would not be involved in the policy-setting of emissions-trading rules, it would offer the government advice on potentially beneficial rule changes and how they should be applied. From an administrative perspective, the authority would be responsible for the management of monitoring, reporting, and verification systems, and the enforcement of compliance regulations.

As for the imposition of emissions targets, once again, the government would assume the policy-setting role of determining emissions targets and the nation's long-term emissions trajectory. It would also determine the nature and timing of any changes to the nation's emissions targets. Notwithstanding this, the nation's emissions targets would be largely determined by a new global climate change protocol of which the government, via international negotiations, would play an active role in establishing. As emissions caps are tightened over time, the central-government authority would be called upon to administer the adjustments, which it would do by altering the quantity of emissions permits it issues.

The precise nature of permit issuance—i.e., whether greenhouse gas emissions permits are auctioned, sold at a fixed price, or issued free of charge—would be determined by the government, although, again, the government's strategy would be heavily influenced by the need to satisfy international standards and link up with the national emissions-trading systems of other nations. Assuming the government opts to sell emissions permits, the authority would, as explained earlier in the chapter, auction the permits on the government's behalf. It would also maintain a register of the ownership of emissions permits; act as a permit clearing-house; and publicise the offer and bid prices for permits.

The revenue raised from the permit auctions conducted by the central-government authority would be distributed in accordance with the government's policy objectives. There would essentially be no international obligations in terms of how the revenue set aside for domestic purposes is redistributed. Conversely, for Annex II and Annex I (non-Annex II) nations, the distribution of the revenue set aside for international purposes would be influenced by the need to meet international funding commitments pertaining to the Green Climate Fund, the Global Climate Change Emergency Fund, and the Carbon Fund of the Forest Carbon Partnership Facility. The same level of commitment would not be required of Annex III nations and LDCs, which, barring contributions to the Global Climate Change Emergency Fund, would be free to distribute permit revenue entirely to meet domestic needs.

In terms of overseeing the operation of the national emissions-trading system, this would not be the responsibility of the government. It would instead be

Table 10.6 Governance of the national emissions-trading system (ETS)

Functions of ETS governance	Government responsibilities (policy-making)	Independent authority (administration and regulation)
ETS rules	• All policy-making, including the coverage of the ETS, point of obligation, acquittal, offset rules and standards, and compliance • Undertake regular reviews of the ETS rules • Lobby for rule changes at UNFCCC meetings	• Advise government on rules and their application • Manage monitoring, reporting, and verification systems • Enforce compliance
Setting emissions targets/limits	• Determine and announce: ⇒ Emissions targets and trajectory ⇒ Nature and timing of changes to the emissions target	• Administer any movement from one target to the next
Permit issuance	• Determine the nature of permit issuance	• Auction permits • Maintain a register of the ownership of emissions permits • Publicise permit price • Act as clearing-house for permits
Use of permit revenue	• Domestically—determine the portion of permit revenue to be redistributed domestically and the manner in which it is spent • Internationally—determine the portion of permit revenue to be distributed to meet international funding commitments under a new global protocol	• Collect permit revenue for redistribution
Market supervision	• Not applicable	• Monitor the integrity of the domestic ETS market
International linking	• Negotiate global agreements (as part of new protocol) to establish uniform standards and conditions for the international trading of emissions permits • Negotiate global agreements, in co-operation with WTO, to establish suitable border-tax adjustments	• Monitor the international trade in emissions units by market participants • Certify that conditions for the international trading of emissions units have been met • In co-operation with customs, apply WTO-sanctioned compensating tariffs • Grant tax rebates to local trade-exposed industries according to an agreed cost-burden threshold test

Source Adapted from Garnaut (2008, Table 14.1)

the responsibility of the central-government authority, which would endeavour to maintain the integrity of the system in order to maximise its effectiveness and ensure it meets international standards.

The final governance area concerns international matters, such as the linking of national emissions-trading systems and border-tax adjustments. For obvious reasons, the government, not the central authority, would be involved in the international negotiations relating to monitoring and reporting standards and the international trading of emissions permits. The government would also negotiate global agreements with respect to permissible border-tax adjustments and the process involved in securing WTO approval for the imposition of compensating tariffs and the payment of tax rebates to trade-exposed domestic producers (e.g., the eligibility criteria and formula used to estimate the magnitude of the adjustments).

As for the central-government authority, its role in international governance matters would involve the monitoring of the international trade in the emissions units of various types (AAU-derived permits, CERs, ERUs, and RMUs) and ongoing assessment to ensure the agreed conditions for the international trading of emissions units are being adequately satisfied (i.e., certification). Last but not least, the government authority would: (i) calculate the size of the border-tax adjustments to be imposed; (ii) assist eligible trade-exposed industries to submit border-tax adjustment applications to the WTO; (iii) apply WTO-sanctioned compensating tariffs; and (iv) grant approved tax rebates to local trade-exposed industries according to an agreed cost-burden threshold test.

10.4 The Separation, Linking, and Gradual Merging of National Emissions-Trading Systems

I have already explained why there are good equity reasons for having a separate emissions-trading system for each of the four UNFCCC groups of nations. Whilst a separation of this sort would restrict some countries from participating in another group-level system, it would not impede a nation's ability to operate within its own group. Indeed, putting aside the quantitative limits imposed on tradeable emissions units to encourage domestic mitigation and exclusion of countries with emissions-trading systems that do not meet UNFCCC standards,[47] there would essentially be no restrictions on the trade in emissions units between countries belonging to a particular group of nations.

It has been suggested that excluding countries with dubious national emissions-trading systems from participating in a larger system is unnecessary. Garnaut (2008), for example, believes that flaws can be dealt with by having national governments set quantitative limits on aggregate permit purchases from external sources. According to Garnaut, a limit could be imposed so that it applies in potentially destabilising situations, but be set high enough to avoid taking effect during a typical trading period.

As feasible as this proposal seems, it would be extremely difficult to estimate what the appropriate limit should be. Given that a limit and what others have recommended effectively amounts to a constraint to thwart the toxic effect of a flawed system, I believe it would be preferable to exclude defective nation-level systems altogether from incorporation in a group-level system. Although this would limit the emissions units available for trading on international carbon markets, it would better protect the integrity of a group-level system and provide greater incentive for a country to ensure its emissions-trading system meets internationally-determined standards.

As for the trade in emissions units across the four group-level emissions-trading systems, countries in a lower-ranked group (e.g., Annex III group) would be able to purchase the AAUs made available for sale by nations belonging to a higher-ranked group (e.g., Annex II group). The converse would not be permitted.[48] Furthermore, the resale of AAUs purchased from a higher-ranked group would be restricted to a country belonging to a group no higher than the group from which the AAUs were initially purchased. For example, an Annex III nation could purchase AAUs from an Annex I (non-Annex II) country and, if it desires, later sell them to the same or a different Annex I (non-Annex II) country. However, the Annex III nation would not be able to sell them to an Annex II nation.

A different situation would apply with respect to domestic offsets, RMU-derived permits, and the CERs, *t*CERs, and ERUs generated through the Clean Development Mechanism and Joint Implementation facility. Upon the establishment of group-level emissions-trading systems for Annex III nations and LDCs, countries from higher-ranked groups would be permitted to purchase the domestic offsets and RMU-derived permits generated by low-GDP nations. This, of course, could be sanctioned in the knowledge that countries from higher-ranked groups would be unable to reduce the AAU-derived permits available for sale to countries in a lower-ranked group. Given the possibility that the RMUs could fetch high prices in international permit markets, unfettered access to RMU-derived permits would encourage Annex III nations and LDCs to engage in domestic action with the capacity to remove greenhouse gases by sinks.

There would also be unfettered market access to CERs, *t*CERs, and ERUs across the four group-level systems, albeit there would be limits on the availability of these emissions in keeping with the restrictions set by central-government authorities. Since all CDM-project investments would be conducted by nations in a group ranked higher than the host countries, there would be no purchases of CERs or *t*CERs by high-ranked nations from low-ranked nations. At the same time, the magnitude of the purchases of CERs, *t*CERs, and ERUs by low-ranked nations from high-ranked nations is likely to be minimal. In addition, virtually all purchases of CERs, *t*CERs, and ERUs would be made within the four group-level systems, rather than across them.

As mentioned earlier in the chapter, as lower-ranked groups graduate to a higher group before graduating to the Annex II group level, the emissions-trading systems of all nations would merge into one. This would result in a single global price for AAU-derived permits and the absence of any restrictions on the ability of a nation to participate in global carbon markets and to purchase the full range of available emissions units (Note: this is provided a nation's emissions-trading systems meet international standards).

Box 10.13 Trade in Emissions Units Across the Four Group-Level Emissions-Trading Systems

- A separate emissions-trading system would exist for each of the four UNFCCC groups of nations
- Besides the limits imposed on tradeable emissions units to encourage domestic mitigation and exclusion of countries with emissions-trading systems that do not meet UNFCCC standards, there would be no restrictions on the trade in emissions units between countries belonging to a particular group-level system
- Inter-group restrictions would apply such that:
  - countries in a higher-ranked group could not purchase the AAUs made available for sale by nations belonging to a lower-ranked group
  - the AAUs purchased from a higher-ranked group could not be resold to a country belonging to a group higher than the group from which the AAUs were initially purchased
- Countries from higher-ranked groups would be permitted to purchase the domestic offsets and RMU-derived permits generated by low-GDP nations
- There would be unfettered market access to CERs, *t*CERs, and ERUs across the four group-level systems, with limits on the availability of these emissions confined to the restrictions set by central-government authorities
- As lower-ranked groups graduate over time, the emissions-trading systems of all nations would eventually be merged into one global emissions-trading system. At this point, there would be:
  - a single global price for AAU-derived permits
  - no restrictions on the ability of nations to participate in global carbon markets
  - the ability of all nations to purchase the entire range of available emissions units

10.5 How Challenging Would It Be to Achieve the Greenhouse Gas Emissions Targets Recommended for the Post-2020 Protocol?

In this final section of the book, I want to allay the fears anyone would have about the difficulty nations might encounter in achieving the greenhouse gas emissions targets recommended in Table 10.1, especially Annex II nations which must make the largest upfront emissions cuts. What I aim to demonstrate is that if all nations are prepared to move to a qualitatively-improving steady-state economy, the task of achieving the required emissions targets will not be unduly arduous.

To justify my sanguinity, we shall consider the mitigation requirements of the four groups of nations between 2015 and 2050—the period in which countries are likely to have the greatest difficulty meeting their mitigation requirements. In order to do this, we first need to consider the factors that determine the ease with which a nation can achieve its emissions targets. Secondly, we need to make assumptions about the change in these factors over the coming 35-year period. Once this is done, we can estimate the ratio of greenhouse gas emissions to energy-resource input (i.e., the *GHG*/*R* ratio), which can indicate the degree of difficulty a nation is likely to have in complying with its emissions obligations.

The *GHG*/*R* ratio is of particular value for two main reasons. Firstly, economic activities are only possible through the expenditure of energy-resources. Secondly, it is because of the use of energy-resources that greenhouse gases are generated. Hence, since a particular quantity of energy-resources must, for given levels of human know-how, be expended to produce a nation's real GDP, the *GHG*/*R* ratio can reveal how much a nation must shift away from greenhouse gas-emitting production techniques and/or increase the rate of its greenhouse gas removals to achieve its emissions targets.

To the first task. As explained at various stages during this book, the difficulty any nation will have in achieving its greenhouse gas emissions targets will depend on three main factors: (i) the rate at which it must cut its greenhouse gas emissions to achieve its emissions targets; (ii) the expected annual rate of technological progress, which determines the quantity of energy needed to produce a given level of real output (real GDP); and (iii) the expected rate of change in real GDP. Clearly, the difficulty increases the greater is the rate at which greenhouse gas emissions must be reduced; the lower is the expected rate of technological progress; and the larger is the expected increase in real GDP. Of course, for an Annex II country that needs to reduce its real GDP to stabilise its per capita GDP at the optimum of Int$15,000 (2004 prices), the difficulty is lessened the more it is required to decrease its real GDP (Daly 1996, 2007; Victor 2008; Jackson 2009).

These three factors can be expressed in the following equation:

$$\frac{GHG_1}{R_1} = \left(\frac{GHG_1}{GHG_0}\right) \times \left(\frac{GDP_0}{GDP_1}\right) \times (1+\beta)^t \tag{10.1}$$

where GHG = net greenhouse gas emissions; R = energy-resource inputs; GDP = real Gross Domestic Product; β = rate of technological progress; and t = length of time between t_0 and t_1, measured in years.

The GHG/R ratio on the left-hand side measures the greenhouse gases emitted per unit of energy-resource input. A reduction in this ratio indicates the extent to which the greenhouse gas emissions associated with a given quantity of energy use must be decreased. A reduction in the ratio can only be achieved by: (i) switching from high-emissions to low-emissions energy-resources (i.e., from fossil fuels to renewable-energy resources); (ii) altering production techniques to reduce the non-CO_2 emissions generated from industrial and agricultural processes; and/or (iii) sequestering more of the greenhouse gases emitted from the use of a given quantity of energy-resources. Quite obviously, the more a nation must reduce this ratio over a given time period, the greater and presumably the more demanding are its mitigation requirements.

The first ratio on the right-hand side of Eq. (10.1) represents the required cuts to net greenhouse gas emissions between t_0 and t_1 (GHG_1/GHG_0). For each of the four groups of nations, a cut of a different magnitude will be required between 2015 and 2050.

The GDP_0/GDP_1 ratio represents the expected change in real GDP between t_0 and t_1. The change in real GDP of each nation will largely depend on its future GDP objectives. Assuming all nations are intent on stabilising their per capita GDP at Int$15,000 (2004 prices), some countries, such as those belonging to the Annex II group, can be expected to reduce their real GDP between 2015 and 2050. Conversely, Annex III nations and LDCs will need to grow their real GDP over coming decades, albeit they will be expected to do this in the most equitable and efficient manner.

The element on the far right-hand side of Eq. (10.1) is the resource-saving technology factor $(1 + \beta)^t$. A crucial mitigation factor in itself, this element represents the reduction in the energy-resources required to produce a given quantity of real output (real GDP). For the purposes of this exercise, it will be assumed that the rate of technological progress for all nations is 1.4 per cent per annum—the same as the rate assumed in the simulation exercises conducted in Chap. 4.[49]

As for the second task, to be consistent with the greenhouse gas emissions cuts recommended in Table 10.1, the following is assumed regarding the emissions reductions of each nation between 2015 and 2050:

- Annex II nations: Net emissions in 2050 will be reduced to 8.6 per cent of 2015 net emissions levels.
- Annex I (non-Annex II) nations: Net emissions in 2050 will be reduced to 13.1 per cent of 2015 net emissions levels.
- Annex III nations: Net emissions in 2050 will be reduced to 39.5 per cent of 2015 net emissions levels.
- LDCs: Net emissions in 2050 will be reduced to 82.2 per cent of 2015 net emissions levels.

Table 10.7 reveals the degree of difficulty faced by each group of nations in achieving their greenhouse gas emissions targets. Consider the Annex II group first. For the very high-GDP countries aiming to halve their real GDP between 2015 and 2050, they must reduce their *GHG/R* ratio by a factor of 3.6.[50] This equates to an average annual decrease of 3.7 per cent per annum. As for the Annex II nations aiming to reduce their real GDP by one-third between 2015 and 2050, they must reduce their *GHG/R* ratio by a factor of 4.7.[51] This equates to a higher annual decrease of 4.5 per cent per annum. What this importantly indicates is that the countries which require a larger reduction in real GDP to move to an optimal macroeconomic scale face an easier task to achieve their emissions targets. This aside, neither a 3.7 per cent nor a 4.5 per cent annual reduction in the *GHG/R* ratio are particularly problematic.

Now contrast these mitigation requirements with the prospect confronting an Annex II nation aiming to double its real GDP between 2015 and 2050.[52] In these circumstances, the *GHG/R* ratio must be reduced by a factor of 14.2, which amounts to an average annual decrease of 7.9 per cent per annum. Compared to the task of achieving emissions targets with a steady-state economic goal in mind, this almost certainly constitutes an unachievable mitigation requirement, especially given that it must be attained on top of the emissions reductions generated by the 1.4 per cent per annum increase in technological progress.

Because many Annex I (non-Annex II) nations are already near the optimal per capita GDP of Int$15,000 (2004 prices), it is assumed that most of them will need to maintain their real GDP at 2015 levels or increase their real GDP between 2015 and 2050 by around 25 per cent.[53] For the former category of Annex I (non-Annex II) countries, they will need to reduce their *GHG/R* ratio by a factor of 4.7 between 2015 and 2050, which equates to an average annual decrease of 4.5 per cent per

Table 10.7 Degree of difficulty in achieving GHG emissions targets as indicated by change in GHG/energy input ratio

Group	Change in real GDP (2015–2050)	Change in GHG/R ratio (2015–2050) (factor change)	Change in GHG/R ratio (2015–2050) (ave. annual change)
Annex II	50 % decrease	3.6-factor cut	3.7 % cut
Annex II	33 % decrease	4.7-factor cut	4.5 % cut
Annex II	2 × increase	14.2-factor cut	7.9 % cut
Annex I (non-AII)	Unchanged	4.7-factor cut	4.5 % cut
Annex I (non-AII)	25 % increase	5.8-factor cut	5.2 % cut
Annex III	2 × increase	3.1-factor cut	3.3 % cut
Annex III	4 × increase	6.2-factor cut	5.3 % cut
LDCs	3 × increase	2.2-factor cut	2.3 % cut
LDCs	5 × increase	3.7-factor cut	3.8 % cut

Notes

- Assumes an average annual rate of technological progress of 1.4 %
- Assumes all nations make the transition to a steady-state economy with per capita GDP stabilising at approximately Int$15,000 (2004 prices)
- GHG denotes greenhouse gas emissions
- R denotes energy-resource input

annum. For the latter category of nations, a 5.8-fold reduction in the *GHG/R* ratio is necessary, which equates to a decrease of 5.2 per cent per annum. Compared to the Annex II group of nations, these reductions are marginally more demanding, albeit the 4.5 per cent annual decrease should be well within their reach. However, should the 5.2 per cent annual decrease be difficult for some Annex I (non-Annex II) nations to achieve, they may need to slow the growth in their real GDP and attain the per capita optimum deeper into the 21st century.

Essentially all Annex III nations require further increases in their real GDP. For those countries aiming to double their real GDP between 2015 and 2050, they will need to reduce their *GHG/R* ratio by a factor of 3.1 or by an average rate of 3.3 per cent per annum. Nonetheless, it is quite conceivable that poorer members of the group will be aiming to quadruple their real GDP between 2015 and 2050.[54] If so, these countries will need to reduce their *GHG/R* ratio by a factor of 6.2, which amounts to an average annual decrease of 5.3 per cent per annum. Whilst the 3.3 per cent annual reduction is feasible, it is questionable whether some Annex III nations would be capable of reducing the *GHG/R* ratio at an average rate of 5.3 per cent per annum. If they cannot, they too may have to reduce the rate of growth in their real GDP and delay the point where they reach the optimal per capita GDP.

The situation facing LDCs is different again, largely because of the less stringent emissions cuts required from them up to 2050. Assuming that LDCs would be aiming to increase their real GDP by a factor between three to five, the need to reduce the *GHG/R* ratio would vary from a factor of 2.2 to 3.7, which equates to average reductions of 2.3–3.8 per cent per annum.[55] Like the Annex III group, it is possible that some LDCs would have difficulty achieving the annual 3.8 per cent requirement. However, rather than have these LDCs slow the growth in their real GDP, it is better that they be given additional financial assistance from high-GDP nations to support their mitigation efforts.

All things considered, it can be seen that the emissions cuts required of all four groups of nations would not be unduly difficult to achieve if countries took the most appropriate course of action with regard to the growth (or de-growth) of their real GDP and its eventual stabilisation at the optimal scale. Nevertheless, the potential difficulty that some low-GDP nations are likely to confront reinforces the need for high-GDP nations to provide them with additional mitigation assistance. As previously suggested, identifying the countries in need should be done through the application of the vulnerability metrics referred to in Chap. 9, whilst the majority of the financial resources should be provided via the Green Climate Fund.

10.6 Concluding Remarks

This book has demonstrated that the climate change crisis is the symptom of a much larger problem that, on the surface, appears intractable given the international collaboration needed to resolve it. Nonetheless, like many global problems linked to humankind's predilection with continued GDP growth, it must be resolved one way or another. Failure is not an option. The post-2020 protocol and emissions-trading framework recommended in this chapter would help overcome the climate change dilemma by contributing towards the sustainable development goals of ecological sustainability, distributional equity (both within and across all nations), and allocative efficiency. I also believe they are politically and administratively feasible.

As important as an effective protocol and emissions-trading framework are, I cannot end this book without reiterating the point that the climate change crisis will not be resolved unless the world's nations work towards the establishment of a steady-state global economy that emphasises equitable qualitative improvement not quantitative growth. As I have stressed numerous times, this will require the world's high-GDP nations to begin the transition to a steady-state domestic economy as soon as possible—a desirable shift given that, for many high-GDP nations, GDP growth has become 'uneconomic'. In the meantime, the world's low-GDP countries need some further GDP growth. However, to be welfare-increasing, it must be growth that is as equitable and efficient as possible. The global protocol and emissions-trading framework presented in this chapter would, for all the reasons given, go a long way towards assisting this adjustment process.

My final point concerns the role of governments. Because resolving the climate change crisis will necessitate strict limits on greenhouse gas emissions, large public investments in energy-efficient and low-emissions infrastructure, and a considerable redistribution of income and wealth from the rich to the poor, more not less government intervention will be required in future. That said, governments need to acknowledge and harness the important efficiency-yielding benefits of markets. What this means is that governments, preferably through democratic sanction, need to establish an institutional framework that allows markets to perform their crucial allocative function whilst confining the resolution of sustainability and distributional matters to non-economic mechanisms and the strict application of ecological and ethical criteria. For this to occur, societies must recognise the need for governments to adopt the stance that governments in high-GDP nations took in the immediate post-World War II period when they exploited their spending and taxation powers to facilitate high rates of GDP growth and generally improved the distribution of income and wealth. The only difference now is that, with the era of desirable growth over for all but impoverished nations, the same interventionist approach is necessary to shift the global economy on a path to sustainable development—the same path that must be travelled on to resolve the climate change crisis.

Notes

1. This would restrict post-2015 global emissions to just under 1,500 $GtCO_2$-e by 2100, which is consistent with the 450 ppm target.
2. To recall from Chap. 4, process-related CO_2 emissions must be cut at the rate of 6.5 per cent annually.

3. Complete de-carbonisation of industrial production is unlikely. As Tables 4.1 and 4.2 both show, a negligible amount of fossil-fuel usage will take place in 2100. There will also be some unavoidable CO_2 emissions from land use, land-use change, and forestry activities, especially from agriculture.
4. The Enforcement Branch would operate in a similar fashion to the Enforcement Branch of the existing Kyoto Protocol.
5. Trade sanctions and tariff penalties should only be applied when a nation: (i) commits a second and subsequent transgression; (ii) refuses to discharge a presently imposed penalty; or (iii) refuses to participate in climate change negotiations and wantonly generates greenhouse gas emissions at levels which undermine the achievement of global emissions targets.
6. The tariff penalties must be set large enough to reduce the post-penalty pay-off that a Party may enjoy from cheating on compliant Parties below that of the pay-off it would receive if it and all other Parties adhered to their emissions targets.
7. Modifying WTO trade rules would involve a separate negotiating process. However, the UNFCCC should participate in these negotiations.
8. The second reason has more to do with realising the goal of allocative efficiency, which is necessary to cost-effectively achieve a safe stabilisation target.
9. There are, of course, other European Union-imposed restrictions on the emissions units generated by forestry activities.
10. In addition to these Kyoto restrictions, there are European Union-imposed restrictions on the types of emissions allowances that can be generated by forestry activities.
11. I say 'potentially' because these conditions would only apply to forestry if it is possible to monitor and attribute greenhouse gas emissions and removals to individual forestry activities.
12. Aesthetic, existence, and wilderness values could also be included here.
13. To recall from Chaps. 3 and 9, the user cost principle should be employed in such a way as to satisfy the condition of 'strong sustainability', which means calculating the user cost on the basis of keeping natural capital—in this case, a forest—intact.
14. There are also a range of European Union-imposed restrictions on the emissions units generated by agricultural activities.
15. Like forestry, there are European Union-imposed restrictions on the types of emissions allowances that can be generated by agricultural activities.
16. Just like forestry, these conditions would only apply to agriculture if it is possible to monitor and attribute greenhouse gas emissions and removals to individual agricultural activities.
17. This sub-section expands on what I briefly alluded to at the beginning of Chap. 9.
18. The inspiration for Table 10.1 (Tables 10.3, 10.4, and 10.5) was drawn from Table 1 in Hamilton et al. (2005).

19. It was shown in Chap. 4 that the real GWP/emissions ratio would need to increase by a factor of 13.6 between 2010 and 2100 to achieve the 450 ppm stabilisation target along with an optimal per capita GWP of Int$15,000 (2004 prices) and a global ecological footprint/biocapacity ratio no greater than 0.9.
20. It is during this time that the Annex III group would graduate all the way to the Annex II group of nations.
21. Exceptions would include LDCs that have yet to reach self-reliance.
22. To recall, this funding commitment would only apply to Annex II and Annex I (non-Annex II) countries with a per capita real GDP of Int$10,000 or more (at 2004 prices).
23. Whether they would fully fund or co-fund such measures would depend on where all nations are with respect to the scheduled timetable of mitigation transfers.
24. Once again, this funding commitment would only apply to Annex II and Annex I (non-Annex II) countries with a per capita real GDP of Int$10,000 or more (at 2004 prices).
25. With a per capita real GDP at or just below the optimum value of Int$15,000 (2004 prices), few if any Annex I (non-Annex II) nations would be required to engage in a de-growth phase.
26. In an indirect way, overseas development assistance would also play a vital adaptation function.
27. Like the mitigation commitment, this funding obligation would only apply to Annex II and Annex I (non-Annex II) countries with a per capita real GDP of Int$10,000 or more (at 2004 prices).
28. Full or co-funding of adaptation measures would, much like mitigation measures, depend on where nations are with respect to the scheduled timetable of adaptation transfers.
29. To recall from Chap. 9, a new global protocol will eventually require non-Annex I nations to meet emissions targets, which would draw them into an emissions-trading system. Because of this, Annex III nations and LDCs would be required to make a contribution to the Global Climate Change Emergency Fund—the former as of 2021; the latter as of 2031. This contribution—equal to 25 per cent of the revenue raised from the auctioning of emissions permits pertaining to emissions-trading systems covering Annex III nations and LDCs—would serve as an 'insurance premium' paid by all countries towards a larger international insurance pool.
30. This would be the result of Annex II nations having a strong capacity to pay for emissions permits coupled with a heavily restricted supply of available permits in the Annex II system owing to the need for Annex II countries to undertake stringent emissions cuts.
31. Restrictions on cross-system participation would also play a part here.
32. Operating independently does not mean that a nation can disregard its emissions targets or other obligations pertaining to a new global protocol.

33. In other words, the number of permits issued would cover the industries where attribution is possible. This does not imply that non-covered industries would escape the need to limit greenhouse gas emissions. It is just that governments would have to rely on other policies and mechanisms to achieve emissions targets in these industries. Given the need to quantitatively limit emissions to meet ecological sustainability targets (see Sect. 3.1), it can be seen why it is so important to develop the institutional and technical capacity to determine attribution and to maximise the coverage of an emissions-trading system.
34. An additional auction could arise if a technological or institutional breakthrough increased the authority's capacity to attribute greenhouse gas emissions to individual activities, in which case a larger proportion of a nation's permissible emissions would be covered by the emissions-trading system.
35. The reason for this is relatively straightforward. If, in the example given, the price of a supplementary-CER permit was $7, the cost of acquiring the right to emit one tonne of CO_2-equivalent greenhouse gases would be $12 if the CER alternative was chosen, or $15 if the AAU alternative is chosen. For obvious reasons, one would take the CER option. There would, as a consequence, be an excessive demand for supplementary-CER permits that would drive up their price. This would continue until the total cost of both options is much the same. The opposite would occur if the price of a supplementary-CER permit was $13. In this case, the lack of demand for supplementary-CER permits (total cost of CER alternative equals $18) would lead to a decrease in their price. Once again, this would continue until the total cost of the two alternatives is much the same.
36. That is, the domestic offsets would be issued to the entities that have created the RMUs.
37. Importantly, domestic offsets would reduce the cost of achieving national emissions targets by allowing some low-cost mitigation activities occurring outside the national emissions-trading system to replace the high-cost mitigation options available to operators in the covered industries. They would do this by enabling the latter group of operators to purchase the offsets generated by the low-cost mitigation activities in non-covered industries, which they presumably would do if the cost of purchasing offsets is less than the cost of undertaking their own mitigation action. Of course, the benefits of domestic offsets would depend on the associated emissions reductions and/or greenhouse gas removals being additional to what would have taken place. This would almost certainly be the case given that the issuance of RMUs requires proof of additionality and the fact that any nation that issues domestic offsets to non-additional activities would simply penalise itself.
38. As indicated in Chap. 7, this would prevent the transgressor from engaging in greenhouse-gas emitting activities.
39. Stern (2007, Chap. 15) has recommended similar guiding principles.

40. In general, setting the point of obligation at the final outputs of any emissions-related activity would be a last-resort measure if all other means of attribution are impractical or too costly. Setting the point of obligation at final outputs would also require a reasonable association to exist between final outputs and the generation of greenhouse gases.
41. The possible exception is *t*CERs given they should be traded in a separate market.
42. Because limiting the throughput of matter-energy requires the imposition of quantitative restrictions (i.e., caps on the rate of resource use and waste generation), there is always the possibility that separate policies would inadequately contain greenhouse gas emissions.
43. It may turn out that, following subtraction of the emissions covered by the permits acquitted by resource-extractors, an electricity-generator has little or no need to acquit emissions permits.
44. Fugitive emissions amount to 5 per cent of total greenhouse gas emissions (WRI 2006).
45. There are already new technologies emerging with the potential to revolutionise the measurement and attribution of fugitive greenhouse gas emissions. One such example is a technology called a 'differential absorption lidar' (DIAL). DIAL can be used to remotely detect and measure the concentration of hydrocarbons in the atmosphere up to 300 metres from a leaky facility. Another possibility exists in the form of portable gas-leak imaging cameras that employ infrared-imaging technology to identify invisible gases escaping from leaky sources. While these cameras cannot measure fugitive emissions, they can be used to detect their sources so that action can be taken to measure them by using appropriate technologies.
46. Much of this sub-section draws from the work of Garnaut (2008).
47. As explained earlier, a country with a national emissions-trading system that does not meet UNFCCC standards would still be able to operate its own system. However, trade in its emissions units—that is, the AAUs it has been initially allocated and any units it may have subsequently generated (e.g., CERs, *t*CERs, ERUs, and RMUs)—would be confined to its emissions-trading system. These emissions units would only be available internally to meet compliance purposes.
48. Because of the higher price of AAUs expected in a higher-ranked group, one would not expect a large quantity of AAUs to be purchased by a nation in a lower-ranked group.
49. For an assumed rate of technological progress of 1.4 per cent per annum, the energy-resource input required to produce one unit of real output in 2050 is 61 per cent lower than in 2015 (i.e., $1/(1.014)^{35} = 0.61$).
50. A halving of real GDP over a 35-year period amounts to an average annual rate of decrease in real GDP of 2 per cent.
51. Reducing real GDP by one-third over a 35-year period amounts to an average annual rate of decrease in real GDP of 1.2 per cent.

52. Doubling real GDP over a 35-year period amounts to an average annual rate of increase in real GDP of 2 per cent.
53. Reducing real GDP by 25 per cent over a 35-year period amounts to an average annual rate of decrease in real GDP of 0.6 per cent.
54. Quadrupling real GDP over a 35-year period amounts to an average annual rate of increase in real GDP of 4 per cent.
55. A 3-fold increase in real GDP over a 35-year period amounts to an average annual rate of increase in real GDP of 3.2 per cent. A 5-fold increase in real GDP over the same period amounts to an average annual rate of increase in real GDP of 4.7 per cent.

Postscript—The UNFCCC Climate Change Conference in Lima, 2014

Various sections of the final stages of this book were written prior to the UNFCCC climate change conference held in December 2014 in Lima (COP-20). Given the importance of this meeting leading up to the crucial Paris conference in late-2015, it is worth considering what was achieved and whether the negotiated agreement has put the global community on track to establish an effective climate change protocol to replace the existing Kyoto Protocol.

Whilst there were many goals that the Lima conference aimed to achieve, two stood out. The first and most obvious goal was an outline text to facilitate the establishment of a new global protocol at the COP-21 Paris conference. Despite the usual concern that the climate change talks could collapse, the first goal was finally achieved in the form of a 37-page text containing a menu of aspirational long-term objectives for possible inclusion in the new protocol. Disconcertingly, the options varied from the seemingly unachievable 'zero net emissions by 2050' to the farcically weak 'deviation from business as usual' and the targetless ramping up of 'low-emissions development strategies'.

The second major goal was to agree on the rules under which the Parties must submit their 'Intended Nationally Determined Contributions' or INDCs—a document outlining a Party's desired emissions targets and the strategies to achieve them. Notwithstanding the shared view that the INDCs should be used to establish the new climate change protocol, it was here that disagreement was greatest. Non-Annex I nations wanted the INDCs to contain plans for climate change adaptation and for high-GDP nations to include their financial support for mitigation action in low-GDP countries. Following a protracted debate, the inclusion of explicit financial commitments was not required. Nor was there any compulsion to include adaptation plans. Not that high-GDP nations got their way with everything. Their call for all Parties to provide standardised information on their emissions targets and strategies was rejected. In addition, their insistence that all Parties' INDCs should be scrutinised was omitted from the conference text.

P. Lawn, *Resolving the Climate Change Crisis*, DOI 10.1007/978-94-017-7502-1

Importantly, while there are no plans to assess national INDCs, the UNFCCC secretariat will be estimating the likely impact that the proposed strategies will have on future global temperatures. The secretariat aims to release its findings just prior to the Paris conference, which many believe will reveal that the proposed collective action is inadequate. According to critics, one of the reasons for the likely shortcoming is that the INDC submission process has given Parties the latitude to set their own commitments free from any standardised guidelines from the UNFCCC. The same critics also fear that the weak starting position now created renders it more difficult for Parties to agree on tougher emissions cuts should it become clear that more stringent emissions targets are required.

Because all nations have been requested to prepare an INDC, many believe the Lima conference has once-and-for-all blurred the distinction between the world's rich and poor nations. To date, of course, only Annex I nations have been subject to notionally binding emissions targets. A number of low-GDP countries (e.g., India) reacted during the conference by insisting on the maintenance of the current distinction fearing they may be required to make emissions cuts similar to those required of high-GDP nations in a post-2020 protocol.

These fears were largely allayed by a reference in the conference text highlighting that Party obligations would be based on "common but differentiated responsibilities and respective capabilities in light of different national circumstances". This reference ensures that past and present emissions, national GDP growth requirements, and the capacity of a country to undertake mitigation and adaptation action will be major factors when final deliberations are made on national emissions targets. Indeed, whilst the text indicates a clear intention to subject all UNFCCC Parties to emissions obligations, it leaves the door open for the creation of different groups of nations with different responsibilities, much like the four groups recommended in Chaps. 9 and 10.

In all, putting aside the concerns regarding the transfer of funds and the flimsy 'firewall' between low-GDP and high-GDP nations, most non-Annex I nations (e.g., China and Brazil) seem relatively comfortable with the Lima agreement. Together with Barak Obama's overt support for stringent emissions cuts, there is reasonable optimism that a new protocol will emerge from the Paris conference that all UNFCCC Parties will be prepared to ratify. However, one crucial question remains: Will it be an effective protocol? With the INDC offerings of most powerful nations likely to be inadequate, I fear not—a sentiment shared by many climate change researchers and echoed in a joint statement by Oxfam, Greenpeace, and Christian Aid. Coupled with the absence of any nation indicating its intention to make the transition to a qualitatively-improving steady-state economy, there appears little hope of preventing the 2 °C 'guard-rail' from being breached. Then again, hope springs eternal. Let us hope that sanity prevails come December 2015 in Paris, since the conference may represent humankind's last chance to prevent catastrophic climate change.

References

Abbott, K., & Gartner, D. (2011). *The green climate fund and the future of environmental governance*. Earth System Governance Working Paper No. 16. http://papers.ssrn.com/sol3/papers.cfm?abstract_id=1931066

Aber, J., Neilson, R., McNulty, S., Lenihan, J., Bachelet, D., & Drapek, R. (2001). Forest processes and global environmental change: Predicting the effects of individual and multiple stressors. *BioScience, 51*, 735–751.

Ackerman, F. (2007). *Debating climate economics: The stern review vs. its critics*. Report to Friends of the Earth England, Wales, and Northern Ireland.

Ackermann, T., Andersson, G., & Söder, L. (2001). Overview of government and market driven programs for the promotion of renewable power generation. *Renewable Energy, 22*(1–3), 197–204.

Adam, D. (2008, May 31). Could US scientist's "CO_2 catcher" help to slow warming? *The Guardian*.

African Development Bank (AfDB), Asian Development Bank (ADB), Department for International Development (DFID), Directorate-General for International Cooperation (DGIS), Directorate General for Development, European Commission (EC), Federal Ministry for Economic Cooperation and Development (BMZ), Organization for Economic Cooperation and Development (OECD), United Nations Development Programme (UNDP), United Nations Environment Programme (UNEP), The World Bank. (2003). *Poverty and climate change: Reducing the vulnerability of the poor through adaptation*. VARG multi-development agency paper, United Nations Development Project (UNDP), United Nations, New York.

African Development Bank (AfDB). (2013). *Progress report on the implementation of the climate change action plan 2011–2015*. Tunisia: African Development Bank Group.

Alchian, A. (1965). The basis of some recent advances in the theory of management of the firm. *Journal of Industrial Economics, 14*(1), 30–41.

Aldred, J. (2002). Cost-benefit analysis, incommensurability, and rough equality. *Environmental Values, 11*, 27–47.

Aldy, J., Barrett, S., & Stavins, R. (2003 December). Thirteen plus one: A comparison of global climate change architectures. *Climate Policy, 3*(4), 373–397.

Alley, R. (2000). The younger Dryas cold interval as viewed from central Greenland. *Quaternary Science Reviews, 19*(1), 213–226.

Alley, R., & Ágústsdóttir, A.-M. (2005). The 8k event: Cause and consequences of a major Holocene abrupt climate change. *Quaternary Science Reviews, 24*(10–11), 1123–1149.

P. Lawn, *Resolving the Climate Change Crisis*, DOI 10.1007/978-94-017-7502-1

Amundsen, H., Berglund, F., & Westskog, H. (2010). Overcoming barriers to climate change adaptation: A question of multilevel governance? *Environment and Planning C: Government and Policy, 28*, 276–289.

Anderson, D. (2006). *Costs and finance of carbon abatement in the energy sector*. Paper prepared for the Stern Review. www.sternreview.org.uk

Anderson, K., & Bows, A. (2008). Reframing the climate change challenge in light of post-2000 emission trends. *Philosophical Transactions of the Royal Society*. doi:10.1098/rsta.2008.0138

Ångström, K. (1900). Ueber die bedeutung des wasserdapfes un der kohlensäure bei der absorption der erdatmosphäre. *Annuls of Physics, 4*, 720–733.

Antle, J., Capalbo, S., Elliott, E., & Paustian, K. (2004). Adaptation, spatial heterogeneity, and the vulnerability of agricultural systems to climate change and CO_2 fertilization: An integrated assessment approach. *Climatic Change, 64*, 289–315.

Archer, D., Winguth, A., Lea, D., & Mahowald, N. (2000). What caused the glacial/interglacial atmospheric pCO_2 cycles? *Reviews of Geophysics, 38*, 159–189.

Arnel, N. (2004). Climate change and global water resources: SRES emissions and socio-economic scenarios. *Global Environmental Change, 14*, 31–52.

Arrhenius, S. (1896). On the influence of carbonic acid in the air upon the temperature of the ground. *Philosophical Magazine and Journal of Science, 41*, 237–276 (Series 5).

Arrow, K. (2007, June). Global climate change: A challenge to policy. *Economists' Voice*. www.bepress.com/ev

Art, R., & Waltz, N. (2009). *The use of force: Military power and international politics*. Lanham: Rowman and Litchfield Publishers.

Arthur, W. (1989). Competing technologies, increasing returns, and lock-in by historical events. *The Economic Journal, 99*, 116–131.

Ashton, P. (2002). Avoiding conflicts over Africa's water resources. *Ambio, 31*, 236–242.

Asian Development Bank (ADB). (2014). *ADB and climate investment funds: Innovation and action on climate change in Asia and the Pacific*. Manila: Asian Development Bank.

Asseng, S., Jamieson, P., Kimball, B., Pinter, P., Sayre, K., Bowden, J., & Howden, S. (2004). Simulated wheat growth affected by rising temperature, increased water deficit, and elevated atmospheric CO_2. *Field Crops Research, 85*, 85–102.

Atkinson, A. (1995). *Public economics in action: The basic income/flat tax proposal*. Oxford, UK: Clarendon Press.

Atkinson, G. (1995). *Measuring sustainable economic welfare: A critique of the UK ISEW*. Working Paper GEC 95-08. Centre for Social and Economic Research on the Global Environment, Norwich and London.

Atteridge, A., Siebert, C. K., Klein, R., Butler, C., & Tella, P. (2009). *Bilateral finance institutions and climate change: A mapping of climate portfolios*. Stockholm Environment Institute Working Paper—2009, Stockholm Environment Institute, Stockholm.

Atwell, B., Henery, M., & Whitehead, D. (2003). Sapwood development in *Pinus radiata* trees grown for three years at ambient and elevated carbon dioxide partial pressures. *Tree Physiology, 23*, 13–21.

Augenstein, D., & Benemann, J. (2007). The cellulosic ethanol delusion. www.aiche-norcal.org/Symposium/Symposium2006/pdfs/EnergySolutions.pdf. Accessed March 30, 2009.

Australian Bureau of Statistics (ABS). (2004). *Australian social capital framework and indicators*. Canberra: Commonwealth of Australia.

Australian Bureau of Statistics (ABS). (2006). *Aspects of social capital*. Canberra: Commonwealth of Australia.

Australian Government. (2012). *An overview of the clean energy legislative package*. www.cleanenergyfuture.gov.au. Accessed December 20, 2012.

Ayres, R. (2005). Resources, scarcity, technology, and growth. In R. Simpson, M. Toman, & R. Ayres (Eds.), *Scarcity and growth revisited* (pp. 142–154). Washington, DC: Resource for the Future.

Ayres, R., & Miller, S. (1980). The role of technological change. *Journal of Environmental Economics and Management, 7*, 353–371.

Ayres, R., & Ayres, L. (1999). *Accounting for resources*. Northampton, MA: Edward Elgar.

Ayres, R., & van den Bergh, J. (2005). A theory of economic growth with material/energy resources and dematerialization: Interaction of three growth mechanisms. *Ecological Economics, 55*(1), 96–118.

Azar, C., & Lindgren, K. (2003). Catastrophic events and stochastic cost-benefit analysis of climate change. *Climate Change, 56*, 245–255.

Baalman, P., & Schlamadinger, B. (2008). *Scaling up AFOLU mitigation activities in non-annex I countries*. Cambridge, UK: Climate Strategies.

Babiker, M. (2005). Climate change policy, market structure and carbon leakage. *Journal of International Economics, 65*(2), 421–445.

Baer, P. (2007). *The worth of an ice-sheet: A critique of the treatment of catastrophic impacts in the Stern Review*. EcoEquity Discussion Paper, January 18, 2007. www.ecoequity.org/docs/WorthOfAnIceSheet.pdf

Baer, P., Harte, J., Haya, B., Herzog, A., Holdren, J., Hultman, N., et al. (2000). Equality and greenhouse gas responsibility. *Science, 289*, 2287.

Baetz, R. (1972). The Nuffield Canadian seminar and after: A personal view. In *Guaranteed annual income: An integrated approach*. Ottawa, Canada: The Canadian Council on Social Development.

Bailey, E. (1981). Contestability and the design of regulatory and antitrust policy. *American Economic Association Paper Proceedings, 71*(2), 178–183.

Barber, D., Dyke, A., Hillaire-Marcel, C., Jennings, A., Andrews, J., Kerwin, M., et al. (1999). Forcing of the cold event 8,200 years ago by catastrophic drainage of Laurentide Lakes. *Nature, 400*, 344–348.

Barnes, P., Costanza, R., Hawken, P., Orr, D., Ostrom, E., Umaña, A., & Young, O. (2008). Creating an earth atmospheric trust: A system to stop global warming and reduce poverty. *Science, 319*, 724.

Barnett, J., & Adger, W. (2003). Climate dangers and atoll countries. *Climatic Change, 61*, 321–337.

Barrett, S. (2003). *Environment and statecraft*. Oxford: Oxford University Press.

Barros, V. (2005). Inundaci6n y cambio climático: costa Argentina del Río de la Plata. In V. Barros, A. Men6ndez, & G. Nagy (Eds.), *El Cambio Climático en el Río de la Plata* (pp. 41–52), Proyectos AIACC.

Barry, J., Buck, K., Lovera, C., Kuhnz, L., Whaling, P., Peltzer, E., et al. (2004). Effects of direct ocean CO_2 injection on deep-sea meiofauna. *Journal of Oceanography, 60*, 759–766.

Batima, P., Ganbaatar, T., Tumerbaatar, D., Erdenetsetseg, B., Bolortsetseg, B., Gantsetseg, B., et al. (2005). Climate change impacts on environment. In P. Batima & B. Bayasgalan (Eds.), *Climate change impacts* (pp. 59–115). Ulaanbaatar: Admon Publishing.

Baumol, W. (1959). *Business behavior. Value, and growth*. New York: Macmillan.

Baumol, W., Panzar, J., & Willig, R. (1982). *Contestable markets and the theory of industry structure*. San Diego: Harcour Brace Jovanovich.

Baylis, K., Fullerton, D., & Karney, D. (2014). Negative leakage. *Journal of the Association of Environment and Resource Economists, 1*(1/2), 51–73.

Becken, S. (2005). Harmonising climate change adaptation and mitigation: The case of tourist resorts in Fiji. *Global Environmental Change, 15*, 381–393.

Beckerman, W. (1997). *Small is stupid*. London: Duckworth.

Beg, N., Morlot, J., Davidson, O., Afrane-Okesse, Y., Tyani, L., Denton, F., et al. (2002). Linkages between climate change and sustainable development. *Climate Policy, 2*(5), 129–144.

Beinhocker, E., Oppenheim, J., Irons, B., Lahti, M., Farrell, D., Nyquist, S., et al. (2008). *The carbon productivity challenge: Curbing climate change and sustaining economic growth*. McKinsey Climate Change Special Initiative, McKinsey Global Institute.

Benedick, R. (1991). *Ozone diplomacy*. Cambridge, MA: Harvard University Press.

Benndorf, R., Federici, S., Forner, C., Pena, N., Rametsteiner, E., Sanz, M. J., & Somogyi, Z. (2007). Including land use, land-use change, and forestry in future climate change agreements: Thinking outside the box. *Environmental Science and Policy, 10*, 283–294.

Berger, A. (1977). Support for the astronomical theory of climate change. *Nature, 269*, 44–45.

Berger, A. (1978). Long-term variations in the caloric insolation resulting from the earth's orbital elements. *Quaternary Research, 9*, 139–167.

Berger, A., Li, X., & Loutré, M. (1999). Modelling northern hemisphere ice volume over the last 3 Ma. *Quaternary Science Reviews, 18*, 1–11.

Berkhout, F., Hertin, J., & Gann, D. (2006). Learning to adapt: Organisational adaptation to climate change impacts. *Climate Change, 78*, 135–156.

Bertram, G., & Simon, T. (2010). *The carbon challenge: New Zealand's emissions trading scheme*. Wellington: Bridget Williams Books.

Bhagwati, J., & Mavroidis, P. (2007). Is action against US exports for failure to sign Kyoto WTO-legal? *World Trade Review, 6*(2), 299–310.

Biddle, D. (2001). Youth unemployment and government policy in Australia. In W. Mitchell & E. Carlson (Eds.), *Unemployment: The tip of the Iceberg* (pp. 117–134). Sydney: Centre of Applied and Economic Research, University of New South Wales Press.

Biermann, F., & Brohm, R. (2003). *Implementing the Kyoto Protocol without the United States: The strategic role of energy tax adjustments at the border*. Global Governance Working Paper No. 5. The Global Governance Project, Potsdam.

Bishop, M., & Thompson, D. (1992). Regulatory reform and productivity growth in the UK's public utilities. *Applied Economics, 24*, 1181–1190.

Bishop, R. (1993, May). Economic efficiency, sustainability, and biodiversity. *Ambio*, 69–73.

Blackman, A., & Harrington, W. (2000). The use of economic incentives in developing countries: Lessons from international experience with industrial air pollution. *Journal of Environment and Development, 9*(1), 5–44.

Blandy, R., & Brummitt, W. (1990). *Labour productivity and living standards*. Sydney: Allen & Unwin.

Block, B. (2008). Hansen to Obama support a carbon tax. Worldwatch Institute. http://www.cworldwatch.org/node/5962. Accessed March 31, 2010.

Blum, H. (1962). *Times arrow and evolution* (3rd ed.). Princeton, NJ: Harper Torchbook.

Blyth, W., & Baron, R. (2003). *Green investment schemes: Options and issues*. Paris: Organization for Economic Cooperation and Development (OECD) and International Energy Agency (IEA).

Bodansky, D., Chou, S., & Jorge-Tresolini, C. (2004, December). *International climate efforts beyond 2012: A survey of approaches*. Arlington, VA: Pew Center on Global Climate Change.

Böhringer, C., & Welsch, H. (2006). Burden sharing in a greenhouse: Egalitarianism and sovereignty reconciled. *Applied Economics, 38*, 981–996.

Bosetti, V., Carraro, C., Massetti, E., & Tavoni, M. (2009). Optimal energy investment and R&D strategies to stabilise greenhouse gas atmospheric concentrations. *Resource and Energy Economics, 31*(2), 123–137.

Bossel, U. (2003). *The world needs a sustainable energy economy, not a hydrogen economy*. European Fuel Cell Forum, Report E09.

Bossel, U. (2004, March–April). The hydrogen illusion: Why electrons are a better energy carrier. *Cogeneration and On-Site Power Production*, 55–59.

Bou-Zeid, E., & El-Fadel, M. (2002). Climate change and water resources in the Middle East: A vulnerability and adaptation assessment. *ASCE, Journal of Water Resources Planning and Management, 128*, 343–355.

Boulding, K. (1964). *The meaning of the twentieth century*. New York: Harper & Row.

Bowen, G., Beerling, D., Koch, P., Zachos, J., & Quattlebaum, T. (2004). A humid climate state during the Paleocene/Eocene thermal maximum. *Nature, 432*, 495–499.

Boyd, E., Hultman, N., Timmons-Roberts, J., Corbera, E., Cole, J., Bozmoski, A., et al. (2009). Reforming the CDM for sustainable development: Lessons learned and policy futures. *Environmental Science and Policy, 12*(7), 820–831.

Bradley, R., & Jones, P. (1993). Little ice age summer temperature variations: Their nature and relevance to recent global warming trends. *The Holocene, 3*, 367–376.

Bradley, R., Hughes, M., & Diaz, H. (2003). Climate in medieval time. *Science, 302*(5644), 404–405.

Bramley, M., Partington, P., & Sawyer, D. (2009). *Linking national cap-and-trade systems in North America*. The Pembina Institute and the International Institute for Sustainable Development.

Broeker, W., & Henderson, G. (1998). The sequence of events surrounding termination II and their implications for the cause of glacial-interglacial CO_2 changes. *Paleoceanography, 13*, 352–364.

Brook, B. (2009). *Carbon tax or cap-and-trade? The debate we never had*. http://bravenewclimate.com/2009/02/14/carbon-tax-or-cap-and-trade-the-debate-we-never-had/

Brook, E. (2008). Windows on the greenhouse. *Nature, 453*(15), 291–292.

Brookes, L. (2000). Energy efficiency fallacies revisited. *Energy Policy, 28*(6/7), 355–366.

Brooks, N., & Adger, W. (2005). Assessing and enhancing adaptive capacity. In B. Lim, E. Spanger-Siegried, I. Burton, E. Malone, & S. Huq (Eds.), *Adaptation policy frameworks for climate change* (pp. 165–182). New York: Cambridge University Press.

Broome, J. (1992). *Counting the cost of global warming*. Cambridge, UK: White Horse Press.

Brovkin, V., Sitch, S., von Bloh, W., Claussen, M., Bauer, E., & Cramer, W. (2004). Role of land cover changes for atmospheric CO_2 increase and climate change during the last 150 years. *Global Change Biology, 10*, 1253–1266.

Brown, A. (1996). *Should Telstra be privatised?* School of Economics. Working Paper Number 8, Griffith University.

Brown, T., Hall, B., & Westerling, A. (2004). The impact of twenty-first century climate change on wildland fire danger in the western United States: A simulation study. *Climatic Change, 62*, 365–388.

Bryan, E., Deressa, T., Gbetibouo, G., & Ringler, C. (2009). Adaptation to climate change in Ethiopia and South Africa: Options and constraints. *Environmental Science and Policy, 12*(4), 413–426.

Buen, J. (2013, February). *CDM criticisms: Don't throw the baby out with the bathwater*. FNI Climate Policy Perspectives 8.

Buesseler, K., Andrews, J., Pike, S., & Charette, M. (2004). The effects of iron fertilisation on carbon sequestration in the Southern Ocean. *Science, 304*, 414–417.

Buesseler, K., Lamborg, C., Boyd, P., Lam, P., Trull, T., Bidigare, R., et al. (2007). Revisiting carbon flux through the ocean's twilight zone. *Science, 316*(5824), 567–570.

Buesseler, K., Doney, S., Karl, D., Boyd, P., Caldeira, K., Chai, F., et al. (2008). Ocean iron fertilization—Moving forward in a sea of uncertainty. *Science, 319*, 162.

Burgess, J., & Mitchell, W. (1998). Employment, unemployment, and the right to work. *Australian Journal of Human Rights, 4*(2), 76–94.

Butler, R., & Laurance, W. (2009). Is oil palm the next emerging threat to the Amazon? *Tropical Conservation Science, 2*(1), 1–10.

Byatt, I., Castles, I., Goklany, I., Henderson, D., Lawson, N., McKitrick, R., et al. (2006). The Stern review: A dual critique—Part II: Economic aspects. *World Economics, 7*(4), 199–229.

Callaghan, T., Björn, L., Chernov, Y., Chapin, F, I. I. I., Christensen, T., Huntley, B., et al. (2005). Arctic tundra and polar ecosystems. *Arctic Climate Impact Assessment. ACIA* (pp. 243–352). Cambridge: Cambridge University Press.

Callendar, C. (1938). The artificial production of carbon dioxide and its influence on temperature. *Quarterly Journal of the Royal Meteorological Society, 64*, 223–237.

Campbell, K. (2008). *Climatic cataclysm: The foreign policy and national security implications of climate change*. Washington, DC: Brookings Institution Press.

Canadell, J., Le Quéré, C., Raupach, M., Field, C., Buitenhuis, E., Ciais, P., Conway, T., et al. (2007). From the cover: Contributions to accelerating atmospheric CO_2 growth from economic activity, carbon intensity, and efficiency of natural sinks. *Proceedings of the National Academy of Sciences, USA, 104*. doi:10.1073/pnas.0702737104

Capra, F. (1982). *The turning point*. London: Fontana.

Carbon-info.org. (2007). Bali: A spectacular failure? http://www.carbon-info.org/carbonnews2/carbonnews2_084.htm. Accessed July 3, 2009.

Carbon Market Watch. (2013, March). *Doha decisions on the Kyoto surplus explained*. Carbon Market Watch Policy Brief.

Carbon Trust. (2009, March). *Global carbon mechanisms: Emerging lessons and implications*. CTC748, Carbon Trust.

Carlsson-Kanyama, A., Carlens, H., & Dreborg, K. (2013). Barriers in municipal climate change adaptation: Results from case studies using backcasting. *Futures, 49*, 9–21.

Carmody, J., & Ritchie, D. (2007). *Investing in clean energy and low carbon alternatives in Asia*. Manila: Asian Development Bank.

Casimiro, E., & Calheiros, J. (2002). Human health. In F. Santos, K. Forbes, & R. Moita (Eds.), *Climate change in Portugal: Scenarios, impacts, and adaptation measures*. Lisbon: Gradiva Publishers.

CDM Policy Dialogue. (2012). *Climate change, carbon markets and the CDM: A call to action*. Report of the High-Level Panel on the CDM Policy Dialogue.

Cerling, T. (1991). Carbon dioxide in the atmosphere: Evidence from Cenozoic and Mesozoic paleosols. *American Journal of Science, 291*, 377–400.

Chamberlin, T. (1897). A group of hypotheses bearing on climate changes. *Journal of Geology, 5*, 653–683.

Chan, Y. (2008, January). Increasing soil organic carbon of agricultural land. *PrimeFact, 735*, 1–5.

Changsheng, L., Frolking, S., & Butterback-Bahl, K. (2005). Carbon sequestration in arable soils is likely to increase nitrous oxide emissions, offsetting reductions in climate radiative forcing. *Climate Change, 72*(3), 321–338.

Charney, J. (1979). *Carbon dioxide and climate: A scientific assessment*. Washington, DC: National Academy of Sciences Press.

Chiew, F., Harrold, T., Siriwardenena, L., Jones, R., & Srikanthan, R. (2003). Simulation of climate change impact on runoff using rainfall scenarios that consider daily patterns of change from GCMs. In D. Post (Ed.), *Proceedings of ODSIM 2003: International Congress on Modelling and Simulation* (pp. 154–159). Modelling and Simulation Society of Australia and New Zealand, Canberra.

Christensen, N., Wood, A., Voisin, N., Lettenmaier, D., & Palmer, R. (2004). The effects of climate change on the hydrology and water resources of the Colorado River basin. *Climatic Change, 62*, 337–363.

Christiansen, L., Ray, A., Smith, J., & Haites, E. (2012). *Assessing international funding for climate change adaptation: A guidebook for developing countries*. TNA Guidebook Series, The Technology Needs Assessment (TNA) Project implemented by the United Nations Environment Programme (UNEP) and the UNEP Risø Centre (URC) on Energy, Climate and Sustainable Development and funded by the Global Environment Facility (GEF), UNEP Risø Centre, Roskilde, Denmark.

Churchill, C. (2007). Insuring the low-income market: Challenges and solutions for commercial insurers. *The Geneva Papers, 32*(3), 401–412.

Ciriacy-Wantrup, S. (1952). *Resource conservation: Economics and policies*. Berkeley, CA: University of California Press.

Clare, R., & Johnston, K. (1992). *Profitability and productivity of government business enterprises*. Economic Planning Advisory Council Research Paper Number 2, EPAC, Canberra.

Clark, C., & Kavanagh, C. (1996). Basic income, inequality, and unemployment: Rethinking the linkage between work and welfare. *Journal of Economic Literature, 30*(2), 399–407.

Clark, P., Pisias, N., Stocker, T., & Weaver, A. (2002). The role of the thermohaline circulation in abrupt climate change. *Nature, 415*, 863–869.

Clark, P., Archer, D., Pollard, D., Blum, J., Rial, J., Brovkin, V., et al. (2006). The middle Pleistocene transition: Characteristics, mechanisms, and implications for long-term changes in atmospheric pCO_2. *Quaternary Science Reviews, 25*, 3150–3184.

Clarke, G., Leverington, D., Teller, J., & Dyke, A. (2004). Palaeohydraulics of the last outburst flood from glacial Lake Agassiz and the 8200 BP cold event. *Quantitative Science Review, 23*, 389–407.

Clarke, M., & Lawn, P. (2008). Is measuring genuine progress at the sub-national level useful? *Ecological Indicators, 8*(5), 573–581.

Clarke, M., & Shaw, J. (2008). Genuine progress in Thailand: A systems-analysis approach. In P. Lawn & M. Clarke (Eds.), *Genuine progress in the Asia-Pacific: Studies using the genuine progress indicator* (pp. 260–298). Cheltenham, UK: Edward Elgar.

Clement, A., & Cane, M. (1999). A role for the tropical Pacific coupled ocean-atmospheric system on Milankovitch and millennial timescales—Part I: A modeling study of tropical Pacific variability. In P. Clark, R. Webb, & L. Keigwin (Eds.), *Mechanisms of global climate change at millennial time scales* (pp. 363–371). Washington, DC: American Geophysical Union.

Climate Policy Initiative. (2013). *The global landscape of climate finance*. CPI Report, October 2013, Climate Policy Initiative.

Cline, W. (1992). *The economics of global warming*. Washington, DC: Institute for International Economics.

Coase, R. (1960). The problem of social cost. *Journal of Law and Economics, 3*, 1–44.

Cole, M., & Fredriksson, P. (2013). Institutionalised pollution havens. In P. Lawn (Ed.), *Globalisation, economic transition, and the environment* (pp. 66–101). UK: Edward Elgar.

Common, M., & Stagl, S. (2005). *Ecological economics: An introduction*. Cambridge: Cambridge University Press.

Conant, R., Ogle, S., Paul, E., & Paustian, K. (2010). Measuring and monitoring soil organic carbon stocks in agricultural lands for climate mitigation. *Frontiers in Ecology and the Environment, 9*(3), 167–173.

Corbera, E., González Soberanis, C., & Brown, K. (2009). Institutional dimensions of payments for ecosystem services: An analysis of Mexico's carbon forestry programme. *Ecological Economics, 68*, 743–761.

Corfee-Morlot, J., & Höhne, N. (2003). Climate change: Long-term targets and short-term commitments. *Global Environmental Change, 13*, 277–293.

Costanza, R. (1980). Embodied energy and economic valuation. *Science, 210*, 1219–1224.

Costanza, R., & Perrings, C. (1990). A flexible assurance boding system for improved environmental management. *Ecological Economics, 2*, 57–76.

Costanza, R., Daly, H., & Bartholomew, J. (1991). Goals, agenda, and policy recommendations for ecological economics. In R. Costanza (Ed.), *Ecological economics: The science and management of sustainability* (pp. 1–20). New York: Columbia University Press.

Cowell, P., & Kench, P. (2001). The morphological response of atoll islands to sea level rise—Part 1: Modifications to the modified shoreface translation model. *Journal of Coastal Research, 34*, 633–644.

Cowling, S., Mitchell, W., & Watts, M. (2006). The right to work versus the right to income. *International Journal of Environment, Workplace and Employment, 2*(1), 89–113.

Cox, P., Betts, R., Jones, C., Spall, S., & Totterdell, I. (2000). Acceleration of global warming due to carbon-cycle feedbacks in a coupled climate model. *Nature, 408*, 184–187.

Cox, P., Huntingford, & Jones, C. (2006). Conditions for sink-to-source transitions and runaway feedbacks from the land carbon-cycle. In H. J. Schellnhuber, W. Cramer, N. Nakicenovic, T. Wigley, & G. Yohe (Eds.), *Avoiding dangerous climate change* (pp. 155–161). Cambridge, UK: Cambridge University Press.

Crowards, T. (1998). Safe minimum standards: Costs and opportunities. *Ecological Economics, 25*, 303–314.

Crowley, T. (2000). Causes of climate change over the past 1000 years. *Science, 289*(5477), 270–277.

Crowley, T., & Lowery, T. (2000). How warm was the medieval period? *Ambio, 29*(1), 51–54.

Crutzen, P. (2006). Albedo enhancement by stratospheric sulphur injections: A contribution to resolve a policy dilemma? *Climate Change, 77*, 211–219.

Cunnold, D., Steele, L., Fraser, P., Simmonds, P., Prinn, R., Weiss, R., Porter, L., et al. (2002). In situ measurements of atmospheric methane at GAGE/AGAGE sites during 1985-2000 and resulting source inferences. *Journal of Geophysical Research, 107*(D14). doi: 10.1029/2001JD001226

Currie, D. (2001). Projected effects of climate change on patterns of vertebrate and tree species in the conterminous United States. *Ecosystems, 4*, 216–225.

Daehler, C. (2005). Upper-montane plant invasions in the Hawaiian Islands: Patters and opportunities. *Perspectives in Plant Ecology, Evolution and Systematics, 7*, 203–216.

Daily, G., & Ehrlich, P. (1992). Population, sustainability, and the Earth's carrying capacity. *BioScience, 42*(10), 761–771.

Daly, H. (1979). Entropy, growth, and the political economy of scarcity. In V. K. Smith (Ed.), *Scarcity and growth reconsidered* (pp. 67–94). Baltimore: John Hopkins University Press.

Daly, H. (Ed.). (1980). *Economics, ecology, and ethics: Essays toward a steady-state economy*. San Francisco: W.H. Freeman.

Daly, H. (1991). *Steady-state economics: Second edition with new essays*. Washington, DC: Island Press.

Daly, H. (1992). Allocation, distribution, and scale: Towards an economics that is efficient, just, and sustainable. *Ecological Economics, 6*, 185–193.

Daly, H. (1993). The perils of free trade. *Scientific American, 269*, 24–29.

Daly, H. (1996). *Beyond growth*. Boston: Beacon Press.

Daly, H. (2000). When smart people make dumb mistakes. *Ecological Economics, 34*, 1–3.

Daly, H. (2007). *Ecological economics and sustainable development: Selected essays of Herman Daly*. USA: Edward Elgar.

Daly, H. (2008). *A steady-state economy: A failed growth economy and a steady-state economy are not the same thing; they are the very different alternatives we face*. Paper produced for the Sustainable Development Commission, UK.

Daly, H. (2013). Globalisation versus internationalisation, and four reasons why internationalisation is better. In P. Lawn (Ed.), *Globalisation, economic transition, and the environment* (pp. 39–49). UK: Edward Elgar.

Daly, H., & Cobb, J. (1989). *For the common good: Redirecting the economy toward community, the environment, and a sustainable future*. Boston: Beacon Press.

Daly, H., & Farley, J. (2004). *Ecological economics: Principles and applications*. Washington, DC: Island Press.

d'Arge, R. (1994). Sustenance and sustainability: how can we preserve and consume without major conflict? In A. Jansson, M. Hamer, C. Folke, & R. Costanza (Eds.), *Investing in natural capital* (pp. 113–127). Washington, DC: Island Press.

Dasgupta, P. (2007). Commentary: The Stern review's economics of climate change. *National Institute Economic Review, 199*, 4–7.

David, P. (1985). Clio and the economics of QWERTY. *American Economic Review, 75*(2), 332–337.

Davis, S., & Caldeira, K. (2010). Consumption-based accounting of CO_2 emissions. *Proceedings of the National Academy of Sciences, USA, 107*(12), 5687–5692.

Davy, R., & Coppin, P. (2003). *South East Australian wind power study*. Canberra: Wind Energy Research Unit, CSIRO.

Deal, T. (2002). *WTO rules and procedures and their implications for the Kyoto Protocol: A background paper*. Washington, DC: US Council for International Business.

Dean, J. (1992). Trade and the environment: A survey of the literature. In P. Low (Ed.), *International trade and the environment*. Washington, DC: World Bank.

DeCanio, S. (1993). Barriers within firms to energy-efficient investments. *Energy Policy, 21*(9), 906–914.
de Klein, C., & Eckard, R. (2008). Targeted technologies for nitrous oxide abatement from animal agriculture. *Australian Journal of Experimental Agriculture, 48*(2), 14–20.
DeLong, B. (2006). Partha Dasgupta makes a mistake in his critique of the Stern Review. http://delong.typepad.com/sdj/2006/11/partha_dasgupta.html
Denniss, R. (2008, November). *Fixing the floor in the ETS: The role of energy efficiency in reducing Australia's emissions.* Australia Institute Research Paper Number 59.
den Elzen, M., & Berk, M. (2004). *Bottom-up approaches for defining future climate mitigation commitments.* RIVM Report 728001029/2004, Bilthoven.
den Elzen, M., & Lucas, P. (2005). The FAIR model: A tool to analyse environmental and costs implications of regimes of future commitments. *Environmental Modeling and Assessment, 10*, 115–134.
den Elzen, M., Schaeffer, M., & Lucas, P. (2005). Differentiating future commitments on the basis of countries' relative historical responsibility for climate change: Uncertainties in the "Brazilian Proposal" in the context of a policy implementation. *Climate Change, 71*, 277–301.
den Elzen, M., & Meinshausen, M. (2006). Multi-gas emission pathways for meeting the EU 2 °C climate target. In H. J. Schellnhuber, W. Cramer, N. Nakicenovic, T. Wigley, & G. Yohe (Eds.), *Avoiding dangerous climate change* (pp. 299–309). Cambridge, UK: Cambridge University Press.
den Elzen, M., & Höhne, N. (2010). Sharing the reduction effort to limit global warming to 2 °C. *Climate Policy, 10*, 249–274.
den Elzen, M., Olivier, J., Höhne, N., & Janssens-Meanhout, G. (2013). Countries' contribution to climate change: Effect of accounting for all greenhouse gases, recent trends, basic needs, and technological progress. *Climatic Change, 121*, 397–412.
Department of Climate Change (DCC). (2008). *Fugitive sector greenhouse gas emissions projections, 2007.* Canberra: Commonwealth of Australia.
Depledge, J. (2000). *Tracing the origins of the Kyoto Protocol: An article-by-article textual history.* Technical Paper prepared under contract to UNFCCC, August 1999/August 2000, United Nations Framework Convention on Climate Change (UNFCCC), New York.
Depledge, J. (2002). Continuing Kyoto: Extending absolute emissions caps to developing countries. In K. Baumert, O. Blanchard, S. Llosa, & J. Perkaus (Eds.), *Building on the Kyoto Protocol: Options for protecting the climate* (pp. 31–60). Washington, DC: World Resources Institute.
Depledge, D., & Feakin, T. (2012). Climate change and international institutions: Implications for security. *Climate Policy, 12*, S73–S84.
DeSerpa, A. (1988). *Microeconomic theory: Issues and applications* (2nd ed.). Needham Heights: Allyn & Bacon.
Dessai, S. (2001). *The climate regime from the Hague to Marrakech: Saving or sinking the Kyoto Protocol?* Tyndall Centre Working Paper 12. Tyndall Centre for Climate Change Research, Norwich.
Diamond, J. (2005). *Collapse: How societies choose to fail or succeed.* New York: Penguin Books.
Dickens, G., Castillo, M., & Walker, J. (1997). A blast of gas in the latest Paleocene: Simulating first-order effects of massive disassociation of oceanic methane hydrate. *Geology, 25*(3), 259–262.
Diefenbacher, H. (1994). The index of sustainable economic welfare in Germany. In C. Cobb & J. Cobb (Eds.), *The green national product.* New York: UPA.
Dietz, S., Hope, C., & Patmore, N. (2007a). Some economics of "dangerous" climate change: Reflections on the *Stern Review*. *Global Environmental Change, 17*, 311–325.
Dietz, S., Hope, C., Stern, N., & Zenghelis, D. (2007b). Reflections on the Stern review (1): A robust case for strong action to reduce the risks of climate change. *World Economics, 8*(1), 121–168.

Dietz, S., Anderson, D., Stern, N., Taylor, C., & Zenghelis, D. (2007c). Right for the right reasons: a final rejoinder on the Stern review. *World Economics, 8*(2), 229–258.

Dietz, S., & Stern, N. (2008). Why economic analysis supports strong action on climate change: A response to the Stern review's critics. *Review of Environmental Economics and Policy, 2*(1), 94–113.

Dinar, A., Larson, D., & Rahman, S. (2013). Clean development mechanism: Past, present, and future. In A. Dinar, D. Larson, & S. Rahman (Eds.), *The clean development mechanism (CDM): An early history of unanticipated outcomes—Volume 1* (pp. 1–8). Singapore: World Scientific Publishing Co.

Domingues, C., Church, J., White, N., Gleckler, P., Wijffels, S., Barker, P., & Dunn, J. (2008). Improved estimates of upper-ocean warming and multi-decadal sea-level rise. *Nature, 453*, 1090–1093.

Donner, S., Skirving, W., Little, C., Oppenheimer, M., & Hoegh-Guldberg, O. (2005). Global assessment of coral bleaching and required rates of adaptation under climate change. *Global Change Biology, 11*, 2251–2265.

Dornbusch, R., & Fischer, S. (1990). *Macroeconomics* (5th ed.). New York: McGraw-Hill.

Dosi, G., & Metcalfe, J. (1991). On some notions of irreversibility in economics. In P. Saviotti & J. Metcalfe (Eds.), *Evolutionary theories of economic and technological change* (pp. 133–159). Reading: Harwood Academic Publishers.

Dowsett, H., & Cronin, T. (1990). High eustatic sea level during the middle Pliocene: Evidence from southeastern U.S. Atlantic coastal plain. *Geology, 18*, 435–438.

Early, J. (1989). Space-based solar shield to offset greenhouse effect. *Journal of the British Interplanetary Society, 42*, 567–569.

Easterlin, R. (1974). Does economic growth improve the human lot? Some empirical evidence. In P. David & M. Reder (Eds.), *Nations and Households in economic growth: Essays in honour of Moses Abramovitz* (pp. 98–125). New York: Academic Press.

Easterlin, R. (1995). Will rising the income for all increase the happiness for all? *Journal of Economic Behaviour and Organization, 27*(1), 35–47.

Ebi, K., Exuzides, K., Lau, E., Kelsh, M., & Branston, A. (2004). Weather changes associated with hospitalizations for cardiovascular diseases and stroke in California, 1983–1998. *International Journal of Biometeorology, 49*(1), 48–58.

Eckholm, N. (1901). On the variations of the climate of the geological and historical past and their causes. *Quarterly Journal of the Royal Meteorological Society, 27*, 61.

Economic Planning and Advisory Commission (EPAC). (1995, July). Microeconomic reform paying off for government business enterprises and the economy generally. Media release.

Edenhofer, O., Knopf, B., Leimbach, M., & Bauer, N. (Eds.). (2009). The economics of low stabilization. *Energy Journal*, Special Issue.

Ehrlich, P. (1993). Biodiversity and ecosystem function: Need we know more? In E.-D. Schulze & H. Mooney (Eds.), *Biodiversity and ecosystem function*. Berlin: Springer.

Ehrlich, P., & Ehrlich, A. (1990). *The population explosion*. New York: Simon & Schuster.

Eitzinger, J., Stastna, M., Zalud, Z., & Dubrovsky, M. (2003). A simulation study of the effect of soil water balance and water stress in winter wheat production under different climate change scenarios. *Agricultural Water Management, 61*, 195–217.

Ekholm, T., Soimakallio, S., Moltmann, S., Höhne, N., Syri, S., & Savolainen, I. (2010). Effort sharing in ambitious, global climate change mitigation scenarios. *Energy Policy, 38*(4), 1791–1810.

Ekins, P., Kesicki, F., & Smith, A. (2011, April). *Marginal abatement cost curves: A call for caution*. A report from the UCL Energy Institute to, and commissioned by, Greenpeace UK.

Eliasch, J. (2008). *Climate change: Financing global forests (The Eliasch Review)*. London: Earthscan.

Ellerman, A., & Buchner, B. (2007). The European Union emissions trading scheme: Origins, allocations, and early results. *Review of Environmental Economics and Policy, 1*(1), 66–87.

Elliott, J., Foster, I., Kortum, S., Munson, T., Pérez Cervantes, F., & Weisbach, D. (2010a). Trade and carbon taxes. *American Economic Review: Papers and Proceedings, 100*(2), 465–469.

Elliott, J., Foster, I., Kortum, S., Munson, T., Pérez Cervantes, F., & Weisbach, D. (2010b). *A quantitative examination of trade and carbon taxes*. ANL/MCS-P1711-0110, MCS Division, Argonne National Laboratory.

Ellis, J., Corfee-Morlot, J., & Winkler, H. (2004). *Taking stock of progress under the clean development mechanism*. OECD/IEA Information Paper for the Annex I Expert Group on the UNFCCC, OECD, Paris.

Ellison, C., Chapman, M., & Hall, I. (2006). Surface and deep ocean interactions during the cold climate event 8,200 years ago. *Science, 312*(5782), 1929–1932.

Ellison, D., Lundblad, M., & Petersson, H. (2014). Reforming the EU approach to LULUCF and the climate policy framework. *Environmental Science and Policy, 40*, 1–15.

El Serafy, S. (1989). The proper calculation of income from depletable natural resources. In Y. Ahmad, S. El Serafy, & E. Lutz (Eds.), *Environmental accounting for sustainable development* (pp. 10–18). Washington, DC: World Bank.

Emori, S., & Brown, S. (2005). Dynamic and thermodynamic changes in mean and extreme precipitation under changed climate. *Geophysical Research Letters, 32*, L17706. doi:10.1029/2005GL023272

Engel, S., Pagiola, S., & Wunder, S. (2008). Designing payments for environmental services in theory and practice: An overview of the issues. *Ecological Economics, 65*, 663–674.

Estrin, S. (1986). *Profit-sharing, motivation and company performance: A survey*. Department of Economics Pamphlet, London School of Economics.

European Commission. (2008). *The EU emissions trading system (EU ETS)*. 2008 Edition. http://ec.europa.eu/environment/climat/emission/index_en.htm. Accessed January 31, 2009.

European Environment Agency (EEA). (2009/2012). *Kyoto burden-sharing targets for EU-15 countries*. http://www.eea.europa.eu. Accessed October 18, 2013.

European Environment Agency (EEA). (2012). *Greenhouse gas emission trends and projections in Europe 2012: Tracking progress towards Kyoto and 2020 Targets*. Copenhagen: European Environment Agency.

European Environment Agency (EEA). (2013). *Trends and projections in Europe 2013: Tracking progress towards Europe's climate and energy targets until 2020*. Copenhagen: European Environment Agency.

Faber, M., & Proops, J. (1990). *Evolution, time, production, and the environment*. Heidelberg: Springer.

Faber, M., Manstetten, R., & Proops, J. (1992). Toward an open future: Ignorance, novelty, and evolution. In R. Costanza, B. Norton, & B. Haskell (Eds.), *Ecosystem health: New goals for environmental management* (pp. 72–96). Washington, DC: Island Press.

Fagan, B. (2001). *The little Ice Age: How climate made history, 1300-1850*. New York: Basic Books.

Fair Climate Network. (2014). Long CERs & temporary CERs. http://www.fairclimate.com/about_cdm/tutorial/lcers.aspx. Accessed November 7, 2014.

Fankhauser, S., & Tol, R. (2005). On climate change and economic growth. *Resource and Energy Economics, 27*(1), 1–17.

Fankhauser, S., & Burton, I. (2011). Spending adaptation money wisely. *Climate Policy, 11*(3), 1037–1049.

Feather, N. (1997). Economic deprivation and the psychological impact of unemployment. *Australian Psychologist, 32*(1), 37–45.

Fiala, N. (2008). Measuring sustainability: Why the ecological footprint is bad economics and bad environmental science. *Ecological Economics, 67*(4), 519–525.

Field, B. (1998). *Environmental economics: An introduction* (2nd ed.). New York: McGraw-Hill.

Filho, L. G. M., & Miguez, J. D. G. (2000). *Technical note on the time-dependent relationship between emissions of greenhouse gases and climate change*. Brazilian Ministry of Science and Technology.

Fischedick, M., Schaeffer, R., Adedoyin, A., Akai, M., Bruckner, T., Clarke, L., et al. (2011). Mitigation potential and costs. In *Special report on renewable energy sources and climate change mitigation*. Prepared by Working Group III of the Intergovernmental Panel on Climate Change. Cambridge: Cambridge University Press.

Fischer, S., Dornbusch, R., & Schmalansee, R. (1988). *Economics*. New York: McGraw-Hill.

Fisher, I. (1906). *Nature of capital and income*. New York: A.M. Kelly.

Fisher, I. (1935). *100% money*. New York: Adelphi.

Fitzherbert, E., Struebig, M., Morel, A., Danielsen, F., Bruhl, Donald, P., & Phalan, B. (2008). How will oil palm expansion affect biodiversity? *Trends in Ecology and Evolution, 23*(10), 538–545.

Flannery, T. (2008). *The weather makers* (2nd ed.). Melbourne: Text Publishers.

Flannigan, M., Logan, K., Amiro, B., Skinner, W., & Stocks, B. (2004). Future area burned in Canada. *Climatic Change, 72*, 1–16.

Fleming, J. (1998). *Historical perspectives on climate change*. New York: Oxford University Press.

Folke, C., Hammer, M., Costanza, R., & Jansson, A. (1994). Investing in natural capital—Why, what, and how? In A. Jansson, M. Hammer, C. Folke, & R. Costanza (Eds.), *Investing in natural capital* (pp. 1–20). Washington, DC: Island Press.

Food and Agriculture Organization (FAO). (2004). *FAO and SIDS: Challenges and emerging issues in agriculture, forestry and fisheries*. Paper prepared by the Food and Agriculture Organization (FAO) on the occasion of the Inter-regional Conference on Small Island Developing States (SIDS), Bahamas, 26–30 January 2004, FAO, Rome.

Forbes, D. (2005). Coastal erosion. In M. Nuttall (Ed.), *Encyclopaedia of the Arctic* (pp. 391–393). New York: Routledge.

Forbes, D., Parkes, G., Manson, G., & Ketch, L. (2004). Storms and shoreline retreat in the southern Gulf of St. Lawrence. *Marine Geology, 210*, 169–204.

Forgie, V., McDonald, G., Zhang, Y., Patterson, M., & Hardy, D. (2008). Calculating the New Zealand genuine progress indicator. In P. Lawn & M. Clarke (Eds.), *Genuine progress in the Asia-Pacific: Studies using the genuine progress indicator* (pp. 126–152). Cheltenham, UK: Edward Elgar.

Fouillet, A., Rey, G., Wagner, V., Laaidi, K., Empereur-Bissonnet, P., Le Terte, A., et al. (2008). Has the impact of heat waves on mortality changed in France since the European heat wave of summer 2003? A study of the 2006 heat wave. *International Journal of Epidemiology, 37*(2), 302–317.

Franke, S. (2005a) *Measurement of social capital: Reference document for public policy research, development, and evaluation*. PRI Project: Social Capital as a Public Policy Tool, Policy Research Initiative, Ottawa.

Frankel, J. (2005b). Climate and trade: Links between the Kyoto Protocol and WTO. *Environment, 47*(7), 8–19.

Frankel, J. (2007). Formulas for quantitative emissions targets. In J. Aldy & R. Stavins (Eds.), *Architectures for agreement: Addressing global climate change in the post-Kyoto World* (pp. 31–56). New York: Cambridge University Press.

Freeman, J., & Kolstad, C. (Eds.). (2006). *Moving to markets: Lessons from thirty years of experience*. New York: Oxford University Press.

Freeman, K., & Hayes, J. (1992). Fractionation of carbon isotopes by photoplankton and estimates of ancient CO_2 levels. *Global Biogeochemical Cycles, 6*, 185–198.

Frenot, Y., Chown, S., Whinam, J., Selkirk, P., Convey, P., Skotnicki, M., & Bergstrom, D. (2005). Biological invasions in the Antarctic: Extent, impacts, and implications. *Biological Reviews, 80*, 45–72.

Friedlingstein, P., Cox, P., Betts, R., Bopp, L., Bloh, W., Brovkin, V., et al. (2006). Climate-carbon cycle feedback analysis: Results from C4MIP model intercomparison. *Journal of Climate, 19*, 3337–3353.

Fryer, D. (1995, June). Labour market disadvantage, deprivation, and mental health: Benefit agency? *The Psychologist*, 265–272.

Funtowitz, S., & Ravetz, J. (1994). The worth of a songbird: Ecological economics as a post-normal science. *Ecological Economics, 10*, 197–207.

Furgal, C., Martin, D., & Gosselin, P. (2002). Climate change and health in Nunavik and Labrador: Lessons from Inuit knowledge. In I. Krupnik & D. Jolly (Eds.), *The earth is faster now: Indigenous observations of Arctic environmental change* (pp. 266–300). Fairbanks, Alaska: Arctic Research Consortium of the United States.

Fyfe, J., & Saenko, O. (2005). Human-induced change in the Antarctic circumpolar current. *Journal of Climate, 18*, 3068–3073.

Fynn, A., Alvarez, P., Brown, J., George, M., Kustin, C., Laca, E., et al. (2009). *Soil carbon sequestration in U.S. Rangelands*. Issues Paper for Protocol Development, Environmental Defense Fund, New York.

Garnaut, R. (2008). *The Garnaut climate change review*. Cambridge: Cambridge University Press.

Garnaut, R. (2011). *The Garnaut climate change review—Update 2011, Update Paper 6: Carbon pricing and reducing Australia's emissions*. www.garnautreview.org.au

Garrod, B. (1998). Are economic globalization and sustainable development compatible? Business strategy and the role of the multinational enterprise. *International Journal of Sustainable Development, 1*, 43–62.

Gedney, N., Cox, P., & Huntingford, C. (2004). Climate feedback from wetland methane emissions. *Geophysical Research Letters, 31*(20), L20503.

Geerts, S., & Raes, D. (2009). Deficit irrigation as an on-farm strategy to maximise crop water productivity in dry areas. *Agricultural Water Management, 96*(9), 1275–1284.

George, H. (1879). *Progress and poverty*. New York: Random House.

Georgescu-Roegen, N. (1971). *The entropy law and the economic process*. Cambridge, MA: Harvard University Press.

Georgescu-Roegen, N. (1978–79). Economic analysis and economic valuation. *Southern Economic Journal, 45*(2), 1023–1058.

Georgescu-Roegen, N. (1979). Comments on the papers by Daly and Stiglitz. In V. K. Smith (Ed.), *Scarcity and growth reconsidered* (pp. 95–105). Baltimore: John Hopkins University Press.

Geres, R., & Michaelowa, A. (2002). A qualitative method to consider leakage effects from CDM and JI projects. *Energy Policy, 30*, 461–463.

Gigli, S., & Agrawala, S. (2007). *Stocktaking of progress on integrating adaptation to climate change into development co-operation activities*. Paris: OECD.

Gillenwater, M., & Seres, S. (2011, March). *The clean development mechanism: A review of the first international offset program*. Pew Center on Global Climate Change, Arlington, VA.

Gilley, B., & Kinsella, D. (2013). *The use of force to achieve climate change goals*'. Paper presented at the International Studies Association. 3–6 April 2013, San Francisco. http://web.pdx.edu/~kinsella/papers/isa13.pdf

Gintis, H. (1997). Review of real freedom for all by Philippe Van Parijs. *Journal of Economic Literature, 35*, 181–182.

Global Commons Institute (GCI). (2008). *The campaign for contractions & convergence*. London: Global Commons Institute.

Global Environment Facility (GEF). (2005). *Operational guidelines for the strategic priority "piloting an operational approach to adaptation" (SPA)*. Global Environment Facility, Washington, DC, GEF/C.27/Inf.10.

Global Environment Facility (GEF). (2006). *Programming paper for funding the implementation of NAPAs under the LDC trust fund*. Global Environment Facility, Washington, DC, GEF/C.28/18.

Global Environment Facility (GEF). (2011, October). *Instrument for the establishment of the restructured global environment facility*. Global Environment Facility, Washington, DC.

Global Environment Facility (GEF). (2012). *Financing adaptation action under the least developed countries fund and special climate change fund*. Global Environment Facility. http://www.thegef.org/gef/sites/thegef.org/files/publication/GEF_FinancingAdaptation-LDCF-SCCF-Oct-31-2012-CRA.pdf. Accessed December 15, 2013.

Global Environment Facility (GEF). (2013a). *Least developed countries fund (LDCF)*. Global Environment Facility. http://www.thegef.org/gef/LDCF. Accessed December 15, 2013.

Global Environment Facility (GEF). (2013b). *Special climate change fund (SCCF)*. Global Environment Facility. http://www.thegef.org/gef/sccf. Accessed December 15, 2013.

Global Footprint Network. (2006). *Ecological footprint and biocapacity*. Oakland, CA: Global Footprint Network.

Global Footprint Network. (2008). *The ecological footprint Atlas 2008: Version 1.0*. Oakland, CA: Global Footprint Network.

Global Wind Energy Council. (2011). *Global wind 2011 report*. Brussels: Global Wind Energy Council.

Goulder, L. (2005). Trade-offs between political feasibility and cost-effectiveness in carbon-emissions permit systems. *NCCR Climate Symposium*. Interlaken, Switzerland, March 3–4, 2005.

Government of Japan. (2008). *Kyoto Protocol target achievement plan (Totally revised 28 March 2008)*. Tokyo: Ministry of the Environment.

Grace, P., Antle, J., Aggarwal, P, Andren, O., Ogle, S., Conant, R., et al. (2004). *Assessment of the costs and enhanced global potential of carbon sequestration in soil*. IEA/CON/03/95.

Graham, N., Wilson, S., Jennings, S., Polunin, N., Bijoux, J., & Robinson, J. (2006). Dynamic fragility of oceanic coral reef and ecosystems. *Proceedings of the National Academy of Sciences, USA, 103*, 8425–8429.

Green Climate Fund (GCF). (2011). *Governing instrument for the green climate fund*. http://www.gcfund.org/about-the-fund/documents. Accessed February 15, 2014.

Green Climate Fund (GCF). (2014). *Status of resources*. http://www.gcfund.org/about-the-fund/meetings. Accessed February 24, 2014).

Groenenberg, H., Phylipsen, D., & Blok, K. (2002). Differentiating commitments worldwide: Global differentiation of GHG emissions reductions based on the Triptych approach—A preliminary assessment. *Energy Policy, 29*(12), 1007–1030.

Groenenberg, H., Blok, K., & van der Sluijs, J. (2004). Global Triptych: a bottom-up approach for the differentiation of commitments under the Climate Convention. *Climate Policy, 4*(2), 153–175.

Grootaert, C., & van Bastelaer, T. (2001). *Understanding and measuring social capital: A synthesis of findings and recommendations from the social capital initiative*. SCI Working Paper No. 250. Washington, DC: The World Bank.

Grubb, M. (2003). The economics of the Kyoto Protocol. *World Economics, 4*(3), 143–189.

Grubb, M., & Neuhoff, K. (2005). Allocation and competitiveness in the EU emissions trading scheme: Policy overview. *Climate Policy, 6*, 7–30.

Grubb, M., Carraro, C., & Schellnhuber, J. (2006, April). Technological change for atmospheric stabilisation: Introductory overview to the Innovation Modelling Comparison Project. *Energy Journal*, Special Issue on Endogenous Technological Change and the Economics of Atmospheric Stabilisation, 1–16.

Grubler, A., & Nakicenovic, N. (1994). *International burden sharing in greenhouse gas reduction*. Austria: IIASA.

Gupta, S., & Bhandari, P. (1999). An effective allocation criterion for CO_2 emissions. *Energy Policy, 27*, 727–736.

Guenno, G., & Tiezzi, S. (1998). *An index of sustainable economic welfare for Italy*. Working Paper 5/98. Fondazione Eni Enrico Mattei, Milan.

Guo, Z. T., Peng, S. Z., Hao, Q. Z., Biscaye, P. E., An, Z. S., & Liu, T. S. (2004). Late miocene-pliocene development of Asian aridification as recorded in the red-earth formation in northern China. *Global Planetary Change, 41*, 135–145.

Gupta, S., & Bhandari, P. (1999). An effective allocation criterion for CO_2 emissions. *Energy Policy, 27*, 727–736.

Haberl, H., Krausman, F., & Gingrich, S. (2006, November 25). Ecological embeddedness of the economy. *Economic and Political Weekly*.

Haberl, H., Erb, K., Krausmann, F., Gaube, V., Bondeau, A., Plutzar, C., et al. (2007). Quantifying and mapping the human appropriation of net primary production in earth's terrestrial ecosystems. *Proceedings of the National Academy of Sciences, USA, 104*(31), 12942–12947.

Haita, C. (2013). *Linking emission trading schemes: Pros and cons*. ICCG Reflection No. 16/2013, International Center for Climate Governance, Venice.

Hall, D., & Lobina, E. (2005). *The relative efficiency of public and private sector water*. Report Commissioned by Public Services International, University of Greenwich, London.

Halloy, S., & Mark, A. (2003). Climate change effects on alpine plant biodiversity: A New Zealand perspective on quantifying the threat. *Arctic, Antarctic, and Alpine Research? 35*, 248–254.

Halsnæs, K. (2002). A review of the literature on climate change and sustainable development. In A. Markandya & K. Halsnæs (Eds.), *Climate change and sustainable development: Prospects for developing countries* (pp. 49–72). London: Earthscan.

Hamilton, C., Sherrard, J., & Tate, A. (2005). *Climate change policy beyond Kyoto: A new global plan*. Australia Institute Discussion Paper Number 75, February 2005.

Hammitt, J., & Shlyakhter, A. (1999). The expected value of information and the probability of surprise. *Risk Analysis, 19*, 135–152.

Hansen, J. (2009, February 25). Carbon tax & 100% dividend vs. tax & trade. Testimony of James E. Hansen to Committee on Ways and Means, United States House of Representatives.

Hansen, J., Sato, M., Kharecha, P., Beerling, D., Masson-Delmotte, V., Pagani, M., et al. (2008). Target atmospheric CO_2: Where should humanity aim? *Open Atmospheric Science Journal, 2*, 217–231.

Hansen, J., Kharecha, P., Sato, M., Masson-Delmotte, V., Ackerman, F., Beerling, D., et al. (2013). Assessing "dangerous climate change": Required reduction of carbon emissions to protect young people, future generations, and nature. *PLOS ONE, 8*(12), e81648.

Hardin, G. (1968). The tragedy of the commons. *Science, 162*, 1243–1248.

Hare, B., & Meinshausen, M. (2005). How much warming are we committed to and how much can be avoided? *Climate Change, 75*, 111–149.

Hare, W. (2003). *Assessment of knowledge on impacts of climate change: Contribution to the specification of article 2 of the UNFCCC*. Berlin: Wiseen.

Harrabin, R. (2009). Harrabin's notes: After Copenhagen. http://news.bbc.co.uk/2/hi/8423822.stm. Accessed September 19, 2013.

Harrington, G., Kemp, S., & Koch, P. (2004a). Paleocene-Eocene paratropical floral change in North America: Responses to climate change and plant immigration. *Journal of the Geological Society of London, 161*, 173–184.

Harrington, W., Morgenstern, R., & Sterner, T. (2004b). Overview: Comparing instrument choices. In W. Harrington, R. Morgenstern, & T. Sterner (Eds.), *Choosing environmental policy: Comparing instruments and outcomes in the United States and Europe* (pp. 1–22). Washington, DC: Resources for the Future Press.

Harris, M. (2007). On income, sustainability, and the microfoundations of the genuine progress indicator. *International Journal of Environment, Workplace, and Employment, 3*(2), 119–131.

Hartmann, J., Ebi, K., McConnell, J., Chan, N., & Weyant, J. (2002). Climate suitability for stable malaria transmission in Zimbabwe under different climate change scenarios. *Global Change and Human Health, 3*, 42–54.

Harvey, L. (2004). Declining temporal effectiveness of carbon sequestration: Implications for compliance with the United National framework convention on climate change. *Climate Change, 63*, 259–290.

Harvey, P. (2000). Combating joblessness: An analysis of the principle strategies that have influenced the development of American employment and social welfare law during the 20th century. *Berkeley Journal of Employment and Labor Law, 21*(2), 677–758.

Hay, J., Mimura, N., Cambell, J., Fifita, S., Koshy, K., McLean, R., et al. (2003). *Climate variability and change and sea level rise in the Pacific Islands Region: A resource book for policy and decision makers, educators and other stakeholders*. Somoa: South Pacific Regional Environment Programme (SPREP), Apia.

Hayden, H. (2003). *The solar fraud*. Pueblo West: Vales Lake Publishing.

Haywood, A., Dekens, P., Ravelo, A., & Williams, M. (2000). Global scale paleoclimate reconstruction of the middle Pliocene climate using the UKMO GCM: Initial results. *Global Planetary Change, 25*, 239–256.

Hazeleger, W. (2005). Can global warming affect tropical ocean heat transport? *Geophysical Research Letters, 32*, L22701. doi:10.1029/2005GL024031

Heer, D. (1975). Marketable licences for babies: Boulding's proposal revisited. *Social Biology, 22*(1), 1–16.

Heitzig, J. (2012). *Bottom-up strategic linking of carbon markets: Which climate coalitions would farsighted players form?*. Potsdam: Potsdam Institute for Climate Impact Research.

Hendricks, C., Graus, W., & van Bergen, F. (2004). Global carbon dioxide storage potential and costs. ECOFYS, Utrecht, www.ecofys.nl. Accessed March 30, 2009.

Henson, R. (2011). *the rough guide to climate change*. London: Rough Guides.

Hepburn, C., Grubb, M., Neuhoff, K., Matthes, F., & Tse, M. (2006). Auctioning of EU ETS phase II allowances: How and why? *Climate Policy, 6*(1), 137–160.

Hicks, J. (1946). *Value and capital* (2nd ed.). London: Clarendon.

Hirsch, F. (1977). *The social limits to growth*. London: Routledge & Kegan Paul.

Hoanh, C., Guttman, H., Droogers, P., & Aerts, J. (2004). Will we produce sufficient food under climate change? Mekong Basin (South-east Asia). In J. Aerts & P. Droogers (Eds.), *Climate change in contrasting river basins: Adaptation strategies for water, food, and environment* (pp. 157–180). Wallington: CABI Publishing.

Hock, R., Jansson, P., & Braun, L. (2005). Modelling the response of mountain glacier discharge to climate warming. In U. Huber, M. Reasoner, & H. Bugmann (Eds.), *Global change series* (pp. 243–252). Dordrecht: Springer.

Hodge, I. (1995). *Environmental economics*. London: Macmillan Press.

Hodgson, G. (1988). *Economics and institutions*. Cambridge: Polity Press.

Hoerner, J. (1998). *The role of border-tax adjustments in environmental taxation: Theory and U.S. experience*. Working Paper. Washington, DC: Center for a Sustainable Economy.

Hoerner, J., & Bosquet, B. (2001). *Environmental tax reform: The European experience*. Washington, DC: Centre for a Sustainable Economy.

Höhne, N., Galleguillos, C., Blok, K., Harnisch, J., & Phylipsen, D. (2003). *Evolution of commitments period of the Kyoto Protocol*. German Federal Environmental Agency, Berlin, 02/05.

Höhne, N., Phylipsen, D., Ullrich, S., & Blok, K. (2005). *Options for the second commitments under the UNFCCC: Involving newly industrialized economies and developing countries*. Research Report 201 41 25, UBA-FB-00412, German Federal Environmental Agency, Berlin.

Höhne, N., den Elzen, M., & Weiss, M. (2006). Common but differentiated convergence (CDC): A new conceptual approach to long-term climate policy. *Climate Policy, 6*, 181–199.

Hope, C. (2006). The marginal impact of CO_2 from PAGE2002: An integrated assessment model incorporating the IPCC's five reasons for concern. *Integrated Assessment Journal, 6*(1), 19–56.

Hotelling, H. (1931). The economics of exhaustible resources. *Journal of Political Economy, 39*, 137–175.

Houghton, J. (1994). *Global warming: The complete briefing*. Oxford, UK: Lion Publishing.

Hovi, J., & Areklett, I. (2004). Enforcing the climate regime: Game theory and the Marrakesh Accords. *International Environmental Agreements: Politics, Law and Economics, 4*, 1–26.

Howarth, R. (2003, February). Climate change. *International Society for Ecological Economics, Internet Encyclopaedia of Ecological Economics*.

Howden, M., & Jones, R. (2004). Risk assessment of climate change impacts on Australia's wheat industry. In T. Fischer, N. Turner, J. Angus, J. McIntyre, L. Robertson, A. Borrell, & D. Lloyd (Eds.), *New directions for a diverse planet: Proceedings of the* 4th *international crop science congress*. Brisbane. http://www.cropscience.org.au/icsc2004/symposia6/2/1848_hodensm.htm

Howden, S., Ash, A., Barlow, E., Booth, T., Charles, S., Cechet, R., et al. (2003). *An overview of the adaptive capacity of the Australian agricultural sector to climate change: Options, costs and benefits*. Canberra: Report to the Australian Greenhouse Office.

Humphries, J. (2007). Exploring a carbon tax for Australia. *Perspectives on Tax Reform*. 14, CIS Policy Monograph 80, The Centre for Independent Studies, St Leonards, NSW.

Humphries, W. (1913). Volcanic dust and other factors in the production of climate changes, and their possible relation to Ice Ages. *Journal of the Franklin Institute, 176*, 132.

Hurrell, J., Kushnir, Y., Ottersen, G., & Visbeck, M. (Eds.). (2003). *The North Atlantic oscillation: Climate significance and environmental impact*. Geophysical Monograph Series 134.

Hutchinson, G. (1991). Efficiency gains through privatisation of UK industries. In A. Ott & K. Hartley (Eds.), *Privatisation and economic efficiency* (pp. 87–107). Aldershot: Edward Elgar.

Iggulden, J. (1996). *Silent lies*. Bellingen: Evandale.

Instanes, A., Anisimov, O., Brigham, L., Goering, D., Ladanyi, B., Larsen, J., & Khrustalev, L. (2005). Infrastructure, buildings, support systems, and industrial facilities. In C. Symon, L. Arris, & B. Heal (Eds.), *Arctic climate impact assessment* (pp. 907–944). Cambridge: Cambridge University Press.

Inter-American Development Bank (IDB). (2013). *Financial instruments and mechanisms for climate programs in Latin America and the Caribbean: A guide for ministries of finance*. Climate Change and Sustainability Division. Policy Brief No. IDB-PB-212, Inter-American Development Bank, Washington, DC.

Intergovernmental Panel on Climate Change (IPCC). (1990). *Climate change: The IPCC scientific assessment*. Cambridge: Cambridge University Press.

Intergovernmental Panel on Climate Change (IPCC). (1996a). *Climate change 1995: The science of climate change*. Contribution of Working Group I to the Second Assessment Report of the Intergovernmental Panel on Climate Change. Cambridge: Cambridge University Press.

Intergovernmental Panel on Climate Change (IPCC). (1996b). *Climate change 1995: Impacts, adaptations and mitigation of climate change: Scientific technical analyses*. Contribution of Working Group II to the Second Assessment Report of the Intergovernmental Panel on Climate Change. Cambridge: Cambridge University Pres.

Intergovernmental Panel on Climate Change (IPCC). (1996c). *Climate change 1995: Economic and social dimensions of climate change*. Contribution of Working Group III to the Second Assessment Report of the Intergovernmental Panel on Climate Change. Cambridge: Cambridge University Press.

Intergovernmental Panel on Climate Change (IPCC). (2000). *Special report on emissions scenarios*. Intergovernmental Panel on Climate Change. Cambridge: Cambridge University Press.

Intergovernmental Panel on Climate Change (IPCC). (2001a). *Third assessment report*. Synthesis and Mitigation Reports. Cambridge: Cambridge University Press.

Intergovernmental Panel on Climate Change (IPCC). (2001b). *Climate change 2001: The scientific basis*. Contribution of Working Group I to the Third Assessment Report of the Intergovernmental Panel on Climate Change. Cambridge: Cambridge University Press.

Intergovernmental Panel on Climate Change (IPCC). (2001c). *Climate change 2001: Impacts, adaptation and vulnerability*. Contribution of Working Group II to the Third Assessment Report of the Intergovernmental Panel on Climate Change. Cambridge: Cambridge University Press.

Intergovernmental Panel on Climate Change (IPCC). (2001d). *Climate change 2001: Mitigation*. Contribution of Working Group III to the Third Assessment Report of the Intergovernmental Panel on Climate Change. Cambridge: Cambridge University Press.

Intergovernmental Panel on Climate Change (IPCC). (2006). *2006 IPCC Guidelines for national greenhouse gas inventories: Volume 2, energy*. Hayama, Japan: Institute for Global Environmental Strategies (IGES).

Intergovernmental Panel on Climate Change (IPCC). (2007a). *Climate change 2007: Synthesis report*. Synthesis of the Fourth Assessment Report, Intergovernmental Panel on Climate Change. Cambridge: Cambridge University Press.

Intergovernmental Panel on Climate Change (IPCC). (2007b). *Climate change 2007: The physical science basis*. Contribution of Working Group I to the Fourth Assessment Report of the Intergovernmental Panel on Climate Change. Cambridge: Cambridge University Press.

Intergovernmental Panel on Climate Change (IPCC). (2007c). *Climate change 2007: Impacts, adaptation and vulnerability*. Contribution of Working Group II to the Fourth Assessment Report of the Intergovernmental Panel on Climate Change. Cambridge: Cambridge University Press.

Intergovernmental Panel on Climate Change (IPCC). (2007d). *Climate change 2007: Mitigation of climate change*. Contribution of Working Group III to the Fourth Assessment Report of the Intergovernmental Panel on Climate Change. Cambridge: Cambridge University Press.

Intergovernmental Panel on Climate Change (IPCC). (2011). *Special report on renewable energy sources and climate change mitigation*. Prepared by Working Group III of the Intergovernmental Panel on Climate Change. Cambridge: Cambridge University Press.

Intergovernmental Panel on Climate Change (IPCC). (2014a). *Summary for policymakers*. Fifth Assessment Report of the Intergovernmental Panel on Climate Change. Cambridge: Cambridge University Press.

Intergovernmental Panel on Climate Change (IPCC). (2014b). *Climate change 2014: The physical science basis*. Contribution of Working Group I to the Fifth Assessment Report of the Intergovernmental Panel on Climate Change. Cambridge: Cambridge University Press.

Intergovernmental Panel on Climate Change (IPCC). (2014c). *Climate change 2014: Impacts, adaptation, and vulnerability*. Contribution of Working Group II to the Fifth Assessment Report of the Intergovernmental Panel on Climate Change. Cambridge: Cambridge University Press.

Intergovernmental Panel on Climate Change (IPCC). (2014d). *Climate change 2014: Mitigation of climate change*. Contribution of Working Group III to the Fifth Assessment Report of the Intergovernmental Panel on Climate Change. Cambridge: Cambridge University Press.

International Alliance of Research Universities (IARU). (2009). *Climate change: Global risks, challenges and decisions—Synthesis report*. Copenhagen: University of Copenhagen.

International Energy Agency (IEA). (2004). *World energy outlook 2004*. Paris: International Energy Agency.

International Energy Agency (IEA). (2006). *World energy outlook*. Paris: OECD/IEA.

International Energy Agency (IEA). (2007). *Reviewing R&D policies: Guidance for IEA review teams*. Paris: OECD/IEA.

International Energy Agency (IEA). (2009). *World energy outlook*. Paris: OECD/IEA.

International Energy Agency (IEA). (2011). *World energy outlook*. Paris: OECD/IEA.

International Energy Agency (IEA). (2012a). *CO_2 emissions from fuel combustion: Highlights* (2012 ed.). Paris: International Energy Agency.

International Energy Agency (IEA). (2012b). *Energy technology perspectives*. Paris: International Energy Agency.

International Energy Agency (IEA). (2014). *IEA energy R&D statistics*. Paris: IEA.

Ishigami, Y., Simizu, Y., & Omasa, K. (2005). Evaluation of the risk to natural vegetation from climate change in Japan. *Journal of Agricultural Meteorology, 61*, 69–75 (text in Japanese with English abstract).

Ishii, M., & Kimoto, M. (2009). Reevaluation of historical ocean heat content variations with time-varying XBT and MBT depth bias corrections. *Journal of Oceanography, 65*(3), 287–299.

Ismer, R., & Neuhoff, K. (2007). Border-tax adjustment: A feasible way to support stringent emission trading. *European Journal of Law and Economics, 24*(2), 137–164.

Jackson, T. (2009). *Prosperity without growth: Economics for a finite planet*. London: Earthscan.

Jackson, T., & Stymne, S. (1996). *Sustainable economic welfare in Sweden: A pilot index 1950–1992*. Stockholm: Stockholm Environment Institute.

Jackson, T., Laing, F., MacGillivray, A., Marks, N., Ralls, J., & Styme, S. (1997). *An index of sustainable economic welfare for the UK, 1950–1996*. Guildford: University of Surrey Centre for Environmental Strategy.

Jaffe, A., Peterson, S., Portney, P., & Stavins, R. (1995). Environmental regulation and the competitiveness of US manufacturing: What does the evidence tell us? *Journal of Economic Literature, 33*, 132–163.

Jakob, M. (2006). Marginal costs and co-benefits of energy-efficiency investments: The case of the Swiss residential sector. *Energy Policy, 34*(2), 172–187.

Jensen, M., & Meckling, W. (1976). Theory of the firm: Managerial behavior, agency costs, and ownership structure. *Journal of Financial Economics, 3*(4), 305–360.

Jiang, D., Wang, H., Ding, Z., Lang, X., & Drange, H. (2005). Modeling the middle Pliocene climate with a global atmospheric general circulation model. *Journal of Geophysical Research, 110*, D14107. doi:10.1029/2004JD005639

Johnstone, N. (2002). The use of tradable permits in combination with other policy instruments: A scoping paper. ENV/EPOC/WPNEP(2002)28, Working Party on National Environmental Policy, OECD, Paris.

Jolley, A. (2006a, March). *Transport technologies*. Climate Change Project Working Paper Number 5, Centre of Strategic Economic Studies, Victoria University.

Jolley, A. (2006b, March). *Technologies for reducing stationary energy use*. Climate Change Project Working Paper Number 6, Centre of Strategic Economic Studies, Victoria University.

Jolley, A. (2006c, March). *Technologies for alternative energy*. Climate Change Project Working Paper Number 7, Centre of Strategic Economic Studies, Victoria University.

Johnstone, N. (2002). *The use of tradable permits in combination with other policy instruments: A scoping paper*. ENV/EPOC/WPNEP(2002)28, Working Party on National Environmental Policy, OECD, Paris.

Jones, C. (2007, March 21–22). *Australian soil carbon accreditation scheme (ASCAS)*. Presentation at the 'Managing the Carbon Cycle Workshop', Katanning, Western Australia.

Jones, C. (2008). Liquid carbon pathway unrecognised. *Australian Farm Journal, 85*(1), 15–17.

Jones, C., Cox, P., & Huntingford, C. (2006). Impact of climate-carbon-cycle feedbacks on emissions scenarios to achieve stabilisation. In H. Schellnhuber, W. Cramer, N. Nakicenovic, T. Wigley, & G. Yohe (Eds.), *Avoiding dangerous climate change* (pp. 323–331). Cambridge, UK: Cambridge University Press.

Jones, P., Briffa, K., Barnett, T., & Tett, S. (1998). High-resolution palaeoclimatic records for the last millennium: Interpretation, integration and comparison with General Circulation Model control run temperatures. *The Holocene, 8*, 455–471.

Jones, R., Bowyer, J., Hoegh-Guldberg, O., & Blackall, L. (2004). Dynamics of a temperature-related coral disease outbreak. *Marine Ecology Progress Series, 281*, 63–77.

Jowit, J., & Wintour, P. (2008, June 26). Cost of tackling global climate change has doubled, warns Stern. *The Guardian*.

Jung, M. (2006). Host country attractiveness for CDM non-sink projects. *Energy Policy, 34*(15), 2173–2184.

Jung, M., Fleck, J., Ellermann, C., Hagemann, M., & Mizerny, J. (Eds.). (2008). *German manual for JI host country approval—Guidance for applicants*. Version 1.1, German Emissions Trading Authority, at the Federal Environment Agency, Berlin.

Kachi, A., Tänzler, D., & Sterk, W. (2014). *The clean development mechanism and emerging offset schemes: Options for reconciliation*. On behalf of the Federal Environment Agency (Germany), Climate Change 06/2014, Berlin.

Kahneman, D., & Tversky, A. (1979). Prospect theory: An analysis of decisions under risk. *Econometrica, 47*(2), 263–291.

Kahneman, D., & Krueger, A. (2006). Developments in the measurement of subjective well-being. *Journal of Economic Perspectives, 20*(1), 3–24.

Kammen, D., & Nemet, G. (2005). Real numbers; Reversing the incredible shrinking energy R&D budget. *Issues in science and technology* (pp. 84–88). National Academy of Sciences, National Academy of Engineering, Institute of Medicine, and the University of Texas at Dallas, Dallas.

Karousakis, K. (2006). *Joint implementation: Current issues and emerging challenges*. Paris: OECD.

Kartha, S., Lazarus, M., & Bosi, M. (2002). *Practical baseline recommendations for greenhouse gas projects in the electric power sector*. Paris: OECD.

Keeling, C. (1960). The concentration and isotopic abundances of carbon dioxide in the atmosphere. *Tellus, 12*, 200–203.

Keith, D. (2009). Why capture CO_2 from the atmosphere? *Science, 325*(5948), 1654–1655.

Kennedy, P., Laplante, B., & Maxwell, J. (1994). Pollution policy: The role for publicly provided information. *Journal of Environmental Economics and Management, 26*, 31–43.

Kennett, J., & Scott, L. (1991). Abrupt deep-sea warming, palaeocanographic changes and benthic extinctions at the end of the Paleocene. *Nature, 353*, 225–229.

Kenny, C. (1999). Does growth cause happiness or does happiness cause growth?'. *Kyklos, 52*, 3–26.

Keynes, J. M. (1936). *The general theory of employment, interest and money*. New York: Harcourt Brace.

Khan, J. (2005). *The economic approach to environmental and natural resources* (3rd ed.). Mason, Ohio: South-Western.

Kahn, J., & Franceschi, D. (2006). Beyond Kyoto: A tax-based system for the global reduction of greenhouse gas emissions. *Ecological Economics, 58*, 778–787.

Kirby, A. (2013, April 3). Least developed countries agree to cut greenhouse gas emissions. *The Guardian*.

Kirch, W., Menne, B., & Bertollini, R. (2005). *Extreme weather events and public health responses*. Heidelberg: Springer.

Klein, R., Nicholls, R., & Thomalla, F. (2002). The resilience of coastal megacities to weather-related hazards. In A. Kreimer, M. Arnold, & A. Carlin (Eds.), *Building safer cities: The future of disaster risk*. Washington, DC: The World Bank Disaster Management Facility, World Bank.

Kleinosky, L., Yarnal, B., & Fisher, A. (2006). Vulnerability of Hampton Roads, Virginia, to storm-surge flooding and sea-level rise. *Natural Hazards, 39*. doi:10.1007/s11069-006-0004-z

Knowlton, B. (2008, November 18). Obama promises action on climate change. *New York Times*. http://www.nytimes.com/2008/11/18/world/americas/18iht-transition.4.17937900.htm. Accessed July 12, 2009.

Knutti, R. Joos, F., Muller, S., Plattner, G.-K., & Stocker, T. (2005). Probabilistic climate change projections for stabilization profiles. *Geophysical Research Letters, 32*, L20707. doi:10.1029/2005GL023294

Kohfeld, K., Le Quéré, C., Harrison, S., & Anderson, R. (2005). Role of marine biology in glacial-interglacial CO_2 cycles. *Science, 308*, 74–78.

Köhl, A., Stammer, D., & Cornuelle, B. (2007). Interannual to decadal changes in the ECCO global synthesis. *Journal of Physical Oceanography, 37*, 313–337.

Köhl, A., & Stammer, D. (2008). Decadal sea level changes in the 50-year GECCO ocean synthesis. *Journal of Climate, 21*, 1876–1889.

Köhler, J., Grubb, M., Popp, D., & Edenhofer, O. (2006). The transition to endogenous technical change in climate-economy models: A technical overview to the innovation modelling comparison project. *Energy Journal, 27*, 17–55.

Korppoo, A. (2003). Forging alliance with Russia: The example of a green investment scheme. *Climate Policy, 103*, 1–10.

Korppoo, A. (2005). Russian energy efficiency projects: Lessons learnt from activities implemented jointly. *Energy Policy, 33*(1), 113–126.

Kosaka, Y., & Xie, S.P. (2013). Recent global-warming hiatus tied to equatorial Pacific surface cooling. *Nature*. doi:10.1038/nature12534

Kovats, S., Campbell-Lendrum, D., & Matthies, F. (2005). Climate change and human health: Estimating avoidable deaths and disease. *Risk Analysis, 25*, 1409–1418.

Krüger, A., Ulbrich, U., & Speth, P. (2002). Groundwater recharge in Northrhine-Westfalia by a statistical model for greenhouse gas scenarios. *Physical Chemistry Earth Part B: Hydrology, Oceans and Atmosphere, 26*, 853–861.

Krupnik, I., & Jolly, D. (Eds.). (2002). *The earth is faster now: Indigenous observations of Arctic environmental change*. Fairbanks, Alaska: Arctic Research Consortium of the United States.

Kubiszewski, I., Costanza, R., Franco, C., Lawn, P., Talberth, J., Jackson, T., & Aylmer, C. (2013). Beyond GDP: Measuring and achieving global genuine progress. *Ecological Economics, 93*, 57–68.

Kuik, O., & Gerlagh, R. (2003). Trade liberalization and carbon leakage. *The Energy Journal, 24*(3), 97–120.

Kunkel, K., & Liang, X. (2005). CGM simulations of the climate in the central United States. *Journal of Climate, 18*, 1016–1031.

Kurtz, A., Kump, L., Arthur, M., Zachos, J., & Paytan, A. (2003). Early Cenozoic decoupling of the global carbon and sulphur cycles. *Paleoceanography, 18*(4), 1090. doi:10.1029/2003PA000908

Kuznets, S. (1974). *Fertility differentials between less developed and developed regions: Components and implications*. Discussion Paper No. 217, New Haven, CT: Yale University Economic Growth Centre.

Laird, D. (2008). The charcoal vision: A win-win-win scenario for simultaneously producing bioenergy, permanently sequestering carbon, while improving soil and water quality. *Agronomy Journal, 100*, 178–181.

Lal, R. (2004). Soil carbon sequestration to mitigate climate change. *Geoderma, 123*(1–2), 1–22.

Lal, R. (2007). Carbon management in agricultural soils. *Mitigation and Adaptation Strategies for Global Climate Change, 12*, 303–322.

Lamb, H. (1982). *Climate, history and the modern world*. London: Routledge.

La Rovere, E., de Macedo, L., & Baumert, K. (2002). The Brazilian proposal on relative responsibility for global warming. In K. Baumert, O. Blanchard, S. Llosa, & J. Perkaus (Eds.), *Building on the Kyoto Protocol: Options for protecting the climate* (pp. 157–174). Washington, DC: World Resources Institute.

Larson, D., Ambrosi, P., Dinar, A., Rahman, S., & Entler, R. (2013). An updated review of carbon markets, institutions, policies, and research. In A. Dinar, D. Larson, & S. Rahman (Eds.), *The clean development mechanism (CDM): An early history of unanticipated outcomes—Volume 1* (Vol. 1, pp. 9–89). Singapore: World Scientific Publishing Co.

Larson, J. (2003, October 9). *Record heat wave in Europe takes 35,000 lives*. Earth Policy Institute. http://www.earth-policy.org

Lasco, R., Habito, C., Delfino, R., Pulhin, F., & Conception, R. (2011). *Climate change adaptation for smallholder farmers in Southeast Asia*. Laguna, Philippines: World Agroforestry Centre.

Laszlo, E. (1972). *The systems view of the world*. New York: Braziller.

Lawn, P. (2000). *Toward sustainable development*. Boca Raton: CRC Press.

Lawn, P. (2003). A theoretical foundation to support the index of sustainable economic welfare (ISEW), genuine progress indicator (GPI), and other related indexes. *Ecological Economics, 44*(1), 105–118.

Lawn, P. (2005). An assessment of the valuation methods used to calculate the Index of Sustainable Economic Welfare (ISEW), Genuine progress indicator (GPI), and sustainable net benefit index (SNBI). *Environment, Development, and Sustainability, 7*(2), 185–208.

Lawn, P. (2006a). Ecological tax reform and the double dividend of ecological sustainability and low unemployment: An empirical assessment. *International Journal of Environment, Workplace, and Employment, 2*(4), 332–358.

Lawn, P. (2006b). Sustainable development: Concept and indicators. In P. Lawn (Ed.), *Sustainable development indicators in ecological economics* (pp. 13–51). UK: Edward Elgar.

Lawn, P. (2007). *Frontier issues in ecological economics*. Cheltenham, UK: Edward Elgar.

Lawn, P. (2008a). Genuine progress in Australia: Time to rethink the growth objective. In P. Lawn & M. Clarke (Eds.), *Genuine progress in the Asia-Pacific: Studies using the genuine progress indicator* (pp. 91–125). Cheltenham, UK: Edward Elgar.

Lawn, P. (2008b). Response to "income, sustainability and the microfoundations of the GPI". *International Journal of Environment, Workplace, and Employment, 4*(1), 59–81.

Lawn, P. (Ed.). (2009). *Environment and employment: A reconciliation*. UK: Routledge.

Lawn, P. (2011). *Wake up economists!—Currency-issuing central governments have no budget constraint*. MPRA Paper No. 28224, Munich Personal RePEc Archive.

Lawn, P. (Ed.). (2013). *Globalisation, economic transition, and the environment*. Cheltenham, UK: Edward Elgar.

Lawn, P., & Clarke, M. (Eds.). (2008). *Sustainable welfare in the Asia-Pacific*. Cheltenham, UK: Edward Elgar.

Lawn, P., & Sanders, R. (1999). Has Australia surpassed its optimal macroeconomic scale? Finding out with the aid of benefit and cost accounts and a sustainable net benefit index. *Ecological Economics, 28*(2), 213–229.

Layard, R. (2005). *Happiness: Lessons from a new science*. London: Allen Lane.

Leemans, R., & Eickhout, B. (2004). Another reason for concern: Regional and global impacts on ecosystems for different levels of climate change. *Global Environmental Change, 14*, 219–228.

Lehmann, J. (2007). A handful of carbon. *Nature, 447*, 143–144.

Leipert, C. (1986). From gross to adjusted national product. In P. Ekins (Ed.), *The living economy: A new economics in the making* (pp. 132–140). London: Routledge & Kegan Paul.

Lenzen, M., & Murray, S. (2001). A modified ecological footprint method and its application to Australia. *Ecological Economics, 37*(2), 229–255.

Leonard, H. (1988). *Pollution and the struggle for the world product: Multinational corporations, environment, and international comparative advantage*. Cambridge, UK: Cambridge University Press.

Le Quéré, C., Rödenbeck, C., Buitenhuis, E., Conway, T., Langenfelds, R., Gomez, A., et al. (2007). Saturation of the Southern ocean CO_2 sink due to recent climate change. *Science, 316*, 1735–1738. doi:10.1126/science.1136188

Lesnikowski, A., Ford, J., Berrang-Ford, L., Barerra, M., Berry, P., Henderson, J., & Heymann, S. (2013). National-level factors affecting planned, public adaptation to health impacts of climate change. *Global Environmental Change, 23*(5), 1153–1163.

Levitus, S., Antonov, J., Boyer, T., & Stephens, C. (2000). Warming of the world ocean. *Science, 287*, 2225–2229.

Levitus, S., Antonov, J., Boyer, T., Locarnini, R., Garcia, H., & Mishonov, A. (2009). Global ocean heat content 1955-2008 in light of recently revealed instrumentation problems. *Geophysical Research Letters, 36*(7), 2009. doi:10.1029/2008GL037155

Li, C., Fan, D., Deng, B., & Korotaev, V. (2004). The coasts of China and issues of sea level rise. *Journal of Coastal Research, 43*, 36–47.

Lisiecki, L., & Raymo, M. (2005). A Pliocene-Pleistocene stack of 57 globally distributed benthic d18O records. *Paleoceanography, 20*, PA1003. doi:10.1029/2004PA001071

Liverman, D. (2009). Conventions of climate change: Constructions of danger and the dispossession of the atmosphere. *Journal of Historical Geography, 35*(2), 279–296.

Lohmann, L. (2006, September). Carbon trading: A critical conversation on climate change, privatization and power. *Development Dialogue, 48*.

Lord, C. (2003). *A citizens income: A foundation for a sustainable world*. Concord: John Carpenter Publishing.

Lovelock, J. (1988). *Ages of Gaia: A biography of our living planet*. New York: Norton & Co.

Lovelock, J. (2006). *The revenge of Gaia*. London: Allen Lane.

Lu, X., & Lu, X. (2003). Climate tendency and analysis of warming and drying in grassland of Northeast Qingzang Plateau of China. *Grassland of China, 24*, 8–13.

Luo, Q., Williams, M., Bellotti, W., & Bryan, B. (2003). Quantitative and visual assessments of climate change impacts on South Australian wheat production. *Agricultural Systems, 77*, 173–186.

Lüthi, D., Le Floch, M., Bereiter, B., Blunier, T., Barnola, J.-M., Siegenthaler, U., et al. (2008). High-resolution carbon dioxide concentration record 650,000-800,000 years before present. *Nature, 453*, 379–382.

MacCracken, C., Edmonds, J., Kim, S., & Sands, R. (1999). *The economics of the Kyoto Protocol* (pp. 25–71). Special Kyoto Issue: Energy Journal.

MacKay, D. (2009). *Sustainable energy—Without the hot air*. Cambridge: UIT.

Maddison, D. (1995). A cost-benefit analysis of slowing climate change. *Energy Policy, 23*(4/5), 337–346.

Maeda, A. (2003). The emergence of market power in emissions rights markets: The role of initial permit distribution. *Journal of Regulatory Economics, 24*(3), 293, 314.

Makino, M. (2008). Genuine progress in Japan and the need for an open economy GPI. In P. Lawn & M. Clarke (Eds.), *Genuine progress in the Asia-Pacific: Studies using the genuine progress indicator* (pp. 153–190). Cheltenham, UK: Edward Elgar.

Malcolm, J., Liu, C., Neilson, R., Hansen, L., & Hannah, L. (2006). Global warming and extinctions of endemic species from biodiversity hotspots. *Conservation Biology, 20*, 538–548.

Manabe, S., & Wetherald, R. (1975). The effects of doubling the CO_2 concentration on the climate of a general circulation model. *Journal of Atmospheric Sciences, 32*, 3–15.

Mangalassery, S., Sjögersten, S., Sparkes, D., Sturrock, C., Craigon, J., & Mooney, S. (2014). To what extent can zero tillage lead to a reduction in greenhouse gas emissions from temperate soils? *Scientific Reports*. *4*, 4586. doi:10.1038/srep04586

Mann, M., Bradley, R., & Hughes, M. (1998). Global-scale temperature patterns and climate forcing over the past six centuries. *Nature, 378*, 266–270.

Manne, A., & Richels, R. (2001). *US rejection of the Kyoto Protocol: The impact on compliance costs and CO_2 emissions*. AEI-Brookings Joint Center for Regulatory Studies, Working Paper October 01–12, 2001.

Maracchi, G., Sirotenko, O., & Bindi, M. (2005). Impacts of present and future climate variability on agriculture and forestry in the temperate regions: Europe. *Climatic Change, 70*, 117–135.

Markelova, K. (2011, January–March). Synthetic tress: A meeting with Klaus Lackner. *The UNESCO Courier, 33*.

Marland, G., Boden, T., & Andres, R. (2006). Global, regional, and national CO_2 emissions. In *Trends: A compendium of data on global change*. Carbon Dioxide Information Analysis Center, Oak Ridge National Laboratory, US Department of Energy, Oak Ridge, TN. http://cdiac.esd.ornl.gov/trends/emis/tre_glob.htm

Marongwe, L., Kwazira, K., Jenrich, M., Thierfelder, C., Kassam, A., & Friedrich, T. (2011). An African success: The case of conservation agriculture in Zimbabwe. *International Journal of Agricultural Sustainability, 9*(1), 153–161.

Martinez, K., & Neuhoff, K. (2005). Allocation of carbon emission certificates in the power sector: How generators profit from grandfathered rights. *Climate Policy, 5*(1), 61–78.

Martinez-Alier, J. (1987). *Ecological economics: Energy*. Blackwell, Oxford: Environment and Society.

Martinez-Alier, J. (2009). Socially sustainable economic de-growth. *Development and Change, 40*(6), 1099–1119.

Martinez-Alier, J., Munda, G., & O'Neill, J. (1998). Weak comparability of values as a foundation for ecological economics. *Ecological Economics, 26*(3), 277–286.

Martinez-Alier, J., & Røpke, I. (Eds.). (2008). *Recent developments in ecological economics*. UK: Edward Elgar.

Maslow, A. (1954). *Motivation and personality*. New York: Harper & Row.

Matthew, R. (2011). Is climate change a national security issue? *Issues in Science & Technology, 27*, 9–60.

Matthews, H., Weaver, A., Meissner, K., Gillett, N., & Eby, M. (2004). Natural and anthropogenic climate change: Incorporating historical land cover change, vegetation dynamics and the global carbon cycle. *Climate Dynamics, 22*, 461–479.

Maunder, W. (1992). *Dictionary of global climate change*. London: Kluwer, UCL Press.

Max-Neef, M. (1995). Economic growth and quality of life. *Ecological Economics, 15*(2), 115–118.

McElwain, J., & Chaloner, W. (1995). Stomatal density and index of fossil plants track atmospheric carbon dioxide in the Palaeocene. *Annuals of Botany, 76*, 389–395.

McGuire, W. (2006). Expert reaction to the Stern Review. BBC, 30 October 2006, Benfield UCL Hazard Research Centre.

McKibbin, W., & Wilcoxen, P. (2002). The role of economics in climate change policy. *Journal of Economic Perspectives, 16*(2), 107–129.

McMichael, A., Woodruff, R., Whetton, P., Hennessy, K., Nicholls, N., Hales, S., et al. (2003). *Human health and climate change in Oceania: A risk assessment 2002*. Canberra: Commonwealth Department of Health and Ageing.

McMichael, A., Campbell-Lendrum, D., Kovats, S., Edwards, S., Wilkinson, P., Wilson, T., et al. (2004). Global climate change. In M. Ezzati, A. Lopez, A. Rodgers, & C. Murray (Eds.), *Comparative quantification of health risks: Global and regional burden of disease due to selected major risk factors* (pp. 1543–1649). Geneva: World Health Organization.

Meade, J. (1952). External economies and diseconomies in a competitive situation. *Economic Journal, 52*, 54–67.

Meadows, D. H., Meadows, D. L., Randers, J., & Brehens, W, I. I. I. (Eds.). (1972). *The limits to growth*. New York: Universe Books.

Measham, T., Preston, B., Smith, T., Brooke, C., Gorddard, R., Withycombe, G., & Morrison, C. (2011). Adapting to climate change through local municipal planning: Barriers and challenges. *Mitigation and Adaptation Strategies for Global Change, 16*(8), 889–909.

Meehl, G., Hu, A., Arblaster, J., Fasullo, J., & Trenberth, K. (2013). Externally forced and internally generated decadal climate variability associated with the Interdecadal Pacific Oscillation. 10.1175/JCLI-D-12-00548.1

Megginson, W., Nash, R., & Van Randenborgh, M. (1994). The financial and operating performance of newly privatised firms: An international empirical analysis. *Journal of Finance, 49*(2), 403–452.

Mehra, R., & Prescott, E. (1985). The equity premium: A puzzle. *Journal of Monetary Economics, 15*(2), 145–161.

Meier, H., Broman, B., & Kjellström, E. (2004). Simulated sea level in past and future climates of the Baltic Sea. *Climate Research, 27*, 59–75.

Meinshausen, M. (2006). What does a 2 °C target mean for greenhouse gas concentrations? A brief analysis based on multi-gas pathways and several climate sensitivity uncertainty estimates. In H. Schellnhuber, W. Cramer, N. Nakicenovic, T. Wigley, & G. Yohe (Eds.), *Avoiding dangerous climate change* (pp. 265–279). Cambridge, UK: Cambridge University Press.

Meinshausen, M., Meinshausen, N., Hare, W., Raper, S., Frieler, K., Knutti, R., et al. (2009). Greenhouse-gas emission targets for limiting global warming to 2 °C. *Nature, 458*(30), 1158–1163.

Menanteau, P., Finon, D., & Lamy, M.-L. (2003). Prices versus quantities: Choosing policies for promoting the development of renewable energy. *Energy Policy, 31*(8), 799–812.

Mendelsohn, R. (2007). A critique of the Stern Report. *Regulation, 29*(4), 40–47.

Mendelsohn, R. (2008). Is the *Stern Review* an economic analysis? *Review of Environmental Economics and Policy, 2*(1), 45–60.

Mendelsohn, R., Dinar, A., & Dalfelt, L. (2000a). *Climate change impacts on African agriculture. Preliminary analysis prepared for the World Bank.* Washington, DC: World Bank.

Mendelsohn, R., Morrison, W., Schlesinger, M., & Andronova, N. (2000b). Country-specific market impacts of climate change. *Climatic Change, 45*(3–4), 553–569.

Menzel, L., & Bürger, G. (2002). Climate change scenarios and runoff response in the Mulde catchment (Southern Elbe, Germany). *Journal of Hydrology, 267*, 53–64.

Merritt, W., Alila, Y., Barton, M., Taylor, B., Cohen, S., & Neilsen, D. (2005). Hydrologic response to scenarios of climate change in sub-watersheds of the Okanagan basin, British Columbia. *Journal of Hydrology, 326*, 79–108.

Metz, B., & van Vuuren, D. (2006). How, and at what costs, can low-level stabilization be achieved—An overview. In H. Schellnhuber, W. Cramer, N. Nakicenovic, T. Wigley, & G. Yohe (Eds.), *Avoiding dangerous climate change* (pp. 337–345). Cambridge, UK: Cambridge University Press.

Meyer, A. (2000). *Contraction and convergence: The global solution to climate change*. London: Greenbooks.

Michaelowa, A., Butzengeiger, S., & Jung, M. (2005a). Graduation and deepening: an ambitious post-2012 climate policy scenario. *International Environmental Agreements, 5*, 25–46.

Michaelowa, A., Tangen, K., & Hasselknippe, H. (2005b). Issues and options for the post-2012 climate architecture—An overview. *International Environmental Agreements, 5*, 5–24.

Michaelowa, A., & Michaelowa, K. (2007). Climate or development: is ODA diverted from its original purpose? *Climate Change, 84*, 5–21.

Miles, L., Grainger, A., & Phillips, O. (2004). The impact of global climate change on tropical forest biodiversity in Amazonia. *Global Ecological Biogeography, 13*, 553–565.

Ministry for the Environment (MfE). (2004). *Coastal hazards and climate change: A guidance manual for local government in New Zealand*. Wellington: Ministry for the Environment.

Mitchell, R. (2005). Flexibility, compliance and norm development in the climate regime. In O. Stokke, J. Ovi, & G. Ulfstein (Eds.), *Implementing the climate regime: International compliance* (pp. 65–83). London: Earthscan Press.

Mitchell, W. (1998). The buffer stock employment model. *Journal of Economic Issues, 32*(2), 547–555.

Mitchell, W. (2000). The job guarantee and inflation control. *Centre of Full Employment and Equity Working Paper* 00-01, January 2000.

Mitchell, W., & Mosler, W. (2002). The imperative of fiscal policy for full employment. *Australian Journal of Labour Economics, 5*(2), 243–259.

Mitchell, W., & Wray, R. (2005). In defence of the employer of last resort. *Journal of Economic Issues, 39*(1), 235–245.

Mitchell, W., & Muysken, J. (2008). *Full employment abandoned: Shifting sands and policy failures*. Cheltenham, UK: Edward Elgar.

Mizerny, J. (2011). Recent changes for joint implementation in Poland—Better late than never. *Joint Implementation Quarterly, 17*(2), 5–7.

Moffat, I., & Wilson, M. (1994). An index of sustainable economic welfare for Scotland, 1980–1991. *International Journal of Sustainable Development and World Ecology, 1*, 264–291.

Möhner, A., & Klein, R. (2007). *The global environment facility: Funding for adaptation or adapting to funds?* Working Paper, Stockholm Environment Institute, Stockholm.

Monbiot, G. (2009). Requiem for a crowded planet. http://www.monbiot.com/2009/12/21/requiem-for-a-crowded-planet/

Montero, J.-P. (2005). Pollution markets with imperfectly observed emissions. *RAND Journal of Economics, 36*(3), 645–660.

Moran, D. (Ed.). (2012). *Climate change and national security: A country-level analysis*. Washington, DC: Georgetown University Press.

Moss, R., Malone, E., Engle, N., de Bremond, A., & Delgado, A. (2012). *Ready or not: Towards a resilience framework for making climate-change adaptation decisions*. Report prepared for the World Bank by the Joint Global Change Research Institute.

Mote, P., Hamlet, A., Clark, M., & Lettenmaier, D. (2005). Declining mountain snowpack in western North America. *Bulletin of the American Meteorological Society, 86*, 39–49.

Mulder, P., & van den Berg, J. (2001). Evolutionary economic theories of sustainable development. *Growth and Change, 32*, 110–134.

Mullan, M., Kingsmill, N., Kramer, A., & Agrawala, S. (2013). *National adaptation planning: Lessons from OECD countries*. OECD Environment Working Papers, No. 54, ENV/WKP(2013)1. Paris: OECD.

Munasinghe, M., & Swart, R. (2005). *Primer on climate change and sustainable development*. Cambridge, UK: Cambridge University Press.

Murase, S. (2002). Conflict of international regimes: Trade and the environment. *Thesaurus Acroasium*, XXXI, Institute of International Public Law and International Relations of Thessaloniki, 297–340.

Murdiyarso, D. (2000). Adaptation to climate variability and change: Asian perspectives on agriculture and food security. *Environmental Monitoring Assessments, 61*, 123–131.

Murphy, J., Solomon, S., Portmann, R., Rosenlof, K., Forster, P., & Wong, T. (2009). An observationally based energy balance for the Earth since 1950. *Journal of Geophysical Research*, 114, D17107. doi:10.1029/2009JD012105

Nakicenovic, N., McGalde, J., Ma, S., Alcamo, J., Bennett, E., Cramer, et al. (2005). Lessons learned from scenario analysis. In S. Carpenter, P. Pingali, E. Bennett, & M. Zurek (Eds.), *Ecosystems and human well-being: Scenarios* (Vol. II, pp. 449–468). Chicago: Island Press.

Nakicenovic, N., Kolp, P., Riahi, K., Kainuma, M., & Hanaoka, T. (2006). Assessment of emissions scenarios revisited. *Environmental Economics and Policy Studies, 7*(3), 137–173.

National Academy of Sciences (NAS). (1979). *Carbon dioxide and climate: A scientific assessment*. Washington, DC.

National Commission on Population. (2000). *National population policy 2000*. New Delhi: Government of India.

National Oceanic and Atmospheric Administration (NOAA). (2013). *Mauna Loa CO_2 monthly mean data*. Earth Systems Research Laboratory. ftp://ftp.cmdl.noaa.gov/ccg/co2/trends/co2_mm_mlo.txt. Accessed September 19, 2013.

National Research Council. (2005). *Radiative forcing of climate change: Expanding the concept and addressing uncertainties*. Washington, DC: National Academy Press.

Nature Code. (2014). *Views on suggested changes to the Modalities and Procedures (M&Ps) for the Clean Development Mechanism (CDM)*. http://carbonmarketwatch.org/wp-content/uploads/2014/04/CMW_Submission_CDMMPs_April2014_FINAL.pdf. Accessed September 21, 2014.

Næss, L., Bang, G., Eriksen, S., & Vevatne, J. (2005). Institutional adaptation to climate change: Flood responses at the municipal level in Norway. *Global Environmental Change, 15*(2), 125–138.

Nauclér, T., & Enkvist, P. (2009). *Pathways to a low-carbon economy—Version 2 of the global greenhouse gas abatement cost curve*. McKinsey and Company.

Neelin, J., Münnich, M., Meyerson, J., & Holloway, C. (2006). Tropical drying trends in global warming models and observations. *Proceedings of the National Academy of Sciences, USA, 103*, 6110–6115.

Nelson, R., Kokic, P., Crimp, S., Martin, P., Meinke, H., & Howden, S. (2010). The vulnerability of Australian rural communities to climate variability and change: Part II—Integrating impacts with adaptive capacity. *Environmental Science & Policy, 13*(1), 18–27.

Nemet, G., & Kammen, D. (2007). U.S. energy research and development: Declining investment, increasing need, and the feasibility of expansion. *Energy Policy, 35*, 746–755.

Nentjes, A., & Klaassen, G. (2004). On the quality of compliance mechanisms in the Kyoto Protocol. *Energy Policy, 32*, 531–544.

Neumayer, E. (1999a). The ISEW—Not an index of sustainable economic welfare. *Social Indicators Research, 48*, 77–101.

Neumayer, E. (1999b). Global warming: Discounting is not the issue, but substitutability is. *Energy Policy, 27*, 33–43.

Neumayer, E. (2000). On the methodology of the ISEW, GPI, and related measures: Some constructive suggestions and some doubt on the threshold hypothesis. *Ecological Economics, 34*, 347–361.

Neumayer, E. (2003). *Weak versus strong sustainability: Exploring the limits of two opposing paradigms* (2nd ed.). Cheltenham and Northampton: Edward Elgar.

Neumayer, E. (2007). A missed opportunity: The Stern review on climate change fails to tackle the issue of the non-substitutable loss of natural capital. *Global Environmental Change, 17*, 297–301.

Nicholls, R. (2004). Coastal flooding and wetland loss in the 21st century: Changes under the CRES climate and socio-economic scenarios. *Global Environmental Change, 14*, 69–86.

Nkomo, J., Nyong, A., & Kulindwa, K. (2006). The impacts of climate change in Africa. Report prepared for the Stern Review on the Economics of Climate Change. http://www.hm-treasury.gov.uk/media/8AD/9E/Chapter_5_The_Impacts_of_Climate_Change_in_Africa-5.pdf

Nobre, C., Assad, E., & Oyama, M. (2005). O impacto do aquecimento global nos ecosistemas da Amazônia e na agricultura. *Scientific American Brasil, 5*, 70–75.

Nordhaus, W. (1991). To slow or not to slow: The economics of the greenhouse effect. *The Economic Journal, 101*, 920–937.

Nordhaus, W. (1993). Rolling the "DICE": An optimal transition path for controlling greenhouse gases. *Resource and Energy Economics, 15*, 27–50.

Nordhaus, W. (2005). Life after Kyoto: Alternative approaches to global warming. *NBER Working Papers*, 11889, National Bureau of Economic Research.

Nordhaus, W. (2007a). A review of the Stern review on the economics of climate change. *Journal of Economic Literature*, XLV(3), 686–702.

Nordhaus, W. (2007b). To tax or not to tax: Alternative approaches to slowing global warming. *Review of Environmental Economics and Policy, 1*(1), 26–44.

Nordhaus, W. (2007c). Critical assumptions in the Stern Review on climate change. *Science, 317*, 201–202.

Nordhaus, W., & Yang, Z. (1996). RICE: A regional dynamic general equilibrium model of optimal climate-change policy. *American Economic Review, 86*(4), 741–765.

Nordhaus, W., & Boyer, J. (2000). *Warming the world: Economic modeling of global warming*. Cambridge, MA: MIT Press.

Norgaard, R. (1985). Environmental economics: An evolutionary critique and a plea for pluralism. *Journal of Environmental Economics and Management, 12*, 382–394.

Norgaard, R. (1988). Sustainable development: A coevolutionary view. *Futures*, December, 606–620.

Norgaard, R. (1990). Economic indicators of resource scarcity: A critical essay. *Journal of Environmental Economics and Management, 19*(1), 19–25.

Norgaard, R. (1994). *Development betrayed: The end of progress and a coevolutionary revisioning*. New York: Routledge.

Norris, K. (1989). *The economics of Australian labour markets* (2nd ed.). Melbourne: Longman Cheshire.

Norton, B. (1986). On the inherent danger of undervaluing species. In B. Norton (Ed.), *The preservation of species* (pp. 110–137). Princeton, NJ: Princeton University Press.

Nusbaumer, J., & Matsumoto, K. (2008). Climate and carbon-cycle changes under the overshoot scenario. *Global Planetary Change, 62*, 164–172.

Nuttall, M., Berkes, F., Forbes, B., Kofinas, G., Vlassova, T., & Wenzel, G. (2005). Hunting, herding, fishing, and gathering: Indigenous peoples and renewable resource use in the Arctic. In C. Symon, L. Arris, & B. Heal (Eds.), *Arctic climate impact assessment* (pp. 649–690). Cambridge: Cambridge University Press.

O'Brien, K., Eriksen, S., Sygna, L., & Naess, L. (2006). Questioning complacency: Climate change impacts, vulnerability, and adaptation in Norway. *Ambio, 35*, 50–56.

O'Brien, S., Mayewski, P., Meeker, L., Meese, D., Twickler, M., & Whitlow, S. (1995). Complexity of Holocene climate as constructed from a Greenland ice core. *Science, 270*, 1962–1964.

O'Connor, M., & Lines, W. (2008). *Overloading Australia: How governments and media dither and deny on population*. NSW: Envirobook.

Olesen, J., Carter, T., Díaz-Ambrona, C., Fronzek, S., Heidmann, T., Hickler, T., et al. (2007). Uncertainties in projected impacts of climate change on European agriculture and terrestrial ecosystems based on scenarios from regional climate models. *Climatic Change, 81*, S123–S143.

Olesen, J., Tranka, M., Kersebaum, K., Skjelvag, A., Seguin, B., Peltonen-Sainio, P., et al. (2011). Impacts and adaptation of European crop production systems to climate change. *European Journal of Agronomy, 34*, 96–112.

Olivier, J., Janssens-Maenhout, G., Peters, J., & Wilson, J. (2011). *Long-term trend in global CO_2 emissions: 2011 report*. PBL Netherlands Environmental Assessment Agency; Institute for Environment and Sustainability (IES) of the European Commission's Joint Research Centre (JCR), The Hague, Netherlands.

Olsen, K. (2007). The clean development mechanism's contribution to sustainable development: A review of the literature. *Climate Change, 84*, 59–73.

Olsen, K., & Fenhann, J. (2008). Sustainable development benefits of Clean Development Mechanism projects: A new methodology for sustainability assessment based on text analysis of the project design documents submitted for validation. *Energy Policy, 36*(8), 2819–2830.

Olson, M. (1965). *The logic of collective action*. Cambridge, MA: Harvard University Press.

O'Neill, B., & Oppenheimer, M. (2004). Climate change impacts sensitive to path to stabilization. *Proceedings of the National Academy of Sciences, USA, 101*, 16411–16416.

Onyx, J., & Bullen, P. (1998). *Measuring social capital in five communities in NSW: An analysis*. Centre for Australian Community Organisations and Management, Working Paper Series No. 41, CACOM, University of Technology, Sydney.

Organisation for Economic Co-operation and Development (OECD), Resources for the Future, and Intergovernmental Panel on Climate Change (IPCC). (2000). *Ancillary benefits and costs of greenhouse-gas mitigation: Workshop on assessing the ancillary benefits and costs of greenhouse-gas mitigation strategies*. Washington, DC: OECD.

Organisation for Economic Co-operation and Development (OECD). (2004). *Sustainable development in OECD countries: Getting the policies right*. Paris: OECD.

Organisation for Economic Co-operation and Development (OECD). (2005/2008). *The paris declaration on aid effectiveness and the accra agenda for action*. Paris: OECD.

Organisation for Economic Co-operation and Development (OECD). (2008). *Climate change mitigation: What do we do?*. Paris: OECD.

Organisation for Economic Co-operation and Development (OECD). *OECD development statistics online*. www.oecd.org/statistics. Accessed October 2, 2009.

Orr, I. (2008). China and the centrality of ecology. *Edinburgh Review*, Autumn 2008 (124).

Ostrom, E. (1990). *Governing the commons: The evolution of institutions for collective action*. Cambridge, MA: Cambridge University Press.

Pacala, S., & Socolow, R. (2004). Stabilization wedges: Solving the climate problem for the next 50 years with current technologies. *Science, 305*(5686), 968–972.

Pagani, M., Zachos, J., Freeman, K., Tipple, B., & Bohaty, S. (2005). Marked decline in atmospheric carbon dioxide concentrations during the Paleogene. *Science, 309*(5734), 600–603.

Pagiola, S., Bishop, J., & Landell-Mills, N. (2002). *Selling forest environmental services: Market-based mechanisms for conservation and development*. London: Earthscan.

Pakrasi, K., & Halder, A. (1981). Fertility in contemporary Calcutta: A biosocial profile. *Genus, 37*(3/4), 201–219.

Paltsev, S. (2001). The Kyoto Protocol: Regional and sectoral contributions to the carbon leakage. *The Energy Journal, 22*(4), 53–79.

Panagariya, A. (2002 April, 24). Stamping in nutrition. *Economic Times*.

Parr, J., & Sullivan, L. (2005). Soil carbon sequestration in phytoliths. *Soil Biology and Biochemistry, 37*, 117–124.

Parry, I. (2003). *Are emissions permits regressive?* Discussion Paper 03-21, June 2003. Washington, DC: Resources for the Future.

Parry, I., Sigman, H., Wallis, M., & Williams, R. (2005a). *The incidence of pollution control policies*. Discussion Paper 05-24, June 2005. Washington, DC: Resources for the Future.

Parry, J.-E., Hammill, A., & Drexhage, J. (2005b). *Climate change and adaptation: A summary*. Winnipeg: International Institute for Sustainable Development (IISD).

Parry, M., Rosenzweig, C., Iglesias, A., Fischer, G., & Livermore, M. (1999). Climate change and world food security: A new assessment. *Global Environmental Change, 9*, 51–67.

Parry, M., Rosenzweig, C., Iglesias, A., Livermore, M., & Fischer, G. (2004). Effects of climate change on global food production under CRES emissions and socio-economic scenarios. *Global Environmental Change, 14*, 53–67.

Parry, M. Arnell, N., Berry, P., Dodman, D., Fankhauser, S., Hope, C., et al. (2009). *Assessing the costs of adaptation to climate change: A review of the UNFCCC and other recent estimates*. International Institute for Environment and Development (IIED) and the Grantham Institute for Climate Change, Imperial College London, London.

Patt, A., & Schröter, D. (2008). Perceptions of climate risk in Mozambique: Implications for the success of adaptation strategies. *Global Environmental Change, 18*(3), 458–467.

Patterson, M. (2006). Selecting headline indicators for tracking progress to sustainability in a nation. In P. Lawn (Ed.), *Sustainable development indicators in ecological economics* (pp. 421–448). Cheltenham, UK: Edward Elgar.

Paulson, H., Darling, A., & Nukaga, F. (2008). Financial bridge from dirty to clean energy. *Financial Times*, February 7, 2008.

Pearce, D. (1991). The role of carbon taxes in adjusting to global warming. *The Economic Journal, 101*, 938–948.

Pearce, D., Markandya, A., & Barbier, E. (1989). *Blueprint for a green economy*. London: Earthscan.

Pearce, D., & Turner, R. (1990). *The economics of natural resources and the environment*. London: Harvester Wheatsheaf.

Pearce, D., & Atkinson, G. (1993). Capital theory and the measurement of sustainable development: An indicator of weak sustainability. *Ecological Economics, 8*, 103–108.

Pearce, D., & Warford, J. (1993). *World without end: Economics, environment, and sustainable development*. New York: Oxford University Press.

Pearce, D., Cline, W., & Achanta, A. (1996). The social costs of climate change: Greenhouse damage and the benefits of control. In J. Bruce, H. Lee, & E. Haites (Eds.), *Climate change 1995: Economic and social dimensions of climate change* (pp. 183–224). Cambridge: Cambridge University Press.

Pearman, G. (2008). *Climate change: Risk in Australia under alternative emissions scenarios*. Graeme Pearman Consulting Pty Ltd, Victoria: Report prepared for the Australian Government Department of Treasury.

Pearson, P., & Palmer, M. (2000). Atmospheric carbon dioxide concentrations over the past 60 million years. *Nature, 406*, 695–669.

Peck, S., & Teisberg, T. (1992). CETA: A model for carbon emissions trajectory assessment. *Energy Journal, 13*(1), 55–77.

Peck, S., & Teisberg, T. (1994). Optimal carbon emissions trajectories when damages depend on the rate or level of global warming. *Climate Change, 28*, 289–314.

Perman, R., Ma, Y., McGilvray, J., & Common, M. (2003). *Natural resource and environmental economics* (3rd ed.). Harlow: Pearson.

Perrings, C. (1986). Conservation of mass and instability in a dynamic economy-environment system. *Journal of Environmental Economics and Management, 13*, 199–211.

Perrings, C. (2003). The economics of abrupt climate change. *Philosophical Transactions of the Royal Society of London, Series A: Mathematical and Physical Sciences, 361*(1810), 2043–2057.

Persson, T., Azar, C., & Lindgren, K. (2006). Allocation of CO_2 permits—Economic incentives for emission reductions in developing countries. *Energy Policy, 34*, 1889–1899.

Peterson, A., Ortega-Huerta, M., Bartley, J., Sánchez-Cordero, V., Soberon, J., Buddemeir, R., & Stockwell, D. (2002). Future projections for Mexican fauna under climate change scenarios. *Nature, 416*, 626–629.

Peterson, W. (1975). *Population*. New York: Macmillan & Co.

Petit, J., Jouzel, J., Raynaud, D., Barkov, N., Barnola, J.-M., Basile, I., et al. (1999). Climate and atmospheric history of the past 420,000 years from the Vostok ice core, Antarctica. *Nature, 399*, 429–436.

Pfister, S., Bayer, P., Koehler, A., & Hellweg, S. (2011). Environmental impacts of water use in global crop production: Hotspots and trade-offs with land use. *Environmental Science and Technology, 45*(13), 5761–5768.

Pigou, A. (1932). *The economics of welfare*. London: Macmillan & Co.

Pitchford, J. (1990). *Australia's foreign debt: Myths and realities*. Sydney: Allen & Unwin.

Plass, G. (1956). Effect of cabon dioxide variations on climate. *American Journal of Physics, 24*, 376–387.

Plimer, I. (2009). *Heaven+Earth—Global warming: The missing science*. Ballan: Connor Court Publishing.

Polimeni, J. (2008). Empirical evidence for the Jevons Paradox. In J. Polimeni, K. Mayumi, M. Giampoetro, & B. Alcott (Eds.), *The Jevons paradox and the myth of resource efficiency improvements*. London: Earthscan.

Polsky, C., & Easterling, W, I. I. I. (2001). Adaptation to climate variability and change in the US Great Plains: A multi-scale analysis of Ricardian climate sensitivities. *Agriculture, Ecosystems and Environment, 85*, 133–144.

Popp, D. (2004). *R&D subsidies and climate policy: Is there a free lunch*? National Bureau of Economic Research, Working Paper No. 11415, Cambridge, MA.

Post, W., & Kwon, K. (2000). Soil carbon sequestration and land-use change: Process and potential. *Global Change Biology, 6*, 317–327.

Postel, S., & Richter, B. (2003). *Rivers for life: Managing water for people and nature*. Washington, DC: Island Press.

Pounds, J., Fogden, M., & Campbell, J. (1999). Biological response to climate change on a tropical mountain. *Nature, 398*, 611–615.

Preston, B., Westaway, R., Dessai, S., & Smith, T. (2009). *Are we adapting to climate change? Research and methods for evaluating progress*. In 89th American meteorological society annual meeting: Fourth symposium on policy and socio-economic research, Phoenix, AZ.

Pulwarty, R., Jocobs, K., & Dole, R. (2005). The hardest working river: Drought and critical water problems in the Colorado River basin. In D. Wilhite (Ed.), *Drought and water crisis—Science, technology and management issues* (pp. 249–286). Boca Raton: CRC Press.

Putman, R. (1993). *Making democracy work: Civic traditions in modern Italy*. Princeton: Princeton University Press.

Putman, R. (2000). *Bowling alone: The collapse and revival of American Community*. New York: Simon and Schuster.

Qin, D. (2002). Assessment of environment change in Western China. In *Prediction of environment change in Western China* (pp. 64–161). Beijing: Science Press.

Quiggin, J. (2002). Privatisation and nationalisation in the 21st century. *Growth, 50*, 66–73.

Quiggin, J. (2008). Stern and his critics on discounting and climate change: An editorial essay. *Climate Change, 89*, 195–205.

Ramanathan, V. (1975). Greenhouse effect due to chlorofluorocarbons: Climatic implications. *Science, 190*, 50–52.

Ramanathan, V., Cicerone, R., Singh, H., & Kiehl, J. (1985). Trace gas trends and their potential role in climate change. *Journal of Geophysical Research: Atmospheres, 90*, 5547–5566.

Ramsey, F. (1928). A mathematical theory of saving. *The Economic Journal, 38*, 543–559.

Ramírez, P. & Brenes, A. (2001). *Informe sobre las condiciones de sequía observadas en el istmo centroamericano en el 2001*. Comité Regional Recursos Hidráulicos/Sistema de la Integración Centroamericana.

Rasch, P., Tilmes, S., Turco, R., Robock, A., Oman, L., Chen, C.-C., et al. (2008). An overview of geoengineering of climate using stratospheric sulphate aerosols. *Philosophical Transactions of the Royal Society A, 366*, 4007–4037.

Ratnayake, R. & Wydeveld, M. (1998). *The multinational corporation and the environment: Testing the pollution haven hypothesis*. University of Auckland, Department of Economics Working Paper No. 179.

Raupach, M., Marland, G., Ciais, P., Le Quéré, C., Canadell, J., Klepper, G., et al. (2007). Global and regional drivers of accelerating CO_2 emissions. *Proceedings of the National Academy of Sciences, USA*, 104, 18 288-18 293 (doi: 10.1073/pnas.0700609104)

Raymo, M., & Rau, G. (1992). Plio-Pleistocene atmospheric CO_2 levels inferred from POMδ^{13}C at DSDP Site 607. *Eos, 73*, 95.

Raymo, M., Grant, B., Horowitz, M., & Rau, G. (1996). Mid-Pliocene warmth: Stronger greenhouse and stronger conveyor. *Marine Micropalaeontology, 27*(1–4), 313–326.

Redefining Progress. (1995). Gross production versus genuine progress. *Excerpt from the genuine progress indicator: Summary of data and methodology*. San Francisco: Redefining Progress.

Rees, W. (2013). Carrying capacity, globalisation, and the unsustainable entanglement of nations. In P. Lawn (Ed.), *Globalisation, economic transition, and the environment* (pp. 50–65). UK: Edward Elgar.

Reist, J., Wrona, F., Prowse, T., Power, M., Dempson, J., Beamish, R., et al. (2006). General effects of climate change on arctic fishes and fish populations. *Ambio, 35*, 370–380.

REN21. (2009). *Renewables 2009: Global Status Report*. Paris, France.

REN21. (2011). *Renewables 2011: Global Status Report*. Paris, France.

REN21. (2014). *Renewables 2014: Global Status Report*. Paris, France.

Retallack, S. (2005). *Setting a long-term climate objective*. A paper for the International Climate Change Taskforce, Institute for Public Policy Research, London.

Revelle, R., & Suess, H. (1957). Carbon dioxide exchange between atmosphere and ocean and the question of an increase in atmospheric CO_2 during the past decades. *Tellus, 9*, 18–27.

Reynolds, D. (1999). The mineral economy: How prices and costs can falsely signal decreasing scarcity. *Ecological Economics, 31*(1), 155–166.

Riahi, K., & Roehrl, R. (2001). *Energy technology strategies for carbon dioxide mitigation and sustainable development*. Laxenburg: International Institute for Applied Systems Analysis.

Riahi, K., Grübler, A., & Nakicenovic, N. (2006). Scenarios of long-term socio-economic and social environmental development under climate stabilization. *Technological Forecasting and Social Change*, 887–935.

Ribes, J., & Nesme-Ribes, E. (1993). The solar sunspot cycle in the Maunder Minimum Ad 1645 to AD 1715. *Astronomy and Astrophysics, 276*, 549–563.

Ricardo, D. (1817). *Principles of political economy and taxation*. Cambridge, UK: Sraffa Edition.

Rietker, M., Dekker, S., de Ruiter, P., & van de Koppel, J. (2004). Self-organized patchiness and catastrophic shifts in ecosystems. *Science, 305*, 1926–1929.

Rignot, E., & Kanagaratnam, P. (2006). Changes in the velocity structure of the Greenland ice sheet. *science, 311*, 986–990.

Roberts, E., Bavishi, R., Lofts, K., Khan, M. H. I., Hasemann, A., Barbosa, A. S., & Huq, S. (2013). Developing an institutional framework to address loss and damage. LDC Paper Series, LDC Group.

Robinson, J. (1962). *Economic Philosophy*. London: C.A. Watts & Co.

Robock, A. (1979). The "Little Ice Age": Northern Hemisphere average observations and model calculations. *Science, 206*(4425), 1402–1404.

Robock, A., Marquardt, A., Kravitz, B., & Stenchikov, G. (2009). Benefits, risks, and costs of stratospheric engineering. *Geophysical Research Letters, 36*, L19703. doi:10.1029/2009GL039209

Rockström, J., Steffen, W., Noone, K., Persson, Å., Chapin, F., Lambin, E., et al. (2009). A safe operating space for humanity. *Nature, 461*, 472–475.

Rohling, E., & Pälike, H. (2005). Centennial-scale climate cooling with a sudden event around 8,200 years ago. *Nature, 434*, 975–979.

Røpke, I. (1994). Trade, environment, and sustainability—A critical assessment of the free trade dogma. *Ecological Economics, 15*, 197–213.

Rosenberg, K. & Oegema, T. (1995). *A pilot ISEW for The Netherlands 1950–1992*. Amsterdam: Instituut Voor Milieu—En Systeemanalyse.

Rosenthal, E. & Lehren, A. (2012). Profits on carbon credits drive output of harmful gases. *The New York Times*, August 8, 2012.

Rosenzweig, C., Iglesias, A., Yang, X., Epstein, P., & Chivian, E. (2001). Climate change and extreme weather events: Implications for food production, plant diseases and pests. *Global Change and Human Health, 2*, 90–104.

Royer, D. (2003). Estimating latest Cretaceous and Tertiary atmospheric CO_2 from stomatal indices. In S. Wing, P. Gingerich, B. Schmitz, & E. Thomas (Eds.), *Causes and consequences of global warm climates in the Early Paleogene* (Vol. 369, pp. 79–93). Boulder, CO: Geological Society of America (Special Paper).

Royer, D., Berner, R., Montañez, I., Tabor, N., & Beerling, D. (2004). CO_2 as a primary driver of Phanerozoic climate. *GSA Today, 14*(3), 4–10.

Rübbelke, D. (2011). International support of climate change policies in developing countries: Strategic, moral and fairness aspects. *Ecological Economics, 70*, 1470–1480.

Runhaar, H., Mees, H., Wardekker, A., van der Sluis, J., & Driessen, P. (2012). Adaptation to climate change-related risks in Dutch urban areas: Stimuli and barriers. *Regional Environmental Change, 12*(4), 777–790.

Ryals, R., Kaiser, M., Torn, M., Berhe, A. A., & Silver, W. (2014). Impacts of organic matter amendments on carbon and nitrogen dynamics in grassland soils. *Soil Biology and Biochemistry, 68*, 52–61.

Sachs, J. (2009). *Eight principles for a global agreement on climate change*. New York: Earth Institute, Columbia University.

Saddler, H., Muller, F., & Cuevas, C. (2006). *Competitiveness and carbon pricing: Border adjustments for greenhouse policies*. Discussion Paper Number 86, April 2006. Canberra: The Australia Institute.

Saenko, O., Fyfe, J., & England, M. (2005). On the response of the ocean wind-driven circulation to atmospheric CO_2 increase. *Climate Dynamics, 25*, 415–426.

Sarmaj-Korjonen, K. (2007). Abrupt and consistent responses of aquatic and terrestrial ecosystems to the 8200 cal. yr cold event: A lacustrine record from Lake Arapisto, Finland. *The Holocene, 17*(4), 457–467.

Sazonova, T., Romanovsky, V., Walsh, J., & Sergueev, D. (2004). Permafrost dynamics in the 2oth and 21st centuries along the East Siberian transect. *Journal of Geophysical Research Atmospheres*, 109. doi:10.1029/2003JD003680

Scheffer, M., Carpenter, S., Foley, J., Folke, C., & Walker, B. (2001). Catastrophic shifts in ecosystems. *Nature, 413*, 591–596.

Schelling, T. (1997). The cost of combating global warming. *Foreign affairs, 76*(6), 8–14.

Schelling, T. (2002). What makes Greenhouse sense? *Foreign Affairs, 81*(3), 2–9.

Schellnhuber, H. (2008). Global warming: Stop worrying, start panicking. *Proceedings of the National Academy of Sciences, USA, 105*(38), 14239–14240.

Scheraga, J., Learly, N., Goettle, R., Jorgenson, D., & Wilkoxen, P. (1993). Macroeconomic modelling and the assessment of climate change impacts. In Y. Kaya, N. Nakicenovic, W. Nordhaus, & F. Toth (Eds.), *Costs, impacts, and benefits of* CO_2 *mitigation* (pp. 107–132). Laxenburg: IIASA.

Schlamadinger, B., Bird, N., Johns, T., Brown, S., Canadell, J., Ciccarese, L., et al. (2007). A synopsis of land use, land-use change and forestry (LULUCF) under the Kyoto Protocol and Marrakech Accords. *Environmental Science and Policy, 10*, 271–282.

Schmidt, J., Helme, N., Lee, J., & Houdashelt, M. (2006). *Sector-based approach to the post-2012 climate change policy architecture*. Future Actions Dialogue Working Paper, Center for Clean Air Policy, Washington, DC.

Schnaiberg, A. (1980). *The environment: From surplus to scarcity*. New York: Oxford University Press.

Schneider, I. (2007). *Is the CDM fulfilling its environmental and sustainable development objectives? An evaluation of the CDM and options for improvement*. Report prepared for the World Wildlife Fund (WWF), Öko-Institut, Berlin.

Schneider, S., & Mastrandrea, M. (2005). Probabilistic assessment of "dangerous" climate change and emission scenarios. *Proceedings of the National Academy of Sciences, USA, 102*, 15728–15735.

Schöb, R. (2005). The double dividend hypothesis of environmental taxes: A survey. In H. Folmer & T. Tietenberg (Eds.), *The international yearbook of environmental and resource economics 2005/2006*. Cheltenham, UK: Edward Elgar.

Schröder, A., Persson, L., & De Roos, A. (2005). Direct experimental evidence for alternative stable states: A review. *Oikos, 110*, 3–19.

Sen, A. (1997). Inequality, unemployment, and contemporary Europe. *International Labour Review, 136*(2), 155–171.

Shackelton, N., Hall, J., & Pate, D. (1995). Pliocene stable isotope stratigraphy of ODP Site 846. In *Proceedings of the Ocean Drilling Program: Scientific Results* (Vol. 138, pp. 337–356). College Station, TX: Ocean Drilling Program.

Shah, P. (2004). *Food stamps: A model for India*. New Delhi: Centre for Civil Society.

Shapiro, R. (2009). Is cap and trade a dead policy walking? http://ndn.org/blog/2009/04/cap-and-trade-dead-policy-walking. Accessed March 26, 2010.

Shepherd, C. (2003). Predicted recurrences of mass coral mortality in the Indian Ocean. *Nature, 425*, 294–297.

Shishlov, I. & Bellassen, V. (2012). *10 Lessons From 10 Years of the CDM*. Climate Report No. 37, October 2012, CDC Climat Research.

Shishlov, I., Bellassen, V., & Leguet, B. (2012). Joint Implementation: A frontier mechanism within the borders of an emissions cap. Climate Report Number 33, February 2012, CDC Climat Research.

Shiyatov, S., Terent'ev, M., & Fomin, F. (2005). Spatiotemporal dynamics of forest-tundra communities in the polar Urals. *Russian Journal of Ecology, 36*, 67–75.

Sigman, D., & Boyle, E. (2000). Glacial/interglacial variations in atmospheric carbon dioxide. *Nature, 407*, 859–869.

Sijm, J. (2005). The interaction between the EU emissions trading scheme and national policy energy policies. *Climate Policy, 1*(4), 481–497.

Silverberg, G. (1988). Modelling economic dynamics and technical change: Mathematical approaches to self-organisation and evolution. In G. Dosi, C. Freeman, R. Nelson, G. Silverberg, & L. Soete (Eds.), *Technical change and economic theory* (pp. 531–559). London: Printer Publishers.

Simpson, G. (1929–30). Past climates. *Manchester literary and philosophical society memoirs*, 74(1), 9–10.

Sindico, F. (2008). The EU and carbon leakage: How to reconcile border adjustments with the WTO? *European Energy and Environmental Law Review, 17*(6), 328–340.

Sinner, J. (2002). *Addressing competitiveness impacts of climate change policies*. Report to the New Zealand Ministry of Economic Development, Wellington.

Skjærseth, J., & Wettestad, J. (2008). Implementing EU emissions trading: Success or failure? *International Environmental Agreements, 8*, 275–290.

Smith, J., Dickinson, T., Donahue, J., Burton, I., Haites, E., Klein, R., & Patwardhan, A. (2011). Development and climate change adaptation funding: Coordination and integration. *Climate Policy, 11*(3), 987–1000.

Smith, R., Fraser, W., Stamnerjohn, S., & Vernet, M. (2003). Palmer long-term ecological research on the Antarctic marine ecosystem. *Antarctic Research Series, 79*, 131–144.

Smith, S., & Swierzbinski, J. (2007). Assessing the performance of the UK emissions trading scheme. *Environmental and Resource Economics, 37*, 131–158.

Soddy, F. (1926). *Wealth, virtual wealth, and debt*. London: Allen & Unwin.

Soden, B., & Held, I. (2005). An assessment of climate feedbacks in coupled ocean-atmospheric models. *Journal of Climate, 19*(14), 3354–3360.

Sorrell, S., & Sijm, J. (2003). Carbon trading in the policy mix. *Oxford Review of Economic Policy, 19*(3), 420–437.

Sorte, C., Jones, S., & Miller, L. (2011). Geographic variation in temperature tolerance as an indicator of potential population responses to climate change. *Journal of Experimental Marine Biology and Ecology, 400*(1), 209–217.

Spash, C. (2002). *Greenhouse economics: Values and ethics*. London: Routledge.

Spash, C. (2007). The economics of climate change impacts à la Stern: Novel and nuanced or rhetorically restricted? *Ecological Economics, 63*(4), 706–713.

Spash, C. (2010). The brave new world of carbon trading. *New Political Economy, 15*(2), 169–195.

Srinivasan, A. & Prabhakar, S. (2009). *Measures of adaptation to climate change and variability (adaptation metrics)*. Final Report to the World Bank.

Steffensen, J., Andersen, K., Bigler, M., Clausen, H., Dahl-Jensen, D., Fischer, H., et al. (2008). High-resolution Greenland ice core data show abrupt climate change happens in few years. *Science, 321*, 680–684.

Stehr, H. (2008). Does the CDM need institutional reform? In K. Olsen & J. Fenhann (Eds.), *A reformed CDM: Including new mechanisms for sustainable development*. UNEP Risoe Centre CD4CDM Perspectives Series 2008, Roskilde.

Stern, N. (2007). *The economics of climate change: The Stern review*. UK: Cambridge University Press.

Stern, N. (2008). *Key elements of a global deal on climate change*. London: London School of Economics and Political Science.

Stern, N. (2009). *A blueprint for a Safer Planet: How to manage climate change and create a new era of progress and prosperity*. London: The Bodley Head.

Stern, N. (2013). The US and China must show leadership on climate change. *The Guardian*. www.theguardian.com/environment/blog. Accessed December 15, 2013.

Sterner, T. (2003). *Policy instruments for environmental and natural resource management*. Washington, DC: Resources for the Future Press.

Sterner, T., & Persson, U. (2008). An even Sterner *Review*: Introducing relative prices into the discounting debate. *Review of Environmental Economics and Policy, 2*(1), 61–76.

Stevens, C. (1993). Do environmental policies affect competitiveness? *OECD Observer, 183*, 22–25.

Stewart, R., & Weiner, J. (2003). *Practical climate change policy. Issues on line in science and technology*. Washington, DC: National Academy of Science.

Stige, L., Stave, J., Chan, K., Ciannelli, L., Pretorelli, N., Glantz, M., et al. (2006). The effect of climate variation on agro-pastoral production in Africa. *Proceedings of the National Academy of Sciences, USA, 103*, 3049–3053.

Stiglitz, J. (2006). A new agenda for global warming. *The Economists' Voice, 3*(7), 1–4.

Stockammer, E., Hochreiter, H., Obermayr, B., & Steiner, K. (1997). The index of sustainable economic welfare (ISEW) as an alternative to GDP in measuring economic welfare: The results of the Austrian (revised) ISEW calculation, 1955–1992. *Ecological Economics, 21*(1), 19–34.

Stone, D., & Fyfe, J. (2005). The effect of ocean mixing parameterisation on the enhanced CO_2 response of the Southern Hemisphere mid-latitude jet. *Geophysical Research Letters, 32*, L06811. doi:10.1029/2004GL022007

Storbjörk, S. (2007). Governing climate adaptation in the local arena: Challenges of risk management and planning in Sweden. *Local Environment, 12*(5), 457–469.

Storm van Leeuwen, J. & Smith, P. (2008). *Nuclear power: The energy balance*. www.stormsmith.nl. Accessed March 30, 2009.

Stouffer, R. (2004). Time scales of climate responses. *Journal of Climate, 17*, 209–217.

Sudo, T. (2006). *Japanese voluntary emissions trading scheme (JVETS)—Overview and analysis*. Tokyo: Institute for Global Environmental Strategies.

Sutter, C., & Parreño, J. (2007). Does the current clean development mechanism (CDM) deliver its sustainable development claim? An analysis of officially registered CDM projects. *Climate Change, 84*, 75–90.

Svensen, H., Plank, S., Malthe-Sørenssen, A., Jamtveit, B., Myklebust, R., Rasmussen Eidem, T., & Rey, S. (2004). Release of methane from a volcanic basin as a mechanism for initial Eocene global warming. *Nature, 429*, 542–545.

Swain, M. (2014). Can palm oil deforestation be stopped? In *The Breakthrough*. Oakland, CA: The Breakthrough Institute.

Swart, R., Robinson, J., & Cohen, S. (2003). Climate change and sustainable development: Expanding the options. *Climate Policy*, 3(S1), S19–S40 (Special Issue on Climate Change and Sustainable Development).

Syri, S., Amann, M., & Capros, P. (2001). Low CO_2 energy pathways and regional air pollution in Europe. *Energy Policy, 29*(11), 871–884.

Tan, J., Zheng, Y., Song, G., Kalkstein, L., & Tang, X. (2007). Heat wave impacts on mortality in Shanghai, 1998–2003. *International Journal of Biometeorology, 51*(3), 193–200.

Tarasov, L., & Peltier, W. (2005). Arctic freshwater forcing of the Younger Dryas cold reversal. *Nature, 435*(7042), 662–665.

Tcherneva, P. (2009). Evaluating the economic and environmental viability of basic income and job guarantees. In P. Lawn (Ed.), *Environment and employment: A reconciliation* (pp. 184–205). UK: Routledge.

Theodossiou, I. (1998). The effects of low-pay and unemployment on psychological well-being: A logistic regression approach. *Journal of Health Economics, 17*(1), 85–104.

Thomas, D., Twyman, C., Osbahr, H., & Hewiston, B. (2007). Adaptation to climate change and variability: Farmer responses to intra-seasonal precipitation trends in South Africa. *Climate Change, 83*(3), 301–322.

Thomas, E. (2003). Extinction and food at the sea floor: A high-resolution benthic foraminiferal record across the Initial Eocene Thermal Maximum, Southern Ocean Site 690. In S. Wing, P. Gingerich, B. Schmitz, & E. Thomas (Eds.), *Causes and consequences of global warm climates in the Early Paleogene* (Vol. 369, pp. 319–332). Special Paper, Geological Society of America, Boulder, CO.

Thompson, L., Mosely-Thompson, E., & Henderson, K. (2000). Ice-core palaeoclimate records in tropical South America since the last glacial maximum. *Journal of Quaternary Science, 15*, 377–394.

Thomson, A., Brown, R., Rosenberg, N., & Izaurralde, R. (2005). Climate change impacts for the conterminous USA: An integrated assessment—Part 5: Irrigated agriculture and national crop production. *Climatic Change, 69*, 89–105.

Thornton, P., Jones, P., Owiyo, T., Kruska, R., Herero, M., Kristjanson,, P., et al. with contributions from Rindi, V., Otiende, B., Ochieng, A., Bhadwal, S., Anantram, K., Nair, S., et al. (2006). *Mapping climate vulnerability and poverty in Africa*. Report to the Department for International Development, ILRI, Nairobi.

Thuiller, W., Lavorel, S., Araújo, M., Sykes, M., & Prentice, I. (2005). Climate change threats to plant diversity in Europe. *Proceedings of the National Academy of Sciences, USA, 102*, 8245–8250.

Thuiller, W., Broennimann, O., Hughes, G., Alkemade, J., Midgley, G., & Corsi, F. (2006). Vulnerability of African mammals to anthropogenic climate change under conservative land transformation assumptions. *Global Change Biology, 12*, 424–440.

Tinbergen, J. (1952). *On the theory of economic policy*. Amsterdam: North-Holland.

Titus, J. (2002). Does sea level rise matter to transportation along the Atlantic coast? In *The potential impacts of climate change on transportation: Workshop summary and proceedings*, Washington, DC.

Titus, J. (2005). Sea-level rise effect. In M. Schwartz (Ed.), *Encyclopaedia of coastal science*. Dordrecht: Springer.

Todaro, M. (1994). *Economic development* (5th ed.). New York: Longman.

Tol, R. (2001). Equitable cost-benefit analysis of climate change. *Ecological Economics, 36*(1), 71–85.

Tol, R. (2002a). Welfare specifications and optimal control of climate change: An application of *FUND*. *Energy Economics, 24*, 367–376.

Tol, R. (2002b). Estimates of the damage costs of climate change. *Environmental and Resource Economics, 21*, 135–160.

Tol, R. & Yohe, G. (2006a). A review of the *Stern Review*. *World Economics*, *7*(4), 233–250.

Tol, R. & Yohe, G. (2006b). Of dangerous climate change and dangerous emission reduction. In H. Schellnhuber, W. Cramer, N. Nakicenovic, T. Wigley, G. Yohe (Eds), *Avoiding Dangerous Climate Change* (pp. 292–298). Cambridge, UK: Cambridge University Press.

Tol, R., & Yohe, G. (2007). The weakest link hypothesis for adaptive capacity: An empirical test. *Global Environmental Change, 17*, 218–227.

Toronto Star (2011). *Canada withdrawing from Kyoto*, December 12, 2011.

Torvanger, A. & Ringius, L. (2000). *Burden differentiation: Criteria for evaluation and development of burden sharing rules*. The Joint CICERO-ECN Project on Sharing of Burden of Greenhouse GAS Reduction Among Countries, Working Paper 2000:1, CICERO, Oslo.

Torvanger, A., Kallbekken, S., & Rypdal, K. (2004). *Prerequisites for geological carbon storage as a climate policy option*. Norway: Centre for the International Climate and Environmental Research.

Trainer, T. (1995). Can renewable energy save industrial society? *Energy Policy, 23*(12), 1009–1026.

Trainer, T. (2005). Can renewable energy sources sustain affluent society? *Energy Policy, 23*(12), 1009–1026.

Trainer, T. (2007). *Renewable energy cannot sustain consumer society*. Dordrecht: Springer.

Trainer, T. (2008). A critique of the Stern and IPCC analyses of CO_2 mitigation: Consumer-capitalist society cannot solve the problem. *The International Journal of Inclusive Democracy, 4*(4), 1–12.

Trainer, T. (2011). *A Critique of the 2011 IPCC report on renewable energy*. http://www.ssis.arts.unsw.edu.au/tsw

Tripati, A., & Elderfield, H. (2004). Abrupt hydrographic changes in the equatorial Pacific and subtropical Atlantic from foraminiferal Mg/Ca indicate greenhouse origin for the thermal maximum at the Paleocene-Eocene Boundary. *Geochemistry, Geophysics, and Geosystems, 5*(2), Q02006. doi:10.1029/2003GC000631

Tuerk, A., Fazekas, D., Schreiber, H., Frieden, D., & Wolf, C. (2013, September). *Green investment schemes: The AAU market between 2008 and 2012*. Joanneum Research and Climate Strategies, Climate Strategies Discussion Paper.

Tversky, A., & Kahneman, D. (1992). Advances in prospect theory: Cumulative representation of uncertainty. *Journal of Risk and Uncertainty, 5*, 297–323.

Tyndall, J. (1861). On the absorption and radiation of heat by gases and vapours, and on the physical connection. *Philosophical Magazine, 22*, 277–302.

Unidad de Cambio Climático (UCC). (2005). Análisis de la Estadística Climática y desarrollo y evaluaci6n de escenarios climáticos de las principales cuecas hidrográficas del Uruguay y de su Zona Costera. Unidad de Cambio Climático, MVOTMA-DINAMA-GEF-UNDP, Montevideo, Uruguay.

United Nations Development Programme (UNDP). (2007). *Human development report 2007/2008—Fighting climate change: Human Solidarity in a divided world*. New York: Palgrave Macmillan.

United Nations Environment Programme (UNEP). (2003). *GEO-LAC 2003—Global environment outlook: Latin America and the Caribbean*. Costa Rica: UNEP.

United Nations Environment Programme (UNEP). (2011). *Bridging the emissions gap: A UNEP synthesis report*. Nairobi, Kenya: UNEP.

United Nations Framework Convention on Climate Change (UNFCCC). (1992). *The United Nations Framework Convention on Climate Change*. New York: UNFCCC.

United Nations Framework Convention on Climate Change (UNFCCC). (1998). *Kyoto Protocol to the United Nations Framework Convention on Climate Change*. New York: UNFCCC.

United Nations Framework Convention on Climate Change (UNFCCC). (2006). Decisions adopted by the Conference of the Parties serving as the meeting of the Parties to the Kyoto Protocol. *Report of the Conference of the Parties serving as the meeting of the Parties to the Kyoto Protocol on its first session, held at Montreal from 28 November to 10 December 2005*. FCCC/FP/CMP/2005/8/Add.1, UNFCCC, New York.

United Nations Framework Convention on Climate Change (UNFCCC). (2007). *Investment and financial flows relevant to the development of an effective and appropriate international response to climate change*. Bonn: UNFCCC.

United Nations Framework Convention on Climate Change (UNFCCC). (2008a). *Clean development mechanism: 2008 in brief*. Bonn: UNFCCC.

United Nations Framework Convention on Climate Change (UNFCCC). (2008b). *Kyoto Protocol reference manual: On accounting and assigned amount*. Bonn: UNFCCC.

United Nations Framework Convention on Climate Change (UNFCCC). (2011). *Benefits of the clean development mechanism*. New York: UNFCCC.

United Nations Framework Convention on Climate Change (UNFCCC). (2012). *Doha amendment to the Kyoto Protocol*. New York: UNFCCC.

United Nations Framework Convention on Climate Change (UNFCCC). (2013a). *Report of the Conference of the Parties on its nineteenth session, held in Warsaw from 11 to 23 November 2013—Part one Proceedings*. New York: UNFCCC.

United Nations Framework Convention on Climate Change (UNFCCC). (2013b). *Report of the Conference of the Parties serving as the meeting of the Parties to the Kyoto Protocol on its eighth session, held in Doha from 26 November to 8 December 2012—Addendum*. New York: UNFCCC.

United Nations Framework Convention on Climate Change (UNFCCC). (2013c). *Executive board annual report 2013: Clean development mechanism*. New York: UNFCCC.

United Nations Framework Convention on Climate Change (UNFCCC). (2013d). *Joint implementation*. UNFCCC. http://unfccc.int/kyoto_protocol/mechanisms/joint_implementation/items/1674.php. Accessed December 15, 2013.

United Nations Framework Convention on Climate Change (UNFCCC) (2013e). *Reporting and accounting of ULUCF activities under the Kyoto Protocol*. UNFCCC. http://unfccc.int/methods/lulucf/items/4129.php. Accessed December 15, 2013.

United Nations Framework Convention on Climate Change (UNFCCC). (2013f). *Emissions from fuel used for international aviation and maritime transport (international bunker fuels)*. UNFCCC, http://unfccc.int/methods_from_intl_transport/items/1057.php. Accessed February 12, 2014.

United Nations Framework Convention on Climate Change (UNFCCC). (2013g). *Compliance: An introduction to the Kyoto Protocol compliance mechanism*. UNFCCC. http://unfccc.int/kyoto_protocol/compliance/items/3024.php. Accessed February 12, 2014.

United Nations Framework Convention on Climate Change (UNFCCC). (2013h). *Least developed countries (LDC) fund*. UNFCCC. http://unfccc.int/cooperation_and_support/financial_mechanism/least_developed_countries_fund/items/4723.php. Accessed December 15, 2013.

United Nations Framework Convention on Climate Change (UNFCCC). (2013i). *National adaptation programmes of action (NAPAs)*. UNFCCC. http://unfccc.int/national_reports/napa/items/2719.php. Accessed February 12, 2014.

United Nations Framework Convention on Climate Change (UNFCCC). (2013j). *The special climate change fund (SCCF)*. UNFCCC. http://unfccc.int/cooperation_and_support/financial_mechanism/special_climate_change_fund/items/3657.php. Accessed December 15, 2013.

United Nations Framework Convention on Climate Change (UNFCCC). (2013k). *Green climate fund*. UNFCCC. http://unfccc.int/cooperation_and_support/financial_mechanism/green_climate_fund/items/5869.php. Accessed December 15, 2013.

United Nations Framework Convention on Climate Change (UNFCCC). (2013l). *Adaptation fund*. UNFCCC. http://unfccc.int/cooperation_and_support/financial_mechanism/adaptation_fund/items/3659.php. Accessed December 15, 2013.

United Nations Framework Convention on Climate Change (UNFCCC). (2013m). *Report on the workshop on the new market-based mechanism*. UNFCCC. http://unfccc.int/resource/docs/2013/sbsta/eng/inf13.pdf. Accessed December 15, 2013.

United Nations Framework Convention on Climate Change (UNFCCC). (2013n). *Report on the workshop on the framework for various approaches*. UNFCCC. http://unfccc.int/resource/docs/2013/sbsta/eng/inf11.pdf. Accessed December 15, 2013.

United Nations Framework Convention on Climate Change (UNFCCC). (2013o). *The mechanisms under the Kyoto Protocol: Emissions trading, the clean development mechanism, and joint implementation*. UNFCCC. http://unfccc.int/kyoto_protcol/mechanisms/items.php. Accessed December 15, 2013.

United Nations Framework Convention on Climate Change (UNFCCC). (2013p). *CDM fact sheet—UNFCCC regional collaborative centres*. UNFCCC, CDM fact sheet—RCC, August 20, 2013. New York: UNFCCC.

United Nations Framework Convention on Climate Change (UNFCCC). (2013q). *CDM Bazaar Newsletter*. CDM Bazaar, UNFCCC, March 21, 2013. http://www.cdmbazaar.net/UserManagement/FileStorage/GTAMKPY3XES07LJHNRQ95D6WOF8CVB. Accessed August 28, 2014.

United Nations Framework Convention on Climate Change. (UNFCCC) (2014a). *Report of the Green Climate Fund to the Conference of the Parties*. FCCC/CP/2014/8, United Nations, New York.

United Nations Framework Convention on Climate Change (UNFCCC). (2014b). *Framework for various approaches Technical paper*. http://unfccc.int/resource.docs/2014/tp/09.pdf

United Nations Framework Convention on Climate Change (UNFCCC). (2014c). *GHG data—UNFCCC, time series—Annex I*. United Nations, New York.

United Nations Framework Convention on Climate Change (UNFCCC). (2014d). *CDM: Project activities*. UNFCCC, http://cdm.unfccc.int/Statistics/Public/CDMinsights/index.htm. Accessed December 2, 2014.

United Nations Framework Convention on Climate Change (UNFCCC). (2014e). *Annual report of the Joint Implementation Supervisory Committee to the Conference of the Parties serving as the meeting of the Parties to the Kyoto Protocol*. UNFCCC, FCCC/KP/CMP/2014/4, 22 October 2014.

US Energy Information Administration (EIA). (2013). *International energy outlook 2013*. Washington, DC: EIA.

Uyarra, M., Cote, I., Gill, J., Tinch, R., Viner, D., & Watkinson, A. (2005). Island-specific preferences of tourists for environmental features: Implications of climate change for tourism-dependent states. *Environmental Conservation, 32*, 11–19.

Valzano, F., Murphy, B., & Koen, T. (2005). *The impact of tillage on changes in soil carbon density with special emphasis on Australian conditions*. Technical Report 43, Australian Government, Canberra.

van Kooten, G. C. (2004). *Climate change economics*. UK: Edward Elgar.

Van Parijs, P. (1991). Why surfers should be fed: The liberal case for an unconditional basic income. *Philosophy and Public Affairs, 20*, 101–131.

Van Parijs, P. (2000). A basic income for all. *Boston Review* (October–November).

Van Parijs, P. (2004). Basic income: A simple and powerful idea for the twenty-first century. *Politics and Society, 32*(1), 7–39.

Vásquez, O. (2004) *El Fenómeno El Niño en Perú y Bolivia: Experiencias en Particpación Local*. Memora del Encuentro Binacional Experiencas de prevención de desastres y manejo de emergencias ante el Fenómeno El Niño, ITDG, Chiclayo, Peru.

Vasquez-Leon, M., West, C., Wolf, B., Moody, J., & Finan, T. (2002). *Vulnerability to climate variability in the faring sector—A case study of groundwater-dependent agriculture in Southeastern Arizona*. The Climate Assessment Project for the South West, Report Series CL 1-02, Institute for the Study of Planet Earth, University of Arizona, Tucson.

Verheyen, R. (2013). *Tackling L&D: A new role for the global climate regime?*. Bonn: Germanwatch.

Victor, D. (1999). Enforcing international law: Implications for an effective global warming regime. *Duke Environmental Law and Policy Forum*, 10 (pp. 147–184). North Carolina: Duke University .

Victor, P. (2008). *Managing without growth: Slower by design not disaster*. Cheltenham, UK and Northampton, USA: Edward Elgar.

Viner, D., Sayer, M., Uyarra, M., & Hodgson, N. (2006). *Climate change and the European countryside: Impacts on land management and response strategies*. Report prepared for the Country Land and Business Association, CLA, UK.

Vining, A., & Boardman, A. (1992). Ownership versus competition: Efficiency in public enterprise. *Public Choice, 73*, 205–239.

Vitousek, P., Ehrlich, P., Ehrlich, A., & Matson, P. (1986). Human appropriation of the products of photosynthesis. *BioScience, 36*, 368–373.

Vivekanandan, J., Mahajan, D., & Vashist, S. (2008). *Optimal scenarios for sustainable development under equity-based approach*. Action for a Global Climate Community, Paper 2, TERI, India.

Voice, M., Harvey, N., & Walsh, K. (Eds.). (2006). *Vulnerability to climate change of Australia's coastal zone: Analysis of gaps in methods, data and system thresholds*. Report to the Australian Greenhouse Office, Department of Environment and Heritage, Canberra.

von Schuckmann, K., Gaillard, F., & Le Traon, P. (2009). Global hydrographic variability patterns during 2003–2008. *Journal of Geophysical Research, 114*, C09007. doi:10.1029/2008JC005237

Wackernagel, M., & Rees, W. (1996). *Our ecological footprint: Reducing human impact on the earth*. Gabriola Island: New Society Publishers.

Wackernagel, M., Onisto, L., Bello, P., Callejas Linares, A., Susana Lopez Falfan, S., Mendez Garcia, J., et al. (1999). National natural capital accounting with the ecological footprint concept. *Ecological Economics, 29*(3), 375–390.

Waddington, C. (1977). *Tools of thought*. St Albans: Paladin.

Wagner, F., Aaby, B., & Visscher, H. (2002). Rapid atmospheric CO_2 changes associated with the 8200-years-B.P. cooling event. *Proceedings of the National Academy of Sciences, USA, 99*(19), 12011–12014.

Wall, E., & Marzall, K. (2006). Adaptive capacity for climate change in Canadian rural communities. *Local Environment, 11*, 373–397.

Walther, G.-R. (2004). Plants in a warmer world. *Perspectives in Plant Ecology, Evolution and Systematics, 6*, 169–185.

Wang, G. (2005). Agricultural drought in a future climate: Results from 15 global climate models participating in the IPCC 4th Assessment. *Climate Dynamics, 25*, 739–753.

Wang, W., Yung, Y., Lacis, A., Mo, T., & Hansen, J. (1976). Greenhouse effects due to man-made perturbations of trace gases. *Science, 194*, 685–690.

Wara, M., & Victor, D. (2008). *A realistic policy on international carbon offsets*. Program on Energy and Sustainable development. Working Paper #74, Stanford University, Stanford, CA.

Warner, K., Ranger, S., Surminski, S., Arnold, M., Linnerooth-Bayer, J., Michel-Kerjan, E, et al. (2009). *Adaptation to climate change: Linking disaster risk reduction and insurance*. Geneva: United Nations International Strategy for Disaster Reduction (UNISDR).

Warren, R., Arnell, N., Nicholls, R., Levy, P., & Price, J. (2006). *Understanding the regional impacts of climate change*. Research Report Prepared for the Stern Review on the Economics of Climate Change, Tyndall Centre Working 90, Tyndall Centre for Climate Change Research, Norwich.

Wassmann, R., Hien, N., Hoanh, C., & Tuong, T. (2004). Sea level rise affecting the Vietnamese Mekong Delta: Water elevation in the flood season and implications for rice production. *Climate Change, 66*, 89–107.

Watkiss, P., Anthoff, D., Downing, T., Hepburn, C., Hope, C., Hunt, A., & Tol, R. (2005). *The social cost of carbon (SCC) review: Methodological approaches for using SCC estimates in policy assessment*. London: DEFRA.

Watts, M., & Mitchell, W. (2000). The costs of unemployment. *Economic and Labour Relations Review, 11*(2), 180–197.

Wätzold, F. (2004). SO_2 emissions in Germany: Regulations to fight Waldsterben. In W. Harrington, R. Morgenstern, & T. Sterner (Eds.), *Choosing environmental policy* (pp. 23–40). Washington, DC: Resources for the Future Press.

Webb, R., Rind, D., Lehman, S., Healy, R., & Sigman, D. (1997). Influence of ocean heat transport on the climate of the Last Glacial Maximum. *Nature, 385*, 695–699.

Weisskopf, W. (1973). Economic growth versus existential balance. In H. Daly (Ed.), *Towards a steady-state economy* (pp. 240–251). San Francisco: W.H. Freeman.

Weiszacker, E., Lovins, A., & Lovins, L. (1998). *Factor four: Doubling wealth*. Earthscan, London: Halving Resource Use.

Weitzman, M. (1974). Prices versus quantities. *Review of Economic Studies, 41*(4), 477–491.

Weitzman, M. (1984). *The share economy*. Cambridge, MA: Harvard University Press.

Weitzman, M. (2007). A review of the Stern review on the economics of climate change. *Journal of Economic Literature*, XLV(3), 703–724.

Weitzman, M. (2009). On modelling and interpreting the economics of catastrophic climate change. *Review of Economics and Statistics, 91*(1), 1–19.

Wen, Z., Yang, Y., & Lawn, P. (2008). From GDP to GPI: Quantifying thirty-five years of development in China. In *Genuine progress in the Asia-Pacific: Studies using the genuine progress indicator* (pp. 228–259). Cheltenham, UK: Edward Elgar.

Weyant, J., de la Chesnaye, F., & Blanford, G. (2006). Overview of EMF-21: Multigas mitigation and climate policy. *The Energy Journal*, Special Issue, 1–32.

Widerquist, K., & Lewis, M. (2009). The basic income guarantee and the goals of equality, efficiency, and environmentalism. In P. Lawn (Ed.), *Environment and employment: A reconciliation* (pp. 163–183). UK: Routledge.

Wilbanks, T. (2003). Integrating climate change and sustainable development in a place-based context. *Climate Policy*, 3(S1), S147–S154 (Special issue on climate change and sustainable development).

Williams, S., Bolitho, E., & Fox, S. (2003). Climate change in Australian tropical forests: An impending environmental catastrophe. *Proceedings of the Royal Society B: Biological Sciences, 270*, 1887–1892.

Williamson, O. (1964). *The economics of discretionary behaviour: Managerial objectives in a theory of the firm*. Englewood Cliffs, NJ: Prentice-Hall.

Wilson, E. O. (2002). *The future of life*. New York: Alfred A. Knopf.

Winkler, H., Spalding-Fecher, R., & Tyani, L. (2002). Comparing developing countries under potential carbon allocation schemes. *Climate Policy, 2*, 303–318.

Wolf, J., Lorenzoni, I., Few, R., Abrahamson, V., & Raine, R. (2009). Conceptual and practical barriers to adaptation: Vulnerability and responses to heat waves in the UK. In N. Adger, I. Lorenzoni, & K. O'Brien (Eds.), *Adapting to climate change: Thresholds, values, governance* (pp. 181–196). Cambridge, UK: Cambridge University Press.

Woodcock, J., Edwards, P., Tonne, C., Armstrong, B., Asiru, O., Banister, D., et al. (2009). Public health benefits of strategies to reduce greenhouse-gas emissions: Urban land transport. *The Lancet, 374*(9705), 1930–1943.

Woodroffe, C., Nicholls, R., Saito, Y. Chen, Z., & Goodbred, S. (2006). Landscape variability and the response of Asian megadeltas to environmental change. In N. Harvey (Ed.), *Global change and integrated coastal management: The Asia-Pacific region* (pp. 277–314). Berlin: Springer.

Woodruff, R., Guest, C., Garner, M., Becker, N., & Lindsay, M. (2006). Early warning of Ross River virus epidemics: Combining surveillance data on climate and mosquitoes. *Epidemiology, 17*, 569–575.

Working Group on Baseline for CDM/JI Project. (2001). *Technical procedures for CDM/JI projects at the planning stage: Interim report to the Ministry of the Environment, Government of Japan*. Tokyo: Ministry of the Environment.

World Bank. (2000). *Cities, seas and storms: Managing change in the Pacific Island economies—Volume IV: Adapting to climate change*. Washington, DC: World Bank.

World Bank. (2006). *Clean energy and development: Towards an investment framework*. Paper Produced for The Development Committee Meeting. Washington, DC: World Bank.

World Bank. (2008). *Carbon finance for sustainable development, 2008*. Washington, DC: Carbon Finance at the World Bank, World Bank.

World Bank. (2011). *Carbon finance—A glossary of terms: Definition of 'Green Investment Scheme' (GIS)*. Carbon Finance Unit (CFU), World Bank Environment Department, Washington, DC.

World Bank. (2012). *State and trends of the carbon market 2012*. Washington, DC: Carbon Finance at the World Bank, World Bank.

World Bank. (2013). *Mapping carbon pricing initiatives: Development and prospects*. Washington, DC: Carbon Finance at the World Bank, World Bank.

World Bank (2014a). *State and trends of carbon pricing*. Carbon Finance at the World Bank. Washington, DC: World Bank.

World Bank (2014b). *Adaptation fund trust fund: Financial report prepared by the Trustee*. Washington, DC: World Bank.

World Commission on Environment and Development (WCED). (1987). *Our common future*. Oxford: Oxford University Press.

World Meteorological Organization (WMO). (1979). *Declaration of world climate conference*. Geneva: WMO.

World Resources Institute (WRI). (2006). Climate analysis indicators tool (CAIT) on-line database version 3.0. Washington, DC: World Resources Institute. http://cait.wri.org

World Resources Institute (WRI). *EarthTrends: The environmental information portal*. http://earthtrends.wri.org

Wray, R. (1998). *Understanding modern money: The key to full employment and price stability*. Cheltenham, UK: Edward Elgar.

Wrona, F., Prowse, T., & Reist, J. (2005). Freshwater ecosystems and fisheries. In C. Symon, L. Arris, & B. Heal (Eds.), *Arctic climate impact assessment* (pp. 353–452). Cambridge: Cambridge University Press.

Ximena, F. (1998). *Contribution to the estimation of countries' inter-dependence in the area of plant genetic resources*. Commission on Genetic Resources for Food and Agriculture, Background Study Paper 7, Rev. 1, Food and Agriculture Organization (FAO), Rome.

Yale Center for Environmental Law and Policy, Center for International Earth Science Information Network, World Economic Forum. (2005). *2005 environmental sustainability index*. New Haven: YCELP.

Yaukey, D., Anderton, D., & Hickes Lundquist, J. (2007). *Demography: The study of human population*. IL: Waveland Press.

Yohe, G. (2006). Some thoughts on the damage estimates presented in the Stern Review—An editorial. *Integrated Assessment Journal, 6*(3), 65–72.

Zachos, J., Pagani, M., Sloan, L., Thomas, E., & Billups, K. (2001). Trends, rhythms, and aberrations in global climate 65 Ma to present. *Science, 292*, 686–693.

Zachos, J., Wara, M., Bohaty, S., Delaney, M., Petrizzo, M.-R., Brill, A., et al. (2003). A transient rise in tropical sea surface temperature during the Paleocene-Eocene thermal maximum. *Science, 302*(5650), 1551–1554.

Zimov, S., Schuur, E., & Chapin, F. (2006). Climate change—Permafrost and the global carbon budget. *Science, 312*(5780), 1612–1613.

Zukewich, N., & Norris, D. (2005). *National experiences and international harmonization in social capital measurement: A beginning*. Draft Paper Submitted at the 9–11 February 2005 Meeting of the Siena Group, Helsinki.

Index

P. Lawn, *Resolving the Climate Change Crisis*, DOI 10.1007/978-94-017-7502-1

D

E

F

M

N

O

P

Q

R

W

Y

MIX
Papier aus verantwortungsvollen Quellen
Paper from responsible sources
FSC® C105338

If you have any concerns about our products,
you can contact us on
ProductSafety@springernature.com

In case Publisher is established outside the EU,
the EU authorized representative is:
Springer Nature Customer Service Center GmbH
Europaplatz 3, 69115 Heidelberg, Germany

Printed by Libri Plureos GmbH
in Hamburg, Germany